Probability Theory

This book introduces Probability Theory with R software and explains abstract concepts in a simple and easy-to-understand way by combining theory and computation. It discusses conceptual and computational examples in detail to provide a thorough understanding of basic techniques and develop an enjoyable read for students seeking suitable material for self-study. It illustrates fundamental concepts including fields, sigma-fields, random variables and their expectations, various modes of convergence of a sequence of random variables, laws of large numbers and the central limit theorem.

- Computational exercises based on R software are included in each Chapter
- Includes a brief introduction to the basic functions of R software for beginners in R and serves as a ready reference
- Includes numerical computations, simulation studies, and visualizations using R software as easy tools to explain abstract concepts
- Provides multiple-choice questions for practice
- Incorporates self-explanatory R codes in every chapter

This textbook is for advanced students, professionals, and academic researchers of Statistics, Biostatistics, Economics and Mathematics.

Probability Theory
An Introduction Using R

Shailaja R. Deshmukh
and
Akanksha S. Kashikar

CRC Press
Taylor & Francis Group
Boca Raton London New York

CRC Press is an imprint of the
Taylor & Francis Group, an **informa** business
A CHAPMAN & HALL BOOK

Designed Cover Image: Shailaja R. Deshmukh, and Akanksha S. Kashikar

First edition published 2025
by CRC Press
2385 NW Executive Center Drive, Suite 320, Boca Raton FL 33431

and by CRC Press
4 Park Square, Milton Park, Abingdon, Oxon, OX14 4RN

CRC Press is an imprint of Taylor & Francis Group, LLC

Library of Congress Cataloging-in-Publication Data
Names: Deshmukh, Shailaja R., author. | Kashikar, Akanksha S., author.
Title: Probability theory : an introduction using R / Shailaja R. Deshmukh
and Akanksha S. Kashikar.
Description: First edition. | Boca Raton : C&H/CRC Press, 2024. | Includes bibliographical references and index. |
Summary: "This book introduces Probability Theory with R software and explains abstract concepts in a
simple and easy-to-understand way by combining theory and computation"-- Provided by publisher.
Identifiers: LCCN 2024010470 (print) | LCCN 2024010471 (ebook) | ISBN
9781032617978 (hbk) | ISBN 9781032619033 (pbk) | ISBN 9781032619057
(ebk)
Subjects: LCSH: Probabilities--Data processing--Textbooks. | R (Computer
program language)--Textbooks.
Classification: LCC QA273.19.E4 D47 2024 (print) | LCC QA273.19.E4
(ebook) | DDC 519.20285/5133--dc23/eng/20240509
LC record available at https://lccn.loc.gov/2024010470
LC ebook record available at https://lccn.loc.gov/2024010471

ISBN: 978-1-032-61797-8 (hbk)
ISBN: 978-1-032-61903-3 (pbk)
ISBN: 978-1-032-61905-7 (ebk)

DOI: 10.1201/9781032619057

Typeset in SFRM1000 font
by KnowledgeWorks Global Ltd.

Dedicated to

Respected Teachers who made us Better Students,

Beloved Students who made us Better Teachers
&
Dear Friends who effortlessly switched between these two roles !

Contents

List of Tables

List of Figures

Preface

The most important questions of life... are indeed for the most part only problems of probability.

– Pierre-Simon Laplace

In his Philosophical essay on Probabilities, Laplace talks about how the theory of probabilities is basically just common sense reduced to calculus and hence has applications in almost every aspect of human knowledge. In today's data-driven world, the importance of probability theory has grown immensely. We have evolved from using conditional probabilities to predict whether an email is spam to creating large language models which are apparently doing a decent job as teachers, diagnosticians and what not! Hence, it is always essential that we don't lose touch with the fundamentals! Probability theory has always served as a foundation of inferential aspect of statistics. In fact, the knowledge of probability theory is what differentiates statisticians from record keepers or accountants. For example, a cricket historian may be able to tell you all about the wins and losses of a particular team. Before a match, he/she may also be able to present you with nice visual displays of the performances of a player in older similar matches. However, a statistician trained in probability theory will be able to account for the variety of factors at play to arrive at a win probability. Similarly using past data, a statistical model based on probability theory can predict whether a recently initiated transaction is a fraud. It can also tell us whether we are in a danger of contracting some disease based on our genetic makeup! Probability theory thus plays the role of a bridge between the exploratory (tabulation, visualization of the given data) and the confirmatory (hypothesis testing for generalizing from samples to populations, predictions, forecasts using sophisticated models) data analysis. The theory of probability is the foundation of statistics as a science.

In the present book, we introduce some basic concepts of the probability theory. We have adopted a rigorous mathematical approach towards the theoretical concepts. To absorb the notions easily, it is simultaneously supported by numerical computations, simulation studies and visualizations, with `R` software. The book begins with a brief introduction to the basic concepts in Chapter 1, such as sequences of sets, their limit, sigma field, Borel field, measurable space, probability measure and probability space. Chapter 2 is devoted to a study of random variables as measurable functions, random vectors, sequence of random variables, induced sigma fields and induced probability measure. In Chapter 3, we discuss the notion of a distribution function and study its properties. Chapter 4 is concerned with the concept of an expectation of a random variable as an integral with respect to the probability measure and related results. A distinguishing and fundamental feature of probability theory is the notion of independence, which has no parallel in measure theory. It is one of the most central concepts of probability theory and a large part of the theory explores the consequences of the presence of this property and the lack of it. Chapter 5 presents the fundamentals of the theory of independence of random variables and some relevant ideas.

Chapters 6 and 7 investigate in detail the various modes of the convergence of a sequence of random variables and the interrelations among these. In probability theory we study limit theorems of two types: (i) strong limit theorems which deal with the almost sure convergence of a sequence of random variables and (ii) weak limit theorems which deal with the convergence in probability of a sequence of random variables as well as the convergence of a sequence of distribution functions. Chapter 8 discusses the convergence of the sequences of expectations of random variables. It includes the monotone convergence theorem, Fatou's lemma and the Lebesgue-dominated convergence theorem. Chapter 9 is concerned with the laws of large numbers which are related with the limiting behaviour of the sequence of partial sums of a sequence of random variables. These state conditions under which the sequence of partial sums converges almost surely and in probability. The weak law of large numbers refers to convergence in probability, whereas the strong law of large numbers refers to the almost sure convergence.

Chapter 10 is devoted to the most important and frequently applied tool in probability theory and it is the central limit theorem. It is concerned with the convergence in distribution of appropriately normed sequence of partial sums. Central limit theorem is one of the oldest and the most useful results in probability theory. It states that the mean of a large number of random variables, no matter what their distribution is, with the only condition on the variance of the random variables, is approximately normally distributed. Empirical findings in applied sciences, dating back to the seventeenth century, showed that the averages of laboratory measurements on various physical quantities tend to have a bell-shaped distribution. The central limit theorem provides a theoretical justification for this observation. Chapters 6 to 10 are the most important chapters and the concerned theory is heavily applied in large sample statistical inference to study the asymptotic optimality properties of the estimators and to derive the asymptotic null distribution of the test statistic.

Chapter 11 presents the solutions of almost all of the conceptual exercises. Solutions to the exercises are important for students who initially do not have the skill in solving these exercises completely. It is expected that students should not be completely driven by the solutions. They should try each exercise first without referring to its solution. Numerous illustrative examples of different difficulty levels are incorporated throughout each chapter to clarify the concepts. These illustrations and several remarks reveal the depth of the incorporated theory. Exercises form an important part of any textbook. For better assimilation of the material, three types of exercises, such as conceptual, computational and MCQs are given at the end of each chapter.

The book has evolved out of the instructional material prepared for teaching a course "Probability Theory" for several years at Savitribai Phule Pune University, formerly known as the University of Pune. To some extent, the topics coincide with what we used to cover in the course. The material covered in the present book is influenced by several standard textbooks, such as Athreya and Lahiri (2006), Bhat B.R. (1999), Billingsley (1986), A. Gut (2005) and Loeve (1978). The novelty and unique feature of our book is augmenting the theory with `R` software as a tool for simulation and computation. Over the years, we have noted that the concepts from probability are difficult for the students to grasp due to their abstract nature. To overcome this difficulty, it is essential that the theory and computations go hand in hand. Hence, in the present book, we have adopted an approach to explain theory along with the computations. The books mentioned above do not address the computational aspect.

We are convinced that the development of computational methods, along with the theory, will greatly contribute to a better understanding of the theory. Hopefully, a better understanding can provide more insights and help students to appreciate the beauty of the subject. We will be deeply rewarded, if the present book helps students to enhance their understanding and to enjoy the subject. The main motive is to provide fairly thorough treatment of basic techniques, theoretically and computationally using `R`, so that the book will be suitable for self-study, particularly in the present era of online education. The style of the book is purposely kept conversational. One more reason to keep the style conversational is to help the students develop an intuition for the subject. In the present era, where students are becoming increasingly reliant on technology for solving their problems, texts which can help develop rigorous mathematical thinking abilities have become a necessity. Thus, we try to explain the thinking behind each and every argument made in a proof, so as to ensure that the students develop the art of problem-solving and logical thinking.

Nowadays `R` is a leading computational tool for statistics and data analysis. It is free and platform-independent and hence can be used on any operating system. Keeping up with the recent trend of using `R` software for statistical computations, we too have used it extensively in this book for illustrating the concepts. Another advantage of using a free software is that a lot of community support is available online for fixing errors in the codes and finding out new commands. Using `R`, we illustrate the concept of limit infimum and limit supremum of the sequence of sets, limits of monotone and non-monotone sequences of sets, various modes of convergence of a sequence of random variables, Lebesgue dominated convergence theorem, laws of large numbers and the central limit theorem. `R` codes are provided in the respective chapters. Computational exercises based on `R` software are included so as to provide a hands-on experience to students.

The mathematical pre-requisites for this book are elementary probability, basic convergence concepts for a sequence of real numbers, familiarity with the properties of standard discrete and continuous distributions. Some background of the measure theory would also be beneficial but is not necessary. In addition, some basic knowledge of `R` software is desirable. In Chapter 1 for ready reference, we have added a section on `R` software, which gives a brief introduction to the basic functions. The intended target audience of the present book is mainly postgraduate students in Statistics, Bio-statistics or Mathematics. It will also provide sufficient background information for studying advanced courses in probability theory and stochastic processes. The book is designed primarily to serve as a textbook for a one-semester introductory course on probability theory in any post-graduate statistics program.

We wish to express our special thanks to all our teachers, who have laid the strong foundation of probability theory and influenced our understanding, appreciation and taste for the subject. The book is essentially based on our class notes when we were students in this course and the notes we prepared as the instructor of this course. We would like to express our sincere thanks to Prof. S. R. Adke and Prof. G. B. Marathe, who taught this course when we were students.

We express our deep gratitude to the `R` core development team and the authors of contributed packages, who have invested a lot of time and effort in creating `R` as it is today. With the help of such a wonderful computational tool, it is possible to showcase the beauty of the theory. We thank Prof. T.V. Ramanathan, Head of the Department of Statistics, Savitribai Phule Pune University, for providing the necessary facilities. We take this opportunity to acknowledge Rajesh Dey, Senior Commissioning Editor (Mathematics, Physics, Statistics and Computer Science), Taylor & Francis India Pvt. Ltd. CRC Press and his team, for providing help from time to time and subsequent processing of the text to its present form. We are deeply grateful to our family members for their constant support and encouragement. We owe profound thanks to all our students whom we have taught during the last several years and who have been the driving force to take up this immense task. Their reactions, positive as well as negative, doubts in the class, and their responses in the answer books urged us to make the theory crystal clear to them, thereby compelling us to pursue this activity and to prepare a variety of illustrations and exercises.

All mistakes and ambiguities in the book are exclusively our responsibility. We would love to know any mistakes that a reader comes across in the book. Feedback in the form of suggestions and comments from colleagues and readers is the most welcome.

December 2023 Shailaja Deshmukh and Akanksha Kashikar

Pune, India
Email Addresses: Shailaja Deshmukh – deshshailaja@gmail.com
Akanksha Kashikar – akanksha.kashikar@gmail.com

Author Biographies

Shailaja Deshmukh is a visiting faculty at the Department of Statistics, Savitribai Phule Pune University, formerly known as University of Pune, India. She retired as a Professor of Statistics from the Savitribai Phule Pune University. She has taught around twenty five different theoretical and applied courses. She worked as a visiting professor at the Department of Statistics, University of Michigan, Ann Arbor, Michigan, during 2009-10 academic year. She has served as an executive editor and as a chief editor of the Journal of Indian Statistical Association. She is an elected member of the International Statistical Institute.
Her areas of interest are inference in stochastic processes, applied probability and analysis of microarray data. She has a number of research publications in various peer-reviewed journals, such as Biometrika, Journal of Multivariate Analysis, J. R. Statist. Society, Australian and New Zealand Journal of Statistics, Journal of Applied Statistics , Environmetrics, J. of Statistical Planning and Inference, Naval Research Logistics, Journal of Translational Medicine, Annals of Institute of Statistical Mathematics. She has published the following six books. (i) Microarray Data: Statistical Analysis Using R, published by Narosa in 2007, (ii) Statistics Using R, published by Narosa in 2008, (iii) Actuarial Statistics: An Introduction Using R, published by Universities Press in 2009, (iv) Multiple Decrement Models in Insurance: An Introduction Using R, published by Springer in 2012 and (v) Asymptotic Statistical Inference: An Introduction Using R, published by Springer in 2021. (vi) Stochastic Processes: An Introduction Using R", published by Springer in 2023.

Akanksha Kashikar is an Assistant Professor at Savitribai Phule Pune University (formerly known as the University of Pune) since 2012, specializing in Statistics. She obtained her masters and PhD in Statistics from the same university. Her research interests include Branching Processes, Count Data Time Series, and Applied Statistics. She has taught around fifteen different theoretical and applied courses. She has published papers in various Sci/Scopus listed journals including the *Journal of Statistical Planning and Inference, Statistics and Probability Letters, Annals of the Institute of Statistical Mathematics, Communication in Statistics, Environmental Science and Pollution Research, Proteome Science, Journal of Astrophysics and Astronomy* etc. She has developed an R Package "Outlier Detection" which has been downloaded more than 15000 times. She is an executive editor for the Journal of Indian Statistical Association. She has received several awards, including the Young

Statistician Award and the Young Scientist Award from Indian statistical societies, and the Best Paper Award from the Indian Society for Medical Statistics. She was granted a research fellowship through the Indo-Japanese Young Researcher Fellowship Programme. She is passionate about science popularization and has written articles in Marathi publications. She has also been recognized for her innovative teaching methods. (Award by SPPU)

1

Sigma Field, Borel Field and Probability Measure

1.1 Introduction

Statistics as a subject comprises the art and science of the collection, interpretation and analysis of data, which enables to draw logical generalities related to the phenomenon under investigation. In view of such essential stages of any scientific method, statistics extensively enters the domain of all scientific investigations. Theory of probability is the foundation of statistics as a science. It forms a strong base for Mathematical statistics. In the present book, we elaborate on some aspects of probability theory.

The main objective of probability theory is to investigate mathematical models of random phenomena from a theoretical point of view. The basis of probability theory is a probability space. To explain this concept properly, we need some definitions and results from measure theory. This chapter provides some principal concepts of probability theory and some important results which form a sound foundation of the theory in subsequent chapters.

As the title of the book conveys, `R` software is used throughout the book to illustrate the concepts with numerical examples and figures. The novel feature of the present book is augmentation of the theory with `R` software (see R [18]) as a tool for simulation, computation and visual presentation. For the last two decades, `R` has been a leading computational tool for statistics and data analysis. It is free and platform-independent and hence can be used on any operating system. Keeping up with the recent trend of using `R` software for statistical computations, we too have used it extensively in this book for illustrating the concepts. In order to facilitate understanding of `R` codes provided in all the chapters, a preliminary introduction to the `R` software is given in Section 2 of this chapter. In Section 3, we review some topics from measure theory, such as set theory, sequence of sets and its convergence. We use `R` to illustrate the concepts of limit supremum, limit infimum and convergence of sequence of sets. `R` codes used in these illustrations are provided.

Borel field plays a major role in the definition of a random variable and a random vector. Hence, in Section 4, we elaborate on the concept of a sigma field of subsets of a non-empty set and study some of its properties. We then define a Borel field and prove various equivalent ways to generate the Borel

DOI: 10.1201/9781032619057-1

field. Section 5 is concerned with the probability measure and some related results. This background material is essential for the solid treatment of the probability theory presented in this book.

One may refer to some standard textbooks on probability theory, such as Ash [2], Athreya and Lahiri [3], Bhat [4], Billingsley [5], Chung [8] and Loeve [16] for more details.

Some readers may be familiar with `R` software as it has been introduced in the curriculum of many undergraduate and post-graduate statistics programs. In the following section, we give a brief introduction to `R`, which will be useful to beginners. We have also tried to make the codes given in all the chapters to be self-explanatory.

1.2 Introduction to `R` Software and Language

In statistical analysis, one needs a good statistical software to carry out computations and draw different types of graphs. There is a variety of software available for the computation such as Excel, Minitab, Matlab, SAS, etc. In the last two decades, `R` software has been strongly advocated and a large proportion of the world's leading statisticians use it for statistical analysis. It is a high-level language and an environment for data analysis and graphics, created by Ross Ihaka and Robert Gentleman in 1996. It is both a software and a programming language considered as a dialect of the S language developed by AT & T Bell Laboratories.

The current `R` software is the result of a collaborative effort with contributions from all over the world. It has become very popular in academics and also in the corporate world for a variety of reasons such as its good computing performance, excellent built-in help system, flexibility in graphical environment, its vast coverage, availability of new, cutting-edge applications in many fields and scripting and interfacing facilities. The most important advantage is that in spite of being the finest integrated software, it is freely available software from the site known as CRAN (Comprehensive R Archive Network) with the address `http://cran.r-project.org/`. From this site one needs to "Download and Install R" by running the appropriate pre-compiled binary distributions. When `R` is installed properly, the `R` icon is displayed on the desktop/laptop. To start `R`, one has to click on the `R` icon. The data analysis in `R` proceeds as an interactive dialogue with the interpreter. As soon as we type a command at the prompt (>) and press the enter key, the interpreter responds by executing the command. The session is ended by typing q(). The latest version of `R` is 4.3.1 released on June 16, 2023. One can type `R` codes in an editor file with "·`R`" extension and save the file for repeated use. A more user-friendly interface for `R` is provided by another software named "`R`"

Studio. It is also available freely for personal use. The latest version available is $2023.06.2 + 561$.

With this non-statistical part of the introduction to R, we now proceed to discuss how it is used for statistical analysis. The discussion is not exhaustive, but we restrict to the functions which are repeatedly used in this book. Like any other programming language, R contains data structures. Vectors are the basic data structures in R. The standard arithmetic functions and operators apply to vectors on an element-wise basis, with usual hierarchy. We specify below some such basic functions which we need to write R codes for understanding the concepts studied in this book. The most useful R command for entering small data sets is the c ("combine") function. This function combines or concatenates terms together. One can use any variable name, but care is to be taken as R is case sensitive. It may be noted that anything that appears after the symbol # is ignored by R. It is only for understanding the code.

Code 1.2.1. This code presents some basic functions.

```
# To input vectors and matrices
x=c(14, 28, 35, 43, 56, 67) # c function to construct a vector
# with given elements
x # displays x, print(x) also displays the object x
length(x) # specifies a number of elements in x
y=1:7; y # constructs a vector with consecutive elements 1 to
# 7 and prints it.
# Two commands can be given on the same line with separator ";"
u=seq(100,250,50); u # sequence function to create a vector
# with first element 100, last element 250 and increment 50
v=c(rep(1,3),rep(2,2),rep(3,5));v # rep function to create a
vector
# where 1 is repeated thrice, 2 twice and 3 five times
m=matrix(c(22,28,33,41,54,67),nrow=2,ncol=3); m # matrix
# with 2 rows and 3 columns, with first two elements
# forming first column, next two second column, output is
> [,1] [,2] [,3]
[1,] 22 33 54
[2,] 28 41 67
m1=matrix(c(22,28,33,41,54,67),nrow=2,ncol=3,byrow=T); m1
# Now elements are row-wise. The output is
     [,1] [,2] [,3]
[1,] 22 28 33
[2,] 41 54 67
# with additional argument byrow=T, we get matrix with 2
```

```
# rows and 3 columns, with first three elements forming
# first row and next three forming second row.
# By default byrow=F, see m matrix defined above
t(m) # function to obtain transpose of matrix m
```

□

`R` has excellent facility to find the values of probability mass function, probability density function and distribution function for specified values, to find quantiles and to draw random samples from some standard discrete and continuous probability distributions. These can be obtained by the functions `d, p, q, r` described below.

(i) The `d` function returns the probability density or probability mass function of the distribution.

(ii) The `p` function gives the value of a distribution function.

(iii) The `q` function gives the quantiles.

(iv) The `r` function returns a random sample from a distribution.

Each family has a name and some parameters. The function name is obtained by combining either `d`, `p`, `q` or `r` with the name for the family. The parameter names vary from family to family but are consistent within the family. Functions `d, p, q and r` are illustrated in the following code for the binomial $B(10, 0.6)$ distribution.

Code 1.2.2.

```
pr=dbinom(c(3,4,5),10,.6); pr1=round(pr,4); pr1
# values in pr rounded to four digits after the decimal point
# 0.0425, 0.1115, 0.2007: probability mass function at 3,4,5
df=pbinom(c(3,4,5),10,.6); df1=round(df,4); df1
# 0.0548, 0.1662, 0.3669: distribution function at 3,4,5
qbinom(c(.25,.5,.75),10,.6)
# 5,6,7: first,second and third quartiles
r=rbinom(8,10,.6); r # random sample of size 8
# 3 6 9 6 5 5 6 6
```

□

If we run the function `rbinom(8,10,.6)` again, we may get a different sample. If we wish to generate the same sample again, we have to use a function `set.seed`. It is used and explained in Code 1.2.4.

We use functions `rexp` and `rnorm` to draw random samples from an exponential distribution and a normal distribution respectively. Thus, we need to change the family name and add appropriate parameters. We can get the names for all probability distributions by following the path on `R` console as help → manuals (in pdf) → An Introduction to `R` → Probability distributions.

The function **round(pr,4)** prints the values of pr, rounded to four digits after the decimal point. It is to be noted that rounding is mainly for printing purpose, original unrounded values of pr are stored in the object pr and used in computations.

There are some useful built-in functions. Apart from many built-in functions, one can write a suitable function as required in a specific situation. In subsequent chapters, you will find many such functions written to serve the specific purpose.

Code 1.2.3. This code illustrates some commonly used functions with a data set stored in variable x.

```
x=rnorm(25,3,2) # random sample of size n=25 from normal
# distribution with mean 3 and standard deviation 2
mean(x); median(x); max(x); min(x); sum(x)
cumsum(x)# cumulative sum
var(x)# returns variance, divisor is (n-1) and not n
quantile(x,c(.25,.5,.75)) # three quartiles
summary(x) # gives minimum, maximum, three quartiles and mean
shapiro.test(x) # Shapiro-Wilk test for normality
# gives value of statistic and p-value
# Partial output
> summary(x)
   Min. 1st Qu. Median Mean 3rd Qu. Max.
 -1.387 2.250 3.426 3.356 4.360 7.978
> shapiro.test(x)
W = 0.97969, p-value = 0.8789
```

□

It is to be noted that we have drawn a random sample from the normal distribution and from the p-value of Shapiro-Wilk test for normality; normality is accepted, as is expected. We can carry out number of test procedures on similar lines; manual from help menu lists some of these. We illustrate one more test procedure after discussing graphical tools of R.

Graphical features of R software have a remarkable variety. Each graphical function has a large number of options making production of graphics very flexible. Graphical device is a graphical window or a file. There are two kinds of graphical functions – the high-level plotting functions, which create a new graph, and the low-level plotting functions, which add elements to an already existing graph. The standard high-level plotting functions are plot() function for scatter plot, hist() function for histogram and boxplot() function for box plot, etc. The lower-level plotting functions are lines() to impose curves on the existing plot, abline() to add a line with a given intercept and slope, `points()` to add points at appropriate places, etc. These functions take extra arguments that control the graphic. The graphs are produced with respect to graphical parameters, which are defined by default and can be modified with the function "par". If we type `?par` on `R` console, we get the description on number of arguments for graphical functions, as documented in `R`. We explain one among these which is frequently used in the book. It is `par(mfrow=c(2,2))` or `par(mfcol=c(2,2))`. This function divides the graphical window invisibly in 2 rows and 2 columns to accommodate 4 graphs. A function legend() is usually added in plots to specify a list of symbols or colours used in the graphs. Following code illustrates one test procedure and `hist` and `lines` functions.

Suppose $\{X_1, X_2, \cdots, X_n\}$ is a random sample from an exponential distribution, with location parameter $\theta \in \mathbb{R}$ and scale parameter 1. It is known that for each n, the distribution of $T_n = n(X_{(1)} - \theta)$ is exponential with mean 1. We verify this result by simulations using the following procedure. Every sample gives us one value of T_n. Hence to study the given distribution of T_n, we generate m samples from exponential distribution, with location parameter $\theta \in \mathbb{R}$ and scale parameter 1 and compute the value of T_n for each of these samples. Thus, we get a sample of size m on T_n. We then examine the distribution of these m values graphically. We draw the histogram of these m values using `hist` function, with relative frequencies on y-axis. Then we impose on it, the graph of the probability density function of the exponential distribution with mean 1, using `lines` function. We observe the closeness between the observed and the expected distributions visually from the graph.

We confirm the visual impression with Karl Pearson's test procedure for testing goodness of fit. Karl Pearson's test statistic is $T = \sum_{i=1}^{k}(o_i - e_i)^2/e_i$, where k denotes the number of classes and o_i and e_i respectively denote the observed frequency and the expected frequency, expected under the exponential distribution with mean 1, of the i-th class. Under the null hypothesis that the observed data are from the exponential distribution with mean 1, T follows chi-square distribution. Using it we find the cut-off point and the p-value to arrive at a decision. In the computation of T, we have not pooled the frequencies which are less than 5. We also use the built-in function `chisq.test(O1,p=ep1)` for goodness of fit test. We note that the two approaches give the same results.

Following code incorporates all these features. In the code we use the function `set.seed`, so that every time we run the code, we get the same sample. In the code the vector "ep" presents the difference between the values of distribution function of the exponential distribution with mean 1 at break points, that is, at class boundaries. Another function `data.frame` is used to create an array-like structure by joining multiple vectors of the same length.

Code 1.2.4.
```
th=1.5; n=60; nsim=2000; t=c()
for(i in 1:nsim)
{
 set.seed(i)
```

```
 x=th+rexp(n,rate=1)
 t[i]=min(x)
}
tn=n*(t-th); summary(tn); v=var(tn)*(nsim-1)/nsim; v
hist(tn,freq=FALSE,main=expression(paste("Histogram of T"[n])),
ylim=c(0,1),xlab=expression(paste("T"[n])),col="light blue")
O=hist(tn,plot=FALSE)$counts; sum(O)
bk=hist(tn,plot=FALSE)$breaks; bk
M=max(bk); u=seq(0,M,.1); y=dexp(u,rate=1)
lines(u,y,"o",pch=20,col="dark blue",lty=1)
legend("topright",pch=20,col="dark blue",lty=1,cex=.7,
legend="pdf of exponential distribution with mean 1")
ep=c()
for(i in 1:(length(bk)-1))
{
 ep[i]=exp(-bk[i])-exp(-bk[i+1])
}
a=1-sum(ep); a; ep1=c(ep,a); ef=sum(O)*ep1; O1=c(O,0)
d=data.frame(O1,round(ef,2)); d
ts=sum((O1-ef)^2/ef);ts
df=length(ef)-1; df; b=qchisq(.95,df); b; p1=1-pchisq(ts,df); p1
chisq.test(O1,p=ep1) # pooling of frequencies less than
#5 is ignored
```

Without pooling the frequencies which are less than 5, $T = 11.956$. The null distribution of the test statistic is χ^2_{10}. The cut-off and p-value are 18.3070 and 0.288 respectively. The built-in function also gives the same value of T and the same p-value. Thus, with both approaches, we arrive at the conclusion that the data may be from an exponential distribution with mean 1. Figure 1.1 also leads to the same conclusion.

Code 1.2.4 illustrates `d,r` functions and a variety of graphical functions. We use `ylim`, `xlab` and `col` arguments in `hist` function to set the limits on y-axis using `ylim=range(c(0,1))`, label the x-axis as required and to set colour of the histogram. Note that from `hist` function, we can extract the observed frequencies and the cut-off points, with the help of $ operator. All these features are used in a variety of graphs drawn in the subsequent chapters.

There are some other functions such as `arrow`, `axis` or many other arguments to the usual functions like `plot`, which may be new to the reader. In such cases, help from `R` or a simple Google search would give us idea about the usage. Another simple way to learn these functions is to experiment by changing the values of these arguments.

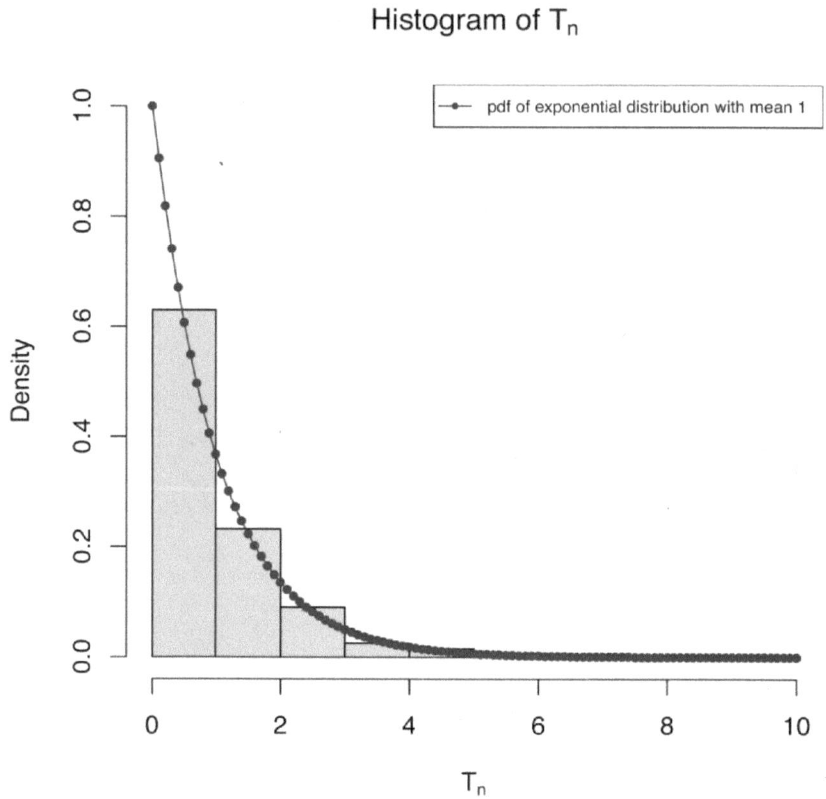

FIGURE 1.1
Exponential Distribution with Mean 1: Observed and Expected Distributions

There are many books on introduction to statistics using `R`, such as Crawley [9], Dalgaard [10], Purohit et al. [17] and Verzani [25]. There is a tremendous amount of information about `R` on the web at `http://cran.r-project.org/` with a variety of `R` manuals. Following are some links useful for beginners to learn `R` software.

1. https://www.datacamp.com/courses/free-introduction-to-r
2. http://www,listendata.com/p/r-programming-tutorials.html
3. http://www.r-tutor.com/r-introduction
4. https://www.r-bloggers.com/list-of-free-online-r-tutorials/
5. https://www.tutorialspoint.com/r/
6. https://www.codeschool.com/courses/try-r

As for any software or programming language, best way to learn R is to use it for understanding the concepts and solving problems. We hope that this brief introduction will definitely be useful to a reader to be comfortable with R and follow the codes written in this and the subsequent chapters.

In the next section, we define a sequence of sets and discuss the concepts related to its convergence.

1.3 Limit of a Sequence of Sets

Suppose Ω is a universal set, which we assume to be non-empty. We begin with the basic concepts from set theory and define union and intersection of two sets, countable and uncountable collections of sets. A countable collection of sets refers to a finite or a countably infinite collection of sets.

Definition 1.3.1. *Union and intersection of two sets: Suppose $A_1, A_2 \subseteq \Omega$. Then union of A_1 and A_2, denoted by $A_1 \cup A_2$ and intersection of A_1 and A_2, denoted by $A_1 \cap A_2$ are defined as,*

$$\begin{aligned} A_1 \cup A_2 &= \{\omega \mid \omega \in A_i, \ \textit{for at least one} \ \ i = 1, 2\} \\ \& \ \ A_1 \cap A_2 &= \{\omega \mid \omega \in A_i, \ \textit{for all} \ \ i = 1, 2\}. \end{aligned}$$

Definition 1.3.2. *Finite union and intersection: Suppose $A_i \subseteq \Omega$, $i = 1, 2, \ldots, k$. Then union and intersection of $A_i, i = 1, 2, \ldots, k$ are respectively defined as,*

$$\begin{aligned} \bigcup_{i=1}^{k} A_i &= \{\omega \mid \omega \in A_i, \ \textit{for at least one} \ \ i = 1, 2, \ldots, k\} \\ \& \ \ \bigcap_{i=1}^{k} A_i &= \{\omega \mid \omega \in A_i, \ \textit{for all} \ \ i = 1, 2, \ldots, k\}. \end{aligned}$$

Definition 1.3.3. *Countable union and intersection: Suppose $\{A_n, n \geq 1\}$ is a countably infinite collection of subsets of Ω. In this case, $\{A_n, n \geq 1\}$ is always referred to as a sequence of subsets of Ω. Then the countable union and countable intersection are respectively defined as,*

$$\begin{aligned} \bigcup_{n \geq 1} A_n &= \{\omega \mid \omega \in A_n, \ \textit{for at least one} \ \ n \geq 1\} \\ \& \ \ \bigcap_{n \geq 1} A_n &= \{\omega \mid \omega \in A_n, \ \textit{for all} \ \ n \geq 1\}. \end{aligned}$$

Definition 1.3.4. *Uncountable union and intersection: Suppose $\{A_t, t \in T\}$ is a collection of subsets of Ω, where T, known as an index set, is uncountable.*

Then uncountable union and uncountable intersection are respectively defined as,

$$\begin{aligned} \bigcup_{t\in T} A_t &= \{\omega \mid \omega \in A_t, \ \textit{for at least one} \ \ t \in T\} \\ \& \ \bigcap_{t\in T} A_t &= \{\omega \mid \omega \in A_t, \ \textit{for all} \ \ t \in T\}. \end{aligned}$$

Remark 1.3.1. (i) These four types of unions and intersections can be summarized as $\bigcup_{t\in T} A_t$ and $\bigcap_{t\in T} A_t$, where $T = \{1,2\}$ or $T = \{1,2,\ldots,k\}$ or $T = \mathbb{N} = \{1,2,\ldots,\}$, a set of natural numbers or T is any uncountable set, such as $T = (0,1)$ or $\mathbb{R}$ or $\mathbb{R}^+$. (ii) If $\{A_t, t \in T\}$ is a collection of disjoint sets in Ω, then $\bigcup_{t\in T} A_t$ is denoted by $\sum_{t\in T} A_t$.

Following examples illustrate countable union and countable intersection of a sequence of sets.

Example 1.3.1. Suppose $\Omega = (0,1)$ and $A_n = (0,1/n), \ n \geq 1$. Then $\bigcup_{n\geq 1} A_n = (0,1)$ and $\bigcap_{n\geq 1} A_n = \emptyset$. If $\Omega = [0,1]$ and $B_n = [0,1/n], \ n \geq 1$, then $\bigcup_{n\geq 1} B_n = [0,1]$ and $\bigcap_{n\geq 1} B_n = \{0\}$. □

Example 1.3.2. Suppose $\Omega = \mathbb{R}$. If $A_n = (a-1/n, a+1/n), \ \ n \geq 1, \ \ a \in \mathbb{R}$, then $\bigcup_{n\geq 1} A_n = (a-1, a+1)$ and $\bigcap_{n\geq 1} A_n = \{a\}$. □

Remark 1.3.2. De Morgan's law related to complement of union/intersection is valid for finite, countable and uncountable unions and intersections. In general,

$$\Big(\bigcup_{t\in T} A_t\Big)^c = \bigcap_{t\in T} A_t^c \ \text{ and } \ \Big(\bigcap_{t\in T} A_t\Big)^c = \bigcup_{t\in T} A_t^c,$$

where $T = \{1,2\}$ or $T = \{1,2,\ldots,k\}$ or $T = \mathbb{N} = \{1,2,\ldots,\}$ or any uncountable set.

As in the setup of a sequence of real numbers, we now discuss the monotone nature of a sequence of sets.

Definition 1.3.5. *Non-decreasing sequence of sets: Suppose $\{A_n, n \geq 1\}$ is a sequence of subsets of Ω. If $A_n \subseteq A_{n+1}, \ \ \forall \ \ n \geq 1$, then $\{A_n, n \geq 1\}$ is called as a non-decreasing sequence of sets.*

Definition 1.3.6. *Non-increasing sequence of sets: Suppose $\{A_n, n \geq 1\}$ is a sequence of subsets of Ω. If $A_n \supseteq A_{n+1}, \ \ \forall \ \ n \geq 1$, then $\{A_n, n \geq 1\}$ is called as a non-increasing sequence of sets.*

In Example 1.3.1 and Example 1.3.2, $\{A_n, n \geq 1\}$ is a non-increasing sequence of sets. If $\{A_n, n \geq 1\}$ is defined by $A_n = (0,n)$ or $A_n = (0, 2^n), n \geq 1$, then it is a non-decreasing sequence of sets. If a sequence of sets is either non-decreasing or non-increasing, then it is referred to as a monotone sequence of sets. If $A_n = (0,1/n)$ if n is odd and $A_n = (-1/n, 1/n)$ if n is even, then the sequence $\{A_n, n \geq 1\}$ is neither increasing nor decreasing and hence is not a monotone sequence.

Once we have defined a sequence of sets, the next natural step is to define a limit of a sequence of sets. However, not every sequence of sets has a limit, as in the case of a sequence of real numbers. The concept of a limit of a sequence of sets is parallel to that for a sequence of real numbers and is defined via a limit supremum and a limit infimum. The limit supremum and the limit infimum for a sequence $\{a_n, n \geq 1\}$ of real numbers is defined as follows.

$$\limsup a_n = \inf_{n\geq 1} \sup_{k\geq n} a_k \quad \text{and} \quad \liminf a_n = \sup_{n\geq 1} \inf_{k\geq n} a_k.$$

If $\limsup a_n = \liminf a_n$, then we say that $\lim_{n\to\infty} a_n$ exists and its value is the common value of $\limsup a_n$ and $\liminf a_n$. If $\{a_n, n \geq 1\}$ is a convergent sequence, then $\lim_{n\to\infty} a_n$ measures roughly the size of a_n for large n. $\limsup a_n$ is a measure of how large a_n can be when n is large and $\liminf a_n$ is a measure of how small a_n can be when n is large.

For a sequence of sets, limit supremum and limit infimum are defined analogously.

Definition 1.3.7. *Limit supremum and limit infimum of a sequence of sets: Suppose $\{A_n, n \geq 1\}$ is a sequence of subsets of Ω. Then the limit supremum and the limit infimum of a sequence of sets, denoted by $\limsup A_n$ and $\liminf A_n$ respectively are defined as,*

$$\limsup A_n = \bigcap_{n\geq 1} \bigcup_{k\geq n} A_k \quad \& \quad \liminf A_n = \bigcup_{n\geq 1} \bigcap_{k\geq n} A_k.$$

Limit supremum and limit inferior of a sequence of sets are also known as a limit superior and a limit inferior respectively of a sequence of sets and are also denoted by $\overline{\lim} A_n$ and $\underline{\lim} A_n$ respectively.

Definition 1.3.8. *Limit of a sequence of sets: Suppose $\{A_n, n \geq 1\}$ is a sequence of subsets of Ω. If $\limsup A_n = \liminf A_n$, then we say that limit of a sequence of sets exists and it is the common set. Thus,*

$$\lim_{n\to\infty} A_n = \limsup A_n = \liminf A_n.$$

Following example illustrates these concepts of a limit of a sequence for a non-decreasing sequence.

Example 1.3.3. We find the limit of the sequence $\{A_n, n \geq 1\}$ where $A_n = (a + 1/n, b - 1/n)$. If a and b are such that $a + 1/n > b - 1/n$, then such A_n sets will be $\emptyset$. Observe that $A_n \subseteq A_{n+1}$ $\forall$ $n \geq 1$, thus $\{A_n, n \geq 1\}$ is a non-decreasing sequence. Further $\forall\, n \geq 1$

$$\begin{aligned} A_1 &\subseteq A_2 \subseteq A_3 \subseteq \cdots \subseteq A_n \\ &\Rightarrow A_1 \cup A_2 \cup \cdots \cup A_n = A_n \;\; \& \;\; A_1 \cap A_2 \cap \cdots \cap A_n = A_1. \end{aligned}$$

By the definition of the limit supremum,

$$\limsup A_n = \bigcap_{n\geq 1}\bigcup_{k\geq n} A_k = \bigcap_{n\geq 1} C_n, \quad \text{where} \quad C_n = \bigcup_{k\geq n} A_k.$$

It is to be noted that

$$C_1 = \bigcup_{k\geq 1} A_k = (a,b) \quad C_2 = \bigcup_{k\geq 2} A_k = \bigcup_{k\geq 1} A_k = (a,b)$$

In general, $C_n = \bigcup_{k\geq n} A_k = \bigcup_{k\geq 1} A_k = (a,b) \;\Rightarrow\; \limsup A_n = \bigcap_{n\geq 1} C_n = (a,b).$

Now, by the definition of the limit infimum,

$$\liminf A_n = \bigcup_{n\geq 1}\bigcap_{k\geq n} A_k = \bigcup_{n\geq 1} B_n \quad \text{where} \quad B_n = \bigcap_{k\geq n} A_k.$$

With $A_n = (a+1/n, b-1/n), \forall\; n \geq 1$

$$\begin{aligned}
B_n &= \bigcap_{k\geq n} A_k = (a+1/n, b-1/n)\cap(a+1/(n+1), b-1/(n+1))\cap\cdots \\
&= (a+1/n, b-1/n) = A_n \\
&\Rightarrow \liminf A_n = \bigcup_{n\geq 1} B_n = \bigcup_{n\geq 1} A_n = (a,b) \\
&\Rightarrow \limsup A_n = \liminf A_n = (a,b) \Rightarrow \lim A_n \text{ exists and is } (a,b).
\end{aligned}$$

If $A_n = [a+1/n, b-1/n]$ or $A_n = (a+1/n, b-1/n]$ or $A_n = [a+1/n, b-1/n)$, then we proceed on similar lines and in all the three cases $\lim A_n$ exists and is (a,b). □

Following theorem proves that the limit of a non-decreasing sequence of sets exists and is a countable union. The proof is similar to the solution of Example 1.3.3.

Theorem 1.3.1. *Suppose $\{A_n, n \geq 1\}$ is a non-decreasing sequence of subsets of Ω. Then $\lim_{n\to\infty} A_n$ exists and is $\bigcup_{n\geq 1} A_n$.*

Proof. By the definition of a limit supremum,

$$\limsup A_n = \bigcap_{n\geq 1}\bigcup_{k\geq n} A_k = \bigcap_{n\geq 1} C_n, \quad \text{where} \quad C_n = \bigcup_{k\geq n} A_k.$$

It is given that $\{A_n, n \geq 1\}$ is a non-decreasing sequence, hence $A_n \subseteq A_{n+1}$, for all $n \geq 1$. In particular, $A_1 \subseteq A_2 \Rightarrow A_1 \cup A_2 = A_2$. Further, $A_1 \subseteq A_2 \subseteq$

$A_3 \subseteq \cdots \subseteq A_n$ implies $A_1 \cup A_2 \cup \cdots \cup A_n = A_n$. As a consequence,

$$\begin{aligned} C_2 &= \bigcup_{k\geq 2} A_k = A_2 \cup A_3 \cup \cdots = A_1 \cup A_2 \cup A_3 \cup \cdots = C_1 \\ C_n &= A_n \bigcup \Big(\bigcup_{k\geq n+1} A_k \Big) = \Big(\bigcup_{k\leq n} A_k \Big) \bigcup \Big(\bigcup_{k\geq n+1} A_k \Big) = C_1, \;\; \forall \;\; n \geq 1 \\ &\Rightarrow \;\; \limsup A_n = \bigcap_{n\geq 1} C_n = C_1 = \bigcup_{n\geq 1} A_n \,. \end{aligned}$$

Now, by the definition of a limit infimum,

$$\liminf A_n = \bigcup_{n\geq 1} \bigcap_{k\geq n} A_k = \bigcup_{n\geq 1} B_n \quad \text{where } B_n = \bigcap_{k\geq n} A_k.$$

Again, in view of the non-decreasing nature of the sequence $\{A_n, n \geq 1\}$ we have,

$$\begin{aligned} A_n \subseteq A_{n+1} \subseteq A_{n+2} \subseteq \cdots \quad &\Rightarrow \quad B_n = \bigcap_{k\geq n} A_k = A_n, \;\; \forall \;\; n \geq 1 \\ &\Rightarrow \quad \liminf A_n = \bigcup_{n\geq 1} B_n = \bigcup_{n\geq 1} A_n \\ &\Rightarrow \quad \limsup A_n = \liminf A_n = \bigcup_{n\geq 1} A_n \\ &\Rightarrow \quad \lim_{n\to\infty} A_n \text{ exists and is } \bigcup_{n\geq 1} A_n. \end{aligned}$$

□

The concepts of limit supremum, limit infimum and limit of a sequence of sets are difficult to grasp due to their abstract nature. To overcome this difficulty, it is essential that the theory and computations go hand in hand. Hence, in the present book we have adopted an approach to explain theory along with the computations and visual presentation of the concepts, wherever possible. In the next example, we demonstrate the limit of a non-decreasing sequence of sets, using R.

Example 1.3.4. In Example 1.3.3, we studied the limit of a non-decreasing sequence $\{A_n, n \geq 1\}$, where $A_n = (a + 1/n, b - 1/n)$. In particular, suppose $a = 2$ and $b = 3$. Note that $A_1 = A_2 = \emptyset$. The limit exists and it is $A = (2, 3)$. We use Code 1.3.1 to demonstrate the limit of a sequence $\{(2 + 1/n, 3 - 1/n), n \geq 1\}$ graphically.

Code 1.3.1. `a=2; b=3`

```
nvec=c(10,25,50,75,100,500,750,1000,2500,5000,10000)
colvec=rainbow(length(nvec),start=0.7,end=0.1)
```

```
nlab=paste("n",nvec,sep="=")
plot(c(a-0.3, b+0.3),c(0,2),type="n",xlab="", ylab="",
ylim=c(0,11),
main="Limiting Behaviour of (2+1/n, 3-1/n)",yaxt="n")
for(i in 1:length(nvec))
{n=nvec[i]
text(x=b+0.2,y=i+0.1,labels=nlab[i])
rect(a+1/n,i,b-1/n,i+0.2,col=colvec[i] )
}
abline(v=c(2,3),col=1)
# function "rect" constructs a rectangle
```

Figure 1.2 clearly depicts the non-decreasing nature of the sequence $\{A_n, n \geq 1\}$ and we note that for $n \geq 500$, A_n is close to $A = (2, 3)$. □

In the next example, we find the limit of a non-increasing sequence of sets.

Example 1.3.5. We find the limit of the sequence $\{A_n, n \geq 1\}$ where, (i) $A_n = (a - 1/n, a + 1/n)$, (ii) $A_n = [a - 1/n, a + 1/n]$, (iii)$A_n = (a - 1/n, a +$

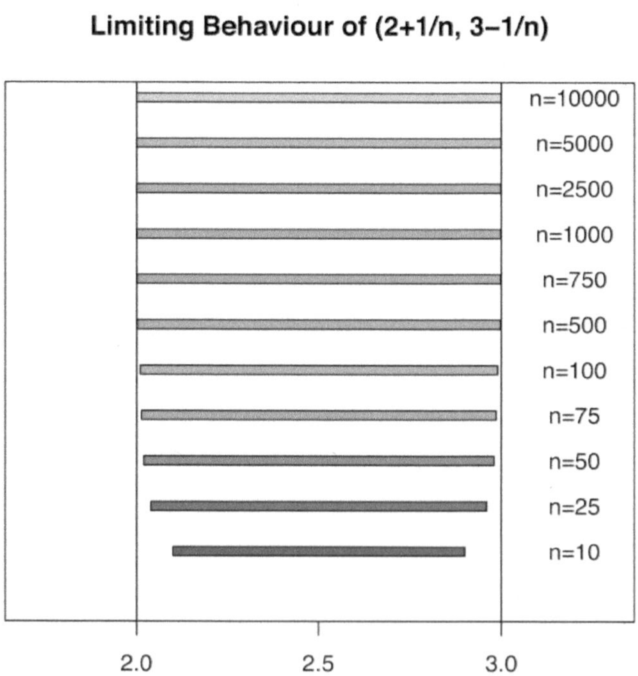

FIGURE 1.2
Limit of a Non-decreasing Sequence

$1/n]$ and (iv) $A_n = [a-1/n, a+1/n)$. By the definition of the limit supremum,

$$\limsup A_n = \bigcap_{n\geq 1}\bigcup_{k\geq n} A_k = \bigcap_{n\geq 1} C_n, \quad \text{where} \quad C_n = \bigcup_{k\geq n} A_k.$$

With $A_n = (a-1/n, a+1/n), C_1 = (a-1, a+1) = A_1, C_2 = (a-1/2, a+1/2) = A_2$ and continuing in this manner $C_n = A_n, \ \forall \ n \geq 1$. Hence, $\limsup A_n = \bigcap_{n\geq 1} A_n = \{a\}$. Now, by the definition of the limit infimum,

$$\liminf A_n = \bigcup_{n\geq 1}\bigcap_{k\geq n} A_k = \bigcup_{n\geq 1} B_n \quad \text{where} \quad B_n = \bigcap_{k\geq n} A_k.$$

With $A_n = (a - 1/n, a + 1/n), \forall \ n \geq 1$,

$$B_n = \bigcap_{k\geq n} A_k = (a - 1/n, a + 1/n) \cap (a - 1/(n+1), a + 1/(n+1)) \cap \cdots = \{a\}.$$

$$\begin{aligned} \text{Hence,} \quad \liminf A_n &= \bigcup_{n\geq 1} B_n = \{a\} \Rightarrow \limsup A_n = \liminf A_n = \{a\} \\ &\Rightarrow \lim A_n \ \text{ exists and is } \{a\}. \end{aligned}$$

For the remaining three sequences, we proceed on similar lines and in all the three cases $\lim A_n$ exists and is $\{a\}$. □

Example 1.3.6. Suppose $\Omega = (0,1)$ and $A_n = (0, 1/n), \ n \geq 1$. By the definition of the limit supremum,

$$\limsup A_n = \bigcap_{n\geq 1}\bigcup_{k\geq n} A_k = \bigcap_{n\geq 1} C_n, \quad \text{where} \quad C_n = \bigcup_{k\geq n} A_k.$$

With $A_n = (0, 1/n), C_1 = (0,1) = A_1, C_2 = (0, 1/2) = A_2$ and continuing in this manner $C_n = (0, 1/n) = A_n, \ \forall \ n \geq 1$. Hence, $\limsup A_n = \bigcap_{n\geq 1} A_n = \emptyset$.

Now, by the definition of limit infimum,

$$\liminf A_n = \bigcup_{n\geq 1}\bigcap_{k\geq n} A_k = \bigcup_{n\geq 1} B_n \quad \text{where} \quad B_n = \bigcap_{k\geq n} A_k.$$

With $A_n = (0, 1/n), B_1 = \emptyset, B_2 = \emptyset$, similarly $B_n = \emptyset, \ \forall \ n \geq 1$ and hence $\liminf A_n = \bigcup_{n\geq 1} B_n = \emptyset$. Thus, $\limsup A_n = \liminf A_n = \emptyset$ hence $\lim A_n$ exists and is $\emptyset$. □

Example 1.3.7. Suppose $\Omega = [0,1]$ and $D_n = [0, 1/n], n \geq 1$. By the definition of the limit supremum,

$$\limsup D_n = \bigcap_{n\geq 1}\bigcup_{k\geq n} D_k = \bigcap_{n\geq 1} C_n, \quad \text{where} \quad C_n = \bigcup_{k\geq n} D_k.$$

On similar lines as in the previous example, we get $C_n = D_n$ and $\limsup D_n = \bigcap_{n\geq 1} D_n = \{0\}$. Further, $B_n = \{0\}, \quad \forall \; n \geq 1$. Hence,

$$\liminf A_n = \bigcup_{n\geq 1} B_n = \{0\} \quad \Rightarrow \quad \limsup D_n = \liminf D_n = \{0\}$$

and hence $\lim D_n$ exists and is $\{0\}$. □

It is to be noted that in all the above three examples, we have non-increasing sequence of subsets of Ω and their limit is the countable intersection. It is true in general for a non-increasing sequence. We prove it in the next theorem.

Theorem 1.3.2. *Suppose $\{A_n, n \geq 1\}$ is a non-increasing sequence of subsets of Ω. Then $\lim_{n\to\infty} A_n$ exists and is $\bigcap_{n\geq 1} A_n$.*

Proof. By the definition of a limit supremum,

$$\limsup A_n = \bigcap_{n\geq 1} \bigcup_{k\geq n} A_k = \bigcap_{n\geq 1} C_n, \quad \text{where} \quad C_n = \bigcup_{k\geq n} A_k.$$

It is given that $\{A_n, n \geq 1\}$ is a non-increasing sequence, hence $A_n \supseteq A_{n+1}$, for all $n \geq 1$, that is, $A_1 \supseteq A_2 \supseteq A_3 \supseteq \cdots$. As a consequence,

$$\begin{aligned} C_1 &= A_1 \cup A_2 \cup A_3 \cup \cdots = A_1, \quad C_2 = A_2 \;\; \& \;\; C_n = A_n, \quad \forall \; n \geq 1 \\ \Rightarrow \quad & \limsup A_n = \bigcap_{n\geq 1} C_n = \bigcap_{n\geq 1} A_n. \end{aligned}$$

Now, by the definition of a limit infimum,

$$\liminf A_n = \bigcup_{n\geq 1} \bigcap_{k\geq n} A_k = \bigcup_{n\geq 1} B_n \quad \text{where} \quad B_n = \bigcap_{k\geq n} A_k.$$

Since $\{A_n, n \geq 1\}$ is a non-increasing sequence, we get $A_1 \supseteq A_2 \supseteq A_3 \supseteq \cdots \supseteq A_n$ which implies $\bigcap_{k\leq n} A_k = A_n, \quad \forall \; n \geq 1$. Hence,

$$\begin{aligned} B_n &= A_n \bigcap \Big(\bigcap_{k\geq n+1} A_k \Big) = \Big(\bigcap_{k\leq n} A_k \Big) \bigcap \Big(\bigcap_{k\geq n+1} A_k \Big) \\ &= \bigcap_{n\geq 1} A_n = B_1, \quad \Rightarrow \quad B_n = B_1, \quad \forall \; n \geq 1 \\ &\Rightarrow \liminf A_n = \bigcup_{n\geq 1} B_n = B_1 = \bigcap_{n\geq 1} A_n \\ &\Rightarrow \limsup A_n = \liminf A_n = \bigcap_{n\geq 1} A_n \Rightarrow \lim_{n\to\infty} A_n \;\; \text{exists and is} \;\; \bigcap_{n\geq 1} A_n. \end{aligned}$$

□

Example 1.3.8. Suppose $\{A_n, n \geq 1\}$ is a sequence of sets where, $A_n = [a, b + 1/n]$, where a and b are such that $a < b + 1/n \;\; \forall \;\; n$. Then $\{A_n, n \geq 1\}$ is a decreasing sequence of sets and hence limit exists and it is $\bigcap_{n\geq 1} A_n = [a, b]$. □

Theorem 1.3.2 can be proved in a different way. It uses Theorem 1.3.1 and the relation among $\limsup A_n$, $\limsup A_n^c$, $\liminf A_n$ and $\liminf A_n^c$. To discuss this approach, we first derive the relations among $\limsup A_n$, $\limsup A_n^c$, $\liminf A_n$ and $\liminf A_n^c$.

Theorem 1.3.3. *Suppose $\{A_n, n \geq 1\}$ is a sequence of subsets of Ω. Then*

$$\limsup A_n^c = (\liminf A_n)^c \quad \& \quad \liminf A_n^c = (\limsup A_n)^c.$$

Proof. By the definition of a limit supremum and a limit infimum we have,

$$\limsup A_n^c = \bigcap_{n\geq 1} \bigcup_{k\geq n} A_k^c = \bigcap_{n\geq 1} \Big(\bigcap_{k\geq n} A_k \Big)^c = \Big(\bigcup_{n\geq 1} \bigcap_{k\geq n} A_k \Big)^c = \Big(\liminf A_n \Big)^c.$$

Similarly,

$$\liminf A_n^c = \bigcup_{n\geq 1} \bigcap_{k\geq n} A_k^c = \bigcup_{n\geq 1} \Big(\bigcup_{k\geq n} A_k \Big)^c = \Big(\bigcap_{n\geq 1} \bigcup_{k\geq n} A_k \Big)^c = \Big(\limsup A_n \Big)^c.$$

□

Following theorem is the same as Theorem 1.3.2, but the proof is given in a different way.

Theorem 1.3.4. *Suppose $\{A_n, n \geq 1\}$ is a non-increasing sequence of subsets of Ω. Then $\lim_{n\to\infty} A_n$ exists and is $\bigcap_{n\geq 1} A_n$.*

Proof. It is given that $\{A_n, n \geq 1\}$ is a non-increasing sequence of sets, hence

$$A_n \supseteq A_{n+1}, \;\; \forall \;\; n \geq 1 \;\; \Rightarrow \;\; A_n^c \subseteq A_{n+1}^c, \;\; \forall \;\; n \geq 1.$$

Thus, $\{A_n^c, n \geq 1\}$ is a non-decreasing sequence. Hence by Theorem 1.3.1, $\lim_{n\to\infty} A_n^c$ exists and it is $\bigcup_{n\geq 1} A_n^c = (\bigcap_{n\geq 1} A_n)^c$. Since, $\lim_{n\to\infty} A_n^c$ exists, we have

$$\begin{aligned} \limsup A_n^c &= \liminf A_n^c = (\bigcap_{n\geq 1} A_n)^c \\ \iff \quad (\liminf A_n)^c &= (\limsup A_n)^c = (\bigcap_{n\geq 1} A_n)^c \\ \iff \quad \liminf A_n &= \limsup A_n = \bigcap_{n\geq 1} A_n. \end{aligned}$$

Thus, $\lim_{n\to\infty} A_n$ exists and is $\bigcap_{n\geq 1} A_n$. □

Using the relations among $\limsup A_n$, $\limsup A_n^c$, $\liminf A_n$ and $\liminf A_n^c$, the next theorem proves that if a sequence $\{A_n, n \geq 1\}$ is a convergent, then so is $\{A_n^c, n \geq 1\}$.

Theorem 1.3.5. *Suppose $\{A_n, n \geq 1\}$ is a convergent sequence of subsets of Ω with limit A. Then $\{A_n^c, n \geq 1\}$ is also a convergent sequence with limit A^c.*

Proof. Since $\{A_n, n \geq 1\}$ is a convergent sequence with limit A, we have,

$$\begin{aligned} \limsup A_n &= \liminf A_n = \lim A_n = A \\ &\Rightarrow (\limsup A_n)^c = (\liminf A_n)^c = (\lim A_n)^c = A^c \\ &\Rightarrow \liminf A_n^c = \limsup A_n^c = A^c \text{ by Theorem 1.3.3} \\ &\Rightarrow \lim A_n^c \text{ exists and is } A^c. \end{aligned}$$

Hence, if $A_n \to A$ then $A_n^c \to A^c$. □

Example 1.3.9. In Example 1.3.5, we discussed the limit of a non-increasing sequence $\{A_n, n \geq 1\}$, where $A_n = (a - 1/n, a + 1/n)$. We have noted that the limit exists and it is $A = \{a\}$. In particular, suppose $a = 2$. We use Code 1.3.2 to present visually the limit of a sequence $\{(2 - 1/n, 2 + 1/n), n \geq 1\}$.

Code 1.3.2.
```
a=2; colvec=rainbow(11,start=0.7, end=0.1)
nvec=c(10,25,50,75,100,500,750,1000,2500,5000,10000)
nlab=paste("n",nvec,sep="=")
for(i in 1:length(nvec))
{
n=nvec[i]
if(i==1) plot(c(a-0.15, a+0.15),c(0, 2),type="n",xlab="",
 ylab="",
ylim=c(0,11),main="Limiting Behaviour of (2-1/n,2+1/n)",yaxt="n")
text(x=a+0.12,y=i+0.1,labels=nlab[i])
rect(a-1/n,i,a+1/n,i+0.2,col=colvec[i])
}
```

From Figure 1.3, we observe that $\{A_n, n \geq 1\}$ is a non-increasing sequence and for $n = 10000$, A_n is almost $A = \{2\}$. □

From Theorem 1.3.1 and Theorem 1.3.2, we note that a monotone sequence of sets is convergent. However, the converse is not true, that is, every convergent sequence need not be monotone, as is clear from the following example.

Example 1.3.10. Suppose $\{A_n, n \geq 1\}$ is a sequence of sets such that $A_n \neq \emptyset$ for at least one $n \geq 1$ and $A_n \cap A_m = \emptyset, \ \forall \ n \neq m$. We examine whether the sequence is monotone and convergent. In the given sequence $\{A_n, n \geq 1\}$ all

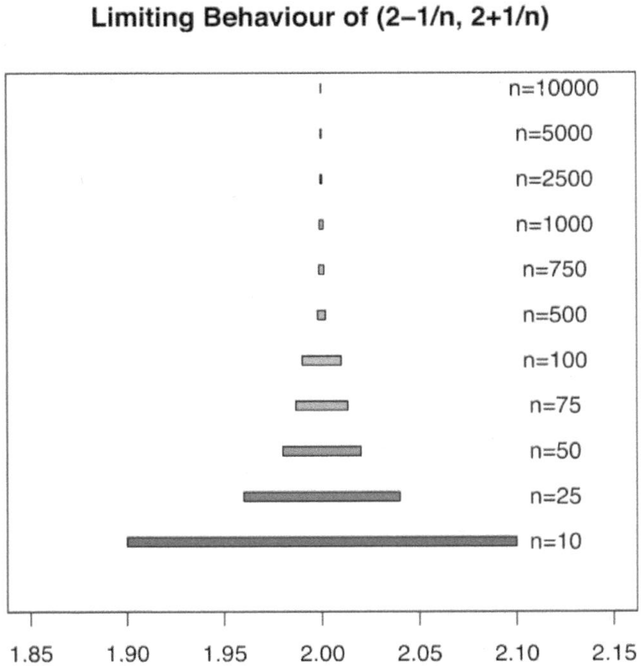

FIGURE 1.3
Limit of a Non-increasing Sequence

the sets are disjoint and hence it is not a monotone sequence. To examine if it is convergent, note that $\liminf A_n = \bigcup_{n\geq 1}\bigcap_{k\geq n} A_k = \bigcup_{n\geq 1}\emptyset = \emptyset$. Further,

$$\limsup A_n = \bigcap_{n\geq 1}\bigcup_{k\geq n} A_k = \bigcap_{n\geq 1} C_n, \quad \text{where} \quad C_n = \bigcup_{k\geq n} A_k.$$

Since $\{A_n, n \geq 1\}$ is a sequence of disjoint sets, $\omega \in \Omega$ is only in one of the sets of the sequence. Suppose $\omega \in A_{k_0}$ and $\omega \notin A_k$ for $k \neq k_0$. Hence,

$$\begin{aligned}
\omega \in C_1 &= \bigcup_{k\geq 1} A_k, \quad \omega \in C_2 = \bigcup_{k\geq 2} A_k, \quad \ldots, \quad \omega \in C_{k_0} = \bigcup_{k\geq k_0} A_k \\
\text{but } \omega &\notin C_{k_0+1} = \bigcup_{k\geq k_0+1} A_k \\
&\Rightarrow \omega \in C_n, \ \forall \ n \leq k_0 \ \text{ but } \ \omega \notin C_n, \ \forall \ n > k_0 \\
&\Rightarrow \limsup A_n = \bigcap_{n\geq 1} C_n = \emptyset = \liminf A_n \\
&\Rightarrow \lim_{n\to\infty} A_n \ \text{ exists and is } \ \emptyset.
\end{aligned}$$

□

Example 1.3.11. Suppose $\{A_n, n \geq 1\}$ is a sequence of sets such that for $n \geq 1$, $A_n = (a - 1/n, b - 1/n)$, $a < b$. We examine whether the sequence is monotone and convergent. Observe that $A_1 = (a-1, b-1)$ and $A_2 = (a-1/2, b-1/2)$, neither $A_1 \subseteq A_2$ nor $A_2 \subseteq A_1$, which implies that $\{A_n, n \geq 1\}$ is not a monotone sequence. To examine whether it is convergent, we find $\limsup A_n$ and $\liminf A_n$. Note that

$$C_n = \bigcup_{k \geq n} A_k = (a - 1/n, b) \Rightarrow \limsup A_n = \bigcap_{n \geq 1} C_n = (a, b).$$

Similarly, $$B_n = \bigcap_{k \geq n} A_k = (a, b - 1/n) \Rightarrow \liminf A_n = \bigcup_{n \geq 1} C_n = (a, b)$$

$$\Rightarrow \limsup A_n = \liminf A_n = (a, b) \Rightarrow \lim A_n = (a, b).$$

Thus, $\{A_n, n \geq 1\}$ is a non-monotone sequence, but it is convergent. □

In the next example using R, we plot a non-monotone convergent sequence.

Example 1.3.12. Suppose $\{A_n, n \geq 1\}$ is a sequence of sets where $A_n = (2 - 1/n, 3 - 1/n)$. From Example 1.3.11, it is clear that it is not a monotone sequence but is convergent with limit $A = (2, 3)$. We use Code 1.3.3 to plot the limit of this sequence.

Code 1.3.3.
```
a=2; b=3
nvec=c(10,25,50,75,100,500,750,1000,2500,5000,10000)
colvec=rainbow(length(nvec),start=0.7,end=0.1)
nlab=paste("n",nvec,sep="=")
plot(c(a-0.5, b+0.5),c(0,2),type="n",xlab="",ylab="",
ylim=c(0,11),main="Limiting Behaviour of(2-1/n,3-1/n)",yaxt="n")
for(i in 1:length(nvec))
{n=nvec[i]
text(x=b+0.2,y=i+0.1,labels=nlab[i])
rect(a-1/n,i,b-1/n,i+0.2, col=colvec[i] )
}
abline(v=c(2,3),col=1)
```

From Figure 1.4, we note that the sequence $\{A_n, n \geq 1\}$ is non-monotone, but for $n \geq 500$, A_n is close to $A = (2, 3)$. □

We now discuss some more examples to illustrate the concept of limit of a non-monotone sequence of sets.

Example 1.3.13. Suppose $\{A_n, n \geq 1\}$ is a sequence of sets where,

$$A_n = \begin{cases} (3 - 1/n, 3 + 1/n), & \text{if } n \text{ is odd}, \\ \emptyset, & \text{if } n \text{ is even.} \end{cases}$$

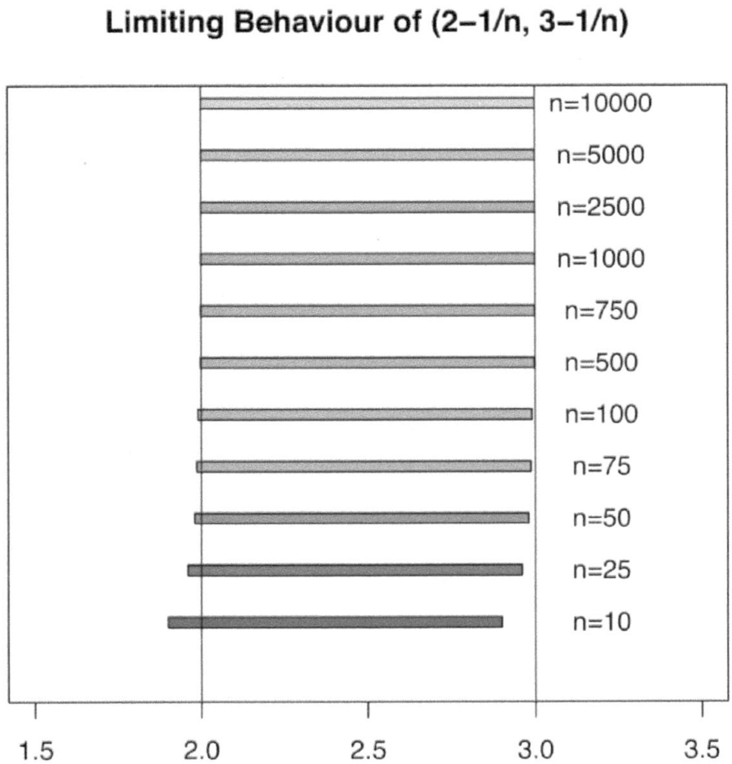

FIGURE 1.4
Non-monotone Convergent Sequence

With $A_n = \emptyset$ for even value of n we have,

$$\limsup A_n = \bigcap_{n\geq 1}\bigcup_{k\geq n} A_k = \bigcap_{n\geq 1}(3-1/n, 3+1/n) \Rightarrow \limsup A_n = \{3\}$$

$$\text{Now,} \quad \liminf A_n = \bigcup_{n\geq 1}\bigcap_{k\geq n} A_k = \bigcup_{n\geq 1} B_n = \emptyset \;\Rightarrow\; \liminf A_n = \emptyset$$

$$\Rightarrow \lim A_n \quad \text{does not exist.}$$

If we define $A_n = (3 - 1/n, 3)$ if n is odd and $A_n = \emptyset$, if n is even, then $C_n = A_n$ and $\limsup A_n = \bigcap_{n\geq 1} C_n = \emptyset$. Thus limit exists in this case and it is an empty set. □

Example 1.3.14. Suppose $\{A_n, n \geq 1\}$ is a sequence of sets where

$$A_{2n} = (0, 1/2n), \quad n \geq 1, \quad A_{2n+1} = [-1, 1/(2n+1)], \quad n \geq 0.$$

We examine whether $\lim A_n$ exists. By the definition,

$$\limsup A_n = \bigcap_{n\geq 1}\bigcup_{k\geq n} A_k = \bigcap_{n\geq 1} C_n \quad \text{where} \quad C_n = \bigcup_{k\geq n} A_k$$

$$\begin{aligned}
\text{For } n = 2m, C_{2m} &= \bigcup_{k\geq 2m} A_k \\
&= (0, 1/(2m)) \cup [-1, 1/(2m+1)] \cup (0, 1/(2m+2)) \cdots \\
&= [-1, 1/(2m)) \\
\text{For } n = 2m-1, C_{2m-1} &= \bigcup_{k\geq 2m-1} A_k = [-1, 1/(2m-1)] \cup (0, 1/(2m)) \cdots \\
&= [-1, 1/(2m-1)] \\
\Rightarrow \limsup A_n &= \bigcap_{n\geq 1} C_n = [-1, 1] \cap [-1, 1/2) \cap [-1, 1/3] \cdots = [-1, 0].
\end{aligned}$$

$$\text{Now,} \quad \liminf A_n = \bigcup_{n\geq 1}\bigcap_{k\geq n} A_k = \bigcup_{n\geq 1} B_n \quad \text{where} \quad B_n = \bigcap_{k\geq n} A_k$$

$$\begin{aligned}
\text{For } n = 2m, B_{2m} &= \bigcap_{k\geq 2m} A_k \\
&= (0, 1/(2m)) \cap [-1, 1/(2m+1)] \cap (0, 1/(2m+2)) \cdots \\
&= \emptyset \\
\text{For } n = 2m-1, B_{2m-1} &= \bigcap_{k\geq 2m-1} A_k = [-1, 1/(2m-1)] \cap (0, 1/(2m)) \cdots \\
&= \emptyset \\
&\Rightarrow \liminf A_n = \emptyset \Rightarrow \lim A_n \text{ does not exist.} \quad \square
\end{aligned}$$

Example 1.3.15. In this example we consider a sequence $\{A_n, n \geq 1\}$, where

$$A_n = \begin{cases} (1 + 1/n, 3 - 1/n), & \text{if } n \text{ is odd }, \\ (2 + 1/n, 4 - 1/n), & \text{if } n \text{ is even.} \end{cases}$$

Proceeding as in Example 1.3.14, it follows that $\limsup A_n = (1, 4)$ and $\liminf A_n = (2, 3)$ and hence it is not a convergent sequence. Using Code 1.3.4, we plot sets A_n for some odd and even values of n, which demonstrates the limiting behaviour of the sequence.

Code 1.3.4.
```
a=1; b=3; c=2; d=4
nvec=c(10,25,50,75,100,505,750,1005,2500,5005,10000)
colvec=rainbow(length(nvec),start=0.7,end=0.1)
nlab=paste("n",nvec,sep="=")
plot(c(a-0.5, d+0.8),c(0,2),type="n", xlab="", ylab="",
```

```
ylim=c(0,11),main="Limiting Behaviour of An",yaxt="n")
for(i in 1:length(nvec))
{
n=nvec[i]
text(x=d+0.5,y=i+0.1,labels=nlab[i])
if(n%%2!=0) rect(a+1/n,i,b-1/n,i+0.2,col=colvec[i])
if(n%%2==0) rect(c+1/n,i,d-1/n,i+0.2,col=colvec[i])
}
abline(v=c(2,3),col=1); abline(v=c(1,4),col=2)
```

From Figure 1.5, it is clear that the sequence is not convergent since for odd and even values of n, sets A_n are different even for large n. With the Code 1.3.5, we plot A_n for $n = 10,000$ to $n = 15,005$.

Code 1.3.5. `n=10000; a=1;b=3;c=2;d=4`

```
nvec=n+c(0,10,25,50,75,100,505,750,1005,2500,5005)
colvec=rainbow(length(nvec),start=0.7,end=0.1)
nlab=paste("n",nvec,sep="=")
plot(c(a-0.5, d+1),c(0,2),type="n", xlab="",
```

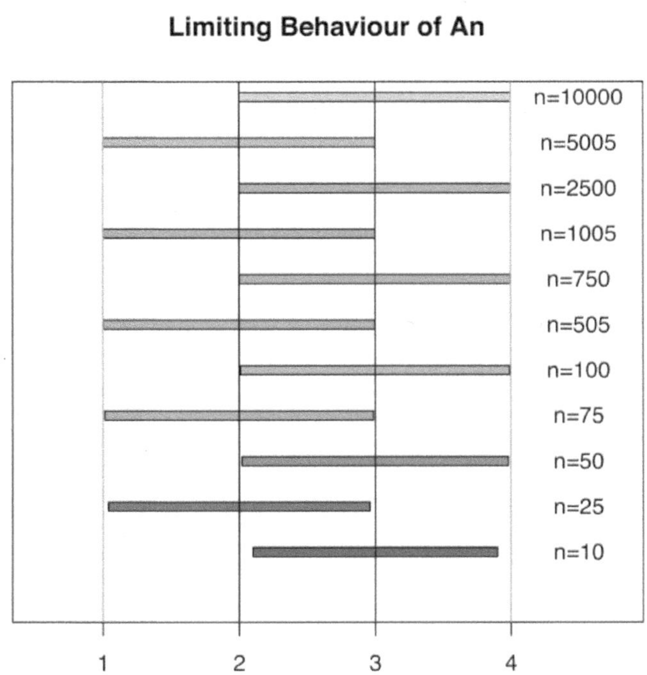

FIGURE 1.5
Non-convergent Sequence

```
ylab="",ylim=c(0,11),
xlim=c(0,5),main="Limsup and Liminf An",yaxt="n")
for(i in 1:length(nvec))
{
n=nvec[i]
text(x=d+0.6,y=i+0.1,labels=nlab[i])
if(n%%2!=0) rect(a+1/n,i,b-1/n,i+0.2,col=colvec[i])
if(n%%2==0) rect(c+1/n,i,d-1/n,i+0.2,col=colvec[i])
}
abline(v=c(2,3),col=1); abline(v=c(1,4),col=2)
```

From Figure 1.6, note that if we take the union of all A_n sets after some n_0, say $n_0 = 10000$, then $C_n = (1,4)$ for all $n \geq n_0$. The countable intersection of C_n is then $(1,4)$ and hence $\limsup A_n = (1,4)$. Similarly, if we take the intersection of all A_n sets after some n_0, then $B_n = (2,3)$ for all $n \geq n_0$. The countable union of B_n is then $(2,3)$ and hence $\liminf A_n = (2,3)$. □

We now investigate the concepts of $\limsup A_n$ and $\liminf A_n$ in more detail and discuss their interpretation. We have defined $\limsup A_n$ as

$$\limsup A_n = \bigcap_{n\geq 1} \bigcup_{k\geq n} A_k = \bigcap_{n\geq 1} C_n, \quad \text{where} \quad C_n = \bigcup_{k\geq n} A_k.$$

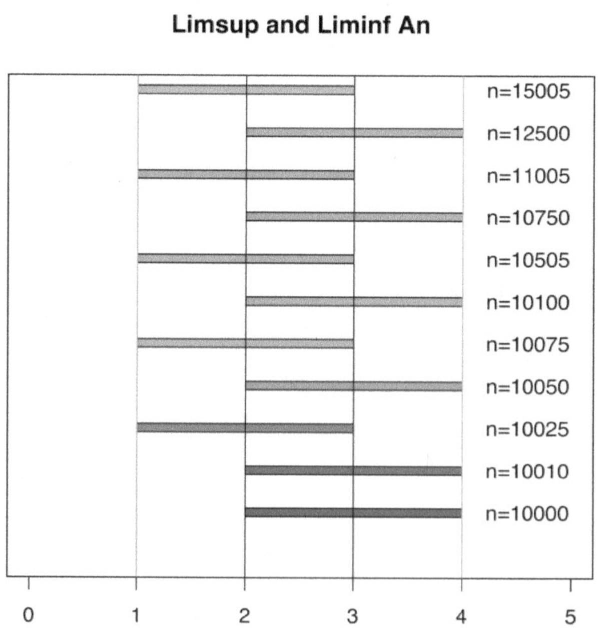

FIGURE 1.6
Limit Supremum and Limit Infimum

C_n is known as $\sup_{k\geq n} A_k$. Now observe that

$$C_n = \bigcup_{k\geq n} A_k = A_n \bigcup C_{n+1} \quad \Rightarrow \quad C_n \supseteq C_{n+1},$$

that is, $\{C_n, n \geq 1\}$ is a non-increasing sequence of sets and hence by Theorem 1.3.2 is convergent with limit $\bigcap_{n\geq 1} C_n$. Thus,

$$\limsup A_n = \bigcap_{n\geq 1} C_n = \lim_{n\to\infty} C_n = \lim_{n\to\infty} \sup_{k\geq n} A_k.$$

Further, $\limsup A_n = \bigcap\limits_{n\geq 1} C_n$ can be interpreted as follows.

$$\begin{aligned}
\omega \in \limsup A_n \quad &\Rightarrow \quad \omega \in \bigcap_{n\geq 1} C_n \Rightarrow \omega \in C_n \ \ \forall \ \ n \geq 1 \\
&\Rightarrow \quad \omega \in \bigcup_{k\geq n} A_k, \ \ \forall \ \ n \geq 1 \\
&\Rightarrow \quad \omega \in A_k, \ \ \text{for at least one } \ k \geq n \ \text{ and } \ \forall \ \ n \geq 1 \\
&\Rightarrow \quad \omega \in A_n \ \ \text{for infinitely many values of } \ n \geq 1.
\end{aligned}$$

Thus, $\limsup A_n$ is a set of points which belong to infinitely many sets of the sequence $\{A_n, n \geq 1\}$. $\liminf A_n$ can also be expressed and interpreted in a similar way. We have defined $\liminf A_n$ as

$$\liminf A_n = \bigcup_{n\geq 1} \bigcap_{k\geq n} A_k = \bigcup_{n\geq 1} B_n \quad \text{where } \ B_n = \bigcap_{k\geq n} A_k.$$

B_n is known as $\inf_{k\geq n} A_k$. Now,

$$B_n = \bigcap_{k\geq n} A_k = A_n \bigcap B_{n+1} \quad \Rightarrow \quad B_n \subseteq B_{n+1},$$

that is, $\{B_n, n \geq 1\}$ is a non-decreasing sequence of sets and hence by Theorem 1.3.1 is convergent with limit $\bigcup_{n\geq 1} B_n$. Thus,

$$\liminf A_n = \bigcup_{n\geq 1} B_n = \lim_{n\to\infty} B_n = \lim_{n\to\infty} \inf_{k\geq n} A_k.$$

Now,

$$\begin{aligned}
\omega \in \liminf A_n \quad &\Rightarrow \quad \omega \in \lim_{n\to\infty} B_n = \bigcup_{n\geq 1} B_n \Rightarrow \omega \in B_n \ \ \text{for at least one } n \geq 1 \\
&\Rightarrow \quad \omega \in \bigcap_{k\geq n} A_k, \ \ \text{for at least one } n \geq 1 \\
&\Rightarrow \quad \omega \in A_k, \ \ \forall \ \ k \geq n \ \text{ and for at least one } \ n \geq 1 \\
&\Rightarrow \quad \omega \in A_n \ \ \text{for all but finitely many values of } n \geq 1.
\end{aligned}$$

Thus, $\liminf A_n$ is a set of points which belong to all but finitely many sets of the sequence $\{A_n, n \geq 1\}$. Note that,

$$\limsup A_n = \inf_{n\geq 1} \sup_{k\geq n} A_k \quad \& \quad \liminf A_n = \sup_{n\geq 1} \inf_{k\geq n} A_k.$$

Thus, the definitions of limit supremum and the limit infimum for a sequence of sets is similar to those for a sequence of real numbers.

As discussed above, $\limsup A_n$ is a set of points which belong to infinitely many sets of the sequence $\{A_n, n \geq 1\}$ and $\liminf A_n$ is a set of points which belong to all but finitely many sets of the sequence $\{A_n, n \geq 1\}$. It indicates that $\liminf A_n \subseteq \limsup A_n$. We prove it in the following theorem.

Theorem 1.3.6. *Suppose $\{A_n, n \geq 1\}$ is a sequence of subsets of Ω. Then*

$$\liminf A_n \subseteq \limsup A_n \ .$$

Proof. By the definitions of $\limsup A_n$ and $\liminf A_n$ we have,

$$\begin{aligned}
\omega \in \liminf A_n \quad &\Rightarrow \quad \omega \in \bigcup_{n\geq 1} \bigcap_{k\geq n} A_k \\
&\Rightarrow \quad \omega \in \bigcap_{k\geq n} A_k, \quad \text{for at least one } n \geq 1, \quad \text{say } n_0 \\
&\Rightarrow \quad \omega \in \bigcap_{k\geq n_0} A_k \quad \Rightarrow \quad \omega \in A_k, \quad \forall \ k \geq n_0 \\
&\Rightarrow \quad \omega \in \bigcup_{k\geq n_0} A_k \quad \Rightarrow \quad \omega \in \bigcup_{k\geq n} A_k \ \ \forall \ \ n \geq 1 \\
&\Rightarrow \quad \omega \in \bigcap_{n\geq 1} \bigcup_{k\geq n} A_k \quad \Rightarrow \quad \omega \in \limsup A_n.
\end{aligned}$$

Thus, $\liminf A_n \subseteq \limsup A_n$. □

We have discussed the interpretation of $\limsup A_n$ and $\liminf A_n$. We illustrate it using R code. A set $\limsup A_n$ contains those points which occur in infinitely many A_n sets. Hence, to check whether, say x belongs to $\limsup A_n$, we adopt the following approach. As per the definition of lim sup, we know that lim sup is a collection of those elements of x which occur in infinitely many A_n sets. Hence, for any x in $\limsup A_n$, if we define a set $T(x)$ as follows,

$$T(x) = \{n \in \mathbb{N} | x \in A_n\},$$

$T(x)$ would be a countably infinite subset of $\mathbb{N}$ and hence an unbounded set. Therefore, any natural number n_0 cannot be an upper bound for $T(x)$. That is, for any natural number n_0, we would find some other natural number $n > n_0$ such that $n \in T(x)$. Therefore, to conclude that x belongs to infinitely may A_ns, given any natural number n_0, we should be able to find at least one $n > n_0$ such that $x \in A_n$. In the next example, we find n for given values of n_0 and x using R. One may try different values of x between $(1, 4)$ and n_0.

Example 1.3.16. In Example 1.3.15, if we consider two separate subsequences corresponding to odd and even values of n respectively, we can see that both these subsequences are increasing. Hence, to prove that x is in infinitely many A_ns, it is enough to find one A_n containing x. It would automatically be in every alternate A_n beyond this n. To find one such n we first locate whether x is closer to 1 or 4. If x is closer to 1, let $\epsilon = x - 1$. Then, we need to find odd n such that $1/n < \epsilon$, so that $1 + 1/n < x$ and hence, $x \in A_n$ for that n. However, in general, we also want to make sure that this n is larger than n_0. Hence we take maximum over the two numbers. If x is closer to 4, we let $\epsilon = 4 - x$, we find even value of n such that $4 - x > 1/n$ and again take $\max\{n, n_0\}$. We use Code 1.3.6.

Code 1.3.6.
```
x=3.1; n0=20000; z=min(c(x-1,4-x)); a=1;b=3;c=2;d=4
{
n1=ifelse(ceiling(1/z)%%2==1,ceiling(1/z),ceiling(1/z)+1)
n2=ifelse(n0%%2==1,n0,n0+1)
n=max(n1,n2)
}
n
nvec=n+c(0,10,25,50,75,100,505,750,1005,2500,5005,10000)
colvec=rainbow(length(nvec),start=0.7,end=0.1)
nlab=paste("n",nvec,sep="=")
plot(c(a-0.5, d+0.5),c(0,2),type="n",xlab="",ylab="",
ylim=c(0,11),xlim=c(0,5),main="Limsup An",yaxt="n")
for(i in 1:length(nvec))
{
n=nvec[i]
text(x=d+0.8,y=i+0.1,labels=nlab[i])
if(n%%2!=0) rect(a+1/n,i,b-1/n,i+0.2,col=colvec[i])
if(n%%2==0) rect(c+1/n,i,d-1/n,i+0.2,col=colvec[i])
}
abline(v=x,col=1,lty=3,lwd=2); axis(1,at=x,labels="x")
```

We observe from Figure 1.7 that $x = 3.1$ is present in all A_n with even values of $n > 20000$. Thus $x \in A_n$ for infinitely many n.

By the definition, $\liminf A_n$ contains those points which occur in all but finitely many A_ns. From the graph in Figure 1.6, it seems that $(2, 3)$ is the $\liminf$. Hence, to check whether, say x belongs to $\liminf A_n$, we need the following approach. Given x, we should be able to find n_0 such that, $x \in A_n$ for every $n \geq n_0$. Hence, if x is closer to 3 than 2, we need to find n such that $x < 3 - 1/n$, i.e., $n > 1/(3 - x)$. Similarly, if x is closer to 2, we will choose $n > 1/(x - 2)$. We use Code 1.3.7 to find such n_0 and draw Figure 1.8.

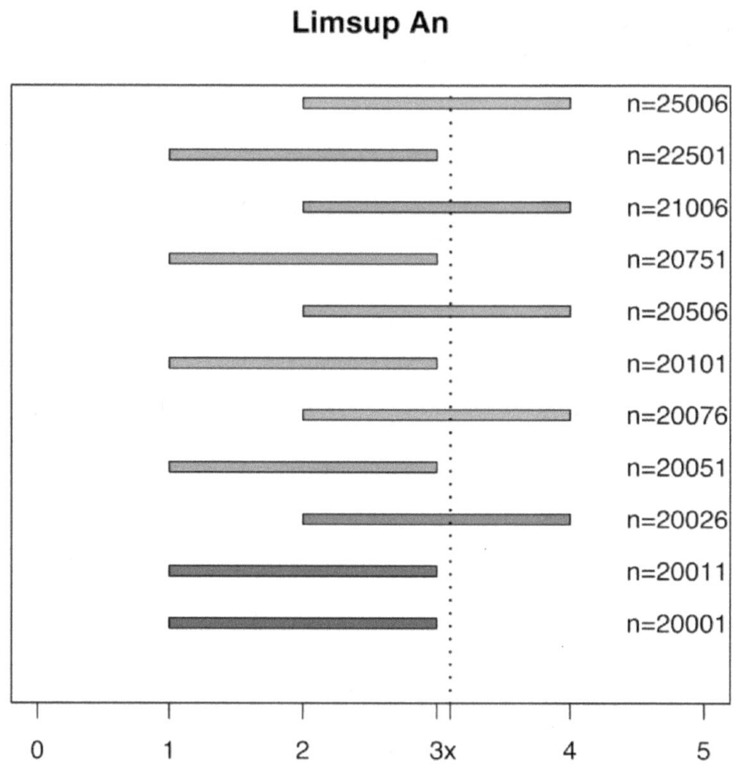

FIGURE 1.7
Limit Supremum

Code 1.3.7.
```
x=2.3; z=min(c(x-2,3-x)); nt=ceiling(1/z); n=nt; n
a=1;b=3;c=2;d=4
nvec=n+c(0,10,25,50,75,100,505,750,1005,2500,5005,10000)
colvec=rainbow(length(nvec),start=0.7,end=0.1)
nlab=paste("n",nvec,sep="=")
plot(c(a-0.5, d+0.5), c(0,2),type="n", xlab="",ylab="",
ylim=c(0,11),xlim=c(0,5),main="Liminf An",yaxt="n")
for(i in 1:length(nvec))
{n=nvec[i]
text(x=d+0.8,y=i+0.1,labels=nlab[i])
if(n%%2!=0) rect(a+1/n,i,b-1/n,i+0.2,col=colvec[i])
if(n%%2==0) rect(c+1/n,i,d-1/n,i+0.2,col=colvec[i])
}
abline(v=x,col=1,lty=3,lwd=2); axis(1,at=x,labels="x")
```

From Figure 1.8, note that the point $x = 2.3 \in \liminf$ is in all the sets after $n = 4$. Thus, the number of sets not containing points in $\liminf A_n$ is finite.

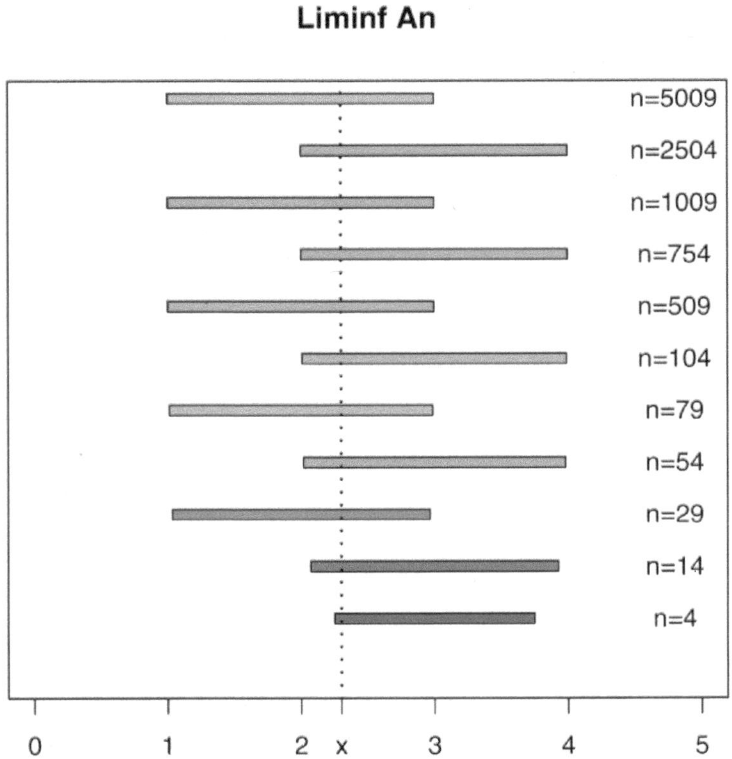

FIGURE 1.8
Limit Infimum

However, in the case of $\limsup A_n$, even though the points occur in infinitely many A_ns, the number of A_n's which do not contain that x is also infinite. For example, in this example $x = 3.1$ does not belong to any A_n where n is odd. Note that $x = 3.1 \notin \liminf A_n$, but all points in $\liminf A_n$ are also in $\limsup A_n$. □

In the next two examples, we examine whether limit supremum and limit infimum are distributive over union and intersection.

Example 1.3.17. In this example we examine whether
(i) $\limsup(A_n \bigcup B_n) = (\limsup A_n) \bigcup (\limsup B_n)$ and
(ii) $\liminf(A_n \bigcap B_n) = (\liminf A_n) \bigcap (\liminf B_n)$.
By the definition of limit supremum of a sequence of sets we have,

$$\begin{aligned}\limsup(A_n \cup B_n) &= \bigcap_{n\geq 1} \bigcup_{k\geq n} (A_k \cup B_k) = \bigcap_{n\geq 1} \left[\left(\bigcup_{k\geq n} A_k\right) \bigcup \left(\bigcup_{k\geq n} B_k\right)\right] \\ &= \bigcap_{n\geq 1} \left[M_n \bigcup N_n\right], \\ \text{where } M_n &= \bigcup_{k\geq n} A_k \ \& \ N_n = \bigcup_{k\geq n} B_k.\end{aligned}$$

Now, $(\limsup A_n)\bigcup(\limsup B_n)$ can be expressed as,

$$\begin{aligned}(\limsup A_n) \cup (\limsup B_n) &= \Big(\bigcap_{n\geq 1}\bigcup_{k\geq n} A_k\Big)\bigcup\Big(\bigcap_{n\geq 1}\bigcup_{k\geq n} B_k\Big)\\ &= \Big(\bigcap_{n\geq 1} M_n\Big)\bigcup\Big(\bigcap_{n\geq 1} N_n\Big)\\ \Rightarrow \quad \limsup(A_n \bigcup B_n) &= (\limsup A_n)\bigcup(\limsup B_n)\\ \text{if } \bigcap_{n\geq 1}\Big(M_n \bigcup N_n\Big) &= \Big(\bigcap_{n\geq 1} M_n\Big)\bigcup\Big(\bigcap_{n\geq 1} N_n\Big),\end{aligned}$$

that is, if the distributive law is valid for countable union and countable intersection. In general, it is not true, but we now show that it is true in this setup, in view of the non-increasing nature of the sequences $\{M_n, n \geq 1\}$ and $\{N_n, n \geq 1\}$. Observe that,

$$\begin{aligned}M_n &= \bigcup_{k\geq n} A_k = A_n \bigcup M_{n+1} \;\;\Rightarrow\;\; M_n \supseteq M_{n+1}\\ N_n &= \bigcup_{k\geq n} B_k = B_n \bigcup N_{n+1} \;\;\Rightarrow\;\; N_n \supseteq N_{n+1},\end{aligned}$$

that is, both $\{M_n, n \geq 1\}$ and $\{N_n, n \geq 1\}$ are non-increasing sequences of sets. We first prove that

$$\begin{aligned}&\bigcap_{n\geq 1}\Big(M_n \bigcup N_n\Big) \subseteq \Big(\bigcap_{n\geq 1} M_n\Big)\bigcup\Big(\bigcap_{n\geq 1} N_n\Big)\\ &\qquad\Longleftrightarrow\; x \in \bigcap_{n\geq 1}\Big(M_n \bigcup N_n\Big) \Rightarrow x \in \Big(\bigcap_{n\geq 1} M_n\Big)\bigcup\Big(\bigcap_{n\geq 1} N_n\Big)\\ &\qquad\Longleftrightarrow\; x \notin \Big(\bigcap_{n\geq 1} M_n\Big)\bigcup\Big(\bigcap_{n\geq 1} N_n\Big) \Rightarrow x \notin \bigcap_{n\geq 1}\Big(M_n \bigcup N_n\Big).\end{aligned}$$

Observe that

$$\begin{aligned}x \;\notin\; & \Big(\bigcap_{n\geq 1} M_n\Big)\bigcup\Big(\bigcap_{n\geq 1} N_n\Big) \;\Rightarrow\; x \notin \Big(\bigcap_{n\geq 1} M_n\Big) \;\&\; x \notin \Big(\bigcap_{n\geq 1} N_n\Big)\\ \Rightarrow\; & \exists\, n_1 \in \mathbb{N} \text{ such that } x \notin M_{n_1} \;\&\; \exists\, n_2 \in \mathbb{N} \text{ such that } x \notin N_{n_2}\\ \Rightarrow\; & x \notin M_{n_0} \;\&\; x \notin N_{n_0} \text{ where } n_0 = \max\{n_1, n_2\}\\ \Rightarrow\; & x \notin M_{n_0} \cup N_{n_0} \;\Rightarrow\; x \notin \bigcap_{n\geq 1}\Big(M_n \bigcup N_n\Big).\end{aligned}$$

In the third step, we use the non-increasing pattern of the sequences $\{M_n, n \geq 1\}$ and $\{N_n, n \geq 1\}$. Thus, $\bigcap_{n\geq 1}\Big(M_n \bigcup N_n\Big) \subseteq \Big(\bigcap_{n\geq 1} M_n\Big)\bigcup\Big(\bigcap_{n\geq 1} N_n\Big)$. Now we prove the inclusion in the reverse direction. Note that

$$\begin{aligned} x \in \Big(\bigcap_{n\geq 1} M_n\Big)\bigcup\Big(\bigcap_{n\geq 1} N_n\Big) &\Rightarrow x \in \Big(\bigcap_{n\geq 1} M_n\Big) \text{ and/or } x \in \Big(\bigcap_{n\geq 1} N_n\Big) \\ &\Rightarrow x \in M_n \ \ \forall\, n \geq 1 \text{ and/or } x \in N_n \ \ \forall\, n\geq 1 \\ &\Rightarrow x \in M_n \bigcup N_n \ \ \forall\, n \geq 1 \\ &\Rightarrow x \in \bigcap_{n\geq 1} \Big(M_n \bigcup N_n\Big). \end{aligned}$$

Thus we have proved that,

$$\begin{aligned} \bigcap_{n\geq 1} \Big(M_n \bigcup N_n\Big) &= \Big(\bigcap_{n\geq 1} M_n\Big)\bigcup\Big(\bigcap_{n\geq 1} N_n\Big) \\ \Rightarrow \quad \limsup(A_n \cup B_n) &= (\limsup A_n) \cup (\limsup B_n). \end{aligned}$$

Now from (i) we have,

$$\begin{aligned} \limsup(A_n^c \cup B_n^c) &= (\limsup A_n^c) \cup (\limsup B_n^c) \\ \Longleftrightarrow \quad (\limsup(A_n^c \cup B_n^c))^c &= ((\limsup A_n^c) \cup (\limsup B_n^c))^c \\ \Longleftrightarrow \quad \liminf(A_n^c \cup B_n^c)^c &= (\limsup A_n^c)^c \cap (\limsup B_n^c)^c \\ \Longleftrightarrow \quad \liminf(A_n \cap B_n) &= (\liminf A_n) \cap (\liminf B_n). \end{aligned}$$

□

Example 1.3.18. In this example we examine whether
(i) $\limsup(A_n \bigcap B_n) = (\limsup A_n) \bigcap (\limsup B_n)$ and
(ii) $\liminf(A_n \bigcup B_n) = (\liminf A_n) \bigcup (\liminf B_n)$.
We define sequences $\{A_n, n \geq 1\}$ and $\{B_n, n \geq 1\}$ as follows.

$$\begin{aligned} A_{2n+1} &= \{2n+1\},\ n \geq 0 \ \ A_{2n} = \{1, 2, \ldots, 2n\},\ n \geq 1 \\ B_{2n+1} &= \{1, 2, \ldots, 2n+1\},\ n \geq 0 \quad \& \ \ B_{2n} = \{2n\}, n \geq 1. \end{aligned}$$

Now by the definition,

$$\limsup A_n = \bigcap_{n\geq 1} \bigcup_{k\geq n} A_k = \bigcap_{n\geq 1} C_n, \quad \text{where} \quad C_n = \bigcup_{k\geq n} A_k.$$

For the given sequence $\{A_n, n \geq 1\}$, observe that,

$$\begin{aligned} C_n = \mathbb{N}, \ \ \forall \ \ n \geq 1 \ &\Rightarrow \ \limsup A_n = \mathbb{N} \\ \text{Similarly,} \ \ \limsup B_n = \mathbb{N} \ &\Rightarrow \ \limsup A_n \bigcap \limsup B_n = \mathbb{N}. \end{aligned}$$

Now, $A_1 \cap B_1 = \{1\}$, $A_2 \cap B_2 = \{2\}$ and in general $M_n = A_n \cap B_n = \{n\}$.

$$\text{Hence,} \ \ \limsup M_n = \bigcap_{n\geq 1} \bigcup_{k\geq n} M_k = \bigcap_{n\geq 1} N_n, \quad \text{where} \quad N_n = \bigcup_{k\geq n} M_k.$$

For the given sequence,

$$\begin{aligned} N_n &= \bigcup_{k\geq n}\{k\} = \{n, n+1, n+2, \ldots\} \Rightarrow \limsup M_n = \bigcap_{n\geq 1} N_n = \emptyset \\ &\Rightarrow \limsup(A_n \bigcap B_n) \neq (\limsup A_n)\bigcap(\limsup B_n), \end{aligned}$$

for the sequences $\{A_n, n \geq 1\}$ and $\{B_n, n \geq 1\}$ as defined above. Thus in general, $\limsup(A_n \bigcap B_n)$ need not be equal to $(\limsup A_n)\bigcap(\limsup B_n)$. We use this result to examine whether (ii) is true. We prove (ii) by contradiction. Thus we assume that $\liminf(A_n \cup B_n) = (\liminf A_n) \cup (\liminf B_n)$ always. Observe that

$$\begin{aligned} \liminf(A_n \cup B_n) &= (\liminf A_n) \cup (\liminf B_n) \\ \Rightarrow \liminf(A_n^c \cup B_n^c) &= (\liminf A_n^c) \cup (\liminf B_n^c) \\ \Rightarrow (\liminf(A_n^c \cup B_n^c))^c &= ((\liminf A_n^c) \cup (\liminf B_n^c))^c \\ \Rightarrow \limsup(A_n^c \cup B_n^c)^c &= (\liminf A_n^c)^c \cap (\liminf B_n^c)^c \\ \Rightarrow \limsup(A_n \cap B_n) &= (\limsup A_n) \cap (\limsup B_n), \end{aligned}$$

but we have proved in (i) that this need not be true. Hence, limit infimum of a sequence of union of two sets need not be equal to union of limit infimums of individual sequences of sets. □

Example 1.3.19. Suppose $\{A_n, n \geq 1\}$ and $\{B_n, n \geq 1\}$ are sequences of sets. If $A_n \to A$ and $B_n \to B$ as $n \to \infty$. We examine whether $A_n \cup B_n \to A \cup B$ and $A_n \cap B_n \to A \cap B$ as $n \to \infty$. By the definition of a limit of a sequence of sets, $A_n \to A$ and $B_n \to B$ implies that

$$\limsup A_n = \liminf A_n = A \quad \text{and} \quad \limsup B_n = \liminf B_n = B.$$

We have proved in Example 1.3.17 that $\limsup(A_n \cup B_n) = (\limsup A_n) \cup (\limsup B_n)$. Using this result we have,

$$\limsup(A_n \cup B_n) = \limsup A_n \cup \ \limsup B_n = A \cup B.$$

Now to find $\liminf(A_n \cup B_n)$ observe that,

$$\omega \in \liminf(A_n \cup B_n) \Rightarrow \omega \in A_k \cup B_k, \quad \forall \ k \geq n_1 \in \mathbb{N}.$$

We make the following three cases.
Case(i): Suppose $\omega \in A_k \ \forall \ k \geq n_1$. Note that

$$\omega \in A_k \ \forall \ k \geq n_1 \Rightarrow \omega \in \liminf A_n = A \Rightarrow \omega \in A \cup B.$$

Case(ii): Suppose $\omega \in B_k \ \forall \ k \geq n_1$. As in Case(i), observe that

$$\omega \in B_k \ \forall \ k \geq n_1 \Rightarrow \omega \in \liminf B_n = B \Rightarrow \omega \in A \cup B.$$

Case(iii) Suppose $\omega \in A_k$ for some $k \geq n_1$ and $\omega \in B_k$ for the remaining $k \geq n_1$. Suppose

$$N_A = \{n \in \mathbb{N} | n \geq n_1 \ \& \ \omega \in A_n\} \quad \& \quad N_B = \{n \in \mathbb{N} | n \geq n_1 \ \& \ \omega \in B_n\}.$$

We know that $N_A \cup N_B = \{n_1, n_1 + 1, n_1 + 2, \ldots, \}$. Hence at least one of the two sets must be countably infinite. Suppose N_A is infinite. Then

$$\begin{aligned} \omega &\in A_n \text{ for infinitely many } n\text{'s} \\ &\Rightarrow \omega \in \limsup A_n = A \quad \Rightarrow \quad \omega \in A \cup B. \end{aligned}$$

If N_B is infinite, then using similar arguments we have $\omega \in A \cup B$. Thus from these three cases, we conclude that $\liminf(A_n \cup B_n) \subseteq A \cup B$. To prove the inclusion in other direction we proceed on similar lines. Observe that

$$\begin{aligned} \omega \in A \cup B &\Rightarrow \omega \in \text{ at least one of } A \ \& \ B \\ \text{Now, } \ \omega \in A = \liminf A_n &\Rightarrow \omega \in A_k \ \ \forall \ k \geq n_2 \in \mathbb{N} \\ \& \text{ or } \ \omega \in B = \liminf B_n &\Rightarrow \omega \in B_k \ \ \forall \ k \geq n_3 \in \mathbb{N} \\ &\Rightarrow \omega \in A_k \ \ \& \text{ or } \omega \in B_k \ \forall \ k \geq n_4 = \max\{n_2, n_3\} \\ &\Rightarrow \omega \in A_k \cup B_k \ \ \forall \ k \geq n_4 \\ &\Rightarrow \omega \in \liminf(A_n \cup B_n) \\ &\Rightarrow A \cup B \subseteq \liminf(A_n \cup B_n). \end{aligned}$$

Consequently,

$$\liminf(A_n \cup B_n) = A \cup B = \limsup(A_n \cup B_n) \quad \Rightarrow \quad A_n \cup B_n \ \to A \cup B.$$

To examine whether $A_n \cap B_n \to A \cap B$, note that by the result in Example 1.3.17, we have

$$\liminf(A_n \cap B_n) = (\liminf A_n) \cap (\liminf B_n) = A \cap B.$$

Now to find $\limsup(A_n \cap B_n)$ observe that

$$\begin{aligned} \omega \in A \cap B \quad &\Rightarrow \quad \omega \in \liminf A_n \cap \liminf B_n \\ &\Rightarrow \quad \omega \in \liminf A_n \ \text{ and } \ \omega \in \liminf B_n \\ &\Rightarrow \quad \omega \in A_n \ \ \forall \ \ n \geq n_5 \in \mathbb{N} \ \& \ \omega \in B_n \ \ \forall \ \ n \geq n_6 \in \mathbb{N} \\ &\Rightarrow \quad \omega \in A_n \cap B_n \ \ \forall \ \ n \geq n_7 = \max\{n_5, n_6\} \\ &\Rightarrow \quad \omega \in \ \ \liminf(A_n \cap B_n) \ \Rightarrow \ \omega \in \ \ \limsup(A_n \cap B_n) \\ &\Rightarrow \quad A \cap B \subseteq \limsup(A_n \cap B_n). \end{aligned}$$

Now to prove that $\limsup(A_n \cap B_n) \subseteq A \cap B$, suppose $A_n \cap B_n = D_n$. By the definition of limit supremum, observe that

$$\begin{aligned}
\omega &\in \limsup D_n \Rightarrow \omega \in \bigcap_{n\geq 1}\bigcup_{k\geq n} D_k \\
&\Rightarrow \omega \in \bigcup_{k\geq n} D_k \ \forall\ n \geq 1 \\
&\Rightarrow \omega \in D_k \text{ for at least one } k \geq n \ \&\ \forall\ n \geq 1 \\
&\Rightarrow \omega \in A_k \cap B_k \text{ for at least one } k \geq n \ \&\ \forall\ n \geq 1 \\
&\Rightarrow \omega \in A_k \text{ for at least one } k \geq n \ \&\ \forall\ n \geq 1 \\
&\ \& \ \omega \in B_k \text{ for at least one } k \geq n \ \&\ \forall\ n \geq 1 \\
&\Rightarrow \omega \in \limsup A_n = A \ \&\ \omega \in \limsup B_n = B \\
&\Rightarrow \omega \in A \cap B \\
&\Rightarrow \limsup(A_n \cap B_n) \subseteq A \cap B.
\end{aligned}$$

Thus, $\limsup(A_n \cap B_n) = A \cap B = \liminf(A_n \cap B_n) \Rightarrow A_n \cap B_n \to A \cap B.$□

The next section is devoted to an important concept of a sigma field of subsets of Ω and a particular sigma field, known as a Borel field.

1.4 Borel Field

Suppose Ω is a non-empty universal set. We now define different classes or collections of subsets of Ω. These classes of sets have certain structures as defined by a set of rules. Different rules yield different types of collections. We begin with two types of collections: a field and a sigma field.

Definition 1.4.1. *Field: A non-empty class $\mathbb{A}$ of subsets of Ω is called a field or an algebra if it is closed under complementation and union of two sets, that is, $\mathbb{A}$ is a field if (i) $A \in \mathbb{A} \Rightarrow A^c \in \mathbb{A}$ and (ii) $A_1, A_2 \in \mathbb{A} \Rightarrow A_1 \cup A_2 \in \mathbb{A}$.*

Following theorem specifies some properties of a field.

Theorem 1.4.1. *(i) The empty set and the universal set are members of a field. (ii) A field is closed under finite unions and finite intersections.*

Proof. (i) Suppose $\mathbb{A}$ is a field. Observe that,

$$A \in \mathbb{A} \Rightarrow A^c \in \mathbb{A} \Rightarrow A \cup A^c = \Omega \in \mathbb{A} \Rightarrow \Omega^c = \emptyset \in \mathbb{A}.$$

(ii) Suppose $A_1, A_2, \ldots A_k \in \mathbb{A}$. By the definition, $B = A_1 \cup A_2 \in \mathbb{A}$, hence, $B \cup A_3 = A_1 \cup A_2 \cup A_3 \in \mathbb{A}$. Continuing in this manner, we get that $A_1 \cup A_2 \cup$

$\cdots \cup A_k \in \mathbb{A}$. Now,

$$\begin{aligned} A_1, A_2, \ldots A_k \in \mathbb{A} &\Rightarrow A_1^c, A_2^c, \ldots A_k^c \in \mathbb{A} \\ &\Rightarrow A_1^c \cup A_2^c \cup \cdots \cup A_k^c \in \mathbb{A} \\ &\Rightarrow (A_1^c \cup A_2^c \cup \cdots \cup A_k^c)^c \in \mathbb{A} \\ &\Rightarrow A_1 \cap A_2 \cap \cdots \cap A_k \in \mathbb{A}. \end{aligned}$$

Thus, a field is closed under finite unions and finite intersections. □

Example 1.4.1. Suppose we want to construct a field containing a subset A of Ω. If a set A is in a field, then, by the definition of a field, A^c has to be in $\mathbb{A}$, union of A and A^c is Ω and its complement is $\emptyset$, thus a field containing the set A or generated by a collection $\{A\}$ is given by, $\mathbb{A} = \{\Omega, A, A^c, \emptyset\}$. □

Example 1.4.2. Suppose $\mathbb{P} = \{A_1, A_2, A_3\}$ is a partition of Ω and we want to find a field generated by the partition $\mathbb{P}$. $\mathbb{P}$ is a partition of Ω implies that A_1, A_2, A_3 are disjoint sets and their union is Ω. If $\mathbb{A}$ denotes the field generated by $\mathbb{P}$, then the empty set and the universal set and A_1, A_2, A_3 are members of $\mathbb{A}$. Further if $A_1 \in \mathbb{A}$, then its complement $A_2 + A_3 \in \mathbb{A}$, note that it is also the union of A_2 and A_3. Continuing in this way we have,

$$\mathbb{A} = \Big\{\Omega, \emptyset, A_1, A_2, A_3, A_1 + A_2, A_1 + A_3, A_2 + A_3\Big\}. \qquad \square$$

Remark 1.4.1. It is to be noted that there are $\binom{3}{0} + \binom{3}{1} + \binom{3}{2} + \binom{3}{3} = 2^3 = 8$ sets in the field. $\binom{3}{0}$ means selecting no set from the partition, which corresponds to an empty set, $\binom{3}{1}$ means there are three ways of selecting 1 set from the partition, which correspond to A_1, A_2, A_3, $\binom{3}{2}$ gives the number of ways in which two sets can be selected from the partition, which correspond to $A_1+A_2, A_1+A_3, A_2+A_3$ and finally $\binom{3}{3}$ corresponds to $A_1+A_2+A_3 = \Omega$. Thus if there are k sets in the partition then the field generated by the partition will consists of 2^k sets corresponding all possible sums of the sets of the partition. Thus, the field $\mathbb{A}$ is given by,

$$\mathbb{A} = \Big\{A \mid A = \sum_{t \in T} A_t \text{ where } T \text{ is the powerset of } \{1, 2, \ldots, k\}\Big\},$$

where summation over empty set is taken as the empty set.

Example 1.4.3. Suppose $\Omega = [0, \infty)$ and $\mathbb{A}$ is a class of all intervals of the type $[a, b)$. It is clear that $[a, b) \in \mathbb{A}$, but $[a, b)^c = [0, a) \cup [b, \infty)$ is not in $\mathbb{A}$. Thus, $\mathbb{A}$ is not closed under complementation and hence it is not a field. □

We now extend the concept of a field of subsets of Ω to a sigma field which is closed under countable unions.

Definition 1.4.2. *Sigma field: A non-empty class $\mathbb{A}$ of subsets of Ω is called a sigma field or a sigma algebra (also written as σ field, or σ algebra), if it is closed under complementation and countable union, that is, $\mathbb{A}$ is a sigma field if (i) $A \in \mathbb{A} \Rightarrow A^c \in \mathbb{A}$, and (ii) $A_n \in \mathbb{A}$, $n \geq 1 \Rightarrow \bigcup_{n \geq 1} A_n \in \mathbb{A}$.*

Some properties of a sigma field are proved in the following theorem.

Theorem 1.4.2. *(i) A sigma field is a field.*

(ii) The empty set and the universal set are members of a sigma field.

(iii) A sigma field is closed under countable intersection.

(iv) A sigma field is closed under limits of monotone sequences.

(v) A sigma field is closed under limit supremum and limit infimum and limit of a sequence of sets, if it exists.

Proof. (i) Suppose $\mathbb{A}$ is a σ field. Then it is closed under complementation. Suppose $A_n \in \mathbb{A}$ are such that $A_n = A_2 \ \forall \ n \geq 2$, then $\bigcup_{n\geq 1} A_n = A_1 \cup A_2 \in \mathbb{A}$. Thus, $\mathbb{A}$ is closed under complementation and finite union, hence it is a field.

(ii) In (i) we have proved that a sigma field is a field, and the empty set and universal set are members of a field and hence of a sigma field. Alternatively, suppose $A_n \in \mathbb{A}, \quad n \geq 1$ are such that $A_n = A_1^c \ \forall \ n \geq 2$. Then

$$\bigcup_{n\geq 1} A_n \in \mathbb{A} \quad \Rightarrow \quad A_1 \cup A_1^c = \Omega \quad \in \mathbb{A} \quad \Rightarrow \quad \Omega^c = \emptyset \quad \in \mathbb{A}\,.$$

Hence, the empty set and the universal set are members of a sigma field.

(iii) Suppose $A_n \in \mathbb{A}, \quad n \geq 1$. Then

$$A_n^c \in \mathbb{A}, \quad n \geq 1 \quad \Rightarrow \quad \bigcup_{n\geq 1} A_n^c \in \mathbb{A} \quad \Rightarrow \quad \Big(\bigcup_{n\geq 1} A_n^c\Big)^c = \bigcap_{n\geq 1} A_n \in \mathbb{A}.$$

Thus, a sigma field is closed under countable intersection.

(iv) It is proved in Theorem 1.3.1 and Theorem 1.3.2 that limit of a monotone sequence is either a countable union or a countable intersection and hence a sigma field is closed under limits of monotone sequences.

(v) A sigma field is closed under countable union and countable intersection and hence $\limsup A_n$ and $\liminf A_n$ are members of a sigma field. If a limit of a sequence $\{A_n, n \geq 1\}$ exists, then $\lim A_n = \limsup A_n = \liminf A_n$. $\limsup A_n$ and $\liminf A_n$ are members of a sigma field and hence a sigma field is closed under limit of a sequence of sets, if it exists. □

Example 1.4.4. In this example we examine whether the given classes are fields and/or sigma fields.

(i) Suppose $\mathbb{A}_1 = \{\Omega, \emptyset\}$. It is obvious that $\mathbb{A}_1$ is a field as $\Omega^c = \emptyset \in \mathbb{A}_1$ and $\emptyset^c = \Omega \in \mathbb{A}_1$. Further, $\Omega \cup \emptyset = \Omega \in \mathbb{A}_1$. Hence, $\mathbb{A}_1$ is a field. A sequence of sets in $\mathbb{A}_1$ will have each set to be either Ω or $\emptyset$. Thus, the countable union will be Ω if all A_n's are not $\emptyset$ and it will be $\emptyset$ if all of A_n's are $\emptyset$. Thus, $\mathbb{A}_1$ is closed under complementation and countable union. Hence, it is a sigma field.

(ii) Suppose $\mathbb{A}_2 = \{\Omega, \emptyset, A, A^c\}$, where $A \subseteq \Omega$. As in (i) we see that $\mathbb{A}_2$ is closed under union and complementation. Again, if we define a sequence of sets in $\mathbb{A}_2$, then the distinct sets are only four, and hence the countable union is just a finite union. Hence, $\mathbb{A}_2$ is a field as well as a sigma field.

(iii) Suppose $\mathbb{A}_3 = \mathbb{A}_2 \cup \mathbb{B}_2$, where $A \subseteq \Omega$ and $\mathbb{B}_2 = \{\Omega, \emptyset, B, B^c\}$ where $B \subseteq \Omega$. Observe that $A, B \in \mathbb{A}_3$, but $A \cup B$ is not in $\mathbb{A}_3$, hence $\mathbb{A}_3$ is neither a field nor a sigma field.

(iv) Suppose $\mathbb{A}_4 = \mathbb{A}_2 \cap \mathbb{A}_3$. Then $\mathbb{A}_4 = \mathbb{A}_2$, hence it is a field as well as a sigma field. □

Remark 1.4.2. (i) We have verified that $\mathbb{A}_1 = \{\Omega, \emptyset\}$ is a sigma field. It is known as a degenerate or a trivial sigma field. Similarly, a powerset $\mathbb{P}(\Omega)$ of Ω is also a sigma field. If $\mathbb{A}$ is any other sigma field of subsets of Ω, then $\mathbb{A}_1 \subset \mathbb{A} \subset \mathbb{P}(\Omega)$. Thus, for any given Ω, there are always two sigma fields. The first one is $\mathbb{A}_1 = \{\Omega, \emptyset\}$, this is the smallest possible sigma field of subsets of Ω. The second one is a powerset $\mathbb{P}(\Omega)$ and is the largest sigma field of subsets of Ω. (ii) It is to be noted that if a class of subsets of Ω is a field and it contains finite number of sets then it is also a sigma field, as countable union essentially reduces to a finite union. A class $\mathbb{A}$ of subsets of Ω is said to be a finite field, if $\mathbb{A}$ is a finite set and is a field. Thus, a finite field is a sigma field.

From (iii) and (iv) of Example 1.4.4, we note that union of two sigma fields may not be a sigma field but intersection of two sigma fields is a sigma field. In the following theorem, we prove that intersection of two sigma fields is always a sigma field.

Theorem 1.4.3. *Intersection of two sigma fields is a sigma field.*

Proof. Suppose $\mathbb{A}_1$ and $\mathbb{A}_2$ are two sigma fields and $\mathbb{A} = \mathbb{A}_1 \cap \mathbb{A}_2$. Being sigma fields, $\{\Omega, \emptyset\} \subseteq \mathbb{A}_1$ and $\{\Omega, \emptyset\} \subseteq \mathbb{A}_2$, thus, $\{\Omega, \emptyset\} \subseteq \mathbb{A}$ implying that $\mathbb{A}$ is a non-empty class of subsets of Ω. Now

$$\begin{aligned} A \in \mathbb{A} \quad &\Rightarrow \quad A \in \mathbb{A}_1 \;\; \& \;\; A \in \mathbb{A}_2 \\ &\Rightarrow \quad A^c \in \mathbb{A}_1 \;\; \& \;\; A^c \in \mathbb{A}_2 \;\; \text{since} \;\; \mathbb{A}_i \;\; \text{are sigma fields } i = 1, 2 \\ &\Rightarrow \quad A^c \in \mathbb{A}_1 \cap \mathbb{A}_2 = \mathbb{A}. \end{aligned}$$

Now consider a sequence $\{A_n, n \geq 1\}$ of sets in $\mathbb{A}$. Observe that,

$$\begin{aligned} \forall \;\; n \geq 1, A_n \in \mathbb{A} \quad &\Rightarrow \quad A_n \in \mathbb{A}_i, \;\; i = 1, 2 \\ &\Rightarrow \quad \bigcup_{n \geq 1} A_n \in \mathbb{A}_i, \;\; i = 1, 2 \;\; \Rightarrow \;\; \bigcup_{n \geq 1} A_n \in \mathbb{A}. \end{aligned}$$

Thus, $\mathbb{A}$ is closed under complementation and countable union and hence is a sigma field. □

Remark 1.4.3. Using similar arguments as in the proof of Theorem 1.4.3, it can be proved that finite intersection and even countable intersection of sigma fields is a sigma field. In the next theorem, we extend it to uncountable intersection.

Theorem 1.4.4. *Suppose* $\{\mathbb{A}_t, t \in T\}$ *is a collection of sigma fields, then* $\mathbb{A} = \bigcap_{t\in T} A_t$ *is a sigma field.*

Proof. Proof is exactly on similar lines as those of Theorem 1.4.3. □

Example 1.4.5. Suppose $\Omega = \mathbb{R}$ and $\mathbb{F}_n$ is the sigma field generated by subsets $\{A_1, A_2, \ldots, A_n\}$ of Ω where $A_k = [k-1, k),\ k = 1, 2, \ldots, n$. We list the sets in $\mathbb{F}_2$. By the definition

$$\begin{aligned}\mathbb{F}_2 &= \sigma(\{A_1, A_2\}) = \sigma(\{[0,1), [1,2)\}) = \Big\{\Omega, \emptyset, [0,1), [1,2),\\ &\quad (-\infty, 0) \cup [1, \infty), (-\infty, 1) \cup [2, \infty), [0, 2), (-\infty, 0) \cup [2, \infty)\Big\}.\end{aligned}$$

We examine whether there is any relation between $\mathbb{F}_n$ and $\mathbb{F}_{n+1}$, such as whether $\mathbb{F}_n \subset \mathbb{F}_{n+1}$. It is to be noted that $\{A_1, A_2, \ldots, A_n\}$ is a collection of disjoint sets. Hence,

$$\begin{aligned}\mathbb{F}_n &= \sigma(\{A_1, A_2, \ldots, A_n\})\\ &= \left\{\Omega, \emptyset, A_1, A_2, \cdots, A_n, A_1^c, A_2^c, \cdots, A_n^c, \sum_T A_i, \Big(\sum_T A_i\Big)^c\right\},\end{aligned}$$

where T denotes all possible subsets of $\{1, 2, \ldots, n\}$ except singletons and the empty set. Similarly,

$$\begin{aligned}\mathbb{F}_{n+1} &= \sigma(\{A_1, A_2, \ldots, A_{n+1}\})\\ &= \left\{\Omega, \emptyset, A_1, A_2, \cdots, A_{n+1}, A_1^c, A_2^c, \cdots, A_{n+1}^c, \sum_T A_i, \Big(\sum_T A_i\Big)^c\right\},\end{aligned}$$

where T denotes all possible subsets of $\{1, 2, \cdots, n+1\}$ except singletons and the empty set. It is clear that $\mathbb{F}_n \subset \mathbb{F}_{n+1}$ for all $n \geq 1$. From $\mathbb{F}_2$ and the sets listed in $\mathbb{F}_n$, it is to be noted that each $\mathbb{F}_n$ consists of finite intervals, finite sums of finite intervals, and sets of the type $(-\infty, a) \cup [b, \infty)$. Hence, $\bigcup_{n\geq1} \mathbb{F}_n$ also consists of finite intervals, finite sums of finite intervals, and sets of the type $(-\infty, a) \cup [b, \infty)$. Thus a set $A_n = [n-1, n) \in \bigcup_{n\geq1} \mathbb{F}_n$ but $\bigcup_{n\geq1} A_n = [0, \infty) \notin \bigcup_{n\geq1} \mathbb{F}_n \;\Rightarrow\; \bigcup_{n\geq1} \mathbb{F}_n$ is not a sigma field. □

Definition 1.4.3. *Co-finite set: A subset* A *of* Ω *is said to be co-finite if its complement is finite.*

For example, suppose $\Omega = \mathbb{N},\ A = \{4, 5, \cdots, \}$ then $A^c = \{1, 2, 3\}$ is finite, hence set A is co-finite.

Example 1.4.6. Suppose Ω is countably infinite and $\mathbb{F}$ is a class of all finite and co-finite subsets of Ω. Suppose $A \in \mathbb{F}$ is finite, that is, $A = (A^c)^c$ is finite, hence A^c is co-finite and hence is in $\mathbb{F}$. Suppose $A \in \mathbb{F}$ is co-finite, then A^c is finite and hence is in $\mathbb{F}$. Thus, $\mathbb{F}$ is closed under complementation. To check whether it is closed under union, we make four cases depending on the nature of the sets.

Case (i): Suppose $A_1, A_2 \in \mathbb{F}$ are both finite, then $A_1 \cup A_2$ is also finite and hence is in $\mathbb{F}$.

Case (ii): Suppose $A_1, A_2 \in \mathbb{F}$ are such that A_1 is co-finite and A_2 is finite, then $A_1^c \cap A_2^c = (A_1 \cup A_2)^c \subseteq A_1^c$ is finite, that is $(A_1 \cup A_2)$ is co-finite and hence is in $\mathbb{F}$.

Case (iii): Suppose $A_1, A_2 \in \mathbb{F}$ are such that A_1 is finite and A_2 is co-finite, then proceeding as in case (ii) we get that $(A_1 \cup A_2)$ is co-finite and hence is in $\mathbb{F}$.

Case (iv): Suppose $A_1, A_2 \in \mathbb{F}$ are both co-finite, that is, both $A_1^c, A_2^c \in \mathbb{F}$ are finite, then $A_1^c \cap A_2^c = (A_1 \cup A_2)^c$ is finite, that is $(A_1 \cup A_2)$ is co-finite and hence is in $\mathbb{F}$. Thus, $\mathbb{F}$ is closed under complementation and finite union, hence it is a field. It will be a sigma field if it is closed under countable union as well. Suppose $\Omega = I^+, A_n = \{2n\} \in \mathbb{F}$, since it is a finite set $\forall\ n \geq 1$. However, $\bigcup_{n\geq 1} A_n = \{2, 4, 6, \cdots\}$ is not in $\mathbb{F}$ as it is neither finite nor co-finite. Thus $\mathbb{F}$ is not a sigma field. □

Remark 1.4.4. In the above example, the universal set Ω is not finite. If it is finite then all the sets and their complements are finite.

Example 1.4.7. Suppose Ω is an uncountable set and $\mathbb{F}$ is a class of all countable subsets of Ω. A set is said to be countable if it is either finite or countably infinite. We examine whether it is a sigma field. Suppose A is a countable set, then $A \in \mathbb{F}$. Since Ω is uncountable, A^c must be uncountable. Hence, $A^c \notin \mathbb{F}$ and hence $\mathbb{F}$ is not a sigma field. For example, suppose $\Omega = \mathbb{R}$ and A is a set of all rational numbers, which is a countable set and hence $A \in \mathbb{F}$. However, $A^c \notin \mathbb{F}$ as it is a set of irrational numbers and is not countable. Thus, $\mathbb{F}$ is not a sigma field. □

In the above example, if $\mathbb{F}$ is a class of all countable and co-countable subsets of Ω, then it is a sigma field as shown in the following example. We first define a co-countable set, it is similar to that of a co-finite set.

Definition 1.4.4. *Co-countable set: A subset A of Ω is said to be co-countable if its complement is countable.*

Example 1.4.8. Suppose Ω is uncountable and $\mathbb{F}$ is a class of all countable and co-countable subsets of Ω. Suppose $A \in \mathbb{F}$ is countable, that is, $A = (A^c)^c$ is countable, hence A^c is co-countable and hence is in $\mathbb{F}$. Suppose $A \in \mathbb{F}$ is co-countable, then A^c is countable and hence is in $\mathbb{F}$. Thus, $\mathbb{F}$ is closed under complementation. To check whether it is closed under union, we make three cases depending on the nature of the sets.

Case (i): Suppose $A_n \ \forall\ n \geq 1$ are countable, then the fact that countable union of countable sets is countable implies that $\bigcup_{n\geq 1} A_n$ is also countable and hence is in $\mathbb{F}$.

Case (ii): Suppose $A_n \ \forall\ n \geq 1$ are co-countable, then A_n^c for all $n \geq 1$ are countable. Then $\bigcap_{n\geq 1} A_n^c$ is atmost countable which implies that $(\bigcap_{n\geq 1} A_n^c)^c = \bigcup_{n\geq 1} A_n$ is co-countable and hence in $\mathbb{F}$.

Case (iii): Suppose some A_n's are countable and remaining in the sequence $\{A_n, n \geq 1\}$ are co-countable. Suppose $n_1 \in \mathbb{N}$ is such that A_{n_1} is co-countable. Note that, A_{n_1} is co-countable and hence $A_{n_1}^c$ is countable. Further,

$$\bigcap_{n\geq 1} A_n^c \subseteq A_{n_1}^c \text{ is countable} \Rightarrow \bigcap_{n\geq 1} A_n^c \in \mathbb{F}$$
$$\Rightarrow (\bigcap_{n\geq 1} A_n^c)^c = \bigcup_{n\geq 1} A_n \in \mathbb{F}.$$

Thus, $\mathbb{F}$ is closed under countable union. From these three cases we conclude that $\mathbb{F}$ is a sigma field. □

Remark 1.4.5.

(i) In the above example, if the universal set Ω is countably infinite, then all the sets and their complements are countably infinite and thus $\mathbb{F}$ is a powerset of Ω. Hence, we have assumed that Ω is uncountable.

(ii) Suppose $\Omega = \mathbb{R}$ and $A = (-\infty, 0]$, which is uncountable. Then $A^c = (0, \infty)$ is also uncountable. Thus neither A nor A^c are in $\mathbb{F}$ defined in Example 1.4.8. It shows that in general, not all subsets of $\Omega = \mathbb{R}$ are members of $\mathbb{F}$.

Suppose Ω is a non-empty universal set and $A \subseteq \Omega$. A sigma field generated by a collection consisting of single set A, is given by $\mathbb{A} = \{\Omega, \emptyset, A, A^c\}$. Similarly, a sigma field $\mathbb{AB}$ generated by a collection $\{A, B\}$ of subsets of Ω, is given by,
$\mathbb{AB} = \Big\{\Omega, \emptyset, A, A^c, B, B^c, A \cup B, A \cup B^c, A^c \cup B, A^c \cup B^c, A \cap B, A \cap B^c,$
$A^c \cap B, A^c \cap B^c, A\Delta B, (A\Delta B)^c\Big\}$, where $A\Delta B$ is known as a symmetric difference and is defined as $A\Delta B = (A^c \cap B) \cup (A \cap B^c)$. $\mathbb{AB}$ can also described

as a sigma field generated by $\{A\}$, but note that $\mathbb{A} \subset \mathbb{AB}$. The power set $\mathbb{P}(\Omega)$ of Ω is a sigma field generated by any class of subsets of Ω. In general, for a given collection of subsets of Ω, there exist many sigma fields covering the collection, so naturally there is a problem of choice and uniqueness. The concept of minimal sigma field, as defined below, provides the solution.

Definition 1.4.5. *Minimal sigma field: Suppose $\mathbb{C}$ is a non-empty class of subsets of Ω. The minimal sigma field covering the class $\mathbb{C}$, or generated by the class $\mathbb{C}$ is denoted by $\sigma(\mathbb{C})$. It is defined as the minimal sigma field if it satisfies the following conditions. (i) $\sigma(\mathbb{C})$ is a sigma field. (ii) $\mathbb{C} \subseteq \sigma(\mathbb{C})$. (iii) If $\sigma^*(\mathbb{C})$ is any sigma field generated by $\mathbb{C}$, then $\sigma^*(\mathbb{C}) \supseteq \sigma(\mathbb{C})$.*

Following theorem establishes the existence and the uniqueness of the minimal sigma field.

Theorem 1.4.5. *There exists a unique minimal sigma field generated by a non-empty class $\mathbb{C}$ of subsets of Ω.*

Proof. Suppose $\{\sigma_t(\mathbb{C}), t \in T\}$ is a collection of sigma fields generated by $\mathbb{C}$. It is known that the powerset $\mathbb{P}(\Omega)$ of Ω is a sigma field generated by any class of subsets of Ω and thus, $\{\sigma_t(\mathbb{C}), t \in T\}$ is a non-empty collection. We define,

$$\sigma(\mathbb{C}) = \bigcap_{t \in T} \sigma_t(\mathbb{C}).$$

From Theorem 1.4.4 it follows that $\sigma(\mathbb{C})$ is a sigma field. Further, $\{\sigma_t(\mathbb{C}), t \in T\}$ is a collection of sigma fields generated by $\mathbb{C}$, hence

$$\mathbb{C} \subseteq \sigma_t(\mathbb{C}) \ \ \forall \ \ t \in T \ \ \Rightarrow \ \ \mathbb{C} \subseteq \bigcap_{t \in T} \sigma_t(\mathbb{C}) = \sigma(\mathbb{C}).$$

It is to be noted that by the definition of $\sigma(\mathbb{C})$, for all $t \in T \ \ \sigma_t(\mathbb{C}) \supseteq \sigma(\mathbb{C})$ and hence, $\sigma(\mathbb{C})$ is a minimal sigma field, which establishes the existence of the minimal sigma field. To prove uniqueness, suppose if possible, $\sigma^*(\mathbb{C})$ is another minimal sigma field generated by $\mathbb{C}$. Then by the definition of minimal sigma field,

$$\sigma^*(\mathbb{C}) \subseteq \sigma(\mathbb{C}) \text{ and } \sigma(\mathbb{C}) \subseteq \sigma^*(\mathbb{C}) \iff \sigma^*(\mathbb{C}) = \sigma(\mathbb{C}).$$

Thus, for any non-empty class of subsets of Ω, there exists a unique minimal sigma field generated by the class. □

Following are some results about the minimal sigma field.

Theorem 1.4.6. *(i) If any non-empty class $\mathbb{C}$ of subsets of Ω, is a sigma field, then $\sigma(\mathbb{C}) = \mathbb{C}$. (ii) Suppose $\mathbb{C}_1$ and $\mathbb{C}_2$ are two classes of sets such that $\mathbb{C}_1 \subseteq \mathbb{C}_2$. Then $\sigma(\mathbb{C}_1) \subseteq \sigma(\mathbb{C}_2)$.*

Proof.

(i) Suppose $\sigma(\mathbb{C})$ is not $\mathbb{C}$, but $\mathbb{F}$. Then by the definition of the minimal sigma field (i) $\mathbb{F}$ is a sigma field. (ii) $\mathbb{C} \subseteq \mathbb{F}$ and (iii) If $\sigma^*(\mathbb{C})$ is any sigma field generated by $\mathbb{C}$, then $\sigma^*(\mathbb{C}) \supseteq \mathbb{F}$. In particular, $\mathbb{C}$ is a sigma field generated by $\mathbb{C}$ and hence $\mathbb{C} \supseteq \mathbb{F}$. Thus,

$$\mathbb{C} \subseteq \mathbb{F} \ \& \ \mathbb{C} \supseteq \mathbb{F} \ \Rightarrow \ \mathbb{F} = \mathbb{C}.$$

(ii) Observe that $\mathbb{C}_1 \subset \mathbb{C}_2 \subset \sigma(\mathbb{C}_2)$. Thus, $\sigma(\mathbb{C}_2)$ is a sigma field covering $\mathbb{C}_1$. Hence, by the definition of the minimal sigma field $\sigma(\mathbb{C}_1) \subset \sigma(\mathbb{C}_2)$.

□

We now proceed to define the most exclusively used minimal sigma field in probability theory and it is the minimal sigma field generated by a collection of any type of intervals of the real line $\mathbb{R}$. Suppose the universal set Ω is the set $\mathbb{R}$ of real numbers. Then an interval of $\mathbb{R}$ can be any one of the following 8 types: (i) (a, b), (ii) $(a, b]$, (iii) $[a, b)$, (iv) $[a, b]$, (v) $(-\infty, a)$, (vi) $(-\infty, a]$, (vii) (a, ∞), (viii) $[a, \infty)$, where $a < b \in \mathbb{R}$.

Definition 1.4.6. *Borel field: Suppose $\mathbb{C}_i$ is the class of intervals of type $i, i = 1, 2, \cdots, 8$. Then the minimal sigma field generated by any one of $\mathbb{C}_i$, denoted by $\mathbb{B}$, is known as a Borel field.*

Definition 1.4.7. *Borel set: A set in the Borel field is known as a Borel set.*

Example 1.4.9. This example shows that almost all the subsets of the real line are Borel sets.

1. A set of real numbers $\mathbb{R}$ is a Borel set, since by the definition, Borel field is the minimal sigma field of subsets of $\mathbb{R}$. Thus, $\Omega = \mathbb{R}$ and hence $\mathbb{R} \in \mathbb{B}$.

2. All eight types of intervals are Borel sets, since by the definition, Borel field is the minimal sigma field generated by classes of any one type of intervals.

3. A singleton set $\{a\}$ can be expressed as $\{a\} = \bigcap_{n \geq 1}(a - 1/n, a + 1/n)$. Thus it is a countable intersection of intervals and hence is in the Borel field. Thus, a singleton set $\{a\}$ is a Borel set.

4. Any finite set can be expressed as a finite union of singleton sets and hence it is a Borel set.

5. Any countable set can be expressed as a countable union of singleton sets and hence it is a Borel set.

6. A set of rational numbers being countable is a Borel set. Similarly a set irrational numbers, being complement of a set of rational numbers is a Borel set.

7. Suppose $A_n = [0, n)$. Then $\bigcup_{n\geq 1} A_n = [0, \infty)$ and hence $\mathbb{R}^+$ is a Borel set. Further, $\mathbb{R}^- = (-\infty, 0)$, being its complement is a Borel set.

8. Any open set can be expressed as a countable union of open intervals and hence it is a Borel set.

9. Any open set is a Borel set and hence a countable union of open sets is a Borel set. Similarly, countable intersection of open sets is a Borel set.

10. Suppose A is a closed set, then A^c is an open set. Since A^c is a Borel set, $A = (A^c)^c$ is also a Borel set. Hence any closed set is a Borel set. Borel field being a sigma field, any countable union of closed sets is in the Borel field and hence is a Borel set. Similarly, countable intersection of closed sets is a Borel set.

11. A set $A = \bigcup_{t\in T}\{a_t\}$, where T is an uncountable set, may not be a Borel set. □

From Definition 1.4.6, we note that there are 8 types of intervals but the minimal sigma field generated by a class of any type of intervals is the same and it is the Borel field. It can be proved that all of these classes of intervals generate the same minimal sigma field. While proving such results, we note that any type of class interval can be expressed in terms of intervals of other types either by countable union or countable intersection or both. Following are some such relations.

1. $(a, b) = \bigcup_{n\geq 1} [a + 1/n, b) = \bigcup_{n\geq 1} (a, b - 1/n] = \bigcup_{n\geq 1} [a + 1/n, b - 1/n]$

2. $[a, b] = \bigcap_{n\geq 1} [a, b + 1/n) = \bigcap_{n\geq 1} (a - 1/n, b] = \bigcap_{n\geq 1} (a - 1/n, b + 1/n)$

3. $[a, b) = \bigcap_{n\geq 1} (a - 1/n, b) = \bigcup_{n\geq 1} [a, b - 1/n] = \bigcap_{n\geq 1} \bigcup_{m\geq 1} (a - 1/n, b - 1/m]$

4. $(a, b] = \bigcap_{n\geq 1} (a, b + 1/n) = \bigcup_{n\geq 1} [a + 1/n, b] = \bigcup_{n\geq 1} \bigcap_{m\geq 1} [a + 1/n, b + 1/m)$

5. $(-\infty, a] = \bigcap_{n\geq 1} (-\infty, a + 1/n)$

6. $(-\infty, a) = \bigcup_{n\geq 1} (-n, a)$

7. $\{a\} = (-\infty, a] \cap [a, \infty)$

In the following theorem, we demonstrate that the minimal sigma field generated by classes of some type of intervals is the same. In the proof, we use some of the relations listed above.

Theorem 1.4.7. *Suppose $\mathbb{C}_1$ is a class of intervals of the type $(a,b), a < b \in \mathbb{R}$, $\mathbb{C}_2$ is a class of intervals of the type $(a,b], a < b \in \mathbb{R}$, $\mathbb{C}_5$ is a class of intervals of the type $(-\infty,a), a \in \mathbb{R}$ and $\mathbb{C}_6$ is a class of intervals of the type $(-\infty,a], a \in \mathbb{R}$. Then (i) $\sigma(\mathbb{C}_1) = \sigma(\mathbb{C}_2)$, (ii) $\sigma(\mathbb{C}_1) = \sigma(\mathbb{C}_5)$, (iii) $\sigma(\mathbb{C}_1) = \sigma(\mathbb{C}_6)$ and hence $\sigma(\mathbb{C}_1) = \sigma(\mathbb{C}_2) = \sigma(\mathbb{C}_5) = \sigma(\mathbb{C}_6)$.*

Proof. (i) Since $\sigma(\mathbb{C}_1)$ is the minimal sigma field generated by $\mathbb{C}_1$,

$$\begin{aligned}\forall\ n \geq 1,\ \ (a, b+1/n) \in \mathbb{C}_1 \ &\Rightarrow\ \forall\ n \geq 1,\ \ (a, b+1/n) \in \sigma(\mathbb{C}_1)\\ &\Rightarrow\ \bigcap_{n\geq 1}(a, b+1/n) = (a,b] \in \sigma(\mathbb{C}_1)\\ &\Rightarrow\ \mathbb{C}_2 \subset \sigma(\mathbb{C}_1)\ \Rightarrow\ \sigma(\mathbb{C}_2) \subseteq \sigma(\mathbb{C}_1).\end{aligned}$$

The last step follows from the following arguments. $\mathbb{C}_2 \subseteq \sigma(\mathbb{C}_1)$ implies that $\sigma(\mathbb{C}_1)$ is a sigma field generated by $\mathbb{C}_2$. Hence, the minimal sigma field generated by $\mathbb{C}_2$ must be included in $\sigma(\mathbb{C}_1)$. We use similar arguments below. Now by the definition of $\mathbb{C}_2$,

$$\begin{aligned}\forall\ n \geq 1,\ \ (a, b-1/n] \in \mathbb{C}_2 \ &\Rightarrow\ \forall\ n \geq 1,\ \ (a, b-1/n] \in \sigma(\mathbb{C}_2)\\ &\Rightarrow\ \bigcup_{n\geq 1}(a, b-1/n] = (a,b) \in \sigma(\mathbb{C}_2)\\ &\Rightarrow\ \mathbb{C}_1 \subseteq \sigma(\mathbb{C}_2)\ \Rightarrow\ \sigma(\mathbb{C}_1) \subseteq \sigma(\mathbb{C}_2).\end{aligned}$$

Thus, $\sigma(\mathbb{C}_1) = \sigma(\mathbb{C}_2)$.

(ii) Since $\sigma(\mathbb{C}_1)$ is the minimal sigma field generated by $\mathbb{C}_1$,

$$\begin{aligned}\forall\ n \geq 1,\ \ (-n, a) \in \mathbb{C}_1 \ &\Rightarrow\ \forall\ n \geq 1,\ \ (-n, a) \in \sigma(\mathbb{C}_1)\\ &\Rightarrow\ \bigcup_{n\geq 1}(-n, a) = (-\infty, a) \in \sigma(\mathbb{C}_1)\\ &\Rightarrow\ \mathbb{C}_5 \subseteq \sigma(\mathbb{C}_1)\ \Rightarrow\ \sigma(\mathbb{C}_5) \subseteq \sigma(\mathbb{C}_1).\end{aligned}$$

Now by the definition of $\mathbb{C}_5$, $(-\infty, b) \in \mathbb{C}_5 \ \Rightarrow\ (-\infty, b) \in \sigma(\mathbb{C}_5)$. Further,

$$\begin{aligned}\forall\ n \geq 1,\ \ (-\infty, a+1/n) \in \mathbb{C}_5 &\Rightarrow \forall\ n \geq 1, (-\infty, a+1/n) \in \sigma(\mathbb{C}_5)\\ &\Rightarrow \bigcap_{n\geq 1}(-\infty, a+1/n) = (-\infty, a] \in \sigma(\mathbb{C}_5)\\ &\Rightarrow (-\infty, a]^c = (a, \infty) \in \sigma(\mathbb{C}_5)\\ &\Rightarrow \text{for}\ \ b > a,\ \ (-\infty, b) \cap (a, \infty)\\ &= (a,b) \in \sigma(\mathbb{C}_5)\\ &\Rightarrow \mathbb{C}_1 \subseteq \sigma(\mathbb{C}_5)\ \Rightarrow\ \sigma(\mathbb{C}_1) \subseteq \sigma(\mathbb{C}_5).\end{aligned}$$

Thus, $\sigma(\mathbb{C}_1) = \sigma(\mathbb{C}_5)$

(iii) Now $\sigma(\mathbb{C}_6)$ is the minimal sigma field generated by $\mathbb{C}_6$. Hence,

$$\forall\ n \geq 1,\ \ (-\infty, b - 1/n] \in \mathbb{C}_6 \Rightarrow \forall\ n \geq 1,\ \ (-\infty, b - 1/n] \in \sigma(\mathbb{C}_6)$$
$$\Rightarrow \bigcup_{n\geq 1} (-\infty, b - 1/n] = (-\infty, b) \in \sigma(\mathbb{C}_6)\ .$$

$$\text{Now,}\ \ (-\infty, a] \in \mathbb{C}_6 \ \Rightarrow\ (-\infty, a] \in \sigma(\mathbb{C}_6) \ \Rightarrow\ (a, \infty) \in \sigma(\mathbb{C}_6)\ .$$

$$\text{Thus,}\ (-\infty, b)\ \ \&\ \ (a, \infty) \in \sigma(\mathbb{C}_6) \Rightarrow (-\infty, b) \cap (a, \infty) = (a, b) \in \sigma(\mathbb{C}_6)$$
$$\Rightarrow \mathbb{C}_1 \subseteq \sigma(\mathbb{C}_6) \ \Rightarrow\ \sigma(\mathbb{C}_1) \subseteq \sigma(\mathbb{C}_6).$$

$$\text{Further,}\ \forall\ n \geq 1,\ \ (a, n) \in \mathbb{C}_1 \Rightarrow \forall\ n \geq 1,\ \ (a, n) \in \sigma(\mathbb{C}_1)$$
$$\Rightarrow \bigcup_{n\geq 1} (a, n) = (a, \infty) \in \sigma(\mathbb{C}_1)$$
$$\Rightarrow (a, \infty)^c = (-\infty, a] \in \sigma(\mathbb{C}_1)$$
$$\Rightarrow \mathbb{C}_6 \subseteq \sigma(\mathbb{C}_1) \ \Rightarrow\ \sigma(\mathbb{C}_6) \subseteq \sigma(\mathbb{C}_1)\ .$$

Hence, $\sigma(\mathbb{C}_6) = \sigma(\mathbb{C}_1)$. We have thus shown that

$$\sigma(\mathbb{C}_1) = \sigma(\mathbb{C}_2) = \sigma(\mathbb{C}_5) = \sigma(\mathbb{C}_6)\ .$$

□

On similar lines one can show that minimal sigma fields generated by a class of any type of intervals coincide and the common sigma field is labeled as the Borel field.

In the following theorem, we show that a sigma field generated by the collection of all open sets is also the Borel-field.

Theorem 1.4.8. *Suppose $\mathbb{O}$ is a class of open sets in $\mathbb{R}$. Then the minimal sigma field generated by $\mathbb{O}$ is the same as the Borel field.*

Proof. By the definition of the Borel field, $\sigma(\mathbb{C}_1) = \mathbb{B}$, where $\mathbb{C}_1$ is a class of intervals of the type $(a, b), a, b \in \mathbb{R}$. Thus, the Borel field $\mathbb{B}$ contains all open intervals. Since any open set can be written as a countable union of open intervals, it follows that $\mathbb{O} \subseteq \sigma(\mathbb{C}_1) = \mathbb{B}$ which further implies that $\sigma(\mathbb{O}) \subseteq \mathbb{B}$. It is known that any open interval is an open set. Hence,

$$\mathbb{C}_1 \subseteq \mathbb{O} \ \Rightarrow\ \mathbb{C}_1 \subseteq \sigma(\mathbb{O}) \ \Rightarrow\ \sigma(\mathbb{C}_1) \subseteq \sigma(\mathbb{O}) \ \Rightarrow\ \mathbb{B} \subseteq \sigma(\mathbb{O}).$$

Thus, we have $\sigma(\mathbb{O}) = \mathbb{B}$. □

Borel field in k dimensions is defined on similar lines as those for one dimension.

Definition 1.4.8. *Borel field in k dimensions: Suppose*

$$I_k = \{(x_1, x_2, \cdots, x_k) | a_i \leq x_i \leq b_i, a_i, b_i, x_i \in \mathbb{R}, i = 1, 2, \cdots, k\}$$

is a k-dimensional rectangle and $\mathbb{C}_k$ *is a collection of rectangles of the type* I_k*. The minimal sigma field generated by* $\mathbb{C}_k$*, denoted by* $\mathbb{B}^k$*, is known as the Borel field in k dimensions.*

As in Theorem 1.4.7, it can be shown that the minimal sigma field generated by rectangles of any type in $\mathbb{R}^k$ coincide and each one is the Borel field in k dimensions. Further, the minimal sigma field generated by a class of open sets in k dimensions is also the Borel field in k dimensions.

Definition 1.4.9. *Measurable space: Suppose* Ω *is a non-empty universal set and* $\mathbb{A}$ *is a sigma field of subsets of* Ω*. Then the set* Ω *along with associated sigma field* $\mathbb{A}$*, that is, a pair* $(\Omega, \mathbb{A})$ *is known as a measurable space.*

For example, $(\mathbb{R}, \mathbb{B})$ is a measurable space, also $(\mathbb{R}^k, \mathbb{B}^k)$ is a measurable space.

Definition 1.4.10. *Event: Suppose* $(\Omega, \mathbb{A})$ *is a measurable space. A set in* $\mathbb{A}$ *is known as an event.*

We have studied two types of classes of subsets of Ω of which sigma field is the most important for theory as well as applications. There are other types of classes also, we define these below.

Definition 1.4.11. *Ring: A non-empty class* $\mathbb{F}$ *of subsets of* Ω *is said to be a ring, if it is closed under finite unions and differences, that is,* $\mathbb{F}$ *is said to be a ring, if* A, $B \in \mathbb{F}$ *implies that* $A \cup B \in \mathbb{F}$ *and* $A - B \in \mathbb{F}$*, where* $A - B = A \cap B^c$*.*

A class of all finite subsets of Ω is a ring. It is closed under finite intersections and symmetric differences, but not under countable unions.

Definition 1.4.12. *Sigma ring: A non-empty class* $\mathbb{F}$ *of subsets of* Ω *is said to be a sigma ring, if it is closed under countable unions and differences.*

Thus, sigma ring is a ring closed under countable unions. Further, every finite ring is a sigma ring. A class of all countable subsets of Ω is a sigma ring.

Definition 1.4.13. *π-class: A non-empty class* $\mathbb{L}$ *of subsets of* Ω *is said to be a π-class or a π-system, if it is closed under finite intersections, that is,* $\mathbb{L}$ *is a π-class if* $A, B \in \mathbb{L} \Rightarrow A \cap B \in \mathbb{L}$*.*

From the definition it follows that a field and a sigma field are π-systems. Further, a class of all intervals in $\mathbb{R}$ is also a π-system, the empty set being taken as (a, a).

Definition 1.4.14. *λ-class: A non-empty class $\mathbb{D}$ of subsets of Ω is said to be a λ-class or a λ system, if $\Omega \in \mathbb{D}$ and if it is closed under complements and countable unions of pairwise disjoint sets. Thus, $\mathbb{D}$ is a λ-class if, (i) $\Omega \in \mathbb{D}$, (ii) $A \in \mathbb{D} \Rightarrow A^c \in \mathbb{D}$ and (iii) $A_n \in \mathbb{D}$ such that $A_i \cap A_j = \emptyset \Rightarrow \bigcup_{n=1}^{\infty} A_n \in \mathbb{D}$.*

Equivalently, a λ-class $\mathbb{D}$ is also defined as a non-empty class of subsets of Ω that satisfies the following conditions. (i) $\Omega \in \mathbb{D}$, (ii) $A, B \in \mathbb{D}$, $B \subseteq A \Rightarrow A - B \in \mathbb{D}$ and (iii) $A_n \in \mathbb{D} \ \forall \ n \geq 1, \ \ A_n \uparrow \ \Rightarrow \ \bigcup_{n=1}^{\infty} A_n \in \mathbb{D}$.

A λ-class is also called as a Dynkin system. In the first set of conditions of λ system, in view of disjointness of sets in condition (iii), λ system is weaker than the sigma field. From the same set, condition (i) and (ii), together imply that $\emptyset$ is a member of λ system. Since $\emptyset$ is a member of λ system, countable union in condition (iii) can be reduced to finite union. Further, λ-class is also a π-class.

Definition 1.4.15. *Monotone class: A non-empty class $\mathbb{M}$ of subsets of Ω is said to be a monotone class, if it is closed under limits of monotone sequences of sets. Thus, $\mathbb{M}$ is a monotone class if, (i) $A_n \in \mathbb{M} \ \forall \ n \geq 1, \ A_n \uparrow$ implies $\bigcup_{n=1}^{\infty} A_n \in \mathbb{M}$ and (ii) $A_n \in \mathbb{M} \ \forall \ n \geq 1, \ A_n \downarrow \ \Rightarrow \ \bigcap_{n=1}^{\infty} A_n \in \mathbb{M}$.*

Definition 1.4.16. *Monotone field: A field which is a monotone class is known as a monotone field.*

The following relations between different collections of sets are obtained by the different properties of set operations, together with De Morgan's rule.

1. Every field is a π-class.

2. Every sigma field is a field.

3. A field is a sigma field if and only if it is a monotone class, that is, a sigma field is a monotone field and monotone field is a sigma field.

4. Every sigma field is a λ-class.

5. A λ-class is a sigma field if and only if it is π-class.

6. Every λ-class is a monotone class.

7. Every sigma field is a monotone class.

8. The power set of any subclass of Ω is a sigma field on that subclass.

9. If $\mathbb{A}$ is a sigma field and $B \subseteq \Omega$, then $B \cap \mathbb{A} = \{B \cap A | A \in \mathbb{A}\}$ is a sigma field of subsets of B.

10. A ring is a sigma ring if and only if it is a monotone class, that is, a sigma ring is a monotone class and a monotone ring is a sigma ring.

11. Every sigma field is a λ-class. However, the converse is not true. For counter example refer to page 5 of Stoyanov [23].

We state below a theorem related to a π-system and a λ-system, which is known as Dynkin's π-λ theorem. It is technical but extremely useful.

Theorem 1.4.9. *Dynkin's $\pi - \lambda$ Theorem: If $\mathbb{P}$ is a π-system and $\mathbb{L}$ is a λ-system, then $\mathbb{P} \subseteq \mathbb{L} \Rightarrow \sigma(\mathbb{P}) \subseteq \mathbb{L}$.*

For proof refer to page 37 of Billingsley [5]. In Chapter 5 while discussing independence of random variables, we discuss one more result about a π-system and a λ-system.

In the next section, we discuss the important concept of a probability measure.

1.5 Probability Space

The basis of probability theory is the probability space. In this section, we discuss the axiomatic theory of probability. Although games of chance have been performed for thousands of years, a mathematically rigorous treatment of the theory of probability was given by the Russian mathematician A.N. Kolmogorov in his fundamental monograph, which appeared in 1933. If a set function P satisfies the three Kolmogorov axioms, as defined below, then P is a probability measure.

Definition 1.5.1. *Probability measure: Suppose $(\Omega, \mathbb{A})$ is a measurable space. A set function P defined on $(\Omega, \mathbb{A})$ is known as a probability measure, if the following conditions are satisfied.*

(i) $P(A) \geq 0, \quad \forall \ A \in \mathbb{A},$ *(non-negativity)*

(ii) $P(\Omega) = 1,$ *(normed measure)*

(iii) If $\{A_n, n \geq 1\}$ is a sequence of disjoint sets in $\mathbb{A}$, then $P(\bigcup_{n\geq 1} A_n) = P(\sum_{n\geq 1} A_n) = \sum_{n\geq 1} P(A_n)$, ($\sigma$- additivity).

Definition 1.5.2. *Probability space: A triplet $(\Omega, \mathbb{A}, P)$ is known as a probability space, where Ω is a non-empty set, $\mathbb{A}$ is a sigma field of subsets of Ω and P is a probability measure on the measurable space $(\Omega, \mathbb{A})$.*

Some important properties of a probability measure are proved in the following theorem.

Theorem 1.5.1. *(i) $P(\emptyset) = 0$. (ii) Probability measure is finitely additive, that is, if $\{A_1, A_2, \cdots, A_k\}$ are disjoint sets from $\mathbb{A}$, then $P(\bigcup_{i=1}^k A_i) = \sum_{i=1}^k P(A_i)$. (iii) $P(A^c) = 1 - P(A)$. (iv) If A and B in $\mathbb{A}$ are such that $A \subseteq B$, then $P(A) \leq P(B)$. (v) $P(A) \leq 1$.*

Proof.

(i) From σ- additivity of the probability measure, if $\{A_n, n \geq 1\}$ is a sequence of disjoint sets in $\mathbb{A}$, then $P(\bigcup_{n\geq 1} A_n) = \sum_{n\geq 1} P(A_n)$. In particular, suppose $A_n = \emptyset, \quad \forall \quad n \geq 2$, then $\bigcup_{n\geq 1} A_n = A_1$. Suppose $P(\emptyset) = p \quad \geq 0$. From σ- additivity,

$$P(\bigcup_{n\geq 1} A_n) = \sum_{n\geq 1} P(A_n) \quad \Rightarrow \quad P(A_1) = P(A_1) + \sum_{n\geq 2} P(\emptyset)$$
$$\Rightarrow \quad \sum_{n\geq 2} p = 0 \quad \Rightarrow \quad p = P(\emptyset) = 0.$$

(ii) Suppose $\{A_n, n \geq 1\}$ is a sequence of disjoint sets in $\mathbb{A}$, such that $A_n = \emptyset$, $\forall \quad n \geq k+1$. Then using the sigma additivity and the property $P(\emptyset) = 0$, the identity $P(\bigcup_{n\geq 1} A_n) = \sum_{n\geq 1} P(A_n)$ reduces to $P(\bigcup_{i=1}^{k} A_i) = \sum_{i=1}^{k} P(A_i)$.

(iii) By finite additivity, $1 = P(\Omega) = P(A \cup A^c) = P(A) + P(A^c)$, hence $P(A^c) = 1 - P(A)$.

(iv) Observe that since $A \subseteq B, \quad B = (A \cap B) \cup (A^c \cap B) = A \cup (A^c \cap B)$. Hence, by finite additivity and non-negativity of probability measure

$$P(B) = P(A \cup (A^c \cap B)) = P(A) + P(A^c \cap B) \geq P(A).$$

(v) Note that, $\forall \quad A \in \mathbb{A}, \quad A \subseteq \Omega \quad \Rightarrow \quad P(A) \leq P(\Omega) = 1$, by property (iv).

□

Remark 1.5.1. From Definition 1.5.2 and Theorem 1.5.1, it is to be noted that $\mathbb{A}$ is a collection of all those subsets of Ω, whose probability can be uniquely determined via relationships for complementation, union and intersection. Hence, $\mathbb{A}$ is a collection of all those subsets of Ω for which probability measure can be assigned. Hence, $(\Omega, \mathbb{A})$ is known as a measurable space.

Example 1.5.1. Suppose $(\Omega, \mathbb{A})$ is a measurable space and a set function P is defined on $(\Omega, \mathbb{A})$ such that the events A, B, C satisfy the following conditions: $P(A) = 0.6, P(B) = 0.8, P(C) = 0.7, P(A \cap B) = 0.5, P(A \cap C) = 0.4$, $P(B \cap C) = 0.4$ and $P(A \cap B \cap C) = 0.1$. We examine whether P can be a probability measure. Suppose P is a probability measure then

$$P(B \cup C) = P(B) + P(C) - P(B \cap C) = 1.5 - 0.4 = 1.1 > 1,$$

which is a contradiction. Hence P cannot be a probability measure.

Example 1.5.2. Suppose $\Omega = \mathbb{W}$, a set of whole numbers, $\mathbb{A} = \mathbb{P}(\Omega)$, power-set of Ω, and a set function P on $(\Omega, \mathbb{P}(\Omega))$ is defined as $P(A) = \sum_{i\in A} e^{-\lambda}\lambda^i/i!$

for $A \in \mathbb{P}(\Omega)$ and λ is a positive real number. We examine whether P is a probability measure on $(\Omega, \mathbb{P}(\Omega))$. By the definition $P(A) \geq 0$. Further, $P(\Omega) = \sum_{i \in \mathbb{W}} e^{-\lambda}\lambda^i/i! = 1$. To examine the third condition of sigma additivity, suppose $\{A_n, n \geq 1\}$ is a sequence of disjoint events. Then

$$P\Big(\bigcup_{n\geq 1} A_n\Big) = \sum_{i \in \bigcup_{n\geq 1} A_n} \frac{e^{-\lambda}\lambda^i}{i!} = \sum_{n\geq 1}\Big(\sum_{i\in A_n} \frac{e^{-\lambda}\lambda^i}{i!}\Big) = \sum_{n\geq 1} P(A_n).$$

Thus, the set function P satisfies all the conditions of a probability measure and hence it is a probability measure on $(\Omega, \mathbb{P}(\Omega))$. $\square$

If Ω is countable then one can usually consider the sigma field as a power set of Ω and a probability measure that assigns a value to every subset of Ω. When Ω is uncountable, it may not be possible to define a reasonable probability measure for every subset of Ω. In view of such a limitation, it is necessary to introduce appropriate sigma fields and assign probability measure for the sets in the sigma field

One of the basic questions is to what extent limits of sets carry over to limits of probabilities of sets, that is, whether probabilities of converging sets converge. We now discuss a continuity theorem for a probability measure, which answers this question. As a pre-requisite, we prove a lemma, known as a disjointification lemma.

Lemma 1.5.1. *Suppose $\{A_n, n \geq 1\}$ is a sequence of sets in $\mathbb{A}$. Then there exists a sequence $\{B_n, n \geq 1\}$ of disjoint sets in $\mathbb{A}$ such that for $m \in \mathbb{N}$,*

$$(i) \bigcup_{n=1}^{m} B_n = \bigcup_{n=1}^{m} A_n \;\; \& \;\; (ii) \bigcup_{n\geq 1} B_n = \bigcup_{n\geq 1} A_n \,.$$

Proof. For a given sequence $\{A_n, n \geq 1\}$ of sets in $\mathbb{A}$, we define another sequence $\{B_n, n \geq 1\}$ as follows.

$$B_1 = A_1 \quad \text{and for} \quad n \geq 2, \quad B_n = A_1^c \cap A_2^c \cap \cdots A_{n-1}^c \cap A_n.$$

Thus $\forall \; n \geq 1, \;\; B_n \in \mathbb{A}$. For $n < m$,

$$B_n = A_1^c \cap A_2^c \cap \cdots A_{n-1}^c \cap A_n \;\; \text{and} \;\; B_m = A_1^c \cap \cdots A_n^c \cap \cdots \cap A_{m-1}^c \cap A_m.$$

Observe that, for $n < m$, $\omega \in B_n \;\Rightarrow\; \omega \in A_n \;\Rightarrow\; \omega \notin A_n^c \;\Rightarrow\; \omega \notin B_m$. Thus, for $n < m, B_n \cap B_m = \emptyset$. Similarly we can prove that for $n > m, B_n \cap B_m = \emptyset$, that is $\{B_n, n \geq 1\}$ is a sequence of disjoint sets in $\mathbb{A}$. Further note that,

$$\omega \in \bigcup_{n=1}^{m} A_n \;\Rightarrow\; \omega \in A_n \text{ for at least one } n, \text{ say } n_0, \text{which is first such } n \leq m$$

$$\Rightarrow\; \omega \in A_1^c \cap A_2^c \cap \cdots A_{n_0-1}^c \cap A_{n_0} = B_{n_0} \;\Rightarrow\; \omega \in \bigcup_{n=1}^{m} B_n.$$

Hence, $\bigcup_{n=1}^{m} A_n \subseteq \bigcup_{n=1}^{m} B_n$. Now suppose,

$$\begin{aligned} \omega \in \bigcup_{n=1}^{m} B_n &\Rightarrow \omega \in B_n, \quad \text{for only one } n, \text{ say } n_0, 1 \le n_0 \le m \\ &\Rightarrow \omega \in B_{n_0} = A_1^c \cap A_2^c \cap \cdots A_{n_0-1}^c \cap A_{n_0} \\ &\Rightarrow \omega \in A_{n_0} \Rightarrow \omega \in \bigcup_{n=1}^{m} A_n. \end{aligned}$$

Thus, $\bigcup_{n=1}^{m} B_n \subseteq \bigcup_{n=1}^{m} A_n$ and hence $\bigcup_{n=1}^{m} B_n = \bigcup_{n=1}^{m} A_n$. We now prove that the relation is true for countable union as well.

$$\begin{aligned} \omega \in \bigcup_{n\ge 1} A_n &\Rightarrow \omega \in A_n \text{ for at least one } n, \text{ say } n_0, \text{ which is first such } n \\ &\Rightarrow \omega \in A_1^c \cap A_2^c \cap \cdots A_{n_0-1}^c \cap A_{n_0} = B_{n_0} \Rightarrow \omega \in \bigcup_{n\ge 1} B_n. \end{aligned}$$

Hence, $\bigcup_{n\ge 1} A_n \subseteq \bigcup_{n\ge 1} B_n$. Now suppose,

$$\begin{aligned} \omega \in \bigcup_{n\ge 1} B_n &\Rightarrow \omega \in B_n, \quad \text{for only one } n, \text{ say } n_0 \\ &\Rightarrow \omega \in B_{n_0} = A_1^c \cap A_2^c \cap \cdots A_{n_0-1}^c \cap A_{n_0} \\ &\Rightarrow \omega \in A_{n_0} \Rightarrow \omega \in \bigcup_{n\ge 1} A_n . \end{aligned}$$

Thus, $\bigcup_{n\ge 1} B_n \subseteq \bigcup_{n\ge 1} A_n$ and hence $\bigcup_{n\ge 1} B_n = \bigcup_{n\ge 1} A_n$. □

Theorem 1.5.2. *Continuity theorem for probability measure: Suppose $(\Omega, \mathbb{A}, P)$ is a probability space.*

(i) If $\{A_n, n \ge 1\}$ is a non-decreasing sequence of sets in $\mathbb{A}$, then $P(A_n) \to P(A)$, as $n \to \infty$ where $A = \lim_{n\to\infty} A_n = \bigcup_{n\ge 1} A_n$.

(ii) If $\{A_n, n \ge 1\}$ is a non-increasing sequence of sets in $\mathbb{A}$, then $P(A_n) \to P(A)$, as $n \to \infty$ where $A = \lim_{n\to\infty} A_n = \bigcap_{n\ge 1} A_n$.

(iii) If $\{A_n, n \ge 1\}$ is a convergent sequence of sets in $\mathbb{A}$, such that $\lim_{n\to\infty} A_n = A$, then

$$\lim_{n\to\infty} P(A_n) = P(A) \iff \lim_{n\to\infty} P(A_n) = P(\lim_{n\to\infty} A_n).$$

Proof.

(i) By Theorem 1.3.1, if $\{A_n, n \ge 1\}$ is a non-decreasing sequence of sets in $\mathbb{A}$, then $\lim_{n\to\infty} A_n = \bigcup_{n\ge 1} A_n = A$. Hence by Lemma 1.5.1 and σ-additivity,

$$
\begin{aligned}
P(A) &= P\Big(\lim_{n\to\infty} A_n\Big) = P\Big(\bigcup_{m\geq 1} A_m\Big) = P\Big(\sum_{m\geq 1} B_m\Big) = \sum_{m\geq 1} P(B_m) \\
&= \lim_{n\to\infty} \sum_{m=1}^{n} P(B_m) = \lim_{n\to\infty} P\Big(\sum_{m=1}^{n} B_m\Big) \quad \text{by finite additivity} \\
&= \lim_{n\to\infty} P\Big(\bigcup_{m=1}^{n} A_m\Big) \quad \text{by Lemma 1.5.1} \\
&= \lim_{n\to\infty} P(A_n), \quad \text{as } \bigcup_{m=1}^{n} A_m = A_n,
\end{aligned}
$$

since $\{A_n, n \geq 1\}$ is a non-decreasing sequence.

(ii) Now $\{A_n, n \geq 1\}$ is a non-increasing sequence of sets in $\mathbb{A}$, hence as proved in Theorem 1.3.2, $\{A_n^c, n \geq 1\}$ is a non-decreasing sequence of sets in $\mathbb{A}$ and it is convergent with limit as $A^c = \bigcup_{n\geq 1} A_n^c = (\bigcap_{n\geq 1} A_n)^c$. Thus by part (i),

$$
\begin{aligned}
\lim_{n\to\infty} P(A_n^c) = P(A^c) &\iff \lim_{n\to\infty} (1 - P(A_n)) = 1 - P(A) \\
&\iff \lim_{n\to\infty} P(A_n) = P(A).
\end{aligned}
$$

(iii) Since $\{A_n, n \geq 1\}$ is a convergent sequence with limit A, we have

$$
\limsup A_n = \liminf A_n = \lim A_n = A,
$$

where, $\limsup A_n$ and $\liminf A_n$ are defined as,

$$
\limsup A_n = \bigcap_{n\geq 1} \bigcup_{k\geq n} A_k = \bigcap_{n\geq 1} C_n, \quad \text{where} \quad C_n = \bigcup_{k\geq n} A_k
$$

$$
\& \quad \liminf A_n = \bigcup_{n\geq 1} \bigcap_{k\geq n} A_k = \bigcup_{n\geq 1} B_n, \quad \text{where} \quad B_n = \bigcap_{k\geq n} A_k.
$$

As shown in Section 1.3, $\{C_n, n \geq 1\}$ is a non-increasing sequence of sets and hence is convergent with

$$
\begin{aligned}
\lim_{n\to\infty} C_n &= \bigcap_{n\geq 1} C_n = \limsup A_n = A \\
\Rightarrow \quad \lim_{n\to\infty} P(C_n) &= P(\lim_{n\to\infty} C_n) = P(\limsup A_n) = P(A),
\end{aligned}
$$

by part(ii). Similarly, $\{B_n, n \geq 1\}$ is a non-decreasing sequence of sets and hence is convergent with

$$
\begin{aligned}
\lim_{n\to\infty} B_n &= \bigcup_{n\geq 1} B_n = \liminf A_n = A \\
\Rightarrow \quad \lim_{n\to\infty} P(B_n) &= P(\lim_{n\to\infty} B_n) = P(\liminf A_n) = P(A),
\end{aligned}
$$

by part (i). Now,

$$
\begin{aligned}
\bigcap_{k\geq n} A_k \subset A_n \subset \bigcup_{k\geq n} A_k \iff & B_n \subset A_n \subset C_n \\
\Rightarrow & P(B_n) \leq P(A_n) \leq P(C_n) \\
\Rightarrow & \lim_{n\to\infty} P(B_n) \leq \lim_{n\to\infty} P(A_n) \leq \lim_{n\to\infty} P(C_n) \\
\Rightarrow & P(\liminf A_n) \leq \lim_{n\to\infty} P(A_n) \leq P(\limsup A_n) \\
\Rightarrow & P(A) \leq \lim_{n\to\infty} P(A_n) \leq P(A) \\
\Rightarrow & \lim_{n\to\infty} P(A_n) = P(A) = P(\lim_{n\to\infty} A_n).
\end{aligned}
$$

□

Remark 1.5.2. (i) Theorem 1.5.2 is labeled as a continuity theorem for probability measures in view of its similarity with the property of a real valued continuous function. If $\{x_n, n \geq 1\}$ is a sequence of real numbers and if g is a continuous function then $x_n \to x$ implies $g(x_n) \to g(x)$. On similar lines, if $A_n \to A$ then $P(A_n) \to P(A)$. (ii) The sigma additivity condition of the probability measure is sometimes referred to as the continuity condition of a probability measure, Bhat [4].

Theorem 1.5.3. *Suppose* $(\Omega, \mathbb{A}, P)$ *is a probability space. If* $\{A_n, n \geq 1\}$ *is a sequence of sets in* $\mathbb{A}$, *then* $P\Big(\bigcup_{n\geq 1} A_n\Big) \leq \sum_{n\geq 1} P(A_n)$.

Proof. As in Lemma 1.5.1, for a given sequence $\{A_n, n \geq 1\}$ of sets in $\mathbb{A}$, we have another sequence of sets $\{B_n, n \geq 1\}$ as,

$$B_1 = A_1 \quad \text{and for} \quad n \geq 2, \quad B_n = A_1^c \cap A_2^c \cap \cdots A_{n-1}^c \cap A_n.$$

It is proved in Lemma 1.5.1 that $\{B_n, n \geq 1\}$ is a sequence of disjoint sets in $\mathbb{A}$ and $\bigcup_{n\geq 1} A_n = \sum_{n\geq 1} B_n$. By the definition of B_n, $B_n \subseteq A_n$, hence $P(B_n) \leq P(A_n)$. Thus,

$$P\Big(\bigcup_{n\geq 1} A_n\Big) = P\Big(\sum_{n\geq 1} B_n\Big) = \sum_{n\geq 1} P(B_n) \leq \sum_{n\geq 1} P(A_n).$$

□

Remark 1.5.3. The inequality proved in Theorem 1.5.3 is known as Boole's inequality and this property of probability measure is known as sub sigma additivity.

Theorem 1.5.4. *Suppose* $(\Omega, \mathbb{A}, P)$ *is a probability space. If* $\{A_n, n \geq 1\}$ *is a sequence of sets in* $\mathbb{A}$, *then* *(i)* $P(\liminf A_n) \leq \liminf P(A_n) \leq \limsup P(A_n) \leq P(\limsup A_n)$.

(ii) If $\{A_n, n \geq 1\}$ *is a convergent sequence of sets in* $\mathbb{A}$*, such that* $\lim_{n\to\infty} A_n = A$*, then* $\lim_{n\to\infty} P(A_n) = P(A)$.

Proof.

(i) By the definition,

$$\liminf A_n = \bigcup_{n\geq 1}\bigcap_{k\geq n} A_k = \bigcup_{n\geq 1} B_n \quad \text{where} \quad B_n = \bigcap_{k\geq n} A_k.$$

Observe that by the definition of B_n, $B_n \subseteq A_n \Rightarrow P(B_n) \leq P(A_n)$. Further, it is known that $\{B_n, n \geq 1\}$ is a non-decreasing sequence of sets and hence is convergent with $\lim_{n\to\infty} B_n = \bigcup_{n\geq 1} B_n = \liminf A_n$. Hence, by continuity theorem for probability measure,

$$\lim_{n\to\infty} B_n = \liminf A_n \quad \Rightarrow \quad \lim_{n\to\infty} P(B_n) = P(\lim_{n\to\infty} B_n) = P(\liminf A_n).$$

Further, $\lim_{n\to\infty} P(B_n)$ exists and hence $\lim_{n\to\infty} P(B_n) = \liminf P(B_n)$. Thus, we have

$$\begin{aligned} P(\liminf A_n) &= P(\lim_{n\to\infty} B_n) = \lim_{n\to\infty} P(B_n) = \liminf P(B_n) \\ &= \lim(\inf_{k\geq n} P(B_k)) \\ &\leq \lim(\inf_{k\geq n} P(A_k)), \quad \text{as} \quad B_k \subseteq A_k, \ \forall \ k \geq n \\ &= \liminf P(A_n). \end{aligned} \tag{1.1}$$

Now using the result $(\limsup A_n)^c = \liminf A_n^c$, we have

$$\begin{aligned} P(\limsup A_n) &= 1 - P((\limsup A_n)^c) = 1 - P(\liminf A_n^c) \\ &\geq 1 - \liminf P(A_n^c), \quad \text{by (1.1)} \\ &= 1 - \liminf(1 - P(A_n)) = 1 - \sup_{n\geq 1}(\inf_{k\geq n}(1 - P(A_k))) \\ &= 1 - \sup_{n\geq 1}((1 - \sup_{k\geq n} P(A_k))) = 1 - (1 - \inf_{n\geq 1}\sup_{k\geq n} P(A_k)) \\ &= \inf_{n\geq 1}\sup_{k\geq n} P(A_k) = \limsup P(A_n) \end{aligned} \tag{1.2}$$

It is well known that for a sequence of real numbers $\{a_n, n \geq 1\}$, $\liminf a_n \leq \limsup a_n$. Hence using this result and combining (1.1) and (1.2), we have

$$P(\liminf A_n) \leq \liminf P(A_n) \leq \limsup P(A_n) \leq P(\limsup A_n).$$

(ii) Since $\{A_n, n \geq 1\}$ is a convergent sequence with limit A, we have $\limsup A_n = \liminf A_n = \lim A_n = A$. Hence by (i) we get

$$\begin{aligned} P(A) &= P(\liminf A_n) \leq \liminf P(A_n) \leq \limsup P(A_n) \\ &\leq P(\limsup A_n) = P(A). \end{aligned}$$

Thus, $\liminf P(A_n) = P(A)$ and $\limsup P(A_n) = P(A)$, hence $\lim P(A_n)$ exists and is equal to $P(A)$.

□

Remark 1.5.4. Result (ii) of Theorem 1.5.4 is the same as result (iii) of Theorem 1.5.2.

Example 1.5.3. Suppose $(\Omega, \mathbb{A}, P)$ is a probability space, where $\Omega = \{\omega_1, \omega_2, \omega_3, \omega_4\}$, $\mathbb{A}$ is a powerset of Ω and probability measure is defined by

$$P(\{\omega_1\}) = 2/9, \quad P(\{\omega_2\}) = 1/9, \quad P(\{\omega_3\}) = 4/9, \quad P(\{\omega_4\}) = 2/9.$$

A set $A_n \in \mathbb{A}$ is defined as $A_n = \{\omega_1, \omega_2\}$ if n is odd and $A_n = \{\omega_2, \omega_3\}$ if n is even. We find $P(\liminf A_n), P(\limsup A_n)$ and $\liminf P(A_n), \limsup P(A_n)$ and examine the relation established in Theorem 1.5.4. By the definition,

$$\begin{aligned}
\liminf A_n &= \bigcup_{n\geq 1}\bigcap_{k\geq n} A_k = \bigcup_{n\geq 1}\{A_n \cap A_{n+1} \cap \cdots\} \\
&= \bigcup_{n\geq 1}\{\{\omega_1, \omega_2\} \cap \{\omega_2, \omega_3\} \cap \{\omega_1, \omega_2\} \cdots\} \quad \text{if } n \text{ is odd} \\
&= \bigcup_{n\geq 1}\{\{\omega_2, \omega_3\} \cap \{\omega_1, \omega_2\} \cap \{\omega_1, \omega_2\} \cdots\} \quad \text{if } n \text{ is even} \\
&= \bigcup_{n\geq 1}\{\omega_2\} = \{\omega_2\}.
\end{aligned}$$

Similarly by the definition,

$$\begin{aligned}
\limsup A_n &= \bigcap_{n\geq 1}\bigcup_{k\geq n} A_k = \bigcup_{n\geq 1}\{A_n \cup A_{n+1} \cup \cdots\} \\
&= \bigcap_{n\geq 1}\{\{\omega_1, \omega_2\} \cup \{\omega_2, \omega_3\} \cup \{\omega_1, \omega_2\} \cdots\} \\
&= \bigcap_{n\geq 1}\{\omega_1, \omega_2, \omega_3\} = \{\omega_1, \omega_2, \omega_3\}.
\end{aligned}$$

Note that $\liminf A_n \subset \limsup A_n$, the two sets are not the same and hence $\lim A_n$ does not exist. Further, $P(\liminf A_n) = 1/9$ and $P(\limsup A_n) = 7/9$. For the given probability measure, $P(A_n) = 3/9$ if n is odd and $5/9$ if n is even. It is then clear that

$$\begin{aligned}
\liminf P(A_n) &= \sup_{n\geq 1}\inf_{k\geq n} P(A_k) = \sup_{n\geq 1} 3/9 = 3/9 \\
\& \quad \limsup P(A_n) &= \inf_{n\geq 1}\sup_{k\geq n} P(A_k) = \inf_{n\geq 1} 5/9 = 5/9.
\end{aligned}$$

Thus, $\liminf P(A_n) \neq \limsup P(A_n)$ hence $\lim P(A_n)$ does not exist. Further,

$$\frac{1}{9} = P(\underline{\lim} A_n) < \underline{\lim} P(A_n) = \frac{3}{9} < \overline{\lim} P(A_n) = \frac{5}{9} < P(\overline{\lim} A_n) = \frac{7}{9}.$$

Thus, the relation established in Theorem 1.5.4 is verified. □

Example 1.5.4. Suppose $(\Omega, \mathbb{A}, P)$ is a probability space and $\{A_n, n \geq 1\}$ is a sequence of disjoint sets from $\mathbb{A}$. We find $\lim_{n\to\infty} P(A_n)$. By the definition,

$$\liminf A_n = \bigcup_{n\geq 1}\bigcap_{k\geq n} A_k = \bigcup_{n\geq 1} \emptyset = \emptyset$$

$$\& \ \limsup A_n = \bigcap_{n\geq 1}\bigcup_{k\geq n} A_k = \bigcap_{n\geq 1} C_n, \ \text{where } C_n = \bigcup_{k\geq n} A_k = A_n \cup A_{n+1} \cup \cdots .$$

Observe that,

$$\begin{aligned} \omega \in A_1 \quad &\Rightarrow \quad \omega \notin A_n \ \text{ for any } \ n \neq 1 \ \Rightarrow \ \omega \notin C_n \ \text{ for any } \ n \neq 1 \\ &\Rightarrow \quad \omega \notin \bigcap_{n\geq 1} C_n = \limsup A_n. \end{aligned}$$

On similar lines if ω is in any one of the sets of the sequence $\{A_n, n \geq 1\}$, then it cannot be in all C_n sets and hence in $\limsup A_n$. It is true for all $\omega \in \Omega$ and hence $\limsup A_n = \emptyset$. Thus $\lim_{n\to\infty} A_n$ exists and it is $\emptyset$. Hence, by the continuity theorem for a probability measure

$$\lim_{n\to\infty} P(A_n) = P(\lim_{n\to\infty} A_n) = P(\emptyset) = 0. \qquad \square$$

Example 1.5.5. Suppose $(\Omega, \mathbb{A}, P)$ is a probability space and $A_n \in \mathbb{A}$ is defined as $A_n = (-\infty, a + 1/n], n \geq 1$. Note that

$$\begin{aligned} A_n \quad \supset \quad & A_{n+1} \ \forall \ n \geq 1 \ \Rightarrow \ \{A_n, n \geq 1\} \ \text{ is a non-increasing sequence} \\ \Rightarrow \quad & \lim_{n\to\infty} A_n = \bigcap_{n\geq 1} A_n = (-\infty, a] = A, \ \text{ say} \\ \Rightarrow \quad & \lim_{n\to\infty} P(A_n) = P(A) \ \text{ by the continuity theorem.} \end{aligned}$$

In general, suppose $\{h_n, n \geq 1\}$ is a non-increasing sequence of positive rational numbers tending to 0 as $n \to \infty$ and A_n is defined as $A_n = (-\infty, a + h_n]$, $n \geq 1$. Then again $\{A_n, n \geq 1\}$ is a non-increasing sequence of sets with limit as $\bigcap_{n\geq 1} A_n = (-\infty, a] = A$ and $P(A_n) \downarrow P(A)$. Now suppose B_n is defined as $B_n = (-\infty, a - h_n], n \geq 1$. Observe that,

$$\begin{aligned} B_n \quad \subset \quad & B_{n+1} \ \forall \ n \geq 1 \ \Rightarrow \ \{B_n, n \geq 1\} \ \text{ is a non-decreasing sequence} \\ \Rightarrow \quad & \lim_{n\to\infty} B_n = \bigcup_{n\geq 1} B_n = (-\infty, a) = B, \ \text{ say} \\ \Rightarrow \quad & \lim_{n\to\infty} P(B_n) = P(B) \ \text{ by the continuity theorem.} \ \square \end{aligned}$$

We will come across sequences of events similar to this example in Chapter 3, while discussing the right continuity and left continuity of a distribution function.

A quick recap of the results discussed in this chapter is given below.

Summary

1. Suppose $\{A_n, n \geq 1\}$ is a sequence of subsets of Ω. Then limit supremum and limit infimum of a sequence of sets are defined as,

$$\limsup A_n = \bigcap_{n\geq 1}\bigcup_{k\geq n} A_k \quad \& \quad \liminf A_n = \bigcup_{n\geq 1}\bigcap_{k\geq n} A_k.$$

2. $\liminf A_n \subseteq \limsup A_n$.

3. If $\{A_n, n \geq 1\}$ is a monotone sequence of subsets of Ω, then $\lim_{n\to\infty} A_n$ exists. If it is non-decreasing sequence, then $\lim_{n\to\infty} A_n = \bigcup_{n\geq 1} A_n$. If it is a non-increasing sequence, then $\lim_{n\to\infty} A_n = \bigcap_{n\geq 1} A_n$.

4. If $\{A_n, n \geq 1\}$ is a non-monotone sequence of subsets of Ω, then $\lim_{n\to\infty} A_n$ exists if $\limsup A_n = \liminf A_n$.

5. A non-empty class $\mathbb{A}$ of subsets of Ω is called a field or an algebra if it is closed under complementation and union of two sets.

6. The empty set and the universal set are members of a field and a field is closed under finite unions and finite intersections.

7. A non-empty class $\mathbb{A}$ of subsets of Ω is called a sigma field or a sigma algebra, if it is closed under complementation and countable union.

8. Properties of a sigma field: (i) A sigma field is a field. (ii) The empty set and the universal set are members of a sigma field. (iii) A sigma field is closed under countable intersection, limits of monotone sequences, limit supremum and limit infimum of sets and limit of a sequence, if it exists. (iv) Any arbitrary intersection of sigma fields is a sigma field.

9. Suppose $\mathbb{C}$ is a non-empty class of subsets of Ω. Then the minimal sigma field $\sigma(\mathbb{C})$ covering the class $\mathbb{C}$, or generated by the class $\mathbb{C}$, satisfies the following conditions: (i) $\sigma(\mathbb{C})$ is a sigma field, (ii) $\mathbb{C} \subseteq \sigma(\mathbb{C})$ and (iii) if $\sigma^*(\mathbb{C})$ is any sigma field generated by $\mathbb{C}$, then $\sigma^*(\mathbb{C}) \supseteq \sigma(\mathbb{C})$.

10. There exists a unique minimal sigma field generated by a non-empty class $\mathbb{C}$ of subsets of Ω.

11. The minimal sigma field generated by a class of any type of intervals of real line is known as a Borel field. A set in the Borel field is known as a Borel set.

12. If $I_k = \{(x_1, x_2, \cdots, x_k) | a_i \leq x_i \leq b_i, a_i, b_i, x_i \in \mathbb{R}, i = 1, 2, \cdots, k\}$ is a k dimensional rectangle and $\mathbb{C}_k$ is the collection of rectangles of the type I_k, then the minimal sigma field generated by $\mathbb{C}_k$, denoted by $\mathbb{B}^k$, is known as the Borel field in k dimensions.

13. Suppose Ω is a non-empty universal set and $\mathbb{A}$ is a sigma field of subsets of Ω. Then a pair $(\Omega, \mathbb{A})$ is known as a measurable space.

14. A set function P defined on $(\Omega, \mathbb{A})$ is known as a probability measure, if the following conditions are satisfied.
(i) $P(A) \geq 0, \ \forall \ A \in \mathbb{A}$, (non-negativity)
(ii) $P(\Omega) = 1$, (normed measure)
(iii) If $\{A_n, n \geq 1\}$ is a sequence of disjoint sets in $\mathbb{A}$, then $P(\bigcup_{n\geq 1} A_n) = \sum_{n\geq 1} P(A_n)$, ($\sigma$- additivity).

15. Properties of a probability measure: (i) $P(\emptyset) = 0$. (ii) Probability measure is finitely additive. (iii) $P(A^c) = 1 - P(A)$. (iv) If $A \subseteq B$, then $P(A) \leq P(B)$. (v) $P(A) \leq 1$.

16. Continuity theorem for probability measures: Suppose $(\Omega, \mathbb{A}, P)$ is a probability space. Then
(i) If $\{A_n, n \geq 1\}$ is a non-decreasing sequence of sets in $\mathbb{A}$, then $P(A_n) \to P(A)$, as $n \to \infty$ where $A = \lim_{n\to\infty} A_n = \bigcup_{n\geq 1} A_n$.
(ii) If $\{A_n, n \geq 1\}$ is a non-increasing sequence of sets in $\mathbb{A}$, then $P(A_n) \to P(A)$, as $n \to \infty$ where $A = \lim_{n\to\infty} A_n = \bigcap_{n\geq 1} A_n$.
(iii) If $\{A_n, n \geq 1\}$ is a convergent sequence of sets in $\mathbb{A}$, such that $\lim_{n\to\infty} A_n = A$, then $\lim_{n\to\infty} P(A_n) = P(A)$.

17. Suppose $(\Omega, \mathbb{A}, P)$ is a probability space. If $\{A_n, n \geq 1\}$ is a sequence of sets in $\mathbb{A}$, then

$$P(\liminf A_n) \leq \liminf P(A_n) \leq \limsup P(A_n) \leq P(\limsup A_n).$$

1.6 Conceptual Exercises

1.6.1 Verify whether the following sequences of the sets are monotone.
(i) $\{(a - \frac{1}{n}, b + \frac{1}{n}), n \geq 1\}$. (ii) $\{(a + \frac{1}{n}, b - \frac{1}{n}), n \geq 1\}$.
(iii) $\{(a + \frac{1}{n}, b + \frac{1}{n}), n \geq 1\}$. (iv) Suppose $\{A_n, n \geq 1\}$ is a sequence of the sets, where $A_n = \{(x, y)|0 \leq x \leq n, 0 \leq y \leq 1/n\}$.

1.6.2 For the following sequences $\{A_n, n \geq 1\}$ of sets examine whether it has a limit. If yes, find it.
(i) $A_n = (a - 1/n, a + 1/n)$, (ii)$A_n = [a - 1/n, a + 1/n]$,
(iii) $A_n = (a - 1/n, a + 1/n]$, (iv) $A_n = [a - 1/n, a + 1/n)$,
(v) $A_n = (a - 1/n, a]$, (vi) $A_n = (a, b + 1/n)$, (vii) $A_n = (a, b + 1/n]$,
(viii) $A_n = [a, b + 1/n)$, (ix) $A_n = [a, b + 1/n]$, (x) $A_n = (a, b - 1/n)$,
(xi) $A_n = (a, b - 1/n]$, (xii) $A_n = [a, b - 1/n)$, (xiii) $A_n = [a, b - 1/n]$,
(xiv) $A_n = (a - 1/n, b + 1/n)$, (xv) $A_n = [a - 1/n, b + 1/n]$,
(xvi) $A_n = (a - 1/n, b + 1/n]$, (xvii) $A_n = [a - 1/n, b + 1/n)$.

1.6.3 For the following sequences $\{A_n, n \geq 1\}$ of sets, examine whether it has a limit. If yes, find it.

$$(i)\ A_n = \begin{cases} (0, 1/n), & \text{if } n \text{ is odd} \\ (-1/n, 1/n), & \text{if } n \text{ is even.} \end{cases}$$

$$(ii)\ A_n = \begin{cases} (0, 1/n), & \text{if } n \text{ is odd} \\ (1 - 1/n, 1), & \text{if } n \text{ is even.} \end{cases}$$

$$(iii)\ A_n = \begin{cases} (0, 1 - 1/n), & \text{if } n \text{ is odd} \\ (1/n, 1), & \text{if } n \text{ is even.} \end{cases}$$

$$(iv)\ A_n = \begin{cases} \left(2 - \frac{1}{n}, 2 + \frac{1}{n}\right), & \text{if } n \text{ is odd} \\ \emptyset, & \text{if } n \text{ is even.} \end{cases}$$

$$(v)\ A_n = \begin{cases} \left(2 - \frac{1}{n}, 2\right), & \text{if } n \text{ is odd} \\ \emptyset, & \text{if } n \text{ is even.} \end{cases}$$

(vi) $A_n = \{0, 1/n, 2/n, \cdots, 1 - 1/n, 1\}$.
(vii) A_n : set of rationals in $\left(1 - \frac{1}{n+1}, 1 + \frac{1}{n}\right]$,
(viii) $A_{2n} = \left(0, \frac{1}{2n}\right), A_{2n+1} = \left[-1, \frac{1}{2n+1}\right], A_1 = [-1, 1]$,
(ix) A_n: set of rationals in $(1 - \frac{1}{n}, 1 + \frac{1}{n})$.

1.6.4 Suppose a sequence $\{A_n, n \geq 1\}$ is defined as follows.

$$A_n = \begin{cases} B, & \text{if } n \text{ is even} \\ C, & \text{if } n \text{ is odd.} \end{cases}$$

Examine whether the sequence $\{A_n, n \geq 1\}$ converges. If yes find the limit. If not, impose a condition on B and C so that it is a convergent sequence.

1.6.5 Examine the limiting behaviour of the following sequence.

$$A_n = \big(-\infty, 0\big) \cup (1/n, \infty), \quad \forall\ n \in \mathbb{N}.$$

1.6.6 Give an example of a sequence of sets $\{A_n, n \geq 1\}$ where $\limsup A_n \neq \liminf A_n$.

1.6.7 Give an example to show that $\liminf A_n \cup \liminf B_n \neq \liminf(A_n \cup B_n)$.

1.6.8 Examine whether (i) $\limsup(A_n \bigcap B_n) = \limsup A_n \bigcap \limsup B_n$ and (ii) $\liminf(A_n \bigcup B_n) = \liminf A_n \bigcup \liminf B_n$ if sequences $\{A_n, n \geq 1\}$ and $\{B_n, n \geq 1\}$ are defined as follows. $A_1 = \{1\}$, $A_{2n} = \{2n\}, A_{2n+1} = [1, 2n+1]$ and $B_1 = \{1\}$, $B_{2n} = [1, 2n]$, $B_{2n+1} = \{2n+1\}$.

1.6.9 Examine whether

$$
\begin{aligned}
(i) \liminf A_n \cup \ \liminf B_n &\subseteq \liminf (A_n \cup B_n) \\
&\subseteq \limsup (A_n \cup B_n) \\
&= \limsup A_n \cup \limsup B_n \\
\& \quad (ii) \liminf A_n \cap \ \liminf B_n &= \liminf (A_n \cap B_n) \\
&\subseteq \limsup (A_n \cap B_n) \\
&\subseteq \limsup A_n \cap \limsup B_n .
\end{aligned}
$$

1.6.10 Give an example where the sequences $\{A_n, n \geq 1\}$ and $\{B_n, n \geq 1\}$ do not converge, but $\{A_n \cap B_n, n \geq 1\}$ and $\{A_n \cup B_n, n \geq 1\}$ converge.

1.6.11 Suppose $A = \limsup A_n$ and $B = \liminf A_n$. Then show that $M - A = \liminf(M - A_n)$ and $M - B = \limsup(M - A_n)$ for any $M \in \Omega$.

1.6.12 Suppose Ω is a finite set. Examine whether (i) a collection of all finite subsets of Ω is a field and (ii) a collection of all singleton subsets of Ω is a field.

1.6.13 Examine whether the following classes of subsets of Ω are field and / or sigma field, where Ω is an uncountable set. Suppose $A \subseteq \Omega$ (i) $\{\Omega, \emptyset\}$ (ii) $\{\Omega, \emptyset, A\}$, (iii) $\{\Omega, \emptyset, A, A^c\}$, (iv) $\mathbb{P}(\Omega)$, (v) the class of all finite subsets of Ω and (vi) the class of all countable subsets of Ω.

1.6.14 Suppose $\mathbb{F}$ is a sigma field of subsets of Ω. Suppose $E \in \mathbb{F}$ and $E \neq \emptyset$. Show that $\mathbb{F}_E = \{A \cap E | A \in \mathbb{F}\}$ is the sigma field of subsets of E.

1.6.15 Suppose $\mathbb{C}_2$ is a class of intervals of the type $(a, b], a < b \in \mathbb{R}$ and $\mathbb{C}_6$ is a class of intervals of the type $(-\infty, a], a \in \mathbb{R}$. Then show that the minimal sigma field generated by $\mathbb{C}_2$ is the same as that by $\mathbb{C}_6$.

1.6.16 Give an example of a class of events, which is closed under finite unions and finite intersections but not under complementation.

1.6.17 Suppose Ω is a finite set and $\mathbb{A}$ is a field of subsets of Ω. Show that it is a sigma field.

1.6.18 Suppose $\{\mathbb{F}_n, n \geq 1\}$ is a sequence of fields of subsets of Ω, such that $\mathbb{F}_n \subseteq \mathbb{F}_{n+1}, \ \forall \ n \geq 1$. Prove that $\bigcup_{n=1}^{\infty} \mathbb{F}_n$ is a field. What can one say if
$\{\mathbb{F}_n, n \geq 1\}$ is a sequence of sigma fields?

1.6.19 Suppose $\Omega = \mathbb{N}$, the set of all natural numbers and $A_n = \{1, 2, \ldots, n\}$, $n \geq 1$. If $\mathbb{A}_n = \sigma(\{A_n\})$, verify whether $\{\mathbb{A}_n, n \geq 1\}$is an increasing sequence.

1.6.20 Suppose $\mathbb{P} = \{A_1, A_2, \ldots, \}$ is a countable partition of Ω and $\mathbb{A} = \{E | E = \emptyset$ or E is a union of sets in $\mathbb{P}\}$. Show that $\mathbb{A}$ is the minimal sigma field containing $\mathbb{P}$.

1.6.21 Suppose $\mathbb{P}_1 = \{A_i, i = 1, 2, \cdots, m\}$ and $\mathbb{P}_2 = \{B_j, j = 1, 2, \cdots, n\}$ are two partitions of Ω. Show that $\sigma(\sigma(\mathbb{P}_1) \cup \sigma(\mathbb{P}_2)) = \sigma(\mathbb{P})$, where $\mathbb{P} = \{A_i \cap B_j, i = 1, 2, \cdots, m, j = 1, 2, \cdots, n\}$ is a partition of Ω.

1.6.22 Suppose $(\Omega, \mathbb{A}, P)$ is a probability space. Prove that for $\{A_1, A_2, \cdots, A_n\}$ in $\mathbb{A}$, $P(\bigcap_{i=1}^{n} A_i) \geq \sum_{i=1}^{n} P(A_i) - (n-1)$.

1.6.23 Suppose $(\Omega, \mathbb{A}, P)$ is a probability space. Suppose $\{A_1, A_2, A_3\} \in \mathbb{A}$ are such that $P(A_1 \cap A_2 \cap A_3) = P(A_1^c \cap A_2^c \cap A_3^c)$. Then show that each is equal to
$(1/2)[1-(P(A_1)+P(A_2)+P(A_3))+P(A_1 \cap A_2)+P(A_1 \cap A_3)+P(A_2 \cap A_3)]$.

1.6.24 Show that if $\sum_{n \geq 1} P(A_n) < \infty$, then $P(\limsup A_n) = 0$. Examine if $P(\liminf A_n) = 0$.

1.6.25 Suppose $\Omega = \mathbb{W}$, a set of whole numbers and $\mathbb{A} = \mathbb{P}(\Omega)$ and a set function P on $(\Omega, \mathbb{P}(\Omega))$ is defined as $P(A) = \sum_{i \in A} p(1-p)^i$ for $A \in \mathbb{P}(\Omega)$ and $p \in (0, 1)$. Examine whether P is a probability measure on $(\Omega, \mathbb{P}(\Omega)$.

1.6.26 Suppose $(\Omega, \mathbb{P}(\Omega), P)$ is a probability space, where $P(\{\omega\}) = p \in (0, 1)$. Show that Ω must be finite. If Ω consists of k elements, then p must be $1/k$.

1.6.27 Suppose $\Omega = \mathbb{N}$, a set of natural numbers and P is a set function defined on $(\Omega, \mathbb{P}(\Omega))$ such that

$$P(A) = \begin{cases} 0, & \text{if } A \text{ is finite} \\ 1, & \text{if } A \text{ is infinite.} \end{cases}$$

Is P a probability measure on $(\Omega, \mathbb{P}(\Omega))$?

1.6.28 Suppose $\Omega = \mathbb{W}$, a set of whole numbers, $\mathbb{A}$ is a power set of Ω. Examine whether set functions P defined below in (i), (ii) and (iii) are probability measures on $(\Omega, \mathbb{A})$. (i) P(A) = number of points in A, (ii) for a non-empty set $A \in \mathbb{A}$, $P(A) = \sum_{x \in A} \mu(\{x\})$, where $\mu(\{x\}) = 1/2^x, x = 1, 2, \cdots$, $\mu(\{0\}) = 1/2$ and $\mu(\emptyset) = 0$ and (iii) $\mu(\{x\}) = 1/2^{x+1}, x \in \Omega$.

1.6.29 Suppose $(\Omega, \mathbb{A}, P)$ is a probability space and $\{A_n, n \geq 1\}$ and $\{B_n, n \geq 1\}$ are sequences of sets from $\mathbb{A}$. Suppose $P(\limsup A_n) = P(\limsup B_n) = 1$. Examine whether (i) $P(\limsup A_n \cap \limsup B_n) = 1$ and (ii) $P(\limsup A_n \cap B_n) = 1$.

1.6.30 Suppose $(\Omega, \mathbb{A})$ is a measurable space and $B \in \mathbb{A}$ is any non empty subset of Ω. Suppose a class is defined as $B \cap \mathbb{A} = \{B \cap A | A \in \mathbb{A}\} = \mathbb{A}_B$ say. Show that $\mathbb{A}_B$ is a sigma field of subsets of B. A function $P_B : (B, \mathbb{A}_B) \to [0, 1]$ is defined as $P_B(A) = P(B \cap A)/P(B) \;\; \forall \; A \in \mathbb{A}$. Show that P_B is a probability measure on $(B, \mathbb{A}_B)$.

1.6.31 Suppose $\{B_n, n \geq 1\}$ is a sequence of sets from $\mathbb{A}$ such that $P(B_n) = 1, \ \forall \ n \geq 1$. Then show that $P(\bigcap_{n\geq 1} B_n) = 1$.

1.7 Computational Exercises

1.7.1 For any non-decreasing sequence $\{A_n, n \geq 1\}$, find its limit and demonstrate the limit of a sequence graphically.

1.7.2 For any non-increasing sequence $\{A_n, n \geq 1\}$, find its limit and demonstrate the limit of a sequence graphically.

1.7.3 Give an example of a non-monotone but convergent sequence of sets. Demonstrate graphically that it is non-monotone but convergent.

1.7.4 Give an example of a non-convergent sequence $\{A_n, n \geq 1\}$ of sets and graphically show that $\liminf A_n \subseteq \limsup A_n$. Further show that the number of sets not containing points from $\liminf A_n$ is finite whereas for $\limsup A_n$ this number is infinite.

1.8 Multiple Choice Questions

Note: In each question, multiple options may be correct. Unless specified otherwise, identify which of the statement(s) is/are correct. Answers are given in Chapter 11, after the solutions of conceptual exercises of Chapter 1.

1.8.1 Which of the following sequences is monotone decreasing?

(a) $A_n = [a + 1/n, b - 1/n]$
(b) $A_n = \{k \in \mathbb{N} | k \geq n\}$
(c) $A_n = \{a\}$
(d) $A_n = [a - 1/n, b - 1/n]$

1.8.2 Which of the following sequences $\{A_n, n \geq 1\}$ does/do NOT converge?

(a) $A_{2n} = C, A_{2n-1} = D, \quad C \neq D$
(b) $A_n = \{k \in \mathbb{N} | k \leq n\}$
(c) $A_n = \{n\} \ \forall \ n \geq 1$
(d) $A_n = A$

1.8.3 Which of the following set operations are commutative?

(a) Union

(b) Intersection
(c) Difference
(d) Symmetric Difference

1.8.4 Which of the following sequences converge to (a, b)?

(a) $[a + 1/n, b)$
(b) $(a, b - 1/n]$
(c) $(a - 1/n, b - 1/n)$
(d) $[a + 1/n, b - 1/n]$

1.8.5 Following are two statements: (I) Monotone sequences of sets are convergent. (II) Convergent sequences of sets are monotone. Then which of the following is correct?

(a) Both (I) and (II) are true
(b) (I) is true but (II) is false
(c) (II) is true but (I) is false.
(d) Both (I) and (II) are false.

1.8.6 Which of the following statements is/are true?

(a) lim sup is distributive over union
(b) lim inf is distributive over union
(c) lim sup is distributive over intersection
(d) lim inf is distributive over intersection

1.8.7 Following are four statements related to a sequence of sets.
(I) lim sup is distributive over union.
(II) lim inf is distributive over union.
(III) lim sup is distributive over intersection.
(IV) lim inf is distributive over intersection
Which of the following statements is true?

(a) Only (I) and (II) are true
(b) Only (I) and (III) are true
(c) Only (I) and (IV) are true
(d) Only (II) and (IV) are true

1.8.8 Which of the following statements is/are false?

(a) $\limsup(A_n \cup B_n) = (\limsup A_n) \cup (\limsup B_n)$
(b) $\liminf(A_n \cap B_n) = (\liminf A_n) \cap (\liminf B_n)$
(c) $\limsup(A_n \cap B_n) = (\limsup A_n) \cap (\limsup B_n)$
(d) $\liminf(A_n \cup B_n) = (\liminf A_n) \cup (\liminf B_n)$.

1.8.9 $\mathbb{F}$ is a field. $A_i \in \mathbb{F}$, $\forall i$. Which of the following belong to $\mathbb{F}$?

(a) A_i^c
(b) $\bigcup_{i=1}^{\infty} A_i$
(c) $\bigcap_{i=1}^{\infty} A_i$
(d) $\bigcap_{i=1}^{k} A_i$.

1.8.10 $\mathbb{F}_1$ and $\mathbb{F}_2$ are two σ-fields. Which of the following is/are σ-field(s)?

(a) $\mathbb{F}_1 \bigcup \mathbb{F}_2$
(b) $\mathbb{F}_1 \bigcap \mathbb{F}_2$
(c) $\sigma(\mathbb{F}_1 \bigcup \mathbb{F}_2)$
(d) $\sigma(\mathbb{F}_1 \bigcap \mathbb{F}_2)$

1.8.11 $\mathbb{F}_1$ and $\mathbb{F}_2$ are two σ-fields with $\mathbb{F}_1 \supseteq \mathbb{F}_2$. Which of the following is/are σ-field(s)?

(a) $\mathbb{F}_1 \bigcup \mathbb{F}_2$
(b) $\mathbb{F}_1 \bigcap \mathbb{F}_2$
(c) $\sigma(\mathbb{F}_1 \bigcup \mathbb{F}_2)$
(d) $\sigma(\mathbb{F}_1 \bigcap \mathbb{F}_2)$

1.8.12 Suppose $\Omega = [0, \infty)$. Which of the following is/are field(s)?

(a) $\mathbb{F}_1 = \{I | I = [a, b) \text{ or } I = [a, \infty), a, b \in \Omega, a < b\}$
(b) $\mathbb{F}_2 =$ Class of all finite unions of intervals in $\mathbb{F}_1$
(c) $\mathbb{F}_3 = \{[0, 3), (3, \infty), \Omega, \phi\}$
(d) $\mathbb{F}_4 = \{(0, 4], (4, \infty), \Omega, \phi\}$

1.8.13 Which of the following sets are measurable with respect to $\mathbb{A}$, if $\mathbb{A} = \sigma(\{A \bigcup B\})$?

(a) A
(b) B^c
(c) $A^c \cap B^c$
(d) $A \cap B$

1.8.14 Which of the following statements is/are true?

(a) A finite field is a σ-field
(b) Every σ-field is a field
(c) A σ-field is closed under a limit of a sequence of sets in it, if it exists
(d) Power set is a σ-field

1.8.15 $\mathbb{E} = \{E_1, E_2, \ldots, E_n\}$ is a partition of Ω. Which of the following is/are field/s?

(a) $\mathbb{G}_1 = \{A | A \text{ is a union of sets in } \mathbb{E}\}$, individual $E_i \notin \mathbb{G}_1$
(b) $\mathbb{G}_2 =$ Power set of $\mathbb{E}$
(c) $\mathbb{G}_3 = \{\Omega,\ \phi,\ \bigcup_{i=2}^{n} E_i,\ E_1\}$
(d) $\mathbb{G}_4 = \mathbb{E}$

1.8.16 $\mathbb{E} = \{E_1, E_2, \ldots, E_n\}$ is a partition of Ω. Which of the following is/are sigma field/s?

(a) $\mathbb{G}_1 = \{A | A \text{ is a union of sets in } \mathbb{E}\}$, individual $E_i \notin \mathbb{G}_1$
(b) $\mathbb{G}_2 =$ Power set of $\mathbb{E}$
(c) $\mathbb{G}_3 = \{\Omega,\ \phi,\ \bigcup_{i=3}^{n} E_i,\ E_1\}$
(d) $\mathbb{G}_4 = \mathbb{E}$

1.8.17 $\mathbb{E} = \{E_1, E_2, \ldots, E_n\}$ is a partition of Ω. Which of the following is/are sigma field/s?

(a) $\mathbb{G}_1 = \{A | A \text{ is a union of sets in } \mathbb{E}\}$, individual $E_i \notin \mathbb{G}_1$
(b) $\mathbb{G}_2 =$ Power set of $\mathbb{E}$
(c) $\mathbb{G}_3 = \{\Omega,\ \phi,\ \bigcup_{i=1}^{n} E_i,\ E_1\}$
(d) $\mathbb{G}_4 = \mathbb{E}$

1.8.18 If $\{A_n, n \geq 1\}$ is a monotone increasing sequence of sets, which of the following is/are true?

(a) $A_n \to \bigcup A_n$
(b) $A_n^c \to \bigcup A_n^c$
(c) $P(A_n) \to P(\bigcup A_n)$
(d) $P(A_n^c) \to P(\bigcap A_n^c)$

1.8.19 Suppose $(\Omega, \mathbb{A}, P)$ is a probability space. Which of the following statements is/are false?

(a) $P(A) \geq 0, \ \ \forall \ \ A \in \mathbb{A}$
(b) $P(\Omega) = 1$
(c) If $\{A_1, A_2, \cdots, A_k\} \in \mathbb{A}$, then $P(\bigcup_{i=1}^{k} A_i) = \sum_{i=1}^{k} P(A_i)$
(d) If $\{A_n, n \geq 1\}$ is a sequence of disjoint sets in $\mathbb{A}$, then $P(\bigcup_{n \geq 1} A_n) = \sum_{n \geq 1} P(A_n)$

1.8.20 Which of the following statements is/are not true?

(a) $P(\liminf A_n) \leq \limsup P(A_n)$
(b) $P(\liminf A_n) \geq \liminf P(A_n)$
(c) $P(\limsup A_n) \geq \limsup P(A_n)$
(d) $P(\limsup A_n) \geq \liminf P(A_n)$

1.8.21 Suppose $\{A_n, n \geq 1\}$ is a sequence of sets such that

$$A_n = \begin{cases} (2 - 1/n, 4 + 1/n), & \text{if } n \text{ is odd} \\ (1 - 1/n, 3 + 1/n), & \text{if } n \text{ is even.} \end{cases}$$

Which of the following points can be in $\limsup A_n$?

(a) 2
(b) 3
(c) 1
(d) 4

1.8.22 Suppose $\{A_n, n \geq 1\}$ is a sequence of sets such that

$$A_n = \begin{cases} (2 - 1/n, 4 + 1/n), & \text{if } n \text{ is odd} \\ (1 - 1/n, 3 + 1/n), & \text{if } n \text{ is even.} \end{cases}$$

Which of the following points can be in $\liminf A_n$?

(a) 2
(b) 3
(c) 1
(d) 4

1.8.23 Suppose $A_n \to A \neq \emptyset$. For $x \in A$, a set N_x is defined as $N_x = \{n \in \mathbb{N} | x \in A_n\}$. Which of the following statements is/are true?

(a) N_x is a countably infinite set
(b) $N_x^c = \{n \in \mathbb{N} | x \notin A_n\}$ is a finite set
(c) N_x can be empty set
(d) $N_x = \mathbb{N}$ always

2

Random Variables and Random Vectors

2.1 Introduction

Any realistic model of a real-world phenomenon has to take into account the possibility of randomness. It is usually achieved by allowing the model to be probabilistic in nature. A starting point in the development of the theory of probability is the concept of a random experiment. An experiment is a systematic method to collect data. Since it is a systematic method, an experiment can be repeated any number of times. Conducting an experiment once is called a trial. An experiment in which a set of possible outcomes of the experiment is known, but an outcome of a given performance of the experiment is not known in advance, is defined as a random experiment. In a random experiment, the outcome is not known at any trial unless the experiment is conducted.

As an illustration, the annual premium of the vehicle insurance is based on the estimate of claims, the vehicle owners will make in a period. The vehicle may or may not meet with an accident. It is a random experiment with two possible outcomes - accident occurs in a given year or accident does not occur in a given period. In case there is no accident, there is no claim. If an accident occurs, it is known before hand that the damage will be between two limits say a and b. However, the actual value of the damage and hence of the claim amount is not known in advance. In the determination of a premium, insurer has to take into account the output of many such repeated cases of accidents and claims, to model the distribution of the number of occurrences of accidents and of the claim amount.

A fundamental concept in probability theory is to model the uncertainty of a random experiment. With each such experiment, we associate a set Ω, a set of all possible outcomes of the experiment. It is known as a sample space associated with the experiment. It is assumed that the sample space is known. Obviously, it is a non-empty set. If it is a singleton set, then the random experiment is said to be deterministic. Thus, for an experiment to be truly a random experiment, Ω must have at least two elements. A sample space plays the role of a universal set as defined in Chapter 1. In modelling a random phenomenon, usually some functions defined on Ω are of more interest than the outcomes of the experiment. In many situations, one is interested in only some aspects of the random experiment. For example, in an experiment of tossing 3 unbiased coins, one is only interested in number of 'heads' turned

DOI: 10.1201/9781032619057-2

up. So our interest is only on certain functions on the sample space. These functions are called random variables, if the image space is the real line. If the image space is k dimensional Euclidean space, these functions are labeled as random vectors and more generally if the image space is any abstract space, then the functions on Ω are labeled as random elements.

In the present chapter, we make these concepts more precise, in which we require the functions to be measurable with respect to a sigma field of subsets of Ω. Towards it, we need to explore the concept of an inverse image of a set and an inverse image of a collection of sets. In Section 2, we define these concepts and study their properties. The main result we prove is that the inverse image of the minimal sigma field covering $\mathbb{C}$ is the minimal sigma field covering the inverse image of $\mathbb{C}$ and it is heavily used in subsequent sections. We define the concept of "random variable" as a measurable function, using two approaches. These lead to descriptive and economical definitions of a random variable. We further elaborate on the concept of sigma field induced by a random variable. We define a Borel function and prove that a Borel function of a random variable is again a random variable, measurable with respect to the same sigma field.

In Section 3, we extend all the results related to random variables, to finite and infinite dimensional random vectors. Using the results about the induced sigma fields, we prove that an infinite dimensional random vector can be labeled as a sequence of random variables. Section 4 is devoted to the study of a simple random variable and two important theorems concerned with a sequence of simple random variables. These are useful in demonstrating that a continuous function is a Borel function and in the definition of expectation of a random variable, discussed in Section 4. Section 5 discusses the probability measure induced by a random variable and a random vector and its connection with the probability measure defined in Chapter 1.

2.2 Random Variables

To define the term "random variable" precisely, we begin with the concept of an inverse image of a set.

Definition 2.2.1. *Inverse image of a set: Suppose Ω_1 and Ω_2 are non empty sets and $X : \Omega_1 \to \Omega_2$ is a function. Suppose $S \subseteq \Omega_2$. Then the inverse image, also known as inverse mapping, of a set S under the function X, denoted by $X^{-1}(S)$, is defined as,*

$$X^{-1}(S) = \{\omega \in \Omega_1 | X(\omega) \in S\}.$$

Thus, the inverse image of a set S is a set of all arguments whose images are in S. It is to be noted that $X^{-1}(S) \subseteq \Omega_1$. Hence, the argument as well as the image of the inverse function are sets.

Example 2.2.1. Suppose $\Omega_1 = \{a, b, c, d, e, f\}$ and Ω_2 is a set of all alphabets. A function X is defined on $\Omega_1 \to \Omega_2$ as

$$X(\omega) \quad = \quad \begin{cases} z, & \text{if} \quad \omega = a, b, c \\ y, & \text{if} \quad \omega = d, e, f. \end{cases}$$

To find the inverse image of any subset S of Ω_2, we divide the class of subsets of Ω_2 in four groups depending on whether y and z are in the set. Suppose $A = \{a, b, c\}$. The inverse image of any subset S is then given by,

$$X^{-1}(S) \quad = \quad \begin{cases} \emptyset, & \text{if} \quad y, z \notin S \\ \Omega_1, & \text{if} \quad y, z \in S \\ A, & \text{if} \quad z \in S, y \notin S \\ A^c, & \text{if} \quad y \in S, z \notin S. \end{cases}$$

Note that $X^{-1}(S) \subseteq \Omega_1$. □

In the definition of "random variable", we assume $\Omega_1 = \Omega$, which is a sample space associated with the given random experiment. Further, instead of having the domain of a function simply as Ω, it is taken as a measurable space $(\Omega, \mathbb{A})$. Similarly, the co-domain is taken as a measurable space $(\mathbb{R}, \mathbb{B})$, where $\mathbb{B}$ is the Borel field defined in Chapter 1.

Definition 2.2.2. *Random variable: Suppose* $X : (\Omega, \mathbb{A}) \to (\mathbb{R}, \mathbb{B})$ *is a function defined on a measurable space* $(\Omega, \mathbb{A})$ *to a measurable space* $(\mathbb{R}, \mathbb{B})$*. A function* X *is a random variable or a measurable function, measurable with respect to the sigma field* $\mathbb{A}$*, if*

$$\forall \ S \in \mathbb{B}, \ \ X^{-1}(S) \in \mathbb{A}.$$

The above definition of a random variable is known as a descriptive definition. If $X : (\Omega, \mathbb{A}) \to ([-\infty, \infty], \mathbb{B}^*)$, where $\mathbb{B}^*$ is the Borel field generated by a class of any type of intervals of $[-\infty, \infty]$, is a $\mathbb{A}$ measurable function, then we call X an extended valued random variable.

In the following examples, we examine whether the given function X is a random variable with respect to some sigma field.

Example 2.2.2. Suppose $\Omega = \{1, 2, 3, 4, 5, 6\}$, $A = \{1, 3, 5\}$ and a function X is defined on $(\Omega, \mathbb{A})$ as

$$X(\omega) \quad = \quad \begin{cases} 0, & \text{if} \quad \omega = 1, 3, 5 \\ 1, & \text{if} \quad \omega = 2, 4, 6. \end{cases}$$

Then the inverse image of any Borel set S is given by,

$$X^{-1}(S) \quad = \quad \begin{cases} \emptyset, & \text{if} \quad 0, 1 \notin S \\ \Omega, & \text{if} \quad 0, 1 \in S \\ A, & \text{if} \quad 0 \in S, 1 \notin S \\ A^c, & \text{if} \quad 1 \in S, 0 \notin S. \end{cases}$$

If we define $\mathbb{A} = \{\Omega, \emptyset, A, A^c\}$, then it has been verified in Section 1.3, that it is a sigma field. Now $\forall \;\; S \in \mathbb{B}, \;\; X^{-1}(S) \in \mathbb{A}$. Hence, X is measurable with respect to $\mathbb{A}$ and is a random variable with respect to $\mathbb{A}$. Further, observe that $\forall \;\; S \in \mathbb{B}, \;\; X^{-1}(S) \in \mathbb{P}(\Omega)$, the powerset of Ω. Thus, X is a random variable with respect to $\mathbb{P}(\Omega)$ also. It is to be noted that $\mathbb{A} \subset \mathbb{P}(\Omega)$. □

Remark 2.2.1. If Ω is finite or countably infinite then the powerset of Ω, is a sigma field and any function $X : (\Omega, \mathbb{A}) \to (\mathbb{R}, \mathbb{B})$ is a measurable function with respect to $\mathbb{P}(\Omega)$ and hence is a random variable with respect to $\mathbb{P}(\Omega)$.

Example 2.2.3. Suppose $(\Omega, \mathbb{A})$ is a measurable space. Suppose $A \in \mathbb{A}$. The indicator function $X \equiv I_A$ is defined as,

$$I_A(\omega) = \begin{cases} 1, & \text{if} \quad \omega \in A \\ 0, & \text{if} \quad \omega \in A^c \end{cases}$$

Suppose $S \in \mathbb{B}$. Then

$$X^{-1}(S) = \begin{cases} \emptyset, & \text{if} \quad 0, 1 \notin S \\ \Omega, & \text{if} \quad 0, 1 \in S \\ A, & \text{if} \quad 1 \in S, 0 \notin S \\ A^c, & \text{if} \quad 0 \in S, 1 \notin S. \end{cases}$$

Observe that if $A \in \mathbb{A}$ then $A^c \in \mathbb{A}$, it being a sigma field. Further Ω and $\emptyset$ are always members of any sigma field. Thus, $\forall \;\; S \in \mathbb{B}, \;\; X^{-1}(S) \in \mathbb{A}$ and I_A is a random variable with respect to $\mathbb{A}$ if $A \in \mathbb{A}$. □

Example 2.2.4. Suppose $\Omega = \{a, b, c, d\}$ and $\mathbb{A} = \{\Omega, \emptyset, A, A^c\}$, where $A = \{a, b\}$. A function X is defined on $(\Omega, \mathbb{A})$ as $X(a) = X(b) = -1, X(c) = 1$ and $X(d) = 2$. Suppose $S = \{2\}$, then $X^{-1}(S) = \{d\} \notin \mathbb{A}$, similarly if $S = \{1\}$, then $X^{-1}(S) = \{c\} \notin \mathbb{A}$ and hence X is not a random variable with respect to $\mathbb{A}$. □

Example 2.2.5. Suppose we want to identify a function on $(\Omega, \mathbb{A})$ which is measurable with respect to $\mathbb{A}$, where $\mathbb{A} = \{\Omega, \emptyset\}$. Since there are only two sets in $\mathbb{A}$, the collection of all Borel sets is partitioned into two classes such that inverse images of Borel sets are either Ω or $\emptyset$. Suppose $X \equiv c$, that is $X(\omega) = c \;\; \forall \;\; \omega \in \Omega$. Then

$$X^{-1}(S) = \begin{cases} \emptyset, & \text{if} \quad c \notin S \\ \Omega, & \text{if} \quad c \in S. \end{cases}$$

Hence, X is a random variable with respect to the sigma field $\mathbb{A} = \{\Omega, \emptyset\}$. We label X as a degenerate random variable, measurable with respect to $\mathbb{A}$. The sigma field $\mathbb{A} = \{\Omega, \emptyset\}$ is known as the trivial sigma field. □

Example 2.2.6. Suppose $\Omega = \{-1, 0, 1\}$ and a function X is defined on $(\Omega, \mathbb{A})$ as $X(\omega) = \omega$. Suppose $\mathbb{A} = \mathbb{P}(\Omega)$. To examine if X is $\mathbb{A}$ measurable random variable, we first find $X^{-1}(S)$ for $S \in \mathbb{B}$. These are given by,

$$X^{-1}(S) = \begin{cases} \emptyset, & \text{if} \quad -1 \notin S, 0 \notin S, 1 \notin S \\ \{-1\}, & \text{if} \quad -1 \in S, 0 \notin S, 1 \notin S \\ \{0\}, & \text{if} \quad -1 \notin S, 0 \in S, 1 \notin S \\ \{1\}, & \text{if} \quad -1 \notin S, 0 \notin S, 1 \in S \\ \{-1, 0\}, & \text{if} \quad -1 \in S, 0 \in S, 1 \notin S \\ \{-1, 1\}, & \text{if} \quad -1 \in S, 0 \notin S, 1 \in S \\ \{0, 1\}, & \text{if} \quad -1 \notin S, 0 \in S, 1 \in S \\ \Omega, & \text{if} \quad -1 \in S, 0 \in S, 1 \in S. \end{cases}$$

Thus, $\forall \; S \in \mathbb{B}, \;\; X^{-1}(S) \in \mathbb{A} = \mathbb{P}(\Omega)$ and X is a random variable with respect to $\mathbb{A}$. □

It is to be noted that for a function X defined on $(\Omega, \mathbb{A})$ to be a random variable, we have to examine if the inverse image of every Borel set is in the sigma field $\mathbb{A}$. It becomes tedious when the number of elements in Ω and in the set of images of X is large. There is an alternative simpler way to examine whether X is a random variable. In this approach, it is not necessary to verify whether inverse image of every Borel set is in the sigma field $\mathbb{A}$, but it is sufficient to examine that inverse image of a set in class $\mathbb{C}$, which generates the Borel field, is in $\mathbb{A}$. Thus, one may restrict to a class of intervals of any one of the eight types as described in Chapter 1, which generates a Borel field. To establish such an equivalence, we need some properties of inverse image of a set and also the properties of inverse image of a collection of sets. We investigate these in the following lemma. It is of interest to note that the inverse mapping preserves all the set operations.

Lemma 2.2.1. *Suppose X is a function defined from Ω_1 to Ω_2.*

(i) *Inverse image of a complement is a complement of the inverse image, that is, for $S \subseteq \Omega_2$, $X^{-1}(S^c) = (X^{-1}(S))^c$.*

(ii) *Inverse image of a union of two sets is a union of inverse images of the sets, that is, if S_1 and S_2 are two subsets of Ω_2, then $X^{-1}(S_1 \cup S_2) = X^{-1}(S_1) \cup X^{-1}(S_2)$.*

(iii) *Inverse image of a countable union is a countable union of inverse images, that is, if $\{S_n, n \geq 1\}$ is a sequence of subsets of Ω_2, then $X^{-1}(\bigcup_{n \geq 1} S_n) = \bigcup_{n \geq 1} X^{-1}(S_n)$.*

(iv) *Inverse image of a countable intersection is a countable intersection of inverse images, that is, if $\{S_n, n \geq 1\}$ is a sequence of subsets of Ω_2, then $X^{-1}(\bigcap_{n \geq 1} S_n) = \bigcap_{n \geq 1} X^{-1}(S_n)$.*

(v) *Inverse images of disjoint sets are disjoint sets, that is, if S_1 and S_2 are two disjoint subsets of Ω_2, then $X^{-1}(S_1)$ and $X^{-1}(S_2)$ are also disjoint subsets of Ω_1.*

(vi) If $S_1 \subseteq S_2 \subseteq \Omega_2$, then $X^{-1}(S_1) \subseteq X^{-1}(S_2) \subseteq \Omega_1$.

(vii) If $X^{-1}(S_1) \subseteq X^{-1}(S_2)$, then $S_1 \subseteq S_2$.

Proof.

(i) By the definition of the inverse image of a set,

$$\begin{aligned} \omega \in X^{-1}(S^c) &\iff X(\omega) \in S^c \iff X(\omega) \notin S \\ &\iff \omega \notin X^{-1}(S) \iff \omega \in (X^{-1}(S))^c. \end{aligned}$$

Hence, $X^{-1}(S^c) = (X^{-1}(S))^c$.

(ii) Suppose S_1 and S_2 are two subsets of Ω_2.

$$\begin{aligned} \omega \in X^{-1}(S_1 \cup S_2) &\iff X(\omega) \in S_1 \cup S_2 \\ &\iff X(\omega) \in S_i \ \text{ for at least one } \ i = 1, 2 \\ &\iff \omega \in X^{-1}(S_i) \ \text{ for at least one } \ i = 1, 2 \\ &\iff \omega \in X^{-1}(S_1) \cup X^{-1}(S_2). \end{aligned}$$

Hence, $X^{-1}(S_1 \cup S_2) = X^{-1}(S_1) \cup X^{-1}(S_2)$. Result (ii) can be extended to a countable union as shown below.

(iii) As in the proof of Result (ii),

$$\begin{aligned} \omega \in X^{-1}\Big(\bigcup_{n\geq 1} S_n\Big) &\iff X(\omega) \in \bigcup_{n\geq 1} S_n \\ &\iff X(\omega) \in S_n \ \text{ for at least one } n \geq 1 \\ &\iff \omega \in X^{-1}(S_n) \ \text{ for at least one } n \geq 1 \\ &\iff \omega \in \bigcup_{n\geq 1} X^{-1}(S_n). \end{aligned}$$

Hence, $X^{-1}(\bigcup_{n\geq 1} S_n) = \bigcup_{n\geq 1} X^{-1}(S_n)$.

(iv) By the definition of the inverse image of a set,

$$\begin{aligned} \omega \in X^{-1}\Big(\bigcap_{n\geq 1} S_n\Big) &\iff X(\omega) \in \bigcap_{n\geq 1} S_n \\ &\iff X(\omega) \in S_n, \ \ \forall\, n \geq 1 \\ &\iff \omega \in X^{-1}(S_n) \ \ \forall\, n \geq 1 \\ &\iff \omega \in \bigcap_{n\geq 1} X^{-1}(S_n). \end{aligned}$$

Hence, $X^{-1}(\bigcap_{n\geq 1} S_n) = \bigcap_{n\geq 1} X^{-1}(S_n)$.

(v) We assume the contrary, that is, $X^{-1}(S_1)$ and $X^{-1}(S_2)$ are not disjoint sets. Then

$$\begin{aligned}X^{-1}(S_1)\cap X^{-1}(S_2)\neq\emptyset &\Rightarrow \exists\ \omega\in\Omega \text{ such that } \omega\in X^{-1}(S_1)\cap X^{-1}(S_2)\\ &\Rightarrow \omega\in X^{-1}(S_1) \text{ and } \omega\in X^{-1}(S_2)\\ &\Rightarrow X(\omega)\in S_1 \text{ and } X(\omega)\in S_2\\ &\Rightarrow X(\omega)\in S_1\cap S_2 \Rightarrow S_1\cap S_2\neq\emptyset,\end{aligned}$$

which is a contradiction to the given statement that S_1 and S_2 are two disjoint sets. Thus, if S_1 and S_2 are two disjoint sets, then $X^{-1}(S_1)$ and $X^{-1}(S_2)$ are also disjoint sets.

(vi) Observe that

$$\begin{aligned}\omega\in X^{-1}(S_1) &\Rightarrow X(\omega)\in S_1 \Rightarrow X(\omega)\in S_2 \text{ as } S_1\subseteq S_2\\ &\Rightarrow \omega\in X^{-1}(S_2) \Rightarrow X^{-1}(S_1)\subset X^{-1}(S_2)\,.\end{aligned}$$

(vii) To prove that if $X^{-1}(S_1)\subseteq X^{-1}(S_2)$, then $S_1\subseteq S_2$, we assume the contrary, that is, S_1 is not a subset of S_2. Then

$$\begin{aligned}\exists\ \omega &\in \Omega_1 \text{ such that } X(\omega)\in S_1 \text{ but } X(\omega)\notin S_2\\ &\Rightarrow \exists\ \omega\in\Omega_1 \text{ such that } \omega\in X^{-1}(S_1) \text{ but } \omega\notin X^{-1}(S_2)\\ &\Rightarrow X^{-1}(S_1) \text{ not a subset of } X^{-1}(S_2),\end{aligned}$$

which is a contradiction. Hence, $X^{-1}(S_1)\subseteq X^{-1}(S_2) \Rightarrow S_1\subseteq S_2$.

□

We now extend the concept of an inverse image of a set to an inverse image of a collection of sets. We define it below and then prove some results regarding it in subsequent lemmas.

Definition 2.2.3. *Inverse image of a collection: Suppose $\mathbb{C}$ is a collection of subsets of Ω_2. Then the inverse image of a collection $\mathbb{C}$ under the function X, denoted by $X^{-1}(\mathbb{C})$, is defined as,*

$$X^{-1}(\mathbb{C})=\{X^{-1}(S)|S\in\mathbb{C}\}.$$

Thus, an inverse image of a collection is basically a collection of inverse images of sets from that collection.

Lemma 2.2.2. *Suppose $X:\Omega_1\to\Omega_2$ is a function. Suppose $\mathbb{C}_1$ and $\mathbb{C}_2$ are classes of subsets of Ω_2. Then*

$$\mathbb{C}_1\subseteq\mathbb{C}_2 \Rightarrow X^{-1}(\mathbb{C}_1)\subseteq X^{-1}(\mathbb{C}_2) \quad \& \quad \sigma[X^{-1}(\mathbb{C}_1)]\subseteq\sigma[X^{-1}(\mathbb{C}_2)].$$

Proof. Suppose $\mathbb{C}_1 \subseteq \mathbb{C}_2$. To prove that $X^{-1}(\mathbb{C}_1) \subseteq X^{-1}(\mathbb{C}_2)$, note that

$$\begin{aligned} A \in X^{-1}(\mathbb{C}_1) &\Rightarrow \exists \ S \in \mathbb{C}_1 \text{ such that } X^{-1}(S) = A \\ &\Rightarrow \exists \ S \in \mathbb{C}_2 \text{ such that } X^{-1}(S) = A \text{ as } \mathbb{C}_1 \subseteq \mathbb{C}_2 \\ &\Rightarrow X^{-1}(S) \in X^{-1}(\mathbb{C}_2) \text{ such that } X^{-1}(S) = A \\ &\Rightarrow A \in X^{-1}(\mathbb{C}_2) \Rightarrow X^{-1}(\mathbb{C}_1) \subseteq X^{-1}(\mathbb{C}_2). \end{aligned}$$

Now, $X^{-1}(\mathbb{C}_1) \subseteq X^{-1}(\mathbb{C}_2) \subseteq \sigma[X^{-1}(\mathbb{C}_2)]$. Thus, $\sigma[X^{-1}(\mathbb{C}_2)]$ is a sigma field covering $X^{-1}(\mathbb{C}_1)$, and hence by the definition of the minimal sigma field $\sigma[X^{-1}(\mathbb{C}_1)] \subseteq \sigma[X^{-1}(\mathbb{C}_2)]$. If $\mathbb{C}_1 = \mathbb{C}_2$, then $\sigma[X^{-1}(\mathbb{C}_1)] = \sigma[X^{-1}(\mathbb{C}_2)]$. □

Lemma 2.2.3. *Suppose $X : \Omega_1 \to \Omega_2$ is a function. If $\mathbb{F}$ is a sigma field of subsets of Ω_2, then $X^{-1}(\mathbb{F})$ is a sigma field of subsets of Ω_1, that is, inverse image of a sigma field is a sigma field.*

Proof. By the definition of an inverse image of a collection of sets, we have

$$X^{-1}(\mathbb{F}) = \{X^{-1}(S) | S \in \mathbb{F}\},$$

which we want to show to be a sigma field. Now,

$$\begin{aligned} A \in X^{-1}(\mathbb{F}) &\Rightarrow \exists \ S \in \mathbb{F} \text{ such that } A = X^{-1}(S) \\ &\Rightarrow S^c \in \mathbb{F} \text{ as } \mathbb{F} \text{ is a sigma field} \\ &\Rightarrow X^{-1}(S^c) \in X^{-1}(\mathbb{F}) \text{ by Definition 2.2.3} \\ &\Rightarrow (X^{-1}(S))^c \in X^{-1}(\mathbb{F}) \text{ by (i) of Lemma 2.2.1} \\ &\Rightarrow A^c \in X^{-1}(\mathbb{F}). \end{aligned} \quad (2.1)$$

Thus, $X^{-1}(\mathbb{F})$ is closed under complementation. Further, consider a sequence $\{A_n, n \geq 1\}$ of sets in $X^{-1}(\mathbb{F})$. Now,

$$\begin{aligned} \forall \ n \geq 1, \ A_n \in X^{-1}(\mathbb{F}) &\Rightarrow \exists \ S_n \in \mathbb{F} \text{ such that } A_n = X^{-1}(S_n) \\ &\Rightarrow \bigcup_{n \geq 1} S_n \in \mathbb{F} \text{ as } \mathbb{F} \text{ is a sigma field} \\ &\Rightarrow X^{-1}(\bigcup_{n \geq 1} S_n) \in X^{-1}(\mathbb{F}) \text{ by Definition 2.2.3} \\ &\Rightarrow \bigcup_{n \geq 1} X^{-1}(S_n) \in X^{-1}(\mathbb{F}) \text{ by (iii) of Lemma 2.2.1} \\ &\Rightarrow \bigcup_{n \geq 1} A_n \in X^{-1}(\mathbb{F}) . \end{aligned} \quad (2.2)$$

Thus, $X^{-1}(\mathbb{F})$ is closed under countable union. Hence by Equation (2.1) and Equation (2.2), it is a sigma field. □

Lemma 2.2.4. *Suppose $X : \Omega_1 \to \Omega_2$ is a function. Suppose $\mathbb{A}$ is a sigma field of subsets of Ω_1. Then, there exists a sigma field $\mathbb{F}$ of subsets of Ω_2 such that $X^{-1}(\mathbb{F}) \subseteq \mathbb{A}$.*

Proof. Suppose a class $\mathbb{F}$ of subsets of Ω_2 is defined as

$$\mathbb{F} = \{S \subset \Omega_2 | X^{-1}(S) \in \mathbb{A}\}.$$

We examine whether $\mathbb{F}$ is a sigma field. Observe that,

$$\begin{aligned} S \in \mathbb{F} \quad &\Rightarrow \quad X^{-1}(S) \in \mathbb{A} \text{ by definition of } \mathbb{F} \\ &\Rightarrow \quad (X^{-1}(S))^c \in \mathbb{A} \text{ as } \mathbb{A} \text{ is a sigma field} \\ &\Rightarrow \quad X^{-1}(S^c) \in \mathbb{A} \text{ by (i) of Lemma 2.2.1} \\ &\Rightarrow \quad S^c \in \mathbb{F} \text{ by definition of } \mathbb{F}. \end{aligned} \tag{2.3}$$

Thus, $\mathbb{F}$ is closed under complementation. Suppose $\{S_n, n \geq 1\}$ is a sequence of sets in $\mathbb{F}$. Note that,

$$\begin{aligned} \forall\ n \geq 1,\ S_n \in \mathbb{F} \quad &\Rightarrow \quad X^{-1}(S_n) \in \mathbb{A},\ \ n \geq 1,\ \text{ by definition of } \mathbb{F} \\ &\Rightarrow \quad \bigcup_{n \geq 1} X^{-1}(S_n) \in \mathbb{A} \text{ as } \mathbb{A} \text{ is a sigma field} \\ &\Rightarrow \quad X^{-1}(\bigcup_{n \geq 1} S_n) \in \mathbb{A} \text{ by (iii) of Lemma 2.2.1} \\ &\Rightarrow \quad \bigcup_{n \geq 1} S_n \in \mathbb{F} \text{ by definition of } \mathbb{F}. \end{aligned} \tag{2.4}$$

Thus, $\mathbb{F}$ is closed under countable union. Hence by Equation (2.3) and Equation (2.4), it is a sigma field.
To examine whether $X^{-1}(\mathbb{F}) \subseteq \mathbb{A}$, note that

$$A \in X^{-1}(\mathbb{F}) = \{X^{-1}(S) | S \in \mathbb{F}\} \ \Rightarrow\ \exists\ S \in \mathbb{F} \text{ such that } A = X^{-1}(S).$$

Now by the definition of $\mathbb{F}$,

$$S \in \mathbb{F} \ \Rightarrow\ X^{-1}(S) \in \mathbb{A} \ \Rightarrow\ A \in \mathbb{A}.$$

Thus, $A \in X^{-1}(\mathbb{F}) \Rightarrow A \in \mathbb{A}$. Hence, $X^{-1}(\mathbb{F}) \subseteq \mathbb{A}$. □

Remark 2.2.2. Note that $X^{-1}(\mathbb{F})$ need not be equal to $\mathbb{A}$. $\mathbb{A}$ can have some additional sets which are not inverse images of any subsets of Ω_2.

Theorem 2.2.1. *Suppose $X : \Omega_1 \to \Omega_2$ is a function. If $\mathbb{C}$ is a class of subsets of Ω_2, then the inverse image of the minimal sigma field covering $\mathbb{C}$ is the minimal sigma field covering the inverse image of $\mathbb{C}$, that is, if $\sigma(\mathbb{C})$ is the minimal sigma field generated by class $\mathbb{C}$ of subsets of Ω_2 then*

$$X^{-1}[\sigma(\mathbb{C})] = \sigma[X^{-1}(\mathbb{C})].$$

Proof. Note that

$$\mathbb{C} \subseteq \sigma(\mathbb{C}) \ \Rightarrow\ X^{-1}(\mathbb{C}) \subseteq X^{-1}[\sigma(\mathbb{C})] \text{ by Lemma 2.2.2.}$$

Observe that $\sigma(\mathbb{C})$ is a sigma field and hence by Lemma 2.2.3, $X^{-1}[\sigma(\mathbb{C})]$ is also a sigma field. Thus, $X^{-1}[\sigma(\mathbb{C})]$ is a sigma field covering the class $X^{-1}(\mathbb{C})$. Hence, the minimal sigma field generated by $X^{-1}(\mathbb{C})$ is included in $X^{-1}[\sigma(\mathbb{C})]$ by the definition of the minimal sigma field, that is,

$$\sigma[X^{-1}(\mathbb{C})] \subseteq X^{-1}[\sigma(\mathbb{C})]. \tag{2.5}$$

To prove that $X^{-1}[\sigma(\mathbb{C})] \subseteq \sigma[X^{-1}(\mathbb{C})]$, suppose a collection $\mathbb{F}$ of subsets of Ω_2 is defined as $\mathbb{F} = \{S \subseteq \Omega_2 | X^{-1}(S) \in \sigma[X^{-1}(\mathbb{C})]\}$. Note that $\sigma[X^{-1}(\mathbb{C})]$ is a sigma field of subsets of Ω_1. Hence by Lemma 2.2.4, $\mathbb{F}$ is sigma field of subsets of Ω_2 and $X^{-1}(\mathbb{F}) \subseteq \sigma[X^{-1}(\mathbb{C})]$. Now by the definition, $X^{-1}(\mathbb{C}) = \{X^{-1}(S) | S \in \mathbb{C}\}$. Note that

$$S \in \mathbb{C} \Rightarrow X^{-1}(S) \in X^{-1}(\mathbb{C}) \subseteq \sigma[X^{-1}(\mathbb{C})] \Rightarrow X^{-1}(S) \in \sigma[X^{-1}(\mathbb{C})] \Rightarrow S \in \mathbb{F},$$

by definition of $\mathbb{F}$. Thus, $\mathbb{C} \subseteq \mathbb{F}$. Now by the definition of the minimal sigma field,

$$\begin{aligned} \mathbb{C} \subseteq \mathbb{F} &\Rightarrow \sigma(\mathbb{C}) \subseteq \mathbb{F} \\ &\Rightarrow X^{-1}[\sigma(\mathbb{C})] \subseteq X^{-1}(\mathbb{F}) \subseteq \sigma[X^{-1}(\mathbb{C})]. \end{aligned} \tag{2.6}$$

From Equation (2.5) and Equation (2.6), we conclude that
$X^{-1}[\sigma(\mathbb{C})] = \sigma[X^{-1}(\mathbb{C})]$. □

Remark 2.2.3. Lemma 2.2.2 conveys that $\mathbb{C}_1 \subseteq \mathbb{C}_2 \Rightarrow X^{-1}(\mathbb{C}_1) \subseteq X^{-1}(\mathbb{C}_2)$. However, its converse is not true, that is, if $X^{-1}(\mathbb{C}_1) \subseteq X^{-1}(\mathbb{C}_2)$, then $\mathbb{C}_1$ may not be included in $\mathbb{C}_2$ as illustrated in the following example

Example 2.2.7. Suppose $X : \Omega_1 \to \Omega_2$ where $\Omega_1 = \{HH, HT, TH, TT\}$, $\Omega_2 = \mathbb{R}^+$ and $X(HH) = 2, X(TH) = X(HT) = 1$ and $X(TT) = 0$. Observe that,

$$\begin{aligned} \mathbb{C}_1 &= \{\{1\},\{2\}\} \Rightarrow X^{-1}(\mathbb{C}_1) = \{\{HT,TH\},\{HH\}\} \\ \mathbb{C}2 &= \{\{0\},(0.5,1.5),\{2\}\} \Rightarrow X^{-1}(\mathbb{C}_2) = \{\{HT,TH\},\{HH\},\{TT\}\}. \end{aligned}$$

Thus, $X^{-1}(\mathbb{C}_1) \subset X^{-1}(\mathbb{C}_2)$. However, $\mathbb{C}_1$ is not included in $\mathbb{C}_2$. □

The next example shows that if a class $\mathbb{C}$ is such that $\sigma(X^{-1}(\mathbb{C})) = X^{-1}(\mathbb{F})$, then $\mathbb{C}$ may or may not be included in $\mathbb{F}$.

Example 2.2.8. Suppose $X : \Omega_1 \to \Omega_2$ where $\Omega_1 = \{HH, HT, TH, TT\}$, $\Omega_2 = \mathbb{R}^+ = [0, \infty)$, $X(HH) = 2, X(TH) = X(HT) = 1$ & $X(TT) = 0$. Suppose $\mathbb{C} = \{\{1\},\{2\}\}$. Then $X^{-1}(\mathbb{C}) = \{\{HT,TH\},\{HH\}\}$. Hence, $\sigma(X^{-1}(\mathbb{C})) = \Big\{\Omega_1, \emptyset, \{HT,TH\}, \{HH\}, \{TT,HH\}, \{HT,TH,HH\}, \{\text{TT}\}, \{HT,TH,TT\}\Big\}$. We now find two different $\mathbb{F}$ such that

$\sigma(X^{-1}(\mathbb{C})) = X^{-1}(\mathbb{F})$. However, $\mathbb{C}$ is included in one but not in the other. Suppose

$$\begin{aligned}\mathbb{F}_1 &= \Big\{\Omega_2, \emptyset, \{1\}, \{2\}, \{1\}^c, \{2\}^c, \{1,2\}, \{1,2\}^c\Big\} \\ &= \Big\{\Omega_2, \emptyset, \{1\}, \{2\}, [0,1) \cup (1,\infty), [0,2) \cup (2,\infty), \{1,2\}, \\ &\quad [0,1) \cup (1,2) \cup (2,\infty)\Big\}.\end{aligned}$$

$$\begin{aligned}\mathbb{F}_2 &= \Big\{\Omega_2, \emptyset, \{2\}, (0.5,1.5), \{2\}^c, (0.5,1.5)^c, (0.5,1.5) \cup \{2\}, \\ &\quad (0.5,1.5)^c \cap \{2\}^c\Big\} \\ &= \Big\{\Omega_2, \emptyset, \{2\}, (0.5,1.5), [0,2) \cup (2,\infty), [0,0.5) \cup (1.5,\infty), \\ &\quad (0.5,1.5) \cup \{2\}, (0.5,1.5)^c \cap \{2\}^c\Big\}.\end{aligned}$$

Note that $X^{-1}(\mathbb{F}_1) = X^{-1}(\mathbb{F}_2) = \sigma(X^{-1}(\mathbb{C}))$ and $\mathbb{C} = \{\{1\},\{2\}\} \subset \mathbb{F}_1$. However $\mathbb{C} = \{\{1\},\{2\}\}$ is not included in $\mathbb{F}_2$. □

It is to be noted that Lemma 2.2.1, Lemma 2.2.2, Lemma 2.2.3 and Lemma 2.2.4 are mainly to prove Theorem 2.2.1. All these results are also valid for a function X from any measurable space to any other measurable space. When we deal with a random variable, we take $(\Omega, \mathbb{A})$ as a domain and $(\mathbb{R}, \mathbb{B})$ as a co-domain. When we deal with a Borel function, both the domain space and the co-domain space are $(\mathbb{R}, \mathbb{B})$ (see Theorem 2.2.3). In the definition of a random vector, the co-domain space is $(\mathbb{R}^k, \mathbb{B}^k)$.

Using Theorem 2.2.1, we prove an important result in the following theorem which leads to a simpler way to verify that a given function is a random variable.

Theorem 2.2.2. *Suppose $X : (\Omega, \mathbb{A}) \to (\mathbb{R}, \mathbb{B})$ is a function. Suppose $\mathbb{C}$ is a class of subsets of $\mathbb{R}$ such that $\sigma(\mathbb{C}) = \mathbb{B}$. Then*

$$\forall\ S \in \mathbb{B}, X^{-1}(S) \in \mathbb{A} \quad \Longleftrightarrow \quad \forall\ S \in \mathbb{C},\ X^{-1}(S) \in \mathbb{A}.$$

Proof. Part(i) - Suppose $\forall\ S \in \mathbb{B}, X^{-1}(S) \in \mathbb{A}$. Now $\mathbb{B} = \sigma(\mathbb{C})$ is the minimal sigma field generated by the class $\mathbb{C}$ and hence by the definition of minimal sigma field, $\mathbb{C} \subseteq \sigma(\mathbb{C}) = \mathbb{B}$. Suppose $S \in \mathbb{C}$. Then,

$$S \in \mathbb{C} \ \Rightarrow\ S \in \mathbb{B} \ \Rightarrow\ X^{-1}(S) \in \mathbb{A}$$

Part(ii) - Suppose $\forall\ S \in \mathbb{C},\ X^{-1}(S) \in \mathbb{A}$. Now

$$\begin{aligned}\forall\ S \in \mathbb{C},\ X^{-1}(S) \in \mathbb{A} \quad &\Rightarrow \quad X^{-1}(\mathbb{C}) \subseteq \mathbb{A}, \ \ (\mathbb{A} \text{ covers } X^{-1}(\mathbb{C})) \\ &\Rightarrow \quad \sigma[X^{-1}(\mathbb{C})] \subseteq \mathbb{A} \\ &\Rightarrow \quad X^{-1}[\sigma(\mathbb{C})] \subseteq \mathbb{A} \ \text{ by Theorem 2.2.1} \\ &\Rightarrow \quad X^{-1}(\mathbb{B}) \subseteq \mathbb{A} \\ &\Rightarrow \quad \forall\ S \in \mathbb{B},\ \ X^{-1}(S) \in \mathbb{A}.\end{aligned}$$

The second step follows by the definition of minimal sigma field. □

From Theorem 2.2.2, we note that

$$\forall \ S \in \mathbb{C}, \ \ X^{-1}(S) \in \mathbb{A} \Rightarrow \forall \ \ S \in \mathbb{B}, \ \ X^{-1}(S) \in \mathbb{A}.$$

Hence we have the following another definition of a random variable.

Definition 2.2.4. *Random variable: Suppose* $X : (\Omega, \mathbb{A}) \to (\mathbb{R}, \mathbb{B})$ *is a function. Suppose* $\mathbb{C}$ *is a class of subsets of* $\mathbb{R}$ *such that* $\sigma(\mathbb{C}) = \mathbb{B}$*. Then* X *is a random variable measurable with respect to* $\mathbb{A}$ *if* $\forall \ S \in \mathbb{C}, \ X^{-1}(S) \in \mathbb{A}$.

Definition 2.2.4 is referred to as an economical definition of a random variable, while Definition 2.2.2 is referred to as a descriptive definition of a random variable. Theorem 2.2.2 proves the equivalence of descriptive and economical definitions of a random variable. With the economical definition, it is simpler to verify that a function defined on the measurable space is a random variable. It is clear from the following examples.

Example 2.2.9. Suppose $\Omega = \{a, b, c, d, e, f\}$ and $\mathbb{A}$ is given by, $\mathbb{A} = \{\Omega, \emptyset, \{a, b\}, \{c, d\}, \{a, b, c, d\}, \{e, f\}, \{c, d, e, f\}, \{a, b, e, f\}\}$. A function X is defined on $(\Omega, \mathbb{A})$ as follows.

$$X(\omega) \quad = \quad \begin{cases} 10, & \text{if} \quad \omega = a \\ 10^{10}, & \text{if} \quad \omega = b, c \\ 10^{20}, & \text{if} \quad \omega = d \\ 10^{30}, & \text{if} \quad \omega = e, f. \end{cases}$$

We use Definition 2.2.4 to examine if X is $\mathbb{A}$ measurable random variable. Suppose $\mathbb{C} = \{(-\infty, x] | x \in \mathbb{R}\}$. Now,

$$X^{-1}((-\infty, x]) \quad = \quad \begin{cases} \emptyset, & \text{if} \quad -\infty < x < 10 \\ \{a\}, & \text{if} \quad 10 \le x < 10^{10}. \end{cases}$$

But $\{a\} \notin \mathbb{A}$, hence X is not $\mathbb{A}$ measurable random variable. □

Example 2.2.10. Suppose $\Omega = \{-1, 0, 1\}$ and a function X is defined on $(\Omega, \mathbb{A})$ as $X(\omega) = \omega$ and $\mathbb{A} = \mathbb{P}(\Omega)$. As in the previous example we use Definition 2.2.4 to examine if X is $\mathbb{A}$ measurable random variable. Suppose $\mathbb{C} = \{(-\infty, x] | x \in \mathbb{R}\}$. Now,

$$X^{-1}((-\infty, x]) \quad = \quad \begin{cases} \emptyset, & \text{if} \quad -\infty < x < -1 \\ \{-1\}, & \text{if} \quad -1 \le x < 0 \\ \{-1, 0\}, & \text{if} \quad 0 \le x < 1 \\ \Omega, & \text{if} \quad x \ge 1. \end{cases}$$

Thus, $X^{-1}((-\infty, x]) \in \mathbb{P}(\Omega), \ \ \forall \ \ x \in \mathbb{R}$ and hence X is a random variable with respect to $\mathbb{P}(\Omega)$. □

Comparison of the solutions of Examples 2.2.6 and 2.2.10 clarifies why Definition 2.2.4 is labeled as the economical definition of a random variable.

Note that $\mathbb{B}$ is a sigma field and hence by Lemma 2.2.3, $X^{-1}(\mathbb{B})$ is also a sigma field. It is referred to by a special term as given in the following definition.

Definition 2.2.5. *Sigma field induced by a random variable X: Suppose X is a random variable defined on $(\Omega, \mathbb{A})$ to $(\mathbb{R}, \mathbb{B})$. Then*

$$X^{-1}(\mathbb{B}) = \{X^{-1}(S) | S \in \mathbb{B}\},$$

also denoted by $\sigma(X)$, is known as a sigma field induced by or generated by a random variable X.

In the following example, we obtain the induced sigma field and also verify Theorem 2.2.1.

Example 2.2.11. Suppose X is a function defined on $(\Omega, \mathbb{A})$ to $(\mathbb{R}, \mathbb{B})$ as

$$X(\omega) = \begin{cases} -1, & \text{if} \quad \omega \in A_1 \\ 0, & \text{if} \quad \omega \in (A_1 \cup A_2)^c \\ 1, & \text{if} \quad \omega \in A_1^c \cap A_2, \end{cases}$$

where A_1 and A_2 are measurable sets, that is, they are in $\mathbb{A}$. We examine whether X is a random variable with respect to $\mathbb{A}$. Suppose $\mathbb{C} = \{(-\infty, x] | x \in \mathbb{R}\}$. Now,

$$X^{-1}((-\infty, x]) = \begin{cases} \emptyset, & \text{if} \quad -\infty < x < -1 \\ A_1, & \text{if} \quad -1 \leq x < 0 \\ A_1 \cup (A_1 \cup A_2)^c = A_1 \cup A_2^c, & \text{if} \quad 0 \leq x < 1 \\ \Omega, & \text{if} \quad x \geq 1. \end{cases}$$

It is given that $A_1, A_2 \in \mathbb{A}$, hence $A_1 \cup A_2^c \in \mathbb{A}$, it being a sigma field. Thus, $X^{-1}((-\infty, x]) \in \mathbb{A}, \ \forall \ x \in \mathbb{R}$ and hence X is a random variable with respect to $\mathbb{A}$. Note that

$$X^{-1}(\mathbb{C}) = \{\Omega, \emptyset, A_1, A_1 \cup A_2^c\}$$

$$\& \ \ \sigma(X^{-1}(\mathbb{C})) = \{\Omega, \emptyset, A_1, A_1 \cup A_2^c, A_1^c, A_1^c \cap A_2, A_1 \cup A_2, A_1^c \cap A_2^c\}.$$

We now find the sigma field induced by X. Thus we need to find $X^{-1}(S) \ \forall \ S \in \mathbb{B}$. We have for $S \in \mathbb{B}$,

$$X^{-1}(S) = \begin{cases} \emptyset, & \text{if} \quad -1 \notin S, 0 \notin S, 1 \notin S \\ A_1, & \text{if} \quad -1 \in S, 0 \notin S, 1 \notin S \\ A_1^c \cap A_2, & \text{if} \quad -1 \notin S, 0 \notin S, 1 \in S \\ A_1^c \cap A_2^c, & \text{if} \quad -1 \notin S, 0 \in S, 1 \notin S \\ A_1 \cup A_2, & \text{if} \quad -1 \in S, 0 \notin S, 1 \in S \\ A_1 \cup A_2^c, & \text{if} \quad -1 \in S, 0 \in S, 1 \notin S \\ A_1^c, & \text{if} \quad -1 \notin S, 0 \in S, 1 \in S \\ \Omega, & \text{if} \quad -1 \in S, 0 \in S, 1 \in S. \end{cases}$$

Thus, the sigma field induced by X is given by,

$$X^{-1}(\mathbb{B}) = \{\Omega, \emptyset, A_1, A_1^c \cap A_2, A_1^c \cap A_2^c, A_1 \cup A_2, A_1 \cup A_2^c, A_1^c\}.$$

Note that $X^{-1}(\mathbb{B}) = X^{-1}(\sigma(\mathbb{C})) = \sigma(X^{-1}(\mathbb{C}))$. □

Remark 2.2.4. Observe that in the above example $X^{-1}(\mathbb{B}) \subseteq \mathbb{A}$ and hence it is a random variable with respect to $\mathbb{A}$. Note that a function X defined on $(\Omega, \mathbb{A})$ to $(\mathbb{R}, \mathbb{B})$ is always a random variable with respect to its induced sigma field. Further note that $X^{-1}(\mathbb{B})$ is a much smaller sigma field than the original sigma field $\mathbb{A}$, which is not even specified in the above example. Often the sigma field $\mathbb{A}$, which may be a powerset, contains too many subsets and we are interested in only some of them. One can then define a random variable X measurable with respect to the induced sigma field containing subsets that are of interest. In general, $X^{-1}(\mathbb{B})$ is between the trivial sigma field and $\mathbb{A}$. In fact, the induced sigma field is the smallest sigma field with respect to which a random variable is measurable.

Example 2.2.12. Suppose X is a function defined on $(\Omega, \mathbb{A})$ to $(\mathbb{R}, \mathbb{B})$ as $X(\omega) = \omega$, where $\Omega = \{-1, 0, 1\}$. We obtain the sigma fields induced by X, $Y = X^2$ and $Z = e^X$ and examine the relation between the sigma fields induced by X and Y as well as between the sigma fields induced by X and Z. To find the sigma field induced by X, for $S \in \mathbb{B}$,

$$X^{-1}(S) = \begin{cases} \emptyset, & \text{if } -1 \notin S, 0 \notin S, 1 \notin S \\ \{-1\}, & \text{if } -1 \in S, 0 \notin S, 1 \notin S \\ \{0\}, & \text{if } -1 \notin S, 0 \in S, 1 \notin S \\ \{1\}, & \text{if } -1 \notin S, 0 \notin S, 1 \in S \\ \{-1, 0\}, & \text{if } -1 \in S, 0 \in S, 1 \notin S \\ \{-1, 1\}, & \text{if } -1 \in S, 0 \notin S, 1 \in S \\ \{0, 1\}, & \text{if } -1 \notin S, 0 \in S, 1 \in S \\ \Omega, & \text{if } -1 \in S, 0 \in S, 1 \in S. \end{cases}$$

Thus, the sigma field induced by X is given by,

$$X^{-1}(\mathbb{B}) = \{\Omega, \emptyset, \{-1\}, \{0\}, \{1\}, \{-1, 0\}, \{-1, 1\}, \{0, 1\}\}$$

and X is a random variable measurable with respect to the induced sigma field. Now $Y = X^2$, hence $Y(-1) = Y(1) = 1$ and $Y(0) = 0$. To find the sigma field induced by Y, we have for $S \in \mathbb{B}$,

$$Y^{-1}(S) = \begin{cases} \emptyset, & \text{if } 0 \notin S, 1 \notin S \\ \{0\}, & \text{if } 0 \in S, 1 \notin S \\ \{-1, 1\}, & \text{if } 0 \notin S, 1 \in S \\ \Omega, & \text{if } 0 \in S, 1 \in S. \end{cases}$$

Thus, the sigma field induced by Y is given by,

$$Y^{-1}(\mathbb{B}) = \{\Omega, \emptyset, \{0\}, \{-1, 1\}\} \quad \text{and} \quad Y^{-1}(\mathbb{B}) \subset X^{-1}(\mathbb{B}).$$

It is to be noted that Y is a random variable measurable with respect to $Y^{-1}(\mathbb{B})$ as well as $X^{-1}(\mathbb{B})$, but $Y^{-1}(\mathbb{B})$ is the minimal sigma field with respect to which Y is measurable. To find the sigma field induced by Z, for $S \in \mathbb{B}$,

$$Z^{-1}(S) = \begin{cases} \emptyset, & \text{if} \quad e^{-1} \notin S, e^0 \notin S, e^1 \notin S \\ \{-1\}, & \text{if} \quad e^{-1} \in S, e^0 \notin S, e^1 \notin S \\ \{0\}, & \text{if} \quad e^{-1} \notin S, e^0 \in S, e^1 \notin S \\ \{1\}, & \text{if} \quad e^{-1} \notin S, e^0 \notin S, e^1 \in S \\ \{-1, 0\}, & \text{if} \quad e^{-1} \in S, e^0 \in S, e^1 \notin S \\ \{-1, 1\}, & \text{if} \quad e^{-1} \in S, e^0 \notin S, e^1 \in S \\ \{0, 1\}, & \text{if} \quad e^{-1} \notin S, e^0 \in S, e^1 \in S \\ \Omega, & \text{if} \quad e^{-1} \in S, e^0 \in S, e^1 \in S. \end{cases}$$

Thus, the sigma field induced by Z is the same as that of X and Z is a random variable measurable with respect to the induced sigma field. It is to be noted that Y is a many-to-one function of X and Z is a one-to-one function of X. In general, all one-to-one functions of X induce the same sigma fields as that of X and the sigma field induced by many-to-one functions of X are included in the sigma field induced by X. Thus, the values of random variables are not important but the partition of the sample space that it induces is important. □

To formalize the results obtained in Example 2.2.12, we define a term "Borel function".

Definition 2.2.6. *Borel function: Suppose $g : (\mathbb{R}, \mathbb{B}) \to (\mathbb{R}, \mathbb{B})$ is a function. If*

$$\forall \ S \in \mathbb{B}, \ \ g^{-1}(S) = \{x \in \mathbb{R} | g(x) \in S\} \in \mathbb{B}$$

then g is measurable with respect to the Borel field and is known as a Borel function.

Just as we have an economical definition of a random variable, we have a following theorem which states that to examine if g is a Borel function, it is enough to examine that inverse images of sets in a class $\mathbb{C}$, which generates $\mathbb{B}$.

Theorem 2.2.3. *Suppose $g : (\mathbb{R}, \mathbb{B}) \to (\mathbb{R}, \mathbb{B})$ is a function. Suppose $\mathbb{C}$ is a class of subsets of $\mathbb{R}$ such that $\sigma(\mathbb{C}) = \mathbb{B}$. Then g is a Borel function if and only if $\forall \ S \in \mathbb{C}, \ g^{-1}(S) \in \mathbb{B}$.*

Proof.

(i) Suppose $g : (\mathbb{R}, \mathbb{B}) \to (\mathbb{R}, \mathbb{B})$ is a Borel function, that is, $\forall \ S \in \mathbb{B}$, $g^{-1}(S) \in \mathbb{B}$. Now $\mathbb{B} = \sigma(\mathbb{C})$ is the minimal sigma field generated by the class $\mathbb{C}$ and hence by the definition of minimal sigma field, $\mathbb{C} \subset \sigma(\mathbb{C}) = \mathbb{B}$. Thus,

$$S \in \mathbb{C} \ \Rightarrow \ S \in \mathbb{B} \ \Rightarrow \ g^{-1}(S) \in \mathbb{B}.$$

(ii) Suppose that

$$\begin{aligned}\forall\ S \in \mathbb{C},\ g^{-1}(S) \in \mathbb{B} &\Rightarrow g^{-1}(\mathbb{C}) \subseteq \mathbb{B}\ (\mathbb{B} \text{ covers } g^{-1}(\mathbb{C}))\\ &\Rightarrow \sigma[g^{-1}(\mathbb{C})] \subseteq \mathbb{B} \text{ by definition of minimal } \sigma \text{ field}\\ &\Rightarrow g^{-1}[\sigma(\mathbb{C})] \subseteq \mathbb{B} \text{ by Theorem 2.2.1}\\ &\Rightarrow g^{-1}(\mathbb{B}) \subseteq \mathbb{B} \Rightarrow \forall\ S \in \mathbb{B},\ \ g^{-1}(S) \in \mathbb{B}.\end{aligned}$$

Hence g is $\mathbb{B}$ measurable function, that is, it is a Borel function.

□

Example 2.2.13. Using the economical definition of a Borel function we examine whether (i) $g(x) = a_0 + a_1x + a_2x^2 + \cdots + a_nx^n,\quad x \in \mathbb{R}$ is a Borel function. Suppose $\mathbb{C}$ is a class of all open intervals of $\mathbb{R}$, then the minimal sigma field generated by $\mathbb{C}$ is the Borel field $\mathbb{B}$. Suppose $S = (a, b) \in \mathbb{C}$, then $g^{-1}(S) = \{x|g(x) = a_0 + a_1x + a_2x^2 + \cdots + a_nx^n \in S\}$ will be an open interval or union of open intervals and hence belongs to $\mathbb{B}$. It follows since the inverse image of an open set under a continuous function is again an open set. Hence g is a Borel function.
(ii) Suppose $g(x) = \sin x$, the range is $(-1, 1)$, hence for any $S = (a, b) \subseteq (-1, 1)$, $g^{-1}(S)$ will be again an union of open intervals in $\mathbb{R}$ and hence a Borel set. For any Borel set S, which is not included in $(-1, 1)$, $g^{-1}(S) = \emptyset$. Thus $g(x) = \sin x$ is a Borel function.
(iii) If $g(x) = \exp(x)$, then inverse image of any open interval of $\mathbb{R}^+ = [0, \infty)$ will be an open interval of $\mathbb{R}$. For any Borel set S, which is included in $\mathbb{R}^-$, $g^{-1}(S) = \emptyset$. For an interval of the type $(-a, a)$, where $a > 0$, $g^{-1}((-a, a)) = g^{-1}((0, a))$. It is again an open interval. Hence $\exp(x)$ is a Borel function. □

Having defined the Borel function, we now prove an important result that measurability of X remains invariant under Borel transformations. The following theorem plays a central role in Statistics.

Theorem 2.2.4. *Suppose $X : (\Omega, \mathbb{A}) \to (\mathbb{R}, \mathbb{B})$ is a random variable measurable with respect to a sigma field $\mathbb{A}$ and $g : (\mathbb{R}, \mathbb{B}) \to (\mathbb{R}, \mathbb{B})$ is a Borel function, then*

(i) $Y = g(X)$ is a random variable, measurable with respect to the sigma field $\mathbb{A}$.

(ii) The sigma field induced by Y is included in a sigma field induced by X.

Proof.

(i) For a Borel set S,

$$\begin{aligned}Y^{-1}(S) &= (g(X))^{-1}(S) = \{\omega|Y(\omega) \in S\} = \{\omega|g(X)(\omega) \in S\}\\ &= \{\omega|g(X(\omega)) \in S\} = \{\omega|X(\omega) \in g^{-1}(S)\}\\ &= X^{-1}(g^{-1}(S))\ .\end{aligned} \tag{2.7}$$

Thus, for a Borel set S, $g(X)^{-1}(S) = X^{-1}(g^{-1}(S))$. Now, $\forall\ S \in \mathbb{B}$,

$$\begin{aligned} S \in \mathbb{B} \quad &\Rightarrow \quad g^{-1}(S) \in \mathbb{B}, \quad \text{g being a Borel function} \\ &\Rightarrow \quad X^{-1}(g^{-1}(S)) \in \mathbb{A}, \quad \text{X being } \mathbb{A} \text{ measurable} \\ &\Rightarrow \quad Y^{-1}(S) = X^{-1}(g^{-1}(S)) \in \mathbb{A}. \end{aligned}$$

Hence, $Y = g(X)$ is a random variable, measurable with respect to the sigma field $\mathbb{A}$.

(ii) To prove that $Y^{-1}(\mathbb{B}) \subseteq X^{-1}(\mathbb{B})$, observe that

$$\begin{aligned} A \in Y^{-1}(\mathbb{B}) \quad &\Rightarrow \quad \exists\ S \in \mathbb{B} \ \text{ such that } \ A = Y^{-1}(S) \\ &\Rightarrow \quad A = Y^{-1}(S) = X^{-1}(g^{-1}(S)) \qquad \text{by } (2.7) \\ &\Rightarrow \quad A = X^{-1}(g^{-1}(S)) \in X^{-1}(\mathbb{B}) \ \text{ as } \ g^{-1}(S) \in \mathbb{B}. \end{aligned}$$

Thus, $A \in Y^{-1}(\mathbb{B}) \ \Rightarrow \ A \in X^{-1}(\mathbb{B})$ and hence $Y^{-1}(\mathbb{B}) \subseteq X^{-1}(\mathbb{B})$.

□

Remark 2.2.5. The result $g(X)^{-1}(S) = X^{-1}(g^{-1}(S))$ is similar to $(AB)^{-1} = B^{-1}A^{-1}$, where A and B are matrices of appropriate order so that matrix product AB is defined.

We prove in Section 4 that every continuous function is a Borel function. Thus, almost any function of a $\mathbb{A}$ measurable random variable X, such as a polynomial function, exponential function, logarithmic function and all trignometric functions, is again a random variable with respect to the same sigma field $\mathbb{A}$. We have verified in Example 2.2.12 that the sigma field induced by a Borel function of X is included in the sigma field induced by X.

Example 2.2.14. In this example, we examine whether $|X|$ is a random variable with respect to $\mathbb{A}$ implies X is a random variable with respect to $\mathbb{A}$. Suppose $\Omega = \{-1, 0, 1\}$ and X be defined on $(\Omega, \mathbb{A})$ as an identity function, that is as $X(i) = i$. Now $Y = |X|$, hence $Y(-1) = Y(1) = 1$ and $Y(0) = 0$. To find the sigma field induced by Y, we have for $S \in \mathbb{B}$,

$$Y^{-1}(S) \quad = \quad \begin{cases} \emptyset, & \text{if } \ 0 \notin S, 1 \notin S \\ \{0\}, & \text{if } \ 0 \in S, 1 \notin S \\ \{-1, 1\}, & \text{if } \ 0 \notin S, 1 \in S \\ \Omega, & \text{if } \ 0 \in S, 1 \in S. \end{cases}$$

Thus, the sigma field induced by Y is given by,

$$Y^{-1}(\mathbb{B}) = \{\Omega, \emptyset, \{0\}, \{-1, 1\}\} = \mathbb{A}, \ \text{ say}$$

and then $Y = |X|$ is a random variable with respect to $\mathbb{A}$. It is to be noted that the sigma field induced by $|X|$ and X^2 is the same. Now for a Borel set S such that $-1 \in S, 0 \notin S, 1 \notin S$, $X^{-1}(S) = \{-1\} \notin \mathbb{A}$, hence X is not a random variable with respect to $\mathbb{A}$. It is to be noted that the correspondence from $|X|$ to X is not a function, it being a one-many correspondence. □

Once we have defined a random variable as a measurable function from $(\Omega, \mathbb{A})$ to $(\mathbb{R}, \mathbb{B})$, we interpret Ω as a set of all possible outcomes of a random experiment and it is known as a sample space and each element in Ω is known as a sample point. Often, we are interested in the occurrence of some outcomes of interest. It is a subset of Ω. If it belongs to $\mathbb{A}$, then it is known as an event. Thus we have a following formal definition.

Definition 2.2.7. *Sample space, sample point and event: Suppose $(\Omega, \mathbb{A})$ is a measurable space. Then the set Ω is a set of all possible outcomes of a random experiment and is known as a sample space. Each element in Ω is known as a sample point and a set in $\mathbb{A}$ is known as an event.*

It should be noted that every subset of Ω may not be an event. An event A is said to occur at a trial of a random experiment, if the outcome $\omega \in A$ at that trial. Otherwise, we say that the event A has not occurred at that trial. In such case, $\omega \in A^c$ and hence A^c has occurred. It may be noted that, Ω is an event that always occurs and the event $\emptyset$ never happens. For this reason Ω is known as a sure event and $\emptyset$ as an impossible event.

In this section, the image space of the function X is the real line and if it is measurable, we have labeled it as a random variable. The image space of the function can be k dimensional Euclidean space or even a sequence of real numbers. In the next section, we extend the theory of univariate random variables, that is random variables with image space as the real line, to multivariate random variable, when image space is k dimensional Euclidean space, also known as random vector and then to a countably infinite collection of random variables, known as a sequence of random variables.

2.3 Random Vectors

Definition of a random vector is similar to that of a random variable.

Definition 2.3.1. *k-Variate random vector: Suppose $\underline{Z} = (X_1, X_2, \cdots, X_k)'$ is a function defined on the measurable space $(\Omega, \mathbb{A})$ to $(\mathbb{R}^k, \mathbb{B}^k)$, where $\mathbb{B}^k$ is the Borel field of subsets of $\mathbb{R}^k$. If*

$$\forall \ S \in \mathbb{B}^k, \ \ \underline{Z}^{-1}(S) \ = \ \{\omega | \ \underline{Z}(\omega) = (X_1(\omega), X_2(\omega), \cdots, X_k(\omega))' \in S\} \in \mathbb{A},$$

then $\underline{Z}$ is a random vector measurable with respect to $\mathbb{A}$.

By Lemma 2.2.3, $\underline{Z}^{-1}(\mathbb{B}^k) = \{\underline{Z}^{-1}(S) | S \in \mathbb{B}^k\}$ is a sigma field and it is known as a sigma field induced by the random vector $\underline{Z}$. It is also denoted by $\sigma(\underline{Z})$. Thus, if the induced sigma field is included in $\mathbb{A}$, then $\underline{Z}$ is a random vector with respect to $\mathbb{A}$. Following example illustrates the concept of a random vector.

Example 2.3.1. Suppose X_1 and X_2 are two functions defined on $(\Omega, \mathbb{A})$ as follows, where $\Omega = \{a, b, c\}$ and $\mathbb{A} = \mathbb{P}(\Omega)$.

$$X_1(\omega) = \begin{cases} 0, & \text{if} \quad \omega = a \\ 1, & \text{if} \quad \omega = b, c \end{cases} \quad \& \quad X_2(\omega) = \begin{cases} 2, & \text{if} \quad \omega = b \\ 3, & \text{if} \quad \omega = a, c. \end{cases}$$

A vector $\underline{Z} = (X_1, X_2)'$ is given by,

$$\underline{Z}(\omega) \quad = \quad \begin{cases} (0,3)', & \text{if} \quad \omega = a, \\ (1,2)', & \text{if} \quad \omega = b, \\ (1,3)', & \text{if} \quad \omega = c. \end{cases}$$

(i) We first examine whether X_1 and X_2 are random variables on $(\Omega, \mathbb{A})$. Suppose $S \in \mathbb{B}$, then

$$X_1^{-1}(S) \quad = \quad \begin{cases} \emptyset, & \text{if} \quad 0 \notin S \;\; \& \;\; 1 \notin S, \\ \{a\}, & \text{if} \quad 0 \in S \;\; \& \;\; 1 \notin S, \\ \{b,c\}, & \text{if} \quad 0 \notin S \;\; \& \;\; 1 \in S, \\ \Omega, & \text{if} \quad 0 \in S \;\; \& \;\; 1 \in S. \end{cases}$$

and

$$X_2^{-1}(S) \quad = \quad \begin{cases} \emptyset, & \text{if} \quad 2 \notin S \;\; \& \;\; 3 \notin S, \\ \{b\}, & \text{if} \quad 2 \in S \;\; \& \;\; 3 \notin S, \\ \{a,c\}, & \text{if} \quad 2 \notin S \;\; \& \;\; 3 \in S, \\ \Omega, & \text{if} \quad 2 \in S \;\; \& \;\; 3 \in S. \end{cases}$$

It is clear that $X_1^{-1}(S) \in \mathbb{A}, \;\; \forall \;\; S \in \mathbb{B}$ and $X_2^{-1}(S) \in \mathbb{A}, \;\; \forall \;\; S \in \mathbb{B}$ and hence X_1 and X_2 and are random variables on $(\Omega, \mathbb{A})$.

(ii) Now using Definition 2.3.1 of a random vector, we examine whether $\underline{Z} = (X_1, X_2)'$ is a random vector on $(\Omega, \mathbb{A})$. Suppose we denote the images $(0,3), (1,2)$ and $(1,3)$ as points as P_1, P_2 and P_3 respectively. For $S \in \mathbb{B}^2$,

$$\underline{Z}^{-1}(S) \quad = \quad \begin{cases} \emptyset, & \text{if} \quad P_1 \notin S, P_2 \notin S, P_3 \notin S,, \\ \{a\}, & \text{if} \quad P_1 \in S, P_i \notin S, i = 2,3, \\ \{b\}, & \text{if} \quad P_2 \in S, P_i \notin S, i = 1,3, \\ \{c\}, & \text{if} \quad P_3 \in S, P_i \notin S, i = 1,2, \\ \{a,b\}, & \text{if} \quad P_1 \in S, P_2 \in S, P_3 \notin S, \\ \{a,c\}, & \text{if} \quad P_1 \in S, P_3 \in S, P_2 \notin S, \\ \{b,c\}, & \text{if} \quad P_1 \notin S, P_2 \in S, P_3 \in S, \\ \Omega, & \text{if} \quad P_1 \in S, P_2 \in S, P_3 \in S. \end{cases}$$

Again it is to be noted that $\underline{Z}^{-1}(S) \in \mathbb{A}, \;\; \forall \;\; S \in \mathbb{B}^2$, hence $\underline{Z} = (X_1, X_2)$ is a random vector on $(\Omega, \mathbb{A})$.

(iii) From the inverse images of Borel sets as obtained above, we have

$$X_1^{-1}(\mathbb{B}) = \{\Omega, \emptyset, \{a\}, \{b, c\}\}, \qquad X_2^{-1}(\mathbb{B}) = \{\Omega, \emptyset, \{b\}, \{a, c\}\}$$
$$\text{and} \quad \underline{Z}^{-1}(\mathbb{B}^2) = \{\Omega, \emptyset, \{a\}, \{b\}, \{c\}, \{a, b\}, \{b, c\}, \{a, c\}\}.$$

Further,

$$X_1^{-1}(\mathbb{B}) \cup X_2^{-1}(\mathbb{B}) = \{\Omega, \emptyset, \{a\}, \{b\}, \{b, c\}, \{a, c\}\}.$$

Observe that $\underline{Z}^{-1}(\mathbb{B}^2) \neq X_1^{-1}(\mathbb{B}) \cup X_2^{-1}(\mathbb{B})$. Further $X_1^{-1}(\mathbb{B}) \cup X_2^{-1}(\mathbb{B})$ is not a sigma field as $\{a\} \cup \{b\} = \{a, b\} \notin X_1^{-1}(\mathbb{B}) \cup X_2^{-1}(\mathbb{B})$. We have noted in Chapter 1 that in general union of two sigma fields is not a sigma field. Also note that $\{b, c\}, \{a, c\} \in X_1^{-1}(\mathbb{B}) \cup X_2^{-1}(\mathbb{B})$, but $\{b, c\} \cap \{a, c\} = \{c\} \notin X_1^{-1}(\mathbb{B}) \cup X_2^{-1}(\mathbb{B})$. Now the minimal sigma field generated by $X_1^{-1}(\mathbb{B}) \cup X_2^{-1}(\mathbb{B})$ is given by

$$\begin{aligned} \sigma(X_1^{-1}(\mathbb{B}) \cup X_2^{-1}(\mathbb{B})) &= \{\Omega, \emptyset, \{a\}, \{b\}, \{c\}, \{a, b\}, \{b, c\}, \{a, c\}\} \\ &= Z^{-1}(\mathbb{B}^2). \end{aligned}$$

(iv) Note that X_1 and X_2 are random variables, with respect to $\mathbb{A}$. Thus,

$$\begin{aligned} X_1^{-1}(\mathbb{B}) \subseteq \mathbb{A} \ \&\ X_2^{-1}(\mathbb{B}) \subseteq \mathbb{A} &\Rightarrow X_1^{-1}(\mathbb{B}) \cup X_2^{-1}(\mathbb{B}) \subseteq \mathbb{A} \\ &\Rightarrow \sigma(X_1^{-1}(\mathbb{B}) \cup X_2^{-1}(\mathbb{B})) \subseteq \mathbb{A} \\ &\Rightarrow \underline{Z}^{-1}(\mathbb{B}^2) \subseteq \mathbb{A}, \end{aligned}$$

as in (iii) it is verified that $\sigma(X_1^{-1}(\mathbb{B}) \cup X_2^{-1}(\mathbb{B})) = Z^{-1}(\mathbb{B}^2)$. Thus this example shows that if components are random variables, then vector of these components is also a random vector. We prove this result in general in this section. □

In general, if $\underline{Z} = (X_1, X_2, \cdots, X_k)'$ is a k-dimensional random vector, then the sigma field induced by $\underline{Z}$ is the same as the minimal sigma field generated by the union of sigma fields, induced by $\{X_1, X_2, \cdots, X_k\}$. Thus,

$$\sigma(\underline{Z}) = \sigma(\{X_1, X_2, \cdots, X_k\}) = \underline{Z}^{-1}(\mathbb{B}^k) = \sigma\Big(\bigcup_{i=1}^{k} X_i^{-1}(\mathbb{B})\Big).$$

It is proved below. Towards it, we need some more results. As in the univariate setup, we have the following theorem which leads to a simpler way to verify that a given function is a random vector.

Theorem 2.3.1. *Suppose $\mathbb{C}^k$ is a class of subsets of $\mathbb{R}^k$ which generates the Borel field $\mathbb{B}^k$. A function $\underline{Z} = (X_1, X_2, \cdots, X_k)'$ is defined on the measurable space $(\Omega, \mathbb{A})$ to $(\mathbb{R}^k, \mathbb{B}^k)$. Then*

$$\forall\ S \in \mathbb{B}^k,\ \underline{Z}^{-1}(S) \in \mathbb{A} \iff \forall\ S \in \mathbb{C}^k,\ \underline{Z}^{-1}(S) \in \mathbb{A}.$$

Proof.

(i) Suppose $\forall\ S \in \mathbb{B}^k,\ \underline{Z}^{-1}(S) \in \mathbb{A}$. By the definition of the minimal sigma field, $\mathbb{C}^k \subseteq \sigma(\mathbb{C}^k) = \mathbb{B}^k$. Hence,

$$S \in \mathbb{C}^k \quad \Rightarrow \quad S \in \mathbb{B}^k \quad \Rightarrow \quad \underline{Z}^{-1}(S) \in \mathbb{A}.$$

(ii) Now suppose that $\forall\ S \in \mathbb{C}^k,\ \underline{Z}^{-1}(S) \in \mathbb{A}$, thus $\underline{Z}^{-1}(\mathbb{C}^k) \subseteq \mathbb{A}$ implying that $\mathbb{A}$ is a sigma field covering $\underline{Z}^{-1}(\mathbb{C}^k)$. Thus,

$$\begin{aligned}\underline{Z}^{-1}(\mathbb{C}^k) \subseteq \mathbb{A} &\Rightarrow \sigma(\underline{Z}^{-1}(\mathbb{C}^k)) \subseteq \mathbb{A} \quad \text{by definition of minimal sigma field}\\ &\Rightarrow \underline{Z}^{-1}(\sigma(\mathbb{C}^k)) \subseteq \mathbb{A} \quad \text{by Theorem 2.2.1}\\ &\Rightarrow \underline{Z}^{-1}(\mathbb{B}^k) \subseteq \mathbb{A} \quad \text{by definition of Borel field}\\ &\Rightarrow \forall\ \ S \in \mathbb{B}^k,\ \ \underline{Z}^{-1}(S)\ \in \mathbb{A}\end{aligned}$$

□

In view of Theorem 2.3.1, we have one more definition for a random vector as given below.

Definition 2.3.2. *k-Variate random vector: Suppose $\underline{Z} = (X_1, X_2, \cdots, X_k)'$ is a function defined on the measurable space $(\Omega, \mathbb{A})$ to $(\mathbb{R}^k, \mathbb{B}^k)$. Suppose $\mathbb{C}^k$ is a class of subsets of $\mathbb{R}^k$ which generates the Borel field $\mathbb{B}^k$. If*

$$\forall\ \ S \in \mathbb{C}^k,\ \ \underline{Z}^{-1}(S)\ = \{\omega\ \mid\ \underline{Z}(\omega) = (X_1(\omega), X_2(\omega), \cdots, X_k(\omega))' \in S\} \in \mathbb{A},$$

then $\underline{Z}$ is a random vector measurable with respect to $\mathbb{A}$.

Definition 2.3.2 is known as an economical definition of a random vector while Definition 2.3.1 is known as a descriptive definition of a random vector. Equivalence of the two definitions is established in Theorem 2.3.1. In the following example, we use the descriptive definition as well as the economical definition of a random vector to verify whether $\underline{Z} = (I_A, I_B)'$ is a random vector.

Example 2.3.2. Suppose $\underline{Z} = (I_A, I_B)$ where $A, B \in \mathbb{A}$. (i) We show that $\underline{Z}$ is a random vector on $(\Omega, \mathbb{A})$ by showing that $\sigma(\underline{Z}) \subseteq \mathbb{A}$. By definition of $\underline{Z}$ we have,

$$\underline{Z}(\omega) \quad = \quad \begin{cases} (1,1), & \text{if} \quad \omega \in A\ \ \&\ \ \omega \in B \quad \Longleftrightarrow \quad \omega \in A \cap B, \\ (1,0), & \text{if} \quad \omega \in A\ \ \&\ \ \omega \notin B \quad \Longleftrightarrow \quad \omega \in A \cap B^c, \\ (0,1), & \text{if} \quad \omega \notin A\ \ \&\ \ \omega \in B \quad \Longleftrightarrow \quad \omega \in A^c \cap B, \\ (0,0), & \text{if} \quad \omega \notin A\ \ \&\ \ \omega \notin B \quad \Longleftrightarrow \quad \omega \in A^c \cap B^c. \end{cases}$$

To find the sigma field induced by $\underline{Z}$, we denote the images as $P_1 = (0,0)$,

$P_2 = (0,1), P_3 = (1,0)$ and $P_4 = (1,1)$. For $S \in \mathbb{B}^2$,

$$\underline{Z}^{-1}(S) = \begin{cases} \emptyset, & \text{if} \quad P_i \notin S, i = 1,2,3,4 \\ A^c \cap B^c, & \text{if} \quad P_1 \in S, P_i \notin S, i = 1,2,3, \\ A^c \cap B, & \text{if} \quad P_2 \in S, P_i \notin S, i = 1,3,4, \\ A \cap B^c, & \text{if} \quad P_3 \in S, P_i \notin S, i = 1,2,4, \\ A \cap B, & \text{if} \quad P_4 \in S, P_i \notin S, i = 1,2,3, \\ A^c, & \text{if} \quad P_1 \in S, P_2 \in S, P_i \notin S, i = 3,4, \\ B^c, & \text{if} \quad P_1 \in S, P_3 \in S, P_i \notin S, i = 2,4, \\ (A\Delta B)^c, & \text{if} \quad P_1 \in S, P_4 \in S, P_i \notin S, i = 2,3, \\ A\Delta B, & \text{if} \quad P_2 \in S, P_3 \in S, P_i \notin S, i = 1,4, \\ B, & \text{if} \quad P_2 \in S, P_4 \in S, P_i \notin S, i = 1,3, \\ A, & \text{if} \quad P_3 \in S, P_4 \in S, P_i \notin S, i = 1,2, \\ (A \cap B)^c, & \text{if} \quad P_1 \in S, P_2 \in S, P_3 \in S, P_4 \notin S, \\ (A \cap B^c)^c, & \text{if} \quad P_1 \in S, P_2 \in S, P_3 \notin S, P_4 \in S, \\ (A^c \cap B)^c, & \text{if} \quad P_1 \in S, P_2 \notin S, P_3 \in S, P_4 \in S, \\ A \cup B, & \text{if} \quad P_1 \notin S, P_2 \in S, P_3 \in S, P_4 \in S, \\ \Omega, & \text{if} \quad P_1 \in S, P_2 \in S, P_3 \in S, P_4 \in S. \end{cases}$$

Thus, $\underline{Z}^{-1}(\mathbb{B}^2) = \{\Omega, \emptyset, A, B, A^c, B^c, A \cap B, A^c \cap B, A \cap B^c, A^c \cap B^c, (A\Delta B)^c, A\Delta B, (A \cap B)^c, (A \cap B^c)^c, (A^c \cap B)^c, A \cup B\}$. It is clear that if $A, B \in \mathbb{A}$, then all sets in $\underline{Z}^{-1}(\mathbb{B}^2)$ are included in $\mathbb{A}$ and hence $\underline{Z}$ is a random vector measurable with respect to $\mathbb{A}$.
(ii) We now examine whether $\underline{Z} = (I_A, I_B)$ is a random vector, using the economical definition, as given in Definition 2.3.2. Suppose for $a, b \in \mathbb{R}$,

$$\mathbb{C}^2 = \{(x,y) | -\infty < x \leq a, -\infty < y \leq b\} = \{S | S = (-\infty, a] \times (-\infty, b]\}.$$

Note that with $S = (-\infty, a] \times (-\infty, b]$,

$$\underline{Z}^{-1}(S) = \begin{cases} \emptyset, & \text{if} \quad a < 0 \ \& \ b < 0, \\ \emptyset, & \text{if} \quad a < 0 \ \& \ b \geq 0, \\ \emptyset, & \text{if} \quad a \geq 0 \ \& \ b < 0, \\ A^c \cap B^c, & \text{if} \quad 0 \leq a < 1 \ \& \ 0 \leq b < 1, \\ A^c, & \text{if} \quad 0 \leq a < 1 \ \& \ 0 \leq b \leq 1, \\ B^c, & \text{if} \quad 0 \leq a \leq 1 \ \& \ 0 \leq b < 1, \\ \Omega, & \text{if} \quad a \geq 1 \ \& \ b \geq 1. \end{cases}$$

It is given that $A, B \in \mathbb{A}$ hence $A^c, B^c, A^c \cap B^c \in \mathbb{A}$, it being a sigma field. Thus, $\underline{Z}^{-1}(S) \in \mathbb{A} \ \forall \ S \in \mathbb{C}^2$ and hence $\underline{Z}$ is a $\mathbb{A}$ measurable random vector. □

In Example 2.3.1, we have noted that $\underline{Z}^{-1}(\mathbb{B}^2) = \sigma(X_1^{-1}(\mathbb{B}) \cup X_2^{-1}(\mathbb{B}))$. It is in general true and we prove this important result below. In the proof, we need the following lemma.

Lemma 2.3.1. *(i) Suppose $\mathbb{C}_1$ and $\mathbb{C}_2$ are two classes of sets such that $\mathbb{C}_1 \subseteq \mathbb{C}_2$. Then $\sigma(\mathbb{C}_1) \subseteq \sigma(\mathbb{C}_2)$. (ii) Suppose $\sigma(\mathbb{C}) \subseteq \mathbb{A}$ where $\mathbb{A}$ is a sigma field, then $\mathbb{C} \subseteq \mathbb{A}$.*

Proof.

(i) As proved in Section 1.4, note that $\mathbb{C}_1 \subseteq \mathbb{C}_2 \subseteq \sigma(\mathbb{C}_2)$. Thus, $\sigma(\mathbb{C}_2)$ is a sigma field covering $\mathbb{C}_1$. Hence, by the definition of the minimal sigma field $\sigma(\mathbb{C}_1) \subseteq \sigma(\mathbb{C}_2)$.

(ii) Note that

$$S \in \mathbb{C} \;\Rightarrow\; S \in \sigma(\mathbb{C}) \;\Rightarrow\; S \in \mathbb{A} \text{ as } \sigma(\mathbb{C}) \subseteq \mathbb{A}.$$

Thus, $\mathbb{C} \subseteq \mathbb{A}$. Note that Result(ii) is valid if $\mathbb{A}$ is any collection of sets, not necessarily a sigma field.

□

Theorem 2.3.2. *Suppose $\underline{Z} = (X_1, X_2)'$ is a random vector defined on a measurable space $(\Omega, \mathbb{A})$ to $(\mathbb{R}^2, \mathbb{B}^2)$. Then*

$$\sigma(\underline{Z}) = \sigma\left((X_1, X_2)'\right) = \underline{Z}^{-1}(\mathbb{B}^2) = \sigma\left(X_1^{-1}(\mathbb{B}) \cup X_2^{-1}(\mathbb{B})\right).$$

Proof. Suppose a_1, b_1, a_2 and b_2 are real numbers and

$$S = \{(x_1, x_2) \in \mathbb{R}^2 \mid a_1 < x_1 < b_1, \quad a_2 < x_2 < b_2\}$$

is a rectangle in $\mathbb{R}^2$. Suppose $\mathbb{C}^1$ is a class of open intervals of $\mathbb{R}$ and $\mathbb{C}^2$ is a class of rectangles of the type S. Then $\sigma(\mathbb{C}^1) = \mathbb{B}$ and $\sigma(\mathbb{C}^2) = \mathbb{B}^2$.

(i) Observe that

$$\begin{aligned}
\underline{Z}^{-1}(S) &= \{\omega \mid a_1 < X_1(\omega) < b_1, \quad a_2 < X_2(\omega) < b_2\} \\
&= \{\omega \mid a_1 < X_1(\omega) < b_1\} \cap \{\omega \mid a_2 < X_2(\omega) < b_2\} \\
&= X_1^{-1}((a_1, b_1)) \cap X_2^{-1}((a_2, b_2)).
\end{aligned}$$

In particular, $(a_2, b_2) = \mathbb{R} \;\Rightarrow\; \underline{Z}^{-1}(S) = X_1^{-1}((a_1, b_1)) \cap \Omega = X_1^{-1}((a_1, b_1))$. Thus, $X_1^{-1}((a_1, b_1))$ can be written as $\underline{Z}^{-1}(S)$ for some S. Hence,

$$\begin{aligned}
\{\underline{Z}^{-1}(S) \mid S \text{ is a rectangle}\} &\supseteq \{X_1^{-1}((a_1, b_1)) \mid a_1, b_1 \in \mathbb{R}\} \\
\Rightarrow \underline{Z}^{-1}(\mathbb{C}^2) &\supseteq X_1^{-1}(\mathbb{C}^1) \\
\Rightarrow \sigma(\underline{Z}^{-1}(\mathbb{C}^2)) &\supseteq \sigma(X_1^{-1}(\mathbb{C}^1)), \text{ by Lemma 2.3.1} \\
\Rightarrow \underline{Z}^{-1}(\sigma(\mathbb{C}^2)) &\supseteq X_1^{-1}(\sigma(\mathbb{C}^1)) \\
\Rightarrow \underline{Z}^{-1}(\mathbb{B}^2) &\supseteq X_1^{-1}(\mathbb{B}).
\end{aligned}$$

Similarly, we have $\underline{Z}^{-1}(\mathbb{B}^2) \supseteq X_2^{-1}(\mathbb{B})$. Hence,

$$X_1^{-1}(\mathbb{B}) \cup X_2^{-1}(\mathbb{B}) \subseteq \underline{Z}^{-1}(\mathbb{B}^2) \;\Rightarrow\; \sigma\left(X_1^{-1}(\mathbb{B}) \cup X_2^{-1}(\mathbb{B})\right) \subseteq \underline{Z}^{-1}(\mathbb{B}^2). \tag{2.8}$$

(ii) Note that $X_1^{-1}((a_1,b_1)) \in X_1^{-1}(\mathbb{B})$ and $X_2^{-1}((a_2,b_2)) \in X_2^{-1}(\mathbb{B})$, however, $X_1^{-1}(a_1,b_1) \cap X_2^{-1}(a_2,b_2)$ may not be in $X_1^{-1}(\mathbb{B}) \cup X_2^{-1}(\mathbb{B})$. On the other hand, it is in $\sigma\left(X_1^{-1}(\mathbb{B}) \cup X_2^{-1}(\mathbb{B})\right)$. Thus, $\forall\ S \in \mathbb{C}^2$

$$\begin{aligned}\underline{Z}^{-1}(S) = X_1^{-1}((a_1,b_1)) \cap X_2^{-1}((a_2,b_2)) &\in \sigma\left(X_1^{-1}(\mathbb{B}) \cup X_2^{-1}(\mathbb{B})\right)\\ \Rightarrow\quad \underline{Z}^{-1}(\mathbb{C}^2) &\subseteq \sigma\left(X_1^{-1}(\mathbb{B}) \cup X_2^{-1}(\mathbb{B})\right)\\ \Rightarrow\quad \sigma(\underline{Z}^{-1}(\mathbb{C}^2)) &\subseteq \sigma\left(X_1^{-1}(\mathbb{B}) \cup X_2^{-1}(\mathbb{B})\right)\\ \Rightarrow\quad \underline{Z}^{-1}(\sigma(\mathbb{C}^2)) &\subseteq \sigma\left(X_1^{-1}(\mathbb{B}) \cup X_2^{-1}(\mathbb{B})\right)\\ \Rightarrow\quad \underline{Z}^{-1}(\mathbb{B}^2) &\subseteq \sigma\left(X_1^{-1}(\mathbb{B}) \cup X_2^{-1}(\mathbb{B})\right). \end{aligned} \tag{2.9}$$

Thus, from (2.8) and (2.9), we have $\underline{Z}^{-1}(\mathbb{B}^2) = \sigma(X_1^{-1}(\mathbb{B}) \cup X_2^{-1}(\mathbb{B}))$.

□

In Example 2.3.1 we have verified that if the components of $\underline{Z}$ are random variables with respect to $\mathbb{A}$, then $\underline{Z}$ is a random vector with respect to $\mathbb{A}$. It is in general true and we prove this important result below. Theorem 2.3.2 is heavily used in the proof.

Theorem 2.3.3. *Suppose $\underline{Z} = (X_1, X_2)'$ is defined on the measurable space $(\Omega, \mathbb{A})$ to $(\mathbb{R}^2, \mathbb{B}^2)$. Then $\underline{Z}$ is a random vector measurable with respect to $\mathbb{A}$ if and only if X_1 and X_2 are random variables measurable with respect to $\mathbb{A}$.*

Proof. In Theorem 2.3.2, it is proved that $\underline{Z}^{-1}(\mathbb{B}^2) = \sigma(X_1^{-1}(\mathbb{B}) \cup X_2^{-1}(\mathbb{B}))$.

(i) Suppose $\underline{Z}$ is a random vector measurable with respect to $\mathbb{A}$. Then

$$\begin{aligned}\underline{Z}^{-1}(\mathbb{B}^2) \subseteq \mathbb{A} &\Rightarrow \sigma(X_1^{-1}(\mathbb{B}) \cup X_2^{-1}(\mathbb{B})) \subseteq \mathbb{A}\\ &\Rightarrow X_1^{-1}(\mathbb{B}) \cup X_2^{-1}(\mathbb{B}) \subseteq \mathbb{A}\\ &\qquad \text{by (ii) of Lemma 2.3.1}\\ \text{Hence,}\quad A \in X_1^{-1}(\mathbb{B}) &\Rightarrow A \in X_1^{-1}(\mathbb{B}) \cup X_2^{-1}(\mathbb{B}) \subseteq \mathbb{A}\\ &\Rightarrow A \in \mathbb{A} \quad\Rightarrow\quad X_1^{-1}(\mathbb{B}) \subseteq \mathbb{A}\\ \text{Similarly,}\quad A \in X_2^{-1}(\mathbb{B}) &\Rightarrow A \in \mathbb{A} \quad\Rightarrow\quad X_2^{-1}(\mathbb{B}) \subseteq \mathbb{A}.\end{aligned}$$

Hence, X_1 and X_2 are random variables measurable with respect to $\mathbb{A}$.

(ii) Suppose X_1 and X_2 are random variables measurable with respect to $\mathbb{A}$. Then

$$\begin{aligned}X_1^{-1}(\mathbb{B}) \subseteq \mathbb{A}\ \ \&\ \ X_2^{-1}(\mathbb{B}) \subseteq \mathbb{A} \quad &\Rightarrow\quad X_1^{-1}(\mathbb{B}) \cup X_2^{-1}(\mathbb{B}) \subseteq \mathbb{A}\\ &\Rightarrow\quad \sigma\left(X_1^{-1}(\mathbb{B}) \cup X_2^{-1}(\mathbb{B})\right) \subseteq \mathbb{A}\\ &\Rightarrow\quad \underline{Z}^{-1}(\mathbb{B}^2) \subseteq \mathbb{A}.\end{aligned}$$

Hence, $\underline{Z}$ is a random vector measurable with respect to $\mathbb{A}$.

□

Remark 2.3.1. Theorem 2.3.2 and Theorem 2.3.3 can be extended to a k dimensional random vector, that is, if $\underline{Z} = (X_1, X_2, \cdots, X_k)'$, then

$$\underline{Z}^{-1}(\mathbb{B}^k) = \sigma\Big(\bigcup_{i=1}^{k} X_i^{-1}(\mathbb{B})\Big).$$

Further, $\underline{Z}$ is a random vector measurable with respect to $\mathbb{A}$ if and only if each of X_i, $i = 1, 2, \cdots, k$ is a random variable measurable with respect to $\mathbb{A}$. It is the most useful result to verify whether $\underline{Z}$ is a random vector, as one can simply proceed marginally and verify whether components are random variables using the economical definition of a random variable.

We apply Theorem 2.3.3 in the next example.

Example 2.3.3. Suppose $\underline{Z} = (I_A, I_B)$ where $A, B \in \mathbb{A}$. In Example 5.3.8, using economical and descriptive definitions of a random vector, it is shown that $\underline{Z}$ is a random vector on $(\Omega, \mathbb{A})$. We now use Theorem 2.3.3. In Example 2.2.3, we have noted that the sigma field induced by I_A is $I_A^{-1}(\mathbb{B}) = \{\Omega, \emptyset, A, A^c\}$ and that of I_B is given by $I_B^{-1}(\mathbb{B}) = \{\Omega, \emptyset, B, B^c\}$. Both are included in $\mathbb{A}$ which implies that I_A and I_B both are random variables with respect to $\mathbb{A}$ and hence by Theorem 2.3.3, $\underline{Z} = (I_A, I_B)$ is $\mathbb{A}$ measurable random vector. It is clear that instead of using the definitions, it is simpler to verify that I_A and I_B are random variables and hence $\underline{Z} = (I_A, I_B)'$ is a random vector. Note that $I_A^{-1}(\mathbb{B}) \cup I_B^{-1}(\mathbb{B}) = \{\Omega, \emptyset, A, A^c, B, B^c\}$ which is not a sigma field but it is included in $\underline{Z}^{-1}(\mathbb{B}^2)$. If we go on adding sets to $I_A^{-1}(\mathbb{B}) \cup I_B^{-1}(\mathbb{B})$ so that it becomes closed under complementation and countable union and hence will be a sigma field, then $\sigma(I_A^{-1}(\mathbb{B}) \cup I_B^{-1}(\mathbb{B})) = \underline{Z}^{-1}(\mathbb{B}^2)$. □

In Theorem 2.2.4 we have proved that measurability of a random variable X remains invariant under Borel transformations and the sigma field induced by any Borel function of X is included in the sigma field induced by X. We prove similar result for a Borel transformation of a random vector in the following Theorem 2.3.5. We first define a Borel function in higher dimensions. The definition is similar to that in one dimension.

Definition 2.3.3. *Borel function: Suppose* $\underline{g} : (\mathbb{R}^k, \mathbb{B}^k) \to (\mathbb{R}^l, \mathbb{B}^l)$ $l \leq k$ *is a function where* $\underline{g} = (g_1, g_2, \cdots, g_l)$. $\underline{g}$ *is a Borel function if* $\forall\ S \in \mathbb{B}^l$, $\underline{g}^{-1}(S) \in \mathbb{B}^k$.

Following theorem gives a necessary and sufficient condition for a function to be a Borel function in higher dimensions.

Theorem 2.3.4. *Suppose* $\underline{g} : (\mathbb{R}^k, \mathbb{B}^k) \to (\mathbb{R}^l, \mathbb{B}^l)$ $l \leq k$ *is a function where* $\underline{g} = (g_1, g_2, \cdots, g_l)$. $\underline{g}$ *is a Borel function if and only if each of* g_i *is a Borel function from* $(\mathbb{R}^k, \mathbb{B}^k) \to (\mathbb{R}, \mathbb{B})$, $i = 1, 2, \cdots, l$.

Proof. Proof follows on similar lines as in Theorem 2.3.3, with $(\Omega, \mathbb{A})$ replaced by $(\mathbb{R}^k, \mathbb{B}^k)$. □

Theorem 2.3.5. *Suppose* $\underline{Z} : (\Omega, \mathbb{A}) \to (\mathbb{R}^k, \mathbb{B}^k)$ *is a random vector, measurable with respect to* $\mathbb{A}$.

(i) Suppose $g : (\mathbb{R}^k, \mathbb{B}^k) \to (\mathbb{R}, \mathbb{B})$ *is a Borel function, then* $Y = g(\underline{Z})$ *is a random variable, measurable with respect to* $\mathbb{A}$.

(ii) $\sigma(Y) \subseteq \sigma(\underline{Z})$.

(iii) Suppose $\underline{g} : (\mathbb{R}^k, \mathbb{B}^k) \to (\mathbb{R}^l, \mathbb{B}^l), l \leq k$ *is a Borel function, then* $\underline{Y} = \underline{g}(\underline{Z})$ *is a random vector, measurable with respect to the sigma field* $\mathbb{A}$.

(iv) $\sigma(\underline{Y}) \subseteq \sigma(\underline{Z})$.

Proof.

(i) For a Borel set $S \in \mathbb{B}$,

$$\begin{aligned} (g(\underline{Z}))^{-1}(S) &= Y^{-1}(S) = \{\omega | Y(\omega) \in S\} = \{\omega | g(\underline{Z})(\omega) \in S\} \\ &= \{\omega | g(\underline{Z}(\omega)) \in S\} = \{\omega | \underline{Z}(\omega) \in g^{-1}(S)\} \\ &= \underline{Z}^{-1}(g^{-1}(S)) \ . \end{aligned} \tag{2.10}$$

Thus, for a Borel set $S \in \mathbb{B}$, $(g(\underline{Z}))^{-1}(S) = \underline{Z}^{-1}(g^{-1}(S))$. Now, $\forall \ S \in \mathbb{B}$,

$$\begin{aligned} S \in \mathbb{B} &\Rightarrow g^{-1}(S) \in \mathbb{B}, \quad \text{g being a Borel function} \\ &\Rightarrow \underline{Z}^{-1}(g^{-1}(S)) \in \mathbb{A}, \quad \underline{Z} \text{ being } \mathbb{A} \text{ measurable} \\ &\Rightarrow (g(\underline{Z}))^{-1}(S) = Y^{-1}(S) \in \mathbb{A}. \end{aligned}$$

Hence, $Y = g(\underline{Z})$ is a random variable, measurable with respect to $\mathbb{A}$.

(ii) To prove that $Y^{-1}(\mathbb{B}) \subseteq \underline{Z}^{-1}(\mathbb{B}^k)$, observe that

$$\begin{aligned} A \in Y^{-1}(\mathbb{B}) &\Rightarrow \exists \ S \in \mathbb{B} \ \text{ such that } \ A = Y^{-1}(S) \\ &\Rightarrow A = Y^{-1}(S) = \underline{Z}^{-1}(g^{-1}(S)) \qquad \text{by (2.10)} \\ &\Rightarrow A = \underline{Z}^{-1}(g^{-1}(S)) \in \underline{Z}^{-1}(\mathbb{B}^k) \ \text{ as } \ g^{-1}(S) \in \mathbb{B}^k \end{aligned}$$

Thus, $A \in Y^{-1}(\mathbb{B}) \Rightarrow A \in \underline{Z}^{-1}(\mathbb{B}^k) \ \Rightarrow \ Y^{-1}(\mathbb{B}) \subseteq \underline{Z}^{-1}(\mathbb{B}^k)$.

(iii) Proof of this part is on similar lines as that of (i). For a Borel set $S \in \mathbb{B}^l$,

$$\begin{aligned} (\underline{g}(\underline{Z}))^{-1}(S) = \underline{Y}^{-1}(S) &= \{\omega | \underline{Y}(\omega) \in S\} = \{\omega | \underline{g}(\underline{Z})(\omega) \in S\} \\ &= \{\omega | \underline{g}(\underline{Z}(\omega)) \in S\} = \{\omega | \underline{Z}(\omega) \in \underline{g}^{-1}(S)\} \\ &= \underline{Z}^{-1}(\underline{g}^{-1}(S)). \end{aligned} \tag{2.11}$$

Thus, for a Borel set $S \in \mathbb{B}^l$, $\underline{g}(\underline{Z})^{-1}(S) = \underline{Z}^{-1}(\underline{g}^{-1}(S))$. Now, $\forall \ S \in \mathbb{B}^l$,

$$\begin{aligned} S \in \mathbb{B}^l &\Rightarrow \underline{g}^{-1}(S) \in \mathbb{B}^k, \quad \text{g being a Borel function} \\ &\Rightarrow \underline{Z}^{-1}(\underline{g}^{-1}(S)) = (\underline{g}(\underline{Z}))^{-1}(S) = \underline{Y}^{-1}(S) \in \mathbb{A}, \end{aligned}$$

$\underline{Z}$ being $\mathbb{A}$ measurable. Hence, $\underline{Y} = \underline{g}(\underline{Z})$ is a random vector, measurable with respect to $\mathbb{A}$. Another approach to prove this result is as follows. $\underline{g} : (\mathbb{R}^k, \mathbb{B}^k) \to (\mathbb{R}^l, \mathbb{B}^l), l \leq k$ is a Borel function implies that each of g_i, $i = 1, 2, \cdots, l$ is a Borel function from $(\mathbb{R}^k, \mathbb{B}^k) \to (\mathbb{R}, \mathbb{B})$. It then follows by part(i) that $Y_i = g_i(\underline{Z})$ is a random variable, measurable with respect to $\mathbb{A}$. Hence, by Theorem 2.3.3, $\underline{Y} = (Y_1, Y_2, \cdots, Y_l) = \underline{g}(\underline{Z})$ is a random vector, measurable with respect to $\mathbb{A}$.

(iv) To prove that $\underline{Y}^{-1}(\mathbb{B}^l) \subseteq \underline{Z}^{-1}(\mathbb{B}^k)$, observe that

$$
\begin{aligned}
A \in \underline{Y}^{-1}(\mathbb{B}^l) &\Rightarrow \exists \ S \in \mathbb{B}^l \text{ such that } A = \underline{Y}^{-1}(S) \\
&\Rightarrow A = \underline{Y}^{-1}(S) = \underline{Z}^{-1}(\underline{g}^{-1}(S)) \qquad \text{by (2.11)} \\
&\Rightarrow A = \underline{Z}^{-1}(\underline{g}^{-1}(S)) \in \underline{Z}^{-1}(\mathbb{B}^k) \text{ as } \underline{g}^{-1}(S) \in \mathbb{B}^k
\end{aligned}
$$

$$\text{Thus,} \quad A \in \underline{Y}^{-1}(\mathbb{B}^l) \Rightarrow A \in \underline{Z}^{-1}(\mathbb{B}^k) \ \Rightarrow \ \underline{Y}^{-1}(\mathbb{B}^l) \subseteq \underline{Z}^{-1}(\mathbb{B}^k).$$

□

Remark 2.3.2. As a consequence of Theorem 2.3.5, the addition, the subtraction and the product of k random variables is again a random variable. Further $Y = \exp(t_1X_1 + t_2X_2 + \cdots + t_kX_k)$, where $t_1, t_2, \cdots, t_k$ are real numbers is also a random variable. Such a result is proved with a different approach in Section 4. Part (iii) and (iv) of Theorem 2.3.5 indicate that results for real valued Borel function can be easily extended to a Borel function $g : (\mathbb{R}^k, \mathbb{B}^k) \to (\mathbb{R}^l, \mathbb{B}^l)$, $l \leq k$. Thus, if $\underline{Z} = (X_1, X_2, X_3)'$ is a random vector, measurable with respect to $\mathbb{A}$, then random vectors such as $\underline{U} = (X_1/X_3, X_2/X_3)'$ or $\underline{V} = (X_1 + X_3, X_2 + X_3)'$ are also $\mathbb{A}$ measurable random vectors.

Example 2.3.4. Suppose $\underline{Z} = (X_1, X_2, \cdots, X_5)'$ is a random vector on $(\Omega, \mathbb{A})$. Suppose $Y = \sum_{i=1}^{4} l_iX_i$, where $l_1, l_2, \cdots, l_4$ are any real numbers. Then Y can be expressed as $Y = g(\underline{Z})$ where $g : \mathbb{R}^5 \to \mathbb{R}$ is defined as $g(x_1, x_2, \cdots, x_5) = \sum_{i=1}^{5} l_ix_i$ with $l_5 = 0$. It is a Borel function and hence by Theorem 2.3.5, Y is a random variable on $(\Omega, \mathbb{A}, P)$. Suppose $\underline{U} = (\sin X_1, X_2^2, X_3^3, e^{X_4}, |X_5|)$. It is to be noted that $\sin X_1$ is a Borel function of X_1, similarly, each of $X_2^2, X_3^3, e^{X_4}, |X_5|$ are Borel functions of X_2, X_3, X_4, X_5 respectively. Thus, each component of $\underline{U}$ is a random variable on $(\Omega, \mathbb{A})$ and hence by Theorem 2.3.3, $\underline{U}$ is a random vector on $(\Omega, \mathbb{A})$. □

Remark 2.3.3. It is known that there is a one-to-one correspondence between all points in $\mathbb{R}^2$ and a set of all complex numbers. Analogously, if $(X_1, X_2)'$ is a random vector, then $Z = X_1 + iX_2$ is a complex valued random variable, where $i = \sqrt{-1}$.

From one dimensional random variable, we proceeded to define a k-dimensional random vector. We now extend it to a countably infinite collection of random variables and define what we mean by a sequence of random variables. Towards it, we need a concept of the Borel field in infinite dimensions. Suppose $\underline{x} = (x_1, x_2, \cdots,)$ is a sequence of real numbers and $\mathbb{R}^\infty$ denotes the collection of all such sequences of real numbers. Thus,

$$\mathbb{R}^\infty = \{\underline{x} = (x_1, x_2, \cdots,) \mid x_i \in \mathbb{R}, \ \forall \ i \geq 1\}.$$

We consider a collection of particular types of subsets of $\mathbb{R}^\infty$ defined as follows.

Definition 2.3.4. *Cylinder set with a finite dimensional base: A subset S of $\mathbb{R}^\infty$ is known as a cylinder set with a finite dimensional base if it is defined as,*

$$S = \{\underline{x} \mid a_{i_j} \leq x_{i_j} \leq b_{i_j}, \ j = 1, 2, \cdots, k, \ \ x_{i_j} \in \mathbb{R}, \ j = k+1, k+2, \cdots\},$$

where $i_1, i_2, \cdots i_k$ is any permutation of positive integers and k is finite.

Definition 2.3.5. *Borel field in infinite dimensions: Suppose $\mathbb{C}^\infty$ is a collection of all cylinder sets with a finite dimensional base, as defined in Definition 2.3.4. Then the minimal sigma field induced by $\mathbb{C}^\infty$ is said to be a Borel field in infinite dimensions and is denoted by $\mathbb{B}^\infty$.*

Having defined the Borel field in infinite dimensions, we now define an infinite dimensional random vector.

Definition 2.3.6. *Infinite dimensional random vector: A function $\underline{Z} = (X_1, X_2, \cdots,)$ defined on $(\Omega, \mathbb{A})$ to $(\mathbb{R}^\infty, \mathbb{B}^\infty)$ is an infinite dimensional random vector if $\forall \ S \in \mathbb{B}^\infty, \ \ \underline{Z}^{-1}(S) \in \mathbb{A}$.*

To prove that $\underline{Z} = (X_1, X_2, \cdots,)$ is an infinite dimensional random vector on $(\Omega, \mathbb{A})$, if and only if each component of $\underline{Z}$ is a random variable on $(\Omega, \mathbb{A})$, we derive a relation between the sigma field induced by $\underline{Z}$ and the sigma fields induced by each of $\{X_n, n \geq 1\}$. We define

$$\mathbb{A}_n = \sigma\{X_1, X_2, \cdots, X_n\} = \sigma\Big(\bigcup_{i=1}^{n} (X_i^{-1}(\mathbb{B}))\Big),$$

which is the minimal sigma field induced by $\{X_1, X_2, \cdots, X_n\}$. Note that

$$\begin{aligned} \bigcup_{i=1}^{n+1} (X_i^{-1}(\mathbb{B})) \quad & \supseteq \quad \bigcup_{i=1}^{n} (X_i^{-1}(\mathbb{B})) \\ \Rightarrow \ \sigma\Big(\bigcup_{i=1}^{n+1} (X_i^{-1}(\mathbb{B}))\Big) \quad & \supseteq \quad \sigma\Big(\bigcup_{i=1}^{n} (X_i^{-1}(\mathbb{B}))\Big) \\ \Rightarrow \ \mathbb{A}_{n+1} \quad & \supseteq \quad \mathbb{A}_n \ \ \forall \ n \geq 1 \,. \end{aligned}$$

Hence, $\{\mathbb{A}_n, n \geq 1\}$ is a monotone increasing sequence of sigma fields and therefore has a limit given by $\bigcup_{n\geq 1} \mathbb{A}_n$, which may not be a sigma field.

Then we find the minimal sigma field generated by $\bigcup_{n\geq 1}\mathbb{A}_n$, denoted by $\sigma(\bigcup_{n\geq 1}\mathbb{A}_n)$. In the following Theorem 2.3.6, we prove that a sigma field $\underline{Z}^{-1}(\mathbb{B}^\infty)$ induced by a sequence $\underline{Z}=(X_1,X_2,\cdots,)$ of random variables is $\sigma(\bigcup_{n\geq 1}\mathbb{A}_n)$. Thus, if

$$\underline{Z}^{-1}(\mathbb{B}^\infty)=\sigma\Big(\bigcup_{n\geq 1}\mathbb{A}_n\Big)=\sigma(\{X_n,n\geq 1\})\subseteq\mathbb{A}$$

then $\{X_n,n\geq 1\}$ is an infinite dimensional random vector, measurable with respect to $\mathbb{A}$.

Theorem 2.3.6. *Suppose an infinite dimensional random vector $\underline{Z}=(X_1,X_2,\cdots,)$ is defined on $(\Omega,\mathbb{A})$ to $(\mathbb{R}^\infty,\mathbb{B}^\infty)$. Then*

$$\underline{Z}^{-1}(\mathbb{B}^\infty)=\sigma\Big((X_1,X_2,\cdots,)\Big)=\sigma\Big(\bigcup_{n\geq 1}\mathbb{A}_n\Big).$$

Proof. Suppose $S=\{\underline{x}\mid a_{i_j}\leq x_{i_j}\leq b_{i_j},\ j=1,2,\cdots,k,\ x_{i_j}\in\mathbb{R}, j=k+1,k+2,\cdots\}$ is a cylinder set, the collection of which generates $\mathbb{B}^\infty$. Then for $S\in\mathbb{B}^\infty$,

$$\underline{Z}^{-1}(S)=\{\omega|\underline{Z}(\omega)\in S\}=\bigcap_{j=1}^k X_{i_j}^{-1}((a_{i_j},b_{i_j})),\quad\text{for any}\quad i_1,i_2,\cdots,i_k\in\mathbb{N}.$$

Then as in Theorem 2.3.2, observe that, for any $j=1,2,\cdots,k$ and $i_1,i_2,\cdots,i_k\in\mathbb{N}$,

$$\begin{aligned}X_{i_j}^{-1}((a_{i_j},b_{i_j}))\in X_{i_j}^{-1}(\mathbb{B})&\Rightarrow\bigcap_{j=1}^k X_{i_j}^{-1}((a_{i_j},b_{i_j}))\in\sigma\Big(\bigcup_{j=1}^k(X_{i_j}^{-1}(\mathbb{B})\Big)\\&\Rightarrow\underline{Z}^{-1}(S)\in\sigma\Big(\bigcup_{j=1}^k(X_{i_j}^{-1}(\mathbb{B})\Big)\\&\Rightarrow\underline{Z}^{-1}(S)\in\sigma\Big(\bigcup_{i=1}^k(X_i^{-1}(\mathbb{B})\Big)=\mathbb{A}_k,\\&\qquad\text{for some}\quad k\geq 1\end{aligned}$$

It further implies that

$$\begin{aligned}\underline{Z}^{-1}(S)\in\bigcup_{k\geq 1}\mathbb{A}_k\quad&\Rightarrow\quad\underline{Z}^{-1}(\mathbb{C}^\infty)\subset\bigcup_{k\geq 1}\mathbb{A}_k\\&\Rightarrow\quad\sigma\Big(\underline{Z}^{-1}(\mathbb{C}^\infty)\Big)\subseteq\sigma\Big(\bigcup_{k\geq 1}\mathbb{A}_k\Big)\\&\Rightarrow\quad\underline{Z}^{-1}(\mathbb{B}^\infty)\subseteq\sigma\Big(\bigcup_{k\geq 1}\mathbb{A}_k\Big).\end{aligned}\tag{2.12}$$

Now to establish the inclusion in the reverse direction, in $\underline{Z}^{-1}(S)$ if we take $(a_{i_j}, b_{i_j}) = \mathbb{R}$ for $j = 2, \cdots, k$, then $\underline{Z}^{-1}(S) = X_{i_1}^{-1}((a_{i_1}, b_{i_1}))$, hence $\underline{Z}^{-1}(\mathbb{C}^\infty) \supseteq X_{i_1}^{-1}(\mathbb{B})$. On similar lines we have, $X_{i_j}^{-1}(\mathbb{B}) \subset \underline{Z}^{-1}(\mathbb{C}^\infty)$ for $j = 2, \cdots, k$ and any permutation of $i_1, i_2, \cdots, i_k \in \mathbb{N}$. In particular, suppose $i_j = j, \quad j = 1, 2, \cdots, k$, then for $i = 1, 2, \cdots, k$

$$
\begin{aligned}
X_i^{-1}(\mathbb{B}) &\subseteq \underline{Z}^{-1}(\mathbb{C}^\infty) \Rightarrow \bigcup_{i=1}^{k} X_i^{-1}(\mathbb{B}) \subseteq \underline{Z}^{-1}(\mathbb{C}^\infty) \\
\Rightarrow \sigma\Big(\bigcup_{i=1}^{k} X_i^{-1}(\mathbb{B})\Big) &\subseteq \sigma\Big(\underline{Z}^{-1}(\mathbb{C}^\infty)\Big) \\
\Rightarrow \mathbb{A}_k &\subseteq \underline{Z}^{-1}(\mathbb{B}^\infty), \quad \text{for any} \quad k \geq 1 \\
\Rightarrow \bigcup_{k\geq 1} \mathbb{A}_k &\subseteq \underline{Z}^{-1}(\mathbb{B}^\infty) \\
\Rightarrow \sigma(\bigcup_{k\geq 1} \mathbb{A}_k) &\subseteq \underline{Z}^{-1}(\mathbb{B}^\infty). \qquad (2.13)
\end{aligned}
$$

From (2.12) and (2.13) we have, $\underline{Z}^{-1}(\mathbb{B}^\infty) \;=\; \sigma(\bigcup_{n\geq 1} \mathbb{A}_n)$. □

Theorem 2.3.3 states that $\underline{Z} = (X_1, X_2, \cdots, X_k)'$ is a k-dimensional random vector, measurable with respect to $\mathbb{A}$, if and only if each of the components $X_1, X_2, \cdots, X_k$ are random variables, measurable with respect to $\mathbb{A}$. On similar lines in the following theorem we prove that $\underline{Z} = (X_1, X_2, \cdots,)$ defined on $(\Omega, \mathbb{A})$ is an infinite dimensional random vector, if and only if for every $n \geq 1$, X_n is a random variable on $(\Omega, \mathbb{A})$. The proof is based on Theorem 2.3.6.

Theorem 2.3.7. *Suppose $\underline{Z} = (X_1, X_2, \cdots,)$ is defined on the measurable space $(\Omega, \mathbb{A})$ to $(\mathbb{R}^\infty, \mathbb{B}^\infty)$. Then $\underline{Z}$ is an infinite dimensional random vector if and only if for every $n \geq 1$, X_n is a random variable on a measurable space $(\Omega, \mathbb{A})$.*

Proof. In Theorem 2.3.6, we have proved that $\underline{Z}^{-1}(\mathbb{B}^\infty) \;=\; \sigma\Big(\bigcup_{n\geq 1} \mathbb{A}_n\Big)$. Suppose for every $i \geq 1$, X_i is a random variable on $(\Omega, \mathbb{A})$. Hence,

$$
\begin{aligned}
X_i^{-1}(\mathbb{B}) \subseteq \mathbb{A}, \ \forall \ i \geq 1 &\Rightarrow \bigcup_{i=1}^{n} X_i^{-1}(\mathbb{B}) \subseteq \mathbb{A} \Rightarrow \sigma\Big(\bigcup_{i=1}^{n} X_i^{-1}(\mathbb{B})\Big) \subseteq \mathbb{A} \\
&\Rightarrow \mathbb{A}_n \subseteq \mathbb{A}, \ \forall \ n \geq 1 \\
&\Rightarrow \bigcup_{n\geq 1} \mathbb{A}_n \subseteq \mathbb{A} \Rightarrow \sigma\Big(\bigcup_{n\geq 1} \mathbb{A}_n\Big) \subseteq \mathbb{A} \\
&\Rightarrow \underline{Z}^{-1}(\mathbb{B}^\infty) \subseteq \mathbb{A} \qquad (2.14)
\end{aligned}
$$

Thus, if for each $i \geq 1$, X_i is a random variable, measurable with respect to $\mathbb{A}$, then $\underline{Z}$ is an infinite dimensional random vector, measurable with respect to $\mathbb{A}$. Now suppose $\underline{Z} = (X_1, X_2, \cdots,)$ is an infinite dimensional random vector, measurable with respect to $\mathbb{A}$. Then

$$\begin{aligned}
\underline{Z}^{-1}(\mathbb{B}^{\infty}) \subseteq \mathbb{A} &\Rightarrow \sigma\Big(\bigcup_{n\geq 1} \mathbb{A}_n\Big) \subseteq \mathbb{A} \\
&\Rightarrow \bigcup_{n\geq 1} \mathbb{A}_n \subseteq \mathbb{A} \Rightarrow \mathbb{A}_n \subseteq \mathbb{A}, \ \forall\ n \geq 1 \\
&\Rightarrow \sigma\Big(\bigcup_{i=1}^{n} X_i^{-1}(\mathbb{B})\Big) \subseteq \mathbb{A}, \ \forall\ n \geq 1 \\
&\Rightarrow \bigcup_{i=1}^{n} X_i^{-1}(\mathbb{B}) \subseteq \mathbb{A}, \ \forall\ n \geq 1 \\
&\Rightarrow X_i^{-1}(\mathbb{B}) \subseteq \mathbb{A}, \ \forall\ i \geq 1 \qquad (2.15)
\end{aligned}$$

Thus for each $i \geq 1$, X_i is a random variable, measurable with respect to $\mathbb{A}$. From (2.14) and (2.15) we conclude that $\underline{Z} = (X_1, X_2, \cdots,)$ defined on $(\Omega, \mathbb{A})$ is an infinite dimensional random vector, if and only if for every $n \geq 1$, X_n is a random variable on $(\Omega, \mathbb{A})$. □

As a consequence of this theorem, an infinite dimensional random vector $\underline{Z}$ will now be referred to as a sequence $\{X_n, n \geq 1\}$ of random variables.

Once we have defined a sequence of random variables, the subsequent natural questions are, how to define its limit, how to find it, whether it is unique and whether it is a random variable. In the following lemmas and theorems and in Chapter 6, we discuss all these issues.

Suppose $\{X_n, n \geq 1\}$ is a sequence of random variables defined on a probability space $(\Omega, \mathbb{A}, P)$. Thus, for every $n \geq 1$, X_n is a measurable function on a measurable space $(\Omega, \mathbb{A})$ with range space as the real line, that is, for fixed $\omega \in \Omega$, $X_n(\omega) = a_n$, a real number. As a consequence, for each fixed $\omega \in \Omega$, $\{X_n(\omega), n \geq 1\} \equiv \{a_n, n \geq 1\}$, a sequence of real numbers and hence all techniques of convergence of a sequence of real numbers can be used to study the convergence of a sequence of random variables. However, note that a sequence $\{X_n, n \geq 1\}$ of random variables is equivalent to a collection of sequences of real numbers, $\{\{X_n(w) = a_n, n \geq 1\}, \omega \in \Omega\}$. This collection is finite, countably infinite or uncountable depending on whether Ω is finite, countably infinite or uncountable respectively. Thus, to discuss convergence of a sequence of random variables, one has to deal with convergence of a collection of sequences of real numbers. There are various modes of convergence of a sequence of random variables, which we will discuss in detail in Chapters 6 and 7. In two approaches, we proceed exactly on the lines of convergence of the sequence of real numbers $\{a_n, n \geq 1\}$, with images $X_n(\omega) = a_n$. Other modes of convergence depend on different approaches to reduce a sequence of random variables to a sequence of real numbers.

We now discuss one mode of convergence of a sequence of random variables, where a sequence of random variables is reduced to a collection of sequences of real numbers corresponding to the images. For a sequence of real numbers, we say that the limit exists, if its limit supremum and limit infimum coincide and value of the limit is the common value of the limit supremum and limit infimum.

We proceed to define a limit supremum and a limit infimum of the sequence of random variables. These are defined on similar lines as for a sequence of real numbers. Suppose

$$Y_n(\omega) = \inf_{k\geq n} X_k(\omega) \quad \& \quad Z_n(\omega) = \sup_{k\geq n} X_k(\omega).$$

From the theory of real numbers, it is known that as $n \to \infty$, $\forall \ \omega \in \Omega$,

$$\begin{aligned} Y_n(\omega) &\uparrow \sup_{n\geq 1} Y_n(\omega) = \sup_{n\geq 1}\{\inf_{k\geq n} X_k(\omega)\} = \liminf X_n(\omega) \\ \& \ Z_n(\omega) &\downarrow \inf_{n\geq 1} Z_n(\omega) = \inf_{n\geq 1}\{\sup_{k\geq n} X_k(\omega)\} = \limsup X_n(\omega). \end{aligned}$$

Thus, for a sequence $\{X_n, n \geq 1\}$ of random variables defined on $(\Omega, \mathbb{A})$, $\liminf X_n$ and $\limsup X_n$ are defined as

$$\liminf_{n\to\infty} X_n = \sup_{n\geq 1} \inf_{k\geq n} X_k \quad \& \quad \limsup_{n\to\infty} X_n = \inf_{n\geq 1} \sup_{k\geq n} X_k \ .$$

As in the setup of real numbers, $\liminf X_n(\omega)$ and $\limsup X_n(\omega)$ always exist, it may happen that for some $\omega \in \Omega$, $\liminf X_n(\omega) = -\infty$ and $\limsup X_n(\omega) = \infty$. A limit of a sequence of random variables, denoted by $\lim_{n\to\infty} X_n$, is defined similar to that for a sequence of real numbers.

Definition 2.3.7. *Limit of a sequence of random variables: Suppose* $\{X_n, \ n \geq 1\}$ *is a sequence of* $\mathbb{A}$ *measurable random variables. If* $\forall \ \omega \in \Omega$, $\liminf X_n(\omega) = \limsup X_n(\omega)$, *then we say that* $\lim_{n\to\infty} X_n$ *exists. Further, if* $\liminf X_n(\omega) = \limsup X_n(\omega) = X(\omega), \ \forall \ \omega \in \Omega$, *then we say that* $\lim_{n\to\infty} X_n = X$ *on* Ω. *If* $X(\omega) < \infty, \ \forall \ \omega \in \Omega$, *then we say that* X_n *converges to* X *and write it as* $X_n \to X$ *on* Ω.

The type of convergence of a sequence of random variables defined above is known as a point-wise convergence or convergence everywhere. If $X_n \to X$ on $A \subseteq \Omega$, then A is said to be a set of convergence of the sequence $\{X_n, n \geq 1\}$. Definition of a point-wise convergence of the sequence $\{X_n, n \geq 1\}$ can also be given as follows.

Definition 2.3.8. *Point-wise convergence of a sequence of random variables: A sequence* $\{X_n, \ n \geq 1\}$ *of random variables is said to converge to a random variable* $X < \infty$ *on* $A \subseteq \Omega$ *if* $\forall \ \omega \in A \ \& \ \forall \ \epsilon > 0, \ \exists \ n_0(\epsilon, \omega)$ *such that* $\forall \ n \geq n_0(\epsilon, \omega)$, $|X_n(\omega) - X(\omega)| < \epsilon$. *If* $A = \Omega$ *then we have the point-wise convergence on* Ω.

Remark 2.3.4. If $n_0(\epsilon, \omega)$ is the same for all $\omega \in A$, then we say that the convergence is uniform on A.

Remark 2.3.5. Since the point-wise convergence of a sequence of random variables is defined in terms of the convergence of a sequence of images, that is real numbers, the limit, whenever exists, is uniquely defined.

Definition 2.3.9. *Almost sure convergence of a sequence of random variables: If a sequence $\{X_n,\ n \geq 1\}$ of random variables converges to a random variable $X < \infty$ on $A \subseteq \Omega$ such that $P(A) = 1$, then we say that $\{X_n,\ n \geq 1\}$ converges to X almost surely.*

A set N such that $P(N) = 0$ is known as a P-null set or P-null event. Thus point-wise convergence on N^c is the almost sure convergence. In Chapters 6 and 7, we study in detail various properties of almost sure convergence.

We now investigate whether the limit of a sequence of random variables is a random variable. In the following theorem, we prove that if $\{X_n, n \geq 1\}$ is a sequence of random variables on a measurable space $(\Omega, \mathbb{A})$, $\liminf X_n$ and $\limsup X_n$ and $\lim X_n$, whenever exists, are random variables, measurable with respect to $\mathbb{A}$. In some cases these may be extended valued random variables.

Theorem 2.3.8. *Suppose $\{X_n, n \geq 1\}$ is a sequence of random variables on a measurable space $(\Omega, \mathbb{A})$. Suppose*

(i) $Y_1 = \max\{X_1, X_2\}$ *and* $Y_2 = \min\{X_1, X_2\}$

(ii) $Z_1 = \sup_{n \geq 1} X_n$ *and* $Z_2 = \inf_{n \geq 1} X_n$

(iii) $U_1 = \liminf X_n$ *and* $U_2 = \limsup X_n$ *and*

(iv) $X = \lim X_n$*, if it exists*

Then $Y_1, Y_2, Z_1, Z_2, U_1, U_2$ and X are random variables, measurable with respect to $\mathbb{A}$.

Proof. Since $\{X_n, n \geq 1\}$ is a sequence of random variables on a measurable space $(\Omega, \mathbb{A})$, for each $n \geq 1, X_n$ is a random variable on a measurable space $(\Omega, \mathbb{A})$, that is, $\forall\ S \in \mathbb{B}, X_n^{-1}(S) \in \mathbb{A}$.

(i) Note that for any $x \in \mathbb{R}$,

$$\begin{aligned} Y_1^{-1}((-\infty, x]) &= \{\omega \mid Y_1(\omega) \leq x\} = \{\omega \mid \max\{X_1, X_2\}(\omega) \leq x\} \\ &= \{\omega \mid \max\{X_1(\omega), X_2(\omega)\} \leq x\} \\ &= \{\omega \mid X_1(\omega) \leq x\} \cap \{\omega \mid X_2(\omega) \leq x\} \\ &= X_1^{-1}((-\infty, x]) \cap X_2^{-1}((-\infty, x]) \in \mathbb{A}, \end{aligned}$$

sigma field being closed under intersection. Hence by the economical definition of the random variable, $Y_1 = \max\{X_1, X_2\}$ is a random variable measurable with respect to $\mathbb{A}$. To prove that Y_2 is a random variable, using arguments similar to those for $Y_1 = \max\{X_1, X_2\}$, we get for any $x \in \mathbb{R}$,

$$\begin{aligned} Y_2^{-1}((-\infty, x]) &= \{\omega \mid \min\{X_1(\omega), X_2(\omega)\} \leq x\} \\ &= \{\omega \mid X_1(\omega) \leq x\} \cup \{\omega \mid X_2(\omega) \leq x\} \\ &= X_1^{-1}((-\infty, x]) \cup X_2^{-1}((-\infty, x]) \in \mathbb{A}. \end{aligned}$$

Thus, $Y_2 = \min\{X_1, X_2\}$ is a $\mathbb{A}$ measurable random variable.

(ii) As in (i), for any $x \in \mathbb{R}$,

$$\begin{aligned} Z_1^{-1}((-\infty, x]) &= \{\omega \mid \sup_{n\geq 1} X_n \leq x\} = \{\omega \mid X_n(\omega) \leq x, \quad \forall \ n \geq 1\} \\ &= \bigcap_{n\geq 1} \{\omega \mid X_n(\omega) \leq x\} = \bigcap_{n\geq 1} X_n^{-1}((-\infty, x]) \in \mathbb{A}, \end{aligned}$$

as sigma field is closed under countable intersection. Thus, $Z_1 = \sup_{n\geq 1} X_n$ is a $\mathbb{A}$ measurable random variable. Now, for any $x \in \mathbb{R}$,

$$\begin{aligned} Z_2^{-1}((-\infty, x]) &= \{\omega \mid \inf_{n\geq 1} X_n \leq x\} \\ &= \{\omega \mid X_n(\omega) \leq x, \quad \text{for at least one} \ \ n \geq 1\} \\ &= \bigcup_{n\geq 1} \{\omega \mid X_n(\omega) \leq x\} = \bigcup_{n\geq 1} X_n^{-1}((-\infty, x]) \in \mathbb{A}, \end{aligned}$$

as sigma field is closed under countable union. Thus, $Z_2 = \inf_{n\geq 1} X_n$ is a $\mathbb{A}$ measurable random variable.

(iii) To prove that $\liminf X_n$ and $\limsup X_n$ are $\mathbb{A}$ measurable random variables, from the definition of $\liminf X_n$ we note that,

$$\liminf X_n = \sup_{n\geq 1} Y_n, \quad \text{where} \quad Y_n = \inf_{k\geq n} X_k.$$

It is proved in (ii) that infimum of $\mathbb{A}$ measurable random variables is again a random variable, measurable with respect to $\mathbb{A}$. Hence, Y_n is a $\mathbb{A}$ measurable random variable. Further, it is also proved that supremum of random variables is again a random variable. Hence, $\liminf X_n = \sup_{n\geq 1} Y_n$ is a $\mathbb{A}$ measurable random variable. By the definition,

$$\limsup X_n = \inf_{n\geq 1} Z_n, \quad \text{where} \quad Z_n = \sup_{k\geq n} X_k.$$

Note that Z_n, being supremum of $\mathbb{A}$ measurable random variables is a $\mathbb{A}$ measurable random variable and $\limsup X_n$ being infimum of Z_n random variables is a $\mathbb{A}$ measurable random variable.

(iv) If $\lim_{n\to\infty} X_n$ exists, then $\liminf X_n = \limsup X_n = \lim X_n$. Further, $\liminf X_n$ and $\limsup X_n$ are $\mathbb{A}$ measurable random variables, hence $\lim_{n\to\infty} X_n$ is a $\mathbb{A}$ measurable random variable.

□

Following example illustrates the point-wise convergence of a sequence of random variables.

Example 2.3.5. Suppose a measurable space $(\Omega, \mathbb{A})$ is defined as $\Omega = (0, 1]$ and $\mathbb{A}$ is a sigma field of subsets of Ω. Suppose a sequence $\{X_n, n \geq 1\}$ of $\mathbb{A}$ measurable random variables and a $\mathbb{A}$ measurable random variable X are defined as follows. $X_n(\omega) = \sin(\omega) + \omega^n,\ n \geq 1\ \ \&\ \ X(\omega) = \sin(\omega)$. It is clear that, $X_n(\omega) = \sin(\omega) + \omega^n \ \rightarrow\ \sin(\omega) = X(\omega),\ \ \forall\ \omega \in (0, 1)$. But at $\omega = 1, X_n(\omega) = \sin(1) + 1\ \forall\ \ n \geq 1$ hence converges to $\sin(1) + 1 \neq X(\omega) = \sin(1)$. Thus the sequence of random variables $\{X_n, n \geq 1\}$ does not converge point-wise to X. However, note that $X_n(\omega) \rightarrow X(\omega)$, except for $\omega = 1$. Thus, if $A = (0, 1)$, then, $X_n \rightarrow X$ on A. To grasp and appreciate the abstract concept of point-wise convergence, we verify the results graphically, using the following R code.

Code 2.3.1. Verification of point-wise convergence:

```
Xnw=function(n,w){return(sin(w)+w^n)}
Xw=function(w){return(sin(w))}
omegavec=seq(0,1,length=100)
nvec=c(1,10,100); nlab=paste("n=",nvec,sep="")
XnMat=matrix(nrow=length(omegavec),ncol=length(nvec))
for(i in 1:nrow(XnMat))
{
omega=omegavec[i]
for(j in 1:ncol(XnMat))
{
n=nvec[j]
XnMat[i,j]=Xnw(n,omega)
}
}
XVec=c()
for(i in 1:length(omegavec))
{
omega=omegavec[i]
XVec[i]=Xw(omega)
}
plot(omegavec,XVec,ylim=c(0,max(XnMat)+0.1),type="l",col=2,
```

```
lty=1,lwd=2,xlim=c(0,1.3),xaxt="n", xlab=expression(omega),
ylab=(expression(Xn(omega)~or~X(omega))),
main=(expression(Xn(omega)~"="~sin(omega)+ omega^n)))
axis(1,at=seq(0,1,by=0.1),labels=seq(0,1,by=0.1))
for(i in 1:ncol(XnMat))
{
lines(omegavec,XnMat[,i],col=1,lty=i+1,type="l",lwd=2)
}
legend("bottomright",legend=c(nlab,expression(X(omega))),
lty=c(2:(ncol(XnMat)+1),1),col=c(rep(1,3),2),lwd=2)
```

In Figure 2.1, the red curve displays values of $X(\omega)$, whereas the black curves give values of $X_n(\omega)$ for different choices of n as indicated in the legend. It can be seen that the black curves approach the red curve as n increases except for the rightmost endpoint, that is, $\omega = 1$. Thus, $X_n \nrightarrow X$ point-wise, but $X_n \to X$ on $A = (0, 1)$. □

The next theorem illustrates the point-wise convergence of a sequence of random variables. In Example 2.2.3, we have noted that if $A \in \mathbb{A}$, then the indicator function I_A is a random variable measurable with respect to $\mathbb{A}$. Thus, if $\{A_n, n \geq 1\}$ is a sequence of sets in $\mathbb{A}$, then I_{A_n} is a random variable

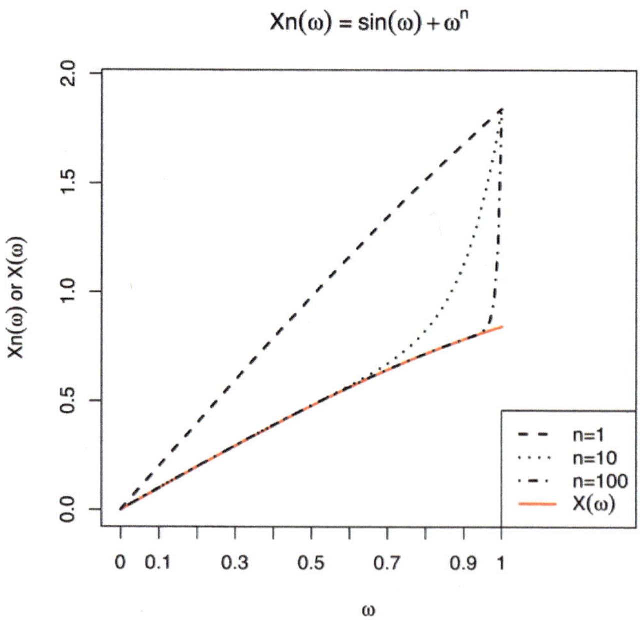

FIGURE 2.1
Point-wise Convergence

measurable with respect to $\mathbb{A}$ $\forall$ $n \geq 1$. Hence by Theorem 2.3.7, $\{I_{A_n}, n \geq 1\}$ is a sequence of random variable measurable with respect to $\mathbb{A}$. Suppose $\{A_n, n \geq 1\}$ is a convergent sequence of sets in $\mathbb{A}$, then $\lim_{n\to\infty} A_n \in \mathbb{A}$. Suppose $\lim_{n\to\infty} A_n = A$. In the next theorem, we prove that $A_n \to A$ if and only if $I_{A_n} \to I_A$ on Ω. The proof is heavily based on the following lemma.

Lemma 2.3.2. *Suppose $\{A_n, n \geq 1\}$ is a sequence of sets in $\mathbb{A}$, then*

$$I_{\liminf A_n} = \liminf I_{A_n} \quad \& \quad I_{\limsup A_n} = \limsup I_{A_n}.$$

Proof. By the definition of the indicator function,

$$\begin{aligned}
I_{\liminf A_n} = 1 \quad &\iff \quad \omega \in \liminf A_n = \bigcup_{n\geq 1}\bigcap_{k\geq n} A_k \\
&\iff \quad \omega \in \bigcup_{n\geq 1} B_n, \quad \text{where } \; B_n = \bigcap_{k\geq n} A_k \\
&\iff \quad \omega \in \bigcap_{k\geq n} A_k, \quad \text{for at least one } \; n \geq 1 \\
&\iff \quad \omega \in A_k, \quad \forall \; k \geq n \quad \text{and for at least one } \; n \geq 1 \\
&\iff \quad I_{A_k} = 1, \quad \forall \; k \geq n \quad \text{and for at least one } \; n \geq 1 \\
&\iff \quad \inf_{k\geq n} I_{A_k} = 1, \quad \text{for at least one } \; n \geq 1 \\
&\iff \quad \sup_{n\geq 1}\inf_{k\geq n} I_{A_k} = 1 \quad \iff \quad \liminf I_{A_n} = 1\,.
\end{aligned}$$

On similar lines we have $I_{\liminf A_n} = 0 \iff \liminf I_{A_n} = 0$. Thus, $I_{\liminf A_n} = \liminf I_{A_n}$. Using similar arguments we examine whether $I_{\limsup A_n} = \limsup I_{A_n}$. Observe that

$$\begin{aligned}
I_{\limsup A_n} = 1 \quad &\iff \quad \omega \in \limsup A_n = \bigcap_{n\geq 1}\bigcup_{k\geq n} A_k \\
&\iff \quad \omega \in \bigcap_{n\geq 1} C_n, \quad \text{where } \; C_n = \bigcup_{k\geq n} A_k \\
&\iff \quad \omega \in \bigcup_{k\geq n} A_k, \quad \forall \; n \geq 1 \\
&\iff \quad \omega \in A_k, \quad \text{for at least one } k \geq n \;\; \& \;\; \forall \; n \geq 1 \\
&\iff \quad I_{A_k} = 1, \quad \text{for at least one } k \geq n \;\; \& \;\; \forall \; n \geq 1 \\
&\iff \quad \sup_{k\geq n} I_{A_k} = 1, \qquad \forall \; n \geq 1 \\
&\iff \quad \inf_{n\geq 1}\sup_{k\geq n} I_{A_k} = 1 \quad \iff \quad \limsup I_{A_n} = 1\,.
\end{aligned}$$

Thus, we have $I_{\limsup A_n} = \limsup I_{A_n}$. □

Theorem 2.3.9. *Suppose $\{A_n, n \geq 1\}$ is a sequence of sets in $\mathbb{A}$, then*

$$A_n \to A \in \mathbb{A} \quad \iff \quad I_{A_n} \to I_A \;\; \textit{on} \;\; \Omega.$$

Proof. By the definition of convergence of sequence of sets and point-wise convergence of sequence of random variables we have,

$$\begin{aligned} A_n \to A &\iff \limsup A_n = \liminf A_n = A \\ &\iff I_{\limsup A_n} = I_{\liminf A_n} = I_A \quad \text{on } \Omega \\ &\iff \limsup I_{A_n} = \liminf I_{A_n} = I_A \quad \text{on } \Omega, \quad \text{by Lemma 2.3.2} \\ &\iff \lim I_{A_n} = I_A \text{ on } \Omega \ . \end{aligned}$$

□

Remark 2.3.6. (i) Observe that in Theorem 2.3.9, $I_A = \lim_{n\to\infty} I_{A_n}$ is $\mathbb{A}$ measurable random variable, when I_{A_n} is $\mathbb{A}$ measurable random variable for each $n \geq 1$. (ii) Note that using Theorem 2.3.9, the convergence of sequence of sets can be defined in terms of the convergence of sequence of indicator functions of sets.

We now define one more sigma field related to a sequence of random variables. It is known as a tail sigma field. Suppose $\{X_n, n \geq 1\}$ is a sequence of random variables. Suppose $\mathbb{C}_n = \sigma\{X_n, X_{n+1}, \cdots\}$. We have proved that

$$\sigma(\{X_n, n \geq 1\}) = \sigma\Big(\bigcup_{n\geq 1} \mathbb{A}_n\Big) \quad \text{where} \quad \mathbb{A}_n = \sigma(\{X_1, X_2, \cdots, X_n\}).$$

Hence by definition,

$$\mathbb{C}_n = \sigma\{X_n, X_{n+1}, \cdots\} = \sigma\Big(\bigcup_{k\geq n} \mathbb{A}_k\Big) \quad \& \quad \mathbb{C}_{n+1} = \sigma\Big(\bigcup_{k\geq n+1} \mathbb{A}_k\Big).$$

Observe that,

$$\begin{aligned} \bigcup_{k\geq n} \mathbb{A}_k \supseteq \bigcup_{k\geq n+1} \mathbb{A}_k \quad &\Rightarrow \quad \sigma\Big(\bigcup_{k\geq n} \mathbb{A}_k\Big) \supseteq \sigma\Big(\bigcup_{k\geq n+1} \mathbb{A}_k\Big) \\ &\Rightarrow \quad \mathbb{C}_n \supseteq \mathbb{C}_{n+1} \ \forall \ n \geq 1. \end{aligned}$$

Thus, $\{\mathbb{C}_n, n \geq 1\}$ is a non-increasing sequence of sigma fields with limit as $\mathbb{T} = \bigcap_{n\geq 1} \mathbb{C}_n$. $\mathbb{T}$ being intersection of sigma fields, is again a sigma field.

Definition 2.3.10. *Tail sigma field and tail event: Suppose $\{X_n, n \geq 1\}$ is a sequence of random variables and $\mathbb{C}_n = \sigma\{X_n, X_{n+1}, \cdots\}$. Then $\mathbb{T} = \bigcap_{n\geq 1} \mathbb{C}_n$ is known as a tail sigma field of the sequence $\{X_n, n \geq 1\}$ and sets in $\mathbb{T}$ are known as tail events.*

If n is interpreted as time, then $\mathbb{C}_n$ contains the information beyond time n and $\mathbb{T}$ contains the information "beyond time $n \ \forall \ n$", that is, loosely speaking, the information at infinity. Another name for the tail sigma field is the sigma algebra of remote events. A random variable X such that $\sigma(X) \subseteq \mathbb{T}$, that is, a random variable which is measurable with respect to a tail sigma field $\mathbb{T}$

is known as a tail random variable. Intuitively, a function measurable with respect to $\mathbb{T}$, does not depend on any finite segment of the sequence $\{X_n, n \geq 1\}$, in a sense that its value is not affected if finitely many X_i's are changed in the sequence $\{X_n, n \geq 1\}$. Similarly, for the tail event, the probability remains unaltered if finitely many X_i's are changed in the sequence. It is thus expected that $\liminf X_n$, $\limsup X_n$ and $\lim X_n$, if it exists, should be measurable with respect to a tail sigma field $\mathbb{T}$. In the following theorem, we prove that it is indeed true.

Theorem 2.3.10. *Suppose $\{X_n, n \geq 1\}$ is a sequence of random variables on a measurable space $(\Omega, \mathbb{A})$. Then $\liminf X_n$, $\limsup X_n$ and $\lim X_n$, if it exists, are measurable with respect to a tail sigma field $\mathbb{T}$.*

Proof.

(i) By definition,

$$\limsup X_n = \inf_{n\geq 1} \sup_{k\geq n} X_k = \inf_{n\geq 1} W_n, \quad \text{where} \quad W_n = \sup_{k\geq n} X_k.$$

Note that if $A \subseteq B$, then $\sup(A) \leq \sup(B)$. Hence $\forall \ n \geq 1$,

$$\begin{aligned} W_n &= \sup\{X_n, X_{n+1}, X_{n+2}, \cdots,\} \geq \sup\{X_{n+1}, X_{n+2}, \cdots,\} = W_{n+1} \\ &\Rightarrow \{W_n, n \geq 1\} \ \text{ is a non-increasing sequence.} \end{aligned}$$

Observe that W_n is a function of $\{X_n, X_{n+1}, \cdots,\}$ and hence it is measurable with respect to $\mathbb{C}_n = \sigma\{X_n, X_{n+1}, \cdots\}$. Further, $\{\mathbb{C}_n, n \geq 1\}$ is a non-increasing sequence of sigma fields. Now, W_1 is $\mathbb{C}_1$ measurable and W_2 is $\mathbb{C}_2$ measurable. However, $\mathbb{C}_2 \subseteq \mathbb{C}_1$ and hence W_2 is $\mathbb{C}_1$ measurable. Using such arguments, we claim that W_n, which is measurable with respect to $\mathbb{C}_n$, is also measurable with respect to $\mathbb{C}_k$, $\forall \ k \leq n$. As a consequence, $\inf_{n\geq 1} W_n$ is measurable with respect to $\mathbb{C}_1$, since each W_n is measurable with respect to $\mathbb{C}_1$. Similarly, $\inf_{n\geq k} W_n$ is measurable with respect to $\mathbb{C}_k$, since each W_n is measurable with respect to $\mathbb{C}_k$. Note that since $\{W_n, n \geq 1\}$ is a non-increasing sequence, $\forall \ k \in \mathbb{N}$, $\inf_{n\geq 1} W_n = \inf_{n\geq k} W_n$. Thus, $\inf_{n\geq 1} W_n$ is measurable with respect to $\mathbb{C}_k$, $\forall \ k \in \mathbb{N}$ implies that $\inf_{n\geq 1} W_n = \limsup X_n$ is measurable with respect to $\bigcap_{k\geq 1} \mathbb{C}_k = \mathbb{T}$.

(ii) We use arguments similar to those in (i) to prove that $\liminf X_n$ is measurable with respect to a tail sigma field $\mathbb{T}$. By definition,

$$\liminf X_n = \sup_{n\geq 1} \inf_{k\geq n} X_k = \inf_{n\geq 1} V_n, \quad \text{where} \quad V_n = \inf_{k\geq n} X_k.$$

Note that if $A \subseteq B$, then $\inf(A) \geq \inf(B)$. Hence $\forall \ n \geq 1$

$$\begin{aligned} V_n &= \inf\{X_n, X_{n+1}, X_{n+2}, \cdots,\} \leq \inf\{X_{n+1}, X_{n+2}, \cdots,\} = V_{n+1} \\ &\Rightarrow \{V_n, n \geq 1\} \ \text{ is a non-decreasing sequence.} \end{aligned}$$

Observe that V_n is a function of $\{X_n, X_{n+1}, \cdots ,\}$ and hence it is measurable with respect to $\mathbb{C}_n = \sigma\{X_n, X_{n+1}, \cdots\}$. Further, $\{\mathbb{C}_n, n \geq 1\}$ is a non-increasing sequence of sigma fields, hence V_n which is measurable with respect to $\mathbb{C}_n$ is also measurable with respect to $\mathbb{C}_k$, $\forall\ k \leq n$. As a consequence, $\sup_{n\geq 1} V_n$ is measurable with respect to $\mathbb{C}_1$, since each V_n is measurable with respect to $\mathbb{C}_1$. Similarly, $\sup_{n\geq k} V_n$ is measurable with respect to $\mathbb{C}_k$, since each V_n is measurable with respect to $\mathbb{C}_k$. Since $\{V_n, n \geq 1\}$ is a non-decreasing sequence, $\forall\ k \in \mathbb{N}$, $\sup_{n\geq 1} V_n = \sup_{n\geq k} W_n$. Thus, $\sup_{n\geq 1} V_n$ is measurable with respect to $\mathbb{C}_k$, $\forall\ k \in \mathbb{N}$ implies that $\sup_{n\geq 1} V_n = \liminf X_n$ is measurable with respect to $\bigcap_{k\geq 1} \mathbb{C}_k = \mathbb{T}$.

(iii) If $\lim X_n$ exists, then $\lim X_n = \liminf X_n = \limsup X_n$ and hence by (i) and (ii) it follows that $\lim X_n$ is measurable with respect to $\mathbb{T}$.

□

Remark 2.3.7. If $\{X_n, n \geq 1\}$ is a sequence of random variables on a measurable space $(\Omega, \mathbb{A})$, then in Theorem 2.3.8, it is proved that $\liminf X_n$, $\limsup X_n$ and $\lim X_n$, if it exists, are measurable with respect to $\mathbb{A}$. In the above Theorem 2.3.10, it is proved that these are measurable with respect to a tail sigma field $\mathbb{T}$. It suggests that there has to be a relation between the two sigma fields $\mathbb{A}$ and $\mathbb{T}$. Observe that the minimal sigma field $\sigma(\{X_n, n \geq 1\})$ induced by the sequence $\{X_n, n \geq 1\}$ is such that $\sigma(\{X_n, n \geq 1\}) \subseteq \mathbb{A}$. From the definition of tail sigma field $\mathbb{C}_1 = \sigma(\{X_n, n \geq 1\})$. Thus, $\mathbb{T} \subseteq \mathbb{C}_1 \subseteq \mathbb{A}$. Thus, Theorem 2.3.10 specifically conveys that $\liminf X_n$, $\limsup X_n$ and $\lim X_n$, if it exists, are measurable with respect to a tail sigma field which is included in $\mathbb{A}$.

In Chapter 5 we study one important result related to the tail sigma field. It is known as the Kolmogorov zero-one law, which states that if $\{X_n, n \geq 1\}$ is a sequence of independent random variables then events in the tail sigma field have probabilities either 0 or 1.

The last result in Theorem 2.3.8 which states that $\lim X_n$, if it exists, is a random variable plays important role, along with the results related to a simple random variable in proving many results in probability theory. It will be clear in the next section. As the nomenclature conveys, a simple random variable is simple in its definition but is extremely useful in establishing number of significant results which play key role in probability theory. The next section is devoted to the theory concerned with a simple random variable.

2.4 Simple Random Variable

The class of simple random variables is obtained by taking linear combinations of indicators of measurable sets. The definition of a simple random variable is based on the concept of a measurable partition of Ω, as defined below.

Definition 2.4.1. *Measurable partition: Suppose* $(\Omega, \mathbb{A})$ *is a measurable space and* $\mathbb{P} = \{A_1, A_2, \cdots, A_k\}$ *is a partition of* Ω*, that is,* $\bigcup_{i=1}^k A_i = \Omega$ *and* $A_i \cap A_j = \emptyset \;\; \forall \;\; i \neq j = 1, 2, \cdots, k$*.* $\mathbb{P}$ *is said to be a measurable finite partition of* Ω*, if* $A_i \in \mathbb{A}, \;\; \forall \;\; i = 1, 2, \cdots, k$*. If* $\mathbb{P} = \{A_1, A_2, \cdots, \}$ *is a countably infinite partition of* Ω *such that* $A_i \in \mathbb{A}, \;\; \forall \;\; i \geq 1$*, then* $\mathbb{P}$ *is said to be a measurable countably infinite partition of* Ω*.*

Definition 2.4.2. *Simple random variable: Suppose a random variable* X *on a measurable space* $(\Omega, \mathbb{A})$ *is defined as follows.*

$$X(\omega) = \sum_{i=1}^{k} a_i I_{A_i}(\omega) \quad \Longleftrightarrow \quad X(\omega) = a_i \;\; \text{if} \;\; \omega \in A_i, \;\; a_i \in \mathbb{R}, \;\; i = 1, 2, \cdots, k,$$

where $\mathbb{P} = \{A_1, A_2, \cdots, A_k\}$ *is a measurable partition of* Ω*. Then* X *is known as a simple random variable, measurable with respect to* $\mathbb{A}$*.*

If X is a simple random variable, then the sigma field $X^{-1}(\mathbb{B})$ induced by X is the sigma field generated by the partition $\mathbb{P} = \{A_1, A_2, \cdots, A_k\} \subset \mathbb{A}$, as illustrated in Example 1.4.2.

Definition 2.4.3. *Elementary random variable: Suppose a random variable* X *on a measurable space* $(\Omega, \mathbb{A})$ *is defined as follows.*

$$X(\omega) = \sum_{i=1}^{\infty} a_i I_{A_i}(\omega) \quad \Longleftrightarrow \quad X(\omega) = a_i \;\; \text{if} \;\; \omega \in A_i, \;\; a_i \in \mathbb{R}, \;\; i \geq 1,$$

where $\mathbb{P} = \{A_1, A_2, \cdots, \}$ *is a measurable countably infinite partition of* Ω*. Then* X *is known as an elementary random variable, measurable with respect to* $\mathbb{A}$*.*

Example 2.4.1. Suppose Ω is defined as,

$$\Omega = \{\omega = (\omega_1, \omega_2, \cdots, \omega_n) | \;\; \omega_j = 0 \;\; \text{or} \;\; 1, \quad j = 1, 2, \cdots, n\},$$

which is a sample space corresponding to a random experiment of tossing a coin n times, where 1 corresponds to head and 0 corresponds to tail. If we are interested in number of ones in each sample point ω, we define

$$A_i = \{\omega = (\omega_1, \omega_2, \cdots, \omega_n) | \sum_{j=1}^{n} \omega_j = i\}, \quad i = 0, 1, 2, \cdots, n.$$

Suppose $\mathbb{A} = \mathbb{P}(\Omega)$, then $A_i \in \mathbb{P}(\Omega), \;\; i = 0, 1, 2, \cdots, n$. Further, $A_i \cap A_j = \emptyset$, $\forall \; i \neq j, \; i, j = 0, 1, 2, \cdots, n$ and $\bigcup_{i=0}^n A_i = \Omega$. Thus, $\{A_0, A_1, A_2, \cdots, A_n\}$ is a measurable partition of Ω. The number of sample points in A_i is $\binom{n}{i}$, $i = 1, 2, \cdots, n$, since A_i a collection of outcomes corresponding to i heads in n tosses. We define a random variable X on $(\Omega, \mathbb{A})$ as follows.

$$X(\omega) = \sum_{i=0}^{n} i I_{A_i}(\omega) \quad \Longleftrightarrow \quad X(\omega) = i \;\; \text{if} \;\; \omega \in A_i, \quad i = 0, 1, 2, \cdots, n.$$

Thus, X is a simple random variable on $(\Omega, \mathbb{A})$. □

In the following theorem, we prove that a class of simple random variables is closed under arithmetic operations.

Theorem 2.4.1. *Suppose X_1 and X_2 are $\mathbb{A}$ measurable simple random variables defined on a measurable space $(\Omega, \mathbb{A})$ to $(\mathbb{R}, \mathbb{B})$. Then*

(i) $X_1 + X_2$, $X_1 - X_2$, $X_1 \times X_2$ and X_1/X_2, provided it is defined, are $\mathbb{A}$ measurable simple random variables on the measurable space $(\Omega, \mathbb{A})$.

(ii) aX_1 is a simple random variable on $(\Omega, \mathbb{A})$ to $(\mathbb{R}, \mathbb{B})$, where $a \in \mathbb{R}$.

(iii) Suppose $g : \mathbb{R} \to \mathbb{R}$ is a function. Then $g(X_1)$ is a simple random variable on $(\Omega, \mathbb{A})$ to $(\mathbb{R}, \mathbb{B})$.

(iv) $\underline{Z} = (X_1, X_2)'$ is a simple random vector on $(\Omega, \mathbb{A})$ to $(\mathbb{R}^2, \mathbb{B}^2)$.

Proof. Suppose $\mathbb{P}_1 = \{A_1, A_2, \cdots, A_k\}$ and $\mathbb{P}_2 = \{B_1, B_2, \cdots, B_l\}$ are measurable partitions of Ω and simple random variables X_1 and X_2 are defined as,

$$X_1(\omega) = \sum_{i=1}^{k} a_i I_{A_i}(\omega) \quad \& \quad X_2(\omega) = \sum_{j=1}^{l} b_j I_{B_j}(\omega)$$

Further, we define $C_{ij} = A_i \cap B_j$, $i = 1, 2, \cdots, k$ and $j = 1, 2, \cdots, l$. Then for $i \neq r$ & $j \neq s$ $C_{ij} \cap C_{rs} = (A_i \cap B_j) \cap (A_r \cap B_s) = (A_i \cap A_r) \cap (B_j \cap B_s) = \emptyset$. Similarly, even if $i = r$ and $j \neq s$ or $i \neq r$ and $j = s$, then $C_{ij} \cap C_{rs} = \emptyset$. Thus,
$C_{ij} = A_i \cap B_j$, $i = 1, 2, \cdots, k$ and $j = 1, 2, \cdots, l$ are all disjoint sets. Further,

$$\bigcup_{i=1}^{k} \bigcup_{j=1}^{l} C_{ij} = \bigcup_{i=1}^{k} \bigcup_{j=1}^{l} (A_i \cap B_j) = \bigcup_{i=1}^{k} \Big(\bigcup_{j=1}^{l} (A_i \cap B_j) \Big) = \bigcup_{i=1}^{k} (A_i \cap \Omega) = \Omega.$$

Note that, $A_i \in \mathbb{A}$ and $B_j \in \mathbb{A}$, hence $C_{ij} = A_i \cap B_j \in \mathbb{A}$, $i = 1, 2, \cdots, k$ and $j = 1, 2, \cdots, l$. Thus, $\{C_{ij}, i = 1, 2, \cdots, k, \ j = 1, 2, \cdots, l\}$ forms a measurable partition of Ω. Observe that,

$$\begin{aligned} \omega \in C_{ij} \Rightarrow \omega \in A_i \quad \& \quad \omega \in B_j \ &\Rightarrow \ X_1(\omega) = a_i, \ \& \ X_2(\omega) = b_j \\ &\Rightarrow X_1(\omega) \pm X_2(\omega) = a_i \pm b_i, \ \ X_1(\omega) \times X_2(\omega) = a_i \times b_j \\ &\& X_1(\omega)/X_2(\omega) = a_i/b_j \\ &\Rightarrow X_1 \pm X_2 = \sum_{i=1}^{k} \sum_{j=1}^{l} (a_i \pm b_j) I_{C_{ij}}, \ \ X_1 \times X_2 = \sum_{i=1}^{k} \sum_{j=1}^{l} (a_i \times b_j) I_{C_{ij}} \end{aligned}$$

$$X_1/X_2 = \sum_{i=1}^{k} \sum_{j=1}^{l} (a_i/b_j) I_{C_{ij}} \ \ \& \ \ \underline{Z}(\omega) = (X_1(\omega), X_2(\omega))' = (a_i, b_j)'.$$

Note that X_1/X_2 is defined if $b_j \neq 0$ for any j. Hence, $X_1 + X_2$, $X_1 - X_2$, $X_1 \times X_2$ and X_1/X_2, provided it is defined, are $\mathbb{A}$ measurable simple random variables on the measurable space $(\Omega, \mathbb{A})$.

(ii) aX_1 is a simple random variable follows from the fact that $X_1 \times X_2$ is a simple random variable in a particular case of $X_2 \equiv a$.

(iii) Note that

$$X_1(\omega) = \sum_{i=1}^{k} a_i I_{A_i}(\omega) \quad \Rightarrow \quad g(X_1)(\omega) = g(X_1(\omega)) = \sum_{i=1}^{k} g(a_i) I_{A_i}(\omega),$$

that is, $g(X_1)(\omega) = g(a_i)$ if $a_i \in A_i, \;\; i = 1, 2, \cdots, k$. Hence, $g(X)$ is a simple random variable on $(\Omega, \mathbb{A})$ to $(\mathbb{R}, \mathbb{B})$.
(iv) Observe that $\underline{Z}$ can be expressed as $\underline{Z} = \sum_{i=1}^{k} \sum_{j=1}^{l} (a_i, b_j)' I_{C_{ij}}$ and hence it is a $\mathbb{A}$ measurable simple random vector. Thus, if the components of a random vector are simple random variables, then the random vector is a simple random vector. □

Following theorem is related to an important result which states that any random variable can be represented as a limit of a sequence of simple random variables. The idea of the proof is similar to the idea of Riemann sum for approximating a Riemann integral. It involves the following steps. (i) We use the fact that a sequence of intervals $\{(-n, n], n \geq 1\}$ converges to $\mathbb{R}$ as $n \to \infty$. Hence it is enough to worry about the values of X_n and X being close to each other when $X(\omega) \in (-n, n]$. As $n \to \infty$, they will be close to each other on almost entire $\mathbb{R}$. (ii) We partition the interval $(-n, n]$ into intervals of length $1/2^n$ or in general $1/\delta^n$ for any $\delta = 2, 3, 4, \cdots$. If $X(\omega)$ belongs to one of these intervals, then $X_n(\omega)$ is defined as upper end point of that interval. Thus, the distance between $X_n(\omega)$ and $X(\omega)$ is less than $1/2^n$. Consequently as $n \to \infty$, the intervals become smaller and the distance between $X_n(\omega)$ and $X(\omega)$ decreases to 0 leading to convergence.

Theorem 2.4.2. *Suppose X is a $\mathbb{A}$ measurable random variable defined on a measurable space $(\Omega, \mathbb{A})$ to $(\mathbb{R}, \mathbb{B})$. Then there exists a sequence $\{X_n, n \geq 1\}$ of $\mathbb{A}$ measurable simple random variables on a measurable space $(\Omega, \mathbb{A})$ such that $X_n \to X$ on Ω as $n \to \infty$.*

Proof. It is given that X is a $\mathbb{A}$ measurable random variable on $(\Omega, \mathbb{A})$ to $(\mathbb{R}, \mathbb{B})$, hence $\forall \;\; S \in \mathbb{B}, \;\; X^{-1}(S) \in \mathbb{A}$. In particular,

$$\begin{aligned} T_1 &= (-\infty, -n], \; T_2 = (n, \infty) \;\; \& \;\; S_k = (k/2^n, (k+1)/2^n], \\ \Rightarrow B_1 &= X^{-1}(T_1), \; B_2 = X^{-1}(T_2) \; \& \; A_k = X^{-1}(S_k) \in \mathbb{A}, \end{aligned}$$

$k = -n2^n, -n2^n+1, \cdots, n2^n-1$. Note that T_1, T_2 and $S_k, \;\; k = -n2^n, -n2^n+1, \cdots, n2^n-1$ are disjoint subsets of $\mathbb{R}$, hence by Lemma 2.2.1, $B_1 = X^{-1}(T_1)$,

$B_2 = X^{-1}(T_2)$ and $A_k = X^{-1}(S_k)$ for all k are also disjoint sets in $\mathbb{A}$. Further,

$$\begin{aligned} T_1 \bigcup \Big(\bigcup_{k=-n2^n}^{n2^n-1} S_k \Big) \bigcup T_2 &= \mathbb{R} \\ \Rightarrow \quad X^{-1}\Big(T_1 \bigcup (\bigcup_{k=-n2^n}^{n2^n-1} S_k) \bigcup T_2 \Big) &= X^{-1}(\mathbb{R}) \\ \Rightarrow \quad X^{-1}(T_1) \bigcup X^{-1}\left(\bigcup_{k=-n2^n}^{n2^n-1} S_k \right) \bigcup X^{-1}(T_2) &= \Omega \\ \Rightarrow \quad B_1 \bigcup \Big(\bigcup_{k=-n2^n}^{n2^n-1} A_k \Big) \bigcup B_2 &= \Omega. \end{aligned}$$

Thus, $\mathbb{P} = \{B_1,\ B_2,\ A_k,\ k = -n2^n, -n2^n+1, \cdots, n2^n - 1\}$ is a measurable partition of Ω. Suppose a function $X_n(\omega)$ on $(\Omega, \mathbb{A})$ is defined as follows.

$$X_n(\omega) = -n\ I_{B_1}(\omega) \ + \sum_{k=-n2^n}^{n2^n-1} ((k+1)/2^n)\ I_{A_k}(\omega) \ + \ n\ I_{B_2}(\omega) \quad (2.16)$$

that is,

$$X_n(\omega) = \begin{cases} -n, & \text{if} \quad X(\omega) \in (-\infty, -n] \iff \omega \in B_1, \\ (k+1)/2^n, & \text{if} \quad X(\omega) \in (k/2^n, (k+1)/2^n] \iff \omega \in A_k, \\ & \qquad k = -n2^n, -n2^n+1, \cdots, n2^n - 1, \\ n, & \text{if} \quad X(\omega) \in (n, \infty) \iff \omega \in B_2. \end{cases}$$

By Definition 2.4.2, it is clear that X_n as defined in Equation (2.16) is a simple random variable on a measurable space $(\Omega, \mathbb{A})$. To prove that $X_n \to X$ on Ω, observe that $\forall\ \omega \in \Omega,\ X(\omega) \in \mathbb{R} \ \Rightarrow\ \exists\ n_1(\omega)$ such that $\forall\ n \geq n_1(\omega)$, $|X(\omega)| < n$. Further, given $\epsilon > 0$, $\exists\ n_2(\epsilon)$ such that $\forall\ n \geq n_2(\epsilon)$, $1/2^n < \epsilon$. If $N(\omega, \epsilon)$ is defined as $N(\omega, \epsilon) = \max\{n_1(\omega), n_2(\epsilon)\}$, then

$$\forall\ n \geq N(\omega, \epsilon),\ |X(\omega)| < n \ \text{ and } \ 1/2^n < \epsilon,\ \forall\ \omega \in \Omega.$$

Since $|X(\omega)| < n$, $X(\omega)$ is in one of the intervals of the type $(k/2^n, (k+1)/2^n]$, each having length $1/2^n$, thus if $X(\omega) \in (k/2^n, (k+1)/2^n]$, then $X_n(\omega) = (k+1)/2^n$, and the difference $|X_n(\omega) - X(\omega)| < 1/2^n$. Thus, $\forall\ \omega \in \Omega$, given $\epsilon > 0$, $\exists\ N(\omega, \epsilon)$ such that $\forall\ n \geq N(\omega, \epsilon)$, $|X_n(\omega) - X(\omega)| < 1/2^n < \epsilon$. Hence, $X_n \to X$ on Ω. □

We illustrate below Theorem 2.4.2 for $X(\omega) = \omega^3$ for $\omega \in [-2, 2]$. Note that $X(\omega) \in [-8, 8]$. We define a simple random variable X_n as defined in Theorem 2.4.2. For example, $X_2(\omega) = -2$ if $X(\omega) \leq -2 \iff \omega \leq (-2)^{(1/3)} = -0.7937$

and $X_2(\omega) = 2$ if $X(\omega) > 2 \iff \omega > 2^{(1/3)} = 0.7937$. When $X(\omega) \in (-2, 2] \iff \omega \in (-0.7937, 0.7937]$, we partition the interval $(-2, 2]$ into 16 intervals, each of length $(1/4)$ and define X_n as the upper boundary of the interval. We use Code 2.4.1 to draw the graph of X and impose on it graphs of X_n for $n = 1, 2, 3, 5, 10$.

Code 2.4.1.
```
a=-2; b=2; omega=seq(from=a,to=b,length=1000)
Xom=omega^3; nvec=c(1,2,3,5,10)
Xnom=matrix(nrow=length(omega),ncol=length(nvec))
for(j in 1:length(nvec))
{
  n=nvec[j]
  for(i in 1:length(omega))
  {
    if(Xom[i]<= -n) Xnom[i,j]=-n
    if(Xom[i]> n) Xnom[i,j]=n
    if(abs(Xom[i])<n)
    {
      k1=ceiling(Xom[i]*(2^n))
      Xnom[i,j]=k1/(2^n)
    }
  }
}
plot(omega,Xom,lwd=3,type="l",xlab=expression(omega),
ylab=expression(X(omega)),ylim=c(-8,8),xaxt="n",
main=expression(X(omega)==omega^3))
omlab1=-round((seq(2,0,by=-0.25))^(1/3),2)
omlab2=round((seq(0,2,by=0.25))^(1/3),2)
omlab=c(omlab1,omlab2)
axis(1,at=omlab,labels=omlab,las=2,cex.axis=0.7)
for(i in 1:length(nvec))
{
  lines(omega,Xnom[,i],lwd=2,lty=i+1,col=i+1)
}
fomlab=(omlab)^3
arrows(x0=omlab,y0=-10,x1=omlab,y1=fomlab,col="black",lty=3,
length=0);T=paste("n",nvec,sep="=")
legend("topleft",legend=T,col=2:(length(nvec)+1),
lty=2:(length(nvec)+1),lwd=2,cex=1.2)
```

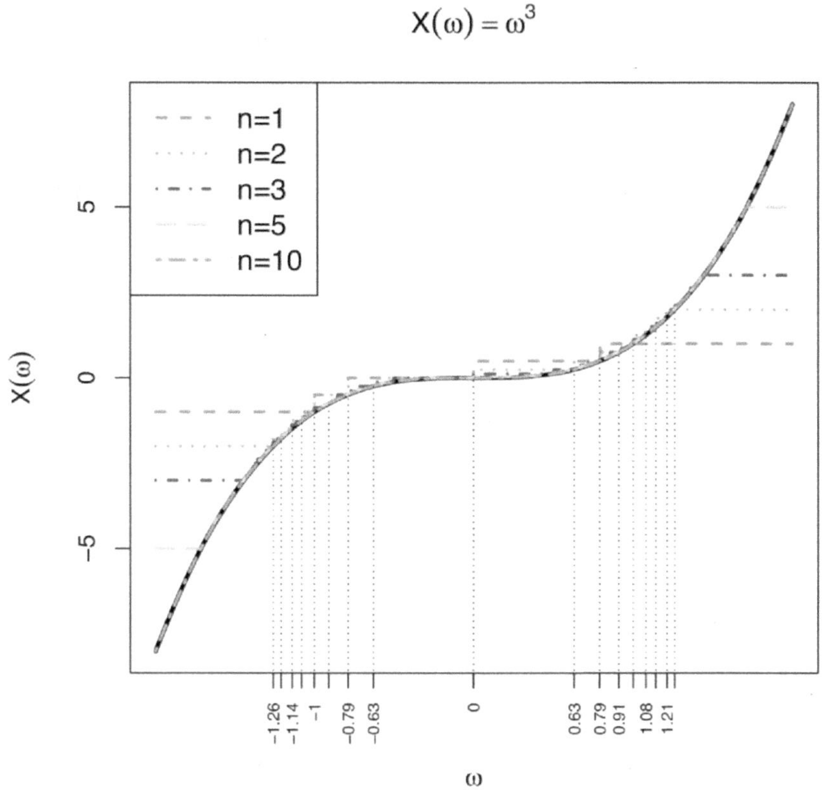

FIGURE 2.2
Visual Illustration of Theorem 2.4.2

From Figure 2.2, we note that as n increases, the graph of X_n approximates graph of X. Note that for $n = 10$, graphs of X and X_n almost coincide, thus the approximation is quite good even for $n = 10$. It is to be noted that the support of X is a bounded interval and hence even for large n, while defining X_n, one has to restrict to a set $\{\omega | X(\omega) \in [-a, a]\}$, where $a = \min\{n, 8\}$. As n increases, the partitions of the intervals become finer and finer and it leads to the better approximation of X by X_n.

The next theorem is similar to Theorem 2.4.2, but we restrict to a non-negative random variable X. As a consequence, the sequence of simple random variables, whose limit is X, is non-decreasing and the simple random variables are non-negative.

Theorem 2.4.3. *Suppose X is a $\mathbb{A}$ measurable non-negative random variable defined on a measurable space $(\Omega, \mathbb{A})$ to $(\mathbb{R}, \mathbb{B})$. Then there exists a non-decreasing sequence $\{X_n, n \geq 1\}$ of $\mathbb{A}$ measurable non-negative simple random variables on a measurable space $(\Omega, \mathbb{A})$ such that $X_n \to X$ on Ω.*

Proof. It is given that X is a non-negative random variable on a measurable space $(\Omega, \mathbb{A})$ to $(\mathbb{R}, \mathbb{B})$, hence $\forall\ \omega \in \Omega, X(\omega) \geq 0$. Since X is $\mathbb{A}$ measurable,

$\forall \;\; S \in \mathbb{B}, \;\; X^{-1}(S) \in \mathbb{A}$. In particular, if $S_k = [k/2^n, (k+1)/2^n), k = 0, 1, \cdots, n2^n - 1$ and $S = [n, \infty)$ then $A_k = X^{-1}(S_k)$, $k = 0, 1, \cdots, n2^n - 1$ and $A = X^{-1}(S)$ are in $\mathbb{A}$. Note that $S_k, \; k = 0, 1, \cdots, n2^n - 1$ and S are disjoint subsets of $\mathbb{R}^+$, hence by Lemma 2.2.1, $A_k = X^{-1}(S_k), k = 0, 1, \cdots, n2^n - 1$ and $A = X^{-1}(S)$ are also disjoint sets in $\mathbb{A}$. As in Theorem 2.4.2,

$$
\begin{aligned}
\left(\bigcup_{k=0}^{n2^n-1} S_k\right)\bigcup S &= \mathbb{R}^+ \;\Rightarrow\; X^{-1}\left((\bigcup_{k=0}^{n2^n-1} S_k)\bigcup S\right) = X^{-1}(\mathbb{R}^+) \\
\Rightarrow \quad \left(\bigcup_{k=0}^{n2^n-1} A_k\right)\bigcup A &= \Omega.
\end{aligned}
$$

Thus, $\mathbb{P} = \{A, \; A_k, k = 0, 1, \cdots, n2^n - 1\}$ is a measurable partition of Ω. Suppose a function $X_n(\omega)$ on $(\Omega, \mathbb{A})$ is defined as follows.

$$
X_n(\omega) = \sum_{k=0}^{n2^n-1} (k/2^n) I_{A_k}(\omega) \quad + \quad n I_A(\omega) \tag{2.17}
$$

that is,

$$
X_n(\omega) \quad = \quad \begin{cases} (k/2^n), & \text{if} \quad \omega \in A_k, \quad \Longleftrightarrow \quad X(\omega) \in [k/2^n, (k+1)/2^n), \\ & \qquad\qquad\qquad\qquad k = 0, 1, \cdots, n2^n - 1, \\ n, & \text{if} \qquad \omega \in A \quad \Longleftrightarrow \quad X(\omega) \geq n. \end{cases}
$$

Then X_n as defined in Equation (2.17) is a non-negative simple random variable on $(\Omega, \mathbb{A})$. To prove that $X_n \to X$ on Ω, observe that, $\forall \;\; \omega \in \Omega, \;\; X(\omega) \in \mathbb{R}^+$ implies that $\exists \;\; n_1(\omega)$ such that $\;\forall \;\; n \geq n_1(\omega), \;\; X(\omega) < n$. Further, given $\epsilon > 0, \;\; \exists \;\; n_2(\epsilon)$ such that $\;\forall \;\; n \geq n_2(\epsilon), \;\; 1/2^n < \epsilon$. If $N(\omega, \epsilon)$ is defined as $N(\omega, \epsilon) = \max\{n_1(\omega), n_2(\epsilon)\}$, then

$$
\forall \;\; n \geq N(\omega, \epsilon), \;\; X(\omega) < n \;\; \forall \;\; \omega \in \Omega \;\; \text{and} \;\; 1/2^n < \epsilon.
$$

Since, $X(\omega) < n, X(\omega)$ is in one of the intervals of the type $\Big[k/2^n, (k+1)/2^n\Big)$, each having length $1/2^n$, thus if $X(\omega) \in [k/2^n, (k+1)/2^n)$, then $X_n(\omega) = k/2^n$, and the difference $|X_n(\omega) - X(\omega)| < 1/2^n$. Thus, $\forall \;\; \omega \in \Omega$, given $\epsilon > 0, \; \exists \; N(\omega, \epsilon)$ such that $\;\forall \;\; n \geq N(\omega, \epsilon), \;\; |X_n(\omega) - X(\omega)| < 1/2^n < \epsilon$. Hence, $X_n \to X$ on Ω. Now to prove that $\{X_n, n \geq 1\}$ is a non-decreasing sequence, we have to show that $X_n(\omega) \leq X_{n+1}(\omega), \;\; \forall \;\; n \geq 1$ and $\;\forall \;\; \omega \in \Omega$. We partition Ω in three sets, as $[X \geq n+1]$, $[n \leq X < n+1)$ and $[X < n]$, as the definitions of $X_{n+1}(\omega)$ and $X_n(\omega)$ differ on these three sets. We have,

$$
X_{n+1}(\omega) \quad = \quad \begin{cases} k/2^{n+1}, & \text{if} \quad X(\omega) \in \left[k/2^{n+1}, (k+1)/2^{n+1}\right), \\ & \qquad\qquad k = 0, 1, \cdots, (n+1)2^{n+1} - 1, \\ n+1, & \text{if} \qquad X(\omega) \in [n+1, \infty). \end{cases}
$$

Case(i) $[X \geq n+1]$: On this partition set, by definition of $X_n(\omega)$ and $X_{n+1}(\omega)$,

$$X_{n+1}(\omega) = n+1 > n = X_n(\omega) \quad \Rightarrow \quad X_n(\omega) < X_{n+1}(\omega), \quad \forall \quad n \geq 1.$$

Case(ii) $[n \leq X < n+1]$: On this partition set, by definition $X_n(\omega) = n$. To find the values of $X_{n+1}(\omega)$, we divide the interval $[n, n+1)$ in 2^{n+1} intervals, each of length $1/2^{n+1}$. By definition of $X_{n+1}(\omega)$, on this partition set, for $k = n2^{n+1}, n2^{n+1}+1, \cdots, (n+1)2^{n+1}-1$ we have,

$$X_{n+1}(\omega) = k/2^{n+1}, \quad \text{if} \quad X(\omega) \in \left[k/2^{n+1}, (k+1)/2^{n+1}\right).$$

Thus, values of $X_{n+1}(\omega)$ are $\left\{n, n+1/2^{n+1}, n+2/2^{n+1}, \cdots, n+1-1/2^{n+1}\right\}$ and hence on the partition set $[n \leq X < n+1]$, $X_n(\omega) \leq X_{n+1}(\omega), \ \forall \ n \geq 1$. Case(iii) $[X < n]$: In this case, $X_n(\omega) = k/2^n$ if $\omega \in A_k, \ k = 0, \cdots, n2^n-1$. To find the values of $X_{n+1}(\omega)$, for each k we divide the interval $[k/2^n, (k+1)/2^n)$ in two equal parts as,

$$\begin{aligned}[k/2^n, (k+1)/2^n) &= \left[2k/2^{n+1}, 2k/2^{n+1} + 1/2^{n+1}\right) \\ &\cup \left[2k/2^{n+1} + 1/2^{n+1}, 2k/2^{n+1} + 2/2^{n+1}\right) = I_1 \cup I_2.\end{aligned}$$

Note that if $X(\omega) \in [k/2^n, (k+1)/2^n)$, then $X(\omega) \in I_1$ or $X(\omega) \in I_2$. Further,

$$X(\omega) \in I_1 \Rightarrow X_{n+1}(\omega) = k/2^n \ \& \ X(\omega) \in I_2 \Rightarrow X_{n+1}(\omega) = k/2^n + 1/2^{n+1}.$$

Thus, on the set $[X < n]$, $X_n(\omega) = k/2^n \leq X_{n+1}(\omega)$, $\forall \ n \geq 1$. Hence, it is proved that $\{X_n, n \geq 1\}$ is a non-decreasing sequence of non-negative simple random variables. □

Theorem 2.4.3 is visually demonstrated below for $X(\omega) = \log(\omega)$ for $\omega \in [1, 8]$. Note that $X(\omega) \in [0, 2.0794]$ and it is a non-negative random variable. We define a simple random variable X_n as defined in Theorem 2.4.3. For example, $X_2(\omega) = 2$ if $X(\omega) \geq 2 \iff \omega \geq \exp(2) = 7.39$. When $X(\omega) \in [0, 2) \iff \omega \in [1, 7.39)$, then we partition the interval $[0, 2)$ into 16 intervals, each of length $(1/4)$ and define X_n as the lower boundary of the interval. We use Code 2.4.2 to draw the graph of X and impose on it graphs of X_n for $n = 1, 2, 3, 5, 10$.

Code 2.4.2. `a=1; b=8; omega=seq(from=a,to=b,length=1000)`

```
Xom=log(omega); nvec=c(1,2,3,5,10)
Xnom=matrix(nrow=length(omega),ncol=length(nvec))
for(j in 1:length(nvec))
{
  n=nvec[j]
  for(i in 1:length(omega))
  {
```

```
    if(Xom[i]> n) Xnom[i,j]=n
    if(abs(Xom[i])<n)
    {
      k1=floor(Xom[i]*(2^n))
      Xnom[i,j]=k1/(2^n)
    }
  }
}
plot(omega,Xom,lwd=3,type="l",xlab=expression(omega),xaxt="n",
ylab=expression(X(omega)),main=expression(X(omega)==log(omega)))
omlab=round(exp(seq(0,2,by=0.25)),2)
axis(1,at=omlab,labels=omlab,las=2,cex.axis=0.7)
for(i in 1:length(nvec))
{
  lines(omega,Xnom[,i],lwd=2,lty=i+1,col=i+1)
}
fomlab=log(omlab)
arrows(x0=omlab,y0=0,x1=omlab,y1=fomlab,col="black",lty=3,
length=0); T=paste("n",nvec,sep="=")
legend("topleft",legend=T,col=2:(length(nvec)+1),
lty=2:(length(nvec)+1),lwd=2,cex=1.2)
```

From Figure 2.3, we note that as n increases, the graph of X_n approximates graph of X, with good approximation for $n = 10$. Further observe that $X_1(\omega) \leq X_2(\omega) \leq X_3(\omega) \leq X_5(\omega) \leq X_{10}(\omega)$ displaying the non-decreasing nature of $\{X_n, n \geq 1\}$ as proved in Theorem 2.4.3. In Figure 2.2, $\{X_n, n \geq 1\}$ is not non-decreasing. As noted in illustration of Theorem 2.4.2, in this case also the support of X is a bounded interval and hence for large n, while defining X_n, one has to restrict to a set $\{\omega | X(\omega) \in [0, a]\}$, where $a = \min\{n, 2\}$. As n increases, the partitions of the intervals become finer and finer which leads to the better approximation of X by X_n.

Theorem 2.4.2 and Theorem 2.4.3 are two widely used results in probability theory. Both the theorems convey that a random variable can be constructed as a limit of a sequence of simple random variables, hence these two theorems are referred to as constructive definitions of a random variable. Theorem 2.4.2 is useful to prove that a continuous function of a random variable is again a random variable and secondly the class of random variables is closed under arithmetic operations. Theorem 2.4.3 plays a fundamental role in the definition of expectation of a random variable as a Lebesgue integral. We first define the expectation of a simple random variable, then use Theorem 2.4.3 to define it for the non-negative random variable. and then extend the definition for an arbitrary random variable. It is discussed in Chapter 4.

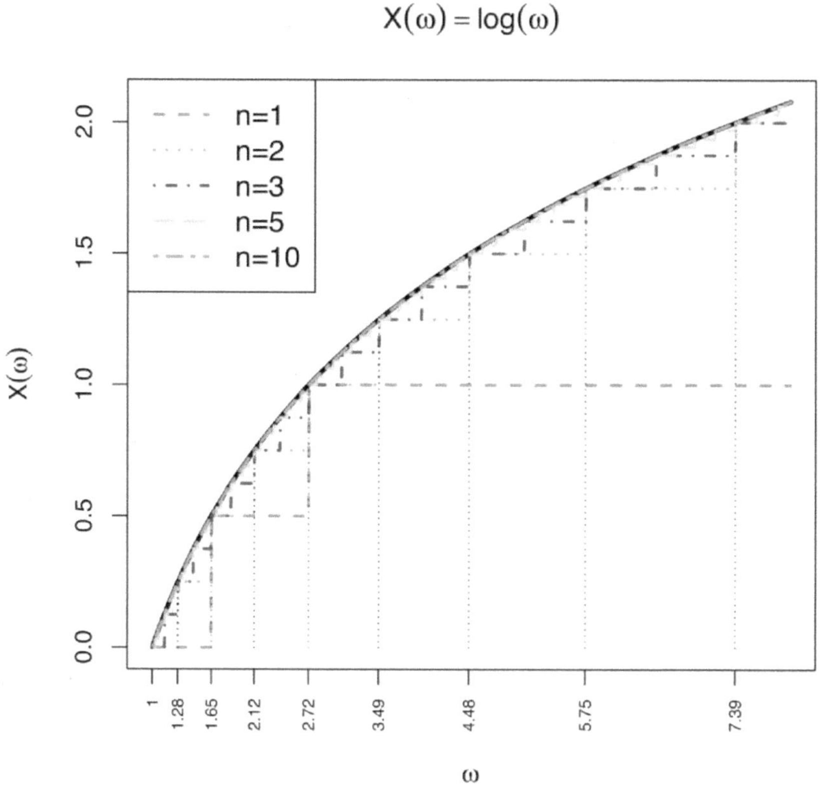

FIGURE 2.3
Visual Illustration of Theorem 2.4.3

Remark 2.4.1. Theorem 2.4.3 states that if X is non-negative $\mathbb{A}$ measurable function on the measurable space $(\Omega, \mathbb{A})$, then there exists a non-decreasing sequence $\{X_n, n \geq 1\}$ of non-negative simple $\mathbb{A}$ measurable functions on a measurable space $(\Omega, \mathbb{A})$, such that $X_n \to X$ on Ω. Using similar arguments it can be proved that if g is non-negative Borel function on the measurable space $(\mathbb{R}, \mathbb{B})$, then there exists a non-decreasing sequence $\{g_n, n \geq 1\}$ of non-negative simple Borel functions on a measurable space $(\mathbb{R}, \mathbb{B})$, such that $g_n \to g$ on $\mathbb{R}$. We need the result in Chapter 4 to derive a formula for $E(g(X))$, as an integral with respect to the induced probability measure.

From Theorem 2.3.5, it follows that the class of random variables is closed under arithmetic operations. We prove the same result below, using a different method. It is based on Theorem 2.4.1 and Theorem 2.4.2.

Theorem 2.4.4. *Suppose X_1 and X_2 are random variables defined on a measurable space $(\Omega, \mathbb{A})$ to $(\mathbb{R}, \mathbb{B})$. Then*

(i) $X_1 + X_2$, $X_1 - X_2$, $X_1 \times X_2$ and X_1/X_2, provided it is defined, are random variables on the measurable space $(\Omega, \mathbb{A})$.

(ii) aX_1 is random variable on $(\Omega, \mathbb{A})$ to $(\mathbb{R}, \mathbb{B})$, where a is any real number.

Proof. If X_1 and X_2 are arbitrary random variables, then by Theorem 2.4.2, there exist sequences $\{X_{1n}, n \geq 1\}$ and $\{X_{2n}, n \geq 1\}$ of simple random variables on a measurable space $(\Omega, \mathbb{A})$, such that $X_{1n} \to X_1$ and $X_{2n} \to X_2$ on Ω. Since $\forall\ n \geq 1, X_{1n}$ and X_{2n} are simple random variables, by Theorem 2.4.1 $X_{1n} \pm X_{2n}, X_{1n} \times X_{2n}$ and X_{1n}/X_{2n} are also simple random variables. Since $X_{1n} \to X_1$ and $X_{2n} \to X_2$ on Ω,

$$\begin{aligned} X_1 \pm X_2 &= \lim(X_{1n} \pm X_{2n}), \quad X_1 \times X_2 = \lim(X_{1n} \times X_{2n}) \\ \text{and } X_1/X_2 &= \lim(X_{1n}/X_{2n}). \end{aligned}$$

We have proved in Theorem 2.3.8, that limit of a sequence of random variables is also a random variable. Hence, $X_1 + X_2$, $X_1 - X_2$, $X_1 \times X_2$ and X_1/X_2 are random variables on the measurable space $(\Omega, \mathbb{A})$. (ii) aX_1 is a random variable follows from the fact that $X_1 \times X_2$ is a random variable when $X_2 \equiv a$. □

With any random variable X defined on $(\Omega, \mathbb{A})$ to $(\mathbb{R}, \mathbb{B})$, we associate two non-negative random variables, known as a positive part and a negative part. A positive part X^+ and a negative part X^- of a random variable X are defined as,

$$X^+ = X\ I_{[X \geq 0]} = \max\{X, 0\}\ \&\ X^- = -X\ I_{[X<0]} = \max\{-X, 0\}.$$

$$\text{Thus, } X^+ = \begin{cases} X, & \text{if } X \geq 0 \\ 0, & \text{if } X < 0 \end{cases} \quad \& \quad X^- = \begin{cases} -X, & \text{if } X \leq 0 \\ 0, & \text{if } X > 0. \end{cases}$$

Note that X^-, although known as a negative part, is a non-negative random variable. An arbitrary random variable X can be expressed as $X = X^+ - X^-$. Using Theorem 2.4.4, we prove that X^+ and X^- are random variables.

Theorem 2.4.5. *Suppose X is a random variable defined on a measurable space $(\Omega, \mathbb{A})$ to $(\mathbb{R}, \mathbb{B})$. Then X^+, X^- and $|X|$ are random variables on a measurable space $(\Omega, \mathbb{A})$.*

Proof. Note that for X defined on $(\Omega, \mathbb{A})$, $[X \geq 0] = X^{-1}([0, \infty)) \in \mathbb{A}$ and $[X < 0] = X^{-1}((-\infty, 0)) \in \mathbb{A}$. Hence, as shown in Example 2.2.3, $I_{[X \geq 0]}$ and $I_{[X<0]}$ are random variables on $(\Omega, \mathbb{A})$. Thus, by Theorem 2.4.4, $X^+ = X\ I_{[X \geq 0]}$ and $X^- = -X\ I_{[X<0]}$ are random variables on $(\Omega, \mathbb{A})$, being product of two random variables. If we use the definition of X^+ and X^- as $X^+ = \max\{X, 0\}$ and $X^- = \max\{-X, 0\}$, then we have one more approach to prove that X^+ and X^- are random variables. In Theorem 2.3.8, it is proved that maximum of two random variables is again a random variable. Further, $Y \equiv 0$ is a random variable with respect to any sigma field, hence,

$$X^+ = \max\{X, 0\} \quad \& \quad X^- = \max\{-X, 0\}$$

are random variables on the same space on which X is defined. Observe that $|X|$ can be expressed as $|X| = X^+ + X^-$ and hence again by Theorem 2.4.4, it is a random variable on $(\Omega, \mathbb{A})$, being addition of two random variables. □

In the next three theorems, we prove that a continuous function of a random variable is a random variable and a continuous function is a Borel function. Theorem 2.4.2 is used heavily.

Following theorem proves that measurability remains invariant under continuous transformations.

Theorem 2.4.6. *Suppose X is a $\mathbb{A}$ measurable random variable defined on a measurable space $(\Omega, \mathbb{A})$ to $(\mathbb{R}, \mathbb{B})$. Suppose $g : \mathbb{R} \to \mathbb{R}$ is a continuous function. Then $g(X)$ is a $\mathbb{A}$ measurable random variable on $(\Omega, \mathbb{A})$ to $(\mathbb{R}, \mathbb{B})$.*

Proof. Suppose $\{a_n, n \geq 1\}$ is a sequence of real numbers such that $a_n \to a$ as $n \to \infty$. Then continuity of g implies that $g(a_n) \to g(a)$ as $n \to \infty$. Since X is a random variable on a measurable space $(\Omega, \mathbb{A})$, by Theorem 2.4.2, there exists a sequence $\{X_n, n \geq 1\}$ of simple random variables on a measurable space $(\Omega, \mathbb{A})$ such that $X_n(\omega) \to X(\omega)\ \forall\ \omega \in \Omega$. Hence, by continuity of g,

$$g(X_n(\omega)) \;\to\; g(X(\omega)) \quad \Longleftrightarrow \quad g(X_n)(\omega) \;\to\; g(X)(\omega)\ \forall\ \omega \in \Omega.$$

By (iii) of Theorem 2.4.1, $\{g(X_n), n \geq 1\}$ is a sequence of simple random variable with pointwise limit $g(X)$. Hence, by Theorem 2.3.8 which states that a limit of a sequence of random variables is again a random variable, we claim that $g(X)$ is a random variable on $(\Omega, \mathbb{A})$. □

Remark 2.4.2. In Theorem 2.2.3, it is proved that a Borel function of a random variable is a random variable. Theorem 2.4.6, proved above, states that a continuous function of a random variable is a random variable. Both these results indicate that there has to be some connection between a Borel function and a continuous function. It is indeed true and it is proved in the following theorem that a continuous function is a Borel function.

Theorem 2.4.7. *Suppose $g : (\mathbb{R}, \mathbb{B}) \to (\mathbb{R}, \mathbb{B})$ is a continuous function. Then g is a Borel function, that is, g is measurable with respect to $\mathbb{B}$.*

Proof. Theorem 2.4.6 states that a continuous function of a random variable is again a random variable with respect to the same measurable space. In particular, suppose X is defined on a measurable space $(\mathbb{R}, \mathbb{B})$ such that $X(\omega) = \omega$, that is, X is an identity function. For the identity function, the inverse image of any Borel set is the set itself, thus the identity function is a Borel measurable function. Since g is a continuous function, by Theorem 2.4.6, $g(X)(\omega) = g(X(\omega)) = g(\omega)$ is a Borel measurable function. Thus a continuous function g on $(\mathbb{R}, \mathbb{B})$ is a Borel function. □

Converse of Theorem 2.4.7 is not true, that is every Borel function need not be continuous. Following are two illustrations.

Example 2.4.2. Suppose Q is a set of rational numbers and $A = (-\infty, 10)$. Suppose functions g & $h : (\mathbb{R}, \mathbb{B}) \to (\mathbb{R}, \mathbb{B})$ are defined as follows.

$$g(x) = \begin{cases} 0, & \text{if} \quad x \in A \\ 1, & \text{if} \quad x \in A^c \end{cases} \quad \& \quad h(x) = \begin{cases} 0, & \text{if} \quad x \in Q^c \\ 1, & \text{if} \quad x \in Q \end{cases}$$

The function h is known as the Dirichlet function. Note that for any Borel set S, inverse images of S under g and h are given by,

$$g^{-1}(S) = \begin{cases} \emptyset, & \text{if} \quad 0, 1 \notin S \\ \Omega, & \text{if} \quad 0, 1 \in S \\ A, & \text{if} \quad 0 \in S, 1 \notin S \\ A^c, & \text{if} \quad 1 \in S, 0 \notin S, \end{cases} \quad \& \quad h^{-1}(S) = \begin{cases} \emptyset, & \text{if} \quad 0, 1 \notin S \\ \Omega, & \text{if} \quad 0, 1 \in S \\ Q^c, & \text{if} \quad 0 \in S, 1 \notin S \\ Q, & \text{if} \quad 1 \in S, 0 \notin S, \end{cases}$$

Thus, $g^{-1}(S)$ & $h^{-1}(S) \in \mathbb{B}, \ \forall \ S \in \mathbb{B}$. Hence both are Borel functions, but g is not continuous, $x = 10$ being a point of discontinuity and the Dirichlet function h is nowhere continuous, refer to page 64 of Kumar and Kumaresan [15]. □

Theorem 2.4.6 and 2.4.7 can be extended to a function g from $\mathbb{R}^k$ to $\mathbb{R}$ and a function $\underline{g}$ from $\mathbb{R}^k$ to $\mathbb{R}^l, \ l \leq k$ and a random vector $\underline{Z}$. It is proved in the next theorem. The proof is on similar lines as that of Theorem 2.4.6 and Theorem 2.4.7.

Theorem 2.4.8. *Suppose $\underline{Z} = (X_1, X_2, \cdots, X_k)'$ is a random vector defined on a measurable space $(\Omega, \mathbb{A})$ to $(\mathbb{R}^k, \mathbb{B}^k)$.*

(i) Suppose $g : \mathbb{R}^k \to \mathbb{R}$ is a continuous function. Then $g(\underline{Z})$ is a random variable on $(\Omega, \mathbb{A})$ to $(\mathbb{R}, \mathbb{B})$ and g is a Borel function.

(ii) Suppose $\underline{g} : \mathbb{R}^k \to \mathbb{R}^l, \ l \leq k$ is a continuous function. Then $\underline{g}(\underline{Z})$ is a random vector on $(\Omega, \mathbb{A})$ to $(\mathbb{R}^l, \mathbb{B}^l)$ and $\underline{g}$ is a Borel function.

Proof.

(i) Since $\underline{Z} = (X_1, X_2, \cdots, X_k)'$ is a random vector on $(\Omega, \mathbb{A})$, each of $X_1, X_2, \cdots, X_k$ are random variables on $(\Omega, \mathbb{A})$. Hence, by Theorem 2.4.2, there exists sequences $\{X_{in}, n \geq 1\}$ of simple random variables on a measurable space $(\Omega, \mathbb{A})$ such that $X_{in}(\omega) \to X_i(\omega) \ \ \forall \ \omega \in \Omega$ as $n \to \infty, \ \forall \ i = 1, 2, \cdots, k$. Suppose $\underline{Z}_n = (X_{1n}, X_{2n}, \cdots, X_{kn})'$, then $\underline{Z}_n \to \underline{Z}$ on Ω which further implies that $g(\underline{Z}_n) \to g(\underline{Z})$ on Ω, in view of continuity of g. In Theorem 2.4.1, it is proved that if components are simple random variables, then the corresponding vector is also a simple random vector. Suppose the simple random variable X_{in} is expressed as

$$\begin{aligned} X_{in} &= \sum_{j=1}^{m} a_{in_j} I_{A_{nj}}, i = 1, 2, \cdots, k \\ \Rightarrow \ \underline{Z}_n &= (a_{1n_j}, a_{2n_j}, \cdots, a_{kn_j})' \ \text{ on } \ A_{nj}, \ j = 1, 2, \cdots, m \\ \Rightarrow \quad g(\underline{Z}_n) &= g((a_{1n_j}, a_{2n_j}, \cdots, a_{kn_j})') \ \text{ on } \ A_{nj}, \ j = 1, 2, \cdots, m \\ \Rightarrow \ \forall \ n &\geq 1, \ \ g(\underline{Z}_n) \ \text{ is a simple random variable,} \end{aligned}$$

measurable with respect to $\mathbb{A}$. Thus, $g(\underline{Z})$ is a limit of a sequence of $\mathbb{A}$ measurable random variables and hence the required result that $g(\underline{Z})$ is a random variable on $(\Omega, \mathbb{A})$ follows from the fact that limit of $\mathbb{A}$ measurable random variable is again $\mathbb{A}$ measurable random variable. The result that g is a Borel function follows by taking $\underline{Z}$ as an identity function on $(\mathbb{R}^k, \mathbb{B}^k)$ to $(\mathbb{R}^k, \mathbb{B}^k)$ and using similar arguments as in Theorem 2.4.7.

(ii) $\underline{g} : \mathbb{R}^k \to \mathbb{R}^l$ is a continuous function hence, if $\underline{g}$ is expressed as

$$\underline{g}(x_1, \cdots, x_k) = (g_1(x_1, \cdots, x_k), g_2(x_1, \cdots, x_k), \cdots, g_l(x_1, \cdots, x_k)),$$

then each of $g_1, g_2, \cdots, g_l$ are continuous functions from $\mathbb{R}^k \to \mathbb{R}$. Thus, by (i) each of $g_i(\underline{Z})$, $i = 1, 2, \cdots, l$ is a random variable on $(\Omega, \mathbb{A})$ to $(\mathbb{R}, \mathbb{B})$ and

$$\underline{g}(\underline{Z}) = (g_1(\underline{Z}), g_2(\underline{Z}), \cdots, g_l(\underline{Z}))$$

is a random vector on $(\Omega, \mathbb{A})$ to $(\mathbb{R}^l, \mathbb{B}^l)$ in view of the fact that components are random variables on $(\Omega, \mathbb{A})$. $\underline{g}$ is a Borel function also follows from the fact that each of $g_1, g_2, \cdots, g_l$ is a Borel function.

□

Remark 2.4.3. If $\underline{Z} = (X_1, X_2, \cdots, X_k)'$ is a random vector on $(\Omega, \mathbb{A})$, then as a consequence of the Theorem 2.4.8, $U = a_1X_1 + a_2X_2 + \cdots + a_kX_k$ or $V = a_1X_1 * a_2X_2 * \cdots * a_kX_k$, where $a_1, a_2, \cdots, a_k$ are any real numbers, are random variables on $(\Omega, \mathbb{A})$, since $U = g(\underline{Z})$ where $g(\underline{x}) = a_1x_1 + a_2x_2 + \cdots + a_kx_k$ is a continuous function from $\mathbb{R}^k \to \mathbb{R}$ and $V = g(\underline{Z})$ where $g(\underline{x}) = a_1x_1 * a_2x_2 * \cdots * a_kx_k$ is a continuous function from $\mathbb{R}^k \to \mathbb{R}$.

Theorem 2.4.5 can be proved using Theorem 2.4.8, as shown below. Suppose functions g^+ and g^- on $(\mathbb{R}, \mathbb{B})$ are defined as follows.

$$g^+(x) = \begin{cases} x, & \text{if } \ x \geq 0 \\ 0, & \text{if } \ x < 0 \end{cases} \quad \& \quad g^-(x) = \begin{cases} -x, & \text{if } \ x \leq 0 \\ 0, & \text{if } \ x > 0. \end{cases}$$

It is clear that g^+ and g^- are continuous functions on $\mathbb{R}$ and hence are Borel functions. Thus, $X^+ = g^+(X)$ and $X^- = g^-(X)$ are Borel functions of $\mathbb{A}$ measurable random variable X and hence are $\mathbb{A}$ measurable random variables. Similarly, $|X| = X^+ + X^- = g(X^+, X^-)$, where $g(x_1, x_2) = x_1 + x_2$ is a continuous and hence Borel function from $\mathbb{R}^2$ to $\mathbb{R}$ and hence is a $\mathbb{A}$ measurable random variable.

Having defined a random variable, a random vector and a sequence of random variables, we now proceed in the next section to define a probability measure associated with these.

2.5 Probability Distribution of a Random Variable

In Section 1.5, we have defined a probability measure on the measurable space $(\Omega, \mathbb{A})$. We now discuss the probability measure associated with a random variable X defined on the measurable space $(\Omega, \mathbb{A})$ to $(\mathbb{R}, \mathbb{B})$, measurable with respect to $\mathbb{A}$. Thus, $\forall\ S \in \mathbb{B}, X^{-1}(S) = A \in \mathbb{A}$. Suppose $(\Omega, \mathbb{A}, P)$ is a probability space. Then,

$$P(A) = P(X^{-1}(S)) = P\{\omega \mid X(\omega) \in S\} = P[X \in S] = P_X(S), \quad \text{say.}$$

The set function $P_X(S)$, related to $P(A)$ in this way, is defined for every Borel set S. Thus, $P_X(S)$ is defined on the measurable space $(\mathbb{R}, \mathbb{B})$. In the following theorem, we verify that a set function $P_X(S)$ defined on $(\mathbb{R}, \mathbb{B})$ satisfies Kolmogorov's three axioms and hence is a probability measure on the measurable space $(\mathbb{R}, \mathbb{B})$.

Theorem 2.5.1. *Suppose X is a random variable defined on a probability space $(\Omega, \mathbb{A}, P)$ to $(\mathbb{R}, \mathbb{B})$. A set function $P_X(S)$ defined on $(\mathbb{R}, \mathbb{B})$ as $P_X(S) = P(X^{-1}(S))$, for $S \in \mathbb{B}$ is a probability measure on the measurable space $(\mathbb{R}, \mathbb{B})$.*

Proof. A set function $P_X(S) = P(X^{-1}(S))$ is a probability measure, if it satisfies Kolmogorov's three axioms as given in Definition 1.5.1. Suppose $S \in \mathbb{B}$ is such that $X^{-1}(S) = A \in \mathbb{A}$. Then

(i) $P_X(S) = P(X^{-1}(S)) = P(A) \geq 0$.

(ii) $P_X(\mathbb{R}) = P(X^{-1}(\mathbb{R})) = P(\Omega) = 1$.

(iii) Suppose $\{S_n, n \geq 1\}$ is a sequence of disjoint Borel sets. Then $\{X^{-1}(S_n), n \geq 1\}$ is also a sequence of disjoint sets in $\mathbb{A}$, by Lemma 2.2.1. Now,

$$\begin{aligned} P_X\Big(\sum_{n\geq 1} S_n\Big) &= P\Big(X^{-1}(\sum_{n\geq 1} S_n)\Big) = P\Big(\sum_{n\geq 1} X^{-1}(S_n)\Big), \\ &\quad \text{by Lemma 2.2.1} \\ &= \sum_{n\geq 1} P\left(X^{-1}(S_n)\right), \\ &\quad \text{by } \sigma\text{-additivity of probability measure } P \\ &= \sum_{n\geq 1} P_X(S_n). \end{aligned}$$

Hence, $P_X(S) = P(X^{-1}(S))$ is a probability measure on the measurable space $(\mathbb{R}, \mathbb{B})$.

□

As a consequence of this theorem, we have the following definitions.

Definition 2.5.1. *Induced probability measure: A probability measure $P_X(S) = P(X^{-1}(S))$ on the measurable space $(\mathbb{R}, \mathbb{B})$ is known as an induced probability measure, induced by a random variable X on $(\Omega, \mathbb{A}, P)$.*

Definition 2.5.2. *Induced probability space: Suppose $P_X(S) = P(X^{-1}(S))$ is a probability measure on the measurable space $(\mathbb{R}, \mathbb{B})$. Then the probability space $(\mathbb{R}, \mathbb{B}, P_X)$ is known as an induced probability space, induced by a random variable X.*

Definition 2.5.3. *Probability distribution of X: Suppose $(\mathbb{R}, \mathbb{B}, P_X)$ is an induced probability space, induced by a random variable X. Then the collection $\{P_X(S), S \in \mathbb{B}\}$ is knows as a probability distribution of X.*

Suppose X is a random variable defined on a probability space $(\Omega, \mathbb{A}, P)$ to $(\mathbb{R}, \mathbb{B}, P_X)$ and g is a Borel function. Then $g(X)$ is also a random variable on the probability space $(\Omega, \mathbb{A}, P)$. The probability measure induced by $g(X)$ is given by,

$$\begin{aligned} P_{g(X)}(S) &= P(g(X)^{-1}(S)) = P((X^{-1}(g^{-1}(S)))) = P((X^{-1}(S_g))) \\ &= P_X(S_g) = P_X(g^{-1}(S)), \text{ where } S_g = g^{-1}(S) \in \mathbb{B}. \end{aligned}$$

Thus, the probability distribution of any Borel function of X is determined by that of X, which in turn is determined by the probability measure P on $(\Omega, \mathbb{A})$. In summary, the knowledge of P is enough to find out the distribution of any random variable defined on $(\Omega, \mathbb{A}, P)$ and also of any Borel function of X. Hence, $(\Omega, \mathbb{A}, P)$ is known as a parent probability space or a fundamental probability space.

Example 2.5.1. Suppose Ω corresponds to the outcome of a random experiment of tossing an unbiased coin twice. Then the sample space Ω is given by $\Omega = \{HH, HT, TH, TT\}$. Suppose $\mathbb{A} = \mathbb{P}(\Omega)$. Since it is given that the coin is unbiased, the probability of getting head is the same as that of tail and it is $1/2$. Thus, for any $A \in \mathbb{A}, P(A) = m/4$, where m is the number of elements in A. Suppose a random variable X is defined as the number of heads in the outcome. Thus,

$$X(\omega) = \begin{cases} 0, & \text{if } \omega = TT \\ 1, & \text{if } \omega = TH, HT \\ 2, & \text{if } \omega = HH. \end{cases}$$

Hence, for a Borel set S,

$$P_X(S) = \begin{cases} 0, & \text{if } 0 \notin S, 1 \notin S, 2 \notin S \\ P(\{TT\}) = 1/4, & \text{if } 0 \in S, 1 \notin S, 2 \notin S \\ P(\{TH, HT\}) = 2/4, & \text{if } 0 \notin S, 1 \in S, 2 \notin S \\ P(\{HH\}) = 1/4, & \text{if } 0 \notin S, 1 \notin S, 2 \in S \\ P(\{TT, TH, HT\} = 3/4, & \text{if } 0 \in S, 1 \in S, 2 \notin S \\ P(\{TT, HH\}) = 2/4, & \text{if } 0 \in S, 1 \notin S, 2 \in S \\ P(\{TH, HT, HH\}) = 3/4, & \text{if } 0 \notin S, 1 \in S, 2 \in S \\ P(\Omega) = 1, & \text{if } 0 \in S, 1 \in S, 2 \in S. \end{cases}$$

This gives us the probability distribution $\{P_X(S), S \in \mathbb{B}\}$ of X. It is usually presented as follows.

$$P[X = 0] = 1/4, \quad P[X = 1] = 1/2 \quad \& \quad P[X = 2] = 1/4. \qquad \square$$

Example 2.5.2. Suppose Ω is a sample space corresponding to a random experiment of tossing a coin n times, independently of each other. For the given random experiment the sample space Ω is given by,

$\{\omega = (\omega_1, \omega_2, \cdots, \omega_n) |\, \omega_j = 0 \text{ or } 1, \; j = 1, 2, \cdots, n\}$ and suppose $\mathbb{A} = \mathbb{P}(\Omega)$. A simple random variable X on $(\Omega, \mathbb{A})$ is defined as follows.

$$X(\omega) = \sum_{i=0}^{n} i I_{A_i}(\omega) \quad \Longleftrightarrow \quad X(\omega) = i \text{ if } \omega \in A_i, \quad i = 0, 1, 2, \cdots, n,$$

where $A_i = \{\omega = (\omega_1, \omega_2, \cdots, \omega_n) \big| \sum_{j=1}^{n} \omega_j = i\}$, $i = 0, 1, 2, \cdots, n$. It is given that probability of a head is p and remains the same $\forall\, n$. By the definition, the probability distribution of X is given by $\{P_X(S), S \in \mathbb{B}\}$, however it is to be noted that by the sigma additivity of the probability measure $P_X(.)$, it is enough to specify the probability assigned to $A_i \equiv [X = i] \;\forall\; i = 1, 2, \cdots, n$. Observe that there are $\binom{n}{i}$ sample points in A_i and for each sample point in A_i there are i heads and $n - i$ tails. Hence

$$P(A_i) = P\{\omega | \omega \in A_i\} = P\{\omega | X(\omega) = i\} = P[X = i] = \binom{n}{i} p^i (i - p)^{n-i},$$

$i = 0, 1, 2, \cdots, n$. Thus, the probability distribution $\{P_X(S), S \in \mathbb{B}\}$ can be specified as $\{P(A_i) = P[X = i] = \binom{n}{i} p^i (i - p)^{n-i}\}$, $i = 0, 1, 2, \cdots, n$. Note that it is the well known binomial distribution. $\square$

Probability distribution of a random vector $\underline{Z}$ is also defined on similar lines as the collection $\{P_{\underline{Z}}(S), S \in \mathbb{B}^k\}$, where $P_{\underline{Z}}(S) = P(\underline{Z}^{-1}(S))$.

Remark 2.5.1. In most of the practical situations, we are usually given the induced probability space. Once it is given, it is not always necessary to have the knowledge of the fundamental probability space $(\Omega, \mathbb{A}, P)$, when we are interested in studying the properties of the given random variables and their Borel functions. However, in order to study the calculus of random variables by introducing new random variables, such as limits of random variables, it is necessary to have the knowledge of the fundamental probability space $(\Omega, \mathbb{A}, P)$.

A quick recap of the results discussed in the present chapter is given below.

Summary

1. Inverse image of a set: Suppose $X : \Omega_1 \to \Omega_2$ is a function and S is a subset of Ω_2. Then the inverse image of a set S under the function X is defined as $X^{-1}(S) = \{\omega \in \Omega_1 | X(\omega) \in S\}$.

2. Properties of the inverse image of a set: (i) Inverse image of a complement is a complement of the inverse image. (ii) Inverse image of a countable union is a countable union of inverse images. (iii) Inverse image of a countable intersection is a countable intersection of inverse images.
   (iv) Inverse images of disjoint sets are also disjoint sets. (v) If $S_1 \subseteq S_2$, then $X^{-1}(S_1) \subseteq X^{-1}(S_2)$. If $X^{-1}(S_1) \subseteq X^{-1}(S_2)$, then $S_1 \subset S_2$.

3. The inverse image of a collection $\mathbb{C}$ of subsets of Ω_2 under the function X is defined as, $X^{-1}(\mathbb{C}) = \{X^{-1}(S) | S \in \mathbb{C}\}$.

4. Suppose $\mathbb{A}$ is a sigma field of subsets of Ω_1, then the class $\mathbb{F}$ of all sets whose inverse images belong to $\mathbb{A}$ is also a sigma field.

5. Inverse image of a sigma field is a sigma field.

6. $\mathbb{C}_1 \subseteq \mathbb{C}_2 \;\Rightarrow\; X^{-1}(\mathbb{C}_1) \subseteq X^{-1}(\mathbb{C}_2)$ and $\sigma(X^{-1}(\mathbb{C}_1)) \subseteq \sigma(X^{-1}(\mathbb{C}_2))$.

7. Inverse image of the minimal sigma field generated by a class is the minimal sigma field of inverse image of the class, that is, $X^{-1}(\sigma(\mathbb{C})) = \sigma(X^{-1}(\mathbb{C}))$.

8. Descriptive definition of a random variable: A function $X : (\Omega, \mathbb{A}) \to (\mathbb{R}, \mathbb{B})$ is said to a random variable or a measurable function with respect to $\mathbb{A}$ if $\forall \; S \in \mathbb{B}, \;\; X^{-1}(S) \in \mathbb{A}$.

9. Economical definition of a random variable: Suppose $X : (\Omega, \mathbb{A}) \to (\mathbb{R}, \mathbb{B})$ is a function and $\mathbb{C}$ is a class of subsets of $\mathbb{R}$ such that the minimal sigma field generated by $\mathbb{C}$ is the Borel field $\mathbb{B}$. Then X is a random variable measurable with respect to $\mathbb{A}$ if $\forall \; S \in \mathbb{C}, \; X^{-1}(S) \in \mathbb{A}$.

10. Suppose X is a random variable defined on $(\Omega, \mathbb{A})$ to $(\mathbb{R}, \mathbb{B})$. Then $X^{-1}(\mathbb{B}) = \{X^{-1}(S) | S \in \mathbb{B}\}$ is a sigma field, being inverse image of a sigma field. It is known as a sigma field induced by or generated by a random variable X.

11. Borel function: Suppose $g : (\mathbb{R}, \mathbb{B}) \to (\mathbb{R}, \mathbb{B})$ is a function. If $\forall \; S \in \mathbb{B}$, $g^{-1}(S) = \{x \in \mathbb{R} | g(x) \in S\} \in \mathbb{B}$, then g is measurable with respect to the Borel field and is known as a Borel function.

12. Suppose $g : (\mathbb{R}, \mathbb{B}) \to (\mathbb{R}, \mathbb{B})$ is a function. Suppose $\mathbb{C}$ is a class of subsets of $\mathbb{R}$ such that the minimal sigma field generated by $\mathbb{C}$ is the Borel field $\mathbb{B}$. Then g is a Borel function if and only if $\forall \; S \in \mathbb{C}, \;\; g^{-1}(S) \in \mathbb{B}$.

13. Suppose $X : (\Omega, \mathbb{A}) \to (\mathbb{R}, \mathbb{B})$ is a random variable measurable with respect to a sigma field $\mathbb{A}$ and $g : (\mathbb{R}, \mathbb{B}) \to (\mathbb{R}, \mathbb{B})$ is a Borel function, then $Y = g(X)$ is a random variable measurable with respect to $\mathbb{A}$ and $\sigma(Y) \subseteq \sigma(X)$.

14. Descriptive definition of a k-variate random vector: Suppose $\underline{Z} = (X_1, X_2, \cdots, X_k)'$ is a function defined on the measurable space $(\Omega, \mathbb{A})$ to $(\mathbb{R}^k, \mathbb{B}^k)$, where $\mathbb{B}^k$ is the Borel field of subsets of $\mathbb{R}^k$. If

    $$\forall \; S \in \mathbb{B}^k, \;\; \underline{Z}^{-1}(S) = \{\omega \mid \underline{Z}(\omega) = (X_1(\omega), X_2(\omega), \cdots, X_k(\omega))' \in S\} \in \mathbb{A},$$

    then $\underline{Z}$ is a random vector measurable with respect to $\mathbb{A}$.

15. Economical definition of a k-variate random vector: Suppose $\underline{Z} = (X_1, X_2, \cdots, X_k)'$ is a function defined on the measurable space $(\Omega, \mathbb{A})$ to $(\mathbb{R}^k, \mathbb{B}^k)$. Suppose $\mathbb{C}^k$ is a class of subsets of $\mathbb{R}^k$ which generates the Borel field $\mathbb{B}^k$. If

$$\forall\, S \in \mathbb{C}^k,\ \underline{Z}^{-1}(S) = \{\omega \mid \underline{Z}(\omega) = (X_1(\omega), X_2(\omega), \cdots, X_k(\omega))' \in S\} \in \mathbb{A},$$

then $\underline{Z}$ is a random vector measurable with respect to $\mathbb{A}$.

16. Suppose $\underline{Z} = (X_1, X_2, \cdots, X_k)'$ is a random vector defined on a measurable space $(\Omega, \mathbb{A})$ to $(\mathbb{R}^k, \mathbb{B}^k)$. Then $\sigma(\underline{Z}) = \underline{Z}^{-1}(\mathbb{B}^k) = \sigma\Big(\bigcup_{i=1}^k X_i^{-1}(\mathbb{B})\Big)$.

17. $\underline{Z} = (X_1, X_2, \cdots, X_k)'$ is a random vector measurable with respect to $\mathbb{A}$ if and only if each of $X_i,\ i = 1, 2, \cdots, k$ is random variable measurable with respect to $\mathbb{A}$.

18. Suppose $\underline{g} : (\mathbb{R}^k, \mathbb{B}^k) \to (\mathbb{R}^l, \mathbb{B}^l)\quad l \leq k$ is a function where $\underline{g} = (g_1, g_2, \cdots, g_l)$. $\underline{g}$ is a Borel function if $\forall\ S \in \mathbb{B}^l,\ \ \underline{g}^{-1}(S) \in \mathbb{B}^k$.

19. Suppose $\underline{g} : (\mathbb{R}^k, \mathbb{B}^k) \to (\mathbb{R}^l, \mathbb{B}^l)\quad l \leq k$ is a function where $\underline{g} = (g_1, g_2, \cdots, g_l)$. $\underline{g}$ is a Borel function if and only if each of g_i is a Borel function from $(\mathbb{R}^k, \mathbb{B}^k) \to (\mathbb{R}, \mathbb{B})$, $i = 1, 2, \cdots, l$.

20. Suppose $\underline{Z} : (\Omega, \mathbb{A}) \to (\mathbb{R}^k, \mathbb{B}^k)$ is a random vector measurable with respect to the sigma field $\mathbb{A}$. (i) Suppose $g : (\mathbb{R}^k, \mathbb{B}^k) \to (\mathbb{R}, \mathbb{B})$ is a Borel function, then $Y = g(\underline{Z})$ is a random variable measurable with respect to the sigma field $\mathbb{A}$ and $\sigma(Y) \subseteq \sigma(X)$. (ii) Suppose $\underline{g} : (\mathbb{R}^k, \mathbb{B}^k) \to (\mathbb{R}^l, \mathbb{B}^l), l \leq k$ is a Borel function, then $\underline{Y} = \underline{g}(\underline{Z})$ is a random vector measurable with respect to the sigma field $\mathbb{A}$ and $\sigma(\underline{Y}) \subseteq \sigma(\underline{Z})$.

21. Borel field in infinite dimensions: Suppose $\mathbb{C}^\infty$ is a collection of cylinder sets S defined as

$$S = \{\underline{x} = (x_1, x_2, \cdots,) | a_{i_j} \leq x_{i_j} \leq b_{i_j}, j = 1, \cdots, k, x_{i_j} \in \mathbb{R},\ j = k+1, \cdots\},$$

where $i_1, i_2, \cdots i_k$ is any permutation of positive integers and k is finite. Then the minimal sigma field induced by $\mathbb{C}^\infty$ is the Borel field in infinite dimensions and is denoted by $\mathbb{B}^\infty$.

22. Infinite dimensional random vector: A function $\underline{Z} = (X_1, X_2, \cdots,)$ defined on $(\Omega, \mathbb{A})$ to $(\mathbb{R}^\infty, \mathbb{B}^\infty)$ is an infinite dimensional random vector if $\forall\ \ S \in \mathbb{B}^\infty,\ \ \underline{Z}^{-1}(S) \in \mathbb{A}$.

23. For an infinite dimensional random vector $\underline{Z} = (X_1, X_2, \cdots,)$,

$$\underline{Z}^{-1}(\mathbb{B}^\infty) = \sigma\Big(\{X_n, n \geq 1\}\Big) = \sigma(\bigcup_{n \geq 1} \mathbb{A}_n)$$

where $\mathbb{A}_n = \sigma\{X_1, \cdots, X_n\} = \sigma\Big(\bigcup_{i=1}^n (X_i^{-1}(\mathbb{B}))\Big)$.

24. Suppose $\underline{Z} = (X_1, X_2, \cdots,)$ is defined on the measurable space $(\Omega, \mathbb{A})$ to $(\mathbb{R}^\infty, \mathbb{B}^\infty)$. Then $\underline{Z}$ is an infinite dimensional random vector if and only if for every $n \geq 1$, X_n is a random variable with respect to $\mathbb{A}$.

25. For a sequence $\{X_n, n \geq 1\}$ of random variables

$$\liminf X_n = \sup_{n\geq 1} \inf_{k\geq n} X_k \quad \& \quad \limsup X_n = \inf_{n\geq 1} \sup_{k\geq n} X_k \ .$$

If $\liminf X_n(\omega) = \limsup X_n(\omega) = X(\omega), \forall \ \omega \in \Omega$, then $\lim_{n\to\infty} X_n$, exists and $\lim_{n\to\infty} X_n = X$. This mode of convergence of a sequence of random variables is known as a point-wise convergence or convergence everywhere.

26. If for each $n \geq 1$, X_n is a random variable measurable with respect to $\mathbb{A}$, then $\liminf X_n$ and $\limsup X_n$, and $\lim X_n$, whenever exists, are random variables measurable with respect to $\mathbb{A}$.

27. Suppose $\{X_n, n \geq 1\}$ is a sequence of random variables and $\mathbb{C}_n = \sigma\{X_n, X_{n+1}, \cdots\}$. Then $\mathbb{T} = \bigcap_{n\geq 1} \mathbb{C}_n$ is known as a tail sigma field of the sequence $\{X_n, n \geq 1\}$ and sets in $\mathbb{T}$ are known as tail events.

28. Suppose a random variable X on a measurable space $(\Omega, \mathbb{A})$ is defined as follows.

$$X(\omega) = \sum_{i=1}^{k} a_i I_{A_i}(\omega) \iff X(\omega) = a_i \text{ if } \omega \in A_i, \ a_i \in \mathbb{R}, \ i = 1, 2, \cdots, k$$

where $\mathbb{P} = \{A_1, A_2, \cdots, A_k\}$ is a measurable partition of Ω. Then X is known as a simple random variable.

29. Suppose X is a random variable defined on a measurable space $(\Omega, \mathbb{A})$ to $(\mathbb{R}, \mathbb{B})$. Then there exists a sequence $\{X_n, n \geq 1\}$ of simple random variables on a measurable space $(\Omega, \mathbb{A})$ such that $X_n \to X$ on Ω as $n \to \infty$.

30. Suppose X is a non-negative random variable defined on a measurable space $(\Omega, \mathbb{A})$ to $(\mathbb{R}, \mathbb{B})$. Then there exists a non-decreasing sequence $\{X_n, n \geq 1\}$ of non-negative simple random variables on a measurable space $(\Omega, \mathbb{A})$ such that $X_n \to X$ on Ω.

31. Suppose X is a simple random variable defined on a measurable space $(\Omega, \mathbb{A})$ to $(\mathbb{R}, \mathbb{B})$. Suppose $g : \mathbb{R} \to \mathbb{R}$ is a function. Then $g(X)$ is a simple random variable on $(\Omega, \mathbb{A})$ to $(\mathbb{R}, \mathbb{B})$.

32. Suppose X is a random variable defined on a measurable space $(\Omega, \mathbb{A})$ to $(\mathbb{R}, \mathbb{B})$. Suppose $g : \mathbb{R} \to \mathbb{R}$ is a continuous function. Then $g(X)$ is a random variable on $(\Omega, \mathbb{A})$ to $(\mathbb{R}, \mathbb{B})$.

33. Suppose $g : (\mathbb{R}, \mathbb{B}) \to (\mathbb{R}, \mathbb{B})$ is a continuous function. Then g is a Borel function.

34. Suppose $\underline{Z} = (X_1, X_2, \cdots, X_k)'$ is a random vector defined on a measurable space $(\Omega, \mathbb{A})$ to $(\mathbb{R}^k, \mathbb{B}^k)$. (i) Suppose $g : \mathbb{R}^k \to \mathbb{R}$ is a continuous function. Then $g(\underline{Z})$ is a random variable on $(\Omega, \mathbb{A})$ to $(\mathbb{R}, \mathbb{B})$ and g is a Borel function. (ii) Suppose $\underline{g} : \mathbb{R}^k \to \mathbb{R}^l, \ \ l \leq k$ is a continuous function. Then $\underline{g}(\underline{Z})$ is a random vector on $(\Omega, \mathbb{A})$ to $(\mathbb{R}^l, \mathbb{B}^l)$ and $\underline{g}$ is a Borel function.

35. Suppose X_1 and X_2 are random variables defined on a measurable space $(\Omega, \mathbb{A})$ to $(\mathbb{R}, \mathbb{B})$. Then (i) $X_1 + X_2$, $X_1 - X_2$, $X_1 X_2$ and X_1/X_2, provided it is defined, are random variables on the measurable space $(\Omega, \mathbb{A})$. (ii) aX_1 is random variable on $(\Omega, \mathbb{A})$ to $(\mathbb{R}, \mathbb{B})$, where a is any real number.

36. Suppose X is a random variable defined on a measurable space $(\Omega, \mathbb{A})$ to $(\mathbb{R}, \mathbb{B})$. Then X^+, X^- and $|X|$ are random variables on a measurable space $(\Omega, \mathbb{A})$.

37. Suppose X is a random variable defined on a probability space $(\Omega, \mathbb{A}, P)$ to $(\mathbb{R}, \mathbb{B})$. A set function $P_X(S)$ defined on $(\mathbb{R}, \mathbb{B})$ as $P_X(S) = P(X^{-1}(S))$, for $S \in \mathbb{B}$ is a probability measure on the measurable space $(\mathbb{R}, \mathbb{B})$. It is known as an induced probability measure, induced by X.

38. The probability space $(\mathbb{R}, \mathbb{B}, P_X)$ is known as an induced probability space.

39. $\{P_X(S), S \in \mathbb{B}\}$ is known as a probability distribution of X.

2.6 Conceptual Exercises

2.6.1 Suppose Ω is a sample space corresponding to the experiment of rolling a die once. Suppose $\mathbb{A} = \{\Omega, \emptyset, \{1, 3, 5\}, \{2, 4, 6\}\}$. Define two functions on $(\Omega, \mathbb{A})$ such that one is a random variable on $(\Omega, \mathbb{A})$ and the other is not a random variable on $(\Omega, \mathbb{A})$.

2.6.2 Suppose $\Omega = (0, 1]$ and $\mathbb{A}$ is a minimal sigma field consisting of all Borel sets which are subsets of Ω. Examine whether a function $X(\omega) = 6\omega + 2$ is a random variable on $(\Omega, \mathbb{A})$. (i) Use the economical definition. (ii) Use the result that a Borel function of a random variable is a random variable.

2.6.3 Suppose X is a random variable, measurable with respect to $\mathbb{A}$. Examine whether (i) $Y_1 = e^X$ and $Y_2 = e^X + X^3 + 5X$ are random variables, measurable with respect to $\mathbb{A}$. (ii) If X is a positive random variable measurable with respect to $\mathbb{A}$, examine whether $Y_3 = \log X$ is a random variable. Is $\underline{Z} = (Y_1, Y_2, Y_3)'$ a random vector measurable with respect to $\mathbb{A}$? Justify.

2.6.4 (i) Suppose A and B are mutually exclusive events. Find the minimal sigma field induced by $I_A + I_B$. (ii) Suppose A and B are not mutually exclusive events. Find the minimal sigma field induced by $I_A + I_B$.

2.6.5 Suppose $\Omega = \{a, b, c\}$ and $\mathbb{A} = \{\Omega, \emptyset, \{a\}, \{b, c\}\}$. Examine whether $\mathbb{A}$ is a sigma field. If yes, specify any non-constant function on $(\Omega, \mathbb{A})$ which is measurable with respect to $\mathbb{A}$.

2.6.6 Suppose X and Y are random variables on $(\Omega, \mathbb{A}, P)$ such that $\sigma(X) = \sigma(Y)$. Show that $\sigma(X + Y) \subseteq \sigma(Y)$.

2.6.7 Suppose X is a random variable. Show that $\sigma(X^2) \subseteq \sigma(X)$.

2.6.8 Give an example of a random variable such that $\sigma(X^2) \neq \sigma(X)$.

2.6.9 Give an example of a random variable such that $\sigma(X^2) = \sigma(X)$.

2.6.10 Suppose $\Omega = \{1, 2, 3, 4\}$, $A = \{1, 2\}$ and $\mathbb{A} = \{\Omega, \emptyset, A, A^c\}$. Functions X, Y and Z on $(\Omega, \mathbb{A})$ are defined as follows. $X(\omega) = 4 \quad \forall \ \omega \in \Omega$, $Y(\omega) = 1$ if $\omega \in A$ and 0 otherwise and $Z(\omega) = \omega \ \ \forall \ \ \omega \in \Omega$. Examine whether $\underline{Z} = (X, Y, Z)$ is random vector on $(\Omega, \mathbb{A})$. If not, modify $\mathbb{A}$, so that $\underline{Z} = (X, Y, W)$ is random vector on $(\Omega, \mathbb{A})$. Find the sigma field induced by $\underline{Z}$.

2.6.11 Suppose $\Omega = \mathbb{N}$, the set of natural numbers, and $\mathbb{A} = \mathbb{P}(\mathbb{N})$, power set of $\mathbb{N}$. Construct one finite and one countably infinite measurable partition of Ω. Give an example of one simple random variable and one elementary random variable defined on $(\mathbb{N}, \mathbb{P}(\mathbb{N}))$.

2.6.12 Prove or disprove: $X^{-1}(\lim S_n) = \lim X^{-1}(S_n)$, where S_n is a Borel set.

2.6.13 Show that a necessary and sufficient condition for a function to be measurable is that its positive and negative parts are measurable.

2.6.14 Suppose f and g are Borel functions on $(\mathbb{R}, \mathbb{B})$. For $A \in \mathbb{B}$, a function h is defined as $h = f$ on A and $h = g$ on A^c. Show that h is a Borel function.

2.6.15 Suppose $|X|$ is $\mathbb{A}$ measurable. Examine whether X^+ and X^- are also $\mathbb{A}$) measurable.

2.6.16 Suppose $(X_1, X_2, \cdots, X_n)$ is a random vector on $(\Omega, \mathbb{A}, P)$. Show that (i) $\sum_{i=1}^{n} l_i X_i$ is real valued random variable where $l_1, l_2, \cdots, l_n$ are any real numbers, (ii) $(\sin X_1, X_2, X_3, \cdots, e^{X_{n-1}}, |X_n|)$ is a random vector and (iii) $(X_2, X_3, \cdots, X_{n-1})$ is a random vector, all defined on the same probability space $(\Omega, \mathbb{A}, P)$.

2.6.17 Suppose a measurable space $(\Omega, \mathbb{A})$ is defined as $\Omega = [0, \infty)$ and $\mathbb{A}$ is a sigma field of subsets of Ω. Suppose a sequence $\{X_n, n \geq 1\}$ of $\mathbb{A}$ measurable random variables and a $\mathbb{A}$ measurable random variable X are defined as follows.

$$X_n(\omega) = \exp(-n\omega), \ n \geq 1 \ \ \& \ \ X(\omega) = 0 \ \ \ \forall \ \ \omega \in \Omega.$$

(i) Examine if $X_n \to X$ on Ω or any subset of Ω. (ii) If not, identify another X such that $X_n \to X$ on Ω.

2.6.18 Suppose a random variable X is defined on $(\Omega, \mathbb{A}, P)$ and $X \sim N(0, 1)$ distribution. Suppose $A_n = [X \leq 1/n]$. Find $\lim_{n\to\infty} P(A_n)$.

2.6.19 Suppose a random variable X is defined on $(\Omega, \mathbb{A}, P)$ and $X \sim \chi^2_5$ and $A_n = [X \leq 1/n]$. Find $\lim_{n\to\infty} P(A_n)$.

2.6.20 Suppose a random variable X is defined on $(\Omega, \mathbb{A}, P)$ and X has exponential distribution with mean 1 and $A_n = [1 - 1/n \leq X \leq 2 + 1/n]$. Find $\lim_{n\to\infty} P(A_n)$. Suppose $B_n = [n < X \leq n + 1]$. Find $\lim_{n\to\infty} P(B_n)$.

2.6.21 Suppose a random variable X is defined on $(\Omega, \mathbb{A}, P)$ and $P[X = n] = a_n$, $n \in \mathbb{N}, a_n \geq 0$ and $\sum_{n\in\mathbb{N}} a_n = 1$. Find $\lim_{n\to\infty} P[X = n]$.

2.7 Computational Exercises

2.7.1 Suppose a measurable space $(\Omega, \mathbb{A})$ is defined as $\Omega = [0, \infty)$ and $\mathbb{A}$ is a sigma field of subsets of Ω. Suppose a sequence $\{X_n, n \geq 1\}$ of $\mathbb{A}$ measurable random variables and a $\mathbb{A}$ measurable random variable X are defined as follows.

$$X_n(\omega) = \exp(-n\omega), \ n \geq 1 \ \ \& \ \ X(\omega) = 0 \ \ \ \forall \ \ \omega \in \Omega.$$

Examine graphically if $X_n \to X$ on Ω or any subset of Ω.

2.7.2 Verify Theorem 2.4.2 graphically.

2.7.3 Verify Theorem 2.4.3 graphically.

2.7.4 Suppose $X \sim N(0, 1)$ and $A_n = [X \leq 1/n]$. Find $\lim_{n\to\infty} P(A_n)$ on the basis of data simulated from the distribution of X.

2.7.5 Suppose $X \sim \chi^2_5$ and $A_n = [X \leq 1/n]$. Using simulation find $\lim_{n\to\infty} P(A_n)$.

2.7.6 Suppose X has exponential distribution with mean 1 and $A_n = [1 - 1/n \leq X \leq 2 + 1/n]$. Find $\lim_{n\to\infty} P(A_n)$ based on simulated data.

2.8 Multiple Choice Questions

Note: In each question, multiple options may be correct. Answers are given in Chapter 11, after the solutions of conceptual exercises of Chapter 2.

2.8.1 Suppose $X : \Omega_1 \to \Omega_2$ is a function and S_1 and S_2 are subsets of Ω_2. Following are two statements: (I) If $S_1 \subseteq S_2$, then $X^{-1}(S_1) \subseteq X^{-1}(S_2)$. (II) If $X^{-1}(S_1) \subset X^{-1}(S_2)$, then $S_1 \subset S_2$. Then which of the following statements is true?

(a) Both (I) and (II) are true
(b) (I) is true but (II) is false
(c) (II) is true but (I) is false
(d) Both (I) and (II) are false

2.8.2 Suppose $X : \Omega_1 \to \Omega_2$ is a function and $\mathbb{C}_1$ and $\mathbb{C}_2$ are classes of subsets of Ω_2. Following are two statements: (I) $\mathbb{C}_1 \subseteq \mathbb{C}_2 \;\; \Rightarrow \;\; X^{-1}(\mathbb{C}_1) \subseteq X^{-1}(\mathbb{C}_2)$. (II) $X^{-1}(\mathbb{C}_1) \subseteq X^{-1}(\mathbb{C}_2) \;\; \Rightarrow \;\; \mathbb{C}_1 \subset \mathbb{C}_2$. Then which of the following statements is true?

(a) Both (I) and (II) are true
(b) (I) is true but (II) is false
(c) (II) is true but (I) is false
(d) Both (I) and (II) are false

2.8.3 Suppose $X : \Omega_1 \to \Omega_2$ is a function and $\mathbb{C}_1$ and $\mathbb{C}_2$ are classes of subsets of Ω_2. Following are two statements: (I) $\mathbb{C}_1 \subseteq \mathbb{C}_2 \;\; \Rightarrow \;\; \sigma[X^{-1}(\mathbb{C}_1)] \subseteq \sigma[X^{-1}(\mathbb{C}_2)]$. (II) $\sigma[X^{-1}(\mathbb{C}_1)] \subseteq \sigma[X^{-1}(\mathbb{C}_2)] \;\; \Rightarrow \;\; \mathbb{C}_1 \subseteq \mathbb{C}_2$. Then which of the following statements is true?

(a) Both (I) and (II) are true
(b) (I) is true but (II) is false
(c) (II) is true but (I) is false
(d) Both (I) and (II) are false

2.8.4 Suppose $X : \Omega_1 \to \Omega_2$ is a function and $\mathbb{C}$ is a sigma field of subsets of Ω_2, then which of the following statements is/are true

(a) $X^{-1}(\mathbb{C}) = \sigma[X^{-1}(\mathbb{C})]$
(b) $X^{-1}[\sigma(\mathbb{C})] = \sigma[X^{-1}(\mathbb{C})]$
(c) $X^{-1}(\mathbb{C}) \subseteq \sigma[X^{-1}(\mathbb{C})]$
(d) $X^{-1}[\sigma(\mathbb{C})] \supseteq \sigma[X^{-1}(\mathbb{C})]$

2.8.5 Which of the following statements is/are true for a random variable X?

(a) X is $\mathbb{A}$ measurable $\Rightarrow$ $|X|$ is $\mathbb{A}$ measurable
(b) X is $\mathbb{A}$ measurable $\Rightarrow$ X^2 is $\mathbb{A}$ measurable
(c) $|X|$ is $\mathbb{A}$ measurable $\Rightarrow$ X is $\mathbb{A}$ measurable
(d) X^2 is $\mathbb{A}$ measurable $\Rightarrow$ X is $\mathbb{A}$ measurable

2.8.6 Which of the following is/are simple random variables with respect to $(\Omega, \mathbb{A})$, if Ω = set of natural numbers and $\mathbb{A}$= powerset of Ω.

(a) $X(\omega) = 0 \ \forall\ \omega \in \Omega$
(b) $X(\omega) = 0$ if ω is odd and $X(\omega) = 1$ if ω is even
(c) $X(\omega) = \omega \ \forall\ \omega \in \Omega$
(d) $X(\omega) = 2\omega \ \ \forall\ \omega \in \Omega$

2.8.7 Suppose $(\Omega, \mathbb{A})$ is a measurable space where Ω = set of natural numbers and $\mathbb{A}$= powerset of Ω. Following are two statements: (I) $X(\omega) = 0 \ \forall\ \omega \in \Omega$ is a simple random variable with respect to $(\Omega, \mathbb{A})$. (II) $X(\omega) = \omega \ \forall\ \omega \in \Omega$ is a simple random variable with respect to $(\Omega, \mathbb{A})$. Then which of the following statements is true?

(a) Both (I) and (II) are true
(b) (I) is true but (II) is false
(c) (II) is true but (I) is false
(d) Both (I) and (II) are false

2.8.8 Which of the following statements is/are NOT true for a random variable X?

(a) X is $\mathbb{A}$ measurable $\Rightarrow$ $|X|$ is $\mathbb{A}$ measurable
(b) X is $\mathbb{A}$ measurable $\Rightarrow$ X^2 is $\mathbb{A}$ measurable
(c) $|X|$ is $\mathbb{A}$ measurable $\Rightarrow$ $\log(|X|)$ is $\mathbb{A}$ measurable
(d) X^2 is $\mathbb{A}$ measurable $\Rightarrow$ X is $\mathbb{A}$ measurable

2.8.9 Suppose $\underline{Z} = (X_1, X_2)'$ is a random vector defined on a measurable space $(\Omega, \mathbb{A})$ to $(\mathbb{R}^2, \mathbb{B}^2)$. Then the sigma field $\underline{Z}^{-1}(\mathbb{B}^2)$ induced by $\underline{Z}$ is given by

(a) $X_1^{-1}(\mathbb{B}) \cup X_2^{-1}(\mathbb{B})$
(b) $\sigma(X_1^{-1}(\mathbb{B}) \cup X_2^{-1}(\mathbb{B}))$
(c) $\sigma(X_1^{-1}(\mathbb{B}) \cap X_2^{-1}(\mathbb{B}))$
(d) $X_1^{-1}(\mathbb{B}) \cap X_2^{-1}(\mathbb{B})$

2.8.10 Suppose an infinite dimensional random vector $\underline{Z} = (X_1, X_2, \cdots,)$ is defined on $(\Omega, \mathbb{A})$ to $(\mathbb{R}^\infty, \mathbb{B}^\infty)$. Suppose $\mathbb{A}_n$ is a sigma field induced by $\{X_1, X_2, \cdots, X_n\}$. Then the sigma field $\underline{Z}^{-1}(\mathbb{B}^\infty)$ induced by $\underline{Z}$ is given by

(a) $\sigma(\bigcap_{n\geq 1} \mathbb{A}_n)$
(b) $\sigma(\bigcup_{n\geq 1} \mathbb{A}_n)$
(c) $\bigcup_{n\geq 1} \mathbb{A}_n$
(d) $\bigcap_{n\geq 1} \mathbb{A}_n$

2.8.11 Suppose $\underline{Z} = (X_1, X_2)'$ is defined on the measurable space $(\Omega, \mathbb{A})$ to $(\mathbb{R}^2, \mathbb{B}^2)$. Following are two statements: (I) $\underline{Z}$ is a random vector measurable with respect to $\mathbb{A}$ implies X_1 and X_2 are random variables measurable with respect to $\mathbb{A}$. (II) X_1 and X_2 are random variables measurable with respect to $\mathbb{A}$, but $\underline{Z} = (X_1, X_2)'$ is not measurable with respect to $\mathbb{A}$. Then which of the following is true?

(a) Both (I) and (II) are true
(b) (I) is true but (II) is false
(c) (II) is true but (I) is false
(d) Both (I) and (II) are false

2.8.12 Suppose $(\sin X_1, e^{X_2})'$ is a random vector measurable with respect to $\mathbb{A}$. Then which of the following is/are NOT always true?

(a) $\sin X_1$ measurable with respect to $\mathbb{A}$
(b) $\cos^2 X_1$ measurable with respect to $\mathbb{A}$
(c) $\cos X_1$ measurable with respect to $\mathbb{A}$
(d) $\cos^2 X_1 e^{7X_2+5}$ measurable with respect to $\mathbb{A}$

2.8.13 Following are two statements: (I) Every continuous function on $\mathbb{R}$ is a Borel function on $(\mathbb{R}, \mathbb{B})$. (II) Every Borel function on $(\mathbb{R}, \mathbb{B})$ is a continuous function on $\mathbb{R}$. Then which of the following is true?

(a) Both (I) and (II) are true
(b) (I) is true but (II) is false
(c) (II) is true but (I) is false
(d) Both (I) and (II) are false

2.8.14 Suppose $\{X_n, n \geq 1\}$ is a sequence of random variables on a measurable space $(\Omega, \mathbb{A})$. Then which of the following is/are NOT true?

(a) $Y_1 = \max\{X_1, X_2\}$ is a $\mathbb{A}$ measurable random variable
(b) $Z_1 = \sup_{n\geq 1} X_n$ is a $\mathbb{A}$ measurable random variable
(c) $U_1 = \limsup X_n$ is a $\mathbb{A}$ measurable random variable
(d) Y_1, Z_1, U_1 are not measurable with respect to $\mathbb{A}$.

2.8.15 Suppose $\underline{Z} = (X_1, X_2, \cdots,)$ is defined on the measurable space $(\Omega, \mathbb{A})$ to $(\mathbb{R}^\infty, \mathbb{B}^\infty)$. Following are two statements.
(I) $\underline{Z}$ is measurable with respect to $\mathbb{A}$ implies that for every $n \geq 1$, X_n is measurable with respect to $\mathbb{A}$.
(II) For every $n \geq 1$, X_n is measurable with respect to $\mathbb{A}$ does not imply that $\underline{Z}$ is measurable with respect to $\mathbb{A}$.
Then which of the following is true?

(a) Both (I) and (II) are true
(b) (I) is true but (II) is false
(c) (II) is true but (I) is false
(d) Both (I) and (II) are false

2.8.16 Suppose $\{X_n, n \geq 1\}$ is a sequence of random variables on a measurable space $(\Omega, \mathbb{A})$. Then which of the following is/are true?

(a) $Y_1 = \max\{X_1, X_2\}$ is a $\mathbb{A}$ measurable random variable
(b) $Z_1 = \sup_{n\geq 1} X_n$ is a $\mathbb{A}$ measurable random variable

(c) $U_1 = \limsup X_n$ is a $\mathbb{A}$ measurable random variable
(d) $U_2 = \liminf X_n$ is a $\mathbb{A}$ measurable random variable

2.8.17 Suppose $\{A_n, n \geq 1\}$ is a sequence of sets in $\mathbb{A}$, then which of the following is/are true?

(a) $I_{\liminf A_n} = \limsup I_{A_n^c}$
(b) $I_{\liminf A_n} = \liminf I_{A_n}$
(c) $I_{\limsup A_n} = \liminf I_{A_n^c}$
(d) $I_{\limsup A_n} = \liminf I_{A_n}$

2.8.18 Which of the following is true?
Suppose $\{X_n, n \geq 1\}$ is a sequence of random variables and $\mathbb{C}_n = \sigma\{X_n, X_{n+1}, \cdots\}$. Then the tail sigma field of the sequence $\{X_n, n \geq 1\}$ is defined as

(a) $\mathbb{T} = \bigcup_{n\geq 1} \mathbb{C}_n$
(b) $\mathbb{T} = \bigcup_{n\geq k} \mathbb{C}_n$
(c) $\mathbb{T} = \bigcap_{n\geq k} \mathbb{C}_n$
(d) $\mathbb{T} = \bigcap_{n\geq 1} \mathbb{C}_n$

2.8.19 Suppose X is a non-negative random variable defined on a measurable space $(\Omega, \mathbb{A})$ to $(\mathbb{R}, \mathbb{B})$. Then which of the following is always true?

(a) There exists a sequence $\{X_n, n \geq 1\}$ of non-negative simple random variables on a measurable space $(\Omega, \mathbb{A})$ such that $X_n \to X$ on Ω
(b) There exists a non-decreasing sequence $\{X_n, n \geq 1\}$ of elementary random variables on a measurable space $(\Omega, \mathbb{A})$ such that $X_n \to X$ on Ω
(c) There exists a non-decreasing sequence $\{X_n, n \geq 1\}$ of non-negative simple random variables on a measurable space $(\Omega, \mathbb{A})$ such that $X_n \to X$ on Ω
(d) There exists a non-increasing sequence $\{X_n, n \geq 1\}$ of non-negative random variables on a measurable space $(\Omega, \mathbb{A})$ such that $X_n \to X$ on Ω

2.8.20 Following are two statements.
(I) A class of random variables is closed under arithmetic operations.
(II) A class of simple random variables is closed under arithmetic operations.
Then which of the following is true?

(a) Both (I) and (II) are true
(b) (I) is true but (II) is false
(c) (II) is true but (I) is false
(d) Both (I) and (II) are false

2.8.21 Which of the following is/are Borel function(s), if $f(x) = \sqrt{x}$?

(a) $f : \mathbb{R} \to \mathbb{R}$

(b) $f : \mathbb{R}^+ \to \mathbb{R}$
(c) $f : \mathbb{R}^+ \to \mathbb{R}^+$
(d) $f : \mathbb{R} \to \mathbb{R}^+$

2.8.22 What is the value of $\lim_{n\to\infty} \exp(-\lambda)\lambda^n/n!$ for $\lambda > 0$ and $n \in \mathbb{N}$?

(a) 0
(b) 1
(c) 0.5
(d) λ

2.8.23 What is the value of $\lim_{n\to\infty}[\exp(-n) - \exp(-(n+1)]$?

(a) 0
(b) 1
(c) $exp(-1)$
(d) $-\infty$

2.8.24 For non-decreasing sequence $\{X_n, n \geq 1\}$, which of the following statements is/are true? (I) $\sup_{n\geq 1} X_n = \sup_{n\geq k} X_n, \ \forall\ k \in \mathbb{N}$. (II) $\inf_{n\geq 1} X_n = \inf_{n\geq k} X_n, \ \forall\ k \in \mathbb{N}$.

(a) Both (I) and (II) are true
(b) Only (I) is true
(c) Only (II) is true
(d) Both (I) and (II) are false

2.8.25 For non-increasing sequence $\{X_n, n \geq 1\}$, which of the following statements is/are true? (I) $\sup_{n\geq 1} X_n = \sup_{n\geq k} X_n, \ \forall\ k \in \mathbb{N}$. (II) $\inf_{n\geq 1} X_n = \inf_{n\geq k} X_n, \ \forall\ k \in \mathbb{N}$.

(a) Both (I) and (II) are true
(b) Only (I) is true
(c) Only (II) is true
(d) Both (I) and (II) are false

3

Distribution Function

3.1 Introduction

In Chapter 2 we introduced the probability distribution of a random variable X defined on the probability space $(\Omega, \mathbb{A}, P)$ to $(\mathbb{R}, \mathbb{B}, P_X)$. It is a collection $\{P_X(S), S \in \mathbb{B}\}$, where $P_X(S)$ is the induced probability measure, induced by the random variable X. Thus, to specify the probability distribution of X, we require the knowledge of $P_X(S)$ for all $S \in \mathbb{B}$. We have a simpler way to specify the probability distribution of X, as we have a simpler way to verify that a function on $(\Omega, \mathbb{A})$ is a random variable via the economical definition. Once again such a simpler approach is based on the fact that a collection of intervals of the type $(-\infty, x]$, $x \in \mathbb{R}$, generates the Borel field and it is enough to specify $P_X(S)$ for $S = (-\infty, x]$, $x \in \mathbb{R}$. It is manifested by introducing the concept of a distribution function in the present chapter.

By definition, the induced probability measure $P_X(S)$ is a set function defined on $(\mathbb{R}, \mathbb{B})$. Using a distribution function, we can specify the probability distribution of a random variable X by a point function defined on $\mathbb{R}$. It is usually easier to work with a point function than a set function and hence in practice the distribution of a random variable is always specified in terms of its distribution function. We define below a distribution function of a random variable and study its properties.

Definition 3.1.1. *Distribution function: Suppose X is a random variable defined on the probability space $(\Omega, \mathbb{A}, P)$. Then the distribution function $F_X(.)$ of X is defined on $\mathbb{R}$ as*

$$F_X(x) = P\left(X^{-1}((-\infty, x])\right) = P\{\omega \mid X(\omega) \le x\} = P[X \le x], \quad x \in \mathbb{R}.$$

Suppose $S = (-\infty, x]$, $x \in \mathbb{R}$, then $P_X(S) = P\left(X^{-1}((-\infty, x]\right) = F_X(x)$. Thus, probability distribution of X completely specifies the distribution function of X. The converse is also true and can be proved using the Caratheodory extension theorem. For proof, see Ash [2] or Athreya and Lahiri [3]. Thus, there is a one to one correspondence between the probability distribution of X and the distribution function of X as stated in the following theorem.

Theorem 3.1.1. *Correspondence theorem: Every probability measure P^* on $(\mathbb{R}, \mathbb{B})$ uniquely determines a distribution function through the correspondence*

DOI: 10.1201/9781032619057-3

$F(x) = P^*((-\infty, x])$, $x \in \mathbb{R}$ with $F(-\infty) = 0$ and $F(\infty) = 1$. Conversely, if $F(x)$ is specified for each $x \in \mathbb{R}$, then the probability measure P^* on $(\mathbb{R}, \mathbb{B})$ is determined uniquely through the relation $P^*((-\infty, x]) = F(x)$.

In view of the correspondence theorem, the terms probability distribution and distribution function of a random variable are used interchangeably. In the next section, we discuss some properties of a distribution function.

3.2 Properties of a Distribution Function

Four important properties of a distribution function are proved in the following theorem. The continuity theorem for probability measure is heavily used in the proof. We also use one property of a dense set, defined below.

Definition 3.2.1. *A set A is said to be dense in $\mathbb{R}$, if every point of $\mathbb{R}$ is either in A or is a limit point of A.*

For example, the set of rational numbers as well as the set of irrational numbers is dense in $\mathbb{R}$.

Theorem 3.2.1. *Suppose $F_X(x)$ is a distribution function of a random variable X defined on the probability space $(\Omega, \mathbb{A}, P)$. Then*

(i) $0 \le F_X(x) \le 1$.

(ii) *$F_X(x)$ is a non-decreasing function on $\mathbb{R}$.*

(iii) *$F_X(x)$ is a right continuous function on $\mathbb{R}$.*

(iv) $\lim_{x\to\infty} F_X(x) = 1$ *and* $\lim_{x\to-\infty} F_X(x) = 0$.

Proof.

(i) By definition, $F_X(x) = P[X \le x]$, $x \in \mathbb{R}$. Hence, $0 \le F_X(x) \le 1$.

(ii) Suppose x_1 and x_2 in $\mathbb{R}$ are such that $x_1 \le x_2$.

$$\begin{aligned} x_1 \le x_2 \Rightarrow (-\infty, x_1] \subseteq (-\infty, x_2] &\Rightarrow X^{-1}((-\infty, x_1]) \subseteq X^{-1}((-\infty, x_2]) \\ &\Rightarrow P(X^{-1}((-\infty, x_1])) \le P(X^{-1}((-\infty, x_2])) \Rightarrow F_X(x_1) \le F_X(x_2). \end{aligned}$$

Thus, $F_X(x)$ is a non-decreasing function on $\mathbb{R}$.

(iii) $F_X(x)$ is a right continuous function on $\mathbb{R}$ if $\lim\limits_{h\to 0} F_X(x+h) = F_X(x)$. To establish this assertion, suppose an event $A_n, n \ge 1$ is defined as $A_n = [X \le x+h_n]$, where $\{h_n, n \ge 1\}$ is a monotone decreasing sequence of positive numbers tending to 0 as $n \to \infty$. Following arguments assure that such a sequence exists. Observe that 0 is a limit point of $[0, \infty)$, since every neighborhood of 0 contains infinitely many points from $[0, \infty)$. Hence by the properties of limit points, there $\exists$ at least one sequence $\{g_n, n \geq 1\}$ from $[0, \infty)$, which converges to 0. Further, every sequence of real numbers has a monotone subsequence (page 55 of Kumar and Kumaresan [15] and if the original sequence is convergent then this subsequence is also converges to the same limit as that of the original sequence. Thus, $\exists$ a monotone subsequence $\{h_n, n \geq 1\}$ of $\{g_n, n \geq 1\}$ such that $h_n \to 0$. Since $h_n \geq 0 \ \forall \ n \geq 1$, it must be a monotone decreasing sequence. Note that $\forall \ n \geq 1$

$$\begin{aligned}
A_n \supset A_{n+1} \ &\Rightarrow \ \{A_n, n \geq 1\} \ \text{ is a non-increasing sequence of events} \\
&\Rightarrow \lim_{n\to\infty} A_n = \bigcap_{n\geq 1} A_n = [X \leq x] = A, \ \ \text{say} \\
&\Rightarrow \lim_{n\to\infty} P(A_n) = P(A) \\
&\quad \text{by continuity theorem for probability measure} \\
&\Rightarrow \lim_{n\to\infty} P[X \leq x + h_n] = P[X \leq x] = F_X(x) \\
&\Rightarrow \lim_{h\to 0} F_X(x+h) = F_X(x) \ \ \Rightarrow \ F_X(x) \text{ is right continuous on } \mathbb{R}.
\end{aligned}$$

(iv) Suppose $\{a_n, n \geq 1\}$ is a non-decreasing sequence of numbers such that $a_n \to \infty$ as $n \to \infty$, for example $a_n = n$. The existence of such a sequence follows using similar arguments as in (iii). Suppose the event A_n is defined as $A_n = [X \leq a_n]$. Observe that $\forall \ \omega \in \Omega$,

$$\begin{aligned}
X(\omega) \leq a_n \ \ &\Rightarrow \ \ X(\omega) \leq a_{n+1} \ \Rightarrow \ A_n \subset A_{n+1} \ \ \forall \ \ n \geq 1 \\
&\Rightarrow \ \ \{A_n, n \geq 1\} \ \text{ is a non-decreasing sequence of events} \\
&\Rightarrow \ \ \lim_{n\to\infty} A_n = \bigcup_{n\geq 1} A_n = \Omega, \ \text{ since } a_n \to \infty \text{ as } n \to \infty.
\end{aligned}$$

Again by the continuity theorem for probability measures, if $A_n \uparrow \Omega$ then $P(A_n) \uparrow P(\Omega) = 1$. Thus,

$$1 = P(\Omega) = P(\lim_{n\to\infty} A_n) = \lim_{n\to\infty} P(A_n) = \lim_{n\to\infty} F_X(a_n) \Rightarrow \lim_{x\to\infty} F_X(x) = 1.$$

To prove $\lim_{x\to-\infty} F_X(x) = 0$, an event B_n is defined as $B_n = [X \leq -a_n]$. Then $\{B_n, n \geq 1\}$ is a non-increasing sequence of events with limit $\bigcap_{n\geq 1} B_n = \emptyset$. Again by the continuity theorem for probability measures, if $B_n \downarrow \emptyset$ then $P(B_n) \downarrow P(\emptyset) = 0$. Thus,

$$\begin{aligned}
0 = P(\emptyset) = P(\lim_{n\to\infty} B_n) &= \lim_{n\to\infty} P(B_n) = \lim_{n\to\infty} F_X(-a_n) \\
\Rightarrow \ \lim_{x\to-\infty} F_X(x) &= 0.
\end{aligned}$$

□

Converse of Theorem 3.2.1 is true and is proved in the following theorem.

Theorem 3.2.2. *Suppose a function $g : \mathbb{R} \to \mathbb{R}$ satisfies the following four properties.*

(i) $0 \leq g(x) \leq 1$.

(ii) $g(x)$ *is a non-decreasing function on* $\mathbb{R}$.

(iii) $g(x)$ *is a right continuous function on* $\mathbb{R}$.

(iv) $\lim_{x\to\infty} g(x) = 1$ *and* $\lim_{x\to-\infty} g(x) = 0$. *Then there exists a probability space on which one can define a random variable, whose distribution function is the given function* g.

Proof. Suppose a function X is defined on a measurable space $(\mathbb{R}, \mathbb{B})$ as $X(y) = y \;\; \forall \;\; y \in \mathbb{R}$, that is X is an identity function. Note that

$$\forall \;\; S \in \mathbb{B}, \;\; X^{-1}(S) = \{y | X(y) \in S\} = \{y | y \in S\} = S \in \mathbb{B}.$$

Thus, X is measurable with respect to $\mathbb{B}$. Suppose $\mathbb{C}$ denotes a collection of all intervals of the form $(a, b], \; a < b \in \mathbb{R}$. Then the minimal sigma field $\sigma(\mathbb{C}) = \mathbb{B}$. We now define an increment function $F((a, b])$ as $F((a, b]) = g(b) - g(a), \; a \leq b \in \mathbb{R}$. Since g is non-decreasing, it follows that $F((a, b]) \geq 0$. Using the increment function, we define a measure P on $\mathbb{F}(\mathbb{C})$, a field generated by $\mathbb{C}$, such that

$$P((-\infty, a]) = F((-\infty, a]) = g(a) - g(-\infty) = g(a) \;\; \forall \;\; a \in \mathbb{R} \text{ as } \lim_{x\to-\infty} g(x) = 0.$$

By the Caratheodory extension theorem, P is further extended to the minimal sigma field $\sigma(\mathbb{C}) = \mathbb{B}$, the Borel field. Note that $\lim_{x\to\infty} g(x) = 1$ implies that $P(\mathbb{R}) = 1$ and the right continuity of g assures that P is a continuous set function. Thus P is a probability measure on $(\mathbb{R}, \mathbb{B})$ and hence there exists a probability space $(\mathbb{R}, \mathbb{B}, P)$. Note that X is a measurable function $(\mathbb{R}, \mathbb{B}, P)$ such that

$$P[X \leq x] = P((-\infty, x]) = g(x) \;\; \forall \;\; x \in \mathbb{R}.$$

Thus, if a function $g : \mathbb{R} \to \mathbb{R}$ satisfies the four properties of a distribution function, then there exists a probability space on which one can define a random variable whose distribution function is the given function g. □

In general, many random variables can have the same distribution function. In such a case we say that the random variables are identically distributed.

Definition 3.2.2. *Identically distributed random variables: Random variables* X *and* Y *are said to be identically distributed if* $F_X(x) = F_Y(x) \;\forall\; x \in \mathbb{R}$.

If two random variables X and Y are identical in the sense that $X(\omega) = Y(\omega) \;\; \forall \;\; \omega \in \Omega$, then X and Y have the same distribution function and the two are identically distributed. However, the converse is not true. Two identically distributed random variables need not be identical as is clear from the following example.

Example 3.2.1. Suppose a probability distribution of a random variable X is given by $P[X = 0] = P[X = 1] = 1/2$. Hence its distribution function is,

$$F(x) = \begin{cases} 0, & \text{if} \quad x \leq 0 \\ 1/2, & \text{if} \quad 0 \leq x < 1 \\ 1, & \text{if} \quad x \geq 1. \end{cases}$$

If a random variable Y is defined as $Y = 1 - X$, then the probability distribution of Y is given by, $P[Y = 0] = P[Y = 1] = 1/2$ and its distribution function is the same as that of X. Hence, X and Y are identically distributed random variables taking values 0 and 1 each with probability $1/2$ but observe that if $X(\omega) = 0$ then $Y(\omega) = 1$ and if $X(\omega) = 1$ then $Y(\omega) = 0$. Thus, $X(\omega) \neq Y(\omega) \ \forall \ \omega \in \Omega$ which implies that X and Y are not identical random variables. □

As proved in Theorem 3.2.1, the distribution function is always right continuous. To study the left hand limit $\lim_{h \to 0} F_X(x - h)$ at x, as in the setup of right continuity, suppose $\{h_n, n \geq 1\}$ is a non-increasing sequence of positive numbers tending to 0 as $n \to \infty$. Suppose an event A_n is defined as $A_n = [X \leq x - h_n], n \geq 1$. Observe that $A_n \subset A_{n+1}, \ \forall \ n \geq 1$ and hence $\{A_n, n \geq 1\}$ is a non-decreasing sequence of events. Further, the sequence $\{(-\infty, x - h_n], n \geq 1\} \to (-\infty, x)$. Hence,

$$\begin{aligned} \lim_{n \to \infty} A_n &= \bigcup_{n \geq 1} A_n = [X < x] \\ \Rightarrow \lim_{n \to \infty} P(A_n) &= P[X < x] \Rightarrow \lim_{h_n \to 0} F_X(x - h_n) = P[X < x] \\ \Rightarrow \lim_{h \to 0} F_X(x - h) &= P[X < x] = F_X(x-), \quad \text{say.} \end{aligned}$$

The second step follows by the continuity theorem of probability measure. Hence, F is continuous from left at x, if $F_X(x-) = F_X(x)$. It may or may not be left continuous at all points in the domain $\mathbb{R}$. We denote the left hand limit $\lim\limits_{h \to 0} F_X(x - h)$ by either $F_X(x-)$ or $F_X(x - 0)$.

Definition 3.2.3. *Point of discontinuity of a distribution Function: If F is not continuous from left at $x \in \mathbb{R}$, that is, if $\lim_{h \to 0} F_X(x - h) = F_X(x-) \neq F_X(x)$, then x is said to be a point of discontinuity of a distribution function F.*

Definition 3.2.4. *Size of discontinuity: Suppose x is a point of discontinuity of F. Then, $p_x = F_X(x) - F_X(x-)$ is known as the size of discontinuity of F at x or the magnitude of the jump at x.*

It is to be noted that in view of the non-decreasing nature of the distribution function, $F_X(x) > F_X(x-)$ and hence $p_x > 0$ for a point x of discontinuity of F. Further,

$$p_x = F_X(x) - F_X(x-) = P[X = x] = P\{\omega \mid X(\omega) = x\}.$$

If x is not a point of discontinuity of F, then $p_x = F_X(x) - F_X(x-) = 0$, that is, $P[X = x] = 0$. Following theorem proves an important result related to the set of points of discontinuity of F.

Theorem 3.2.3. *Suppose F is a distribution function of a random variable X defined on a probability space $(\Omega, \mathbb{A}, P)$. Then the set D of points of discontinuity of F is at most countable set.*

Proof. Suppose x is a point of discontinuity of F, with $p_x > 0$ as the size of the discontinuity. Suppose D_n is a set of points x of discontinuity of F, such that the size of discontinuity $p_x \geq 1/n$. Thus, $D_n = \{x \in \mathbb{R} | p_x \geq 1/n\}$. Now,

$$\begin{aligned} x \in D &\iff p_x > 0 \iff p_x \geq 1/n, \quad \text{for at least one } n \geq 1 \\ &\iff x \in D_n, \quad \text{for at least one } n \geq 1 \\ &\iff x \in \bigcup_{n \geq 1} D_n. \end{aligned}$$

Thus, $D = \bigcup_{n \geq 1} D_n$. Suppose D_n contains more than n points. In particular, suppose $x_1, x_2, \cdots, x_{n+k}$, $k \geq 1$ are $n + k$ distinct points in D_n. Hence, the singleton sets $\{x_i\}, i = 1, 2, \cdots, n + k$ are disjoint Borel sets, which further implies that the inverse images

$$X^{-1}(\{x_i\}) = \{\omega \mid X(\omega) = x_i\} = [X = x_i], \quad i = 1, 2, \cdots, n + k$$

are also disjoint subsets of $\mathbb{A}$. Further, size of discontinuity $p_{x_i} \geq 1/n$ at x_i, $i = 1, 2, \cdots, n + k$. Hence, by finite additivity of the probability measure P,

$$P\Big[\sum_{i=1}^{n+k} [X = x_i]\Big] = \sum_{i=1}^{n+k} P[X = x_i] = \sum_{i=1}^{n+k} p_{x_i} \geq \frac{1}{n}(n + k) = 1 + \frac{k}{n} > 1,$$

which is a contradiction. Thus the assumption that D_n contains more than n points is not correct and we conclude that the number of points in D_n is $\leq n$, that is, the set D_n is finite for all $n \geq 1$. Now D is a countable union of finite sets D_n and hence D is at most countable. □

Suppose C denotes a set of points of continuities of a distribution function F, then $C = D^c$. Since D is at most countable, it follows that C is either $\mathbb{R}$ or an uncountable subset of $\mathbb{R}$. In the following lemma, we prove that C is dense in $\mathbb{R}$.

Lemma 3.2.1. *The set C of points of continuity of F is dense in $\mathbb{R}$.*

Proof. We prove the result by contradiction. Suppose $C = D^c$ is not dense in $\mathbb{R}$. Hence $\exists \; x \in \mathbb{R}$ such that $x \notin D^c$ and x is not a limit point of D^c. It implies that $\exists \; \epsilon > 0$ such that $N_\epsilon(x) = (x - \epsilon, x + \epsilon)$ does not contain any point from D^c, which further implies that $N_\epsilon(x) \subseteq D$. It is a contradiction, since $N_\epsilon(x)$ is an uncountable set while D is at most countable. □

Using this property and the right continuity of F, it is proved in the following theorem that F is completely determined by its values at all the points of continuity.

Theorem 3.2.4. *Suppose F and G are two distribution functions with $C(F)$ and $C(G)$ as the sets of points of continuity respectively. Then*

(i) $C(F) \cap C(G) \neq \emptyset$.

(ii) $C(F) \cap C(G)$ *is dense in* $\mathbb{R}$.

(iii) If $F(x) = G(x) \;\; \forall \;\; x \in C(F) \cap C(G)$, *then* $F(x) = G(x) \;\; \forall \;\; x \in \mathbb{R}$.

Proof.

(i) To prove that $C(F)\cap C(G) \neq \emptyset$, we assume the contrary. Suppose $C(F)\cap C(G) = \emptyset$. Then,

$$(C(F) \cap C(G))^c = D(F) \cup D(G) = \emptyset^c = \mathbb{R},$$

where $D(F)$ and $D(G)$ are sets of points of discontinuity of F and G respectively. In Theorem 3.2.3, it is proved that the set of points of discontinuity of a distribution function is at most countable. Further the union of two at most countable sets is at most countable. Hence, $(D(F) \cup D(G))$ is at most countable and it cannot be equal to $\mathbb{R}$. Thus, our assumption that $C(F) \cap C(G) = \emptyset$ is wrong and $C(F) \cap C(F) \neq \emptyset$.

(ii) As in Lemma 3.2.1, we assume the contrary that $C(F) \cap C(G)$ is not dense in $\mathbb{R}$. Hence $\exists \;\; x \in \mathbb{R}$ such that $x \notin C(F) \cap C(G)$ and x is not a limit point of $C(F) \cap C(G)$. It implies that $\exists \;\; \epsilon > 0$ such that $N_\epsilon(x) = (x - \epsilon, x + \epsilon)$ does not contain any point from $C(F) \cap C(G)$, which further implies that $N_\epsilon(x) \subseteq (C(F) \cap C(G))^c = D(F) \cup D(G)$. It is a cotradiction, since $N_\epsilon(x)$ is an uncountable set while $D(F) \cup D(G)$ is atmost countable.

(iii) Suppose $x \in (C(F) \cap C(G))^c$. It is proved in (ii) that $C(F) \cap C(G)$ is dense in $\mathbb{R}$. As a consequence, for any real number x, there exists a sequence $\{x + r_n, n \geq 1\}$ of points in $C(F)\cap C(G)$ such that $x + r_n \to x$, as $r_n \downarrow 0$. It is given that $F(x) = G(x) \;\; \forall \;\; x \in C(F) \cap C(G)$. Hence, $F(x + r_n) = G(x + r_n)$. Note that being distribution functions, F and G are right continuous. Hence, for $x \in (C(F) \cap C(G))^c$,

$$\begin{aligned} F(x) &= \lim_{h \downarrow 0} F(x + h) = \lim_{r_n \downarrow 0} F(x + r_n) = \lim_{r_n \downarrow 0} G(x + r_n) \\ &= \lim_{h \downarrow 0} G(x + h) = G(x) \;\; \Rightarrow \;\; F(x) = G(x) \;\; \forall \;\; x \in \mathbb{R}. \end{aligned}$$

□

We use Theorem 3.2.4 in Section 7.3, to prove that limit random variables in convergence in distribution are identically distributed. In fact, if two distribution functions agree on a dense subset of $\mathbb{R}$, then these agree on $\mathbb{R}$, the result is stated below. For proof refer to Gut [13].

Theorem 3.2.5. *Suppose F and G are two distribution functions. If $F = G$ on a dense subset of $\mathbb{R}$, then $F = G$ on $\mathbb{R}$.*

The next example illustrates how to identify the points of discontinuity and their respective sizes.

Example 3.2.2. Suppose $\Omega = \{1, 2, 3\}$, $\mathbb{A} = \mathbb{P}(\Omega)$ and the probability measure P is specified by $P(\{1\}) = 1/4, P(\{2\}) = 1/2, P(\{3\}) = 1/4$. Suppose a random variable X on $(\Omega, \mathbb{A})$ is defined as $X(i) = i$. Then the distribution function $F_X(x) = P\{\omega \mid X(\omega) \leq x\} = P[X \leq x]$, for $x \in \mathbb{R}$ is given by,

$$F_X(x) \quad = \quad \begin{cases} 0, & \text{if} \quad x < 1, \\ 1/4, & \text{if} \quad 1 \leq x < 2, \\ 3/4, & \text{if} \quad 2 \leq x < 3, \\ 1, & \text{if} \quad x \geq 3. \end{cases}$$

To find the points of discontinuities of F, observe that $\forall\ x < 1,\ F_X(x) = 0$, thus it is a constant function and hence a continuous function on $(-\infty, 1)$, hence all points in this interval are points of continuity. At $x = 1$,

$$F_X(1) = 1/4\ \neq 0 = F_X(1-) = \lim_{h \to 0} F_X(1-h)\ \&\ p_1 = F_X(1) - F_X(1-) = 1/4.$$

Hence $x = 1$ is a point of discontinuity with size of discontinuity $1/4$. Further, $\forall\ x \in (1, 2),\ F_X(x) = 1/4$ is a constant function and hence a continuous function on $(1, 2)$, hence all points in this interval are points of continuity. At $x = 2$,

$$F_X(2) = 3/4\ \neq 1/4 = F_X(2-) = \lim_{h \to 0} F_X(2-h)\ \&\ p_2 = F_X(2) - F_X(2-) = 1/2.$$

Hence $x = 2$ is a point of discontinuity with size of discontinuity $1/2$. Note that, $\forall\ x \in (2, 3),\ F_X(x) = 3/4$, thus it is again a constant function and hence a continuous function on $(2, 3)$, hence all points in this interval are points of continuity. At $x = 3$,

$$F_X(3) = 1\ \ \neq 3/4 = F_X(3-) = \lim_{h \to 0} F_X(3-h)\ \&\ p_3 = F_X(3) - F_X(3-) = 1/4.$$

Hence $x = 3$ is a point of discontinuity with size of discontinuity $1/4$. Now, $\forall\ x \geq 3,\ F_X(x) = 1$ is a constant function and hence a continuous function on $[3, \infty)$, hence all points in $(3, \infty)$ are points of continuity. Thus, the set D of points of discontinuity of $F_X(x)$ is $\{1, 2, 3\}$, with size of jump at point of discontinuity being $p_1 = 1/4,\ \ p_2 = 1/2$ and $p_3 = 1/4$. It is to be noted that sum of sizes of jump at points of discontinuity is 1. □

Having defined a distribution function, the next natural question is to determine how many different types of distribution functions exist. To define these types, we now introduce a concept of a point of increase of a distribution function. It is useful to define a discrete and a continuous type distribution function.

Definition 3.2.5. *Point of increase of a distribution function: A point $x \in \mathbb{R}$ is said to be a point of increase of a distribution function F if*

$$\forall \quad \epsilon > 0, \qquad F_X(x+\epsilon) - F_X(x-\epsilon) > 0.$$

Example 3.2.3. We find the points of increase for the distribution function derived in Example 3.2.2. Note that

$$\forall \quad x < 1 \ , \ \forall \quad \epsilon \in (0, 1-x) \quad F_X(x+\epsilon) - F_X(x-\epsilon) = 0.$$

Hence, no point in this interval is a point of increase. At

$$x = 1, \ \forall \quad \epsilon > 0, \quad F_X(x+\epsilon) - F_X(x-\epsilon) = 1/4 \ \text{ or } \ 3/4 \ \text{ or } \ 1,$$

depending on the values of ϵ, thus, $x = 1$ is a point of increase. For $x \in (1, 2)$, if
$\epsilon = (1/2)\min\{2-x, x-1\}$, then for all such ϵ, $F_X(x+\epsilon) - F_X(x-\epsilon) = 0$ but for other ϵ, $F_X(x+\epsilon) - F_X(x-\epsilon) > 0$ and hence no $x \in (1, 2)$ is a point of increase. At

$$x = 2, \ \forall \quad \epsilon > 0, \quad F_X(x+\epsilon) - F_X(x-\epsilon) > 0,$$

thus, $x = 2$ is a point of increase. For $x \in (2, 3)$, using similar arguments as for the case $x \in (1, 2)$, no point is a point of increase. For

$$x = 3, \ \forall \quad \epsilon > 0, \quad F_X(x+\epsilon) - F_X(x-\epsilon) > 0,$$

thus, $x = 3$ is a point of increase. For all $x \geq 3, F_X(x+\epsilon) - F_X(x-\epsilon) = 0$. Hence, no point in this interval is a point of increase. Thus, a set E of points of increase of $F_X(\cdot)$ is $E = \{1, 2, 3\}$. □

Remark 3.2.1. From Example 3.2.2 and Example 3.2.3, we note that the set of points of discontinuities of F is the same as the set of points of increase of F. From the definitions of point of discontinuity and of point of increase, it is clear that a point of discontinuity of F is always a point of increase of F. But the converse is not true. Suppose a distribution function F is continuous and strictly increasing on $\mathbb{R}$, thus the set of points of discontinuities of F is the empty set, but

$$\forall \quad x \in \mathbb{R} \ \text{ and } \quad \forall \quad \epsilon > 0, \quad F_X(x+\epsilon) - F_X(x-\epsilon) > 0$$

implying that all $x \in \mathbb{R}$ are points of increase. This observation leads to the following definitions of a discrete and a continuous random variable.

Definition 3.2.6. *Discrete random variable or discrete distribution function: A random variable X or its distribution function F is said to be discrete if all points of increase of F are its points of discontinuities. Alternatively, X or its distribution function F is said to be discrete, if for some countable set of real numbers $\{x_j\}$ and $\{p_{x_j} \in (0,1)\}$, $F_X(x) = \sum_{x_j \leq x} p_{x_j}, \quad \forall \; x \in \mathbb{R}$.*

Definition 3.2.7. *Continuous and absolutely continuous random variable or continuous and absolutely continuous distribution function: A random variable X or its distribution function F is said to be continuous if the set of points of discontinuity of F is an empty set, that is, if the distribution function F is continuous on $\mathbb{R}$. A continuous random variable X or its distribution function F is said to be absolutely continuous, if there exists a non-negative function f such that $F_X(b) - F_X(a) = \int_a^b f(x)dx$, for all $a < b$.*

Definition 3.2.8. *Singular distribution function: A distribution function F is said to be singular if $F \neq 0$, but $\frac{d}{dx}F_X(x)$ exists and is 0 almost everywhere.*

Definition 3.2.9. *Support of a distribution function: A set of values of $x \in \mathbb{R}$ for which $F(x)$ increases is known as a support of a distribution function.*

If F is a discrete distribution function, then its support is the set D of points of discontinuity of F, which is the same as the set of points of increase of F. If F is a continuous distribution function, then its support is the set of points of increase of F.

Definition 3.2.10. *Probability mass function of a discrete random variable: Suppose D is a set of points of discontinuity of F and $p_x, x \in D$ is the size of discontinuity at x. Then the set $\{(x, p_x) \mid x \in D\}$ is a probability mass function of a discrete random variable X.*

Definition 3.2.11. *Probability density function of an absolutely continuous random variable: Suppose a random variable X is absolutely continuous with distribution function F. Hence, there exists a non-negative function f such that $F_X(b) - F_X(a) = \int_a^b f(x)dx$, for all $a < b$. The function f is known as a probability density function of an absolutely continuous random variable X. If F is differentiable on $\mathbb{R}$, then $f(x) = \frac{d}{dx}F_X(x)$.*

Remark 3.2.2. Probability mass function of a discrete random variable X completely specifies the probability distribution of X, because when the set $\{(x, p_x) \mid x \in D\}$ is known, then by the finite additivity or sigma additivity of the probability measure P, $P[X \in S]$ is known for every Borel set S and $P[X \in S] = \sum_{x \in S} p_x$. Similarly, probability density function of an absolutely continuous random variable X completely specifies the probability distribution of X as $P[X \in S]$ is given by $P[X \in S] = \int_{x \in S} f(x)\, dx$. In particular, $F_X(x) = P[X \leq x] = \int_{-\infty}^{x} f(x)dx$. Thus, there is one to one correspondence between the probability density function and the distribution function. Note that the probability density function of an absolutely continuous random variable or the probability mass function of a discrete random variable are completely determined by the probability measure P of the probability space $(\Omega, \mathbb{A}, P)$.

A distribution function can be neither discrete nor continuous. It may be a mixture of discrete and continuous distribution functions. In the next section, we discuss how such a distribution function can be decomposed.

3.3 Decomposition of a Distribution Function

Suppose X denotes the waiting time at a traffic signal. It is 0, if the signal light is green upon arrival. If the signal light is red, then waiting time is a continuous random variable. We have a similar scenario, if X denotes the waiting time at a bill counter at the mall. It is 0, if there is no customer at the counter, otherwise it is a continuous random variable. In such cases the distribution function is a mixture of discrete and continuous distribution functions. Every distribution function can always be decomposed in terms of discrete and continuous distribution functions. The following theorem, known as the Jordan decomposition theorem, conveys this concept more precisely.

Theorem 3.3.1. *Jordan decomposition theorem: Every distribution function F can be uniquely expressed as,*

$$F = l_1F_1 + l_2F_2, \quad \textit{where} \quad l_1, l_2 \geq 0 \quad \textit{and} \;\; l_1 + l_2 = 1,$$

where F_1 is a discrete distribution function and F_2 is a continuous distribution function.

Proof. It is proved in Theorem 3.2.3 that the set of points of discontinuity of F is at most countable. Suppose $\{x_n, n \geq 1\}$ is a sequence of points of discontinuity of F. Thus, the sequence $\{x_n, n \geq 1\}$ consists of finite or countably infinite number of points or no points at all. The size of discontinuity defined as $p_{x_n} = F(x_n) - F(x_n-) > 0 \; \forall \; n \geq 1$. Suppose a function G on $\mathbb{R}$ is defined as

$$G(x) = \sum_{x_n \leq x} p_{x_n}, \quad x \in \mathbb{R}.$$

By the definition of G it is clear that it satisfies the following properties. (i) $0 \leq G(x) \leq 1$ for all $x \in \mathbb{R}$, (ii) G is non-decreasing and right continuous on $\mathbb{R}$ and (iii) $G(-\infty) = 0, G(\infty) \leq 1$. Another function $H(x)$ based on F and G is defined as $H(x) = F(x) - G(x)$, $x \in \mathbb{R}$ and it satisfies the following properties. (i) $G(x) \leq F(x) \quad \Rightarrow \quad H(x) \geq 0$. (ii) Since $F(x) \leq 1$ and $G(x)$

may be 0 for some or all x, $H(x) \leq 1$. (iii) $H(-\infty) = 0$. (iv) Observe that for

$$\begin{aligned} x < y, \quad H(y) - H(x) &= F(y) - F(x) - \sum_{x_n \leq y} p_{x_n} + \sum_{x_n \leq x} p_{x_n} \\ &= P[x < X \leq y] - \sum_{x < x_n \leq y} p_{x_n} \geq 0 \end{aligned}$$

and hence H is non-decreasing on $\mathbb{R}$. (v) Since F and G are right continuous, H is also right continuous on $\mathbb{R}$. To examine the left continuity, observe that if x is a point of continuity of F, then for $h > 0$

$$H(x) - H(x-h) = F(x) - F(x-h) - \sum_{x-h<x_n\leq x} p_{x_n} \quad \to \quad 0 \text{ as } h \to 0.$$

Further, if x is a point of discontinuity of F, then for $h > 0$

$$H(x)-H(x-h) = F(x)-F(x-h)- \sum_{x-h<x_n\leq x} p_{x_n} \quad \to \quad p_x - p_x = 0 \text{ as } h \to 0.$$

It thus follows that H is continuous on $\mathbb{R}$. Suppose $l_1 = G(\infty) < 1$. Since for all $x \in \mathbb{R}$, $F(x) = H(x) + G(x)$, $H(\infty) = 1 - G(\infty) = l_2$ say. Suppose F_1 and F_2 are defined as

$$F_1 = (1/l_1)G \;\;\&\;\; F_2 = (1/l_2)H \quad \Rightarrow \quad F_1(\infty) = 1 \;\;\&\;\; F_2(\infty) = 1.$$

From the properties of G and H and the fact that $F_1(\infty) = F_2(\infty) = 1$, it is clear that F_1 and F_2 are distribution functions. Thus,

$$F(x) = G(x) + H(x) = l_1 F_1(x) + l_2 F_2(x),$$

where F_1 is of discrete type while F_2 is continuous. Note that the decomposition remains valid even if either l_1 or l_2 is 0. To establish the uniqueness, suppose if possible, $F = G + H = G^* + H^*$ which implies $G - G^* = H - H^*$, however G and G^* are discrete distribution functions and $G - G^*$ is not a continuous function, but $H - H^*$ is a continuous function. Hence $G - G^* = H - H^*$ is possible if and only if each is 0, that is, if and only if $G = G^*$ and $H = H^*$. □

Remark 3.3.1. In the decomposition $F = l_1F_1 + l_2F_2$, if $l_1 = 0$, then F is a continuous distribution function and if $l_2 = 0$, then F is a discrete distribution function. If both l_1 and l_2 are positive then F is neither discrete nor continuous. A discrete distribution function is always a step function while graph of a continuous distribution function is a non-decreasing continuous function, tending to 0 as $x \to -\infty$ and tending to 1 as $x \to \infty$.

In Jordan decomposition theorem, F is decomposed as $F = l_1F_1 + l_2F_2$. The continuous distribution function F_2 can be further decomposed into absolutely continuous and singular components. Thus, F can be decomposed in three components. We state this result in the following theorem. For proof one may refer to Athreya and Lahiri [3].

Theorem 3.3.2. *Decomposition theorem: Every distribution function can be uniquely expressed as,*

$$F = l_1F_1 + l_2F_2 + l_3F_3, \quad \textit{where} \quad l_1, l_2, l_3 \geq 0 \quad \textit{and} \quad l_1 + l_2 + l_3 = 1,$$

where F_1 is a discrete distribution function, F_2 is an absolutely continuous distribution function and F_3 is a singular distribution function.

In the following example, we adopt the procedure outlined in Theorem 3.3.1 to decompose a distribution function F into discrete and continuous components.

Example 3.3.1. Suppose a distribution function $F : \mathbb{R} \to [0, 1]$ is defined as follows.

$$F(x) = \begin{cases} 0, & \text{if} \quad x < -2, \\ 1/3, & \text{if} \quad -2 \leq x < 0, \\ 1/2, & \text{if} \quad 0 \leq x < 5, \\ 1/2 + (x-5)^2/2, & \text{if} \quad 5 \leq x < 6, \\ 1, & \text{if} \quad x \geq 6. \end{cases}$$

We first find a set of points of discontinuity and corresponding size of the discontinuity. Observe that at $F(-2-) = 0 \neq 1/3 = F(-2)$, implying that -2 is a point of discontinuity with size of discontinuity $1/3$. Similarly, $F(0-) = 1/3 \neq 1/2 = F(0)$, hence 0 is a point of discontinuity with size of discontinuity $1/2 - 1/3 = 1/6$. No other point is a point of discontinuity. Hence, as in Theorem 3.3.1, we have $G(x) = \sum_{x_n \leq x} p_{x_n}$ and $H(x) = F(x) - G(x)$, $x \in \mathbb{R}$. These are given by,

$$G(x) = \begin{cases} 0, & \text{if} \quad x < -2, \\ 1/3, & \text{if} \quad -2 \leq x < 0, \\ 1/2, & \text{if} \quad x \geq 0 \end{cases} \quad \& \quad H(x) = \begin{cases} 0, & \text{if} \quad x < 5, \\ (x-5)^2/2, & \text{if} \quad 5 \leq x < 6, \\ 1/2, & \text{if} \quad x \geq 6. \end{cases}$$

Note that $l_1 = G(\infty) = 1/2 < 1$ and $H(\infty) = 1 - G(\infty) = l_2 = 1/2$. Suppose F_1 and F_2 are defined as $F_1 = (1/l_1)G$ and $F_2 = (1/l_2)H$. Hence, $F_1(\infty) = 1$ and $F_2(\infty) = 1$. Thus, F_1 and F_2 are given by,

$$F_1(x) = \begin{cases} 0, & \text{if} \quad x < -2, \\ 2/3, & \text{if} \quad -2 \leq x < 0, \\ 1, & \text{if} \quad x \geq 0 \end{cases} \quad \& \quad F_2(x) = \begin{cases} 0, & \text{if} \quad x < 5, \\ (x-5)^2, & \text{if} \quad 5 \leq x < 6, \\ 1, & \text{if} \quad x \geq 6. \end{cases}$$

It is easy to verify that $F(x) = l_1F_1(x) + l_2F_2(x)$ for all $x \in \mathbb{R}$. F_1 is a discrete distribution function and corresponding probability mass function is given by, $P[X = -2] = 2/3$ and $P[X = 0] = 1/3$. F_1 is a continuous distribution function and it is differentiable with corresponding probability density function $f(x) = 2(x-5)$ if $5 \leq x < 6$ and 0 otherwise. □

In some cases it is possible to decompose F into discrete and continuous components by observing the given F. We adopt this approach in the following example.

Example 3.3.2. Suppose a function $F : \mathbb{R} \to [0, 1]$ is defined as follows.

$$F(x) = \begin{cases} 0, & \text{if} \quad x < 0, \\ 1/4, & \text{if} \quad 0 \le x < 1, \\ 1/2, & \text{if} \quad 1 \le x < 2, \\ 1/2 + (x-2)/2, & \text{if} \quad 2 \le x < 3, \\ 1, & \text{if} \quad x \ge 3. \end{cases}$$

We first examine whether F is a distribution function. It will be a distribution function, if it satisfies the four conditions given in Theorem 3.2.1. By the definition of F, (i) it is clear that $0 \le F(x) \le 1 \;\; \forall \;\; x \in \mathbb{R}$. (ii) F is non-decreasing on $\mathbb{R}$, if $\forall \; x_1, x_2 \in \mathbb{R}, x_1 \le x_2$ implies $F(x_1) \le F(x_2)$. Suppose $x_1 < x_2$ are in $(-\infty, 0)$. For all $x \in (-\infty, 0), F(x) = 0$, hence $x_1 \le x_2 \Rightarrow F(x_1) \le F(x_2)$. For $x_1 < 0, \;\; F(x_1) = 0$, and for $x_2 \in [0,1), F(x_2) = 1/4$, for $x_2 \in [1,2), F(x_2) = 1/2$, for $x_2 \in [2,3), F(x_2) = 1/2 + (x_2 - 2)/2$ and for $x_2 \in [3, \infty), F(x_2) = 1$. Thus, for all such x_1 and x_2, $\;F(x_1) \le F(x_2)$. Now suppose both $x_1, x_2 \in [0,1)$, then $F(x_1) = F(x_2) = 1/4$. For $x_2 \in [1,2), F(x_2) = 1/2$, for $x_2 \in [2,3), F(x_2) = 1/2 + (x_2 - 2)/2$, for $x_2 \in [3, \infty)$, $F(x_2) = 1$. Thus, for all such x_1 and x_2, $\;F(x_1) \le F(x_2)$. For both $x_1, x_2 \in [1,2)$, $F(x_1) = F(x_2) = 1/2$. For $x_2 \in [2,3), F(x_2) = 1/2 + (x_2 - 2)/2$ and for $x_2 \in [3, \infty)$, $F(x_2) = 1$.Thus, for all such x_1 and x_2, $\;F(x_1) \le F(x_2)$. Further, assume that both $x_1 < x_2 \in [2,3)$, then $F(x_1) = 1/2 + (x_1 - 2)/2 \le 1/2 + (x_2 - 2)/2 = F(x_2)$ and for $x_2 \in [3, \infty)$, $F(x_2) = 1$. Thus, for all such x_1 and x_2, $\;F(x_1) \le F(x_2)$. Now suppose both $x_1, x_2 \in [3, \infty)$, then $F(x_1) = F(x_2) = 1$. Hence, we have noted that $\;\forall \;\; x_1, x_2 \in \mathbb{R}, \; x_1 \le x_2$ implies that $F(x_1) \le F(x_2)$. (iii) Note that for $x \in (-\infty, 0)$, $x \in (0,1)$, $x \in (1,2)$ and $x \in (3, \infty)$, F(x) is a constant function, hence continuous and hence right continuous function on all these intervals of the domain. For $x \in (2,3), F(x) = 1/2 + (x-2)/2$ is a polynomial of degree 1 and hence is continuous and hence right continuous function. We now have to check right continuity at $x = 0, 1, 2, 3$. Observe that,

$$\lim_{h \to 0} F(0+h) = 1/4 = F(0), \quad \lim_{h \to 0} F(1+h) = 1/2 = F(1),$$

$$\lim_{h \to 0} F(2+h) = \lim_{h \to 0} (1/2 + (2+h-2)/2) = 1/2 = F(2), \lim_{h \to 0} F(3+h) = 1 = F(3).$$

Thus F is right continuous at $x = 0, 1, 2, 3$. (iv) Note that

$$\forall \, x < 0, \; F(x) = 0 \Rightarrow \lim_{x \to -\infty} F(x) = 0 \;\&\; \forall \, x > 3, \; F(x) = 1 \Rightarrow \lim_{x \to \infty} F(x) = 1.$$

Thus, the function F satisfies the four conditions for it to qualify as a distribution function. Observe that F can be expressed as,

$$F(x) = \begin{cases} 0, & \text{if} \quad x < 0, \\ \frac{1}{4} = \frac{1}{2} * \frac{1}{2}, & \text{if} \quad 0 \le x < 1, \\ \frac{1}{2} = \frac{1}{2} * 1, & \text{if} \quad 1 \le x < 2, \\ \frac{1}{2} + \frac{(x-2)}{2} = \frac{1}{2} * 1 + \frac{1}{2}(x-2), & \text{if} \quad 2 \le x < 3, \\ 1 = \frac{1}{2} * 1 + \frac{1}{2} * 1, & \text{if} \quad x \ge 3. \end{cases}$$

This expression of F suggests that F can be decomposed as $F = l_1F_1 + l_2F_2$, where $l_1 = 1/2$, $l_2 = 1/2$ and F_1 and F_2 are

$$F_1(x) = \begin{cases} 0, & \text{if} \quad x < 0, \\ 1/2, & \text{if} \quad 0 \le x < 1, \\ 1, & \text{if} \quad x \ge 1 \end{cases} \quad \& \quad F_2(x) = \begin{cases} 0, & \text{if} \quad x < 2, \\ x-2, & \text{if} \quad 2 \le x < 3, \\ 1, & \text{if} \quad x \ge 3. \end{cases}$$

Observe that F_1 is a discrete distribution function with set $D = \{0, 1\}$ as a set of points of discontinuity, with jump size 1/2 at each jump. It is the same as the set of points of increase. Thus, F_1 is a distribution function of a discrete random variable X_1 with probability mass function $P[X_1 = 0] = P[X_1 = 1] = 1/2$. Further, F_2 is a continuous distribution function and it is a distribution function of an absolutely continuous random variable X_2 having uniform $U(2, 3)$ distribution with probability density function $f(x)$ given by,

$$f(x) = \begin{cases} 1, & \text{if} \quad 2 \le x \le 3, \\ 0, & \text{otherwise.} \end{cases}$$

□

Example 3.3.3. In this example we draw four graphs corresponding to different types of distribution functions. We use Code 3.3.1 to draw these graphs. Figure 3.1 displays these graphs.

Code 3.3.1. Graphs of distribution functions:

```
par(mfrow=c(2,2),mai=c(0.5,0.5,0.5,0.1)); n=1000
#Part I: Discrete uniform on {0,1,2,3}
x=seq(from=-1,to=4,length=n)
F3=function(z)
{
  if(z<0) Fz=0
  if(z>=0 && z<1) Fz=1/4
  if(z>=1 && z<2) Fz=1/2
  if(z>=2 && z<3) Fz=3/4
  if(z>=3) Fz=1
  return(Fz)
}
y=c()
for(i in 1:n)
{
  y[i]=F3(x[i])
}
plot(x,y,type="l",col="dark blue",lwd=2,xlab="x",ylab="F3(x)",
```

```
main="Discrete Uniform on {0,1,2,3}")
lines(x=c(0,1,2,3),y=c(F3(0),F3(1),F3(2),F3(3)),type="p",pch=16,
col=2,cex=2)
#Part II: Uniform distribution on (0,3)
a=0; b=3
F2=function(z)
{
  if(z<a) Fz=0
  if(z>=a && z<b) Fz=(z-a)/(b-a)
  if(z>b) Fz=1
  return(Fz)
}
y=c()
for(i in 1:n)
{
  y[i]=F2(x[i])
}
plot(x,y,type="l",col="dark blue",lwd=2,xlab="x",ylab="F2(x)",
main="Uniform Distribution on (0,3)")
#Part III: Mixture distribution from Ex 3.3.1
x=seq(from=-3,to=7,length=n)
F1=function(z)
{
  Fz=c()
  if(z< -2) Fz=0
  if(z>=-2 && z<0) Fz=1/3
  if(z>=0 && z<5) Fz=0.5
  if(z>=5 && z<6) Fz=0.5+(z-5)^2/2
  if(z>=6) Fz=1
  return(Fz)
}
y=c()
for(i in 1:n)
{
y[i]=F1(x[i])
}
plot(x,y,type="l",col="dark blue",lwd=2,xlab="x",ylab="F1(x)",
main="DF: Ex 3.3.1")
lines(x=c(-2,0),y=c(F1(-2),F1(0)),type="p",pch=16,col=2,cex=2)
```

```
#Part IV: Mixture distribution from Ex 3.3.2 and its components
x=seq(from=-1,to=4,length=n)
F1=function(z)
{
  Fz=c()
  if(z<0) Fz=0
  if(z>=0 && z<1) Fz=0.25
  if(z>=1 && z<2) Fz=0.5
  if(z>=2 && z<3) Fz=0.5+(z-2)/2
  if(z>=3) Fz=1
  return(Fz)
}
F2=function(z)
{
  Fz=c()
  if(z<0) Fz=0
  if(z>=0 && z<1) Fz=0.5
  if(z>=1 ) Fz=1
  return(Fz)
}
F3=function(z)
{
  Fz=c()
  if(z<2) Fz=0
  if(z>=2 && z<3) Fz=z-2
  if(z>=3 ) Fz=1
  return(Fz)
}
y1=y2=y3=c()
for(i in 1:n)
{
  y1[i]=F1(x[i])
  y2[i]=F2(x[i])
  y3[i]=F3(x[i])
}
plot(x,y1,type="l",col="dark blue",lwd=2,xlab="x",ylab="F1(x)",
main="DF & its components:Ex 3.3.2")
lines(x,y2,type="l",col="dark red",lwd=2,lty=2)
lines(x,y3,type="l",col="dark green",lwd=2,lty=3)
```

```
lines(x=c(0,1),y=c(F1(0),F1(1)),type="p",pch=16,col=2,cex=2)
legend("topleft",legend=c("Mixture","Discrete","Continuous"),
col=c("dark blue","dark red", "dark green"),lty=1:3,
lwd=c(1.5,2,2),bty="n")
```

From Figure 3.1, we note the nature of the following different types of distribution functions. (i) The upper left panel of Figure 3.1 shows the graph of a distribution function of the discrete uniform distribution with support $\{0, 1, 2, 3\}$. Observe that it is a step function with jump of the same size 1/4 at each of the four points of discontinuity. (ii) The upper right panel shows the graph of a distribution function of the continuous uniform distribution with support $(0, 3)$. On the support, the graph is a straight line with slope 1/3. (iii) The lower left panel shows the graph of a distribution function discussed in Example 3.3.1, which is neither discrete nor continuous. (iv) The lower right panel shows the graph of a distribution function discussed in Example 3.3.2, which is also neither discrete nor continuous. In Example 3.3.2, we have decomposed it into discrete and continuous components. In this panel, the graphs of discrete and continuous components are imposed on the graph of the distribution function.

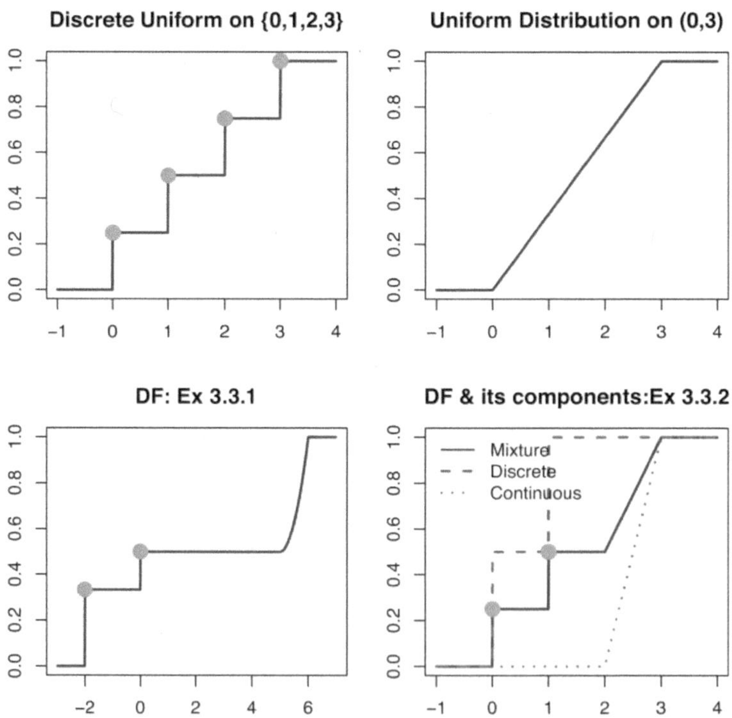

FIGURE 3.1
Distribution Functions

□

Example 3.3.4. Suppose F is a distribution function of a random variable X. In this example we express the distribution functions of (i) X^+, (ii) X^-, (iii) $|X|$ and (iv) $-X$ in terms of F. Suppose $P[X > 0] = p \geq 0$ and $P[X \leq 0] = 1 - p = q = F(0) \geq 0$.
(i) By the definition of X^+, it is a non-negative random variable defined as,

$$X^+ = \begin{cases} X, & \text{if } X > 0, \\ 0, & \text{if } X \leq 0. \end{cases}$$

For $x < 0$, $F_{X^+}(x) = P[X^+ \leq x] = 0$, for $x = 0$, $F_{X^+}(0) = P[X^+ = 0] = P[X \leq 0] = q = F(0)$ and for $x > 0$

$$\begin{aligned} F_{X^+}(x) &= P[X^+ \leq x] = P[X^+ = 0] + P[0 < X^+ \leq x] \\ &= P[X \leq 0] + P[0 < X^+ \leq x, X > 0] + P[0 < X^+ \leq x, X \leq 0] \\ &= q + P[0 < X \leq x] + 0 = q + (F(x) - F(0)) = F(x). \end{aligned}$$

Thus, $F_{X^+}(x)$ is given by,

$$F_{X^+}(x) = \begin{cases} 0, & \text{if } x < 0, \\ q = F(0), & \text{if } x = 0, \\ q + (F(x) - F(0)) = F(x), & \text{if } x > 0. \end{cases}$$

That is,

$$F_{X^+}(x) = \begin{cases} 0, & \text{if } x < 0, \\ F(x), & \text{if } x \geq 0. \end{cases}$$

Observe that when $q > 0$, 0 is a point of discontinuity, with size of discontinuity to be q. Further, right hand limit at 0 is q, since F is right continuous at 0. Further, $\lim_{x \to \infty} F_{X^+}(x) = 1$. Thus, if $q > 0$, then the distribution function of X^+ is neither discrete nor continuous. If $q = 0$, then X^+ distributed as X.
(ii) By the definition of X^-,

$$X^- = \begin{cases} -X, & \text{if } X \leq 0, \\ 0, & \text{if } X > 0. \end{cases}$$

Note that X^- is a non-negative random variable. Hence, for $x < 0$, $F_{X^-}(x) = 0$. For $x = 0$, $F_{X^-}(0) = P[X^- = 0] = P[X \geq 0] = 1 - F(0-)$. Now for $x > 0$,

$$\begin{aligned} F_{X^-}(x) &= P[X^- \leq x] = P[X^- \leq 0] + P[0 < X^- \leq x] \\ &= 1 - F(0-) + P[0 < X^- \leq x, X \leq 0] + P[0 < X^- \leq x, X > 0] \\ &= 1 - F(0-) + P[0 < -X \leq x] + 0 \\ &= 1 - F(0-) + P[-x \leq X < 0] \\ &= 1 - F(0-) + P[X < 0] - P[X < -x] \\ &= 1 - F(0-) + F(0-) - F(-x-) = 1 - F(-x-) \end{aligned}$$

Thus, $F_{X^-}(x)$ is given by,

$$F_{X^-}(x) = \begin{cases} 0, & \text{if } x < 0, \\ 1 - F(-x-), & \text{if } x \geq 0. \end{cases}$$

Observe that 0 is a point of discontinuity with size $1 - F(0-)$, that is, $1 - P[X < 0] = P[X \geq 0]$. Further, right hand limit at 0 is $1 - F(0-)$. Note that, $\lim_{x \to \infty} F_{X^-}(x) = 1$. Thus, if $1 - F(0-) > 0$, then the distribution function of X^- is neither discrete nor continuous.
(iii) Since $|X|$ is a non-negative random variable, for $x < 0$, $F_{|X|}(x) = 0$. If $x = 0$, then $F_{|X|}(0) = P[|X| \leq 0] = P[X = 0]$. For $x > 0$,

$$F_{|X|}(x) = P[|X| \leq x] = P[-x \leq X \leq x] = F(x) - F(-x-).$$

(iv) It is clear that for $x \in \mathbb{R}$,

$$F_{-X}(x) = P[-X \leq x] = P[X \geq -x] = 1 - P[X < -x] = 1 - F(-x-). \quad \square$$

Example 3.3.5. Suppose a random variable X follows either a Cauchy $C(0,1)$ distribution or the Laplace distribution with location parameter 0 and scale parameter 1, or the standard normal distribution. All the three distributions are symmetric around 0 and for each, the value of p and q as defined in Example 3.3.4, is 1/2. Thus, the distribution functions of X^+ and of X^- are neither discrete nor continuous. In all the cases 0 is a point of discontinuity with size 1/2 and the distributions of X^+ and of X^- are the same. In the next example, we demonstrate it for the Cauchy distribution. $\square$

Example 3.3.6. In this example we plot the distribution functions of X, X^+, X^- and $|X|$ in two cases: (i) when X follows the Cauchy $C(0,1)$ distribution, which is symmetric around 0 and (ii) when X is discrete random variable with finite support and asymmetric probability distribution.

(i) When $X \sim C(0,1)$, its distribution function F is $F(x) = 1/2 + \tan^{-1}(x)/\pi$,
$x \in \mathbb{R}$. For $x < 0$, the values of the distribution function of X^+ and X^- are 0, there is a discontinuity at $x = 0$ with size 1/2, while for $x > 0$, these are respectively given by $F(x)$ and $1 - F(-x) = F(x)$, due to symmetry. Thus, for $x > 0$, the values of the distribution function of X, X^+ and X^- are the same. Again it is a consequence of the symmetry of the Cauchy $C(0,1)$ distribution around 0. The values of the distribution function of $|X|$ are 0 for $x < 0$ and $F(x) - F(-x)$ for $x \geq 0$. At $x = 0$ it is 0. Thus, the distribution function of $|X|$ is a continuous function. We note these features in the graph in the left panel of Figure 3.2.

(ii) Suppose X is a discrete random variable with support $\{-2, -1, 0, 1, 2, 3\}$ with respective probabilities $\{0.2, 0.3, 0.1, 0.1, 0.2, 0.1\}$. Note that the distribution of X is not symmetric. Further, X^+ is a discrete random variable with support $\{0, 1, 2, 3\}$ with respective probabilities

$\{0.6, 0.1, 0.2, 0.1\}$, X^- is a discrete random variable with support $\{0, 1, 2\}$ with respective probabilities $\{0.5, 0.3, 0.2\}$. Observe that the distributions of X^+ and X^- are different. Note that $|X|$ is a discrete random variable with support $\{0, 1, 2, 3\}$ with respective probabilities $\{0.1, 0.4, 0.4, 0.1\}$ and its distribution is different from that of X^+. The right panel of Figure 3.2 displays the distribution functions corresponding to these four random variables. We use the following code to draw these graphs.

Code 3.3.2. Graphs of distribution functions of X, X^+, X^- and $|X|$:

```
par(mfrow=c(1,2),mai=c(0.5,0.5,0.5,0.1)); n=1000
x=seq(from=-5,to=5,length=n)
y=pcauchy(x) # DF of X
x1=seq(from=-5,to=0,length=n/2); y1=rep(0,n/2)
x2=seq(from=0,to=5,length=n/2)
y2=pcauchy(x2); x3=c(x1,x2);y3=c(y1,y2) #DF of X+
y4=1-pcauchy(-x2);y5=c(y1,y4) # DF of X-
y6=pcauchy(x2)-pcauchy(-x2); y7=c(y1,y6) # Df of |X|
plot(x,y,type="l",col="black",xlab="x",lty=1,ylim=c(0,1),
main="Symmetric Distribution",lwd=2)
lines(x3,y3,type="l",col="blue",xlab="x",lty=2,lwd=2)
lines(x,y5,type="l",col="dark red",xlab="x",lty=5,lwd=2)
lines(x,y7,type="l",col="dark green",xlab="x",lty=6,lwd=2)
y8=rep(0.5,n/2); lines(x1,y8)
xp=expression(paste(X^"+")); xm=expression(paste(X^"-"))
legend("topleft",legend=c("X",xp,xm,"|X|"),lty=c(1,2,5,6),
lwd=2,
col=c("black","blue","dark red","dark green"))
## Discrete random variable
Fx=function(z)
{
  if(z< -2) Fz=0
  if(z>=-2 && z< -1) Fz=0.2
  if(z>=-1 && z<0) Fz=0.5
  if(z>=0 && z<1) Fz=0.6
  if(z>=1 && z<2) Fz=0.7
  if(z>=2 && z<3) Fz=0.9
  if(z>=3) Fz=1
  return(Fz)
}
Fxp=function(z)
```

```
{
  if(z<0) Fz=0
  if(z>=0&& z< 1) Fz=0.6
  if(z>=1 && z<2) Fz=0.7
  if(z>=2 && z<3) Fz=0.9
  if(z>=3) Fz=1
  return(Fz)
}
Fxm=function(z)
{
  if(z<0) Fz=0
  if(z>=0&& z< 1) Fz=0.5
  if(z>=1 && z<2) Fz=0.8
  if(z>=2) Fz=1
  return(Fz)
}
Fmodx=function(z)
{
  if(z<0) Fz=0
  if(z>=0&& z< 1) Fz=0.1
  if(z>=1 && z<2) Fz=0.5
  if(z>=2 && z<3) Fz=0.9
  if(z>=3) Fz=1
  return(Fz)
}
a=-3; b=4; n=1000; y1=y2=y3=y4=c()
x=seq(from=a,to=b,length=n)
for(i in 1:n)
{
  y1[i]=Fx(x[i])
  y2[i]=Fxp(x[i])
  y3[i]=Fxm(x[i])
  y4[i]=Fmodx(x[i])
}
plot(x,y1,type="l",col="black",lwd=2,xlab="x",lty=1,
main="Asymmetric Distribution")
points(x=c(-2,-1,0,1,2,3),y=c(.2,.5,.6,.7,.9,1),type="p",
pch=20,col="black",cex=2)
lines(x,y2,type="l",col="blue",lty=2,lwd=2)
```

```
points(x=c(0,1,2,3),y=c(.6,.7,.9,1),type="p",pch=20,
col="blue",cex=2)
lines(x,y3,type="l",col="dark red",lty=5,lwd=2)
points(x=c(0,1,2),y=c(.5,.8,1),type="p",pch=20,
col="dark red",cex=2)
lines(x,y4,type="l",col="dark green",lty=6,lwd=2)
points(x=c(0,1,2,3),y=c(.1,.5,.9,1),type="p",pch=20,
col="dark green",cex=2)
legend("topleft",legend=c("X",xp,xm,"|X|"),lty=c(1,2,5,6),
lwd=2,col=c("black","blue","dark red","dark green"))
```

Observe that in the left panel of the Figure 3.2, there is a discontinuity at $x = 0$ with size 1/2 in the graphs corresponding to the distribution functions of X^+ and X^-. Further for $x > 0$, the graphs of the distribution function of X, X^+ and X^- coincide. The right panel of the Figure 3.2 displays the graphs of four discrete distribution functions. These are step functions with size of the jump at discontinuity points to be the probability attached with the discontinuity point. □

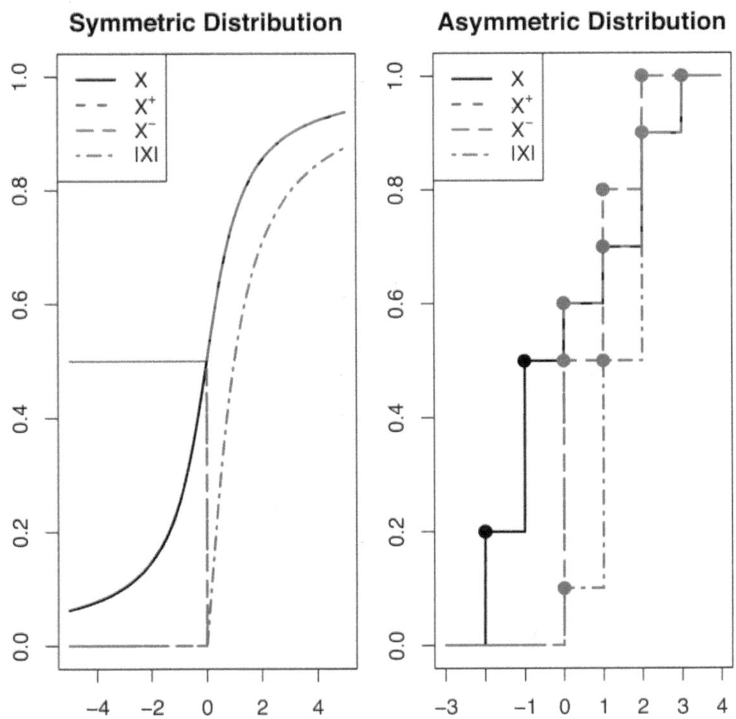

FIGURE 3.2
Distribution Functions of X, X^+, X^- and $|X|$

In the next section, we now extend the concept of distribution function of a random variable to a distribution function of a random vector.

3.4 Distribution Function of a Random Vector

We begin with a bivariate random vector.

Definition 3.4.1. *Distribution function of a bivariate random vector: Suppose a random vector $\underline{Z} = (X_1, X_2)'$ is defined on a probability space $(\Omega, \mathbb{A}, P)$ to a probability space $(\mathbb{R}^2, \mathbb{B}^2, P_{\underline{Z}})$. Then the distribution function $F_{\underline{Z}}(.)$ of $\underline{Z}$ is defined on $\mathbb{R}^2$ as follows. Suppose $\underline{z} = (x_1, x_2)' \in \mathbb{R}^2$.*

$$\begin{aligned} F_{\underline{Z}}(\underline{z}) &= P\{\omega \mid X_1(\omega) \leq x_1, X_2(\omega) \leq x_2\} \\ &= P\left[X_1^{-1}((-\infty, x_1)) \cap X_2^{-1}((-\infty, x_2))\right]. \end{aligned}$$

As in the setup of a random variable, $F_{\underline{Z}}(\underline{z})$ is a point function with domain as $\mathbb{R}^2$ and range $[0,1]$. Observe that if $x_2 \to \infty$ in $F_{\underline{Z}}(\underline{z})$, then $X_2^{-1}((-\infty, \infty)) = \Omega$ and hence

$$\lim_{x_2 \to \infty} F_{\underline{Z}}(x_1, x_2) = P\left[X_1^{-1}((-\infty, x_1)) \cap \Omega\right] = P\{\omega \mid X_1(\omega) \leq x_1\} = F_{X_1}(x_1).$$

Similarly, if $x_1 \to \infty$ in $F_{\underline{Z}}(\underline{z})$, we get

$$\lim_{x_1 \to \infty} F_{\underline{Z}}(x_1, x_2) = P\left[\Omega \cap X_2^{-1}((-\infty, x_2))\right] = P\{\omega \mid X_2(\omega) \leq x_2\} = F_{X_2}(x_2).$$

We state below the properties of $F_{\underline{Z}}(\underline{z})$.

(i) $\lim_{x_1 \to \infty} F_{\underline{Z}}(x_1, x_2) = F_{X_2}(x_2)$.

(ii) $\lim_{x_2 \to \infty} F_{\underline{Z}}(x_1, x_2) = F_{X_1}(x_1)$.

(iii) $\lim_{x_1 \to -\infty} F_{\underline{Z}}(x_1, x_2) = \lim_{x_2 \to -\infty} F_{\underline{Z}}(x_1, x_2) = 0$.

(iv) $\lim_{x_1, x_2 \to \infty} F_{\underline{Z}}(x_1, x_2) = 1$.

(v) If a_1, a_2, b_1, b_2 are real numbers such that $a_1 \leq a_2$ and $b_1 \leq b_2$, then

$$F_{\underline{Z}}(a_2, b_2) - F_{\underline{Z}}(a_1, b_2) - F_{\underline{Z}}(a_2, b_1) + F_{\underline{Z}}(a_1, b_1) \geq 0.$$

The function $F_{\underline{Z}}(.)$ is known as a joint distribution function of $(X_1, X_2)'$, while $F_{X_1}(.)$ and $F_{X_2}(.)$ are known as the marginal distribution functions to emphasize that it is a distribution function of a component variable in a bivariate setup. The property (v) is analogous to the non-decreasing property of univariate distribution function. This expression gives $P[a_1 < X_1 \leq a_2, b_1 < X_2 \leq b_2]$ and hence it must be non-negative. Distribution function of a k-variate random vector is defined analogously.

Definition 3.4.2. *Distribution function of a k-variate random vector: Suppose a random vector* $\underline{Z} = (X_1, X_2, \cdots, X_k)'$ *is defined on a probability space* $(\Omega, \mathbb{A}, P)$ *to a probability space* $(\mathbb{R}^k, \mathbb{B}^k, P_{\underline{Z}})$. *Then the distribution function,* $F_{\underline{Z}}(.)$ *of* $\underline{Z}$ *is defined on* $\mathbb{R}^k$ *as follows. Suppose* $\underline{z} = (x_1, x_2, \cdots, x_k)' \in \mathbb{R}^k$.

$$F_{\underline{Z}}(\underline{z}) = P\{\omega \mid X_i(\omega) \leq x_i, i = 1, 2, \cdots, k\} = P\Big[\bigcap_{i=1}^{k} X_i^{-1}((-\infty, x_i])\Big].$$

Properties of a Distribution Function of a Random Vector

(i) $\lim_{x_1 \to \infty} F_{\underline{Z}}(x_1, x_2, \cdots, x_k) = F_{X_2, X_3, \cdots, X_k}(x_2, x_3, \cdots, x_k)$.
If further we allow $x_2 \to \infty$, then we get the distribution function of $(X_3, X_4, \cdots, X_k)'$. In this manner distribution function of a random vector of lower dimensions can be derived.

(ii) $\lim_{x_i \to -\infty} F_{\underline{Z}}(x_1, x_2, \cdots, x_k) = 0$, for any $i = 1, 2, \cdots, k$.

(iii) $\lim_{x_i \to \infty \ \forall \ i} F_{\underline{Z}}(x_1, x_2, \cdots, x_k) = 1$.

The quick recap of the results discussed in this chapter is given below.

Summary

1. Suppose X is a random variable defined on the probability space $(\Omega, \mathbb{A}, P)$. Then the distribution function $F_X(.)$ of X is defined on $\mathbb{R}$ as

$$F_X(x) = P\left(X^{-1}((-\infty, x]\right) = P\{\omega \mid X(\omega) \leq x\} = P[X \leq x], \quad x \in \mathbb{R}.$$

2. Correspondence theorem: Every probability measure P^* on $(\mathbb{R}, \mathbb{B})$ uniquely determines a distribution function through the correspondence $F(x) = P^*((-\infty, x]), \ x \in \mathbb{R}$ with $F(-\infty) = 0$ and $F(\infty) = 1$. Conversely, if $F(x)$ is specified for each $x \in \mathbb{R}$, then the probability measure P^* on $(\mathbb{R}, \mathbb{B})$ is determined uniquely through the relation $P^*((-\infty, x]) = F(x)$.

3. A distribution function $F_X(x)$ of a random variable X satisfies the following properties. (i) $0 \leq F_X(x) \leq 1$. (ii) $F_X(x)$ is a non-decreasing function on $\mathbb{R}$. (iii) $F_X(x)$ is a right continuous function on $\mathbb{R}$ and (iv) $\lim_{x \to \infty} F_X(x) = 1$ and $\lim_{x \to -\infty} F_X(x) = 0$. Conversely, if a function $g : \mathbb{R} \to \mathbb{R}$ satisfies these four properties, then there exists a probability space on which one can define a random variable whose distribution function is the given function g.

4. A set of values of $x \in \mathbb{R}$ for which $F(x)$ increases is known as a support of a distribution function.

5. If $F(x)$ is not continuous from left at $x \in \mathbb{R}$, that is, if $\lim_{h \to 0} F_X(x-h) = F(x-) \neq F(x)$, then x is said to be a point of discontinuity of $F(x)$ and $p_x = F(x) - F(x-)$ is the size of discontinuity of F at x or the magnitude of the jump at x.

6. A set of points of discontinuity of F is at most countable.

7. A point $x \in \mathbb{R}$ is said to be a point of increase of $F_X(x)$ if $\forall\ \epsilon > 0$, $F_X(x+\epsilon) - F_X(x-\epsilon) > 0$.

8. A random variable X or its distribution function F is said to be discrete, if all points of increase of F are its points of discontinuities. Suppose D is a set of points of discontinuity of F and $p_x, x \in D$ is the size of discontinuity at x. Then the set $\{(x, p_x) \mid x \in D\}$ is a probability mass function of a discrete random variable X.

9. A random variable X or its distribution function F is said to be continuous, if the set of points of discontinuity of F is an empty set, that is, if the distribution function F is continuous on $\mathbb{R}$. A continuous random variable X or its distribution function F is said to be absolutely continuous, if there exists a non-negative function f such that $F_X(b) - F_X(a) = \int_a^b f(x)dx$, for all $a < b$. The function f is known as a probability density function of an absolutely continuous random variable X. If F is differentiable on $\mathbb{R}$, then $f(x) = \frac{d}{dx} F_X(x)$.

10. Every distribution function can be uniquely expressed as,

$$F = l_1 F_1 + l_2 F_2, \quad \text{where} \quad l_1, l_2 \geq 0 \ \text{ and } \ l_1 + l_2 = 1,$$

where F_1 is a discrete distribution function and F_2 is a continuous distribution function.

11. Suppose a random vector $\underline{Z} = (X_1, X_2, \cdots, X_k)'$ is defined on a probability space $(\Omega, \mathbb{A}, P)$ to a probability space $(\mathbb{R}^k, \mathbb{B}^k, P_{\underline{Z}})$. Then the distribution function, $F_{\underline{Z}}(.)$ of $\underline{Z}$ is defined on $\mathbb{R}^k$ as follows. Suppose $\underline{z} = (x_1, x_2, \cdots, x_k)' \in \mathbb{R}^k$.

$$F_{\underline{Z}}(\underline{z}) = P\{\omega \mid X_i(\omega) \leq x_i, i = 1, 2, \cdots, k\} = P\Big[\bigcap_{i=1}^{k} X_i^{-1}((-\infty, x_i])\Big].$$

3.5 Conceptual Exercises

3.5.1 Suppose functions F are defined as follows. In each case examine whether F is a distribution function. If yes, find the set of points of discontinuities and points of increase. Decompose the distribution function and decide the type of distribution function F.

$$(i)\ \ F(x)=\begin{cases}0, & \text{if} \quad x<0,\\ x^2/4, & \text{if} \quad 0\le x<2,\\ 1, & \text{if} \quad x\ge 2.\end{cases} \qquad (ii)\ \ F(x)=\begin{cases}0, & \text{if} \quad x<0,\\ \frac{x}{2}, & \text{if} \quad 0\le x\le \frac{2}{3},\\ 1, & \text{if} \quad x>\frac{2}{3}.\end{cases}$$

$$(iii) F(x)=\begin{cases}0, & \text{if} \quad x\le 1,\\ 1-\frac{1}{x^2}, & \text{if} \quad x>1.\end{cases} \qquad (iv) F(x)=\begin{cases}0, & \text{if} \quad x\le 0,\\ \frac{1}{2}+\frac{1}{2}e^{-x}, & \text{if} \quad x>0.\end{cases}$$

$$(v) F(x)=\begin{cases}0, & \text{if} \quad x<0,\\ 1-\frac{1}{2}e^{-x}, & \text{if} \quad x\ge 0.\end{cases} \qquad (vi) F(x)=\begin{cases}0, & \text{if} \quad x<0,\\ \sin x, & \text{if} \quad 0\le x<\frac{\pi}{2},\\ 1, & \text{if} \quad x\ge \frac{\pi}{2}.\end{cases}$$

(vii) $F(x)=\sin x, \quad 0<x<\pi/2.$
(viii) $F(x)=\sin x, \quad x\in\mathbb{R}.$

$$(ix) F(x)=\begin{cases}0, & \text{if} \quad x<0,\\ x/2, & \text{if} \quad 0\le x<1,\\ 1, & \text{if} \quad x\ge 1.\end{cases} \qquad (x)\ F(x)=\begin{cases}0, & \text{if} \quad x<2,\\ x/3, & \text{if} \quad 2\le x<3,\\ 1, & \text{if} \quad x\ge 3.\end{cases}$$

3.5.2 If D is a set of points of discontinuity of a distribution function, give an example of each of the following types of distribution functions. (i) D is empty. (ii) D is singleton. (iii) D is finite, but not singleton. (iv) D is countably infinite. Write the distribution function or the probability density function or the probability mass function.

3.5.3 A function $F:\mathbb{R}\to\mathbb{R}$ is defined as $F(x)=0$ if $x\le 0$ and $F(x)=\alpha+ke^{-x^2/2}$ if $x>0$. Determine values of α and k so that F is a distribution function.

3.5.4 Examine whether a convex combination of k distribution functions is a distribution function.

3.5.5 Does there exist a distribution function F such that $\forall\ x,y\in\mathbb{R}$ $F(x+y)=F(x)+F(y)$? Justify your answer.

3.5.6 Suppose F and G are distribution functions of X and Y respectively. If $X>Y$ on Ω, show that $F(x)\le G(x)\ \forall\ x\in\mathbb{R}$. Show that the converse is not true.

3.5.7 Suppose a distribution function $F:\mathbb{R}\to[0,1]$ is defined as follows.

$$F(x) = \begin{cases}0, & \text{if} \quad x<-2,\\ 0.2, & \text{if} \quad -2\le x<1,\\ 0.2+x^2/10, & \text{if} \quad 1\le x<2,\\ 0.6+x/10, & \text{if} \quad 2\le x<3,\\ 1, & \text{if} \quad x\ge 3.\end{cases}$$

Decompose F into discrete and continuous components. Identify the distributions in the decomposition.

3.5.8 Suppose a distribution function $F : \mathbb{R} \to [0, 1]$ is defined as follows.

$$F(x) = \begin{cases} 0, & \text{if} \quad x < 0, \\ (pr + (1-p)x)/5, & \text{if} \quad r \le x < r+1, r = 0, 1, \cdots, 4, \\ 1, & \text{if} \quad x \ge 5, \end{cases}$$

$0 < p < 1$. Examine whether F is a distribution function. If yes, decompose F into discrete and continuous components. Identify the distributions in the decomposition.

3.6 Computational Exercises

3.6.1 Suppose a distribution function $F : \mathbb{R} \to [0, 1]$ is defined as follows.

$$F(x) = \begin{cases} 0, & \text{if} \quad x < -2, \\ 0.2, & \text{if} \quad -2 \le x < 1, \\ 0.2 + x^2/10, & \text{if} \quad 1 \le x < 2, \\ 0.6 + x/10, & \text{if} \quad 2 \le x < 3, \\ 1, & \text{if} \quad x \ge 3. \end{cases}$$

Draw the graph of the distribution function. On the same graph impose the graphs of discrete and continuous components.

3.6.2 Suppose $X \sim N(0, 1)$ Plot the distribution functions of X, X^+, X^- and $|X|$.

3.6.3 Suppose X follows a discrete distribution with with support $\{-3, -2, -1, 0, 1, 2, 3, 4\}$ with respective probabilities $\{0.1, 0.1, 0.2, 0.1, 0.1, 0.2, 0.1, 0.1\}$. Plot the distribution functions of X, X^+, X^- and $|X|$.

3.7 Multiple Choice Questions

Note: In each question, multiple options may be correct. Answers are given in Chapter 11, after the solutions of conceptual exercises of Chapter 3.

3.7.1 Suppose $F_X(x)$ is a distribution function of a random variable X. Then which of the following is NOT always true?

(a) $F_X(x)$ is a non-decreasing function on $\mathbb{R}$
(b) $F_X(x)$ is a right continuous function on $\mathbb{R}$
(c) $F_X(x)$ is a continuous function on $\mathbb{R}$
(d) $0 \le F_X(x) \le 1$

3.7.2 Following are two statements.
(I) X and Y are identically distributed random variables implies X and Y are identical random variables.
(II) X and Y are identical random variables implies X and Y are identically distributed random variables.
Then which of the following is true?

(a) Both (I) and (II) are true
(b) (I) is true but (II) is false
(c) (II) is true but (I) is false
(d) Both (I) and (II) are false

3.7.3 For a distribution function, which of the following is NOT true?

(a) Every point of discontinuity is a point of increase
(b) A set of points of discontinuity is at most countable
(c) A set of points of increase is at most countable
(d) Every point of increase is need not be a point of discontinuity

3.7.4 For a distribution function, which of the following need NOT be true?

(a) Every point of discontinuity is a point of increase
(b) A set of points of discontinuity is at most countable
(c) A set of points of increase can be uncountable
(d) Every point of increase is a point of discontinuity

3.7.5 For a distribution function, which of the following is/are always true?

(a) Every point of discontinuity is a point of increase
(b) A set of points of discontinuity is at most countable
(c) A set of points of increase is always uncountable
(d) Every point of increase need not be a point of discontinuity

3.7.6 Which of the following sets can not be a set of discontinuity points of a distribution function?

(a) $D = \emptyset$
(b) $D = \mathbb{N}$
(c) $D = \{0, 1, 2, 3\}$
(d) $D = (0, \infty)$

3.7.7 Which of the following statements is/are always true?

(a) Set of points of increase of a distribution function is dense in $\mathbb{R}$
(b) Set of points of continuity of a distribution function is dense in $\mathbb{R}$
(c) Set of points of discontinuity of a distribution function is dense in $\mathbb{R}$
(d) None of the above

3.7.8 If $F_1(x)$ and $F_2(x)$ are distribution functions, which of the following is/are distribution functions?

(a) $F(x) = (F_1(x) + F_2(x))/2$
(b) $F(x) = F_1(x) + F_2(x)$
(c) $F(x) = F_1(x) - F_2(x)$
(d) None of the above

3.7.9 Which of the following functions is/are distribution functions?

(a)

$$F(x) = \begin{cases} 0, & x < 0 \\ x/2, & 0 \leq x \leq \frac{2}{3} \\ 1, & x > \frac{2}{3} \end{cases}$$

(b)

$$F(x) = \begin{cases} 0, & x < 0 \\ \sin x, & 0 \leq x < \frac{\pi}{2} \\ 1, & x \geq \frac{\pi}{2} \end{cases}$$

(c)

$$F(x) = \begin{cases} 0, & x \leq 1 \\ 1 - 1/x^2, & x > 1 \end{cases}$$

(d)

$$F(x) = \begin{cases} 0, & x \leq 0 \\ 1/2 + e^{-x}/2, & x > 0 \end{cases}$$

3.7.10 If $X \sim B(1, 0.5)$, which of the following random variables is/are NOT distributed as X?

(a) $Y_1 = 1 - X$
(b) $Y_2 =$ No. of heads in a single toss of a fair coin
(c) $Y_3 = X^2$
(d) e^X

4

Expectation and Characteristic Function

4.1 Introduction

In Chapter 2 we defined a random variable and its probability distribution. Probability distribution of a random variable can also be specified in terms of its distribution function, as noted in Chapter 3. It is of interest to find some characteristics of a random variable which will summarize its probability distribution. Many such characteristics are based on an expectation of a random variable or expectation of a particular Borel function of a random variable. We discuss the notion of expectation in this chapter. Mathematically, expectations are integrals with respect to an arbitrary finite measure in measure theory, with respect to a probability measure in probability theory or Riemann Stieltjes integral with respect to the distribution function.

Suppose a random variable X is defined on a probability space $(\Omega, \mathbb{A}, P)$ to $(\mathbb{R}, \mathbb{B}, P_X)$. Expectation of a random variable X, denoted by, $E(X)$ or $\int_\Omega XdP$ is defined as an integral of X with respect to the probability measure P. Expectation of a random variable is also known as a mean of a random variable. Expected value of a random variable is a probabilistic version of the centre of gravity of a physical body.

Differentiation and integration are two main components of calculus. It is also true in measure theory or probability theory. We briefly introduce the concept of an integral with respect to a measure and the theory of Lebesgue integration. The classical integral is the Riemann integral, which is generalized to the Riemann-Stieltjes integral. However, it has certain deficiencies which are overcome by the Lebesgue integral. The Lebesgue integral is an integral with respect to a Lebesgue measure which is defined below.

Definition 4.1.1. *Lebesgue measure: The Lebesgue measure λ is a measure on the measurable space $(\mathbb{R}, \mathbb{B})$ defined as*

$$\lambda((a, b]) = b - a \qquad \forall \;\; a < b \in \mathbb{R}.$$

The Lebesgue integral of a Borel function is defined in three steps, first for a simple Borel function, then for a non-negative Borel function and for an arbitrary Borel function. Analogous to a simple random variable, a simple Borel function is defined as follows.

DOI: 10.1201/9781032619057-4

Definition 4.1.2. *Simple Borel function: A function g on a measurable space $(\mathbb{R}, \mathbb{B})$ to $(\mathbb{R}, \mathbb{B})$ defined as $g(x) = \sum_{i=1}^{k} x_i I_{S_i}(x)$ where $x_i \in \mathbb{R}, \ \ i = 1, 2, \cdots, k$ and $\{S_1, S_2, \cdots, S_k\}$ is a measurable partition of $\mathbb{R}$, is known as a simple Borel function.*

We also have an analogue of Theorem 2.4.3 for a non-negative Borel function. It is stated in the following theorem.

Theorem 4.1.1. *Suppose g is a non-negative Borel function defined on a measurable space $(\mathbb{R}, \mathbb{B})$ to $(\mathbb{R}, \mathbb{B})$. Then there exists a non-decreasing sequence $\{g_n, n \geq 1\}$ of non-negative simple Borel functions on a measurable space $(\mathbb{R}, \mathbb{B})$ such that $g_n \to g$ on $\mathbb{R}$.*

We now define the Lebesgue integral $\int_{\mathbb{R}} g \, d\lambda$ of a Borel function with respect to a Lebesgue measure. The definition proceeds in three steps as follows.

Definition 4.1.3. *Lebesgue integral: Suppose g is a Borel function on a measurable space $(\mathbb{R}, \mathbb{B})$ and λ is a Lebesgue measure.*

i *For a simple Borel function given by $g = \sum_{i=1}^{k} x_i I_{S_i}$, the Lebesgue integral $\int_{\mathbb{R}} g \, d\lambda$ is defined as $\int_{\mathbb{R}} g \, d\lambda = \sum_{i=1}^{k} x_i \lambda(S_i)$.*

ii *If g is a non-negative Borel function, then $\int_{\mathbb{R}} g \, d\lambda = \lim_{n \to \infty} \int_{\mathbb{R}} g_n \, d\lambda$ where $\{g_n, n \geq 1\}$ is a non-decreasing sequence of non-negative simple Borel functions such that $g_n \to g$ on $\mathbb{R}$.*

iii *For arbitrary Borel function g, $\int_{\mathbb{R}} g \, d\lambda = \int_{\mathbb{R}} g^+ \, d\lambda - \int_{\mathbb{R}} g^- \, d\lambda$ provided at least one of the integrals on the right hand side is finite.*

Remark 4.1.1. If g is a Borel function defined on a measurable space $(\mathbb{R}, \mathbb{B})$ to $(\mathbb{R}, \mathbb{B})$, then its integral with respect to any measure μ on $(\mathbb{R}, \mathbb{B})$ is defined on similar lines. In Section 3, we come across with the integral of g with respect to the induced probability measure.

Observe that an integral of a simple function is a weighted average of x_i's with weights as $\lambda(S_i), \ i = 1, 2, \cdots, k$. The Lebesgue integral and the Riemann-Stieltjes integral are related to each other in the sense that if the Riemann-Stieltjes integral of a function exists, then the Lebesgue integral also exists and both are the same.

Using this concept of Lebesgue integration we now define the expectation of a random variable as an integral with respect to a probability measure. We proceed exactly on the same three steps, first we define it for a simple random variable in Section 2, then for a non-negative random variable and for an arbitrary random variable in Section 3. We discuss a variety of properties of the expectation of a random variable. Some of these seem to be obvious or are already known. However, we have now defined expectation as an integral with respect to a probability measure and hence we have a different approach to prove these properties. Section 4 is devoted to the important concept of a characteristic function and some related results, including inversion theorem, its versions and uniqueness theorem. Section 5 presents some inequalities concerning moments, which are the expectations of Borel functions of a random variable. These inequalities are heavily used in the study of convergence of a sequence of random variables discussed in Chapters 6 and 7.

4.2 Expectation of a Simple Random Variable

We have defined a simple random variable in Section 2.4. Its expectation is defined as follows.

Definition 4.2.1. *Expectation of a simple random variable: Suppose a simple random variable X on a probability space $(\Omega, \mathbb{A}, P)$ to $(\mathbb{R}, \mathbb{B}, P_X)$ is defined as, $X = \sum_{i=1}^{k} a_i I_{A_i}$, where $\{A_1, A_2, \cdots, A_k\}$ is a measurable partition of Ω and $a_1, a_2, \cdots, a_k$ are real numbers. Then $E(X)$ is defined as,*

$$E(X) = \int_{\Omega} X dP = \sum_{i=1}^{k} a_i P(A_i).$$

Note that expectation of a simple random variable may be viewed as a weighted average of the possible values of X, where weight attached to a_i is $P(A_i)$. It is also a centre of gravity of the values of X with mass associated with a_i as $P(A_i)$. If $P(A_i) = 1/k$, then $E(X) = (1/k)\sum_{i=1}^{k} a_i$ is the arithmetic mean of $a_1, a_2, \cdots, a_k$. $E(X)$ gives us an idea of the central value around which the values of X are spread. Hence, it is known as the measure of central tendency of X and is termed as the location parameter of the probability distribution of X.

In the following theorem, we discuss the properties of expectation of a simple random variable, some of these are required in the definition of expectation of a non-negative random variable.

Theorem 4.2.1. *Suppose X_1 and X_2 are simple random variables defined on a probability space $(\Omega, \mathbb{A}, P)$. Then*

(i) $E(X_1 \pm X_2) = E(X_1) \pm E(X_2)$.

(ii) $E(aX_1) = aE(X_1)$, *where a is any real number.*

(iii) $E(aX_1 + bX_2) = aE(X_1) + bE(X_2)$, *where a and b are any real numbers.*

(iv) If $X_1 = X_2$ a.s., then $E(X_1) = E(X_2)$ (Equivalence).

(v) If $X_1 \geq X_2$ on Ω, then $E(X_1) \geq E(X_2)$ (Dominance).

(vi) If $X_1 \geq 0$ a.s., then $E(X_1) \geq 0$.

(vii) If $X_1 \geq X_2$ a.s. then $E(X_1) \geq E(X_2)$ (Dominance).

Proof.

(i) Suppose $\mathbb{P}_1 = \{A_1, A_2, \cdots, A_k\}$ and $\mathbb{P}_2 = \{B_1, B_2, \cdots, B_l\}$ are measurable partitions of Ω and simple random variables X_1 and X_2 on $(\Omega, \mathbb{A}, P)$ are defined as follows.

$$X_1(\omega) = \sum_{i=1}^{k} a_i I_{A_i}(\omega) \quad \& \quad X_2(\omega) = \sum_{j=1}^{l} b_j I_{B_j}(\omega),$$

where $a_1, a_2, \cdots, a_k$ and $b_1, b_2, \cdots, b_l$ are any real numbers. Suppose $C_{ij} = A_i \cap B_j, i = 1, 2, \cdots, k$ and $j = 1, 2, \cdots, l$. In Theorem 2.4.7 it is proved that $\{C_{ij}, i = 1, 2, \cdots, k, \; j = 1, 2, \cdots, l\}$ forms a measurable partition of Ω and $X_1 \pm X_2$ is again a simple random variable given by,

$$X_1 \pm X_2 = \sum_{i=1}^{k} \sum_{j=1}^{l} (a_i \pm b_j) I_{C_{ij}}.$$

Further, note that for any $i = 1, 2, \cdots, k$,

$$\begin{aligned} A_i &= A_i \cap \Omega = A_i \cap \Big(\sum_{j=1}^{l} B_j \Big) = \sum_{j=1}^{l} (A_i \cap B_j) \\ \Rightarrow \quad P(A_i) &= P\Big(\sum_{j=1}^{l} (A_i \cap B_j) \Big) = \sum_{j=1}^{l} P(A_i \cap B_j), \end{aligned}$$

the last equality follows from the finite additivity of the probability measure P. Similarly, for any $j = 1, 2, \cdots, l$,

$$B_j = \sum_{i=1}^{k} (A_i \cap B_j) \quad \Rightarrow \quad P(B_j) = \sum_{i=1}^{k} P(A_i \cap B_j).$$

Now by the definition of expectation of a simple random variable, we have

$$\begin{aligned} E(X_1 \pm X_2) &= \sum_{i=1}^{k} \sum_{j=1}^{l} (a_i \pm b_j) P(A_i \cap B_j) \\ &= \sum_{i=1}^{k} \sum_{j=1}^{l} a_i P(A_i \cap B_j) \pm \sum_{i=1}^{k} \sum_{j=1}^{l} b_j P(A_i \cap B_j) \\ &= \sum_{i=1}^{k} a_i \Big(\sum_{j=1}^{l} P(A_i \cap B_j) \Big) \pm \sum_{j=1}^{l} b_j \Big(\sum_{i=1}^{k} P(A_i \cap B_j) \Big) \\ &= \sum_{i=1}^{k} a_i P(A_i) \pm \sum_{j=1}^{l} b_j P(B_j) = E(X_1) \pm E(X_2). \end{aligned}$$

(ii) Since X_1 is a simple random variable,

$$X_1(\omega) = \sum_{i=1}^{k} a_i I_{A_i}(\omega) \quad \Rightarrow \quad aX_1(\omega) = \sum_{i=1}^{k} aa_i I_{A_i}(\omega),$$

for any real number a. Thus, aX_1 is also a simple random variable. Hence,

$$E(aX_1) = \sum_{i=1}^{k} aa_i P(A_i) = a \sum_{i=1}^{k} a_i P(A_i) = aE(X_1).$$

(iii) Observe that

$$\begin{aligned} E(aX_1 + bX_2) &= E(aX_1) + E(bX_2) \text{ by (i)} \\ &= aE(X_1) + bE(X_2) \text{ by (ii).} \end{aligned}$$

(iv) As stated in (i), $X_1 - X_2 = \sum_{i=1}^{k}\sum_{j=1}^{l}(a_i - b_j)I_{C_{ij}}$ is a simple random variable. Note that $X_1 = X_2$ a.s. implies that $X_1(\omega) = X_2(\omega)\ \forall\ \omega \in N^c$ such that $P(N) = 0$. Thus, $a_i - b_j = 0$ on some partition sets C_{ij} and union of such sets is N^c. We define sets A and B as $A = \{(i,j) | a_i - b_j = 0\}$ and
$B = \{(i,j) | a_i - b_j \neq 0\}$, where the number of elements in B is $r < kl$. Suppose events C_{ij} are labeled as $D_1 D_2, \cdots, D_r$ corresponding to $(i,j) \in B$. Similarly, $a_i - b_j$ such that $(i,j) \in B$, are labeled as $d_m, m = 1, 2, \cdots, r$. Suppose $N = \bigcup_{m=1}^{r} D_m$. Since $P(N) = 0$, $P(D_m) = 0\ \forall\ m = 1, \cdots, r$. Hence,

$$\begin{aligned} E(X_1 - X_2) &= \sum_{i=1}^{k}\sum_{j=1}^{l}(a_i - b_j)P(C_{ij}) \\ &= \sum_{(i,j)\in A}(a_i - b_j)P(C_{ij}) + \sum_{(i,j)\in B}(a_i - b_j)P(C_{ij}) \\ &= \sum_{m=1}^{r} d_m P(D_m) = 0 \quad \Rightarrow \quad E(X_1) = E(X_2). \end{aligned}$$

(v) As shown in (i) $X_1 - X_2 = \sum_{i=1}^{k}\sum_{j=1}^{l}(a_i - b_j)I_{C_{ij}}$ is a simple random variable, with $(a_i - b_j) \geq 0\ \ \forall\ \ i\ \&\ j$ since it is given that $X_1 \geq X_2$ on Ω. By definition of expectation,

$$\begin{aligned} E(X_1 - X_2) &= \sum_{i=1}^{k}\sum_{j=1}^{l}(a_i - b_j)P(C_{ij}) \\ &\geq 0 \text{ since } (a_i - b_j)\ \&\ P(C_{ij}) \geq 0\ \ \forall\ \ i\ \ \&\ \ j \\ \Rightarrow E(X_1) &\geq E(X_2). \end{aligned}$$

(vi) $X_1 \geq 0 \;\; a.s. \;\; \Rightarrow \;\; X_1(\omega) \geq 0, \;\; \forall \;\; \omega \in N^c$ such that $P(N) = 0$. Thus if, $X_1(\omega) = \sum_{i=1}^{k} a_i I_{A_i}(\omega)$, some a_i's may be negative on some partition sets $A_i, \; i = 1, 2, \cdots, k$, but for such sets $P(A_i) = 0$. Hence,

$$E(X_1) = \sum_{i=1}^{k} a_i P(A_i) = \sum_{i:a_i \geq 0} a_i P(A_i) + \sum_{i:a_i < 0} a_i P(A_i),$$

where the first summation is over i for which $a_i \geq 0$ and the second summation is over i for which $a_i < 0$. Hence, $E(X_1) \geq 0$, the first term being non-negative and the second term being 0.

(vii) Note that

$$X_1 \geq X_2 \; a.s. \Rightarrow X_1 - X_2 \geq 0 \; a.s. \Rightarrow E(X_1 - X_2) \geq 0 \;\; \Rightarrow \;\; E(X_1) \geq E(X_2).$$

□

We use these properties and Theorem 2.4.3 to define the expectation of a non-negative and an arbitrary random variable in the next section.

Suppose X is a non-negative simple random variable. Then we have some more results related to $E(X)$. These are proved in the following theorem.

Theorem 4.2.2. *Suppose X is a non-negative simple random variable defined on a probability space $(\Omega, \mathbb{A}, P)$.*

(i) If $X = 0$ a.s., then $E(X) = 0$.

(ii) If $E(X) > 0$, then $P[X > 0] > 0$.

(iii) If $E(X) = 0$, then $X = 0 \quad a.s.$

(iv) $E(XI_{[X>0]}) = E(X)$.

Proof. X is a non-negative simple random variable. Hence, X can be expressed as $X(\omega) = \sum_{i=1}^{k} a_i I_{A_i}(\omega)$, where $\mathbb{P}_1 = \{A_1, A_2, \cdots, A_k\}$ is a measurable partition of Ω and $a_i \geq 0$ are real numbers.
(i) $X = 0 \;\; a.s. \;\; \Rightarrow \;\; X(\omega) = 0, \;\; \forall \;\; \omega \in N^c$ such that $P(N) = 0$. Thus in $X(\omega) = \sum_{i=1}^{k} a_i I_{A_i}(\omega)$, some a_i's may be positive on some partition sets $A_i, \; i = 1, 2, \cdots, k$, but for such sets $P(A_i) = 0$. Hence,

$$E(X) = \sum_{i=1}^{k} a_i P(A_i) = \sum_{i:a_i = 0} a_i P(A_i) + \sum_{i:a_i > 0} a_i P(A_i),$$

where the first summation is over i for which $a_i = 0$ and the second summation is over i for which $a_i > 0$. Hence, $E(X_1) = 0$, the first term being 0 since $a_i = 0$

and the second term being 0 since $P(A_i) = 0$.
(ii) Since $X \geq 0$, $E(X) \geq 0$. From (i)

$$\begin{aligned} X = 0 \text{ a.s.} &\Rightarrow E(X) = 0 \\ \text{Hence, } E(X) > 0 &\Rightarrow X \neq 0 \text{ a.s.} \\ &\Rightarrow P[X = 0] \neq 1 \Rightarrow P[X > 0] > 0. \end{aligned}$$

(iii) We prove $X = 0$ a.s. by contradiction. Suppose X takes non-zero values with positive probability, that is, $\exists\ j \in \{1, 2, \cdots, k\}$ such that $a_j > 0$ and $P(A_j) > 0$. Hence $E(X) = \sum_{i=1}^{k} a_i P(A_i) > a_j P(A_j) > 0$, which is a contradiction to the fact that $E(X) = 0$. Hence $X = 0$ a.s.
(iv) Observe that $X = XI_{[X>0]} + XI_{[X=0]}$. Hence by (i) of Theorem 4.2.1,

$$E(X) = E(XI_{[X>0]}) + E(XI_{[X=0]}) = E(XI_{[X>0]}) + 0 = E(XI_{[X>0]}).$$

□

4.3 Expectation of Non-Negative and Arbitrary Random Variables

Suppose X is a non-negative random variable defined on a probability space $(\Omega, \mathbb{A}, P)$ to $(\mathbb{R}, \mathbb{B}, P_X)$. Then by Theorem 2.4.3 there exists a non-decreasing sequence $\{X_n, n \geq 1\}$ of non-negative simple random variables such that $X_n \to X$ on Ω. Since $\{X_n, n \geq 1\}$ is a non-decreasing sequence of simple random variables, it follows from result (v) of Theorem 4.2.1, that $\{E(X_n), n \geq 1\}$ is also a non-decreasing sequence of non-negative real numbers and therefore its limit exists, it may be ∞. Hence, expectation of a non-negative random variable X is defined as follows.

Definition 4.3.1. *Expectation of a non-negative random variable: Suppose X is a non-negative random variable, then*

$$E(X) = \int_{\Omega} X dP = \lim_{n \to \infty} E(X_n),$$

where $\{X_n, n \geq 1\}$ is a non-decreasing sequence of non-negative simple random variables such that $X_n \to X$ on Ω and $E(X_n)$ is as in Definition 4.2.1.

Remark 4.3.1. It is possible that there exist more than one non-decreasing sequences $\{X_n, n \geq 1\}$ of non-negative simple random variables converging to X on Ω. However, it has been proved in Bhat [4] that for all such sequences, $\lim_{n\to\infty} E(X_n)$ is the same and hence $E(X)$ is uniquely defined,

Suppose X is an arbitrary random variable with $X^+$and X^- as its positive and negative parts respectively. Then X can be expressed as $X = X^+ - X^-$. By definition, X^+ and X^- are non-negative random variables and their expectations can be defined as in Definition 4.3.1. Hence, expectation of an arbitrary random variable X is defined as follows.

Definition 4.3.2. *Expectation of an arbitrary random variable: Suppose X is an arbitrary random variable with $X^+$and X^- as its positive and negative parts respectively. Then $E(X)$ is defined as $E(X) = \int_\Omega X dP = E(X^+) - E(X^-)$, provided at least one of the two $E(X^+)$ and $E(X^-)$ is finite.*

If both $E(X^+)$ and $E(X^-)$ are infinite then $E(X)$ does not exist as it is of the form $\infty - \infty$. If both $E(X^+)$ and $E(X^-)$ are finite then $E(X)$ is finite. If $E(X^+)$ is infinite then $E(X)$ exists and is ∞. If $E(X^-)$ is infinite then $E(X)$ exists and is $-\infty$.

Definition 4.3.3. *Integrable random variable: A random variable X is said to be an integrable random variable if $E(X)$ is finite.*

Theorem 4.3.1. *An arbitrary random variable X is integrable if and only if both $X^+$and X^- are integrable.*

Proof. Observe that

$$\begin{aligned} X \text{ is integrable} &\Rightarrow E(X) < \infty \Rightarrow E(X^+) < \infty \ \& \ E(X^-) < \infty \\ &\Rightarrow X^+ \ \& \ X^- \text{are integrable, conversely,} \\ X^+ \ \& \ X^- \text{are integrable} &\Rightarrow E(X^+) < \infty \ \& \ E(X^-) < \infty \\ &\Rightarrow E(X) = E(X^+) - E(X^-) < \infty \\ &\Rightarrow X \text{ is integrable.} \end{aligned}$$

□

In the following theorem, we prove some properties of expectation of non-negative and arbitrary random variables.

Theorem 4.3.2. *Suppose X_1 and X_2 are random variables defined on a probability space $(\Omega, \mathbb{A}, P)$ such that $E(X_1), E(X_2)$ and $E(X_1) + E(X_2)$ exist.*

(i) If $X_1 \geq 0$ a.s., then $E(X_1) \geq 0$ (Non-negativity).

(ii) $E(cX_1) = cE(X_1)$, where c is any real number.

(iii) If $X_1 = 0$ a.s., then $E(X_1) = 0$.

(iv) $E(X_1 + X_2) = E(X_1) + E(X_2)$ (Linearity).

(v) $E(aX_1 + bX_2) = aE(X_1) + bE(X_2)$, where $a, b \in \mathbb{R}$.

(vi) If $X_1 \geq X_2$ a.s., then $E(X_1) \geq E(X_2)$.

(vii) X_1 *is integrable if and only if* $|X_1|$ *is integrable.*

(viii) If $|X_1| \leq Y$, *where* Y *is an integrable random variable then* X_1 *is integrable.*

Proof. (i) $X_1 \geq 0$ a.s. implies that $X_1(\omega) \geq 0, \;\; \forall \;\; \omega \in N^c$ where $P(N) = 0$. Further, by Theorem 2.4.3, there exists a non-decreasing sequence $\{X_{1n}, n \geq 1\}$ of almost surely non-negative simple random variables such that $X_{1n} \to X_1$ on N^c. It is proved in Theorem 4.2.1 that if $X_{1n} \geq 0$ a.s. then $E(X_{1n}) \geq 0$. Hence,

$$E(X_1) = \lim_{n\to\infty} E(X_{1n}) \geq 0.$$

(ii) Suppose $X_1 \geq 0$ and $c \geq 0$. Since X_1 is a non-negative random variable, there exists a non-decreasing sequence $\{X_{1n}, n \geq 1\}$ of non-negative simple random variables such that $X_{1n} \to X_1$ on Ω. For $c \geq 0$, cX_{1n} is a non-negative simple random variable for all $n \geq 1$. Thus, $\{cX_{1n}, n \geq 1\}$ is a non-decreasing sequence of non-negative simple random variables such that $cX_{1n} \to cX_1$ on Ω. Hence,

$$E(cX_1) = \lim_{n\to\infty} E(cX_{1n}) = \lim_{n\to\infty} cE(X_{1n}) = c \lim_{n\to\infty} E(X_{1n}) = cE(X_1), \quad (4.1)$$

by result (ii) of Theorem 4.2.1. For arbitrary X_1 and $c \geq 0$, $cX_1 = (cX_1)^+ - (cX_1)^-$. Since $c > 0$, Observe that

$$(cX_1)^+ \quad = \quad \begin{cases} cX_1, & \text{if} \quad cX_1 \geq 0 \quad \Longleftrightarrow \quad X_1 \geq 0 \\ 0, & \text{if} \quad cX_1 < 0 \quad \Longleftrightarrow \quad X_1 < 0 \end{cases}$$

Hence, $(cX_1)^+ = cX_1^+$. Similarly, it can be shown that $(cX_1)^- = cX_1^-$. Thus, by definition

$$\begin{aligned} E(cX_1) &= E(cX_1)^+ - E(cX_1)^- = cE(X_1^+) - cE(X_1^-) \\ &= c(E(X_1^+) - E(X_1^-)) = cE(X_1). \end{aligned}$$

To establish such a relation for non-negative and arbitrary X and for $c < 0$, we first prove that $E(-X) = -E(X)$. By definition of a positive part and a negative part of a random variable, it is noted that,

$$\begin{aligned} (-X)^+ &= (-X)I_{[-X\geq 0]} = -XI_{[X\leq 0]} = X^- \\ \text{and } (-X)^- &= -(-X)I_{[-X<0]} = XI_{[X>0]} = X^+, \end{aligned}$$

which is logically true. Now, by definition of the expectation of an arbitrary random variable, we have,

$$\begin{aligned} E(-X) &= \int_\Omega (-X)^+ dP - \int_\Omega (-X)^- dP = \int_\Omega X^- dP - \int_\Omega X^+ dP \\ &= -\Big(\int_\Omega X^+ dP - \int_\Omega X^- dP\Big) = -E(X). \end{aligned} \quad (4.2)$$

Thus, for any random variable X, $E(-X) = -E(X)$. Now suppose $X_1 \geq 0$ and $c < 0$. To show that $E(cX_1) = cE(X_1)$, suppose $c = -a$, where $a > 0$ and $Y = aX_1$. Then

$$E(cX_1) = E(-aX_1) = E(-Y) = -E(Y) = -E(aX_1) = -aE(X_1) = cE(X_1),$$

where $E(aX_1) = aE(X_1)$ follows from (4.1). If X_1 is arbitrary and $c < 0$, we prove the result $E(cX_1) = cE(X_1)$ as follows. Note that

$$\begin{aligned} E(cX_1) &= E(-aX_1) = E(-Y) = -E(Y) \text{ by Equation (4.2)} \\ &= -(E(Y^+) - E(Y^-)) = -(E((aX_1)^+) - E((aX_1)^-)) \\ &= -(aE(X_1^+) - aE(X_1^-)) = -aE(X_1) = cE(X_1). \end{aligned}$$

Thus, for non-negative and arbitrary random variables $E(cX_1) = cE(X_1)$, where c is any real number.
(iii) Observe that,

$$\begin{aligned} X_1 = 0 \ a.s. \ &\Rightarrow \ X_1 \geq 0 \ a.s. \ \& \ -X_1 \geq 0 \ a.s. \\ &\Rightarrow \ E(X_1) \geq 0 \ \& \ -E(X_1) \geq 0 \ \Rightarrow \ E(X_1) = 0. \end{aligned}$$

(iv) We first prove the linearity property of expectations for the non-negative random variables. Since X_1 and X_2 are non-negative random variables, by Theorem 2.4.3, there exists a non-decreasing sequence $\{X_{1n}, n \geq 1\}$ of non-negative simple random variables such that $X_{1n} \to X_1$ on Ω and a non-decreasing sequence $\{X_{2n}, n \geq 1\}$ of non-negative simple random variables such that $X_{2n} \to X_2$ on Ω. Note that,

(a) $X_{1n} \to X_1, \ \& \ X_{2n} \to X_2$ on $\Omega \ \Rightarrow \ X_{1n} + X_{2n} \to X_1 + X_2$ on Ω,

(b) $X_{1n} \leq X_{1(n+1)} \ \& \ X_{2n} \leq X_{2(n+1)} \ \Rightarrow \ X_{1n} + X_{2n} \leq X_{1(n+1)} + X_{2(n+1)}$,

(c) $X_{1n} \geq 0 \ \& \ X_{2n} \geq 0 \ \Rightarrow \ X_{1n} + X_{2n} \geq 0, \ \forall \ n \geq 1$

(d) X_{1n} & X_{2n} are simple $\Rightarrow$ $X_{1n} + X_{2n}$ are simple

random variables for all $n \geq 1$ by Theorem 2.4.4. From $(a), (b), (c)$ and (d), it follows that $\{X_{1n} + X_{2n}, n \geq 1\}$ is a non-decreasing sequence of non-negative simple random variables such that $X_{1n} + X_{2n} \to X_1 + X_2$ on Ω. Hence, by the definition of expectation of a non-negative random variable,

$$\begin{aligned} E(X_1 + X_2) &= \lim_{n \to \infty} E(X_{1n} + X_{2n}) = \lim_{n \to \infty} E(X_{1n}) + \lim_{n \to \infty} E(X_{2n}) \\ &= E(X_1) + E(X_2). \end{aligned}$$

To prove linearity for arbitrary random variables, note that for any $A \in \mathbb{A}$ and $X \geq 0 \ a.s.$,

$$\begin{aligned} X &= XI_\Omega = X(I_A + I_{A^c}) = XI_A + XI_{A^c} \\ \Rightarrow \ E(X) &= E(XI_A + XI_{A^c}) = E(XI_A) + E(XI_{A^c}), \end{aligned}$$

by linearity of expectation of non-negative random variables. Since $XI_A \geq 0$ *a.s.* and $XI_{A^c} \geq 0$ *a.s.*, it implies that $E(X)$ is finite if $E(XI_A)$ and $E(XI_{A^c})$ are finite. This result can be extended to a finite partition of Ω. Suppose Ω is partitioned into $\{A_1, A_2, \cdots, A_6\}$ as follows.

$$A_1 = [X_1 \geq 0, X_2 < 0, X_1 + X_2 \geq 0], \quad A_2 = [X_1 < 0, X_2 \geq 0, X_1 + X_2 \geq 0],$$
$$A_3 = [X_1 \geq 0, X_2 \geq 0, X_1 + X_2 \geq 0], \quad A_4 = [X_1 < 0, X_2 \geq 0, X_1 + X_2 < 0],$$
$$A_5 = [X_1 \geq 0, X_2 < 0, X_1 + X_2 < 0], \quad A_6 = [X_1 < 0, X_2 < 0, X_1 + X_2 < 0] .$$

Since $E(X_1), E(X_2)$ and $E(X_1)+E(X_2)$ exist, we can assume that at least one of the random variables is integrable, so we assume that X_2 is integrable. Now on A_1, $X_1 + X_2 \geq 0$ and $-X_2 > 0$. Hence by linearity property of expectation of non-negative random variables, we have,

$$\begin{aligned} E((X_1 + X_2)I_{A_1}) + E(-X_2 I_{A_1}) &= E(X_1 I_{A_1}) \\ \Rightarrow \quad E((X_1 + X_2)I_{A_1}) + E(-X_2 I_{A_1}) + E(X_2 I_{A_1}) &= E(X_1 I_{A_1}) + E(X_2 I_{A_1}) \\ \Rightarrow \quad E((X_1 + X_2)I_{A_1}) &= E(X_1 I_{A_1}) + E(X_2 I_{A_1}). \end{aligned}$$

On A_2, $X_1 + X_2 \geq 0$ and $-X_1 > 0$, hence

$$E((X_1 + X_2)I_{A_2}) + E(-X_1 I_{A_2}) = E(X_2 I_{A_1}) < \infty \quad \Rightarrow \quad E(X_1 I_{A_2}) < \infty.$$

Adding $E(X_1 I_{A_2})$ to both sides we have,

$$E((X_1 + X_2)I_{A_2}) = E(X_1 I_{A_2}) + E(X_2 I_{A_2}).$$

On $A_3, X_1 \geq 0, X_2 \geq 0$, hence by linearity property of expectation of non-negative random variables,

$$E((X_1 + X_2)I_{A_3}) = E(X_1 I_{A_3}) + E(X_2 I_{A_3}).$$

On $A_4, -(X_1 + X_2) \geq 0, X_2 \geq 0$. Hence,

$$E(-(X_1 + X_2)I_{A_4}) + E(X_2 I_{A_4}) = E(-X_1 I_{A_4}).$$

Subtracting $E(X_2 I_{A_4})$ and changing signs on both sides, we get,

$$E((X_1 + X_2)I_{A_4}) = E(X_1 I_{A_4}) + E(X_2 I_{A_4}).$$

Using similar arguments we have,

$$\begin{aligned} E((X_1 + X_2)I_{A_5}) &= E(X_1 I_{A_5}) + E(X_2 I_{A_5}) \\ \text{and } E((X_1 + X_2)I_{A_6}) &= E(X_1 I_{A_6}) + E(X_2 I_{A_6}). \end{aligned}$$

Thus we have,

$$E(X_1 I_{A_i}) + E(X_2 I_{A_i}) = E((X_1 + X_2)I_{A_i}), \quad i = 1, 2, \cdots, 6.$$

From the definition of sets $\{A_1, A_2, \cdots, A_6\}$, we note that,

$$E(X_1(I_{A_1} + I_{A_3} + I_{A_5})) = E(X_1 I_{A_1} + X_1 I_{A_3} + X_1 I_{A_5}) = E(X_1^+)$$

$$\text{and } E(-X_1(I_{A_2} + I_{A_4} + I_{A_6})) = E(-X_1 I_{A_2} - X_1 I_{A_4} - X_1 I_{A_6}) = E(X_1^-),$$

so that $\sum_{i=1}^{6} E(X_1 I_{A_i}) = E(X_1)$. Similarly,

$$E(X_2(I_{A_2} + I_{A_3} + I_{A_4})) = E(X_2^+), \quad E(-X_2(I_{A_1} + I_{A_5} + I_{A_6})) = E(X_2^-),$$

and hence, $\sum_{i=1}^{6} E(X_2 I_{A_i}) = E(X_2)$. Further, on similar lines we get, $\sum_{i=1}^{6} E((X_1+X_2)I_{A_i}) = E(X_1+X_2)$ and hence, $E(X_1+X_2) = E(X_1)+E(X_2)$.

(v) Combining (ii) and (iv) we get, $E(aX_1+bX_2) = aE(X_1)+bE(X_2)$, where a and b are any real numbers. The linearity property is valid for a finite collection of random variables. It is valid for a countable collection of non-negative random variables, which follows from the monotone convergence theorem, which we prove in Chapter 8.

(vi) Observe that

$$\begin{aligned} X_1 \geq X_2 \ \ a.s. \quad &\Rightarrow \quad X_1 - X_2 \geq 0 \ \ a.s. \quad \Rightarrow \quad E(X_1 - X_2) \geq 0 \\ &\Rightarrow \quad E(X_1) \geq E(X_2). \end{aligned}$$

Thus, if $X_1 = X_2 \ \ a.s.$ then $E(X_1) = E(X_2)$.

(vii) It is given that X_1 is integrable, hence $E(X_1^+)$ and $E(X_1^-)$ are finite. As a consequence, using linearity of expectation

$$E(|X_1|) = E(X_1^+ + X_1^-) = E(X_1^+) + E(X_1^-) < \infty \quad \Rightarrow \quad |X_1| \text{ is integrable.}$$

Conversely, suppose $|X_1|$ is integrable, then

$$X_1 \leq |X_1| \quad \Rightarrow \quad E(X_1) \leq E(|X_1|) < \infty \quad \Rightarrow \quad X_1 \text{ is integrable.}$$

Alternatively, $|X_1|$ is integrable implies

$$\begin{aligned} E(|X_1|) < \infty \quad &\Rightarrow \quad E(X_1^+) < \infty \ \ \& \ \ E(X_1^-) < \infty \\ &\Rightarrow \quad E(X_1) = E(X_1^+) - E(X_1^-) < \infty \\ &\Rightarrow \quad E(X_1) \ \text{ is integrable.} \end{aligned}$$

(viii) $|X_1| \leq Y \quad \Rightarrow \quad E(|X_1|) < E(Y) < \infty \quad \Rightarrow \quad X_1$ is integrable. □

As a consequence of linearity of expectation proved in (iv), we have the following lemma, which is useful in many derivations.

Lemma 4.3.1. *Suppose X is an integrable random variable defined on $(\Omega, \mathbb{A}, P)$ and $A \in \mathbb{A}$. Then $E(X) = \int_\Omega X \, dP = \int_A X \, dP + \int_{A^c} X \, dP$.*

Proof. From the definition of an indicator function, $XI_A(\omega) = X(\omega)$ if $\omega \in A$ and 0 if $\omega \in A^c$. Similarly, $XI_{A^c}(\omega) = X(\omega)$ if $\omega \in A^c$ and 0 if $\omega \in A$. Since $A \cup A^c = \Omega, \ \forall \ \omega \in \Omega, I_A(\omega) + I_{A^c}(\omega) = 1$. Hence,

$$\begin{aligned} E(X) &= \int_\Omega X \, dP = \int_\Omega X \, (I_A + I_{A^c})dP = \int_\Omega (XI_A + XI_{A^c})dP \\ &= \int_\Omega XI_A dP + \int_\Omega XI_{A^c} dP \text{ by linearity of expectation} \\ &= \int_A XdP + \int_{A^c} XdP. \end{aligned}$$

□

Theorem 4.3.3. *Suppose X is an integrable random variable defined on $(\Omega, \mathbb{A}, P)$ and $A \in \mathbb{A}$ is such that $P(A) = 0$. Then $\int_A X \, dP = 0$.*

Proof. As in Lemma 4.3.1, $XI_A(\omega) = X(\omega)$ if $\omega \in A$ and it is 0 if $\omega \in A^c$. By the definition of expectation,

$$E(XI_A) = \int_\Omega XI_A \, dP = \int_A XI_A \, dP + \int_{A^c} XI_A \, dP = \int_A XI_A \, dP = \int_A X \, dP \, .$$

We prove the result $\int_A X \, dP = E(XI_A) = 0$ in three steps as we have defined $E(X)$ in three steps.

Case(i) Suppose X is a simple random variable given by $X(\omega) = \sum_{i=1}^k a_i I_{A_i}(\omega)$, where $\{A_1, A_2, \cdots, A_k\}$ is a measurable partitions of Ω. Then XI_A can be expressed as $XI_A(\omega) = \sum_{i=1}^k a_i I_{A_i \cap A}(\omega)$. Thus, XI_A is also a simple random variable and hence by the definition of the expectation of a simple random variable, we have

$$\int_A X \, dP = E(XI_A) = \sum_{i=1}^k a_i P(A_i \cap A) = 0 \ ,$$

since for each $i = 1, 2 \cdots, k, P(A_i \cap A) \leq P(A) = 0$.

Case(ii) Suppose X is a non-negative random variable, then there exists a non-decreasing sequence $\{X_n, n \geq 1\}$ of non-negative simple random variables such that $X_n \to X$ on Ω. As noted in case (i) for each $n \geq 1, X_n I_A$ is a simple random variable. Further, $X_n \leq X_{n+1} \ \Rightarrow \ X_n I_A \leq X_{n+1} I_A \ \forall \ n \geq 1$. Now $X_n \to X$ on Ω implies that for each $\omega \in \Omega$, given $\epsilon > 0 \ \exists \ n_0(\epsilon, \omega)$ such that $|X_n - X| < \epsilon$. Hence,

$$|X_n I_A - XI_A| = |X_n - X||I_A| \leq |X_n - X| < \epsilon \ \ \forall \ n_0(\epsilon, \omega) \ \Rightarrow \ X_n I_A \to XI_A$$

on Ω. Thus, $\{X_n I_A, n \geq 1\}$ is a non-decreasing sequence of non-negative simple random variables converging to XI_A on Ω and hence by the definition of expectation of a non-negative random variable,

$$\int_A X \, dP = E(XI_A) = \lim_{n \to \infty} E(X_n I_A) = 0 \ ,$$

by case(i).
Case(iii) Suppose X is an arbitrary random variable. Then

$$X = X^+ - X^- \quad \Rightarrow \quad XI_A = (X^+ - X^-)I_A = X^+I_A - X^-I_A.$$

Hence by linearity property of expectation and case(ii),

$$\int_A X\, dP = E(XI_A) = E(X^+I_A - X^-I_A) = E(X^+I_A) - E(X^-I_A) = 0.$$

□

We will prove the same result in Chapter 5, using another method. The result is parallel to the result that the Riemann integral $\int_a^a h(x)\, dx = 0$. In this case the Lebesgue measure of (a, a) is 0. Similarly if the probability measure of the event A is 0, then $\int_A X\, dP = 0$.

In Chapter 2, we have proved that Borel function of a random variable is again a random variable and thus we can find its expectation as an integral over Ω with respect to probability measure P. In the next theorem, we discuss how to find the expectation of a Borel function of a random variable, using the definition of integral of a Borel function as stated in Remark 4.1.1, with respect to a probability measure $P_X(\cdot)$ on $\mathbb{R}$.

Theorem 4.3.4. *Suppose X is a random variable defined on a probability space $(\Omega, \mathbb{A}, P)$ to $(\mathbb{R}, \mathbb{B}, P_X)$. Suppose $g : (\mathbb{R}, \mathbb{B}) \to (\mathbb{R}, \mathbb{B})$ is a Borel function. Then,*

$$E(g(X)) = \int_\Omega g(X)dP = \int_\mathbb{R} g(x)dP_X.$$

Proof. The result is proved in three steps as follows.
Case(i) Suppose g is a simple Borel function given by $g(x) = \sum_{j=1}^m a_j I_{S_j}(x)$, $x, a_j \in \mathbb{R}$, $S_j \in \mathbb{B}$, $j = 1, 2, \cdots, m$, where $\{S_1, S_2, \cdots, S_m\}$ is a measurable partition of $\mathbb{R}$. Hence as stated in Remark 4.1.1,

$$\int_\mathbb{R} g(x)dP_X = \sum_{j=1}^m a_j P_X(S_j). \tag{4.3}$$

Now to find $\int_\Omega g(X)dP$, we have to find out the nature of $g(X)$. Observe that,

$$\begin{aligned} g(X)(\omega) &= g(X(\omega)) = a_j \text{ if } X(\omega) \in S_j \iff \omega \in X^{-1}(S_j) \\ \iff g(X) &= \sum_{j=1}^m a_j I_{X^{-1}(S_j)}, \end{aligned}$$

where $\{X^{-1}(S_1), X^{-1}(S_2), \cdots, X^{-1}(S_m)\}$ is a measurable partition of Ω. Thus, $g(X)$ is a simple random variable on $(\Omega, \mathbb{A}, P)$, hence by the definition of expectation,

$$E(g(X)) = \int_\Omega g(X)dP = \sum_{j=1}^m a_j P(X^{-1}(S_j)) = \sum_{j=1}^m a_j P_X(S_j) = \int_\mathbb{R} g(x)dP_X,$$

the last equality follows from (4.3).
Case(ii) Suppose g is a non-negative Borel function. Then there exists a non-decreasing sequence $\{g_n, n \geq 1\}$ of non-negative simple Borel functions such that $g_n \to g$ on $\mathbb{R}$. Hence,

$$\int_{\mathbb{R}} g(x)dP_X = \lim_{n\to\infty} \int_{\mathbb{R}} g_n(x)dP_X = \lim_{n\to\infty} \int_{\Omega} g_n(X)dP, \tag{4.4}$$

by Case(i). Now observe that,

(i) $g(X)(\omega) = g(X(\omega)) \geq 0, \ \forall \ X(\omega) \in \mathbb{R} \iff \forall \ \omega \in X^{-1}(\mathbb{R}) = \Omega$. Thus, $g(X)$ is a non-negative random variable on $(\Omega, \mathbb{A}, P)$.

(ii) $g_n \to g$ on $\mathbb{R} \Rightarrow g_n(X(\omega)) \to g(X(\omega))$ on Ω .

(iii) For each $n \geq 1$, g_n is a simple Borel function, hence as shown in Case(i), for each n, $g_n(X)$ is a simple random variable on $(\Omega, \mathbb{A}, P)$.

(iv) Further the sequence $\{g_n, n \geq 1\}$ is non-decreasing implies that $\{g_n(X), n \geq 1\}$ is also a non-decreasing sequence of simple random variables.

Hence by the definition of expectation of a non-negative random variable and (4.4),

$$E(g(X)) = \int_{\Omega} g(X)dP = \lim_{n\to\infty} \int_{\Omega} g_n(X)dP = \lim_{n\to\infty} \int_{\mathbb{R}} g_n(x)dP_X = \int_{\mathbb{R}} g(x)dP_X \,.$$

Case(iii) Suppose g is an arbitrary function. Then $g = g^+ - g^-$ where g^+ and g^- are positive and negative parts of g respectively. Further both these are non-negative Borel functions. Hence, by the definition of expectation and Case(ii) we have,

$$\int_{\mathbb{R}} g(x)dP_X = \int_{\mathbb{R}} g^+(x)dP_X - \int_{\mathbb{R}} g^-(x)dP_X = \int_{\Omega} g^+(X)dP - \int_{\Omega} g^-(X)dP. \tag{4.5}$$

Now, observe that

$$(g(X))^+(\omega) = \begin{cases} g(X)(\omega) = g(X(\omega)), & \text{if} \quad g(X(\omega)) \geq 0, \\ 0, & \text{otherwise.} \end{cases}$$

Now, for $\omega \in \Omega, X(\omega) = x \in \mathbb{R}$ and hence

$$(g(X))^+(\omega) = \begin{cases} g(x), & \text{if} \quad g(x) \geq 0, \\ 0, & \text{otherwise.} \end{cases}$$

Thus,

$$(g(X))^+(\omega) = g^+(x) = g^+(X(\omega)), \ \forall \ \omega \in \Omega \iff (g(X))^+ = g^+(X).$$

On similar lines we get $(g(X))^- = g^-(X)$. Hence, by (4.5)

$$\begin{aligned} E(g(X)) &= \int_\Omega g(X)dP = \int_\Omega (g(X))^+ dP - \int_\Omega (g(X))^- dP \\ &= \int_\Omega g^+(X)dP - \int_\Omega g^-(X)dP \\ &= \int_\mathbb{R} g^+(x)dP_X - \int_\mathbb{R} g^-(x)dP_X \quad \text{by Case(ii)} \\ &= \int_\mathbb{R} g(x)dP_X. \end{aligned}$$

□

Remark 4.3.2. In Section 2.4, we have proved that if X is a simple random variable, then for any function g, $g(X)$ is also a simple random variable. From Case(i) of Theorem 4.3.4, we note that if g is a simple Borel function then for any random variable X, $g(X)$ is a simple random variable.

Theorem 4.3.4 states that expectation of a random variable can be obtained as an integral with respect to probability measure P or with respect to induced probability measure P_X on Ω and $\mathbb{R}$ respectively. Further note that to compute $E(g(X))$, it is not necessary to know the distribution of $g(X)$. For example, suppose $X \sim U(0,1)$, then $E(\sin(X)) = \int_0^1 \sin x \, dx = 1 - \cos(1)$ and it is not necessary to compute the distribution of $\sin(X)$.

In the correspondence theorem, Theorem 3.1.1, it is proved that there is a one to one correspondence between the distribution function F and the probability measure P. Hence, $E(X)$ can also be defined in terms of the distribution function of X as follows.

$$E(g(X)) = \int_\Omega g(X)dP = \int_\mathbb{R} g(x)dP_X = \int_\mathbb{R} g(x)dF(x),$$

where the last integral is the Riemann-Stieltjes integral.

Depending on the nature of F, the Riemann-Stieltjes integral can be obtained as follows.

(i) Suppose F is a discrete distribution function with set $D = \{x_i, i \geq 1\}$ as a set of points of discontinuity with p_{x_i} as the size of jump at $x_i, i \geq 1$, that is, X is a discrete random variable with probability mass function as $\{(x_i, p_{x_i}), i \geq 1\}$. In such a setup F is a step function with steps at x_i with jump size $p_{x_i}, i \geq 1$. From the definition of the Riemann-Stieltjes integral for a step function F, the expectation of any Borel function g of X is given by,

$$E(g(X)) = \int_\mathbb{R} g(x)dF(x) = \sum_{i\geq 1} g(x_i)p_{x_i}.$$

Note that $g(X)$ is integrable if $|g(X)|$ is integrable, that is, if the series $\sum_{i\geq 1} |g(x_i)|p_{x_i}$ is convergent.

(ii) Suppose F is a continuous distribution function and in addition if it is differentiable function with derivative $\frac{d}{dx}F(x) = f(x)$, then from the definition of the Riemann-Stieltjes integral for a differentiable function F, we have

$$E(g(X)) = \int_{\mathbb{R}} g(x)dF(x) = \int_{\mathbb{R}} g(x)f(x)dx.$$

Using this formula in the next example it is shown that the mean of a Cauchy distribution does not exist.

Example 4.3.1. Suppose a random variable X follows Cauchy $C(0, 1)$ distribution with location parameter 0, scale parameter 1 and the probability density function $f(x) = (1/\pi)(1/(1 + x^2))$, $x \in \mathbb{R}$. We examine whether $E(X^+)$ is finite using the following result. Suppose $g_r(x)$ and $h_s(x)$ are polynomial functions of degree r and s respectively. Then $\int_0^\infty g_r(x)/h_s(x)\ dx$ is convergent if and only if $s - r \geq 2$. Now $E(X^+)$ is given by, $E(X^+) = \int_0^\infty (1/\pi)(x/(1 + x^2))\ dx$. Observe that it is a divergent integral as $s - r = 1$. On similar lines, it follows that $E(X^-)$ is also divergent. Hence, the mean of the Cauchy $C(0, 1)$ distribution does not exist. □

If F is neither discrete nor continuous, then we find $E(X)$ as $E(X) = \int_{\mathbb{R}} xdF(x)$ using the definition of the Riemann-Stieltjes integral. Following example illustrates the computation of $E(X)$ in this setup.

Example 4.3.2. Suppose a distribution function $F : \mathbb{R} \to [0, 1]$ is defined as follows.

$$F(x) = \begin{cases} 0, & \text{if} \quad x < 0, \\ 1/4, & \text{if} \quad 0 \leq x < 1, \\ 1/2, & \text{if} \quad 1 \leq x < 2, \\ 1/2 + (x-2)/2, & \text{if} \quad 2 \leq x < 3, \\ 1, & \text{if} \quad x \geq 3. \end{cases}$$

In Example 3.3.2, we have noted that F is neither discrete nor continuous. Thus, we find $E(X)$ using the formula $E(X) = \int_{\mathbb{R}} g(x)dF(x)$ as a Riemann-Stieltjes integral. Thus, $E(X) = \sum xp_x + \int xF'(x)\ dx$, where x in summation denotes the point of discontinuity and corresponding p_x denotes the size of jump at the point of discontinuity. Further integral is over the range when F is continuous and differentiable with F' denoting its derivative. Thus,

$$E(X) = 0 \times \frac{1}{4} + 1 \times \frac{1}{4} + \frac{1}{2}\int_2^3 x\ dx = \frac{1}{4} + \frac{5}{4} = \frac{3}{2}\ .$$

In Example 3.3.2, it is shown that $F = (1/2)F_1 + (1/2)F_2$ where F_1 is a distribution function of a discrete random variable X_1 with probability mass function given by, $P[X_1 = 0] = 1/2 = P[X_1 = 1]$ and its mean is $(1/2)$. F_2 is a distribution function of uniform $U(2, 3)$ distribution and its mean is $(5/2)$. Using this approach $E(X) = (1/2)(1/2) + (1/2)(5/2) = 3/2$. □

Different choices of g in $E(g(X))$ correspond to various characteristics of a random variable X. We list these below.

(i) If g is an identity function, then $E(X)$ is the expectation of X, also known as the mean of X.

(ii) If $g(x) = x^r$, $r > 0$ then $E(X^r)$ is the r-th raw moment of X.

(iii) If $g(x) = (x - E(X))^r$, $r > 0$, then $E(X - E(X))^r$ is the r-th central moment of X.

(iv) If $g(x) = (x - a)^r$, $r > 0$, then $E(X - a)^r$ is the r-th moment of X around a.

(v) By taking $g(x) = \exp(tx)$, we get $E(g(X)) = E(\exp(tX)) = M_X(t)$, where the expectation exists for some values of t, which depends on the distribution of X. $M_X(t)$ is known as a moment generating function of X.

(vi) Suppose X is a discrete random variable with support as the set W of whole numbers. Then taking $g(x) = s^x$, we get $E(g(X)) = E(s^X) = P_X(s)$, where the expectation exists when $|s| \leq 1$. $P_X(s)$ is known as a probability generating function of X.

The existence of a mean clearly depends on how quickly tails decay. Following two theorems support this assertion. For a discrete random variable with support as the set W of whole numbers and for a continuous random variable, we have one more procedure to find $E(X)$. It is in terms of right tail probabilities. We derive these formulae in the following theorems.

Theorem 4.3.5. *Suppose X is a discrete random variable with support as the set W of whole numbers and suppose $P[X = i] = p_i, i \in W$. Then $E(X) = \sum_{n \geq 1} P[X \geq n]$.*

Proof. We have

$$\begin{aligned} E(X) &= \int_{\mathbb{R}} x dF(x) = \sum_{i \geq 0} i p_i = \sum_{i \geq 1} i p_i = p_1 + 2p_2 + 3p_3 + \cdots, \\ &= (p_1 + p_2 + p_3 + \cdots,) + (p_2 + p_3 + \cdots,) + (p_3 + p_4 + \cdots,) + \cdots \\ &= P[X \geq 1] + P[X \geq 2] + P[X \geq 3] + \cdots = \sum_{n \geq 1} P[X \geq n]. \end{aligned}$$

$$\begin{aligned} \text{Alternatively, } E(X) &= \int_{\mathbb{R}} x dF(x) = \sum_{i \geq 0} i p_i = \sum_{i \geq 1} i p_i = \sum_{i=1}^{\infty} \sum_{n=1}^{i} p_i \\ &= \sum_{n=1}^{\infty} \sum_{i=n}^{\infty} p_i = \sum_{n \geq 1} P[X \geq n]. \end{aligned}$$

□

Theorem 4.3.6. *Suppose X is a continuous random variable with distribution function F. Suppose F is differentiable with derivative $f(x)$. Then*

$$E(X) = \begin{cases} \int\limits_0^\infty [1 - F(x)]dx, & \text{if} \quad X \geq 0, \\ \int\limits_0^\infty [1 - F(x)]dx - \int\limits_{-\infty}^0 F(x)dx, & \text{if} \quad X \text{ is arbitrary.} \end{cases}$$

Proof. Suppose $X \geq 0$, then using integration by parts we have,

$$\begin{aligned} \int_0^\infty [1 - F(x)]dx &= \int_0^\infty [1 - F(x)]\frac{d}{dx}x\ dx = [x(1 - F(x))]_0^\infty + \int_0^\infty xf(x)\ dx \\ &= \int_0^\infty xf(x)\ dx = E(X), \end{aligned}$$

the term $[x(1 - F(x))]_0^\infty$ is 0 in view of the fact that $(1 - F(x)) \to 0$ faster than $x \to \infty$. Thus, for non-negative X,

$$E(X) = \int_0^\infty [1 - F(x)]\ dx = \int_0^\infty S(x)\ dx = \int_0^\infty P[X \geq x]\ dx,$$

where $1 - F(x) = S(x)$ is known as a survival function. Now suppose X is an arbitrary random variable. Consider

$$\begin{aligned} -\int_{-\infty}^0 F(x)\ dx &= -\int_{-\infty}^0 F(x)\frac{d}{dx}x\ dx = -[xF(x)]_{-\infty}^0 + \int_{-\infty}^0 xf(x)\ dx \\ &= 0 + \int_{-\infty}^0 xf(x)\ dx. \end{aligned}$$

Hence,

$$\begin{aligned} -\int_{\infty}^0 F(x)\ dx + \int_0^\infty [1 - F(x)]dx &= \int_{-\infty}^0 xf(x)\ dx + \int_0^\infty xf(x)\ dx \\ &= \int_{-\infty}^\infty xf(x)\ dx = E(X). \end{aligned}$$

□

Example 4.3.3. Suppose X follows exponential distribution with scale parameter θ. We find $E(X)$ using the formula in terms of its distribution function. The distribution function of X is given by,

$$F(x) = \begin{cases} 0 & \text{if} \quad x \leq 0, \\ 1 - e^{-\theta x}, & \text{if} \quad x \geq 0. \end{cases}$$

$$\text{Hence,} \quad E(X) = \int_0^\infty [1 - F(x)]dx = \int_0^\infty e^{-\theta x}dx = 1/\theta. \qquad \square$$

Example 4.3.4. Suppose X is a discrete random variable such that $P[X = i] = c/i^{r+2}$, $i = 1, 2, \cdots$, $r \geq 1$ and c is a norming constant. We have $E(X^k) = \sum_{i\geq 1} i^k(c/i^{r+2}) = \sum_{i\geq 1} c/i^{r-k+2}$ and the series is convergent for all k such that $r - k + 2 \geq 2$, that is, $k \leq r$ and divergent for all $k > r$. Hence, moments up to order r are finite, but moments of order higher than r are infinite. □

Example 4.3.5. Suppose X is a discrete random variable such that $P[X = 2^i] = 2^{-i}$, $i = 1, 2, \cdots$. Observe that

$$P[X < \infty] = \sum_{i\geq 1} P[X = 2^i] = \sum_{i\geq 1} 2^{-i} = 1.$$

But $E(X) = \sum_{i\geq 1} 2^i 2^{-i} = \infty$. Thus, even if X is finite, $E(X)$ is infinite. □

So far we discussed the expectation of a random variable. We now define the expectation of a random vector.

Definition 4.3.4. *Expectation of a random vector: Suppose* $\underline{Z} = (X_1, X_2, \cdots, X_k)'$ *is a k-variate random vector defined on a probability space* $(\Omega, \mathbb{A}, P)$ *to* $(\mathbb{R}^k, \mathbb{B}^k, P_{\underline{Z}})$. *Then expectation of* $\underline{Z}$ *is defined as* $E(\underline{Z}) = (E(X_1), E(X_1), \cdots, E(X_k))'$.

Definition 4.3.5. *Expectation of a sequence of random variable: Suppose* $\underline{Z} = \{X_n, n \geq 1\}$ *is a sequence of random variables defined on a probability space* $(\Omega, \mathbb{A}, P)$ *to* $(\mathbb{R}^\infty, \mathbb{B}^\infty, P_{\underline{Z}})$. *Then expectation of a sequence of random variables is defined as a sequence of expectations of component random variables and it is given by* $\{E(X_n), n \geq 1\}$.

Thus, the expectation of a random vector is basically defined in terms of the expectations of component random variables and hence all the results developed above for univariate setup are applicable for vector setup as well.

It is known that there is an equivalence between $\mathbb{R}^2$ and a class of complex numbers. This fact is used to define a complex valued random variable and its expectation.

Definition 4.3.6. *Complex valued random variable and its expectation: Suppose* $\underline{Z} = (X, Y)'$ *is a bivariate random vector defined on a probability space* $(\Omega, \mathbb{A}, P)$ *to* $(\mathbb{R}^2, \mathbb{B}^2, P_{\underline{Z}})$. *Then* $W = X + iY$ *is a complex valued random variable and its expectation is defined as* $E(W) = E(X) + iE(Y)$.

A complex valued random variable $W = X + iY$ is said to be integrable if and only if both X and Y are integrable. In the following theorem, we prove that $|E(W)| \leq E(|W|)$.

Theorem 4.3.7. *Suppose* $W = X + iY$ *is a complex valued random variable, then*

$$|E(W)| \leq E(|W|).$$

Proof. Using polar co-ordinates, a complex valued random variable W can be expressed as

$$W = re^{i\theta} \quad \text{where} \quad r = |W| = (X^2 + Y^2)^{1/2} \quad \& \quad \theta = \tan^{-1}(Y/X).$$

Suppose $E(W) = E(X) + iE(Y) = \alpha e^{i\beta}$. Thus,

$$\begin{aligned} |E(W)| &= \alpha = e^{-i\beta}E(W) = e^{-i\beta}E(re^{i\theta}) = E(re^{i(\theta-\beta)}) \\ &= E(r\cos(\theta-\beta)) \quad \text{complex part is 0 as left hand side is real} \\ &\leq E(r) = E(|W|). \end{aligned}$$

□

In the next section, we discuss a special type of complex valued random variable and its expectation. It is known as a characteristic function of a random variable and it is heavily used in probability theory to derive a variety of results. We use it in all the subsequent chapters. It is similar to the moment generating function or probability generating function as defined in this section. But we note in the next section that a characteristic function has some special features and hence it is preferred over other generating functions.

4.4 Characteristic Function

We begin with a definition of a characteristic function.

Definition 4.4.1. *Characteristic function of a random variable: The characteristic function $\phi_X(t)$ of a random variable X for $t \in \mathbb{R}$ is defined as*

$$\begin{aligned} \phi_X(t) &= E(\exp(itX)) = E(\cos tX + i\sin tX) = E(\cos tX) + iE(\sin tX) \\ &= \int_{-\infty}^{\infty} \exp(itx)dF(x). \end{aligned}$$

Note that for any $t \in \mathbb{R}$,

$$|\cos tX| \leq 1 \;\; \& \;\; |\sin tX| \leq 1 \;\; \Rightarrow \;\; E(\cos tX) \leq 1 \;\; \& \;\; E(\sin tX) \leq 1.$$

Thus, real and complex parts are integrable random variables and hence $\exp(itX)$ is an integrable random variable. Consequently, the domain of definition of $\phi_X(t) = E(\exp(itX))$ is $\mathbb{R}$. It is an important property of a characteristic function and hence while proving a variety of results in probability theory, a characteristic function is preferred over other generating functions. Observe that the moment generating function may not exist for all real numbers. For example, if X has exponential distribution with scale parameter θ, then its moment generating function is given by $(1 - t/\theta)^{-1}$ and it is defined for all $t < \theta$. On the other hand the characteristic function is $(1-it/\theta)^{-1}$ and it is defined for all $t \in \mathbb{R}$.

We now discuss some properties of a characteristic function.

(i) If $Y = aX + b$ where $a \neq 0$ and b are any real numbers, then a characteristic function of Y is given by,

$$\phi_Y(t) = E(e^{itY}) = E(e^{it(aX+b)}) = e^{itb}\phi_X(at).$$

In particular, if $a = -1$ and $b = 0$, then

$$\begin{aligned} \phi_{-X}(t) &= \phi_X(-t) = E(\cos(-tX)) + iE(\sin(-tX)) \\ &= E(\cos tX) - iE(\sin(tX)) = \overline{\phi_X(t)}, \end{aligned}$$

a complex conjugate of $\phi_X(t)$.

(ii) Suppose the distribution of X is symmetric around 0, then X and$-X$ are identically distributed, hence they have the same characteristic function. Thus,

$$\phi_X(t) = \phi_{-X}(t) = \phi_X(-t) = \overline{\phi_X(t)} \iff \phi_X(t) \quad \text{is a real valued}$$

function for the distribution symmetric around 0.

(iii) If F is defined as $F = \sum_{i=1}^k \alpha_i F_i, \alpha_i \geq 0$ & $\sum_{i=1}^k \alpha_i = 1$, then it is a convex combination of distribution functions. Its characteristic function $\phi(t)$ is again a convex combination of the characteristic functions $\phi_i(t)$ of component distribution functions and is given by $\phi(t) = \sum_{i=1}^k \alpha_i \phi_i(t)$. This result can also be extended to convex combination of countably many distribution functions.

Table 4.1 lists the characteristic functions of some standard distributions.

To study some more properties of a characteristic function, we first introduce a concept of a random variable and a sequence of random variables being bounded in probability.

TABLE 4.1
Characteristic Functions of Standard Distributions

Distribution	Characteristic Function
Degenerate at a	e^{ita}
Binomial $B(n,p)$	$(1-p+pe^{it})^n$
Geometric $Ge(p)$	$p/(1-(1-p)e^{it})$
Poisson $P(\lambda)$	$e^{\lambda(e^{it}-1)}$
Uniform $U(a,b)$	$(e^{itb}-e^{ita})(it(b-a))$
Uniform $U(-1,1)$	$\sin t/t$
Exponential $Exp(\alpha)$	$(1-it/\alpha)^{-1}$
Gamma $G(\alpha,\lambda)$	$(1-it/\alpha)^{-\lambda}$
Normal $N(\mu,\sigma^2)$	$e^{i\mu t-t^2\sigma^2/2}$
Cauchy $C(0,1)$	$e^{-\lvert t\rvert}$

Definition 4.4.2. *A random variables X is said to be bounded in probability if given $\epsilon > 0 \;\exists\; M > 0$ such that $P[|X| > M] < \epsilon$.*

Definition 4.4.3. *A sequence of random variables $\{X_n, n \geq 1\}$ is said to be bounded in probability if given $\epsilon > 0$, there exists a constant K and an integer n_0 such that,*

$$P[|X_n| \leq K] \geq 1 - \epsilon \iff P[|X_n| > K] < \epsilon, \quad \forall \quad n \geq n_0.$$

If a sequence $\{X_n, n \geq 1\}$ is bounded in probability, then we write $X_n = O_p(1)$. More generally, if $a_n X_n$ is bounded in probability, then we write $X_n = O_p(1/a_n)$. In the following example, the sequence is not bounded in probability.

Example 4.4.1. Suppose $\{X_n, n \geq 1\}$ is a sequence of random variables such that $X_n \sim U(n, n+1)$ distribution, $n \geq 1$. Hence for any,

$$K > 0, \quad P[|X_n| > K] = P[X_n > K] = 1, \quad \forall \quad n > K .$$

Hence the sequence $\{X_n, n \geq 1\}$ is not bounded in probability. □

Following lemma proves that a real random variable is always bounded in probability. The proof is based on the property of a distribution function which states that $F(x) \to 1$ as $x \to \infty$ and $F(x) \to 0$ as $x \to -\infty$.

Lemma 4.4.1. *Suppose X is a random variable defined on a probability space $(\Omega, \mathbb{A}, P)$ to $(\mathbb{R}, \mathbb{B}, P_X)$. Then X is bounded in probability.*

Proof. Suppose F denotes the distribution function of X, then $F(x) \to 1$ as $x \to \infty$ and $F(x) \to 0$ as $x \to -\infty$. Hence, as $x \to \infty$

$$\begin{aligned} P[|X| > x] &= P[X > x] + P[X < -x] = 1 - F(x) - F(-x-) \to 0, \\ &\Rightarrow \text{given } \epsilon > 0 \;\exists\; M > 0, \text{ such that } P[|X| > M] < \epsilon . \end{aligned}$$

Thus, X is bounded in probability. □

We use Lemma 4.4.1 in the following theorem to establish the uniform continuity of a characteristic function.

Theorem 4.4.1. *Suppose $\phi_X(t)$ is a characteristic function of a random variable X. Then (i) $|\phi_X(t)| \leq 1$, (ii) $\phi_X(0) = 1$ and (iii) $\phi_X(\cdot)$ is uniformly continuous on $\mathbb{R}$.*

Proof. A characteristic function of a random variable X is given by $\phi_X(t) = E(\exp(itX))$ for $t \in \mathbb{R}$.
(i) Using Theorem 4.3.7, we have

$$\begin{aligned} |\phi_X(t)| &= |E(\cos tX) + iE(\sin tX)| \leq E(|\cos tX + i \sin tX|) \\ &= E((\cos^2 tX + \sin^2 tX)^{1/2}) = 1. \end{aligned}$$

(ii) Using the result that $\cos(0) = 1$ and $\sin(0) = 0$, we get $\phi_X(0) = 1$.
(iii) To prove the uniform continuity of a characteristic function, observe that for t & $h \in \mathbb{R}$,

$$\begin{aligned} |\phi_X(t+h) - \phi_X(t)| &= \left|\int_{-\infty}^{\infty} e^{i(t+h)u} dF(u) - \int_{-\infty}^{\infty} e^{itu} dF(u)\right| \\ &\leq \int_{-\infty}^{\infty} |e^{itu}|\ |e^{ihu} - 1| dF(u) \\ &= \int_{-\infty}^{\infty} |e^{ihu} - 1| dF(u) \ \text{ as } \ |e^{itu}| = 1. \end{aligned}$$

Now X is a real random variable and hence it is bounded in probability, that is, given $\epsilon > 0\ \exists\ M = M(\epsilon) > 0$ such that $P[|X| > M] < \epsilon$. Note that for $|u| \leq M$,

$$|e^{ihu} - 1| = [(\cos(hu) - 1)^2 + \sin^2(hu)]^{1/2} \ \to\ 0 \ \text{ as } \ h \ \to\ 0,$$

that is, for $|u| \leq M$, given $\epsilon > 0$ we can find $h_0(\epsilon)$ such that $|e^{ihu} - 1| < \epsilon$ for all $|h| < h_0(\epsilon)$. Using these results in the above derivation we have,

$$\begin{aligned} |\phi_X(t+h) - \phi_X(t)| &\leq \int_{-\infty}^{\infty} |e^{ihu} - 1| dF(u) \\ &= \int_{|u|\leq M} |e^{ihu} - 1| dF(u) + \int_{|u|>M} |e^{ihu} - 1| dF(u) \\ &\leq \epsilon P[|X| \leq M] + 2P[|X| > M] \\ &\leq \epsilon + 2\epsilon \quad \forall \ |h| < h_0(\epsilon). \end{aligned}$$

In the third step, we use the result that $|e^{ihu} - 1| \ \leq |e^{ihu}| + |1| \leq 2$. Thus, $|\phi_X(t_1) - \phi_X(t)| < 3\epsilon \ \forall\ t_1 \in N_h(t)$ and h does not depend on t. Hence, $\phi_X(\cdot)$ is uniformly continuous on $\mathbb{R}$. □

We now discuss the connection between the existence of moments of X and differentiability of its characteristic function in the following theorems. We state a theorem below and refer to Loeve [16] for its proof.

Theorem 4.4.2. *If the characteristic function of a random variable X is differentiable n times and has derivatives at $t = 0$, then all moments up to order n exist and are finite if n is even and all moments up to order $n - 1$ exist and are finite if n is odd.*

In the following two theorems, the scenario is opposite to the above theorem. We first state a result (see p. 136 of Rao [19]), which specifies a sufficient condition for the interchange of integration and differentiation. It is to be noted that differentiation is basically a limit. Hence the lemma is also useful to interchange limit and integration. We prove similar result in Chapter 8, which allows interchange of limit and expectation, that is, limit and integration. This lemma is used in the proof of the next theorem and in some more theorems in this section.

Lemma 4.4.2. *Suppose g is a measurable function such that $\frac{d}{dt}g(t,x)$ exists for t in some set D. If $|\frac{d}{dt}g(t,x)| < G(x)$ where G is integrable, then $\frac{d}{dt}\int_{\mathbb{R}} g(t,x)dF(x) = \int_{\mathbb{R}} \frac{d}{dt}g(t,x)dF(x)$.*

Theorem 4.4.3. *Suppose $\phi_X(t)$ is a characteristic function of a random variable X. If $E(|X|^n) < \infty$ where n is a positive integer, then $\phi_X(t)$ is differentiable n times for all $t \in \mathbb{R}$ and n-th derivative $\phi_X^{(n)}(t)$ is given by,*

$$\phi_X^{(n)}(t) = \frac{d^n}{dt^n}\phi_X(t) = \int_{-\infty}^{\infty} (ix)^n e^{itx} dF(x) \quad \& \quad \phi_X^{(n)}(0) = i^n E(X^n).$$

Proof. It is given that $E(|X|^n) < \infty$ and hence $E(X^n) < \infty$. Using the fact that $|e^{itx}| \leq 1$ and $|i^n| = 1$ we note that,

$$\begin{aligned}\left|\int_{-\infty}^{\infty} (ix)^n e^{itx} dF(x)\right| &\leq \int_{-\infty}^{\infty} |(ix)^n||e^{itx}|dF(x)\\ &\leq \int_{-\infty}^{\infty} |x|^n dF(x) = E(|X|^n) < \infty.\end{aligned}$$

Further, $\int_{-\infty}^{\infty}(ix)^n e^{itx}dF(x) = \int_{-\infty}^{\infty} \frac{d^n}{dt^n} e^{itx}dF(x)$. Since $\left|\int_{-\infty}^{\infty}(ix)^n e^{itx}dF(x)\right|$ is finite, by Lemma 4.4.2, the integration and differentiation can be interchanged. Hence we have,

$$\begin{aligned}\int_{-\infty}^{\infty} (ix)^n e^{itx} dF(x) &= \int_{-\infty}^{\infty} \frac{d^n}{dt^n} e^{itx} dF(x) = \frac{d^n}{dt^n}\int_{-\infty}^{\infty} e^{itx} dF(x) = \frac{d^n}{dt^n}\phi_X(t)\\ \Rightarrow \quad \frac{d^n}{dt^n}\phi_X(t)\Big|_{t=0} &= \int_{-\infty}^{\infty} (ix)^n dF(x) = i^n E(X^n).\end{aligned}$$

□

Example 4.4.2. Suppose $\phi(t) = 1/(1+t^2)$ is a characteristic function. Then

$$\frac{d}{dt}\frac{1}{1+t^2} = -\frac{2t}{(1+t^2)^2} \quad \& \quad \frac{d^2}{dt^2}\frac{1}{1+t^2} = \frac{-2(1+t^2)^2 + 8t^2(1+t^2)}{(1+t^2)^4}.$$

Hence, $\mu_1' = 0, \mu_2' = 2 = \mu_2$. □

The next theorem proves that if X has finite moments up to certain order then its characteristic function can be expressed in terms of moments. It is a sort of Taylor's series expansion for a characteristic function around $t = 0$ with a remainder term.

Theorem 4.4.4. *Suppose $\phi_X(t)$ is a characteristic function of a random variable X. If $E(|X|^k) < \infty$ where k is a positive integer, then $\phi_X(t)$ can be expressed,*

$$\phi_X(t) = \sum_{s=0}^{k-1} \frac{(it)^s}{s!}E(X^s) + t^k \int_0^1 \frac{(1-u)^{k-1}}{(k-1)!}\phi_X^{(k)}(tu)\, du,$$

where $\phi_X^{(k)}(tu)$ is the k-th derivative of $\phi_X(t)$ at tu.

Proof. By the definition of a characteristic function for $t \in \mathbb{R}$,

$$\phi_X(t) = \int_{-\infty}^{\infty} e^{itx} dF(x) \quad \Rightarrow \quad \frac{d^k}{dt^k}\phi_X(t) = \int_{-\infty}^{\infty} (ix)^k e^{itx} dF(x).$$

Hence, k-th derivative of $\phi_X(t)$ at tu is $\int_{-\infty}^{\infty}(ix)^k e^{itux} dF(x)$. Using integration by parts repeatedly we simplify the integral $I_k = \int_0^1 \frac{(1-u)^{k-1}}{(k-1)!} e^{iutx}\, du$ as follows.

$$I_k = \int_0^1 \frac{(1-u)^{k-1}}{(k-1)!} \frac{d}{du}\frac{e^{iutx}}{itx}\, du = \left[\frac{(1-u)^{k-1}}{(k-1)!}\frac{e^{iutx}}{itx}\right]_0^1 + \int_0^1 \frac{(1-u)^{k-2}}{(k-2)!}\frac{e^{iutx}}{itx}\, du$$

$$= \frac{-1}{itx(k-1)!} + \frac{1}{itx}\int_0^1 \frac{(1-u)^{k-2}}{(k-2)!} e^{iutx}\, du$$

$$= -\frac{1}{itx(k-1)!} - \frac{1}{(itx)^2(k-2)!} + \frac{1}{(itx)^2}\int_0^1 \frac{(1-u)^{k-3}}{(k-3)!} e^{iutx}\, du$$

$$\vdots$$

$$= -\frac{1}{itx(k-1)!} - \frac{1}{(itx)^2(k-2)!} - \cdots - \frac{1}{(itx)^{k-1}(k-(k-1))!}$$
$$+ \frac{1}{(itx)^{k-1}}\int_0^1 \frac{(1-u)^{k-k}}{(k-k)!} e^{iutx}\, du.$$

It further simplifies to

$$I_k = -\frac{1}{itx(k-1)!} - \frac{1}{(itx)^2(k-2)!} - \cdots - \frac{1}{(itx)^{k-1}(k-(k-1))!} + \frac{1}{(itx)^{k-1}}\left[\frac{e^{itx}-1}{itx}\right].$$

Hence, $\frac{e^{itx}}{(itx)^k}$ can be expressed as follows.

$$\frac{e^{itx}}{(itx)^k} = \frac{1}{itx(k-1)!} + \frac{1}{(itx)^2(k-2)!} + \cdots + \frac{1}{(itx)^k} + \int_0^1 \frac{(1-u)^{k-1}}{(k-1)!} e^{iutx}\, du$$

$$\Rightarrow \quad e^{itx} = \sum_{s=0}^{k-1} \frac{(itx)^s}{s!} + (itx)^k \int_0^1 \frac{(1-u)^{k-1}}{(k-1)!} e^{iutx}\, du.$$

Hence, $\phi_X(t) = \int_{-\infty}^{\infty} e^{itx} dF(x)$ can be written as follows.

$$\begin{aligned}\phi_X(t) &= \sum_{s=0}^{k-1} \frac{(it)^s}{s!} \int_{-\infty}^{\infty} x^s dF(x) + \int_0^1 \frac{(1-u)^{k-1}}{(k-1)!} \left[\int_{-\infty}^{\infty} (itx)^k e^{iutx} dF(x)\right] du \\ &= \sum_{s=0}^{k-1} \frac{(it)^s}{s!} E(X^s) + t^k \int_0^1 \frac{(1-u)^{k-1}}{(k-1)!} \left[\int_{-\infty}^{\infty} (ix)^k e^{iutx} dF(x)\right] du \\ &= \sum_{s=0}^{k-1} \frac{(it)^s}{s!} E(X^s) + t^k \int_0^1 \frac{(1-u)^{k-1}}{(k-1)!} \phi_X^{(k)}(tu)\, du.\end{aligned}$$

□

The next corollary follows immediately from this theorem.

Corollary 4.4.1. *Suppose $\phi_X(t)$ is a characteristic function of a random variable X. If moments of all order of X are finite, then $\phi_X(t)$ can be expressed as,*

$$\phi_X(t) = \sum_{s=0}^{\infty} \frac{(it)^s}{s!} E(X^s).$$

We have noted above that an important property of a characteristic function is that its domain is $\mathbb{R}$. Another important property is a uniqueness theorem which states that if two random variables have the same characteristic function then they also have the same distribution. This theorem is heavily used to identify the limit laws in convergence in distribution in Chapter 7. We first prove an inversion theorem, from which the uniqueness theorem follows. In addition, an inversion theorem provides a formula for explicitly computing the distribution function from the characteristic function. By definition, given a distribution function F, we can find the corresponding characteristic function and using the inversion theorem, we can find a distribution function from the characteristic function. Thus, there is a one-to-one correspondence between a distribution function and a characteristic function and every random variable possesses a unique characteristic function. In other words, the characteristic function characterizes the distribution uniquely and hence it is labeled as a characteristic function.

We state below a lemma which is needed to prove the inversion theorem. For proof we refer to p. 557 of Gut [13].

Lemma 4.4.3. *For $\alpha > 0$,*

$$(i) \int_0^c \frac{\sin(\alpha t)}{t} dt \le \pi \;\forall\; c > 0 \;\;\&\;\; (ii) \int_0^c \frac{\sin(\alpha t)}{t}\, dt \to \frac{\pi}{2} \quad as \;\; c \to \infty.$$

As a consequence of the second result of this lemma we have

$$\frac{1}{\pi}\int_0^{\infty} \frac{\sin(\alpha t)}{t}\, dt = \lim_{c\to\infty} \frac{1}{\pi}\int_0^c \frac{\sin(\alpha t)}{t}\, dt \;=\; \begin{cases} \frac{1}{2} & \text{if} \quad \alpha > 0, \\ -\frac{1}{2} & \text{if} \quad \alpha < 0, \\ 0, & \text{if} \quad \alpha = 0. \end{cases} \quad (4.6)$$

Suppose $\alpha < 0$ and $\alpha = -\delta$ where $\delta > 0$. Then from (i) of the above lemma, we have $\forall\ c > 0\ \&\ \forall\ \alpha \in \mathbb{R}$,

$$\int_0^c \frac{\sin(\alpha t)}{t}\, dt = -\int_0^c \frac{\sin(\delta t)}{t}\, dt \ \geq\ -\pi \Rightarrow \quad \left|\frac{1}{\pi}\int_0^c \frac{\sin(\alpha t)}{t}\, dt\right| \leq 1.$$

Theorem 4.4.5. *Inversion theorem: Suppose $\phi_X(t)$ is the characteristic function of a random variable X with distribution function F_X. Then*

$$\begin{aligned} &F_X(b) - F_X(a) + \frac{1}{2}P[X = a] - \frac{1}{2}P[X = b] \\ &= \frac{1}{2\pi}\lim_{c\to\infty}\int_{-c}^{c}\left[\frac{e^{-ita} - e^{-itb}}{it}\right]\phi_X(t)\, dt. \end{aligned}$$

If $a < b \in C(F_X)$, where $C(F_X)$ is a set of points of continuity of F, then

$$F_X(b) - F_X(a) = \frac{1}{2\pi}\lim_{c\to\infty}\int_{-c}^{c}\left[\frac{e^{-ita} - e^{-itb}}{it}\right]\phi_X(t)\, dt.$$

The value of the integrand at $t = 0$ is determined by continuity, that is, by using L'Hospital's rule and it is $b - a$.

Proof. Suppose the integral $\frac{1}{2\pi}\int_{-c}^{c}\left[\frac{e^{-ita}-e^{-itb}}{it}\right]\phi_X(t)\, dt$ is denoted by J_c.

$$\begin{aligned} \text{Then } J_c &= \frac{1}{2\pi}\int_{-c}^{c}\left[\frac{e^{-ita} - e^{-itb}}{it}\right]\phi_X(t)\, dt \\ &= \frac{1}{2\pi}\int_{-c}^{c}\left[\frac{e^{-ita} - e^{-itb}}{it}\right]\left[\int_{-\infty}^{\infty} e^{itu} dF_X(u)\right] dt \\ &= \frac{1}{2\pi}\int_{-c}^{c}\int_{-\infty}^{\infty}\left[\frac{e^{-it(a-u)} - e^{-it(b-u)}}{it}\right] dF_X(u)\, dt \\ &= \frac{1}{2\pi}\int_{-\infty}^{\infty}\left[\int_{-c}^{c}\left[\frac{e^{-it(a-u)} - e^{-it(b-u)}}{it}\right] dt\right] dF_X(u). \quad (4.7) \end{aligned}$$

It is to be noted that in the last step the order of integration can be interchanged as the integral over $(-\infty, \infty)$ is absolutely convergent and the integral over $(-c, c)$ is over a finite interval. Now we simplify the integral $\int_{-c}^{c}\frac{e^{-it(a-u)}}{it}\, dt$ as follows.

$$\begin{aligned} \int_{-c}^{c}\frac{e^{-it(a-u)}}{it}\, dt &= \int_{-c}^{0}\frac{e^{-it(a-u)}}{it}\, dt + \int_{0}^{c}\frac{e^{-it(a-u)}}{it}\, dt \\ &= -\int_{0}^{c}\frac{e^{it(a-u)}}{it}\, dt + \int_{0}^{c}\frac{e^{-it(a-u)}}{it}\, dt \\ &= \int_{0}^{c}\frac{e^{-it(a-u)} - e^{it(a-u)}}{it}\, dt. \end{aligned}$$

Similarly we have $\int_{-c}^{c} \frac{e^{-it(b-u)}}{it}\, dt = \int_0^c \frac{e^{-it(b-u)} - e^{it(b-u)}}{it}\, dt$. Substituting these expressions in Equation (4.7) and using the fact that $e^{i\alpha} - e^{-i\alpha} = 2i\sin(\alpha)$, we get

$$\begin{aligned}
J_c &= \frac{1}{2\pi}\int_{-\infty}^{\infty}\left[\int_0^c \left[\frac{e^{-it(a-u)} - e^{it(a-u)} - e^{-it(b-u)} + e^{it(b-u)}}{it}\right] dt\right] dF_X(u) \\
&= \frac{1}{2\pi}\int_{-\infty}^{\infty}\left[\int_0^c \left[\frac{e^{it(u-a)} - e^{-it(u-a)} - e^{it(u-b)} + e^{-it(u-b)}}{it}\right] dt\right] dF_X(u) \\
&= \frac{1}{\pi}\int_{-\infty}^{\infty}\left[\int_0^c \left[\frac{\sin(t(u-a)) - \sin(t(u-b))}{t}\right] dt\right] dF_X(u) \\
&= \int_{-\infty}^{\infty} g(c,u) dF_X(u), \text{ where } g(c,u) = \frac{1}{\pi}\int_0^c \left[\frac{\sin(t(u-a)) - \sin(t(u-b))}{t}\right] dt.
\end{aligned}$$

Thus, J_c can be written as

$$J_c = \int_{-\infty}^{a} g(c,u) dF_X(u) + \int_a^b g(c,u) dF_X(u) + \int_b^{\infty} g(c,u) dF_X(u).$$

Now to find $\lim_{c\to\infty} J_c$ we first find $\lim_{c\to\infty} g(c,u)$ in the following five cases: (i) $u < a$, (ii) $u = a$, (iii) $a < u < b$, (iv) $u = b$ and (v) $u > b$.

(i) Suppose $u < a$. Then $u - a < 0$ and $u - b < u - a < 0$. Thus, using the result in Equation (4.6) we have,

$$\lim_{c\to\infty} \frac{1}{\pi}\int_0^c \left[\frac{\sin(t(u-a)) - \sin(t(u-b))}{t}\right] dt = \left(-\frac{1}{2} - \frac{-1}{2}\right) = 0.$$

(ii) Suppose $u = a$. Then $u - a = 0$ and $u - b = a - b < 0$. Again using the result in Equation (4.6), we have,

$$\lim_{c\to\infty} \frac{1}{\pi}\int_0^c \left[\frac{\sin(t(u-a)) - \sin(t(u-b))}{t}\right] dt = \left(0 - \frac{-1}{2}\right) = \frac{1}{2}.$$

(iii) Suppose $a < u < b$. Then $u - a > 0$ and $u - b < 0$. By Equation (4.6), we have,

$$\lim_{c\to\infty} \frac{1}{\pi}\int_0^c \left[\frac{\sin(t(u-a)) - \sin(t(u-b))}{t}\right] dt = \left(\frac{1}{2} - \frac{-1}{2}\right) = 1.$$

(iv) Suppose $u = b$. Then $u - a = b - a > 0$ and $u - b = 0$. By Equation (4.6), we have,

$$\lim_{c\to\infty} \frac{1}{\pi}\int_0^c \left[\frac{\sin(t(u-a)) - \sin(t(u-b))}{t}\right] dt = \left(\frac{1}{2} - 0\right) = \frac{1}{2}.$$

(v) Suppose $u > b$. Then $u - a > b - a > 0$ and $u - b > 0$. Hence,

$$\lim_{c\to\infty} \frac{1}{\pi}\int_0^c \left[\frac{\sin(t(u-a)) - \sin(t(u-b))}{t}\right] dt = \left(\frac{1}{2} - \frac{1}{2}\right) = 0.$$

Thus,

$$g(u) = \lim_{c\to\infty} g(c,u) = \begin{cases} 0 & \text{if} \quad u < a, \\ 1/2 & \text{if} \quad u = a, \\ 1 & \text{if} \quad a < u < b, \\ 1/2 & \text{if} \quad u = b, \\ 0, & \text{if} \quad u > b. \end{cases} \tag{4.8}$$

By Lemma 4.4.3 we have noted that $\frac{1}{\pi}|\int_0^c \frac{\sin(\alpha t)}{t}| \leq 1 \ \forall \ c > 0$ and for all $\alpha \in \mathbb{R}$, which implies that

$$|g(c,u)| \leq \frac{1}{\pi}\left|\int_0^c \frac{\sin(t(u-a))}{t}\right| + \frac{1}{\pi}\left|\int_0^c \frac{\sin(t(u-b))}{t}\right| \leq 2.$$

Further $\lim_{c\to\infty} g(c,u) = g(u)$ exists for all $u \in \mathbb{R}$. Thus, $g(c,u) \to g(u)$ on $\mathbb{R}$ and $g(c,u)$ is bounded uniformly by 2. Hence, by Lemma 4.4.2 we can interchange limit and integration in $\lim_{c\to\infty} \int_{-\infty}^{\infty} g(c,u)dF_X(u)$. We thus have,

$$\begin{aligned}
\lim_{c\to\infty} J_c &= \lim_{c\to\infty} \int_{-\infty}^{a} g(c,u)dF_X(u) + \lim_{c\to\infty} g(c,a)P[X=a] \\
&+ \lim_{c\to\infty} \int_{a}^{b} g(c,u)dF_X(u) + \lim_{c\to\infty} g(c,b)P[X=b] \\
&+ \lim_{c\to\infty} \int_{b}^{\infty} g(c,u)dF_X(u) \\
&= \int_{-\infty}^{a} \lim_{c\to\infty} g(c,u)dF_X(u) + \lim_{c\to\infty} g(c,a)P[X=a] \\
&+ \int_{a}^{b} \lim_{c\to\infty} g(c,u)dF_X(u) + \lim_{c\to\infty} g(c,b)P[X=b] \\
&+ \int_{b}^{\infty} \lim_{c\to\infty} g(c,u)dF_X(u) \\
&= (1/2)P[X=a] + P[a < X < b] + (1/2)P[X=b] \qquad \text{by (4.8)} \\
&= (1/2)P[X=a] + P[a < X \leq b] - P[X=b] + (1/2)P[X=b] \\
&= F_X(b) - F_X(a) + (1/2)P[X=a] - (1/2)P[X=b].
\end{aligned}$$

If a and b are points of continuity, then $P[X=a] = P[X=b] = 0$ and hence

$$\lim_{c\to\infty} J_c = F_X(b) - F_X(a).$$

□

The inversion theorem leads to an important result that a characteristic function uniquely determines the distribution of a random variable. Such a uniqueness property of a characteristic function is heavily used in the proof of the central limit theorem, to be discussed in Chapter 10 and in determining the limiting distribution of an estimator or of a test statistic in large sample inference. It is illustrated in Chapter 7.

Theorem 4.4.6. *Uniqueness theorem: Suppose X and Y are random variables with characteristic functions ϕ_X and ϕ_Y respectively.*

$$\phi_X(t) = \phi_Y(t) \ \forall \ t \in \mathbb{R} \quad \Longleftrightarrow \quad X \stackrel{d}{=} Y.$$

Proof. Suppose F_X and F_Y are distribution functions of X and Y respectively. Suppose X and Y are identically distributed, hence $F_X(x) = F_Y(x) \ \forall \ x \in \mathbb{R}$. By the definition of a characteristic function, for $t \in \mathbb{R}$

$$\phi_X(t) = \int_{\mathbb{R}} e^{itx} dF_X(x) = \int_{\mathbb{R}} e^{itx} dF_Y(x) = \phi_Y(t) \ \forall \ t \in \mathbb{R}.$$

Suppose, $\phi_X(t) = \phi_Y(t) \ \forall \ t \in \mathbb{R}$. As proved in Theorem 3.2.4, $C = C(F_X) \cap C(F_Y) \neq \emptyset$, where $C(F_X)$ and $C(F_Y)$ are the sets of points of continuity of F_X and F_Y respectively. Suppose $a, b \in C$. Then by the inversion theorem,

$$\begin{aligned} F_X(b) - F_X(a) &= \frac{1}{2\pi} \lim_{c \to \infty} \int_{-c}^{c} \left[\frac{e^{-ita} - e^{-itb}}{it} \right] \phi_X(t) \ dt \\ &= \frac{1}{2\pi} \lim_{c \to \infty} \int_{-c}^{c} \left[\frac{e^{-ita} - e^{-itb}}{it} \right] \phi_Y(t) \ dt = F_Y(b) - F_Y(a). \end{aligned}$$

Thus, $\forall \ a, b \in C$,

$$F_X(b) - F_X(a) = F_Y(b) - F_Y(a) \quad \Longleftrightarrow \quad F_X(b) - F_Y(b) = F_X(a) - F_Y(a).$$

Allowing b to vary, for fixed a, we have $F_X(b) - F_Y(b) = F_X(a) - F_Y(a) = d$, where d is a constant. To find d, suppose $b \uparrow \infty$ through the points in C, then we get $F_X(\infty) - F_Y(\infty) = 0$ implies $d = 0$ which further implies that $F_X(b) = F_Y(b) \ \forall \ b \in C$. Using arguments same as in Theorem 3.2.4, if $F_X(x) = F_Y(x) \ \forall \ x \in C$, then $F_X(x) = F_Y(x) \ \forall \ x \in \mathbb{R}$, which implies that X and Y are identically distributed. $\square$

In particular, uniqueness theorem of a characteristic function implies that $\phi(t) \equiv 1$ must be a characteristic function of a distribution which is degenerate at zero and $\phi(t) = \exp(-t^2/2)$ must be a characteristic function of the standard normal distribution.

Using the links between moments and a characteristic function and the uniqueness theorem, in the following example we examine whether a given function is a characteristic function.

Example 4.4.3. Suppose $\phi(t) = \exp(-t^4/2)$, $t \in \mathbb{R}$. We examine whether it is a characteristic function. Suppose $\phi(t) = \exp(-t^4/2)$ is a characteristic function. Then it is a differentiable function and derivatives of all order at $t = 0$ exists. Hence by Theorem 4.4.2, moments of all order exist. Further, $\phi(t) = \exp(-t^4/2)$ can be expressed as $\phi(t) = \exp(-t^4/2) = 1 - t^4/2 + \cdots$. Thus by Corollary 4.4.1, $\mu_1' = \mu_2' = 0$ which implies that $\phi(t)$ is a characteristic function of a random variable X which is degenerate at 0. However, if X is degenerate at 0, then its characteristic function is $E(\exp(itX)) = 1$ which is not the same as $\phi(t) = \exp(-t^4/2)$. Hence our assumption that $\phi(t) = \exp(-t^4/2)$ is a characteristic function is wrong and it is not a characteristic function. □

We have already noted at the beginning of this section that if the distribution of X is symmetric around 0, then its characteristic function is a real valued function. Using uniqueness theorem, we now prove that the converse is also true.

Theorem 4.4.7. *If a characteristic function $\phi_X(t)$ of a random variable X is a real valued function, then its distribution is symmetric around* 0.

Proof. Observe that $\phi_X(t)$ is a real valued function implies that $\forall\ t \in \mathbb{R}$,

$$\phi_X(t) = \overline{\phi_X(t)} \quad \Rightarrow \quad \phi_X(t) = \phi_X(-t) = \phi_{-X}(t) \quad \Rightarrow \quad X \stackrel{d}{=} -X$$

and hence the distribution of X is symmetric around 0. □

In the following corollaries, we discuss some particular cases of the inversion theorem, depending on the nature of the distribution function F.

Corollary 4.4.2. *(i) If F is differentiable at x with $F'(x) = f(x)$, then*

$$f(x) = \lim_{h \to 0} \lim_{c \to \infty} \frac{1}{2\pi} \int_{-c}^{c} \left(\frac{1 - e^{-ith}}{ith} \right) e^{-itx} \phi_X(t)\, dt,$$

if and only if the limits on the right hand side exist.
(ii) If ϕ is Riemann integrable on $\mathbb{R}$ and if $F'(x) = f(x)$ exists $\forall\ x \in \mathbb{R}$, then

$$f(x) = \frac{1}{2\pi} \int_{\mathbb{R}} e^{-itx} \phi_X(t)\, dt$$

and $f(x)$ is continuous on $\mathbb{R}$.

Proof. Suppose $J_c = \frac{1}{2\pi}\int_{-c}^{c}\left[\frac{e^{-ita}-e^{-itb}}{it}\right]\phi_X(t)\,dt$. From Theorem 4.4.5,

$$\begin{aligned}
\lim_{c\to\infty} J_c &= \lim_{c\to\infty}\frac{1}{2\pi}\int_{-c}^{c}\left[\frac{e^{-ita}-e^{-itb}}{it}\right]\phi_X(t)\,dt \\
&= (1/2)P[X=a]+P[a<X<b]+(1/2)P[X=b] \\
&= F_X(b-0)-F_X(a+0) \\
&+ (1/2)[F_X(a+0)-F_X(a-0)+F_X(b+0)-F_X(b-0)] \\
&= (1/2)[F_X(b+0)+F_X(b-0)]-(1/2)[F_X(a+0)+F_X(a-0)] \\
&= (1/2)[F_X(x+h+0)+F_X(x+h-0)] \\
&- (1/2)[F_X(x+0)+F_X(x-0)],
\end{aligned}$$

with $b=x+h$ & $a=x$. Since F is differentiable, F is continuous and hence

$$F(x+h)=F(x+h+0)=F(x+h-0), \quad F(x)=F(x+0)=F(x-0).$$

By definition

$$\begin{aligned}
f(x) &= \lim_{h\to 0}\frac{F(x+h)-F(x)}{h} \\
&= \lim_{h\to 0}\left[\frac{F(x+h+0)+F(x+h-0)}{2h}-\frac{F(x+0)+F(x-0)}{2h}\right] \\
&= \lim_{h\to 0}\frac{1}{h}\left[\lim_{c\to\infty}\frac{1}{2\pi}\int_{-c}^{c}\left[\frac{e^{-itx}-e^{-it(x+h)}}{it}\right]\phi_X(t)\,dt\right] \\
&= \lim_{h\to 0}\left[\lim_{c\to\infty}\frac{1}{2\pi}\int_{-c}^{c}\left[\frac{e^{-itx}(1-e^{-ith})}{ith}\right]\phi_X(t)\,dt\right] \\
&= \lim_{h\to 0}\lim_{c\to\infty}\frac{1}{2\pi}\int_{-c}^{c}\left(\frac{1-e^{-ith}}{ith}\right)e^{-itx}\phi_X(t)\,dt,
\end{aligned}$$

provided the limits on the right hand side exist.
(ii) It is given that ϕ is Riemann integrable on $\mathbb{R}$, hence by Theorem 4.3.7, $|\phi|$ is integrable. Note that for any $h>0$ and $t\in\mathbb{R}$

$$\begin{aligned}
\frac{1-e^{-ith}}{ith} &= \frac{1-e^{-ith}}{ith}\,\frac{ith}{ith} = \frac{ith-ithe^{-ith}}{-th} = -i+ie^{-ith} \\
&= -i+i(\cos(-th)+i\sin(-th)) = i(\cos(th)-1)-\sin(th)
\end{aligned}$$

$$\begin{aligned}
\Rightarrow \quad \left|\frac{1-e^{-ith}}{ith}\right|^2 &= \cos^2(th)-2\cos(th)+1+\sin^2(th) \\
&= 2(1-\cos(th)) = 2\times 2\sin^2(th/2)\le 4.
\end{aligned}$$

Hence, for any c,h and $x\in\mathbb{R}$,

$$\left|\int_{-c}^{c}\left(\frac{1-e^{-ith}}{ith}\right)e^{-itx}\phi(t)\,dt\right| \le 2\int_{-c}^{c}|\phi(t)|\,dt \le 2\int_{\mathbb{R}}|\phi(t)|\,dt<\infty.$$

Hence, by Lemma 4.4.2, $\lim_{h\to 0}$ can be taken inside the integral. Thus,

$$\begin{aligned} f(x) &= \lim_{h\to 0}\lim_{c\to\infty}\frac{1}{2\pi}\int_{-c}^{c}\left(\frac{1-e^{-ith}}{ith}\right)e^{-itx}\phi(t)\,dt \\ &= \lim_{h\to 0}\frac{1}{2\pi}\int_{\mathbb{R}}\left(\frac{1-e^{-ith}}{ith}\right)e^{-itx}\phi(t)\,dt \\ &= \frac{1}{2\pi}\int_{\mathbb{R}}\lim_{h\to 0}\left(\frac{1-e^{-ith}}{ith}\right)e^{-itx}\phi(t)\,dt = \frac{1}{2\pi}\int_{\mathbb{R}}e^{-itx}\phi(t)\,dt, \end{aligned}$$

since $\lim_{h\to 0}(1-e^{-ith})/ith = 1$. To establish continuity of f, we note that for $x\in\mathbb{R}$,

$$\begin{aligned} |f(x+\delta)-f(x)| &= \frac{1}{2\pi}\left|\int_{\mathbb{R}}(e^{-it(x+\delta)}-e^{-itx})\phi(t)\,dt\right| \\ &\le \frac{1}{2\pi}\int_{\mathbb{R}}|(e^{-it\delta}-1)||\phi(t)|\,dt \;\le\; 2\frac{1}{2\pi}\int_{\mathbb{R}}|\phi(t)|\,dt<\infty, \end{aligned}$$

since $|(e^{-it\delta}-1)|^2 = 2-2\cos(t\delta) = 2\times 2\sin^2(t\delta/2)\le 4$, as shown above. Further, $e^{-it\delta}-1\to 0$ as $\delta\to 0$. Hence, again by Lemma 4.4.2, $\lim_{\delta\to 0}$ can be taken inside the integral. Thus,

$$\lim_{\delta\to 0}|f(x+\delta)-f(x)| = \frac{1}{2\pi}\left|\int_{\mathbb{R}}\lim_{\delta\to 0}(e^{-it(x+\delta)}-e^{-itx})\phi(t)\,dt\right| = 0$$

which proves that f is continuous on $\mathbb{R}$. □

Following examples illustrate the inversion theorem.

Example 4.4.4. Suppose X follows Cauchy $C(0,1)$ distribution. Then its characteristic function is $\phi_X(t)=\exp(-|t|)$. It is Riemann integrable. Hence using Corollary 4.4.2, we find the probability density function of X as follows.

$$\begin{aligned} f(x) &= \frac{1}{2\pi}\int_{\mathbb{R}}e^{-itx}\phi(t)\,dt = \frac{1}{2\pi}\int_{\mathbb{R}}e^{-itx}e^{-|t|}\,dt \\ &= \frac{1}{2\pi}\int_{\mathbb{R}}(\cos(-tx)+i\sin(-tx))e^{-|t|}\,dt = \frac{1}{\pi}\int_0^{\infty}\cos(tx)e^{-t}\,dt \\ &= \frac{1}{\pi}\int_0^{\infty}\cos(tx)\frac{d}{dt}[-e^{-t}]\,dt \\ &= \frac{1}{\pi}\left[[-e^{-t}\cos(tx)]_0^{\infty}+\int_0^{\infty}x\sin(tx)e^{-t}\,dt\right] \\ &= \frac{1}{\pi}+\frac{x}{\pi}\left[\int_0^{\infty}\sin(tx)\frac{d}{dt}[-e^{-t}]\,dt\right] \\ &= \frac{1}{\pi}+\frac{x}{\pi}\left[[-e^{-t}\sin(tx)]_0^{\infty}-x\int_0^{\infty}\cos(tx)e^{-t}\,dt\right] \\ &= \frac{1}{\pi}+\frac{x}{\pi}\left[-x\int_0^{\infty}\cos(tx)e^{-t}\,dt\right] = \frac{1}{\pi}-x^2f(x) \;\;\Rightarrow\;\; f(x)=\frac{1}{\pi}\frac{1}{1+x^2}, \end{aligned}$$

for $x \in \mathbb{R}$. In the second step, we use the result that sine is an odd function while cosine is an even function. Note that $f(x)$ is the probability density function of X. $\square$

Example 4.4.5. Suppose $X \sim N(0,1)$. Then its characteristic function is given by,

$$\begin{aligned}\phi_X(t) = E(e^{itX}) &= \int_{\mathbb{R}} \frac{1}{\sqrt{2\pi}} e^{itx} e^{-x^2/2}\, dx = \int_{\mathbb{R}} \frac{1}{\sqrt{2\pi}} e^{-(x^2-2itx+i^2t^2-i^2t^2)/2}\, dx \\ &= e^{-t^2/2} \int_{\mathbb{R}} \frac{1}{\sqrt{2\pi}} e^{-(x-it)^2/2}\, dx = e^{-t^2/2}.\end{aligned}$$

Hence, if $X \sim N(\mu, \sigma^2)$. Then $\phi_X(t) = \exp(i\mu t - t^2\sigma^2/2)$. Now suppose $\phi_X(t) = \exp(-t^2/2)$, it is Riemann integrable. Hence by Corollary 4.4.2, the probability density function is given by,

$$\begin{aligned}f(x) &= \frac{1}{2\pi} \int\limits_{\mathbb{R}} e^{-itx}\phi(t)dt = \frac{1}{2\pi} \int\limits_{\mathbb{R}} e^{-itx} e^{-\frac{t^2}{2}}\, dt = \frac{1}{2\pi} \int\limits_{\mathbb{R}} e^{-\frac{x^2}{2}} e^{-\frac{t^2+2itx+i^2x^2}{2}}\, dt \\ &= e^{-\frac{x^2}{2}} \frac{1}{2\pi} \int_{\mathbb{R}} e^{-\frac{(t+ix)^2}{2}}\, dt = e^{-\frac{x^2}{2}} \frac{1}{2\pi} \sqrt{2\pi} = \frac{1}{\sqrt{2\pi}} e^{-\frac{x^2}{2}} \quad x \in \mathbb{R}. \quad \square\end{aligned}$$

In both the above examples we note a typical link between the characteristic function and the corresponding probability density function. More precisely, we observe that in Example 4.4.5, the characteristic function is $\phi_X(t) = \exp(-t^2/2)$ and the probability density function is $(1/\sqrt{2\pi})\exp(-x^2/2)$. In Example 4.4.4, characteristic function is $\phi_X(t) = \exp(-|t|)$ and the probability density function is $f(x) = (1/\pi)(1/(1+x^2))$, $x \in \mathbb{R}$. It can be shown that if X follows Laplace distribution, then its characteristic function is $\phi_X(t) = 1/(1+t^2)$ and its probability density function is given by $f(x) = (1/2)\exp(-|x|)$, $x \in \mathbb{R}$. Thus, except for a multiplicative constant, the density of one of them equals the characteristic function of the other. If a pair of distributions are related in such a manner, then these distributions are known as a conjugate pair of distributions. We thus note that Laplace and Cauchy distributions are conjugate to each other while normal distribution is conjugate to itself.

Following corollary shows how to obtain the magnitude of jump at a point of discontinuity of F, from the characteristic function.

Corollary 4.4.3. *If $P[X = a] > 0$, then*

$$P[X = a] = F_X(a) - F_X(a-0) = \lim_{c\to\infty} \frac{1}{2c} \int_{-c}^{c} e^{-ita}\phi_X(t)\, dt.$$

Proof. Suppose $I_c = \frac{1}{2c}\int_{-c}^{c} e^{-ita}\phi_X(t)\, dt$. As in Theorem 4.4.5, interchanging the order of integration we have

$$
\begin{aligned}
I_c &= \frac{1}{2c}\int_{-c}^{c} e^{-ita}\left[\int_{-\infty}^{\infty} e^{itu}dF_X(u)\right]\,dt = \frac{1}{2c}\int_{-\infty}^{\infty}\left[\int_{-c}^{c} e^{it(u-a)}\,dt\right]dF_X(u) \\
&= \frac{1}{2c}\int_{-\infty}^{\infty}\left[\int_{-c}^{c}(\cos(t(u-a)) + i\sin(t(u-a)))dt\right]dF_X(u) \\
&= \frac{1}{2c}\int_{-\infty}^{\infty}\left[\int_{-c}^{c}\cos(t(u-a))\,dt\right]dF_X(u) \ \text{ sine being an odd function} \\
&= \frac{2}{2c}\int_{-\infty}^{\infty}\left[\int_{0}^{c}\cos(t(u-a))\,dt\right]dF_X(u) \ \text{ cosine being an even function} \\
&= \frac{1}{c}\int_{\mathbb{R}-\{a\}}\left[\frac{\sin(c(u-a))}{(u-a)}\right]dF_X(u) + \frac{1}{c}\,cP[X=a] \ \text{ as } \ \cos(0)=1 \\
&= \int_{\mathbb{R}-\{a\}}\frac{\sin(c(u-a))}{c(u-a)}dF_X(u) + P[X=a],
\end{aligned}
$$

since at $u = a$, $\int_0^c \cos(t(u-a))\,dt = c$. Now $|\sin(c(u-a))| \le 1$ and hence for $u \ne a$, $\lim_{c\to\infty}\frac{\sin(c(u-a))}{c(u-a)} = 0$. Further, for any $u \in \mathbb{R}$, $|\frac{\sin(c(u-a))}{c(u-a)}|$ is bounded. Hence, by Lemma 4.4.2,

$$
\begin{aligned}
\lim_{c\to\infty} I_c &= \int_{\mathbb{R}-\{a\}}\left[\lim_{c\to\infty}\frac{\sin(c(u-a))}{c(u-a)}\right]dF_X(u) + P[X=a] \\
&= 0 + P[X=a] = P[X=a].
\end{aligned}
$$

□

If X is a discrete random variable, then the above corollary conveys how the probability mass function can be obtained from its characteristic function. For a discrete distribution the support is a countable set. As a particular case, if the support is

$$S = \{a + xd | x = 0, \pm 1, \pm 2, \cdots,\}, \quad d > 0 \ \& \ a \in \mathbb{R}$$

then the discrete distribution is known as a lattice distribution and smallest such d is known as a span of the distribution. The characteristic function for a lattice distribution with support S and $p_x = P[X = a + xd]$, $x \in I$ and the corresponding inversion formula are given by,

$$\phi_X(t) = \sum_{x\in I} p_x e^{it(a+xd)} \quad \& \quad p_x = \frac{d}{2\pi}\int_{-\pi/d}^{\pi/d} e^{-it(a+xd)}\phi_X(t)\,dt.$$

In particular, when $a = 0$ & $d = 1$, then

$$\phi_X(t) = \sum_{x\in I} p_x e^{itx} \quad \& \quad p_x = P[X = x] = \frac{1}{2\pi}\int_{-\pi}^{\pi} e^{-itx}\phi_X(t)\,dt. \tag{4.9}$$

Following examples illustrate these results.

Example 4.4.6. Suppose X is a degenerate random variable, degenerate at C. Then its characteristic function is $\phi_X(t) = E(\exp(itX)) = \exp(itC)$. By the inversion formula, the probability mass function is given by,

$$\begin{aligned}
p_x &= P[X = x] = \frac{1}{2\pi}\int_{-\pi}^{\pi} e^{-itx}e^{itC}\ dt \\
&= \frac{1}{2\pi}\int_{-\pi}^{\pi}(\cos((C-x)t) + i\sin((C-x)t))\ dt \\
&= \frac{1}{\pi}\int_0^{\pi}\cos((C-x)t) = \frac{1}{\pi}(\sin(C-x)t/(C-x))\Big|_0^{\pi} = 0 \ \text{ if } x \neq C.
\end{aligned}$$

$$\text{At } x = C, \ p_x = \frac{1}{2\pi}\int_{-\pi}^{\pi} e^{-itx}e^{itC}\ dt = \frac{1}{2\pi}\int_{-\pi}^{\pi} 1\ dt = 1\ .$$

Thus, $P[X = C] = 1$, which implies that the distribution of X is degenerate at C. □

Example 4.4.7. Suppose X follows geometric distribution with probability mass function $p_x = P[X = x] = pq^x, \quad x = 0, 1, 2, \cdots, \quad q = 1 - p$. Then its characteristic function is given by,

$$\phi_X(t) = E(e^{itX}) = \sum_{x=0}^{\infty} e^{itx}pq^x = \sum_{x=0}^{\infty}(qe^{it})^x p = p(1 - qe^{it})^{-1}.$$

Now, by the inversion formula given in equation (4.9), the probability mass function from its characteristic function is given by,

$$\begin{aligned}
p_x &= P[X = x] = \frac{1}{2\pi}\int_{-\pi}^{\pi} e^{-itx}\phi_X(t)\ dt = \frac{p}{2\pi}\int_{-\pi}^{\pi} e^{-itx}(1 - qe^{it})^{-1}\ dt \\
&= \frac{p}{2\pi}\int_{-\pi}^{\pi} e^{-itx}\sum_{r=0}^{\infty}(qe^{it})^r\ dt = \frac{p}{2\pi}\sum_{r=0}^{\infty} q^r\int_{-\pi}^{\pi} e^{-itx+itr}\ dt \\
&= \frac{p}{2\pi}\sum_{r=0}^{\infty} q^r\int_{-\pi}^{\pi}(\cos((r-x)t) + i\sin((r-x)t))\ dt \\
&= \frac{p}{\pi}\sum_{r=0, r\neq x}^{\infty} q^r\int_0^{\pi}\cos((r-x)t)\ dt + \frac{p}{\pi}q^x\int_0^{\pi}\cos 0\ dt \\
&= \frac{p}{\pi}\sum_{r=0, r\neq x}^{\infty} q^r(\sin(r-x)t/(r-x))\Big|_0^{\pi} + \frac{p}{\pi}q^x\int_0^{\pi} 1\ dt \\
&= 0 + pq^x, \qquad x = 0, 1, 2, \cdots,
\end{aligned}$$

as for $r \neq x$, $\sin(r-x)\pi = 0 = \sin(0)$. □

Example 4.4.8. Suppose X follows binomial distribution with probability mass function $p_x = P[X = x] = \binom{n}{x} p^x q^{n-x}, \quad x = 0, 1, 2, \cdots, n, \quad q = 1 - p.$ Then its characteristic function is given by,

$$\phi_X(t) = E(e^{itX}) = \sum_{x=0}^{n} e^{itx} \binom{n}{x} p^x q^{n-x} = \sum_{x=0}^{n} (pe^{it})^x q^{n-x} = (pe^{it} + q)^n.$$

By the inversion formula given in equation (4.9), the probability mass function from its characteristic function is given by,

$$\begin{aligned}
p_x = P[X = x] &= \frac{1}{2\pi} \int_{-\pi}^{\pi} e^{-itx} \phi_X(t) \, dt = \frac{1}{2\pi} \int_{-\pi}^{\pi} e^{-itx} (pe^{it} + q)^n \, dt \\
&= \frac{1}{2\pi} \int_{-\pi}^{\pi} e^{-itx} \sum_{r=0}^{n} \binom{n}{r} (pe^{it})^r q^{n-r} \, dt = \frac{1}{2\pi} \sum_{r=0}^{n} \binom{n}{r} p^r q^{n-r} \int_{-\pi}^{\pi} e^{-itx+itr} \, dt \\
&= \frac{1}{2\pi} \sum_{r=0, r \neq x}^{n} \binom{n}{r} p^r q^{n-r} \int_{-\pi}^{\pi} e^{-itx+itr} \, dt + \frac{1}{2\pi} \binom{n}{x} p^x q^{n-x} \int_{-\pi}^{\pi} 1 \, dt \\
&= \frac{1}{\pi} \sum_{r=0, r \neq x}^{n} \binom{n}{r} p^r q^{n-r} (\sin(r - x)t/(r - x))|_0^{\pi} + \frac{1}{2\pi} \binom{n}{x} p^x q^{n-x} \, 2\pi \\
&= 0 + \binom{n}{x} p^x q^{n-x}, \qquad x = 0, 1, 2, \cdots, n.
\end{aligned}$$

Last two steps follow from the same arguments as in Example 4.4.7. □

We now briefly discuss the characteristic function of a random vector. We first define it.

Definition 4.4.4. *Characteristic function of a random vector: A characteristic function $\phi_{\underline{Z}}(\underline{t})$ of a random vector $\underline{Z} = (X_1, X_2, \cdots, X_k)'$, where $\underline{t} = (t_1, t_2, \cdots, t_k)'$, $t_i \in \mathbb{R}$, $i = 1, 2, \cdots, k$, is defined as*

$$\phi_{\underline{Z}}(\underline{t}) = \phi_{\underline{Z}}(t_1, t_2, \cdots, t_k) = E(\exp(i(t_1X_1 + t_2X_2 + \cdots + t_kX_k))) = E(\exp(\underline{t}'\underline{Z})).$$

In particular, if $\underline{t} = (t, t, \cdots, t)'$, then $\phi_{\underline{Z}}(\underline{t}) = E(\exp(it(X_1 + X_2 + \cdots + X_k))$, which is a characteristic function of $X_1 + X_2 + \cdots + X_k$. From the characteristic function $\phi_{\underline{Z}}(\underline{t})$ of $\underline{Z}$ we can obtain the characteristic function of the random vectors of lower order by substituting corresponding $t_i = 0$. For example, the characteristic function of the random vector $(X_1, X_2, X_3)'$ is $\phi_{\underline{Z}}(t_1, t_2, t_3, 0, \cdots, 0)$. In particular, the characteristic function of X_1 is $\phi_{\underline{Z}}(t_1, 0, 0, \cdots, 0)$, it is known as a marginal characteristic function.

Suppose a random vector $\underline{Z}$ has a multivariate normal distribution with mean vector $\underline{\mu}$ and dispersion matrix Σ, then its characteristic function is given by $\phi_{\underline{Z}}(\underline{t}) = \exp(\underline{t}'\underline{\mu} + \underline{t}'\Sigma\underline{t}/2)$. If Σ is singular and positive semi-definite, then one cannot have the probability density function of $\underline{Z}$, but its characteristic function exists.

Some of the most useful tools in probability theory and statistics for providing finiteness or convergence of sums and integrals are probability inequalities and moment inequalities. In the next section, we study some moment inequalities.

4.5 Moment Inequalities

Moment inequalities play an important role in probability theory and statistics as many times we need to determine or approximate probabilities of certain events in terms of the other probabilities. These are also useful in the determination of moments of sum or in finding bounds on them by the sum of moments. In this section we discuss some inequalities which are frequently used in subsequent chapters.

Lemma 4.5.1. *If $E(|X|^r) < \infty$, then $E(|X|^{r'}) < \infty$, for $0 < r' \leq r$ and $E(X^k)$ exists and is finite for $k < r$, if k is an integer.*

Proof. For $0 < r' \leq r$ and for any $\omega \in \Omega$,

$$|X(\omega)|^{r'} \leq \begin{cases} 1 & \text{if } |X(\omega)| \leq 1, \\ |X(\omega)|^r, & \text{if } |X(\omega)| > 1. \end{cases}$$

Therefore, for all ω and for $0 < r' \leq r$,

$$\begin{aligned} |X(\omega)|^{r'} &\leq 1 + |X(\omega)|^r \Rightarrow E(|X|^{r'}) \leq 1 + E(|X|^r) \\ \text{Hence } E(|X|^r) &< \infty \Rightarrow E(|X|^{r'}) < \infty. \end{aligned}$$

For an integer $k < r$ we have,

$$|X(\omega)|^k = |X(\omega)||X(\omega)| \cdots k \text{ times} = |X(\omega)X(\omega) \cdots k \text{ times}| = |X^k(\omega)|.$$

Thus, $|X^k| = |X|^k$ and $|X|^k$ is integrable for $k < r$. It is known that X^k is integrable if and only if $|X^k|$ is integrable. Thus, $E(X^k)$ exists and is finite. □

Remark 4.5.1. The space of all random variables such that $E(|X|^r) < \infty$ is known as L_r space. Lemma 4.5.1 states that if a moment of a certain order is finite, then the moments of lower order are also finite, that is, if $X \in L_r$ then $X \in L_s$, $\forall\ s < r$. However, the converse is not true, as shown in the following examples.

Example 4.5.1. Suppose the probability density function of a random variable X is given by,

$$f(x) = 2a^2/(x+a)^3 \ , \quad x \geq 0, \quad a \geq 0.$$

Here, $E(X) < \infty$, but $E(X^2)$ is not finite. □

Example 4.5.2. Suppose X is a discrete random variable with $P[X = j] = c/j^{k+2}$, $j = 1, 2, \cdots$, where $c = (\sum_{j=1}^{\infty} 1/j^{k+2})^{-1}$, $k \in \mathbb{N}$. Note that,

$$E(X^k) = c\sum_{j=1}^{\infty} j^k/j^{k+2} = c\sum_{j=1}^{\infty} 1/j^2 < \infty \text{ but } E(X^{k+1}) = c\sum_{j=1}^{\infty} 1/j = \infty. \ \square$$

In the next lemma, we derive C_r inequality, which has many applications.

Lemma 4.5.2. *C_r inequality: Suppose $X \in L_r$ and $Y \in L_r$, then for any $r > 0$, $E(|X+Y|^r) \leq C_r\left(E(|X|^r) + E(|Y|^r)\right)$, where,*

$$C_r = \begin{cases} 1, & \text{if } r \leq 1, \\ 2^{r-1}, & \text{if } r > 1. \end{cases}$$

Proof. For any two positive real numbers a and b and $\forall\ r \leq 1$,

$$\begin{aligned}\frac{a}{a+b} &< 1\ \&\ \frac{b}{a+b} < 1 \Rightarrow \left(\frac{a}{a+b}\right)^r \geq \frac{a}{a+b}\ \&\ \left(\frac{b}{a+b}\right)^r \geq \frac{b}{a+b} \\ \Rightarrow\ \left(\frac{a}{a+b}\right)^r + \left(\frac{b}{a+b}\right)^r &\geq \frac{a}{a+b} + \frac{b}{a+b} = 1 \\ \Rightarrow\ (a^r + b^r) &\geq (a+b)^r.\end{aligned}$$

Hence, substituting $a = |X(\omega)|$ and $b = |Y(\omega)|$ and using triangle inequality, we get, for all ω and for $r \leq 1$,

$$\begin{aligned}|X(\omega)|^r + |Y(\omega)|^r &\geq (|X(\omega)| + |Y(\omega)|)^r \geq |X(\omega) + Y(\omega)|^r \\ \Rightarrow\ E(|X+Y|^r) &\leq E(|X|^r) + E(|Y|^r),\quad r \leq 1\,. \end{aligned} \tag{4.10}$$

For $r > 1$, we consider a function $\phi(p) = p^r + (1-p)^r$, where $0 < p < 1$. Taking derivative with respect to p, we get,

$$\frac{d\phi(p)}{dp} = rp^{r-1} - r(1-p)^{r-1}.$$

Equating the derivative to zero yields, $p = 1 - p$. Hence $p = 1/2$. Second derivative of $\phi(p)$ is,

$$\frac{d^2\phi(p)}{dp^2} = r(r-1)p^{r-2} + r(r-1)(1-p)^{r-2} > 0, \quad \forall\ p \in (0,1).$$

Thus, $\phi(p)$ has minimum at $p = 1/2$. Therefore, using the fact that $0 < a/(a+b) < 1$, we get

$$\begin{aligned}\phi\left(\frac{a}{a+b}\right) &\geq \phi\left(\frac{1}{2}\right) \\ \Rightarrow\ \left(\frac{a}{a+b}\right)^r + \left(\frac{b}{a+b}\right)^r &\geq \left(\frac{1}{2}\right)^r + \left(\frac{1}{2}\right)^r = 2^{-(r-1)} \\ \Rightarrow\ 2^{r-1}(a^r + b^r) &\geq (a+b)^r.\end{aligned}$$

Again substituting $a = |X(\omega)|$ and $b = |Y(\omega)|$ and using triangle inequality, we get, for $r > 1$,

$$E(|X+Y|^r) \leq E((|X|+|Y|)^r) \leq 2^{r-1}\left(E(|X|^r) + E(|Y|^r)\right). \quad (4.11)$$

Thus, combining equations (4.10) and (4.11), we get the C_r inequality. □

C_r inequality implies that if the r-th absolute moments of X and Y exist and are finite, then r-th absolute moment of $(X+Y)$ also exists and is finite. Thus, the class of random variables, whose r-th absolute moment is finite is closed under addition. In this inequality the constant C depends on r, it is 1 if $0 < r \leq 1$ and is 2^{r-1} if $r > 1$. Hence this inequality is known as C_r inequality.

The following lemma specifies Holder's inequality, which is about the expectation of the product of two random variables.

Lemma 4.5.3. *Holder's inequality: For* $r > 1$ *and* $1/r + 1/s = 1$,

$$E(|XY|) \leq E^{1/r}(|X|^r)E^{1/s}(|Y|^s).$$

Proof. Using the condition that $1/r + 1/s = 1$, we get $s = r/(r-1) > 1$. Further, $(r+s)/s = r$ and $(r+s)/r = s$. For $p > 0$, we consider a function,

$$\begin{aligned}
\phi(p) &= p^r/r + p^{-s}/s \Rightarrow \phi(1) = 1 \\
\& \quad \frac{d\phi(p)}{dp} &= p^{r-1} - p^{-s-1} = 0 \Rightarrow p^{r+s} = 1 \Rightarrow p = 1 \\
\text{Further,} \quad \frac{d^2\phi(p)}{dp^2} &= (r-1)p^{r-2} - (-s-1)p^{-s-2} \\
\text{and at } p &= 1, \quad \frac{d^2\phi(p)}{dp^2} = r + s > 0 \\
&\Rightarrow \phi(p) \text{ has minimum at } p = 1.
\end{aligned}$$

Now suppose $p_0 = a^{1/s}/b^{1/r}$, for some positive real numbers a, b. note that

$$\begin{aligned}
\phi(p_0) &\geq \phi(1) = 1 \\
&\Rightarrow a^{r/s}/rb + b^{s/r}/sa = (br)^{-1}a^{r/s} + (as)^{-1}b^{s/r} \geq 1 \\
&\Rightarrow ab[(br)^{-1}a^{r/s} + (as)^{-1}b^{s/r}] \geq ab \\
&\Rightarrow r^{-1}a^{r/s+1} + s^{-1}b^{s/r+1} \geq ab \\
&\Rightarrow r^{-1}a^{(r+s)/s} + s^{-1}b^{(s+r)/r} \geq ab \\
&\Rightarrow r^{-1}a^r + s^{-1}b^s = (sa^r + rb^s)/rs \geq ab.
\end{aligned}$$

If $E(|XY|) = 0$ then $P[|X| = 0] = 1$ or $P[|Y| = 0] = 1$ which further implies $E(|X|^r) = 0$ or $E(|Y|^s) = 0$. Thus Holder's inequality is trivially true. Thus we assume that $E(|X|^r) > 0$ and $E(|Y|^s) > 0$. Substituting $a = |X|/E^{1/r}(|X|^r)$ and $b = |Y|/E^{1/s}(|Y|^s)$ in the above inequality we get,

$$\frac{s\frac{|X|^r}{E(|X|^r)} + r\frac{|Y|^s}{E(|Y|^s)}}{rs} \geq \frac{|X||Y|}{[E^{1/r}(|X|^r)][E^{1/s}(|Y|^s)]}.$$

Taking expectation on both sides and using monotonicity of expectations, we get,

$$\frac{s+r}{rs} \geq \frac{E(|X||Y|)}{[E^{1/r}(|X|^r)][E^{1/s}(|Y|^s)]}.$$

But, $(s+r)/rs = 1$. Hence, we get,

$$E(|X||Y|) = E(|XY|) \quad \leq \quad E^{1/r}(|X|^r)E^{1/s}(|Y|^s).$$

□

Following two inequalities, Lyapounov's inequality and Schwarz inequality, follow from Holder's inequality.

Lemma 4.5.4. *Lyapounov's inequality:*

$$E^{1/s}(|X|^s) \quad \leq \quad E^{1/t}(|X|^t) \qquad \textit{for} \quad s < t.$$

Proof. By Holder's inequality, for $r > 1$ and $1/r + 1/s = 1$ we have,

$$\begin{aligned} E(|XY|) &\leq E^{1/r}(|X|^r)E^{1/s}(|Y|^s) \\ \Rightarrow \quad E(|X|) &\leq E^{1/r}(|X|^r) \quad \text{with } Y \equiv 1 \\ \Rightarrow \quad E(|X|^s) &\leq E^{1/r}(|X|^{rs}) \quad \text{by replacing } |X| \text{ by } |X|^s \\ \Rightarrow \quad E^{1/s}(|X|^s) &\leq E^{1/rs}(|X|^{rs}) \\ \Rightarrow \quad E^{1/s}(|X|^s) &\leq E^{1/t}(|X|^t) \quad \text{with } rs = t > s. \end{aligned}$$

□

Lemma 4.5.5. *Schwarz inequality :*

$$E(|XY|) \leq \sqrt{E(|X|^2)E(|Y|^2)}.$$

Proof. In Holder's inequality if we take $r = s = 2$.

$$E(|XY|) \quad \leq \quad E^{1/2}(|X|^2)E^{1/2}(|Y|^2) \quad \Longleftrightarrow \quad E(|XY|) \leq \sqrt{E(|X|^2)E(|Y|^2)}.$$

□

Schwarz inequality is also known as the Cauchy-Schwarz inequality. Using this inequality we show that the correlation coefficient between any two random variables is always between -1 and 1. Note that $XY \leq |XY|$ and $-XY \leq |XY|$. Thus, using monotonicity of expectations we get,

$$\begin{aligned} E(XY) &\leq E(|XY|) \quad \text{and} \quad E(-XY) = -E(XY) \leq E(|XY|) \\ \Rightarrow \quad |E(XY)| &\leq E(|XY|). \end{aligned}$$

Further, using Schwarz inequality, we have, $|E(XY)| \leq \sqrt{E(|X|^2)E(|Y|^2)}$.

Substituting $X - E(X)$ and $Y - E(Y)$ in place of X and Y respectively, we get,

$$\begin{aligned} |E[(X - E(X))(Y - E(Y))]| &\leq \sqrt{E(|X - E(X)|^2)E(|Y - E(Y)|^2)} \\ \Rightarrow \quad |Cov(X,Y)| &\leq \sqrt{Var(X)Var(Y)} \\ \Rightarrow \quad |Corr(X,Y)| &\leq 1 \ . \end{aligned}$$

Holder's inequality is concerned with the moments of product of random variables, while the following Minkowski's inequality is concerned with the moment of the sum of random variables.

Lemma 4.5.6. *Minkowski's inequality: For $r \geq 1$,*

$$E^{1/r}(|X + Y|^r) \quad \leq \quad E^{1/r}(|X|^r) + E^{1/r}(|Y|^r).$$

Proof. If $r = 1$ then Minkowski's inequality is trivially true. Hence we assume $r > 1$. For $r > 1$, we can write,

$$\begin{aligned} E(|X + Y|^r) &= E(|X + Y||X + Y|^{r-1}) \\ &\leq E[(|X| + |Y|)|X + Y|^{r-1}] \\ &= E(|X||X + Y|^{r-1}) + E(|Y||X + Y|^{r-1}). \end{aligned} \quad (4.12)$$

We know that in Holder's inequality, $s = r/(r - 1)$. For the first term on the right hand side of equation (4.12) with X as first random variable and $|X + Y|^{r-1}$ as second random variable, applying Holder's inequality, we get,

$$E(|X||X + Y|^{r-1}) \quad \leq \quad E^{1/r}(|X|^r)E^{1-1/r}(|X + Y|^r).$$

Using similar adjustments for the second term, we get,

$$E(|X + Y|^r) \quad \leq \quad \left[E^{1/r}(|X|^r) + E^{1/r}(|Y|^r)\right] E^{1-1/r}(|X + Y|^r).$$

If $E(|X+Y|^r) = 0$ then Minkowski's inequality is trivially true. Hence assuming $E(|X + Y|^r) > 0$ and dividing by it on both the sides of above inequality we get

$$E^{1/r}(|X + Y|^r) \quad \leq \quad E^{1/r}(|X|^r) + E^{1/r}(|Y|^r).$$

□

We now derive Jensen's inequality for an expectation of a convex and a concave function, which is frequently used in mathematical statistics. We first define a convex and a concave function.

Definition 4.5.1. *Convex and concave functions: A real valued Borel function f defined on an interval I of $\mathbb{R}$, which may be finite or infinite, is said to be a convex function if*

$$\forall \ x_1, x_2 \in I, \ \ f((x_1 + x_2)/2) \leq (1/2)f(x_1) + (1/2)f(x_2).$$

Alternatively, for every $x_0 \in I$, there corresponds a number $\lambda(x_0)$ such that

$$\forall \ x \in I, \ \ f(x_0) + \lambda(x_0)(x - x_0) \ \ \leq \ \ f(x). \tag{4.13}$$

If for every pair $x_1, x_2 \in I$, $f((x_1 + x_2)/2) \geq (1/2)f(x_1) + (1/2)f(x_2)$, then the function f is a concave function.

In Equation (4.13), the left hand side of the inequality may be interpreted as a tangent through x_0. If it exists, then the inequality implies that all the points of the curve $f(x)$ are above this tangent line. If f is twice differentiable then the necessary and sufficient condition for f to be convex is $f''(x) > 0$ and for f to be concave is $f''(x) < 0$.

Lemma 4.5.7. *Jensen's inequality: If $f(\cdot)$ is a convex function and if $E(X)$ is finite, then*

$$f(E(X)) \leq E(f(X)).$$

If $f(\cdot)$ is a concave function and if $E(X)$ is finite, then

$$f(E(X)) \geq E(f(X)).$$

Proof. Suppose $f(\cdot)$ is a convex function and I is any interval in the domain of $f(\cdot)$. By the property of convex function, for every $x_0 \in I$, there corresponds a number $\lambda(x_0)$ such that for all $x \in I$,

$$\lambda(x_0)(x - x_0) \ \ \leq \ \ f(x) - f(x_0) \tag{4.14}$$

Suppose X is a random variable with support I. Replacing x_0 by $E(X)$ and x by X in (4.14), we have,

$$\lambda(E(X))(X - E(X)) \ \ \leq \ \ f(X) - f(E(X)).$$

Taking expectation on both the sides of the above inequality, we get

$$0 \leq E(f(X)) - f(E(X)) \quad \iff \quad f(E(X)) \ \ \leq \ \ E(f(X)).$$

Proof for the concave function follows on similar lines. □

If $f(\cdot)$ is convex, monotone increasing and has an inverse $f^{-1}(\cdot)$, we get,

$$E(X) \ \ \leq \ \ f^{-1}(E(f(X))).$$

For example, if $f(x) = x^r$, for $r \geq 1, \ \ x \geq 0$, we have, $E(X) \ \ \leq \ \ E^{1/r}(X^r)$. If $r = 2$, then Jensen's inequality conveys the well known fact that variance of X is always non-negative.

Lemma 4.5.8. *Basic inequality: Suppose X is an arbitrary random variable and $g(\cdot)$ is a non-negative Borel function on $\mathbb{R}$. If $g(\cdot)$ is even and non-decreasing on $[0, \infty)$, then for every $a > 0$,*

$$\frac{E(g(X)) - g(a)}{a.s. \sup g(X)} \ \ \leq \ \ P[|X| \geq a] \ \ \leq \ \ \frac{E(g(X))}{g(a)}.$$

Proof. Since $g(\cdot)$ is a Borel function on $\mathbb{R}$, it follows that $g(X)$ is a random variable. Further, $g(X)$ is non-negative, hence there exists a non-decreasing sequence of non-negative simple random variables $g_n(X)$ such that $g_n(X(\omega)) \to g(X(\omega))\ \forall\ \omega \in \Omega$. Note that $E[g_n(X)]$ is a non-decreasing sequence of non-negative real numbers, hence it converges. Thus, $E[g(X)] = \lim_{n\to\infty} E[g_n(X)]$ exists. Therefore we can write,

$$E[g(X)] = \int_A g(X)dP + \int_{A^c} g(X)dP, \tag{4.15}$$

where, $A = [|X| \geq a]$. We are given that $g(\cdot)$ is non-decreasing on $[0, \infty)$. Therefore, on A, $a \leq |X| \Rightarrow g(a) \leq g(|X|) = g(X)$, in view of the fact that $g(X)$ is an even function. Thus, $g(a) \leq g(X) \leq a.s. \sup g(X)$. Using monotonicity of integrals, we have,

$$\begin{aligned} \int_A g(a)dP &\leq \int_A g(X)dP \leq \int_A a.s. \sup g(X)dP \\ \Rightarrow g(a)P(A) &\leq \int_A g(X)dP \leq [a.s. \sup g(X)]P(A). \end{aligned} \tag{4.16}$$

On A^c, $|X| < a$. Therefore, using the same arguments as above and non-negative nature of $g(X)$, we get $0 \leq g(X) \leq g(a)$ and hence integrating each of these terms over A^c, we get,

$$0 \leq \int_{A^c} g(X)dP \leq g(a)P(A^c).$$

However, $P(A^c) \leq 1$. Hence, we get,

$$0 \leq \int_{A^c} g(X)dP \leq g(a). \tag{4.17}$$

Adding inequalities in (4.16) and (4.17) and using (4.15), we get,

$$g(a)P(A) \leq E[g(X)] \leq [a.s. \sup g(X)]P(A) + g(a) . \tag{4.18}$$

From the right hand side of the inequality in (4.18), we get

$$\frac{E(g(X)) - g(a)}{a.s. \sup g(X)} \leq P(A) = P[|X| \geq a].$$

Using the left hand side of the inequality in (4.18), we get

$$P[|X| \geq a] = P(A) \leq \frac{E[g(X)]}{g(a)}.$$

□

With specific choices of g in basic inequality we get famous Markov inequality and Chebyshev's inequality.

Lemma 4.5.9. *Markov inequality: For $r > 0$,*

$$P(|X| \geq a) \quad \leq \quad \frac{E(|X|^r)}{a^r}.$$

Proof. It is to be noted that a function $g(x) = |x|^r$ is non-negative and even as $g(-x) = |-x|^r = |x|^r = g(x)$. Also, for any two positive real numbers x_1, x_2, $x_1 \leq x_2$ implies $x_1^r \leq x_2^r$. Thus, $g(x)$ is non-decreasing on $[0, \infty)$. Hence, using this function in basic inequality we get,

$$\frac{E(|X|^r) - a^r}{a.s. \sup |X|^r} \quad \leq \quad P[|X| \geq a] \quad \leq \quad \frac{E(|X|^r)}{a^r}.$$

The right hand side of the above inequality is the Markov inequality. □

Lemma 4.5.10. *Chebychev's inequality*

$$P(|X| \geq a) \quad \leq \quad \frac{E(X^2)}{a^2} \qquad \Longleftrightarrow \quad P[|X| < a] \;\geq\; 1 - \frac{E(X^2)}{a^2}\;.$$

Proof. Substituting $r = 2$ in Markov inequality, we get Chebyshev's inequality. □

From Chebyshev's inequality we note that $P[|X - E(X)| < a] \;\geq\; 1 - \frac{Var(X)}{a^2}$. It shows that if a random variable X has a small variance, then the distribution of X is tightly concentrated about $E(X)$, however the converse is not true. The fact that X is tightly concentrated about $E(X)$ tells us little about the variance or other moments of X.

A quick recap of the results discussed in this chapter is given below.

Summary

1. For a simple random variable $X = \sum_{i=1}^{k} a_i I_{A_i}$, expectation $E(X)$ is defined as $E(X) = \int_\Omega X dP = \sum_{i=1}^{k} a_i P(A_i)$.

2. Suppose X_1 and X_2 are simple random variables. Then (i) $E(X_1 \pm X_2) = E(X_1) \pm E(X_2)$. (ii) $E(aX_1) = aE(X_1)$, where a is any real number. (iii) $E(aX_1 + bX_2) = aE(X_1) + bE(X_2)$ where a and b are any real numbers. (iv) If $X_1 \geq X_2$ on Ω, then $E(X_1) \geq E(X_2)$.
(v) If $X_1 \geq 0$ *a.s.*, then $E(X_1) \geq 0$.

3. Suppose X is a non-negative random variable, then $E(X) = \int_\Omega X dP = \lim_{n\to\infty} E(X_n)$, where $\{X_n, n \geq 1\}$ is a non-decreasing sequence of non-negative simple random variables such that $X_n \to X$ on Ω.

4. Suppose X is an arbitrary random variable with $X^+$and X^- as its positive and negative parts respectively. Then $E(X)$ is defined as $E(X) = \int_\Omega X dP = E(X^+) - E(X^-)$, provided at least one of the two $E(X^+)$ and $E(X^-)$ is finite.

5. A random variable X is said to be integrable if $E(X)$ is finite. An arbitrary random variable X is integrable if and only if both X^+ and X^- are integrable.

6. Suppose X is an integrable random variable defined on $(\Omega, \mathbb{A}, P)$ and $A \in \mathbb{A}$ is such that $P(A) = 0$. Then $\int_A X\, dP = 0$.

7. Suppose X_1 and X_2 are random variables such that $E(X_1), E(X_2)$ and $E(X_1) + E(X_2)$ exist. (i) If $X_1 \geq 0 \quad a.s.$, then $E(X_1) \geq 0$ (Non-negativity). (ii) $E(cX_1) = cE(X_1)$, where c is any real number. (iii) If $X_1 = 0 \quad a.s.$, then $E(X_1) = 0$. (iv) $E(X_1 + X_2) = E(X_1) + E(X_2)$ (Linearity). (v) If $X_1 \geq X_2 \quad a.s.$, then $E(X_1) \geq E(X_2)$. (vi) X_1 is integrable if and only if $|X_1|$ is integrable. (vii) If $|X_1| \leq Y$, where Y is an integrable random variable then X_1 is integrable.

8. Suppose X is random variable defined on a probability space $(\Omega, \mathbb{A}, P)$ to $(\mathbb{R}, \mathbb{B}, P_X)$. Suppose $g : (\mathbb{R}, \mathbb{B}) \to (\mathbb{R}, \mathbb{B})$ is a Borel function. Then, $E(g(X)) = \int_\Omega g(X) dP = \int_\mathbb{R} g(x) dP_X = \int_\mathbb{R} g(x) dF(x)$.

9. Suppose X is a discrete random variable with $P[X = i] = p_i, i \in W$, the set of whole numbers . Then $E(X) = \sum_{n \geq 1} P[X \geq n]$.

10. Suppose X is a continuous random variable with distribution function F and suppose F is differentiable. Then

$$E(X) = \begin{cases} \int\limits_0^\infty [1 - F(x)] dx, & \text{if} \quad X \geq 0, \\ -\int\limits_{-\infty}^0 F(x) dx + \int\limits_0^\infty [1 - F(x)] dx, & \text{if} \quad X \text{ is arbitrary.} \end{cases}$$

Characteristic Function

1. Suppose $\underline{Z} = (X, Y)'$ is a bivariate random vector defined on a probability space $(\Omega, \mathbb{A}, P)$ to $(\mathbb{R}^2, \mathbb{B}^2, P_Z)$. Then $W = X + iY$ is a complex valued random variable and its expectation is defined as $E(W) = E(X) + iE(Y)$. A complex valued random variable $W = X + iY$ is said to be integrable if and only if both X and Y are integrable. Further, $|E(W)| \leq E(|W|)$.

2. The characteristic function $\phi_X(t)$ of a random variable X is defined as

$$\phi_X(t) = E(\exp(itX)) = E(\cos tX + i \sin tX) = E(\cos tX) + iE(\sin tX).$$

It is defined for all $t \in \mathbb{R}$.

3. If $Y = aX+b$ where $a \neq 0$ and b are any real numbers, then a characteristic function of Y is given by,
$$\phi_Y(t) = E(\exp(itY)) = E(\exp(it(aX+b))) = \exp(itb)\phi_X(at).$$
4. $\phi_{-X}(t) = \phi_X(-t) = \overline{\phi_X(t)}$, a complex conjugate of $\phi_X(t)$.
5. If the distribution of X is symmetric around 0, then $\phi_X(t)$ is a real valued function.
6. If $F = \sum_{i=1}^{k} \alpha_i F_i, \quad \alpha_i \geq 0 \ \& \ \sum_{i=1}^{k} \alpha_i = 1$ is a convex combination of distribution functions, then the characteristic function $\phi(t)$ corresponding to F is again a convex combination of the characteristic functions $\phi_i(t)$ of component distribution functions and is given by $\phi(t) = \sum_{i=1}^{k} \alpha\phi_i(t)$.
7. A sequence of random variables $\{X_n, n \geq 1\}$ is said to be bounded in probability if for a given $\epsilon > 0$, there exists a constant K and an integer n_0 such that,
$$P[|X_n| \leq K] \geq 1-\epsilon \quad \Longleftrightarrow \quad P[|X_n| > K] < \epsilon, \quad \forall \quad n \geq n_0.$$
8. A real random variable is bounded in probability.
9. (i) $|\phi_X(t)| \leq 1$. (ii) $\phi_X(0) = 1$. (iii) $\phi_X(\cdot)$ is uniformly continuous on $\mathbb{R}$.
10. If $E(|X|^n) < \infty$ where n is a positive integer, then $\phi_X(t)$ is differentiable n times for all $t \in \mathbb{R}$ and n-th derivative $\phi_X^{(n)}(t)$ is given by,
$$\phi_X^{(n)}(t) = \frac{d^n}{dt^n}\phi_X(t) = \int_{-\infty}^{\infty} (ix)^n e^{itx} dF(x) \ \& \ \phi_X^{(n)}(0) = i^n E(X^n).$$
11. If $E(|X|^k) < \infty$ where k is a positive integer, then $\phi_X(t)$ can be expressed,
$$\phi_X(t) = \sum_{s=0}^{k-1} \frac{(it)^s}{s!} E(X^s) + t^k \int_0^1 \frac{(1-u)^{k-1}}{(k-1)!}\phi_X^{(k)}(tu)\, du,$$
where $\phi_X^{(k)}(tu)$ is the k-th derivative of $\phi_X(t)$ at tu.
12. If moments of all order of X are finite, then $\phi_X(t)$ can be expressed,
$$\phi_X(t) = \sum_{s=0}^{\infty} \frac{(it)^s}{s!} E(X^s).$$
13. Inversion Theorem: Suppose $\phi_X(t)$ is a characteristic function of a random variable X with distribution function F_X. Then
$$\begin{aligned} F_X(b) - F_X(a) &+ (1/2)P[X=a] - (1/2)P[X=b] \\ &= \frac{1}{2\pi}\lim_{c\to\infty}\int_{-c}^{c}\left[\frac{e^{-ita}-e^{-itb}}{it}\right]\phi_X(t)\, dt. \end{aligned}$$

If $a < b \in C(F_X)$, then

$$F_X(b) - F_X(a) = \frac{1}{2\pi} \lim_{c \to \infty} \int_{-c}^{c} \left[\frac{e^{-ita} - e^{-itb}}{it} \right] \phi_X(t)\, dt.$$

The value of the integrand at $t = 0$ is determined by continuity, that is, by using L'Hospital's rule and it is $b - a$.

14. Uniqueness theorem: $\phi_X(t) = \phi_Y(t)\ \forall\, t \in \mathbb{R} \iff X \overset{d}{=} Y$.

15. If F is differentiable at x with $F'(x) = f(x)$, then

$$f(x) = \lim_{h \to 0} \lim_{c \to \infty} \frac{1}{2\pi} \int_{-c}^{c} \left(\frac{1 - e^{-ith}}{ith} \right) e^{-itx} \phi(t)\, dt,$$

if and only if the limits on the right hand side exist.
(ii) If ϕ is Riemann integrable on $\mathbb{R}$ and if $F'(x) = f(x)$ exists $\forall\ x \in \mathbb{R}$, then

$$f(x) = \frac{1}{2\pi} \int_{\mathbb{R}} e^{-itx} \phi(t)\, dt$$

and $f(x)$ is continuous on $\mathbb{R}$.

16. If $P[X = a] > 0$, then

$$P[X = a] = F_X(a + 0) - F_X(a - 0) = \lim_{c \to \infty} \frac{1}{2c} \int_{-c}^{c} e^{-ita} \phi_X(t)\, dt.$$

17. If X is a discrete random variable with support I, then

$$\phi_X(t) = \sum_{x \in I} p_x e^{itx} \quad \& \quad p_x = P[X = x] = \frac{1}{2\pi} \int_{-\pi}^{\pi} e^{-itx} \phi_X(t)\, dt.$$

18. A characteristic function $\phi_{\underline{Z}}(\underline{t})$ of a random vector $\underline{Z} = (X_1, X_2, \cdots, X_k)'$, where $\underline{t} = (t_1, t_2, \cdots, t_k)'$, $t_i \in \mathbb{R}$, $i = 1, 2, \cdots, k$, is defined as

$$\begin{aligned} \phi_{\underline{Z}}(\underline{t}) &= \phi_{\underline{Z}}(t_1, t_2, \cdots, t_k) = E(\exp(i(t_1 X_1 + t_2 X_2 + \cdots + t_k X_k)) \\ &= E(\exp(\underline{t}' \underline{Z})). \end{aligned}$$

Moment Inequalities

1. If $E(|X|^r) < \infty$, then $E(|X|^{r'}) < \infty$, for $0 < r' \le r$ and $E(X^k)$ exists and is finite for $k < r$, if k is an integer.

2. C_r inequality: Suppose $X \in L_r$ and $Y \in L_r$, then for any $r > 0$, $E(|X + Y|^r) \le C_r \left(E(|X|^r) + E(|Y|^r) \right)$, where,

$$C_r = \begin{cases} 1, & \text{if } r \le 1, \\ 2^{r-1}, & \text{if } r > 1. \end{cases}$$

3. Holder's inequality: $E(|XY|) \leq E^{1/r}(|X|^r)E^{1/s}(|Y|^s)$, for $r > 1$ and $1/r + 1/s = 1$. With $r = s = 2$ we get $E(|XY|) \leq \sqrt{E(|X|^2)E(|Y|^2)}$ which is known as Schwarz inequality or Cauchy-Schwarz inequality.

4. Minkowski inequality: $E^{1/r}(|X + Y|^r) \leq E^{1/r}(|X|^r) + E^{1/r}(|Y|^r)$ for $r \geq 1$.

5. Jensen's inequality: If $f(\cdot)$ is a convex function and if $E(X)$ is finite, then $f(E(X)) \leq E(f(X))$. If $f(\cdot)$ is a concave function and if $E(X)$ is finite, then $f(E(X)) \geq E(f(X))$.

6. Basic inequality: Suppose X is an arbitrary random variable and $g(\cdot)$ is a non-negative Borel function on $\mathbb{R}$. If $g(\cdot)$ is even and non-decreasing on $[0, \infty)$, then for every $a > 0$,

$$\frac{E(g(X)) - g(a)}{a.s. \sup g(X)} \leq P[|X| \geq a] \leq \frac{E(g(X))}{g(a)}.$$

7. Markov inequality: For $r > 0$, $P[|X| \geq a] \leq E(|X|^r)/a^r$.

8. Chebychev inequality: $P[|X| \geq a] \leq E(X^2)/a^2$ which is equivalent to $P[|X| < a] \geq 1 - E(X^2)/a^2$.

4.6 Conceptual Exercises

4.6.1 Give an example of a discrete random variable with infinite mean.

4.6.2 Give an example of a discrete random variable with finite mean but with infinite variance.

4.6.3 Suppose X and Y are random variables such that $P[|X - Y| \leq M] = 1$ for some $M > 0$. Show that Y has finite mean if X has finite mean.

4.6.4 Show that for any non-negative random variable X,
(i) $E(X) + E(1/X) \geq 2$ and (ii) $E(\max\{X, 1/X\}) \geq 1$.

4.6.5 Prove that for a non-negative random variable X, $E(X) = 0$ implies $X = 0$ a.s.

4.6.6 Prove that if $Var(X) = 0$, then $X = E(X)$ a.s.

4.6.7 Suppose X is an integrable random variable and $A\Delta B$ is a null event. Show that $\int_A X\, dP = \int_B X\, dP$.

4.6.8 Prove or disprove: $X \equiv 0$ implies $\int_\Omega X dP = 0$, but $\int_\Omega X dP = 0$ may not imply $X \equiv 0$.

4.6.9 Suppose $\Omega = \{-2, -1, 3, 7\}$ and $\mathbb{A}$ is a powerset of Ω. A measure μ on $(\Omega, \mathbb{A})$ is defined as $\mu(\{-2\}) = 2, \mu(\{-1\}) = 1, \mu(\{3\}) = 3$ and $\mu(\{7\}) = 7$. Obtain a probability measure P from μ. Suppose a random variable X is defined on $(\Omega, \mathbb{A}, P)$ as follows.

$$X(\omega) = \begin{cases} 1, & \text{if} \quad \omega = 3 \text{ or } 7 \\ -1, & \text{if} \quad \omega = -2 \text{ or } -1 \,. \end{cases}$$

Find $\int_\Omega X \, dP$.

4.6.10 Suppose $(\Omega, \mathbb{A}, P)$ is a probability space where $\mathbb{A}$ is a sigma field generated by a countable partition $\{A_n, n \geq 1\}$ of Ω. Suppose a random variable X is defined on $(\Omega, \mathbb{A}, P)$ as $X = i$ on A_i, $i \geq 1$. Find $\int_\Omega X \, dP$ if the probability measure P is defined as (i) $P(A_i) = c/i^3$, where c is such that $\sum_{i \geq 1} c/i^3 = 1$, (ii) $P(A_i) = c/i^2$, where c is such that $\sum_{i \geq 1} c/i^2 = 1$ and (iii) $P(A_i) = 1/2^i$. In all the three cases, find $\int_\Omega Y \, dP$, when $Y = X^2$ and $Y = e^X$.

4.6.11 Suppose a distribution function $F : \mathbb{R} \to [0, 1]$ is defined as follows.

$$F(x) = \begin{cases} 0, & \text{if} \quad x < 0, \\ 0.2, & \text{if} \quad 0 \leq x < 1, \\ 0.5, & \text{if} \quad 1 \leq x < 2, \\ 0.8, & \text{if} \quad 2 \leq x < 4, \\ 1, & \text{if} \quad x \geq 4. \end{cases}$$

Compute $E(X)$ from $F(x)$, without finding the probability mass function.

4.6.12 Suppose a distribution function $F : \mathbb{R} \to [0, 1]$ is defined as follows.

$$F(x) = \begin{cases} 0, & \text{if} \quad x < -2, \\ 1/3, & \text{if} \quad -2 \leq x < 0, \\ 1/2, & \text{if} \quad 0 \leq x < 5, \\ 1/2 + (x-5)^2/2, & \text{if} \quad 5 \leq x < 6, \\ 1, & \text{if} \quad x \geq 6. \end{cases}$$

Find (i) $E(X)$ from $F(x)$. (ii) Decompose F into discrete and continuous components. (iii) Identify the distributions in the decomposition and hence find $E(X)$.

4.6.13 Suppose a distribution function $F : \mathbb{R} \to [0, 1]$ is defined as follows.

$$F(x) = \begin{cases} 0, & \text{if} \quad x < -2, \\ 0.2, & \text{if} \quad -2 \leq x < 1, \\ 0.2 + x^2/10, & \text{if} \quad 1 \leq x < 2, \\ 0.6 + x/10, & \text{if} \quad 2 \leq x < 3, \\ 1, & \text{if} \quad x \geq 3. \end{cases}$$

Find (i) $E(X)$ from $F(x)$. (ii) Decompose F into discrete and continuous components. (iii) Identify the distributions in the decomposition and hence find $E(X)$.

4.6.14 Suppose a distribution function $F : \mathbb{R} \to [0, 1]$ is defined as follows.

$$F(x) = \begin{cases} 0 , & \text{if} \quad x < 2 \\ (2/3)x - 1 , & \text{if} \quad 2 \leq x < 3 \\ 1 , & \text{if} \quad x \geq 3. \end{cases}$$

Find (i) $E(X)$ from $F(x)$. (ii) Decompose F into discrete and continuous components. (iii) Identify the distributions in the decomposition and hence find $E(X)$.

4.6.15 Suppose a distribution function $F : \mathbb{R} \to [0, 1]$ is defined as follows.

$$F(x) = \begin{cases} 0 , & \text{if} \quad x < 0 \\ 1/4 , & \text{if} \quad 0 \leq x < 1 \\ 1 - \frac{1}{2}e^{-(x-1)}, & \text{if} \quad x \geq 1. \end{cases}$$

(i) Find $E(X)$ from $F(x)$. (ii) Decompose F into discrete and continuous components. (iii) Identify the distributions in the decomposition and hence find $E(X)$.

4.6.16 Suppose a distribution function $F : \mathbb{R} \to [0, 1]$ is defined as follows.

$$F(x) = \begin{cases} 0, & \text{if} \quad x < 0, \\ (pr + (1-p)x)/k, & \text{if} \quad r \leq x < r+1, r = 0, 1, \cdots, k-1, \\ 1, & \text{if} \quad x \geq k, \end{cases}$$

$0 < p < 1$. (i) Find $E(X)$ from $F(x)$. (ii) Decompose F into discrete and continuous components. (iii) Identify the distributions in the decomposition and hence find $E(X)$.

4.6.17 Prove that for a non-negative random variable X, $E(X) = 0$ implies $X = 0$ a.s.

4.6.18 If $X \geq 0$ on A and $\int_A X \, dP = 0$ then show that $X = 0$ a.s. on A.

4.6.19 If $a \leq X \leq b$ a.s. on A then show that $aP(A) \leq \int_A X \, dP \leq bP(A)$.

4.6.20 Examine whether following functions are characteristic functions.
(i) $\phi_1(t) = \phi(t/a) \exp(ibt)$, $a \neq 0, b \in \mathbb{R}$ and ϕ is a characteristic function,
(ii) $\phi_1(t) = \phi(-t)$, where ϕ is a characteristic function,
(iii) $\phi(t) = \exp\{-|t|\}$, $t \in \mathbb{R}$,
(iv) $\phi(t) = \exp\{\lambda(\psi(t-1))\}$ where ψ is a characteristic function and $\lambda > 0$,
(v) $\phi(t) = \sum_{j=0}^{\infty} a_j \phi_j(t)$, $a_j \geq 0$ and $\sum_{j=0}^{\infty} a_j = 1$ and $\phi_j(t)$ are characteristic functions for all $j = 0, 1, \cdots$ and
(vi) $\phi(t) = 1/(1 + t^4)$ where $t \in \mathbb{R}$, (vii) $\phi(t) = |\cos t|$.

4.6.21 Suppose a discrete random variable has finite second moment. Show that it has finite mean.

4.6.22 Suppose $L_r = \{Z|E(|Z|^r) < \infty\}$. Prove that the set L_r is closed under addition.

4.6.23 Suppose $X \sim U(0,2)$ distribution then without evaluating, show that $E(X\ \log X) \geq 0$.

4.7 Multiple Choice Questions

Note: In each question, multiple options may be correct. Answers are given in Chapter 11, after the solutions of conceptual exercises of Chapter 4.

4.7.1 If X is a real valued random variable, which of the following quantities will always be finite?

(a) $E(X)$
(b) $E(X^+X^-)$
(c) $E(X^+ - X^-)$
(d) None of the above.

4.7.2 In which of the following cases, expectation may not exist?

(a) X is a degenerate random variable
(b) X is a simple random variable
(c) X is a non-negative random variable
(d) Support of X is $\mathbb{R}$

4.7.3 Which of the following need not be always true?

(a) $E(X+Y) = E(X) + E(Y)$
(b) $E(X-Y) = E(X) - E(Y)$
(c) $E(aX+bY) = aE(X) + bE(Y)$
(d) $E(XY) = E(X)E(Y)$

4.7.4 Suppose X is a discrete random variable with support as the set W of whole numbers and suppose $P[X=i] = p_i, i \in W$. Then which of the following is not always true?

(a) $E(X) = \sum_{i\geq 0} ip_i$
(b) $E(X) = \sum_{n\geq 1} P[X \geq n]$
(c) $E(X) = \sum_{n>1} P[X \geq n]$
(d) $E(X) = \sum_{i\geq 1} ip_i$

4.7.5 If X is a non-negative random variable, which of the following is false?

(a) $E(X^2) \geq [E(X)]^2$
(b) $E(X^3) \geq [E(X)]^3$
(c) $E(X^{-2}) \geq [E(X)]^{-2}$
(d) $E(X^4) \leq [E(X)]^4$

4.7.6 Which of the following is/are not always true, if X is an arbitrary random variable?

(a) $E(X^2) \geq [E(X)]^2$
(b) $E(X^3) \geq [E(X)]^3$
(c) $E(X^{-2}) \geq [E(X)]^{-2}$
(d) $E(X^4) \leq [E(X)]^4$

4.7.7 If $E(X^2)$ is finite, which of the following need not be finite?

(a) $E(X)$
(b) $E(|X|)$
(c) $E(X^{3/2})$
(d) $E(X^3)$

4.7.8 Which of the following statements is/are false?

(a) X is integrable implies $|X|$ is integrable
(b) $|X|$ is integrable implies X is integrable
(c) X is integrable implies X^2 is integrable
(d) X^2 is integrable implies X is integrable

4.7.9 Following are four statements: (I) X is integrable implies $|X|$ is integrable. (II) $|X|$ is integrable implies X is integrable. (III) X is integrable implies X^2 is integrable. (IV) X^2 is integrable implies X is integrable. Then which of the following is/are true?

(a) Both (I) and (II) are true
(b) Both (III) and (IV) are true
(c) Both (II) and (IV) are true
(d) Both (I) and (III) are true

4.7.10 If support of a random variable X is denoted by S, in which of the following cases, $E(X)$ may not exist?

(a) S is a singleton set.
(b) $S = (0, \infty)$
(c) S is a finite set having positive as well as negative elements
(d) $S = \mathbb{R}$

4.7.11 Suppose X and Y are two random variables whose third order moments exist. Then which of the following statements is correct?

(a) $\left(E(|X+Y|)^3\right)^{1/3} \leq \left(E(|X|)^3\right)^{1/3} \times \left(E(|Y|)^3\right)^{1/3}$
(b) $\left(E(|X+Y|)^3\right)^{1/3} \leq \left(E(|X|)^3\right)^{1/3} + \left(E(|Y|)^3\right)^{1/3}$
(c) $\left(E(|X+Y|)^3\right)^{1/3} \leq (E(|X|) + E(|Y|))^{1/3}$

(d) $\left(E(|X+Y|)^3\right)^{1/3} \leq (E(|X|) \times E(|Y|))^{1/3}$

4.7.12 Suppose X is a non-negative random variable whose mean exists. Then which of the following statements is correct?

(a) $E(\log X) \leq \log E(X)$
(b) $E(\log X) \geq \log E(X)$
(c) $E(\log X) = \log E(X)$
(d) None of (a), (b) and (c) is true

4.7.13 Suppose X is an arbitrary random variable and $g(\cdot)$ is a non-negative Borel function on $\mathbb{R}$. If $g(\cdot)$ is even and non-decreasing on $[0, \infty)$, then which of the following statements is correct? $\forall \;\; a > 0$

(a) $P[|X| \leq a] \leq E(g(X))/g(a)$
(b) $P[|X| \leq a] \geq E(g(X))/g(a)$
(c) $P[|X| \geq a] \leq E(g(X))/g(a)$
(d) $P[|X| \geq a] \geq E(g(X))/g(a)$

4.7.14 Following are two statements. (I) If the distribution of X is symmetric around 0, then its characteristic function is a real valued function. (II) If the characteristic function of X is a real valued function, then the distribution of X is symmetric around 0. Which of the following options is correct?

(a) Both (I) and (II) are true
(b) Both (I) and (II) are false
(c) (I) is true but (II) is false
(d) (I) is false but (II) is true

4.7.15 The characteristic function of $X \sim B(n, p)$ is

(a) $(p + (1-p)e^{it})^n$
(b) $(1 - p + pe^{it})^{-n}$
(c) $(1 - p + pe^{it})^n$
(d) $(p + (1-p)e^{it})^{-n}$

4.7.16 The characteristic function of $X \sim P(\lambda)$ is

(a) $e^{-\lambda(e^{it}-1)}$
(b) $e^{\lambda(e^{it}+1)}$
(c) $e^{\lambda(e^{-it}-1)}$
(d) $e^{\lambda(e^{it}-1)}$

4.7.17 The domain of the characteristic function of any random variable X is

(a) $\mathbb{R}$
(b) $(0, \infty)$
(c) $(0, 1)$
(d) depends on the distribution of X.

4.7.18 Suppose $\phi(t) = 1/(1+t^2)$ is a characteristic function of a random variable X. Then $E(X)$ and $Var(X)$ are respectively given by

(a) $\{1/2, 1/8\}$
(b) $\{0, 2\}$
(c) $\{0, 1\}$
(d) $\{1/2, 1/16\}$

4.7.19 Suppose $\phi(t) = e^{-|t|}$ is a characteristic function of a random variable X. Then the distribution of X is

(a) normal
(b) Laplace
(c) Students t with 1 degree of freedom
(d) Cauchy

4.7.20 Following are two statements. (I) if X is a simple random variable, then for any function g, $g(X)$ is also a simple random variable. (II) If g is a simple Borel function, then for any random variable X, $g(X)$ is a simple random variable. Which of the following options is correct?

(a) Both (I) and (II) are true
(b) Both (I) and (II) are false
(c) (I) is true but (II) is false
(d) (I) is false but (II) is true

5

Independence

5.1 Introduction

There are some similarities between probability theory and measure theory. A probability measure is essentially a normed measure with the measure of the sample space equal to one. A probability space is a measure space where the measure is replaced by a probability measure. A random variable in probability theory is nothing but a measurable function in measure theory, the concept of expectation in probability theory parallels the concept of integration in measure theory. Some modes of convergence of a sequence of random variables are similar to the convergence of a sequence of measurable functions. For example, almost sure convergence from probability theory is the same as the almost everywhere convergence in measure theory, convergence in probability is the same as the convergence in measure. In spite of such similarities, probability theory is not merely a branch of measure theory.

A distinguishing and fundamental feature of probability theory is the notion of independence, which has no parallel in measure theory. It is one of the most central concepts of probability theory and a large part of the theory explores the consequences of the presence of this property and the lack of it. The present chapter is devoted to the study of the fundamentals of the theory of independence of random variables and some related ideas.

In simple words, independence of trials of the experiment means that the outcome in one trial of the experiment does not influence the chance of occurrence of a particular outcome in the next trial, that is, the probabilities of occurrence of events in future does not depend on the occurrence of events in the past, that knowledge of the outcomes of trials so far does not provide any information about the chance of outcomes of the future trials. The classical examples are coin-tossing and throwing dice where successive outcomes are independent. In sampling without replacement from a finite population, successive outcomes are not independent, the typical example being drawing cards from a deck of cards. We now proceed for the rigorous discussion of the concept of independence.

In the next section, we discuss the concept of independence of events and independence of classes of events. Section 3 is concerned with the independence of random variables and some related results. Section 4 presents

DOI: 10.1201/9781032619057-5

Kolmogorov zero-one law about tail events, which is concerned with a sequence of independent random variables.

5.2 Independence of Events and Classes of Events

Suppose $(\Omega, \mathbb{A}, P)$ is a probability space. We begin with the basic definition of independence of two events.

Definition 5.2.1. *Independence of two events: Two events A_1 and A_2 in $\mathbb{A}$ are said to be independent if $P(A_1 \cap A_2) = P(A_1)P(A_2)$.*

As a consequence of this definition,

$$\begin{aligned} P(A_1|A_2) &= P(A_1 \cap A_2)/P(A_2) = P(A_1) \\ \& \ P(A_2|A_1) &= P(A_1 \cap A_2)/P(A_1) = P(A_2). \end{aligned}$$

Thus, the knowledge of occurrence of A_2 does not affect the probability of occurrence of A_1 and the knowledge of occurrence of A_1 does not affect the probability of occurrence of A_2. The conditional probability $P(A_1|A_2)$ is the same as the unconditional probability $P(A_1)$. Further, A_1 and A_2 are independent implies that

$$\begin{aligned} P(A_1 \cap A_2) &= P(A_1)P(A_2) \\ \Longleftrightarrow \quad P(A_1 \cap A_2)/P(A_1) &= P(A_2) = P(A_2)/P(\Omega) \\ \& \ P(A_1 \cap A_2)/P(A_2) &= P(A_1) = P(A_1)/P(\Omega). \end{aligned}$$

Note that $P(A_1 \cap A_2)/P(A_1) = P(A_2)/P(\Omega)$ implies that the relative probability measure of $A_1 \cap A_2$ relative to that of A_1 is the same as the relative probability measure of A_2 relative to that of Ω when A_1 and A_2 are independent. Similarly, $P(A_1 \cap A_2)/P(A_2) = P(A_1) = P(A_1)/P(\Omega)$ implies that the relative probability measure of $A_1 \cap A_2$ relative to that of A_2 is the same as the relative probability measure of A_1 relative to that of Ω.

An important point to be noted in the concept of independence is that it is not a property of events, but it is a property of underlying probability measure. The two events may be independent with respect one probability measure but may not be with respect to another probability measure. This feature is illustrated in the following example.

Example 5.2.1. Suppose $(\Omega, \mathbb{A})$ is a measurable space, where $\Omega = \{1, 2, 3, 4, 5, 6\}$, $\mathbb{A} = \mathbb{P}(\Omega)$. Two probability measures P and P^* on $(\Omega, \mathbb{A})$ are defined as follows.

$$P(\omega) = 1/6 \ \ \forall \ \ \omega \in \Omega \ \ \& \ \ P^*(\{1\}) = P^*(\{3\}) = P^*(\{6\}) = 1/4,$$

$$P^*(\{2\}) = P^*(\{4\}) = 1/8 \;\;\&\;\; P^*(\{5\}) = 0.$$

Suppose $A = \{1, 3, 5\}$ and $B = \{2, 3, 4\}$. Observe that,

$$\begin{aligned} P(A) &= P(B) = 1/2 \;\;\&\;\; P(A \cap B) = P(\{3\}) = 1/6 \\ \Rightarrow \;\; P(A)P(B) &= 1/4 \neq P(A \cap B) = 1/6 \\ P^*(A) &= P^*(B) = 1/2 \;\;\&\;\; P^*(A \cap B) = P(\{3\}) = 1/4 \\ \Rightarrow \;\; P^*(A)P^*(B) &= 1/4 = P^*(A \cap B). \end{aligned}$$

Thus A is not independent of B with respect to the probability measure P, but A and B are independent with respect to the probability measure P^*. □

Example 5.2.1 conveys that whenever we discuss independence of events or independence of random variables, underlying probability space must be the same. Next theorem addresses some simple but frequently used results related to independence of events.

Theorem 5.2.1. *Suppose $(\Omega, \mathbb{A}, P)$ is a probability space.*

(i) An event $A \in \mathbb{A}$ is independent of itself if and only if its probability measure is 0 or 1.

(ii) An event $A \in \mathbb{A}$ is independent of A^c if and only if $P(A)$ is 0 or 1.

(iii) Any event $A \in \mathbb{A}$ is independent of sure event Ω and null event $\emptyset$.

(iv) Suppose $N \in \mathbb{A}$ is a P-null event. Then any event $A \in \mathbb{A}$ is independent of N and N^c.

(v) Suppose $N \in \mathbb{A}$ is a P-null event. Then N is independent of N^c.

(vi) Suppose $A_1, A_2 \in \mathbb{A}$. If A_1 and A_2 are independent then A_1 and A_2^c, A_1^c and A_2, A_1^c and A_2^c are also independent.

Proof. (i) By the definition of independence, A is independent of itself if,

$$\begin{aligned} P(A \cap A) = P(A)P(A) \;\;&\Rightarrow\;\; P(A) = (P(A))^2 \;\;\Rightarrow\;\; P(A)(1 - P(A)) = 0 \\ &\Rightarrow\;\; P(A) = 0 \;\text{ or }\; 1. \end{aligned}$$

Suppose $P(A) = 0$ or 1. To examine if A is independent of itself, observe that

$$\begin{aligned} P(A) = 0 \;\;&\Rightarrow\;\; P(A \cap A) = P(A) = 0 = P(A)P(A) \\ \&\;\; P(A) = 1 \;\;&\Rightarrow\;\; P(A \cap A) = P(A) = 1 = P(A)P(A) \\ &\Rightarrow\;\; A \text{ is independent of itself.} \end{aligned}$$

Thus, A is independent of itself if and only if its probability measure is 0 or 1, that is, it is either a P-null event or a complement of P-null event.
(ii) Note that $P(A \cap A^c) = P(\emptyset) = 0 = P(A)P(A^c)$ if and only if $P(A)$ is 0

or 1.
(iii) It is clear that

$$P(A \cap \Omega) = P(A) = P(A)P(\Omega) \quad \& \quad P(A \cap \emptyset) = P(\emptyset) = 0 = P(A)P(\emptyset).$$

Thus, any event A is independent of sure event Ω and impossible event $\emptyset$.
(iv) Note that

$$P(N) = 0 \;\Rightarrow\; P(A \cap N) \leq P(N) = 0 \;\Rightarrow\; P(A \cap N) = 0 = P(A)P(N).$$

Hence, any event $A \in \mathbb{A}$ is independent of N. Further,

$$P(N^c) = 1 \;\Rightarrow\; P(A \cap N^c) = P(A) - P(A \cap N) = P(A) = P(A)P(N^c).$$

Hence, any event A is independent of N^c.
(v) It is clear that

$$P(N \cap N^c) = P(\emptyset) = 0 = P(N)P(N^c).$$

Hence, P-null event N is independent of N^c.
(vi) Observe that

$$\begin{aligned}
A_1 &= (A_1 \cap A_2) \cup (A_1 \cap A_2^c) \;\Rightarrow\; P(A_1) = P(A_1 \cap A_2) + P(A_1 \cap A_2^c) \\
&\Rightarrow P(A_1 \cap A_2^c) = P(A_1) - P(A_1 \cap A_2) \\
&\Rightarrow P(A_1 \cap A_2^c) = P(A_1) - P(A_1)P(A_2) = P(A_1)P(A_2^c) \\
&\Rightarrow A_1 \;\&\; A_2^c \;\text{ are independent.}
\end{aligned}$$

Now

$$\begin{aligned}
A_2 &= (A_1 \cap A_2) \cup (A_1^c \cap A_2) \Rightarrow P(A_2) = P(A_1 \cap A_2) + P(A_1^c \cap A_2) \\
&\Rightarrow P(A_1^c \cap A_2) = P(A_2) - P(A_1 \cap A_2) \\
&\Rightarrow P(A_1^c \cap A_2) = P(A_2) - P(A_1)P(A_2) = P(A_1^c)P(A_2) \\
&\Rightarrow A_1^c \;\&\; A_2 \;\text{ are independent.} \\
A_1^c \cap A_2^c &= (A_1 \cup A_2)^c \;\Rightarrow\; P(A_1^c \cap A_2^c) = 1 - P(A_1 \cup A_2) \\
&\Rightarrow P(A_1^c \cap A_2^c) = 1 - P(A_1) - P(A_2) + P(A_1)P(A_2) \\
&\Rightarrow P(A_1^c \cap A_2^c) = (1 - P(A_1))(1 - P(A_2)) = P(A_1^c)P(A_2^c) \\
&\Rightarrow A_1^c \;\&\; A_2^c \;\text{ are independent.}
\end{aligned}$$

□

We now extend the concept of independence of two events to a collection of countable number of events.

Definition 5.2.2. *Independence of k events: Suppose $\{A_1, A_2, \cdots, A_k\}$ is a finite collection of events from $\mathbb{A}$. We say that these k events are independent if,*

$$P(A_{i_1} \cap A_{i_2} \cap \cdots \cap A_{i_r}) = \prod_{l=1}^{r} P(A_{i_l}), \tag{5.1}$$

for every possible combination $\{i_1, i_2, \cdots, i_r\}$ of $\{1, 2, \cdots, k\}, 2 \leq r \leq k$.

To check the independence of k events, we need to verify,

$$\binom{k}{2} + \binom{k}{3} + \cdots + \binom{k}{k} = 2^k - \binom{k}{1} - \binom{k}{0} = 2^k - k - 1$$

conditions as specified in Equation (5.1).

It is to be noted that independence of $\{A_1, A_2, A_3\}$ implies that for three possible pairs, $P(A_i \cap A_j) = P(A_i)P(A_j)$ and $P(A_1 \cap A_2 \cap A_3) = P(A_1)P(A_2)P(A_3)$. The concept of pair-wise independence, as defined below, is slightly weaker than the concept of independence.

Definition 5.2.3. *Pair-wise independence: Suppose $\{A_1, A_2, \cdots, A_k\}$ is a finite collection of events from $\mathbb{A}$. We say that these k events are pair-wise independent if,*

$$P(A_i \cap A_j) = P(A_i)P(A_j), \quad \forall \ \ i \neq j = 1, 2, \cdots, k. \tag{5.2}$$

A collection of independent events is obviously a collection of pair-wise independent events. However, the converse is not true as shown in the following example.

Example 5.2.2. Suppose a probability space $(\Omega, \mathbb{A}, P)$ is defined as follows. $\Omega = \{(1,0,0), (0,1,0), (0,0,1), (1,1,1)\}$, $\mathbb{A} = \mathbb{P}(\Omega)$ and $P(\{\omega\}) = 1/4 \ \forall \ \omega \in \Omega$. Suppose an event A_i is defined as $A_i = \{\omega | i\text{-th coordinate is } 1\}$, $i = 1, 2, 3$. Then it is clear that for $i = 1, 2, 3$,

$$P(A_i) = 1/2, \quad P(A_i \cap A_j) = 1/4 \ \ i \neq j = 1, 2, 3 \ \ \& \ \ P(A_1 \cap A_2 \cap A_3) = 1/4.$$

Observe that $\forall \ \ i \neq j = 1, 2, 3, \ \ P(A_i \cap A_j) = P(A_i)P(A_j)$, however $P(A_1 \cap A_2 \cap A_3) \neq P(A_1)P(A_2)P(A_3)$. Thus, A_1, A_2, A_3 are pair-wise independent but not independent, which is usually referred to as A_1, A_2, A_3 are not mutually independent. □

Example 5.2.3. Suppose a probability space $(\Omega, \mathbb{A}, P)$ is defined as follows.

$$\Omega = \{\underline{\omega} = (a_1, a_2, \cdots, a_n) | a_i = 0, 1, i = 1, 2, \cdots, n\}, \mathbb{A} = \mathbb{P}(\Omega) \ \& \ P(\underline{\omega}) = 1/2^n.$$

An event A_i is defined as $A_i = \{(a_1, a_2, \cdots, a_n) | a_i = 1\}$, $i = 1, 2, \cdots, n$. We examine whether the collection $\{A_1, A_2, \cdots, A_n\}$ is a collection of independent events. By the definition of A_i, i-th component is fixed at 1 and remaining $n-1$ components can be either 0 or 1. Thus, the number of elements in A_i is $2^{(n-1)}$ and each has the same probability. Hence, for the given probability measure, $P(A_i) = 2^{n-1}/2^n = 1/2 \ \ \forall \ \ i = 1, 2, \cdots, n$. Suppose $\{i_1, i_2, \cdots, i_m\}$ is any combination of $\{1, 2, \cdots, n\}$ for $2 \leq m \leq n$. Then

$$\begin{aligned} A_{i_1} \cap A_{i_2} \cap \cdots \cap A_{i_m} &= \{(a_1, a_2, \cdots, a_n) | a_{i_j} = 1, j = 1, 2, \cdots, m\} \\ \Rightarrow \ P(A_{i_1} \cap A_{i_2} \cap \cdots \cap A_{i_m}) &= 2^{n-m}/2^n = 1/2^m \\ \Rightarrow \ P(A_{i_1} \cap A_{i_2} \cap \cdots \cap A_{i_m}) &= P(A_{i_1})P(A_{i_2}) \cdots P(A_{i_m}) \, . \end{aligned}$$

Thus, every sub-collection of the collection $\{A_1, A_2, \cdots, A_n\}$ is a collection of independent events. Hence, we claim that the collection $\{A_1, A_2, \cdots, A_n\}$ is a collection of independent events. □

Definition 5.2.4. *Sequence of independent events: Suppose $\{A_n, n \geq 1\}$ is a sequence of events from $\mathbb{A}$. We say that $\{A_n, n \geq 1\}$ is a sequence of independent events if every finite sub-collection of $\{A_n, n \geq 1\}$ is a collection of independent events.*

We now extend the definition of independence of events to independence among collections of events, in particular sigma fields of events.

Definition 5.2.5. *Independence of two collections of events: Suppose $\mathbb{C}_1$ and $\mathbb{C}_2$ are two collections of events in $\mathbb{A}$. $\mathbb{C}_1$ and $\mathbb{C}_2$ are said to be independent classes if*

$$\forall \;\; A_1 \in \mathbb{C}_1 \;\; \textit{and} \;\; \forall \;\; A_2 \in \mathbb{C}_2, \;\; P(A_1 \cap A_2) = P(A_1)P(A_2).$$

Example 5.2.4. Suppose $\mathbb{C}_1 = \{A, A^c, \emptyset\}$ and $\mathbb{C}_2 = \{B, \Omega\}$ are collections of events. If A and B are independent events, then from the results in Theorem 5.2.1, it immediately follows that A and Ω, A^c and B, A^c and Ω, $\emptyset$ and B, $\emptyset$ and Ω are independent events. Thus,

$$\forall \;\; A_1 \in \mathbb{C}_1 \;\; \text{and} \;\; \forall \;\; A_2 \in \mathbb{C}_2, \;\; P(A_1 \cap A_2) = P(A_1)P(A_2).$$

Hence $\mathbb{C}_1$ and $\mathbb{C}_2$ are independent classes. □

Definition 5.2.6. *Independence of k collections of events: Suppose $\mathbb{C}_1, \mathbb{C}_2, \cdots, \mathbb{C}_k$ are finite number of collections of events in $\mathbb{A}$. Then $\mathbb{C}_1, \mathbb{C}_2, \cdots, \mathbb{C}_k$ are said to be independent collections if $\forall \;\; A_1 \in \mathbb{C}_1, A_2 \in \mathbb{C}_2, \cdots, A_k \in \mathbb{C}_k$, $\{A_1, A_2, \cdots, A_k\}$ is a collection of independent events.*

In particular, suppose $\mathbb{C}_i = \sigma(X_i) = \{X_i^{-1}(S) | S \in \mathbb{B}\}$ is a sigma field induced by X_i, $i = 1, 2, \cdots, k$, where $\{X_1, X_2, \cdots, X_k\}$ are random variables defined on $(\Omega, \mathbb{A}, P)$ and $\mathbb{B}$ is the Borel field.

Definition 5.2.7. *Sequence of independent collections: Suppose $\{\mathbb{C}_n, n \geq 1\}$ is a sequence of collections of events from $\mathbb{A}$. We say that $\{\mathbb{C}_n, n \geq 1\}$ is a sequence of independent collections, if $\forall \;\; A_n \in \mathbb{C}_n, n \geq 1$, $\{A_n, n \geq 1\}$ is a sequence of independent events.*

Remark 5.2.1. From the above definitions it follows that

(i) if every collection of events in Definition 5.2.7 contains exactly one event, then this definition reduces to Definition 5.2.4.

(ii) an infinite collection of events is independent if and only if every finite sub-collection is independent.

(iii) disjoint sub-collections of a sequence of independent events are independent.

Example 5.2.5. Suppose $\{A_n, n \geq 1\}$ is a sequence of independent events. Thus, every finite sub collection of $\{A_n, n \geq 1\}$ is a collection of independent events. Suppose we have a sequence of collections of events defined as $\mathbb{C}_n = \mathbb{A}_n = \{\{\Omega, \emptyset, A_n, A_n^c\}, n \geq 1\}$. It is a sequence of sigma fields. We examine whether it is a sequence of independent sigma fields, given that $\{A_n, n \geq 1\}$ is a sequence of independent events. In particular, $\mathbb{A}_1 = \{\Omega, \emptyset, A_1, A_1^c\}$ and $\mathbb{A}_2 = \{\Omega, \emptyset, A_2, A_2^c\}$. It is proved in Theorem 5.2.1, that any event is independent of Ω and $\emptyset$. Further, A_1 is independent of A_2 and also of A_2^c. Similarly, A_1^c is independent of A_2 and also of A_2^c. Thus, $\mathbb{A}_1$ and $\mathbb{A}_1$ are independent sigma fields. On similar lines we can prove that every finite sub collection of $\{\mathbb{A}_n, n \geq 1\}$ is a collection of independent sigma fields and hence $\{\mathbb{A}_n = \{\Omega, \emptyset, A_n, A_n^c\}, n \geq 1\}$ is a sequence of independent sigma fields. Note that $\mathbb{A}_n$ is a σ-field induced by $X_n = I_{A_n}, n \geq 1$. □

Using these concepts of independence of sequence of events and independence of collections of events, we define independence of random variables in the next section.

5.3 Independence of Random Variables

Suppose $\{X_1, X_2, \cdots, X_k\}$ are random variables defined on $(\Omega, \mathbb{A}, P)$ with $\sigma(X_i)$ as the sigma field induced by X_i, $i = 1, 2, \cdots, k$. The induced sigma field contains all the information about the random variable. Hence, it is appropriate to define independence of random variables in terms of induced sigma fields.

Definition 5.3.1. *Independence of random variables: Suppose $\{X_1, \cdots, X_k\}$ are random variables defined on $(\Omega, \mathbb{A}, P)$ with $\sigma(X_i)$ as the sigma field induced by X_i, $i = 1, 2, \cdots, k$. If $\sigma(X_i)$ $i = 1, 2, \cdots, k$ are independent sigma fields, then $X_1, X_2, \cdots, X_k$ are said to be independent random variables.*

We now demonstrate how this definition leads to the well known definitions of independence in terms of distribution functions or density functions. Suppose X_1 and X_2 are independent random variables. Thus, the induced sigma fields $\sigma(X_1)$ and $\sigma(X_2)$ are independent collections of events in $\mathbb{A}$. By Definition 5.2.5, $\forall\ A_1 \in \sigma(X_1)$ and $\forall\ A_2 \in \sigma(X_2)$, $P(A_1 \cap A_2) = P(A_1)P(A_2)$. Suppose S_1 and S_2 in $\mathbb{B}$ are such that $X_1^{-1}(S_1) = A_1$ and $X_2^{-1}(S_2) = A_2$. Now,

$$\begin{aligned}
& P(A_1 \cap A_2) = P(A_1)P(A_2) \\
\iff & P(X_1^{-1}(S_1) \cap X_2^{-1}(S_2)) = P(X_1^{-1}(S_1))P(X_2^{-1}(S_2)) \\
\iff & P\{\omega | \omega \in X_1^{-1}(S_1) \cap X_2^{-1}(S_2)\} = P\{\omega | \omega \in X_1^{-1}(S_1)\}P\{\omega | \omega \in X_2^{-1}(S_2)\} \\
\iff & P\{\omega | X_1(\omega) \in S_1, X_2(\omega) \in S_2\} = P\{\omega | X_1(\omega) \in S_1\}P\{\omega | X_2(\omega) \in S_2\} \\
\iff & P[X_1 \in S_1, X_2 \in S_2] = P[X_1 \in S_1]P[X_2 \in S_2]
\end{aligned}$$

Thus, $\forall\ A_1 \in \sigma(X_1)$ and $\forall\ A_2 \in \sigma(X_2)$

$$\begin{aligned} P(A_1 \cap A_2) &= P(A_1)P(A_2) \\ \iff \quad P[X_1 \in S_1, X_2 \in S_2] &= P[X_1 \in S_1]P[X_2 \in S_2], \qquad (5.3) \end{aligned}$$

$\forall\ S_1, S_2 \in \mathbb{B}$. Equation (5.3) asserts that the joint probability distribution of X_1 and X_2 is the product of marginal probability distributions of X_1 and X_2. In particular, suppose $S_1 = (-\infty, x_1]$ and $S_2 = (-\infty, x_2]$, $x_1, x_2 \in \mathbb{R}$. Then Equation (5.3) reduces to

$$\begin{aligned} P[X_1 \leq x_1, X_2 \leq x_2] &= P[X_1 \leq x_1]P[X_2 \leq x_2] \\ \iff \quad F_{X_1,X_2}(x_1, x_2) &= F_{X_1}(x_1)F_{X_2}(x_2)\ \ \forall\ \ x_1, x_2 \in \mathbb{R}. \end{aligned}$$

Thus, the joint distribution function of X_1 and X_2 is the product of marginal distribution functions of X_1 and X_2. Further if $F_{X_1,X_2}(x_1, x_2)$ is totally differentiable, then the condition of independence in terms of distribution function reduces to that in terms of probability density function as given by,

$$f_{X_1,X_2}(x_1, x_2) = f_{X_1}(x_1)f_{X_2}(x_2).$$

If X_1 and X_2 are discrete random variables, then in view of sigma additivity of probability measure, it is enough to take $S_1 = \{x_1\}$ and $S_2 = \{x_2\}$. Then Equation (5.3) reduces to,

$$P[X_1 = x_1, X_2 = x_2] = P[X_1 = x_1]P[X_2 = x_2]\ \ \forall\ \ x_1, x_2 \in \mathbb{R}.$$

Thus, if X_1 and X_2 are independent discrete random variables, then the joint probability mass function is the product of marginal probability mass functions.

Example 5.3.1. Suppose X is a random variable defined on $(\Omega, \mathbb{A}, P)$ and $Y \equiv C$ is a degenerate random variable. The sigma field generated by $Y \equiv C$ is $\{\Omega, \emptyset\}$. For any set $A \in \sigma(X)$,

$$P(A \cap \Omega) = P(A) = P(A)P(\Omega) \ \text{ and } \ P(A \cap \emptyset) = P(\emptyset) = 0 = P(A)P(\emptyset).$$

Thus, $\sigma(X)$ and $\sigma(Y) = \{\Omega, \emptyset\}$ are independent sigma fields and hence X and $Y \equiv C$ are independent random variables. Thus, any random variable is always independent of a degenerate random variable. □

In particular a degenerate random variable is also independent of a degenerate random variable, as shown in the next example.

Example 5.3.2. Suppose $X \equiv C$ and $Y \equiv D$ are degenerate random variable defined on $(\Omega, \mathbb{A}, P)$. The sigma field generated by both is the same and is $\{\Omega, \emptyset\}$. It is known that Ω is independent of itself and of $\emptyset$. Hence X and Y are independent random variables. □

Example 5.3.3. Suppose $\mathbb{P}_1 = \{A_1, A_2\}$ and $\mathbb{P}_2 = \{B_1, B_2, B_3\}$ form measurable partitions of Ω. Then

$$\sigma(\mathbb{P}_1) = \{\Omega, \emptyset, A_1, A_2\} \ \& \ \sigma(\mathbb{P}_2) = \{\Omega, \emptyset, B_1, B_2, B_3, B_1 \cup B_2, B_1 \cup B_3, B_2 \cup B_3\}.$$

Suppose $\mathbb{P}_1$ and $\mathbb{P}_2$ are independent of each other. We examine whether $\sigma(\mathbb{P}_1)$ is independent of $\sigma(\mathbb{P}_2)$. Since $\mathbb{P}_1$ and $\mathbb{P}_2$ are independent of each other, $A_i, i = 1, 2$ are independent of $B_j, j = 1, 2, 3$. By Theorem 5.2.1, $A_i, i = 1, 2$ are independent of Ω and $\emptyset$, similarly $B_j, j = 1, 2, 3$ are independent of Ω and $\emptyset$. By the same theorem $A_i, i = 1, 2$ are independent of $B_j^c, j = 1, 2, 3$. Note that $B_1 \cup B_2 = B_3^c, B_1 \cup B_3 = B_2^c$ and $B_2 \cup B_3 = B_1^c$. Independence of A_1 with $B_1 \cup B_2$ can also be proved as follows.

$$\begin{aligned} P(A_1 \cap (B_1 \cup B_2)) &= P(A_1 \cap B_1) + P(A_1 \cap B_2) \\ &= P(A_1)P(B_1) + P(A_1)P(B_2) \\ &= P(A_1)(P(B_1) + P(B_2)) = P(A_1)P(B_1 \cup B_2). \end{aligned}$$

Thus, A_1 and $B_1 \cup B_2$ are independent events. Thus we have noted that every event in $\sigma(\mathbb{P}_1)$ is independent of every event in $\sigma(\mathbb{P}_2)$. Hence, if $\mathbb{P}_1$ and $\mathbb{P}_2$ are independent of each other, then $\sigma(\mathbb{P}_1)$ is independent of $\sigma(\mathbb{P}_2)$. □

Example 5.3.4. Suppose X and Y are simple random variables defined on $(\Omega, \mathbb{A}, P)$ as

$$X = \sum_{i=1}^{m} a_i I_{A_i} \quad \text{and} \quad Y = \sum_{j=1}^{n} b_j I_{B_j}$$

where $\{A_1, A_2, \cdots, A_m\}$ and $\{B_1, B_2, \cdots, B_n\}$ form measurable partitions of Ω and $\{a_1, a_2, \cdots, a_m\}$ and $\{b_1, b_2, \cdots, b_n\}$ are real numbers. Since X is a simple random variable, $\sigma(X)$ includes an empty set and all sets of the form $A = \sum_{s \in T} A_s$, where $T \subseteq \{1, 2, \cdots, m\}$. Similarly, $\sigma(Y)$ includes an empty set and all sets of the form $B = \sum_{r \in M} B_r$, where $M \subseteq \{1, 2, \cdots, n\}$. Suppose $\{A_1, A_2, \cdots, A_m\}$ are independent of $\{B_1, B_2, \cdots, B_n\}$. Note that

$$\begin{aligned} P(A \cap B) &= P\Big(\sum_{s \in T} A_s \cap \sum_{r \in M} B_r\Big) = P\Big(\sum_{s \in T} \sum_{r \in M} A_s \cap B_r\Big) \\ &= \sum_{s \in T} \sum_{r \in M} P(A_s \cap B_r) \ \text{ by finite additivity} \\ &= \sum_{s \in T} \sum_{r \in M} P(A_s)P(B_r) = \sum_{s \in T} P(A_s) \sum_{r \in M} P(B_r) \\ &= P\Big(\sum_{s \in T} A_s\Big) P\Big(\sum_{r \in M} B_r\Big) = P(A)P(B) \end{aligned}$$

It thus follows that any $A \in \sigma(X)$ is independent of $B \in \sigma(Y)$. Hence, induced sigma fields $\sigma(X)$ and $\sigma(Y)$ are independent and hence X and Y are independent random variables. Thus, if the measurable partitions are independent, then the associated simple random variables are independent. □

Remark 5.3.1. For simple random variables X and Y defined in Example 5.3.4, note that

$$\begin{aligned} P[X = a_i] &= P(A_i), \quad P[Y = b_j] = P(B_j) \\ \& \quad P[X = a_i, Y = b_j] &= P(A_i \cap B_j) \\ P(A_i \cap B_j) &= P(A_i)P(B_j) \quad \text{if } A_i \text{ and } B_j \text{ are independent} \\ \Rightarrow \quad P[X = a_i, Y = b_j] &= P[X = a_i]P[Y = b_j] \quad \text{if } X \text{ and } Y \text{ are independent.} \end{aligned}$$

Thus, if X and Y are independent, the joint probability mass function of $(X, Y)'$ is the product of marginal probability mass functions of X and Y.

We have noted in Example 5.2.1 that the concept of independence is not a property of events, but it is a property of underlying probability measure. Thus the two events may be independent with respect one probability measure but not with respect to another probability measure. It is applicable for independence of two random variables as well. The next example illustrates this important feature.

Example 5.3.5. Suppose $X = I_A$ and $Y = I_B$ are two random variables defined on $(\Omega, \mathbb{A}, P)$, where $A, B \in \mathbb{A}$. Suppose the probability measure P is such that$P(A) = P(B) = 1/2$ and $P(A \cap B) = 1/4$. The sigma fields induced by X and Y are

$$\sigma(X) = \{\Omega, \emptyset, A, A^c\} \quad \& \quad \sigma(Y) = \{\Omega, \emptyset, B, B^c\}.$$

We know that Ω and $\emptyset$ are independent of each other and also with any event. Further, since $P(A) = P(B) = 1/2$ and $P(A \cap B) = 1/4$, $P(A^c \cap B) = 1/4$, $P(A \cap B^c) = 1/4$ and $P(A^c \cap B^c) = 1/4$. Thus any event in $\sigma(X)$ is independent of any event in $\sigma(Y)$. Hence, $X = I_A$ and $Y = I_B$ are independent random variables corresponding to the given probability measure P. Now suppose the probability measure P^* is such that, $P^*(A) = P^*(B) = 1/3$ and $P^*(A \cap B) = 1/4$. Then $P^*(A)P^*(B) = 1/9 \neq P^*(A \cap B) = 1/4$. Hence, A and B are not independent with respect to P^*. As a consequence, $X = I_A$ and $Y = I_B$ are not independent random variables corresponding to the probability measure P^*. □

We now discuss the most heavily used result regarding independence of two random variables, which states that independence remains valid under Borel transformations of X and Y.

Theorem 5.3.1. *Suppose X and Y are independent random variables defined on $(\Omega, \mathbb{A}, P)$. If g and h are Borel functions, then $g(X)$ and $h(Y)$ are independent random variables.*

Proof. It is proved in Section 2.3 that if g and h are Borel functions, then $g(X)$ and $h(Y)$ are random variables on $(\Omega, \mathbb{A}, P)$ and

$$\sigma(g(X)) \subseteq \sigma(X) \quad \& \quad \sigma(h(Y)) \subseteq \sigma(Y).$$

To prove that, $g(X)$ and $h(Y)$ are independent is equivalent to prove that $\sigma(g(X))$ and $\sigma(h(Y))$ are independent sigma fields. Hence, observe that $\forall\ A \in \sigma(g(X)) \subseteq \sigma(X)$ & $\forall\ B \in \sigma(h(Y)) \subseteq \sigma(Y)$

$$\begin{aligned} X \ \&\ Y \ \text{ are independent} \quad &\Rightarrow \quad \sigma(X) \ \&\ \sigma(Y) \ \text{ are independent} \\ &\Rightarrow \quad P(A \cap B) = P(A)P(B) \\ &\Rightarrow \quad \sigma(g(X)) \ \&\ \sigma(h(Y)) \ \text{ are independent} \\ &\Rightarrow \quad g(X) \ \&\ h(Y) \ \text{ are independent.} \end{aligned}$$

□

As a consequence of this theorem, if X and Y are independent, then X^r and Y^s are independent random variables for any r, s such that X^r and Y^s are defined. Similarly, $\exp(X)$ and $\exp(Y)$ are also independent, any trignometric function of X will be independent of any trignometric function of Y. Note that establishing independence using this theorem is much simpler than proving it via distribution functions or density functions or probability mass functions.

It is to be noted that Theorem 5.3.1 states that if X and Y are independent random variables, then $g(X)$ and $h(Y)$ are independent random variables. However, if X and Y are not independent random variables, then we cannot conclude that $g(X)$ and $h(Y)$ are also not independent random variables. They may be independent or may not be independent as shown in the next two examples.

Example 5.3.6. Suppose X_1 and X_2 are two random variables defined on $(\Omega, \mathbb{A}, P)$ as follows, where $\Omega = \{a, b, c\}$, $\mathbb{A} = \mathbb{P}(\Omega)$ and P is defined as $\forall\ \omega \in \Omega$, $P(\{\omega\}) = 1/3$.

$$X_1(\omega) = \begin{cases} -1, & \text{if} \quad \omega = a \\ 1, & \text{if} \quad \omega = b, c \end{cases} \quad \& \quad X_2(\omega) = \begin{cases} 5, & \text{if} \quad \omega = b \\ 10, & \text{if} \quad \omega = a, c. \end{cases}$$

Note that $X_1^{-1}(\mathbb{B}) = \{\Omega, \emptyset, \{a\}, \{b, c\}\}$ & $X_2^{-1}(\mathbb{B}) = \{\Omega, \emptyset, \{b\}, \{a, c\}\}$. Observe that $P(\{a\} \cap \{b\}) = P(\emptyset) = 0 \neq P(\{a\})P(\{b\})$. Hence, X_1 and X_2 are not independent. Suppose $Y_1 = g(X_1) = X_1^2$ and $Y_2 = h(X_2) = \log X_2$. Observe that Y_1 is degenrate at 1 and Y_2 is a one-to-one function of X_2. Hence,

$$Y_1^{-1}(\mathbb{B}) = \{\Omega, \emptyset\} \ \&\ Y_2^{-1}(\mathbb{B}) = X_2^{-1}(\mathbb{B}) = \{\Omega, \emptyset, \{b\}, \{a, c\}\}.$$

In Example 5.3.1, we have shown that any random variable is always independent of a degenerate random variable. Thus, X_1 and X_2 are not independent but $g(X_1)$ and $h(X_2)$ are independent random variables. □

Example 5.3.7. Suppose X_1 and X_2 are two random variables defined on $(\Omega, \mathbb{A}, P)$ as follows, where $\Omega = \{a, b, c\}$, $\mathbb{A} = \mathbb{P}(\Omega)$ and P is defined as $\forall\ \omega \in \Omega$, $P(\{\omega\}) = 1/3$.

$$X_1(\omega) = \begin{cases} -1, & \text{if} \quad \omega = a \\ 0, & \text{if} \quad \omega = b \\ 1, & \text{if} \quad \omega = c. \end{cases} \quad \& \quad X_2(\omega) = \begin{cases} 50, & \text{if} \quad \omega = b \\ 100, & \text{if} \quad \omega = a, c. \end{cases}$$

Note that $X_1^{-1}(\mathbb{B}) = \mathbb{P}(\Omega)$ & $X_2^{-1}(\mathbb{B}) = \{\Omega, \emptyset, \{b\}, \{a,c\}\}$. Observe that for $\{a\} \in X_1^{-1}(\mathbb{B})$ and $\{b\} \in X_2^{-1}(\mathbb{B})$, $P(\{a\} \cap \{b\}) = P(\emptyset) = 0$ which is not equal to $P(\{a\})P(\{b\}) = 1/9$. Hence, X_1 and X_2 are not independent. Suppose $Y_1 = g(X_1) = X_1^2$ and $Y_2 = h(X_2) = 3X_2^3 + 4X_2^2 + 7$. Observe that

$$Y_1(\omega) \quad = \quad \begin{cases} 1, & \text{if} \quad \omega = a, c \\ 0, & \text{if} \quad \omega = b. \end{cases}$$

Further, Y_2 is a one-to-one function of X_2. Hence,

$$Y_1^{-1}(\mathbb{B}) = \{\Omega, \emptyset, \{b\}, \{a,c\}\} \ \ \& \ Y_2^{-1}(\mathbb{B}) = X_2^{-1}(\mathbb{B}) = \{\Omega, \emptyset, \{b\}, \{a,c\}\}.$$

Thus, the sigma fields induced Y_1 and Y_2 are the same. These are independent if and only if every set in the sigma fields have probability 0 or 1. However, $P(\{b\}) = 1/3$ and $P(\{a,c\}) = 2/3$. Hence, $Y_1 = g(X_1)$ and $Y_2 = h(X_2)$ are not independent random variables. Thus, X_1 and X_2 are not independent random variables, as well as $g(X_1)$ and $h(X_2)$ are also not independent random variables. □

In Example 5.3.1, we have shown that any random variable is always independent of a degenerate random variable. In Example 5.3.2, it shown that a degenerate random variable is independent of itself. Both these results follow as a particular case of Theorem 5.3.1 as shown in the following theorem.

Theorem 5.3.2. *(i) Any random variable is always independent of a degenerate random variable. (ii) A degenerate random variable is independent of itself.*

Proof. In Theorem 5.3.1 we have proved that if X and Y are independent random variables defined on $(\Omega, \mathbb{A}, P)$ and if g and h are Borel functions, then $g(X)$ and $h(Y)$ are independent random variables.
(i) In particular, suppose g and h are Borel functions such that $g(X) = X$ and $h(Y) \equiv C$. Then it follows that X and a random variable degenerate at C are independent random variables.
(ii) With $g(X) \equiv C$ and $h(Y) \equiv C$, we conclude that a random variable degenerate at C is always independent of itself. □

From Theorem 5.3.1 it is known that if X and Y are independent random variables their Borel functions are also independent. Following example shows that X and Y are not independent but X^2 and Y^2 are independent random variables.

Example 5.3.8. Suppose X is a discrete random variable with $P[X = \pm 1] = 1/2$ and a random variable Y is defined as $Y = -X$. Since Y is one-to-one function of X, it follows that the sigma fields induced by X and Y are the same and it is given by $\{\Omega, \emptyset, A, A^c\}$ where $A = \{\omega | X(\omega) = 1\}$ with probability 1/2. Thus, X and Y are not independent random variables. Now observe that $P[X^2 = 1] = P[Y^2 = 1] = 1$, which implies that both X^2 and Y^2 are degenerate random variables and hence by Theroem 5.3.2, are independent random variables. Thus, X^2 and Y^2 are independent random variables, but X and Y are not independent random variables. □

Remark 5.3.2. Apparently the result from Example 5.3.8 may seem contradictory to Theorem 5.3.1. However, it is to be noted in this example that X^2 and Y^2 are independent random variables, but X is not a Borel function of X^2 or Y is not a Borel function of Y^2, in fact it is not a function, in view of a one-many correspondence from X^2 to X. In Example 5.3.8, X and Y would have been independent if $P(A) = 0$ or 1, which amounts to assuming that X and hence Y are degenerate random variables.

In Chapter 4, we have noted that the expectation of sum of two random variables is the sum of their expectations. However, expectation of product of two random variables is in general not the product of their expectations. The result is true if the underlying random variables are independent. We prove it in the following theorem. The proof proceeds in six steps as the definition of the expectation is different for a simple random variable, a non-negative random variable and for an arbitrary random variable.

Theorem 5.3.3. *If X and Y are independent random variables then,*

$$E(XY) = E(X)E(Y).$$

Proof. Case(i): Suppose X and Y are simple random variables defined on $(\Omega, \mathbb{A}, P)$ as

$$X = \sum_{i=1}^{m} a_i I_{A_i} \quad \text{and} \quad Y = \sum_{j=1}^{n} b_j I_{B_j}$$

where $\{a_1, a_2, \cdots, a_m\}$ and $\{b_1, b_2, \cdots, b_n\}$ are real numbers and $\{A_1, A_2, \cdots, A_m\}$ and $\{B_1, B_2, \cdots, B_n\}$ are measurable partitions of Ω. It is proved in Theorem 2.4.1 that when $\{A_1, A_2, \cdots, A_m\}$ and $\{B_1, B_2, \cdots, B_n\}$ are measurable partitions of Ω, $\{A_i \cap B_j, i = 1, 2, \cdots, m,\ j = 1, 2, \cdots, n\}$ is also a measurable partition of Ω. As proved in Theorem 2.4.1, XY is also a simple random variable and is given by $XY = \sum_{i=1}^{m} \sum_{j=1}^{n} a_i b_j I_{A_i \cap B_j}$. Since X is a simple random variable, as discussed in Example 5.3.4, $\sigma(X)$ includes an empty set and all sets of the form $A = \sum_{r \in T} A_r$, where $T \subseteq \{1, 2, \cdots, m\}$. Thus, $\{A_1, A_2, \cdots, A_m\}$ are in $\sigma(X)$. Similarly, $\{B_1, B_2, \cdots, B_n\}$ are in $\sigma(Y)$. Now by the definition of expectation of a simple random variable,

$$E(X) = \sum_{i=1}^{m} a_i P(A_i), \ \ E(Y) = \sum_{j=1}^{n} b_j P(B_j) \ \&\ E(XY) = \sum_{i=1}^{m} \sum_{j=1}^{n} a_i b_j P(A_i \cap B_j).$$

It is given that X and Y are independent, that is $\sigma(X)$ and $\sigma(Y)$ are independent sigma fields. Hence, $P(A_i \cap B_j) = P(A_i)P(B_j)$ for all $i = 1, 2, \cdots, m$, $j = 1, 2, \cdots, n$. With this substitution in $E(XY)$ we get,

$$E(XY) = \sum_{i=1}^{m} \sum_{j=1}^{n} a_i b_j P(A_i) P(B_j) = \sum_{i=1}^{m} a_i P(A_i) \sum_{j=1}^{n} b_j P(B_j)) = E(X)E(Y).$$

Case(ii): Suppose X and Y are non-negative random variables defined on $(\Omega, \mathbb{A}, P)$. It is proved in Theorem 2.4.3 that there exists a non-decreasing sequence $\{X_n, n \geq 1\}$ of non-negative simple random variables such that $X_n \uparrow X$ on Ω and $E(X) = \lim_{n\to\infty} E(X_n)$. Similarly, there exists a non-decreasing sequence $\{Y_n, n \geq 1\}$ of non-negative simple random variables such that $Y_n \uparrow Y$ on Ω and $E(Y) = \lim_{n\to\infty} E(Y_n)$. Observe that as $X_n \uparrow X$ and $Y_n \uparrow Y$ on Ω, $X_nY_n \uparrow XY$ on Ω. Further as proved in case(i), if X_n and Y_n are simple random variables, then X_nY_n is also a simple random variable. As discussed in Theorem 2.4.3, X_n can be expressed as,

$$X_n \quad = \quad \sum_{k=1}^{n2^n} \frac{k-1}{2^n} I_{\left[\frac{k-1}{2^n} \leq X < \frac{k}{2^n}\right]} \quad + \quad nI_{[X \geq n]} \quad = \quad g(X),$$

where g is a Borel function. Similarly,

$$Y_n \quad = \quad \sum_{k=1}^{n2^n} \frac{k-1}{2^n} I_{\left[\frac{k-1}{2^n} \leq Y < \frac{k}{2^n}\right]} \quad + \quad nI_{[Y \geq n]} \quad = \quad g(Y).$$

It is given that X and Y are independent random variables, hence $X_n = g(X)$ and $Y_n = g(Y)$ are also independent random variables $\forall \; n \geq 1$. Thus,

$$E(XY) = \lim_{n\to\infty} E(X_nY_n) = \lim_{n\to\infty} E(X_n) \lim_{n\to\infty} E(Y_n) = E(X)E(Y).$$

Case(iii): Suppose X and Y are arbitrary random variables defined on $(\Omega, \mathbb{A}, P)$. Hence, $X = X^+ - X^-$ and $Y = Y^+ - Y^-$, where $X^+ = XI_{[X\geq 0]} = g(X)$ and $X^- = -XI_{[X\geq 0]} = h(X)$. Y^+ and Y^- are defined on similar lines. Thus, $Y^+ = g(Y)$ and $Y^- = h(Y)$. Note that both g and h are Borel functions. Since it is given that X and Y are independent, the collection $\{g(X), h(X)\} \equiv \{X^+, X^-\}$ and $\{g(Y), h(Y)\} \equiv \{Y^+, Y^-\}$ are independent collections of random variables, that is, X^+ is independent of $\{Y^+, Y^-\}$ and X^- is independent of $\{Y^+, Y^-\}$. Hence,

$$\begin{aligned}
E(XY) &= E(X^+ - X^-)(Y^+ - Y^-) \\
&= E(X^+Y^+) - E(X^+Y^-) - E(X^-Y^+) + E(X^-Y^-) \\
&= E(X^+)E(Y^+) - E(X^+)E(Y^-) - E(X^-)E(Y^+) + E(X^-)E(Y^-) \\
&= E(X^+)(E(Y^+) - E(Y^-)) - E(X^-)(E(Y^+) - E(Y^-)) \\
&= (E(X^+) - E(X^-))(E(Y^+) - E(Y^-)) = E(X)E(Y).
\end{aligned}$$

Case(iv): Suppose X is a non-negative simple random variable and Y is non-negative random variable. Since $Y \geq 0$, there exists a non-decreasing sequence $\{Y_n, n \geq 1\}$ of non-negative simple random variables such that $Y_n \uparrow Y$ on Ω and $E(Y) = \lim_{n\to\infty} E(Y_n)$. It is to be noted that $XY_n \geq 0$, $\{XY_n, n \geq 1\}$ is a non-decreasing sequence, $XY_n \uparrow XY$ and as proved in case(i), XY_n is also a simple random variable. As in case (ii) $Y_n = g(Y)$, hence independence of

X and Y implies independence of X and Y_n for each $n \geq 1$. Thus, by the definition of expectation,

$$\begin{aligned} E(XY) &= \lim_{n\to\infty} E(XY_n) = \lim_{n\to\infty} E(X)E(Y_n) \\ &= E(X) \lim_{n\to\infty} E(Y_n) = E(X)E(Y). \end{aligned}$$

Case(v): Suppose X is a non-negative simple random variable and Y is an arbitrary random variable. As in case (iii) Y^+ and Y^- are Borel functions of Y, hence independence of X and Y implies independence of X and Y^+ and of X and Y^-. Further X, Y^+, Y^- are non-negative random variables. Thus, by the definition of expectation,

$$\begin{aligned} E(XY) &= E(X(Y^+ - Y^-)) = E(XY^+) - E(XY^-) \\ &= E(X)E(Y^+) - E(X)E(Y^-) \text{ by case (iv)} \\ &= E(X)(E(Y^+) - E(Y^-)) = E(X)E(Y). \end{aligned}$$

Case(vi): Suppose X is a non-negative random variable and Y is an arbitrary random variable. As in case (iii) Y^+ and Y^- are Borel functions of Y, hence independence of X and Y implies independence of X and Y^+ and of X and Y^-. Further X, Y^+, Y^- are non-negative random variables. Thus, by the definition of expectation,

$$\begin{aligned} E(XY) &= E(X(Y^+ - Y^-)) = E(XY^+) - E(XY^-) \\ &= E(X)E(Y^+) - E(X)E(Y^-) \text{ by case (ii)} \\ &= E(X)(E(Y^+) - E(Y^-)) = E(X)E(Y). \end{aligned}$$

Thus, it is proved that if X and Y are independent random variables then, $E(XY) = E(X)E(Y)$. □

Remark 5.3.3. Theorem 5.3.3 can be extended to the expectation of product of more than two random variables.

Example 5.3.9. Suppose X is an integrable random variable on $(\Omega, \mathbb{A}, P)$ and $A \in \mathbb{A}$ is such that $P(A) = 0$. The sigma field induced by I_A is $\{\emptyset, \Omega, A, A^c\}$ and probability attached to these four events is either 0 or 1. Hence from result (iii) of Theorem 5.2.1, any $B \in \sigma(X)$ is independent of any $D \in \sigma(I_A)$, which implies that X and I_A are independent. Hence,

$$\int_A X \, dP = \int_\Omega X I_A \, dP = E(XI_A) = E(X)E(I_A) = E(X)P(A) = 0 \, .$$

It then follows that

$$\int_{A^c} X \, dP = \int_{A^c} X \, dP + \int_A X \, dP = \int_{A \cup A^c} X \, dP = \int_\Omega X \, dP = E(X) \, . \quad \square$$

Theorem 5.3.3 along with Theorem 5.3.1 immediately implies that if X and Y are independent random variables, then the joint moment generating function is the product of marginal moment generating functions, moment generating function of $X+Y$ is the product of marginal moment generating functions. Similar results are true for the probability generating function and the characteristic function. In the following theorem, we prove it for the moment generating function, provided it exists.

Theorem 5.3.4. *Suppose X and Y are independent random variables defined on $(\Omega, \mathbb{A}, P)$. Then (i) joint moment generating function of X and Y is the product of marginal moment generating functions of X and Y and (ii) the moment generating function of $X+Y$ is the product of marginal moment generating functions of X and Y, provided these exist.*

Proof. The joint moment generating function of X and Y is defined as, $M_{X,Y}(t_1, t_2) = E(\exp(t_1 X + t_2 Y))$. Note that $\exp(x)$ and $\exp(y)$ are continuous functions and hence are Borel functions. Thus, if X and Y are independent random variables, then by Theorem 5.3.1 $\exp(t_1 X)$ and $\exp(t_2 Y)$ are also independent random variables. Hence,

$$\begin{aligned} M_{X,Y}(t_1, t_2) &= E(\exp(t_1 X + t_2 Y)) = E(\exp(t_1 X)\exp(t_2 Y)) \\ &= E(\exp(t_1 X))E(\exp(t_2 Y)) = M_X(t_1)M_Y(t_2). \end{aligned}$$

On similar lines it follows that the moment generating function of $X+Y$ is the product of marginal moment generating functions. The moment generating function of $X+Y$ can be expressed as,

$$\begin{aligned} M_{X+Y}(t) &= E(\exp(t(X+Y))) = E(\exp(tX)\exp(tY)) \\ &= E(\exp(tX))E(\exp(tY)) = M_X(t)M_Y(t). \end{aligned}$$

□

Similar result is true for a characteristic function. We prove it below.

Theorem 5.3.5. *Suppose X and Y are independent random variables defined on $(\Omega, \mathbb{A}, P)$. Then (i) joint characteristic function of X and Y is the product of characteristic functions of X and Y and (ii) the characteristic function of $X+Y$ is the product of characteristic functions of X and Y.*

Proof. Since X and Y are independent random variables, by Theorem 5.3.1, $\cos(t_1 X)$ is independent of $\cos(t_2 Y)$ and $\sin(t_2 Y)$, similarly $\sin(t_1 X)$ is independent of $\cos(t_2 Y)$ and $\sin(t_2 Y)$. We use Theorem 5.3.3 for the expectation of product of these random variables. By the definition of the joint characteristic function of X and Y

$$\begin{aligned} \phi_{X,Y}(t_1, t_2) &= E(\exp(it_1 X + it_2 Y)) = E(\exp(it_1 X)exp(it_2 Y)) \\ &= E\{(\cos(t_1 X) + i\sin(t_1 X))(\cos(t_2 Y) + i\sin(t_2 Y))\} \\ &= E(\cos(t_1 X)\cos(t_2 Y) + i\cos(t_1 X)\sin(t_2 Y)) \\ &+ E(i\sin(t_1 X)\cos(t_2 Y) + i^2\sin(t_1 X)\sin(t_2 Y)) \\ &= E(\cos(t_1 X))E(\cos(t_2 Y)) + iE(\cos(t_1 X))E(\sin(t_2 Y)) \\ &+ iE(\sin(t_1 X)E(\cos(t_2 Y) + i^2 E(\sin(t_1 X))E(\sin(t_2 Y)). \end{aligned}$$

It further simplifies as follows.

$$\begin{aligned}\phi_{X,Y}(t_1,t_2) &= E(\cos(t_1X))\{E(\cos(t_2Y)) + iE(\sin(t_2Y))\}\\ &+ iE(\sin(t_1X)\{E(\cos(t_2Y) + iE(\sin(t_2Y))\}\\ &= \{E(\cos(t_1X)) + iE(\sin(t_1X)\}\{E(\cos(t_2Y)) + iE(\sin(t_2Y))\}\\ &= \phi_X(t_1)\phi_Y(t_2)\end{aligned}$$

and (i) follows. Using similar arguments with $t_1 = t_2 = t$, (ii) follows. □

Using Theorem 5.3.5 and the inversion theorem, we now derive the characteristic function of the Cauchy $C(0,1)$ distribution in the following example.

Example 5.3.10. Suppose X and Y are independent and identically distributed random variables, each having exponential distribution with scale parameter 1. Using the convolution formula, one can show that $Z = X - Y$ follows a Laplace distribution with the probability density function $f(x) = (1/2)e^{-|x|}$, $x \in \mathbb{R}$. The characteristic function of X and Y is the same and is given by $\phi(t) = (1 - it)^{-1}$. Hence, using Theorem 5.3.5, the characteristic function of Z is given by
$\phi_Z(t) = (1-it)^{-1} \times (1+it)^{-1} = (1+t^2)^{-2}$. Since this characteristic function is integrable, using inversion formula as derived in Theorem 4.4.2 we have

$$\frac{1}{2}e^{-|x|} = \frac{1}{2\pi}\int_{\mathbb{R}} e^{itx}\frac{1}{1+t^2}dt \iff e^{-|t|} = \frac{1}{\pi}\int_{\mathbb{R}} e^{itx}\frac{1}{1+x^2}dt$$

by interchanging t and x. We note that the right-hand side is the characteristic function of the Cauchy $C(0,1)$ distribution and is given by $\exp\{-|t|\}, t \in \mathbb{R}$. Since Cauchy $C(0,1)$ distribution is symmetric around 0, the characteristic function is a real valued function. □

In the following theorem, we derive three equivalent conditions for independence of two random variables, which can be extended to a finite collection of random variables. We state one result about independence of π-system in the following theorem. It is needed to establish the equivalence of conditions for independence.

Theorem 5.3.6. *Suppose* $\mathbb{P}_i$, $i = 1, 2, \cdots, k$ *are independent* π*-systems. Then* $\sigma(\mathbb{P}_i)$, $i = 1, 2, \cdots, k$ *are independent sigma fields.*

For proof we refer to p.50 of Billingsley [5] or p.41 of Shao [21]. The proof is based on Dynkin's π-λ theorem. This theorem is also known as an extension theorem, (p.246, Loeve [16]).

Theorem 5.3.7. *Suppose* X *and* Y *are random variables defined on* $(\Omega, \mathbb{A}, P)$. *Then* X *and* Y *are independent random variables if any one of the following three conditions are satisfied.*
(i) $\sigma(X)$ *and* $\sigma(Y)$ *are independent sigma fields, that is*

$$\forall\ S_1, S_2 \in \mathbb{B},\quad P[X \in S_1, Y \in S_2] = P[X \in S_1]P[Y \in S_2].$$

(ii) $\forall\ x, y \in \mathbb{R},\quad F_{X,Y}(x,y) = F_X(x)F_Y(y)$.
(iii) $\forall\ t_1, t_2 \in \mathbb{R},\quad \phi_{X,Y}(t_1,t_2) = \phi_X(t_1)\phi_Y(t_2)$.

Proof. We prove the equivalence of three conditions as follows. Suppose (i) is true. Thus, $X^{-1}(\mathbb{B})$ and $Y^{-1}(\mathbb{B})$ are independent sigma fields which further implies for all $S_1, S_2 \in \mathbb{B}$, $P[X \in S_1, Y \in S_2] = P[X \in S_1]P[Y \in S_2]$. In particular, if we take $S_1 = (-\infty, x]$ and $S_2 = (-\infty, y]$ for $x, y \in \mathbb{R}$, then we get (ii). Thus, (i) implies (ii). Further, (i) implies (iii) follows from Theorem 5.3.5. Suppose now (ii) is true. We define $\mathbb{C} = \{(-\infty, x] | x \in \mathbb{R}\}$. Observe that,

$$\begin{aligned} S_1 &= (-\infty, a] \in \mathbb{C}\ \&\ S_2 = (-\infty, b] \in \mathbb{C} \\ &\Rightarrow S_1 \cap S_2 = (-\infty,\ d] \in \mathbb{C} \text{ where } d = \min\{a, b\} \\ &\Rightarrow \mathbb{C} \text{ is a } \pi \text{ system.} \end{aligned}$$

Suppose $\mathbb{A}_1 = X^{-1}(\mathbb{C})$ and $\mathbb{A}_2 = Y^{-1}(\mathbb{C})$. Suppose $A_1, A_2 \in \mathbb{A}_1$ are such that $A_1 = X^{-1}(S_1) = \{\omega | X(\omega) \leq a\}$, $A_2 = X^{-1}(S_2) = \{\omega | X(\omega) \leq b\}$. Then $A_1 \cap A_2 = \{\omega | X(\omega) \leq d\}$. Now

$$(-\infty,\ d] \in \mathbb{C} \quad \Rightarrow \quad X^{-1}(-\infty,\ d]) = A_1 \cap A_2 \in \mathbb{A}_1 \quad \Rightarrow \quad \mathbb{A}_1 \text{ is a } \pi \text{ system.}$$

On similar lines we get that $\mathbb{A}_2$ is a π-system. From (ii) we have $\forall\ x, y \in \mathbb{R}$,

$$\begin{aligned} F_{X,Y}(x,y) &= F_X(x)F_Y(y) \\ \iff P[X \leq x, Y \leq y] &= P[X \leq x]P[Y \leq y] \\ \iff P(A_x \cap B_y) &= P(A_x)P(B_y), \end{aligned}$$

for all $A_x \in \mathbb{A}_1$ and $B_y \in \mathbb{A}_2$, where $A_x = X^{-1}((-\infty, x])$ and $B_y = Y^{-1}((-\infty, y])$. It shows that $\mathbb{A}_1$ and $\mathbb{A}_2$ are independent π-systems. Hence by Theorem 5.3.6, $\sigma(\mathbb{A}_1)$ and $\sigma(\mathbb{A}_2)$ are independent classes. However, $\sigma(\mathbb{A}_1) = \sigma(X^{-1}(\mathbb{C})) = X^{-1}(\sigma(\mathbb{C})) = X^{-1}(\mathbb{B})$ and $\sigma(\mathbb{A}_2) = \sigma(Y^{-1}(\mathbb{C})) = Y^{-1}(\sigma(\mathbb{C})) = Y^{-1}(\mathbb{B})$, that is, sigma fields induced by X and Y are independent and thus (i) follows from (ii). Hence, *(ii)* $\Rightarrow$ *(i)* $\Rightarrow$ *(iii)*, hence, *(ii)* $\Rightarrow$ *(iii)*. Now we show that (iii) implies (i) using inversion formula which gives the distribution function corresponding to a given characteristic function. Suppose B is a two dimensional rectangle defined as $B = \{(x, y) | a_1 < x \leq a_1 + h_1, a_2 < y \leq a_2 + h_2\}$. Then by the multivariate inversion formula (Billingsley [5]) $P[(X, Y) \in B]$ is given by,

$$\begin{aligned} P[(X,Y) \in B] &= \lim_{T\to\infty} \frac{1}{4\pi^2} \int_{-T}^{T} \int_{-T}^{T} \left(\frac{e^{-ia_1t_1} - e^{-i(a_1+h_1)t_1}}{it_1} \right) \\ &\times \left(\frac{e^{-ia_2t_2} - e^{-i(a_2+h_2)t_2}}{it_2} \phi_{X,Y}(t_1,t_2) \right) dt_1 dt_2 \\ &= \lim_{T\to\infty} \frac{1}{2\pi} \int_{-T}^{T} \frac{e^{-ia_1t_1} - e^{-i(a_1+h_1)t_1}}{it_1} \phi_X(t_1)\ dt_1 \\ &\times \lim_{T\to\infty} \frac{1}{2\pi} \int_{-T}^{T} \frac{e^{-ia_2t_2} - e^{-i(a_2+h_2)t_2}}{it_2} \phi_Y(t_2)\ dt_2 \quad \text{by (iii)} \\ &= P[a_1 < X \leq a_1 + h_1]P[a_2 < Y \leq a_2 + h_2]. \end{aligned}$$

Thus, for any $a_1, a_2, h_1, h_2 \in \mathbb{R}$, we have

$$\begin{aligned} P[(X,Y) \in B] &= P[a_1 < X \leq a_1 + h_1]P[a_2 < Y \leq a_2 + h_2] \\ &\Longleftrightarrow P[X^{-1}((a_1, a_1 + h_1]) \cap Y^{-1}((a_2, a_2 + h_2])] \\ &= P[X^{-1}((a_1, a_1 + h_1])]P[Y^{-1}((a_2, a_2 + h_2])]. \quad (5.4) \end{aligned}$$

We now define $\mathbb{C}_1 = \{(a,b] | a \leq b \in \mathbb{R}\}$. If $a = b$, we define $(a,b] = \emptyset$. Observe that,

$$\begin{aligned} S_1 = (a_1, a_2] \in \mathbb{C}_1 \;\; \& \;\; S_2 = (b_1, b_2] \in \mathbb{C}_1 \;\Rightarrow\; S_1 \cap S_2 = (d_1, d_2] \in \mathbb{C}_1 \\ \Rightarrow \mathbb{C}_1 \text{ is a } \pi \text{ system where, } d_1 = \max\{a_1, b_1\} \;\& d_2 = \min\{a_2, b_2\}. \end{aligned}$$

In some cases, $S_1 \cap S_2 = \emptyset$. Suppose $\mathbb{F}_1 = X^{-1}(\mathbb{C}_1)$ and $\mathbb{F}_2 = Y^{-1}(\mathbb{C}_1)$. Suppose $A_1, A_2 \in \mathbb{F}_1$ are such that

$$A_1 = X^{-1}(S_1) \;\; \& \;\; A_2 = X^{-1}(S_2) \;\; \text{then} \;\; A_1 \cap A_2 = \{\omega | d_1 < X(\omega) \leq d_2\}.$$

Now

$$(d_1, d_2] \in \mathbb{C}_1 \quad \Rightarrow \quad X^{-1}((d_1, d_2]) = A_1 \cap A_2 \in \mathbb{F}_1 \quad \Rightarrow \quad \mathbb{F}_1 \text{ is a } \pi \text{ system.}$$

On similar lines we get that $\mathbb{F}_2$ is a π-system. From Equation (5.4), we claim that $\mathbb{F}_1$ and $\mathbb{F}_2$ are independent π-systems. Hence by Theorem 5.3.6, $\sigma(\mathbb{F}_1)$ and $\sigma(\mathbb{F}_2)$ are independent classes. However, as shown above $\sigma(\mathbb{F}_1) = X^{-1}(\mathbb{B})$ and $\sigma(\mathbb{F}_2) = Y^{-1}(\mathbb{B})$, that is, sigma fields induced by X and Y are independent and thus (i) follows from (iii). Thus all the three conditions are equivalent and the proof is complete. □

So far we discussed the concept of independence of a collection of finitely many random variables. We now extend it to a sequence of random variables and any collection of random variables.

Definition 5.3.2. *Sequence of independent random variables: A sequence $\{X_n, n \geq 1\}$ is a sequence of independent random variables if and only if any finite collection of the sequence is a collection of independent random variables.*

Definition 5.3.3. *Class of independent random variables: Suppose $\{X_t, t \in T\}$ is a class of random variables, where T is any non-empty index set. It is said to a class of independent random variables if and only if any finite sub class of this class is a class of independent random variables.*

In view of these definitions, to examine independence of a sequence of random variables or an indexed class of random variables, all the tools developed for a finite collection of random variables are applicable.

Independence of random vectors and a class of independent random vectors are defined analogously.

Definition 5.3.4. *Independence of random vectors: Two random vectors $\underline{X}$ and $\underline{Y}$ defined on the same probability space are independent random vectors if the induced sigma fields are independent.*

In the following theorem, we prove that if two random vectors are independent then each component random variable of one random vector is independent of each component random variable of the second random vector.

Theorem 5.3.8. *Suppose $\underline{Z}_1 = (X_1, X_2, \cdots, X_k)$ and $\underline{Z}_2 = (Y_1, Y_2, \cdots, Y_l)$ are random vectors defined on the same probability space. Then $\underline{Z}_1$ and $\underline{Z}_2$ are independent implies each component of $\underline{Z}_1$ is independent of each component of $\underline{Z}_2$.*

Proof. Note that the sigma fields induced by $\underline{Z}_1$ and $\underline{Z}_2$ are given by,

$$\sigma(\underline{Z}_1) = \sigma\Big(\bigcup_{i=1}^{k} \sigma(X_i)\Big) \quad \& \quad \sigma(\underline{Z}_2) = \sigma\Big(\bigcup_{j=1}^{l} \sigma(Y_j)\Big).$$

Suppose $\underline{Z}_1$ and $\underline{Z}_2$ are independent. Then the corresponding sigma fields are independent. Hence,

$$\forall \ A_1 \in \sigma(\underline{Z}_1) \ \& \ \forall \ A_2 \in \sigma(\underline{Z}_2), \ \ P(A_1 \cap A_2) = P(A_1)P(A_2).$$

To prove that X_i is independent of Y_j, we examine whether the corresponding sigma fields are independent. Observe that $\forall \ i = 1, 2, \cdots, k$ and $j = 1, 2, \cdots, l$,

$$\begin{aligned} A_1 &\in \sigma(X_i) \ \& \ A_2 \in \sigma(Y_j) \ \Rightarrow \ A_1 \in \sigma(\underline{Z}_1) \ \& \ A_2 \in \sigma(\underline{Z}_2) \\ &\Rightarrow \ P(A_1 \cap A_2) = P(A_1)P(A_2) \\ &\Rightarrow \ \sigma(X_i) \ \& \ \sigma(Y_j) \ \text{ are independent sigma fields} \\ &\Rightarrow \ X_i \ \& \ Y_j \ \text{ are independent.} \end{aligned}$$

□

Following example illustrates that the converse of Theorem 5.3.8 is not true, that is, if the component random variables of $\underline{Z}_1$ are independent of the component random variables $\underline{Z}_2$ then $\underline{Z}_1$ and $\underline{Z}_2$ may not be independent random vectors.

Example 5.3.11. Suppose $(\Omega, \mathbb{A}, P)$ is a probability space, where $\Omega = \{a, b, c, d\}$, $\mathbb{A} = \mathbb{P}(\Omega)$ and $P(\{\omega\}) = 1/4 \ \forall \ \omega \in \Omega$. Suppose events $A, B, C, D \in \mathbb{A}$ are defined as follows.

$$A = \{a, b\}, \quad B = \{a, c\}, \quad C = \{b, c\} \ \ \& \ \ D = \{a, d\}.$$

We define $\underline{Z}_1 = (I_A, I_B)'$ and $\underline{Z}_2 = (I_C, I_D)'$. The sigma fields induced by $I_A, I_B, I_C, I_D, \underline{Z}_1$ and $\underline{Z}_2$ are respectively given by,

$$\begin{aligned}
I_A^{-1}(\mathbb{B}) &= \{\Omega, \emptyset, A, A^c\} \quad I_B^{-1}(\mathbb{B}) = \{\Omega, \emptyset, B, B^c\} \\
I_C^{-1}(\mathbb{B}) &= \{\Omega, \emptyset, C, C^c\} \quad I_D^{-1}(\mathbb{B}) = \{\Omega, \emptyset, D, D^c\} \\
\underline{Z}_1^{-1}(\mathbb{B}) &= \{\Omega, \emptyset, A, A^c, B, B^c, A \cup B, A \cup B^c, A^c \cup B, A^c \cup B^c, A^c \cap B^c, \\
& \quad A^c \cap B, A \cap B^c, A \cap B, (A \cup B^c) \cup (A^c \cup B), \\
& \quad ((A \cup B^c) \cup (A^c \cup B))^c\} \\
\underline{Z}_2^{-1}(\mathbb{B}) &= \{\Omega, \emptyset, C, C^c, D, D^c, C \cup D, C \cup D^c, C^c \cup D, C^c \cup D^c, C^c \cap D^c, \\
& \quad C^c \cap D, C \cap D^c, C \cap D, (C \cup D^c) \cup (C^c \cup D), \\
& \quad ((C \cup D^c) \cup (C^c \cup D))^c\}.
\end{aligned}$$

Now note that

$$P(A \cap B) = P(\{a\}) = 1/4 = P(A)P(B) = (1/2)(1/2).$$

Thus, A is independent of B and hence $I_A^{-1}(\mathbb{B})$ is independent of $I_B^{-1}(\mathbb{B})$, implying that I_A and I_B are independent random variables. Further,

$$P(C \cap D) = P(\emptyset) = 0 \neq P(C)P(D) = (1/2)(1/2).$$

Thus, I_C and I_D are not independent random variables. Further,

$$\begin{aligned}
P(A \cap C) &= P(\{b\}) = 1/4 = P(A)P(C) = (1/2)(1/2) \\
P(A \cap D) &= P(\{a\}) = 1/4 = P(A)P(D) = (1/2)(1/2) \\
P(B \cap C) &= P(\{c\}) = 1/4 = P(B)P(C) = (1/2)(1/2) \\
P(B \cap D) &= P(\{a\}) = 1/4 = P(B)P(D) = (1/2)(1/2).
\end{aligned}$$

Thus, A is independent of C and D, which implies that $I_A^{-1}(\mathbb{B})$ is independent of $I_C^{-1}(\mathbb{B})$ and $I_D^{-1}(\mathbb{B})$. Thus, I_A is independent of I_C and I_D. Similarly, B is independent of C and D, which implies that I_B is independent of I_C and I_D. Thus, the component random variables of $\underline{Z}_1$ are independent of the component random variables $\underline{Z}_2$. We now examine whether $\underline{Z}_1$ and $\underline{Z}_2$ are independent random vectors, by verifying whether the induced sigma fields are independent. Note that A and B are independent of C and D implies that A and B are independent of C^c and D^c. Similarly, A^c and B^c are independent of C and D. However, observe that for $A \cap B \in \underline{Z}_1^{-1}(\mathbb{B})$ and $D \in \underline{Z}_2^{-1}(\mathbb{B})$, $P(A \cap B \cap D) = 1/4 \neq P(A \cap B)P(D) = (1/4)(1/2)$. Similarly, for $A \cap B \in \underline{Z}_1^{-1}(\mathbb{B})$ and $C \in \underline{Z}_2^{-1}(\mathbb{B})$, $P(A \cap B \cap C) = 0 \neq P(A \cap B)P(C) = (1/4)(1/2)$. Hence, $\underline{Z}_1$ is not independent of $\underline{Z}_2$. Thus, components of $\underline{Z}_1$ are independent of $\underline{Z}_2$ but the random vector $\underline{Z}_1$ is not independent of $\underline{Z}_2$. □

In Theorem 5.3.1 it is proved that Borel functions of independent random variables are also independent random variables. The result remains valid even for random vectors, as proved below.

Theorem 5.3.9. *Suppose $\underline{X}$ and $\underline{Y}$ are independent random vectors defined on $(\Omega, \mathbb{A}, P)$ to $(\mathbb{R}^k, \mathbb{B}^k, P_{\underline{X}})$ and $(\mathbb{R}^l, \mathbb{B}^l, P_{\underline{Y}})$ respectively.*
(i) Suppose $g : (\mathbb{R}^k, \mathbb{B}^k) \to (\mathbb{R}, \mathbb{B})$ is a Borel function and $h : (\mathbb{R}^l, \mathbb{B}^l) \to (\mathbb{R}, \mathbb{B})$ is a Borel function. Then $g(\underline{X})$ and $h(\underline{Y})$ are independent random variables.
(ii) Suppose $\underline{g} : (\mathbb{R}^k, \mathbb{B}^k) \to (\mathbb{R}^m, \mathbb{B}^m), m \le k$ is a Borel function and $\underline{h} : (\mathbb{R}^l, \mathbb{B}^l) \to (\mathbb{R}^s, \mathbb{B}^s), s \le l$ is a Borel function. Then $\underline{g}(\underline{X})$ and $\underline{h}(\underline{Y})$ are independent random vectors

Proof. In Theorem 2.3.5, it is proved that the sigma field induced by the Borel function of a random vector is included in the sigma field induced by the random vector. Hence,

$$\sigma(g(\underline{X})) \subseteq \sigma(\underline{X}) \quad \& \quad \sigma(h(\underline{Y})) \subseteq \sigma(\underline{X}).$$

(i) To prove that, $g(\underline{X})$ and $h(\underline{Y})$ are independent is equivalent to prove that $\sigma(g(\underline{X}))$ and $\sigma(h(\underline{Y}))$ are independent sigma fields. Hence observe that,

$$\begin{aligned} A \in \sigma(g(\underline{X})) \ \& \ B \in \sigma(h(\underline{Y})) &\Rightarrow A \in \sigma(\underline{X}) \ \& \ B \in \sigma(\underline{Y}) \\ &\Rightarrow P(A \cap B) = P(A)P(B) \\ &\Rightarrow \sigma(g(\underline{X})) \ \& \ \sigma(h(\underline{Y})) \ \text{ are independent} \\ &\Rightarrow g(\underline{X}) \ \& \ h(\underline{Y}) \ \text{ are independent} \end{aligned}$$

random variables. (ii) follows on similar lines. □

In particular, suppose $\underline{X} = (X_1, X_2)'$ and $\underline{Y} = (Y_1, Y_2)'$. Then $X_1 + X_2$ is independent of $Y_1 + Y_2$, Y_1Y_2, $e^{Y_1+Y_2}$. $(X_1 + X_2, X_1X_2)'$ is independent of $(Y_1 + Y_2, Y_1Y_2)'$, $e^{X_1+X_2}$ is independent of $e^{Y_1+Y_2}$ and so on.

We now proceed to one important result in probability theory, concerned with a sequence of independent random variables. It deals with a situation in which the probability of an event can only be 0 or 1. Theorems that provide criteria for such a situation are called zero-one laws. The next section presents one such law, known as Kolmogorov zero-one law. The other, known as Borel zero-one law, is discussed in Section 6.4.

5.4 Kolmogorov Zero-One Law

In Section 2.3, we have introduced the concept of a tail sigma field generated by a sequence $\{X_n, n \ge 1\}$ of random variables. In addition, if $\{X_n, n \ge 1\}$ is a sequence of independent random variables, then the tail sigma field has a peculiar nature and it is investigated in the Kolmogorov zero-one law. It states that if $\{X_n, n \ge 1\}$ is a sequence of independent random variables, then the tail sigma field generated by the sequence contains sets of probability measure 0 or 1 only. We state below a lemma which is needed in the proof of Kolmogorov zero-one law. For proof we refer to p.62 of Chow and Teicher [7].

Lemma 5.4.1. *If $\mathbb{C}$ and $\mathbb{D}$ are independent classes of events and if $\mathbb{D}$ is a π-system, then $\mathbb{C}$ and $\sigma(\mathbb{D})$ are independent classes.*

One more result related to a sequence of independent random variables is needed in the proof of Kolmogorov zero-one law. We prove it below, which is based on the extension theorem stated in Theorem 5.3.6.

Lemma 5.4.2. *Suppose $\{X_n, n \geq 1\}$ is a sequence of independent random variables. If T_1 and T_2 are non-empty disjoint subsets of $\{1, 2, \cdots, \}$, then the minimal sigma fields $\sigma(\{X_n, n \in T_1\})$ and $\sigma(\{X_n, n \in T_2\})$ are independent sigma fields.*

Proof. Since $\{X_n, n \geq 1\}$ is a sequence of independent random variables, $\{\sigma(X_n), n \geq 1\}$ is a sequence of independent sigma fields. Suppose $\mathbb{A}_1 = \bigcup_{n \in T_1} \sigma(X_n)$. Now,

$$\begin{aligned} A \in \mathbb{A}_1 \quad &\Rightarrow \quad A \in \sigma(X_n) \text{ for at least one } n \in T_1 \\ &\Rightarrow \quad A^c \in \sigma(X_n) \text{ for at least one } n \in T_1 \\ &\Rightarrow \quad A^c \in \bigcup_{n \in T_1} \sigma(X_n) = \mathbb{A}_1. \end{aligned}$$

Thus, $\mathbb{A}_1$ is closed under complementation. Further,

$$\begin{aligned} A,\ B \in \mathbb{A}_1 \quad &\Rightarrow \quad A,\ B \in \sigma(X_n) \text{ for at least one } n \in T_1 \\ &\Rightarrow \quad A \in \sigma(X_r) \text{ for at least one } r \in T_1 \\ \&\ B \quad &\in \quad \sigma(X_s) \text{ for at least one } s \geq r \in T_1 \end{aligned}$$

Thus,

$$\begin{aligned} A \quad &\in \quad \bigcup_{i \in T_1, i \leq r} \sigma(X_i) \ \ \&\ \ B \in \bigcup_{j \in T_1, j \leq s} \sigma(X_j) \\ &\Rightarrow \quad A \cup B \in \bigcup_{j \in T_1, j \leq s} \sigma(X_j) \ \Rightarrow \ A \cup B \in \bigcup_{n \in T_1} \sigma(X_n) = \mathbb{A}_1. \end{aligned}$$

Thus, $\mathbb{A}_1$ is closed under finite union and hence it is a field. Similarly, $\mathbb{A}_2 = \bigcup_{n \in T_2} \sigma(X_n)$ is also a field and hence both $\mathbb{A}_1$ and $\mathbb{A}_2$ are π- systems. Since T_1 and T_2 are non-empty disjoint classes and $\sigma(\{X_n, n \geq 1\})$ are independent sigma fields, it follows that $\mathbb{A}_1$ and $\mathbb{A}_2$ are independent π- systems. Further, by Theorem 5.3.6, $\sigma(\mathbb{A}_1) = \sigma(\{X_n, n \in T_1\})$ and $\sigma(\mathbb{A}_2) = \sigma(\{X_n, n \in T_2\})$ are independent sigma fields. □

Theorem 5.4.1. *Kolmogorov zero-one law: Suppose $\{X_n, n \geq 1\}$ is a sequence of independent random variables. Then the probability of tail events is either 0 or 1.*

Proof. It is given that $\{X_n, n \geq 1\}$ is a sequence of independent random variables. Suppose $T_1 = \{1, 2, \cdots, n\}$ and $T_2 = \{n+1, n+2, \cdots\}$. Hence by Lemma 5.4.2 $\forall\ n \geq 1$,

$$\mathbb{A}_n = \sigma\{X_1, \cdots, X_n\} \ \&\ \mathbb{C}_{n+1} = \sigma\{X_{n+1}, X_{n+2}, \cdots, \} \text{ are independent.} \tag{5.5}$$

As shown in Section 2.3, $\forall\, n \geq 1\ \mathbb{A}_n \subset \mathbb{A}_{n+1}$ which implies that $\{\mathbb{A}_n, n \geq 1\}$ is a non-decreasing sequence of sigma fields. Hence, its limit exists and is $\mathbb{A} = \bigcup_{n \geq 1} \mathbb{A}_n$. As discussed in Lemma 5.4.2, $\mathbb{A}$ is a field, hence a π class. By definition, the tail sigma field $\mathbb{T} = \bigcap_{n \geq 1} \mathbb{C}_n$. Now

$$
\begin{aligned}
\mathbb{T} = \bigcap_{n \geq 1} \mathbb{C}_n \Rightarrow \mathbb{T} \subset \mathbb{C}_n\ \ \forall\ n \geq 1, &\quad \text{hence } A \in \mathbb{T} \Rightarrow A \in \mathbb{C}_n\ \ \forall\ n \geq 1 \\
&\Rightarrow \forall\ A \in \mathbb{T}\ \subset \mathbb{C}_{n+1}\ \&\ \forall\ B \in \mathbb{A}_n,\ \ P(A \cap B) = P(A)P(B) \\
&\Rightarrow \mathbb{T}\ \ \&\ \ \mathbb{A}_n \text{ are independent sigma fields } \forall\ n \geq 1 \\
\text{Further, } B \in \mathbb{A} \Rightarrow B \in \mathbb{A}_n &\text{ for at least one } n \geq 1 \\
&\Rightarrow \text{For } A \in \mathbb{T}\ \&\ B \in \mathbb{A},\ P(A \cap B) = P(A)P(B) \\
&\Rightarrow \mathbb{T}\ \ \&\ \ \mathbb{A} \text{ are independent} \\
&\Rightarrow \mathbb{T}\ \ \&\ \ \sigma(\mathbb{A}) \text{ are independent, by Lemma 5.4.1} \\
&\Rightarrow \mathbb{T}\ \ \&\ \ \sigma(\{X_n, n \geq 1\}) \text{ are independent, by Theorem 2.3.6} \\
&\Rightarrow \mathbb{T}\ \ \&\ \ \mathbb{C}_1 \text{ are independent, by definition of } \mathbb{C}_1 \\
&\Rightarrow \forall\ A \in \mathbb{T}\ \&\ \forall\ B \in \mathbb{C}_1,\ \ P(A \cap B) = P(A)P(B) \\
&\Rightarrow \forall\ A \in \mathbb{T}\ \& A \in \mathbb{C}_1\ \ P(A \cap A) = P(A)P(A) \text{ as } \mathbb{T} \subset \mathbb{C}_1 \\
&\Rightarrow \forall\ A \in \mathbb{T},\ \ P(A) = 0 \text{ or } P(A) = 1.
\end{aligned}
$$

The second step follows by Equation (5.5) and the fifth step follows since $\mathbb{T}$ and $\mathbb{A}_n$ are independent for all $n \geq 1$. In other words, $\mathbb{T}$ is independent of $\mathbb{T}$ itself which means that every event $A \in \mathbb{T}$ is independent of itself implying that $P(A)$ is either 0 or 1. □

Tail events are determined by the behaviour of the sequence $\{X_n, n \geq 1\}$ for large n and they remain unchanged if any finite sub collection of the X_n's are dropped or replaced by another finite set of random variables. Hence, events such as $[\limsup_{n \to \infty} X_n < x]$ or $[\lim_{n \to \infty} X_n = X]$ are tail events. In fact in Theorem 2.3.10, it is proved that $\liminf X_n, \limsup X_n$ and $\lim X_n$, if it exists, are measurable with respect to tail sigma field $\mathbb{T}$. Kolmogorov zero one law implies that if $\{X_n, n \geq 1\}$ is a sequence of independent random variables, then $\liminf X_n, \limsup X_n$ and $\lim X_n$, if it exists, are almost surely degenerate random variables, as all these are measurable with respect to the tail sigma field. More precisely, suppose a random variable X is measurable with respect to $\mathbb{T}$. Then for each $x \in \mathbb{R}$, $A = [X \leq x] \in \mathbb{T}$. Further, by Kolmogorov zero-one law $P(A) = 0$ or 1. Since, $P[X \leq x]$ increases as x increases, it is 0 for $x < x_0$ and is 1 for all $x \geq x_0$ for some $x_0 \in \mathbb{R}$, which shows that X is degenerate at x_0.

In the following example, we note that the limit random variable is degenerate at ∞.

Example 5.4.1. Suppose $\{X_n, n \geq 1\}$ is a sequence of random variables where for each $n \geq 1$, $X_n \equiv n$. Thus, X_n is degenerate at n for each $n \geq 1$ and hence it is a sequence of independent random variables with $\lim_{n\to\infty} X_n = X = \infty$. The limit random variable is also a degenerate random variable. Further, $\sigma(X_n) = \{\Omega, \emptyset\}\ \forall\ n \geq 1$ implies $\mathbb{T} = \{\Omega, \emptyset\}$ and the tail events have probability either 0 or 1. □

A quick recap of the results discussed in this chapter is given below.

Summary

1. A finite collection $\{A_1, A_2, \cdots, A_k\}$ of events from $\mathbb{A}$ is said to be a collection of independent events if, $P(A_{i_1} \cap A_{i_2} \cap \cdots \cap A_{i_r}) = \prod_{l=1}^{r} P(A_{i_l})$, for every possible sub collection $\{i_1, i_2, \cdots, i_r\}$ of $\{1, 2, \cdots, k\}$, $2 \leq r \leq k$.

2. A sequence $\{A_n, n \geq 1\}$ of events from $\mathbb{A}$ is a sequence of independent events if every finite sub-collection of $\{A_n, n \geq 1\}$ is a collection of independent events.

3. Suppose $\mathbb{C}_1, \mathbb{C}_2, \cdots, \mathbb{C}_k$ are collections of events in $\mathbb{A}$. Then these are independent collections if $\forall\ A_1 \in \mathbb{C}_1, A_2 \in \mathbb{C}_2, \cdots, A_k \in \mathbb{C}_k$, $\{A_1, A_2, \cdots, A_k\}$ is a collection of independent events.

4. Suppose $\{\mathbb{C}_n, n \geq 1\}$ is a sequence of collections of events from $\mathbb{A}$. It is a sequence of independent collections if $\{A_n, n \geq 1\}$ is a sequence of independent events for every $A_n \in \mathbb{C}_n, n \geq 1$.

5. Random variables $X_1, X_2, \cdots, X_k$ are independent if $\sigma(X_i)$ are independent sigma fields, $i = 1, 2, \cdots, k$.

6. Suppose X and Y are independent random variables and g and h are Borel functions, then $g(X)$ and $h(Y)$ are independent random variables.

7. Any random variable is always independent of a degenerate random variable.

8. If X and Y are independent random variables then, $E(XY) = E(X)E(Y)$.

9. Suppose X and Y are independent random variables. Then (i) joint characteristic function of X and Y is the product of characteristic functions of X and Y and (ii) the characteristic function of $X + Y$ is the product of characteristic functions of X and Y.

10. Two random variables X and Y are independent random variables if any one of the following three conditions are satisfied.
(i) $\sigma(X)$ and $\sigma(Y)$ are independent sigma fields, that is

$$\forall\ S_1, S_2 \in \mathbb{B}, \quad P[X \in S_1, Y \in S_2] = P[X \in S_1]P[Y \in S_2].$$

(ii) $\forall\ x, y \in \mathbb{R}, \quad F_{X,Y}(x, y) = F_X(x)F_Y(y)$.
(iii) $\forall\ t_1, t_2 \in \mathbb{R}, \quad \phi_{X,Y}(t_1, t_2) = \phi_X(t_1)\phi_Y(t_2)$.

11. A sequence $\{X_n, n \geq 1\}$ is a sequence of independent random variables if and only if any finite collection of the sequence is a collection of independent random variables. A class $\{X_t, t \in T\}$ is a class of independent random variables if and only if any finite sub class of this class is a class of independent random variables.

12. Two random vectors $\underline{X}$ and $\underline{Y}$ defined on the same probability space are independent random vectors if the induced sigma fields are independent.

13. If two random vectors are independent, then each component random variable of one random vector is independent of each component random variable of the second random vector.

14. Borel functions of independent random vectors are also independent random vectors.

15. Kolmogorov zero-one law: If $\{X_n, n \geq 1\}$ is a sequence of independent random variables, then the tail events have probability either 0 or 1.

5.5 Conceptual Exercises

5.5.1 Prove or disprove: A P-null event is independent of itself.

5.5.2 Suppose N is a P-null event. Examine whether N^c is independent of itself.

5.5.3 Suppose $P(C|A \cap B) = P(C|B)$ and $P(A \cap B \cap C) > 0$. Show that $P(A|B \cap C) = P(A|B)$.

5.5.4 Prove or disprove: (i) Mutually exclusive events are independent.
(ii) Independent events are mutually exclusive.

5.5.5 Show that two events A and B are independent if and only if $\sigma(\{A\})$ and $\sigma(\{B\})$ are independent. Show further that it is equivalent to I_A and I_B are independent random variables.

5.5.6 Give an example of events A_1, A_2, A_3 which are pair-wise independent but not independent.

5.5.7 Suppose $\{A_n, n \geq 1\}$ is a sequence of independent events. Show that $P\Big(\bigcup_{r=1}^{k} A_{i_r}\Big) = 1-\prod_{r=1}^{k}(1-P(A_{i_r}))$, for any combination $i_1, i_2, \cdots, i_k$ of $1, 2, \cdots, k$ and for finite $k \geq 2$.

5.5.8 Suppose X_1 and X_2 are random variables defined on $(\Omega, \mathbb{A}, P)$ as follows, where $\Omega = \{a, b, c\}$, $\mathbb{A} = \mathbb{P}(\Omega)$ and P is defined as $P(\{\omega\}) = 1/3 \ \forall \ \omega \in \Omega$.

$$X_1(\omega) = \begin{cases} 0, & \text{if} \quad \omega = a \\ 1, & \text{if} \quad \omega = b, c. \end{cases} \quad \& \quad X_2(\omega) = \begin{cases} 2, & \text{if} \quad \omega = b \\ 3, & \text{if} \quad \omega = a, c. \end{cases}$$

(i) Examine whether X_1 and X_2 are independent random variables on $(\Omega, \mathbb{A}, P)$. (ii) Examine whether $5X_1^2+4$ and $3X_2^3+e^{X_2}$ are independent random variables on $(\Omega, \mathbb{A}, P)$.

5.5.9 Suppose $(\Omega, \mathbb{A}, P)$ is a probability space, where $\Omega = \{a, b, c, d\}$, $\mathbb{A} = \mathbb{P}(\Omega)$ and $P(\{\omega\}) = 1/4 \ \forall \ \omega \in \Omega$. Suppose X_1 and X_2 are random variables defined on $(\Omega, \mathbb{A}, P)$ as follows.

$$X_1(\omega) = \begin{cases} 10, & \text{if} \quad \omega = a, b \\ 20, & \text{if} \quad \omega = c, d. \end{cases} \quad \& \quad X_2(\omega) = \begin{cases} 30, & \text{if} \quad \omega = a, c \\ 40, & \text{if} \quad \omega = b, d. \end{cases}$$

Examine whether $\exp(5X_1-7)$ and $\log(4X_2+9)$ are independent random variables.

5.5.10 Suppose $(\Omega, \mathbb{A}, P)$ is a probability space and $\{A_1, A_2, A_3\}$ is a measurable partition of Ω. Suppose $B \in \mathbb{A}$. Show that a random variable X defined as $X = \sum_{i=1}^{3} iI_{A_i}$ and I_B are independent if and only if B is independent of each of $\{A_1, A_2, A_3\}$.

5.5.11 Suppose $(\Omega, \mathbb{A}, P)$ is a probability space and $A, B, C \in \mathbb{A}$ are independent events. Examine whether C is independent of an event obtained from A and B by complementation, union and intersection.

5.5.12 Suppose $(\Omega, \mathbb{A}, P)$ is a probability space and $A, B, C \in \mathbb{A}$ are independent events. Can we claim that I_A+3I_B and I_C are independent random variables? Justify your answer.

5.5.13 Suppose $(\Omega, \mathbb{A}, P)$ is a probability space. Suppose $A, C \in \mathbb{A}$ are independent events and $B, C \in \mathbb{A}$ are independent events. Can we claim that $I_A + I_B$ and I_C are independent random variables? Justify your answer.

5.5.14 Give two examples in which a random variable is independent of itself.

5.5.15 Give an example of random variables X_1, X_2, X_3 which are pairwise independent but not independent.

5.5.16 Give an example of random variables X_1, X_2, X_3, X_4 which are independent.

5.5.17 Prove or disprove: If F is a distribution function, then (i) F^4 is also a distribution function, (ii) $2F - F^2$ is also a distribution function.

5.5.18 Suppose $\{X_n, n \geq 1\}$ is a sequence of independent random variables and $\{f_n, n \geq 1\}$ is a sequence of Borel functions. Show that $\{f_n(X_n), n \geq 1\}$ is a sequence of independent random variables.

5.5.19 If X and Y are independent and X is integrable, show that $\int_{[Y \in B]} X \, dP = E(X)P[Y \in B]$ for any Borel set B.

5.5.20 Give an example of two random variables X and Y such that X and Y are not independent but the characteristic function of their sum is the product of their characteristic functions.

5.5.21 Suppose X and Y are independent random variables. Show that if X and $X - Y$ are independent, then X must be degenerate.

5.6 Multiple Choice Questions

Note: In each question, multiple options may be correct. Unless specified otherwise, identify which of the statement(s) is/are correct. Answers are given in Chapter 11, after the solutions of conceptual exercises of Chapter 5.

5.6.1 A and B are two independent events with $0 < P(A), P(B) < 1$. Which of the following is/are false?

(a) $P(A^c \bigcap B) = P(B) - P(B)P(A)$
(b) $P(A \bigcap B) = P(A)P(B)$
(c) A and B are mutually exclusive
(d) $P(A \bigcap B^c) = P(B) - P(B)P(A)$

5.6.2 Following are two statements: (I) Mutually exclusive events are independent. (II) Independent events are mutually exclusive. Then

(a) Both (I) and (II) are true
(b) (I) is true but (II) is false
(c) (II) is true but (I) is false
(d) Both (I) and (II) are false

5.6.3 Suppose A and B are independent events. Then

(a) $P(A^c \cap B) = P(B) - P(B)P(A)$
(b) $P(A \cap B) = 0$

(c) A and B are mutually exclusive events
(d) $P(A \cup B) = P(A) + P(B)$.

5.6.4 Suppose $(\Omega, \mathbb{A}, P)$ is a probability space. Then an event $A \in \mathbb{A}$ is

(a) always independent of itself
(b) independent of itself if $P(A) = 0$
(c) independent of itself if $P(A) = 1$
(d) independent of a sure event

5.6.5 Suppose $(\Omega, \mathbb{A}, P)$ is a probability space. Which of the following statements is false?

(a) An event $A \in \mathbb{A}$ is always independent of itself
(b) An event $A \in \mathbb{A}$ is independent of itself if $P(A) = 0$
(c) An event $A \in \mathbb{A}$ is independent of itself if $P(A) = 1$
(d) An event $A \in \mathbb{A}$ is independent of a sure event

5.6.6 Suppose $(\Omega, \mathbb{A}, P)$ is a probability space. Following are two statements about an event $A \in \mathbb{A}$. (I) A is independent of sure event Ω. (II) A is independent of null event $\emptyset$.

(a) Both (I) and (II) are true
(b) (I) is true but (II) is false
(c) (II) is true but (I) is false
(d) Both (I) and (II) are false

5.6.7 Two random variables X_1 and X_2 are independent if

(a) $\sigma(X_1) \subset \sigma(X_2)$
(b) $\sigma(X_1) \supset \sigma(X_2)$
(c) $\sigma(X_1) = \sigma(X_2)$
(d) $\sigma(X_1)$ and $\sigma(X_2)$ are independent sigma fields.

5.6.8 Suppose X and Y are independent random variables. Following are two statements: (I) If g and h are Borel functions, then $g(X)$ and $h(Y)$ are independent random variables. (II) If g and h are continuous functions, then $g(X)$ and $h(Y)$ are independent random variables.

(a) Both (I) and (II) are true
(b) (I) is true but (II) is false
(c) (II) is true but (I) is false
(d) Both (I) and (II) are false

5.6.9 Following are two statements: (I) Any random variable is always independent of a degenerate random variable. (II) Any random variable is always independent of itself.

(a) Both (I) and (II) are true
(b) (I) is true but (II) is false
(c) (II) is true but (I) is false

(d) Both (I) and (II) are false

5.6.10 Following are two statements: (I) A random variable is independent of itself. (II) A random variable is independent of itself if and only if it is a degenerate random variable.

(a) Both (I) and (II) are true
(b) (I) is true but (II) is false
(c) (II) is true but (I) is false
(d) Both (I) and (II) are false

5.6.11 Following are two statements: (I) Suppose $\{X_n, n \geq 1\}$ is a sequence of independent random variables. Then the tail events have probability either 0 or 1. (II) Suppose $\{X_n, n \geq 1\}$ is a sequence of random variables. Then the tail events have probability either 0 or 1.

(a) Both (I) and (II) are true
(b) (I) is true but (II) is false
(c) (II) is true but (I) is false
(d) Both (I) and (II) are false

5.6.12 Suppose g and h are Borel functions and X and Y are random variables. Following are two statements: (I) $g(X)$ and $h(Y)$ are independent random variables implies X and Y are independent random variables. (II) $g(X)$ and $h(Y)$ are not independent random variables for some functions g and h implies X and Y are not independent random variables.

(a) Both (I) and (II) are true
(b) (I) is true but (II) is false
(c) (II) is true but (I) is false
(d) Both (I) and (II) are false

5.6.13 Following are two statements: (I) If two random vectors are independent, then each component random variable of one random vector is independent of each component random variable of the second random vector. (II) If each component random variable of one random vector is independent of each component random variable of the second random vector, then two random vectors are independent.

(a) Both (I) and (II) are true
(b) (I) is true but (II) is false
(c) (II) is true but (I) is false
(d) Both (I) and (II) are false

5.6.14 Suppose X and Y are independent random variables and the expectations in the following statements are finite. Following are three statements:
(I) $E(X^3 + \tan Y) = E(X^3) + E(\tan Y)$.
(II) $E((\exp(5X^3 + 10))(\cos(4Y - 6))) = E((\exp(5X^3 + 10))E((\cos(4Y - 6))$.

(III) $E((3X^3 + 6X + 9)(4Y^4 - 6Y)) \neq E((3X^3 + 6X + 9)E(4Y^4 - 6Y)$ always.

(a) Only (I) is true
(b) Only (II) is true
(c) Both (I) and (II) are true
(d) (III) is false

5.6.15 Suppose $\underline{Z}_1 = (X_1, X_2)'$ and $\underline{Z}_2 = (Y_1, Y_2)'$. Following are two statements: (I) If $\underline{Z}_1$ and $\underline{Z}_2$ are independent, then X_i is independent of Y_j $\forall\ i, j = 1, 2$. (II) If X_i is independent of Y_j $\forall\ i, j = 1, 2$, then $\underline{Z}_1$ and $\underline{Z}_2$ are independent.

(a) Both (I) and (II) are true
(b) (I) is true but (II) is false
(c) (II) is true but (I) is false
(d) Both (I) and (II) are false

6

Almost Sure Convergence and Borel Zero-One Law

6.1 Introduction

Statistics as a scientific discipline deals with various methods of collection of data, a variety of tools for their analysis and interpretation of the results of the analysis. Numerous techniques of analysis and their optimality properties are discussed in statistical inference in its three branches - point estimation, interval estimation and tests of hypotheses. These differ as the size of the data differs. In some cases if the data size is small, optimal solution may not exist. However, scenario changes as sample size increases. Statistics is concerned with gathering data, hence it is naturally of interest to judge what happens when we get more and more data. It is investigated in large sample or asymptotic inference theory. Two most desirable properties of a point estimator in asymptotic inference setup are consistency and asymptotic normality with suitable normalization. An estimator is a Borel function of sample observations and hence is again a random variable. The optimality properties of the estimator for finite sample size, such as unbiasedness and variance, are studied from its distribution, known as the sampling distribution. When the sample size is large, the optimality properties such as consistency and asymptotic normality are based on the limiting behaviour of a sequence of estimators. Hence, the principal probability tool in large sample inference is the convergence of a sequence of random variables. It is the most important concept from probability theory and is heavily applied in statistical analysis. This chapter and Chapters 7 to 10 are devoted to an elaborate discussion of this notion.

Suppose $\{X_n, n \geq 1\}$ is a sequence of random variables defined on a probability space $(\Omega, \mathbb{A}, P)$. As discussed in Chapter 2, a sequence $\{X_n, n \geq 1\}$ of random variables is equivalent to a collection of sequences of real numbers, $\{\{X_n(w) = a_n, n \geq 1\}, \omega \in \Omega\}$. Hence, to discuss the convergence of a sequence of random variables, one has to deal with the convergence of a collection of sequences of real numbers. Using this approach in Chapter 2 we have studied point-wise convergence. Almost sure convergence, which we discuss in Section 2, also adopts the similar approach. Other approaches to reduce a sequence of random variables to a collection of sequences of real numbers lead to other modes of convergence, namely, convergence in probability, convergence

DOI: 10.1201/9781032619057-6

in law and convergence in r-th mean. We define all these modes of convergence in the next section and study their properties in subsequent sections and Chapter 7. In the present chapter, we concentrate on the almost sure convergence and Borel zero-one law, which is also concerned with the almost sure convergence. The other modes of convergence, namely, convergence in probability, convergence in law and convergence in r-th mean are studied in Chapter 7.

6.2 Definitions of Various Modes of Convergence

In all the modes of convergence, except convergence in law, it is assumed that a sequence $\{X_n, n \geq 1\}$ of random variables and the limit random variable X are defined on the same probability space $(\Omega, \mathbb{A}, P)$. We begin with point-wise convergence, denoted by $X_n \to X$ on Ω, which has been already introduced in Chapter 2.

Definition 6.2.1. *Point-wise convergence: A sequence $\{X_n, n \geq 1\}$ is said to converge point-wise to a random variable X, if $X_n(\omega) \to X(\omega), \ \forall \ \omega \in \Omega$. More precisely, $\forall \ \omega \in \Omega$ and $\forall \ \epsilon > 0, \ \exists$ an integer $n_0(\epsilon, \omega)$, such that $\forall \ n \geq n_0(\epsilon, \omega), \ |X_n(\omega) - X(\omega)| < \epsilon$.*

A point-wise convergence of a sequence of random variables is denoted by, $X_n \to X$ on Ω. The set B on which $X_n(\omega) \to X(\omega)$, can be expressed as follows.

$$B = \bigcap_{\epsilon>0} \bigcup_{n\geq 1} \bigcap_{m\geq n} \{\omega | |X_m(\omega) - X(\omega)| < \epsilon\}.$$

The set B is referred to as a set of convergence. It is to be noted that $\bigcap_{\epsilon>0}$ is an uncountable intersection, but it can be replaced by a countable intersection, since for every $\epsilon > 0$ there exists a positive integer k, such that $1/k < \epsilon$. Thus, the set B can be represented as

$$B = \bigcap_{k\geq 1} \bigcup_{n\geq 1} \bigcap_{m\geq n} \{\omega | |X_m(\omega) - X(\omega)| < 1/k\}.$$

The set B expressed in this way is a measurable set, so one can assign probability to B. In point-wise convergence $B = \Omega$.

If $X_n(\omega) \to X(\omega)$ where $X(\omega)$ is finite, then $X_n(\omega) - X_m(\omega) \to 0$ as $m, n \to \infty$ or equivalently $X_{m+n}(\omega) - X_n(\omega) \to 0$ as $n \to \infty$ uniformly in m and conversely. This is known as a Cauchy criterion of convergence. If a sequence converges in a Cauchy sense then it is said to converge mutually. If B_c denotes the set of mutual convergence, then it is given by,

$$B_c = \bigcap_{k\geq 1} \bigcup_{n\geq 1} \bigcap_{m\geq 1} \{\omega | |X_{m+n}(\omega) - X_n(\omega)| < 1/k\}.$$

Since mutual convergence implies convergence, $B_c \subset B$. If X is finite on B, then $B_c = B$.

In practice it is very rare to have a point-wise convergence of a sequence of random variables, in view of the fact that in most of the situations Ω is countably infinite or uncountable. In the next mode of convergence, the almost sure convergence, the requirement of convergence for all sample points is slightly relaxed. Almost sure convergence, also known as almost everywhere convergence, is similar to the point-wise convergence of a sequence of random variables, except that the convergence need not occur on a P-null set. Following is the definition of almost sure convergence of a sequence of random variables.

Definition 6.2.2. *Almost sure convergence: A sequence $\{X_n, n \geq 1\}$ is said to converge almost surely to a random variable X, if and only if $X_n(\omega) \to X(\omega)\ \ \forall\ \omega \in N^c$, such that $P(N) = 0$. More precisely, $\forall\ \ \omega \in N^c$ and $\forall\ \epsilon > 0$, $\exists$ an integer $n_0(\epsilon, \omega)$, such that $\forall\ n \geq n_0(\epsilon, \omega)$, $|X_n(\omega) - X(\omega)| < \epsilon$.*

If $n_0(\epsilon, \omega)$ does not depend on ω, then we have uniform almost sure convergence. The above definition can be expressed as $P[\lim_{n\to\infty} X_n = X] = 1$. Hence, almost sure convergence is also known as convergence with probability 1. Almost sure convergence of X_n to X is denoted by $X_n \overset{a.s.}{\to} X$. The set N is a P-null set. In almost sure convergence, the set N^c of convergence is the same as the set B of convergence defined above, with $P(N^c) = 1$.

In point-wise convergence and in almost sure convergence, a sequence of random variables is reduced to a sequence of real numbers by considering images $X_n(\omega)$. In the next mode of convergence, convergence in probability, the closeness of X_n and X is measured in terms of the underlying probability measure and in this way the convergence of sequence of random variables is reduced to convergence of sequence of real numbers. More precisely, suppose $p_n(\epsilon) = P[|X_n - X| < \epsilon]$. Thus, $\{p_n(\epsilon), n \geq 1\}$ is a sequence of real numbers and we examine whether it converges to 1 $\forall\ \epsilon > 0$ in convergence in probability. We define this mode of convergence below.

Definition 6.2.3. *Convergence in probability: A sequence $\{X_n, n \geq 1\}$ is said to converge in probability to a random variable X, if and only if*

$$\forall\ \epsilon > 0,\quad \lim_{n\to\infty} P[|X_n - X| < \epsilon] = 1 \quad \Longleftrightarrow \quad \lim_{n\to\infty} P[|X_n - X| \geq \epsilon] = 0.$$

Convergence in probability can be viewed as a stochastic analog of the convergence of a sequence of real numbers. It is denoted by $X_n \overset{P}{\to} X$. Observe that from the definition

$$X_n \overset{P}{\to} X \quad \Longleftrightarrow \quad X_n - X \overset{P}{\to} 0.$$

If $X_n \overset{P}{\to} 0$, then we write $X_n = o_p(1)$. If more generally, $a_n X_n \overset{P}{\to} 0$, then we write $X_n = o_p(1/a_n)$. From the definitions of almost sure convergence and convergence in probability, observe that

$$X_n \xrightarrow{P} X, \text{ if } \lim_{n\to\infty} P[|X_n - X| < \epsilon] = 1, \forall\ \epsilon > 0$$
$$\text{and } X_n \xrightarrow{a.s.} X, \text{ if } P[\lim_{n\to\infty} X_n = X] = 1.$$

Thus, the two modes are quite different. In the next section, we prove that almost sure convergence is stronger than the convergence in probability.

The frequent use of normal distribution in a variety of areas is based on the fact that the sample mean is approximately normal when the sample size is large. The large sample inference theory mainly deals with approximations of probability distributions and with the limit theorems underlying these approximations for large sample size. Such limit theorems are based on a concept of convergence of a sequence of distributions. Probability distributions can be characterized in different ways, for example, by their distribution functions or their probability mass functions or probability density functions. The convergence of a sequence of probability distributions is defined in terms of convergence of their distribution functions, and this mode of convergence is known as convergence in law or convergence in distribution. Thus in this mode we study the behaviour of a sequence $\{F_n(x), n \geq 1\}$ of real numbers for some values of $x \in \mathbb{R}$.

Definition 6.2.4. *Convergence in law: Suppose $\{X_n, n \geq 1\}$ is a sequence of random variables defined on a probability space $(\Omega, \mathbb{A}, P)$. Suppose $F_n(\cdot)$ is a distribution function of X_n and $F_X(\cdot)$ is a distribution function of X. A sequence $\{X_n, n \geq 1\}$ is said to converge in law to a random variable X, not necessarily defined on the same probability space, if and only if*

$$F_n(x) = P[X_n \leq x] \to P[X \leq x] = F_X(x), \quad \forall\ x \in C(F_X),$$

where $C(F_X)$ is a set of points of continuity of the distribution function F_X.

Convergence in law is denoted by $X_n \xrightarrow{L} X$ or $X_n \xrightarrow{d} X$. From the definition of convergence in law, it immediately follows that for any interval $I = (a, b]$, such that $a, b \in C(F_X)$, as $n \to \infty$

$$P[a < X_n \leq b] = F_{X_n}(b) - F_{X_n}(a) \to F_X(b) - F_X(a) = P[a < X \leq b]\ .$$

It is to be noted that in the definition of convergence in law, it is required that $F_n(x) \to F_X(x)$ for all x which are the points of continuity of $F_X(x)$. One might be concerned about such a restriction. However, it is enough to restrict to a set $C(F_X)$ in view of the fact that the set of points of continuity is dense in $\mathbb{R}$. Further, it is shown in Section 3.2, that the distribution function is completely determined by its values at all continuity points. It is to be noted that in convergence in law, a sequence of random variables $\{X_n, n \geq 1\}$ and the limit random variable is X need not be defined on the same probability space.

In convergence in law, a sequence of random variables $\{X_n, n \geq 1\}$ is reduced to a sequence of real numbers by considering the sequence of distribution functions, which is equivalent to a sequence of real numbers in $[0, 1]$. In the next mode of convergence, convergence in r-th mean, a sequence of random variables $\{X_n, n \geq 1\}$ is reduced to a sequence of real numbers by considering the sequence of expectations of the deviation $|X_n - X|$. This mode of convergence is defined when both X_n and X belong to L_r space, as defined in Chapter 4. If X_n and X belong to L_r space, by the C_r inequality, $E(|X_n - X|^r) < \infty$, $r > 0$ and hence $X_n - X$ also belongs to L_r space.

Definition 6.2.5. *Convergence in r-th mean: Suppose X_n for each $n \geq 1$ and X belong to L_r space. A sequence $\{X_n, n \geq 1\}$ is said to converge in r-th mean to a random variable X, if and only if*

$$E(|X_n - X|^r) \to 0 \quad as \quad n \to \infty, \quad r > 0.$$

If $r = 2$ then the convergence is referred to as a convergence in quadratic mean.

Convergence in r-th mean is denoted by $X_n \xrightarrow{r} X$ and convergence in quadratic mean is denoted by $X_n \xrightarrow{q.m.} X$. Convergence in r-th mean is also known as L_r convergence.

It is to be noted that the concept of convergence in probability, convergence in r-th mean or almost sure convergence represents a sense in which, for n sufficiently large, X_n and X approximate each other as functions on the original probability space. The concept of convergence in distribution however, depends only on the closeness of distribution functions F_{X_n} and F_X and does not require that X_n and X are close in any other sense; in fact, X and X_n's may not be defined on the same probability space. Among all these modes of convergence, point-wise and uniformly point-wise convergence are from calculus theory, almost sure convergence, convergence in probability and convergence in r-th mean are unique to measure theory while convergence in distribution is unique to probability theory.

In this and the next chapter we study all these modes of convergence in detail. We investigate which properties of convergence of a sequence of real numbers get extended to a convergence of a sequence of random variables. We discuss some aspects listed below.

(i) Whether the limit is unique and in what sense?

(ii) Whether the convergence is closed under arithmetic operations?

(iii) Whether the convergence is invariant under continuous transforms?

(iv) Whether the convergence and the corresponding Cauchy convergence are equivalent?

(v) What are the links among these modes of convergence?

In Section 3, we investigate the properties of the almost sure convergence. Section 4 presents Borel zero-one law, which is mainly related to the almost sure convergence of a sequence of independent random variables. We illustrate all the results with a variety of theoretical examples. The concepts related to various modes of convergence are difficult to grasp due to their abstract nature. To absorb the notions easily, theoretical results are simultaneously supported by the numerical computations, graphs and the simulation studies, with `R` software. The `R` codes are given for some examples. Interested readers may refer to Serfling [20] and van der Vaart [24] for some more theoretical results.

6.3 Almost Sure Convergence

In this section we discuss various aspects of the almost sure convergence of a sequence of random variables. We begin with an example illustrating the almost sure convergence and the point-wise convergence. One illustration is already discussed in Section 2.3.

Example 6.3.1. Suppose the probability space $(\Omega, \mathbb{A}, P)$ is defined as follows. $\Omega = [0,1]$ and $\mathbb{A}$ is a sigma field of subsets of Ω. Suppose for $A = (a,b) \subset [0,1]$, $P(A) = b - a$, the Lebesgue measure of A. In such a setup $(\Omega, \mathbb{A}, P)$ is known as a unit interval probability space. Definition of P(A) for $A = (a,b)$ will define P(A) for all A in $\mathbb{A}$ using the properties of probability measure. As in Example 2.3.5, a sequence $\{X_n, n \geq 1\}$ of $\mathbb{A}$ measurable random variables and a $\mathbb{A}$ measurable random variable X are defined as follows.

$$X_n(\omega) = e^{\omega} + \omega^n, \; n \geq 1 \quad \& \quad X(\omega) = e^{\omega}.$$

It is clear that,

$$X_n(\omega) = e^{\omega} + \omega^n \quad \to \quad e^{\omega} = X(\omega), \;\; \forall \;\; \omega \in [0,1).$$

But at $\omega = 1, X_n(\omega) = e+1 \neq e = X(\omega)$. Thus $X_n \nrightarrow X$ point-wise. However, note that $X_n(\omega) \to X(\omega)$, except for $\omega = 1$. Thus, if $N^c = [0,1)$, then given $\epsilon > 0$

$$\forall \;\; \omega \in N^c, \;\; \exists \;\; n_0(\omega, \epsilon) \;\; \text{such that} \;\; |X_n(\omega) - X(\omega)| < \epsilon.$$

Observe that $P(N^c) = 1$ and hence, $X_n \stackrel{a.s.}{\to} X$ with $N = \{1\}$ as the P-null set. □

To understand the abstract concept of almost sure convergence, we verify the results of Example 6.3.1, graphically using `R` software in the next example.

Example 6.3.2. Suppose a sequence $\{X_n, n \geq 1\}$ of random variables and X are as defined in Example 6.3.1. We use the following code for verification of the results obtained in Example 6.3.1.

Code 6.3.1. Verification of almost sure convergence:

```
Xnw=function(n,w){return(exp(w)+w^n)}
Xw=function(w){return(exp(w))}
omegavec=seq(0,1,length=100)
nvec=c(1,10,100); nlab=paste("n=",nvec,sep="")
XnMat=matrix(nrow=length(omegavec),ncol=length(nvec))
for(i in 1:nrow(XnMat))
{
omega=omegavec[i]
for(j in 1:ncol(XnMat))
{
n=nvec[j]
XnMat[i,j]=Xnw(n,omega)
}
}
XVec=c()
for(i in 1:length(omegavec))
{
omega=omegavec[i]
XVec[i]=Xw(omega)
}
plot(omegavec,XVec,ylim=c(1,max(XnMat)+0.1),type="l",col=2,
lty=1,lwd=2,xlim=c(0,1.5),xaxt="n", xlab=expression(omega),
ylab=(expression(Xn(omega)~or~X(omega))),
main=(expression(Xn(omega)~"="~exp(omega)+ omega^n)))
axis(1,at=seq(0,1,by=0.1),labels=seq(0,1,by=0.1))
for(i in 1:ncol(XnMat))
{
lines(omegavec,XnMat[,i],col=1,lty=i+1,type="l",lwd=2)
}
legend("bottomright",legend=c(nlab,expression(X(omega))),
lty=c(2:(ncol(XnMat)+1),1),col=c(rep(1,3),2),lwd=2)
```

In Figure 6.1, the red curve displays the values of $X(\omega)$, whereas the black curves present the values of $X_n(\omega)$ for different choices of n as indicated in the legend. It can be seen that the black curves approach the red curve as n increases except for the rightmost endpoint, that is, $\omega = 1$. Thus, $X_n \nrightarrow X$ point-wise, but $X_n \overset{a.s.}{\rightarrow} X$. □

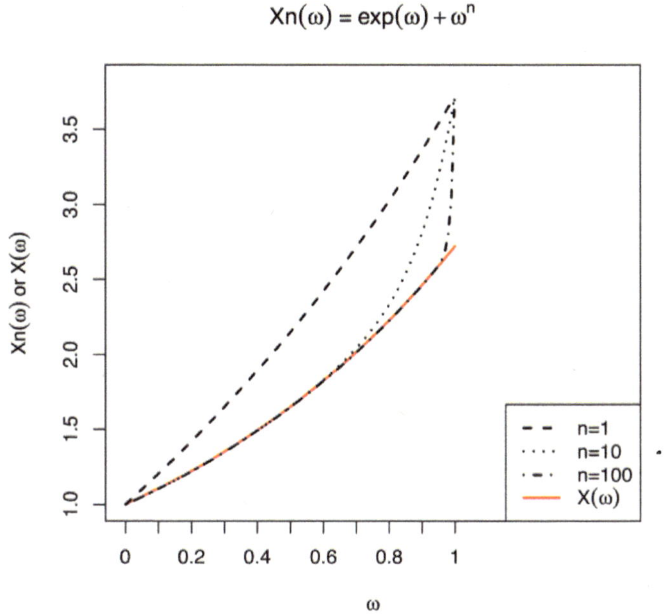

FIGURE 6.1
Almost Sure Convergence

It is of interest to see that even if ω is very close to 1, corresponding to given $\epsilon > 0$, we can find $n_0(\omega, \epsilon)$ such that $|X_n(\omega) - X(\omega)| < \epsilon$. We demonstrate it in the following example.

Example 6.3.3. Suppose a sequence $\{X_n, n \geq 1\}$ of random variables and X are as defined in Example 6.3.1. Using the following code, we find the values of n_0 for different values of ϵ and of ω, which are close to 1.

Code 6.3.2. Computation of $n_0(\omega, \epsilon)$ in almost sure convergence:

```
Xnw=function(n,w){return(exp(w)+w^n)}
Xw=function(w){return(exp(w))}
omegavec=seq(0.999,0.9999, length=1000)
epsilon=c(0.1,0.5)
nvec=c(10,100,1000,5000,10000,50000)
nlab=paste("n=",nvec,sep="")
XnMat=matrix(nrow=length(omegavec),ncol=length(nvec))
for(i in 1:nrow(XnMat))
{
omega=omegavec[i]
for(j in 1:ncol(XnMat))
{
n=nvec[j]
XnMat[i,j]=Xnw(n,omega)
```

```
}
}
XVec=c()
for(i in 1:length(omegavec))
{
omega=omegavec[i]
XVec[i]=Xw(omega)
}
plot(omegavec,XVec,ylim=c(XVec[1]-0.7,max(XnMat)+0.1),type="l",
col=2,lty=1,lwd=2,xlim=c(0.999,1),xaxt="n",
xlab=expression(omega),ylab=(expression(Xn(omega)~or~X(omega))),
main=(expression(Xn(omega)~"="~exp(omega)+ omega^n)))
text(x=0.999,y=XVec[1]-0.05,labels=expression(X(omega)),col=2)
axis(1,at=c(seq(0.999,1,by=0.0005),0.9999),
labels=c(seq(0.999,1,by=0.0005),0.9999))
for(i in 1:ncol(XnMat))
{
lines(omegavec,XnMat[,i],col=1,lty=i+1,type="l",lwd=2)
text(x=0.99995,y=XnMat[nrow(XnMat),i]-0.01*(-1)^i,labels=
nlab[i])
}
abline(h=XVec[1]+epsilon[1],col=2,lty=2,lwd=1.5)
abline(h=XVec[1]-epsilon[1],col=2,lty=2,lwd=1.5)
text(x=0.9995,y=XVec[1]+0.15,labels=expression
(X(omega)~"+"~0.1),
col=2)
text(x=0.9995,y=XVec[1]-0.15,labels=expression
(X(omega)~"-"~0.1),
col=2)
abline(h=XVec[1]-epsilon[2],col=4,lty=3,lwd=1.5)
abline(h=XVec[1]+epsilon[2],col=4,lty=3,lwd=1.5)
text(x=0.9995,y=XVec[1]+0.55,labels=expression
(X(omega)~"+"~0.5),
col=4)
text(x=0.9995,y=XVec[1]-0.55,labels=expression
(X(omega)~"-"~0.5),
col=4)
lines(1,Xw(1),col=2,type="p",pch=15)
lines(1,Xnw(100,1),col=1,type="p",pch=16)
```

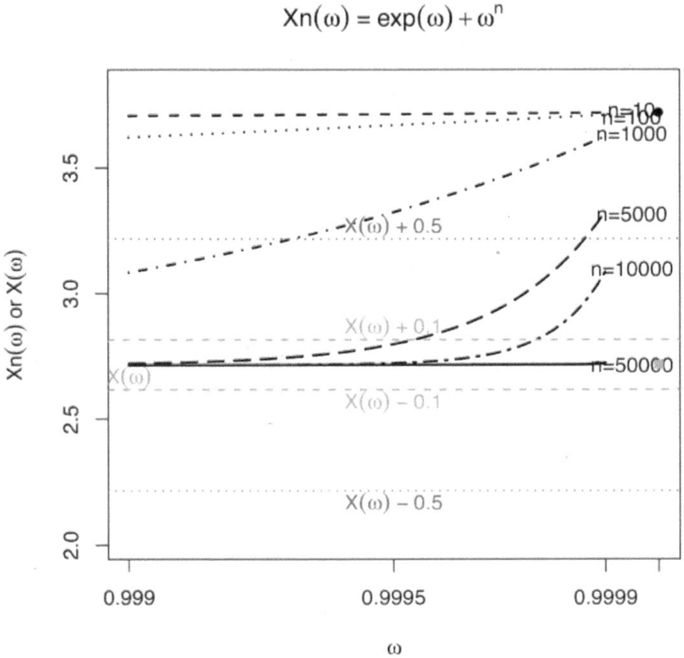

FIGURE 6.2
Almost Sure Convergence: n_0

Figure 6.2 displays the curves of $X_n(\omega)$ for ω values which are very close to 1. More precisely, in Figure 6.2, we focus on ω values between 0.999 and 0.9999. It can be seen that even for ω values which are so close to 1, we can always find n_0 for each ϵ. Since the black curves approach the red curves as seen in Figure 6.1, we can see that any n for which $X_n(\omega)$ lies within $(X(\omega) - \epsilon, X(\omega) + \epsilon)$ can work as n_0. Note that n_0 is not unique. Once we find some n_0, any n larger than it, can work as another n_0. Looking at the graph, some suggested values of n_0 for different values ω and ϵ are given in Table 6.1.

From Table 6.1, we note that for $\omega = 0.9999$ and $\epsilon = 0.1$, $n_0 = 50000$ is very large. □

It is to be noted that to verify the almost sure convergence graphically, we need to know Ω and the functional form of X_n and X. In many cases we know

TABLE 6.1
Almost Sure Convergence: $n_0(\epsilon, \omega)$

ϵ	0.1	0.1	0.5	0.5
ω	0.9990	0.9999	0.9990	0.9999
n_0	5000	50000	1000	10000

only the probability distribution of X_n. To verify almost sure convergence in such a setup, we have to use the definition or some necessary and sufficient conditions for the almost sure convergence. We now proceed to derive such conditions and illustrate its use by examples.

As defined in Section 2, $X_n \stackrel{a.s.}{\to} X \quad \Rightarrow \quad X_n(\omega) \to X(\omega)\ \forall\ \omega \in N^c$, such that $P(N) = 0$. Thus, the set N^c of convergence is given by,

$$N^c = \bigcap_{k\geq 1}\bigcup_{n\geq 1}\bigcap_{m\geq n}\{\omega||X_m(\omega) - X(\omega)| < 1/k\} \quad \text{with} \quad P(N^c) = 1\ .$$

Suppose an event D_k is defined as

$$\begin{aligned} D_k &= \bigcap_{n\geq 1}\bigcup_{m\geq n}\{\omega||X_m(\omega) - X(\omega)| \geq 1/k\} &(6.1)\\ &= [\limsup[|X_n - X| \geq 1/k]] = [|X_n - X| \geq 1/k, \quad \text{infinitely often}], \end{aligned}$$

by the definition of the limit superior of a set. Thus,

$$\begin{aligned} X_n \stackrel{a.s.}{\to} X \quad &\Rightarrow \quad P(N) = 0\\ &\Rightarrow \quad P\Big[\bigcup_{k\geq 1}\bigcap_{n\geq 1}\bigcup_{m\geq n}\{\omega||X_m(\omega) - X(\omega)| \geq 1/k\}\Big] = 0\\ &\Rightarrow \quad P\Big(\bigcup_{k\geq 1} D_k\Big) = 0\\ &\Rightarrow \quad P(D_k) = 0, \quad \forall\ k \geq 1 \quad \text{since}\ D_k \subset \bigcup_{k\geq 1} D_k\\ &\Rightarrow \quad P[\limsup[|X_n - X| \geq 1/k]] = 0\\ &\Longleftrightarrow \quad P[\liminf[|X_n - X| < 1/k]] = 1\\ &\Longleftrightarrow \quad P[|X_n - X| < 1/k, \quad \text{except for finitely many } n] = 1. \end{aligned}$$

Thus, $X_n \stackrel{a.s.}{\to} X$ implies that the probability of infinitely many deviations are large is 0. In other words, with probability 1, most of the deviations are small. Using these arguments, a necessary and sufficient condition for the almost sure convergence can be derived as follows.

Theorem 6.3.1. *Suppose a sequence of random variables* $\{X_n, n \geq 1\}$ *and* X *are defined on the same probability space* $(\Omega, \mathbb{A}, P)$. *Then* $\forall\ k \geq 1$,

$$\begin{aligned} X_n \stackrel{a.s.}{\to} X \quad &\Longleftrightarrow \quad \lim_{n\to\infty} P\Big[\bigcup_{m\geq n}[|X_m - X| \geq 1/k]\Big] = 0\\ &\Longleftrightarrow \quad \lim_{n\to\infty}\lim_{N\to\infty} P\Big[\bigcup_{m=n}^{N}[|X_m - X| \geq 1/k]\Big] = 0. &(6.2) \end{aligned}$$

Proof. Suppose an event D_k is as defined in Equation (6.1). Thus,

$$\begin{aligned} D_k &= \bigcap_{n\geq 1}\bigcup_{m\geq n}\{\omega||X_m(\omega)-X(\omega)|\geq 1/k\} \\ &= \bigcap_{n\geq 1} B_{nk} \text{ where } B_{nk}=\bigcup_{m\geq n}\{\omega||X_m(\omega)-X(\omega)|\geq 1/k\}. \end{aligned}$$

Observe that $\forall\ k\geq 1$,

$$\begin{aligned} B_{nk}\supset B_{(n+1)k} &\Rightarrow \{B_{nk}, n\geq 1\} \text{ is a decreasing sequence of sets} \\ &\Rightarrow \lim_{n\to\infty} B_{nk} \text{ exists} \\ \& \lim_{n\to\infty} B_{nk} &= \bigcap_{n\geq 1} B_{nk} = D_k \\ \text{Further, } B_{nk} &= \bigcup_{m\geq n}\{\omega||X_m(\omega)-X(\omega)|\geq 1/k\} \\ &= \lim_{N\to\infty}\Big[\bigcup_{m=n}^{N}[|X_m-X|\geq 1/k]\Big]. \end{aligned}$$

Now from the definition of almost sure convergence, we have

$$\begin{aligned} X_n \stackrel{a.s.}{\to} X &\iff P[X_n \nrightarrow X]=0 \\ &\iff P\Big[\bigcup_{k\geq 1}\bigcap_{n\geq 1}\bigcup_{m\geq n}[|X_m-X|\geq 1/k]\Big]=P\Big[\bigcup_{k\geq 1}D_k\Big]=0 \\ &\iff \forall\ k\geq 1\ \ P(D_k)=0 \\ &\iff \forall\ k\geq 1\ \ P[\lim_{n\to\infty}B_{nk}]=0 \\ &\iff \forall\ k\geq 1\ \ \lim_{n\to\infty}P(B_{nk})=0 \\ &\iff \forall\ k\geq 1\ \ \lim_{n\to\infty}P\Big[\lim_{N\to\infty}\bigcup_{m=n}^{N}[|X_m-X|\geq 1/k]\Big]=0 \\ &\iff \forall\ k\geq 1\ \ \lim_{n\to\infty}\lim_{N\to\infty}P\Big[\bigcup_{m=n}^{N}[|X_m-X|\geq 1/k]\Big]=0. \end{aligned}$$

Note that

$$\begin{aligned} X_n \stackrel{a.s.}{\to} X &\iff \forall\ k\geq 1\ \ P(D_k)=0 \\ &\iff \forall\ k\geq 1\ \ P[\limsup[|X_n-X|\geq 1/k]]=0 \quad (6.3) \end{aligned}$$

□

Following example illustrates Theorem 6.3.1.

Example 6.3.4. Suppose $\{X_n, n \geq 1\}$ is a sequence of random variables with $P[X_n = \pm 1/n] = 1/2$. Note that,

$$\begin{aligned} j < k \quad &\Rightarrow \quad |X_j| > |X_k| \text{ a.s. } \Rightarrow A_k = [|X_k| > \epsilon] \subset A_j = [|X_j| > \epsilon] \\ &\Rightarrow \quad A_n \supset A_{n+1} \supset A_{n+2} \cdots \\ &\Rightarrow \quad \bigcup_{m \geq n} [|X_m| > \epsilon] = [|X_n| > \epsilon] \\ &\Rightarrow \quad P[\bigcup_{m \geq n} [|X_m| > \epsilon]] = P[|X_n| > \epsilon] \leq P[|X_n| > 1/n] = 0, \end{aligned}$$

since $\forall \;\; \epsilon > 0 \;\; \exists \;\; n \geq 1$ such that $1/n < \epsilon$. Hence, by the necessary and sufficient condition for the almost sure convergence, $X_n \overset{a.s.}{\to} 0$. □

Following corollary gives a sufficient condition for the almost sure convergence.

Corollary 6.3.1. *Suppose a sequence $\{X_n, n \geq 1\}$ of random variables and X are defined on the same probability space $(\Omega, \mathbb{A}, P)$.*

(i) If $\sum_{n=1}^{\infty} P[|X_n - X| \geq 1/k] < \infty \;\; \forall \;\; k \geq 1$, then $X_n \overset{a.s.}{\to} X$.

(ii) If $\forall \;\; n \geq 1$, X_n and X belong to $\mathbb{L}_r$ and $\sum_{n=1}^{\infty} E(X_n - X)^r < \infty$, for some $r > 0$, then $X_n \overset{a.s.}{\to} X$.

Proof. By Theorem 6.3.1,

$$\forall \;\; k \geq 1 \quad \lim_{n \to \infty} P\Big[\bigcup_{m \geq n} [|X_m - X| > 1/k]\Big] = 0 \quad \Rightarrow \quad X_n \overset{a.s.}{\to} X.$$

(i) Note that $\;\; \forall \;\; k \geq 1$,

$$\lim_{n \to \infty} P\Big[\bigcup_{m \geq n} [|X_m - X| > 1/k]\Big] \leq \lim_{n \to \infty} \sum_{m \geq n} P[|X_m - X| > 1/k] = 0,$$

it being a remainder term of the convergent series $\sum_{n=1}^{\infty} P[|X_n - X| \geq 1/k]$. Hence,

$$\sum_{n=1}^{\infty} P[|X_n - X| \geq 1/k] < \infty \;\; \forall \;\; k \geq 1 \quad \Rightarrow \quad X_n \overset{a.s.}{\to} X.$$

(ii) By Markov's inequality,

$$\begin{aligned} P[|X_n - X| > \epsilon] \quad &\leq \quad E(X_n - X)^r/\epsilon^r \\ \Rightarrow \quad \sum_{n=1}^{\infty} P[|X_n - X| \geq \epsilon] \quad &\leq \quad \sum_{n=1}^{\infty} E(X_n - X)^r/\epsilon^r < \infty \;\; \forall \;\; \epsilon > 0, \end{aligned}$$

and the result follows from (i). □

Remark 6.3.1. If $\sum_{n=1}^{\infty} P[|X_n - X| \geq \epsilon] < \infty$ for all $\epsilon > 0$, then the sequence $\{X_n, n \geq 1\}$ is said to converge completely to a random variable X. From Corollary 6.3.1, it follows that the complete convergence is stronger than the almost sure convergence.

Example 6.3.5. As in Example 6.3.4, suppose $\{X_n, n \geq 1\}$ is a sequence of random variables with $P[X_n = \pm 1/n] = 1/2$. Thus, $|X_n| \equiv 1/n$ and hence $\sum_{n=1}^{\infty} E(X_n^2) = \sum_{n=1}^{\infty} 1/n^2 < \infty$. Hence, by Corollary 6.3.1 for the almost sure convergence, $X_n \stackrel{a.s.}{\to} 0$. □

In Example 6.3.8 also, we illustrate Corollary 6.3.1. In Section 4, we discuss some more illustrative examples and a different proof of the same corollary as an application of the Borel-Cantelli lemma.

Example 6.3.6. Suppose $\{X_n, n \geq 0\}$ is a branching process with $X_0 = 1$ and offspring mean $\lambda > 0$. Then $E(X_n) = \lambda^n$. Note that

$$\lambda < 1 \quad \Rightarrow \quad \sum_{n=1}^{\infty} E(|X_n|) = \sum_{n=1}^{\infty} \lambda^n < \infty \quad \Rightarrow \quad X_n \stackrel{a.s.}{\to} 0 \text{ by Corollary 6.3.1.}$$

Thus, if the offspring mean is less than 1, then $X_n \stackrel{a.s.}{\to} 0$. □

It is well known that the limit of a convergent sequence of real numbers is unique. On similar lines, the limit random variable in almost sure convergence is almost surely unique. It is proved in the following theorem.

Theorem 6.3.2. *If $X_n \stackrel{a.s.}{\to} X$ and $X_n \stackrel{a.s.}{\to} Y$, then X and Y are equivalent random variables, that is, $P[X = Y] = 1$.*

Proof. By the definition

$$\begin{aligned} X_n \stackrel{a.s.}{\to} X \quad &\Rightarrow \quad X_n(\omega) \to X(\omega) \;\; \forall \;\; \omega \in N_1^c \;\text{ where }\; P(N_1) = 0 \\ \& \;\; X_n \stackrel{a.s.}{\to} Y \quad &\Rightarrow \quad X_n(\omega) \to Y(\omega) \;\; \forall \;\; \omega \in N_2^c \;\text{ where }\; P(N_2) = 0. \end{aligned}$$

If $N = N_1 \cup N_2$, then $P(N) = 0$. Suppose $\omega \in N^c = N_1^c \cap N_2^c$. Then

$$\begin{aligned} \forall \;\; \epsilon_1 > 0 \;\; \exists \;\; n_1 \in \mathbb{N} \text{ such that } |X_n(\omega) - X(\omega)| &< \epsilon_1, \;\; \forall \;\; n \geq n_1 \\ \& \;\; \forall \;\; \epsilon_2 > 0 \;\; \exists \;\; n_2 \in \mathbb{N} \text{ such that } |X_n(\omega) - Y(\omega)| &< \epsilon_2, \;\; \forall \;\; n \geq n_2. \end{aligned}$$

Suppose $n_0 = \max\{n_1, n_2\}$. Then $\forall \;\; n \geq n_0$ and $\forall \;\; \omega \in N^c$,

$$\begin{aligned} |X(\omega) - Y(\omega)| &= |X(\omega) - X_n(\omega) + X_n(\omega) - Y(\omega)| \\ &\leq |X_n(\omega) - X(\omega)| + |X_n(\omega) - Y(\omega)| < \epsilon_1 + \epsilon_2 \,. \end{aligned}$$

It is true for all ϵ_1 & $\epsilon_2 > 0$. Hence, $\forall \;\; \omega \in N$, $P[|X - Y| > 1/k] = 0 \; \forall \; k \geq 1$ as $n \to \infty$. It then follows that as $n \to \infty$,

$$P[X \neq Y] = P\Big[\bigcup_{k \geq 1} \{\omega | |X(\omega) - Y(\omega)| > 1/k\}\Big] \leq \sum_{k \geq 1} P[|X - Y| > 1/k] = 0,$$

$$\Rightarrow P[X = Y] = 1 \quad \Rightarrow \quad X = Y \;\text{ a.s.}$$

□

As noted in Section 2, we now prove that almost sure convergence implies convergence in probability.

Theorem 6.3.3. *For a sequence $\{X_n, n \geq 1\}$ of random variables,*

$$X_n \xrightarrow{a.s.} X \;\Rightarrow\; X_n \xrightarrow{P} X.$$

Proof. Suppose $X_n \xrightarrow{a.s.} X$. By the definition of almost sure convergence,

$$X_n \xrightarrow{a.s.} X \;\Rightarrow\; X_n(\omega) \to X(\omega), \quad \forall\, \omega \in N^c,$$

such that $P(N) = P(\bigcup_{k\geq 1} D_k) = 0$, where $D_k = \bigcap_{n\geq 1} \bigcup_{m\geq n} [|X_m - X| \geq 1/k]$. As discussed at the beginning of this section,

$$\begin{aligned}
D_k &= \bigcap_{n\geq 1} \bigcup_{m\geq n} \{\omega | |X_m(\omega) - X(\omega)| \geq 1/k\} \\
&= \bigcap_{n\geq 1} B_{nk} \quad \text{where} \quad B_{nk} = \bigcup_{m\geq n} \{\omega||X_m(\omega) - X(\omega)| \geq 1/k\} \\
&= \lim_{n\to\infty} B_{nk}, \qquad \text{as } \{B_{nk}, n \geq 1\} \text{ is a decreasing sequence} \\
\Rightarrow \quad P(D_k) &= P(\lim_{n\to\infty} B_{nk}) \\
&= \lim_{n\to\infty} P(B_{nk}) \quad \text{by continuity theorem for probability measures.}
\end{aligned}$$

Further note that $[|X_n - X| \geq 1/k] \subset B_{nk}$. As a consequence,

$$\begin{aligned}
X_n \xrightarrow{a.s.} X &\Rightarrow P(\bigcup_{k\geq 1} D_k) = 0 \\
&\Rightarrow P(D_k) = 0 \;\; \forall\, k \geq 1 \\
&\Rightarrow P(\lim_{n\to\infty} B_{nk}) = 0, \;\; \forall\, k \geq 1 \\
&\Rightarrow \lim_{n\to\infty} P(B_{nk}) = 0, \;\; \forall\, k \geq 1 \\
&\Rightarrow \lim_{n\to\infty} P[|X_n - X| \geq 1/k] = 0, \;\; \forall\, k \geq 1 \\
&\Rightarrow X_n \xrightarrow{P} X.
\end{aligned}$$

Thus almost sure convergence implies convergence in probability. □

Remark 6.3.2. In general convergence in probability does not imply almost sure convergence. It is illustrated in Example 6.4.5, after the discussion of Borel zero-one law. However, the implication is valid if the sequence $\{X_n, n \geq 1\}$ is a monotone sequence of random variables (Gut [13]).

Following example illustrates that for a monotone sequence of random variables, convergence in probability implies almost sure convergence.

Example 6.3.7. Suppose $(\Omega, \mathbb{A}, P)$ is a probability space, where $\Omega = [0,1]$, $\mathbb{A}$ is a sigma field of subsets of Ω and P is a Lebesgue measure.Thus, $P(A_n) = 1/n$ for $A_n = [0, 1/n)$. Suppose $X_n = I_{A_n}, n \geq 1$. Now $\forall \quad \epsilon > 0$,

$$P[|X_n| > \epsilon] = P[I_{A_n} > \epsilon] = P[I_{A_n} = 1] = P(A_n) = \frac{1}{n} \to 0 \Rightarrow X_n \xrightarrow{P} X \equiv 0.$$

It is clear that $\{A_n, n \geq 1\}$ is a decreasing sequence of sets and hence is convergent with $\lim_{n\to\infty} A_n = \bigcap_{n\geq 1} A_n = \emptyset$. Hence by Theorem 2.3.9

$$I_{A_n} \to I_\emptyset \text{ on } \Omega \quad \iff \quad X_n \to X \equiv 0 \text{ on } \Omega$$

Thus, X_n converges to 0 point-wise and hence almost surely. It is to be noted that $\{X_n, n \geq 1\}$ is a decreasing sequence. Further observe that for any $r > 0$, $E(X_n^r) = 1/n \to 0 \Rightarrow X_n \xrightarrow{r} X \equiv 0$. Note that the distribution function F_n of X_n, given by,

$$F_n(x) = \begin{cases} 0, & \text{if} \quad x < 0 \\ 1 - 1/n, & \text{if} \quad 0 \leq x < 1 \\ 1, & \text{if} \quad x \geq 1. \end{cases} \quad \to \quad F_X(x) = \begin{cases} 0, & \text{if} \quad x < 0 \\ 1, & \text{if} \quad x \geq 0. \end{cases}$$

Thus $F_n(x) \to F_X(x)$ where F_X is the distribution function of $X \equiv 0$. Hence, $X_n \xrightarrow{L} X \equiv 0$. Thus, in this example we have convergence in probability, in law, in r-th mean, point-wise convergence and hence almost sure convergence. □

Almost sure convergence of a sequence of random variables is closed under arithmetic operations in view of the fact that almost sure convergence is a point-wise convergence on N^c. It is proved in the following theorem.

Theorem 6.3.4. *Suppose $\{X_n, n \geq 1\}$, X, $\{Y_n, n \geq 1\}$ and Y are defined on the same probability space $(\Omega, \mathbb{A}, P)$. If $X_n \xrightarrow{a.s.} X$ and $Y_n \xrightarrow{a.s.} Y$, then (i) $X_n \pm Y_n \xrightarrow{a.s.} X \pm Y$. (ii) $X_nY_n \xrightarrow{a.s.} XY$. (iii) $X_n/Y_n \xrightarrow{a.s.} X/Y$, provided X_n/Y_n and X/Y are defined.*

Proof. By definition

$$\begin{aligned} X_n \xrightarrow{a.s.} X \quad &\Rightarrow \quad X_n(\omega) \to X(\omega) \;\; \forall \;\; \omega \in N_1^c \text{ where } P(N_1) = 0 \\ \text{and } Y_n \xrightarrow{a.s.} Y \quad &\Rightarrow \quad Y_n(\omega) \to Y(\omega) \;\; \forall \;\; \omega \in N_2^c \text{ where } P(N_2) = 0. \end{aligned}$$

Suppose $N = N_1 \cup N_2$, then $P(N) = 0$. If $\omega \in N^c = N_1^c \cap N_2^c$, then $\omega \in N_1^c$ and $\omega \in N_2^c$. Hence for $\omega \in N^c$

$$\begin{aligned} X_n(\omega) \pm Y_n(\omega) \quad &\to \quad X(\omega) \pm Y(\omega), \quad X_n(\omega) \times Y_n(\omega) \to X(\omega) \times Y(\omega) \\ \& \; X_n(\omega) \,/ Y_n(\omega) \quad &\to \quad X(\omega)/Y(\omega) \,. \end{aligned}$$

Thus, the almost sure convergence is closed under arithmetic operations. □

The next theorem proves that almost sure convergence is closed under continuous transformation.

Theorem 6.3.5. *Suppose $X_n \stackrel{a.s.}{\to} X$ and $g : \mathbb{R} \to \mathbb{R}$ is a continuous function on $\mathbb{R}$. Then $g(X_n) \stackrel{a.s.}{\to} g(X)$.*

Proof. By the definition, $X_n \stackrel{a.s.}{\to} X \Rightarrow P[X_n \to X] = 1$. Since g is a continuous function on $\mathbb{R}$, it is a Borel function and hence $g(X_n)$ and $g(X)$ are random variables on the same probability space. Continuity of g at every point $a \in \mathbb{R}$ implies that for every sequence $\{x_n, n \geq 1\}$ of real numbers, as $n \to \infty$,

$$\begin{aligned} x_n \to a \quad & \Rightarrow \quad g(x_n) \to g(a) \\ & \Rightarrow \quad \{\omega | g(X_n(\omega)) \to g(X(\omega))\} \supset \{\omega | X_n(\omega) \to X(\omega)\} \\ & \Rightarrow \quad P\{\omega | g(X_n(\omega)) \to g(X(\omega))\} \geq P\{\omega | X_n(\omega) \to X(\omega)\} = 1 \\ & \Rightarrow \quad P\{\omega | g(X_n(\omega)) \to g(X(\omega))\} = 1 \quad \Rightarrow \quad g(X_n) \stackrel{a.s.}{\to} g(X). \end{aligned}$$

□

Following example illustrates Corollary 6.3.1, Theorem 6.3.4 and Theorem 6.3.5.

Example 6.3.8. Suppose $X_n \sim B(n, p)$, $0 < p < 1$ and a random variable Y_n is defined as,

$$Y_n = \begin{cases} X_n/n, & \text{if} \quad X_n \geq 1 \\ 0.001, & \text{if} \quad X_n = 0. \end{cases}$$

Note that Y_n can be expressed as

$$Y_n = (X_n/n)I_{[X_n \neq 0]} + 0.001 \times I_{[X_n = 0]} = (X_n/n)(1 - U_n) + 0.001 \times U_n.$$

where $U_n = I_{[X_n = 0]}$. Now for $0 < \epsilon < 1$,

$$\begin{aligned} P[|U_n| > \epsilon] \quad & = \quad P[U_n = 1] = P[X_n = 0] = q^n, \quad q = 1 - p \\ & \Rightarrow \quad \sum_{n \geq 1} P[|U_n| > \epsilon] = \sum_{n \geq 1} q^n < \infty \\ \text{For } \epsilon \geq 1, P[|U_n| > \epsilon] = 0 \quad & \Rightarrow \quad \sum_{n \geq 1} P[|U_n| > \epsilon] = 0 < \infty \\ & \Rightarrow \quad U_n = I_{[X_n = 0]} \stackrel{a.s.}{\to} 0 \text{ by Corollary 6.3.1} \\ & \Rightarrow \quad 1 - U_n = I_{[X_n \neq 0]} \stackrel{a.s.}{\to} 1 \text{ by Theorem 6.3.5.} \end{aligned}$$

Further, if $X_n \sim B(n, p)$, then the fourth central moment μ_4 is given by,

$\mu_4 = 3(npq)^2 + npq(1-6pq)$ (Johnson et al. [14]). Hence,

$$\begin{aligned}
X_n \sim B(n,p) \quad &\Rightarrow \quad E\left(\frac{X_n - np}{n}\right)^4 = \frac{3(npq)^2 + npq(1-6pq)}{n^4} \\
&\Rightarrow \quad \sum_{n=1}^{\infty} E\left(\frac{X_n - np}{n}\right)^4 = \sum_{n=1}^{\infty}\left[\frac{3(pq)^2}{n^2} + \frac{pq(1-6pq)}{n^3}\right] < \infty \\
&\Rightarrow \quad X_n/n \xrightarrow{a.s.} p \text{ by Corollary 6.3.1} \\
&\Rightarrow \quad (X_n/n)(1-U_n) \xrightarrow{a.s.} p \text{ by Theorem 6.3.4} \\
&\Rightarrow \quad (X_n/n)(1-U_n) + 0.001 \times U_n \xrightarrow{a.s.} p \text{ by Theorem 6.3.4} \\
&\Rightarrow \quad Y_n \xrightarrow{a.s.} p.
\end{aligned}$$

Suppose Z_n is defined as $Z_n = -\log Y_n$. Then by Theorem 6.3.5, $Z_n \xrightarrow{a.s.} -\log p$. If g is any continuous function, then by Theorem 6.3.5, $g(Y_n) \xrightarrow{a.s.} g(p)$. □

Remark 6.3.3. It is to be noted that Y_n defined in Example 6.3.8, can be considered as an estimator of p. In Example 6.3.8, it is proved that Y_n is a strongly consistent estimator of p. Further, $g(Y_n)$ is also a strongly consistent estimator of $g(p)$, where g a continuous function.

As in the setup of sequences of real numbers, we have a concept of Cauchy convergence for a sequence of random variables. We now discuss the Cauchy criterion of almost sure convergence.

Definition 6.3.1. *Almost sure Cauchy convergence: A sequence $\{X_n, n \geq 1\}$ is said to be Cauchy (fundamental) almost surely, if*

$$\forall \ \omega \in N^c, \ \ |X_n(\omega) - X_m(\omega)| \to 0 \quad \text{as} \ \ m,n \to \infty, \text{where} \ \ P(N^c) = 1.$$

In the following theorem, we prove that almost sure convergence and almost sure Cauchy convergence are equivalent.

Theorem 6.3.6. *Suppose a sequence $\{X_n, n \geq 1\}$ of random variables and X are defined on the probability space $(\Omega, \mathbb{A}, P)$. Then*

$$X_n \xrightarrow{a.s.} X \quad \iff \quad \{X_n, n \geq 1\} \quad \text{is Cauchy a.s.}$$

Proof. Suppose $X_n \xrightarrow{a.s.} X$, then there exists a set N with $P(N) = 0$ such that $\forall \ \omega \in N^c$, $X_n(\omega) \to X(\omega)$ as $n \to \infty$. Hence, $\forall \ \omega \in N^c$ and arbitrary positive numbers m and n,

$$|X_n(\omega) - X_m(\omega)| \leq |X_n(\omega) - X(\omega)| + |X_m(\omega) - X(\omega)| \to 0, \quad \text{as} \ \ n,m \to \infty.$$

Consequently, the sequence $\{X_n, n \geq 1\}$ is Cauchy almost surely.
Conversely, assume that the sequence $\{X_n, n \geq 1\}$ is Cauchy almost surely. Thus $\forall \ \omega \in N^c$, $\{X_n(\omega), n \geq 1\}$ is Cauchy sequence of real numbers. Hence, $\exists$ a unique real number $X(\omega)$ such that $X(\omega) = \lim_{n\to\infty} X_n(\omega)$, $\forall \ \omega \in N^c$, which implies that $X_n \xrightarrow{a.s.} X$. □

We now derive a necessary and sufficient condition for the Cauchy almost sure convergence. It is similar to that for almost sure convergence as derived in Equation (6.2).

Theorem 6.3.7. *Suppose $\{X_n, n \geq 1\}$ is a sequence of random variables defined on the probability space $(\Omega, \mathbb{A}, P)$. Then $\{X_n, n \geq 1\}$ is Cauchy a.s. if and only if*

$$\lim_{n\to\infty} P\Big[\bigcup_{m\geq n} [|X_m - X_n| > 1/k]\Big] = 0 \ \ \forall \ \ k \geq 1 \ . \tag{6.4}$$

Proof. By triangular inequality, $|X_m - X_n| \leq |X_m - X| + |X_n - X|$. Hence,

$$\forall \ \ k \geq 1, |X_m - X| \leq 1/2k \ \ \& \ \ |X_n - X| \leq 1/2k \ \ \Rightarrow \ \ |X_m - X_n| \leq 1/k.$$

Thus $\forall \ \ k \geq 1$ and $\forall \ \ m \geq n$,

$$\begin{aligned} \bigcap_{m\geq n} [|X_m - X| \leq 1/2k] \ &\subset \ \bigcap_{m\geq n} [|X_m - X_n| \leq 1/k] \\ \Rightarrow \ P\Big[\bigcap_{m\geq n} [|X_m - X| \leq 1/2k]\Big] \ &\leq \ P\Big[\bigcap_{m\geq n} [|X_m - X_n| \leq 1/k]\Big] \ . \end{aligned}$$

Now,

$$\begin{aligned} \{X_n, n \geq 1\} \ \text{ is Cauchy a.s.} \ &\Longleftrightarrow \ X_n \stackrel{a.s.}{\to} X \\ &\Longleftrightarrow \ \lim_{n\to\infty} P\Big[\bigcap_{m\geq n} [|X_m - X| \leq 1/2k]\Big] = 1 \\ &\Longleftrightarrow \ \lim_{n\to\infty} P\Big[\bigcap_{m\geq n} [|X_m - X_n| \leq 1/k]\Big] = 1 \\ &\Longleftrightarrow \ \lim_{n\to\infty} P\Big[\bigcup_{m\geq n} [|X_m - X_n| > 1/k]\Big] = 0, \end{aligned}$$

$\forall \ \ k \geq 1$. The second step follows by (6.2). □

In the next section, we revisit almost sure convergence when $\{X_n, n \geq 1\}$ is a sequence of independent random variables. In Corollary 6.3.1 we have proved that $X_n \stackrel{a.s.}{\to} X$, if $\forall \ k \geq 1, \sum_{n=1}^{\infty} P[|X_n - X| > 1/k] < \infty$. If $\{X_n, n \geq 1\}$ is a sequence of independent random variables, converse of this result is true. It is a consequence of Borel zero-one law, to be discussed in the next section.

6.4 Borel Zero-One Law

Suppose $\{X_n, n \geq 1\}$ is a sequence of random variables and $A_n \in \sigma(X_n), n \geq 1$. If $\{X_n, n \geq 1\}$ is a sequence of independent random variables, then

$\{A_n, n \geq 1\}$ is a sequence of independent events. Borel zero-one law states that for a sequence $\{A_n, n \geq 1\}$ of independent events, $P(\limsup A_n)$ is 0 or 1, depending on whether $\sum_{n=1}^{\infty} P(A_n)$ is a convergent series or a divergent series. Condition of independence is not needed to prove that $\sum_{n=1}^{\infty} P(A_n) < \infty$ implies that $P(\limsup A_n) = 0$. This result is the famous Borel-Cantelli lemma. Both Borel zero-one law and Borel-Cantelli lemma are stated in terms of a sequence $\{A_n, n \geq 1\}$ of events. For a sequence of random variables $\{X_n, n \geq 1\}$, event A_n is defined appropriately depending on a particular requirement. We begin with a simple but frequently used Borel-Cantelli lemma.

Lemma 6.4.1. *Borel-Cantelli Lemma: Suppose* $\{A_n, n \geq 1\}$ *is a sequence of events defined on* $(\Omega, \mathbb{A}, P)$. *If* $\sum_{n=1}^{\infty} P(A_n) < \infty$, *then* $P(\limsup A_n) = 0$.

Proof. By the definition of a limit supremum of a sequence of sets we have,

$$\limsup A_n = \bigcap_{n\geq 1} \bigcup_{k\geq n} A_k = \bigcap_{n\geq 1} C_n, \quad \text{where } C_n = \bigcup_{k\geq n} A_k$$

$$\text{Further, } C_n = \bigcup_{k\geq n} A_k = A_n \bigcup C_{n+1} \Rightarrow C_n \supseteq C_{n+1} \Rightarrow \lim_{n\to\infty} C_n = \bigcap_{n=1}^{\infty} C_n,$$

$\{C_n, n \geq 1\}$ being a non-increasing sequence of events in $\mathbb{A}$. Now using the result that a tail of the convergent series converges to 0 we have,

$$\begin{aligned} P(\limsup A_n) &= P(\bigcap_{n=1}^{\infty} \bigcup_{k\geq n} A_k) = P(\bigcap_{n=1}^{\infty} C_n) = P(\lim_{n\to\infty} C_n) = \lim_{n\to\infty} P(C_n) \\ &= \lim_{n\to\infty} P(\bigcup_{k\geq n} A_k) \leq \lim_{n\to\infty} \sum_{k\geq n} P(A_k) = 0, \end{aligned}$$

$\sum_{k\geq n} P(A_k)$ being a tail of the convergent series $\sum_{n=1}^{\infty} P(A_n)$. Thus, $\sum_{n=1}^{\infty} P(A_n) < \infty \;\Rightarrow\; P(\limsup A_n) = 0$. □

Converse of this result is not true in general, it is evident from the following example.

Example 6.4.1. Suppose the probability space $(\Omega, \mathbb{A}, P)$ is defined as $\Omega = [0, 1]$, $\mathbb{A}$ is a Borel field of subsets of $[0, 1]$ and P is a Lebesgue measure on $(\Omega, \mathbb{A})$. A sequence $\{A_n, n \geq 1\}$ of sets in $\mathbb{A}$ is defined as $A_n = [0, 1/n]$. It is clear that $\{A_n, n \geq 1\}$ is a decreasing sequence of events, hence $\lim_{n\to\infty} A_n$ exists and is given by,

$$\lim_{n\to\infty} A_n = \limsup_{n\to\infty} A_n = \bigcap_{n=1}^{\infty} A_n = \{0\} \quad \Rightarrow \quad P(\limsup A_n) = P(\{0\}) = 0,$$

but $\sum_{n=1}^{\infty} P(A_n) = \sum_{n=1}^{\infty} 1/n$ is divergent. Thus, $P(\limsup A_n) = 0$, but $\sum_{n=1}^{\infty} P(A_n)$ is not finite. □

In Section 3, we have derived two sufficient conditions for the almost sure convergence. We prove below the same results, using the Borel-Cantelii lemma.

Theorem 6.4.1. *Suppose* $\{X_n, n \geq 1\}$ *is a sequence of random variables on* $(\Omega, \mathbb{A}, P)$. *Then* $X_n \stackrel{a.s.}{\to} X$, *if (i)* $\forall\ \epsilon > 0$,

$$\sum_{n=1}^{\infty} P[|X_n - X| > \epsilon] < \infty \iff \forall\ k \geq 1,\ \sum_{n=1}^{\infty} P[|X_n - X| > 1/k] < \infty.$$

$$(ii) \quad \sum_{n=1}^{\infty} E(|X_n - X|^r) < \infty \ \textit{ for some } r > 0.$$

Proof. (i) By the Borel-Cantelli lemma,

$$\sum_{n=1}^{\infty} P[|X_n - X| > 1/k] < \infty,\ \forall\ k \geq 1 \ \Rightarrow\ P[\limsup[|X_n - X| > 1/k] = 0.$$

Thus by the definition of a limit supremum, $\forall\ \ k \geq 1$,

$$\begin{aligned} P(D_k) &= P\Big(\bigcap_{m\geq 1}\bigcup_{n\geq m}\{\omega||X_n(\omega) - X(\omega)| > 1/k\}\Big) = 0. \\ \Rightarrow &\quad P(\bigcup_{k\geq 1} D_k) \leq \sum_{k\geq 1} P(D_k) = 0. \end{aligned}$$

Hence, $\forall\ \ k \geq 1$

$$\begin{aligned} P(D_k) = 0 \quad &\Rightarrow \quad P\Big[\bigcup_{k\geq 1}\bigcap_{m\geq 1}\bigcup_{n\geq m}\{\omega||X_n(\omega)X(\omega)| > 1/k\}\Big] = 0 \\ &\iff \quad P\Big[\bigcap_{k\geq 1}\bigcup_{m\geq 1}\bigcap_{n\geq m}\{\omega||X_n(\omega) - X(\omega)| < 1/k\}\Big] = 1 \\ &\Rightarrow \quad X_n \stackrel{a.s.}{\to} X. \end{aligned}$$

(ii) By Markov's inequality,

$$\begin{aligned} P[|X_n - X| > \epsilon] &\leq E(|X_n - X|^r)/\epsilon^r \\ \Rightarrow \sum_{n=1}^{\infty} P[|X_n - X| > \epsilon] &\leq \sum_{n=1}^{\infty} E(|X_n - X|^r)/\epsilon^r < \infty \\ \Rightarrow X_n &\stackrel{a.s.}{\to} X \text{ by part (i).} \end{aligned}$$

□

Following two examples illustrate how these conditions are useful to establish almost sure convergence.

Example 6.4.2. Suppose $\{Y_n, n \geq 1\}$ is a sequence of independent and identically distributed random variables each having uniform $U(0, \theta)$ distribution.

Hence the distribution function $F_n(y)$ of $X_n = Y_{(n)} = \max\{Y_1, Y_2, \cdots, Y_n\}$ is given by,

$$F_n(y) = (F(y))^n \quad = \quad \begin{cases} 0, & \text{if} \quad y < 0 \\ (y/\theta)^n, & \text{if} \quad 0 \le y < \theta \\ 1, & \text{if} \quad y \ge \theta. \end{cases}$$

Note that if $\epsilon \ge \theta$, then $P[|X_n - \theta| > \epsilon] = 0$. Suppose $\epsilon < \theta$, then

$$\begin{aligned} P[|X_n - \theta| > \epsilon] &= 1 - P[\theta - \epsilon < X_n < \theta + \epsilon] \\ &= 1 - P[\theta - \epsilon < X_n < \theta] \\ &= 1 - [1 - (\theta - \epsilon)^n/\theta^n] \\ &= (1 - \epsilon/\theta)^n \to 0 \quad \text{as} \quad n \to \infty \Rightarrow X_n \xrightarrow{P} \theta \\ \text{Further} \ \sum_{n=1}^{\infty} P[|X_n - \theta| > \epsilon] &= \sum_{n=1}^{\infty} (1 - \epsilon/\theta)^n < \infty, \quad \forall \, \epsilon < \theta \\ \& \ \sum_{n=1}^{\infty} P[|X_n - \theta| > \epsilon] &= 0 \quad \forall \, \epsilon \ge \theta \Rightarrow X_n \xrightarrow{a.s.} \theta. \quad \square \end{aligned}$$

Example 6.4.3. Suppose $\{X_n, n \ge 1\}$ is a sequence of independent and identically distributed random variables each following normal $N(\mu, \sigma^2)$ distribution. Then $\overline{X}_n \sim N(\mu, \sigma^2/n)$ distribution. Hence,

$$\begin{aligned} Y_n &= \left(\sqrt{n}(\overline{X}_n - \mu)/\sigma\right)^2 \sim \chi_1^2 \\ &\Rightarrow E(Y_n^2) = Var(Y_n) + (E(Y_n))^2 = 2 + 1 = 3 \\ &\Rightarrow \sum_{n=1}^{\infty} E(|\overline{X}_n - \mu|^4) = \sum_{n=1}^{\infty} 3\sigma^4/n^2 < \infty \Rightarrow \overline{X}_n \xrightarrow{a.s.} \mu. \end{aligned}$$

Another approach to show that $E(Y_n^2) = 3$ is based on the cumulant generating function. Note that $U_n = \sqrt{n}(\overline{X}_n - \mu)/\sigma \sim N(0, 1)$. Hence, its moment generating function is $\exp(t^2/2)$. As a consequence, its cumulant generating function is $t^2/2$ which implies the second cumulant k_2 is 1 and the fourth cumulant k_4 is 0. Hence μ_4 of U_n is $\mu_4 = k_4 + 3k_2^2 = 3$. Thus, $E(Y_n^2) = E(U_n^4) = 3$. $\square$

Remark 6.4.1. It is to be noted that in Example 6.4.3, if we take $r = 2$, then $\sum_{n=1}^{\infty} E(|\overline{X}_n - \mu|^2) = \sum_{n=1}^{\infty} \sigma^2/n$, it is not convergent. However, the sufficient condition states that it is enough to have $\sum_{n=1}^{\infty} E(|X_n - x|^r) < \infty$ for some $r > 0$. Here $r = 2$ does not work but $r = 4$ works.

The Borel-Cantelli lemma states that if $\sum_{n=1}^{\infty} P(A_n) < \infty$, then $P(\limsup A_n)$ is 0. We have seen in Example 6.4.1 that the converse is not true in general, but if we impose some additional conditions on A_n's, we get that $\sum_{n=1}^{\infty} P(A_n) = \infty$ implies that $P(\limsup A_n) = 1$. The additional condition requires that the sequence $\{A_n, n \geq 1\}$ is a sequence of independent events. In Example 6.4.1,

$$\begin{aligned} A_2 &= [0,1/2], A_3 = [0,1/3] \ \& \ A_2 \cap A_3 = A_3 \\ \text{Further, } P(A_2 \cap A_3) &= P(A_3) = 1/3 \ \& \ P(A_2)P(A_3) = 1/6 \\ &\Rightarrow A_2 \ \& \ A_3 \text{ are not independent.} \end{aligned}$$

In general,

$$m > n \Rightarrow A_n \cap A_m = [0, 1/m] \Rightarrow P(A_n \cap A_m) = \frac{1}{m} \neq \frac{1}{n}\frac{1}{m} = P(A_n)P(A_m).$$

Hence, the sequence $\{A_n, n \geq 1\}$ is not a sequence of independent events.

In the next example, we note one more scenario of $P(\limsup A_n)$ and $\sum_{n=1}^{\infty} P(A_n)$.

Example 6.4.4. Suppose $\{A_n, n \geq 1\}$ is a sequence of events such that $A_n = A$ for all $n \geq 1$ with $P(A_n) = p, \ 0 < p < 1$. Then

$$\lim_{n\to\infty} A_n = \limsup_{n\to\infty} A_n = A \Rightarrow P(\limsup A_n) = P(A) = p \ \& \sum_{n=1}^{\infty} P(A_n) = \infty.$$

Thus, $P(\limsup A_n) > 0 \ \& \ \sum_{n=1}^{\infty} P(A_n) = \infty$. Note that the sequence $\{A_n, n \geq 1\}$ is not a sequence of independent events, as all A_n's are A and $P(A) \in (0,1)$. □

In the next theorem, known as Borel zero-one law, it is proved that if $\{A_n, n \geq 1\}$ is a sequence of independent events then $\sum_{n=1}^{\infty} P(A_n) = \infty$ implies $P(\limsup A_n) = 1$. We require the following two results in the proof.

Result 6.4.1. *If $\{A_n, n \geq 1\}$ is a sequence of independent events, then a sequence obtained by replacing some or all A_n's by their complements is also a sequence of independent events.*

Result 6.4.2. *For all $x \in [0,1]$, $1 - x \leq e^{-x}$. To prove this result, we define a function $f(x)$ as $f(x) = 1 - x - e^{-x}$. Note that $\frac{d}{dx} f(x) = -1 + e^{-x}$. At $x = 0$ it is 0 and for $x \in (0,1]$, $e^{-x} < 1$. Thus, $\frac{d}{dx} f(x) \leq 0$ for $x \in [0,1]$ and hence $f(x)$ is a decreasing function. Hence, $x > 0 \Rightarrow f(x) \leq f(0) = 0 \Rightarrow 1 - x \leq e^{-x}$.*

Theorem 6.4.2. *Borel zero-one law: Suppose $\{A_n, n \geq 1\}$ is a sequence of independent events. Then $\sum_{n=1}^{\infty} P(A_n) < \infty \Rightarrow P(\limsup A_n) = 0$ and $\sum_{n=1}^{\infty} P(A_n) = \infty \Rightarrow P(\limsup A_n) = 1$.*

Proof. Part (i) is the Borel-Cantelli lemma, which we have already proved and it does not require the assumption of independence. To prove part(ii), note that from the proof of Borel-Cantelli lemma,

$$\begin{aligned} P(\limsup A_n) &= P\big(\bigcap_{n\geq 1}\bigcup_{k\geq n} A_k\big) = P\big(\bigcap_{n\geq 1} C_n\big) = P\big(\lim_{n\to\infty} C_n\big) \\ &= \lim_{n\to\infty} P(C_n) = \lim_{n\to\infty} P\big(\bigcup_{k\geq n} A_k\big) \end{aligned}$$

$$\begin{aligned} \text{Further } \omega \in \bigcup_{k\geq n} A_k &\iff \omega \in A_m, \ \text{ for at least one } \ m \geq n \\ &\iff \omega \in \bigcup_{k=n}^{m} A_k, \ \text{ for at least one } \ m \geq n \\ &\iff \omega \in \bigcup_{m=n}^{\infty}\bigcup_{k=n}^{m} A_k \ \Rightarrow \ \bigcup_{k\geq n} A_k = \bigcup_{m=n}^{\infty}\bigcup_{k=n}^{m} A_k. \end{aligned}$$

$$\text{Observe that } \bigcup_{k=n}^{m+1} A_k = \bigcup_{k=n}^{m} A_k \bigcup A_{m+1} \Rightarrow \Big\{\bigcup_{k=n}^{m} A_k, m \geq n\Big\}$$

is a non-decreasing sequence of events and hence its limit exists. Hence,

$$\lim_{m\to\infty} \bigcup_{k=n}^{m} A_k = \bigcup_{k=n}^{\infty} A_k = \bigcup_{m=n}^{\infty}\bigcup_{k=n}^{m} A_k.$$

As a consequence,

$$\begin{aligned} P(\limsup A_n) &= \lim_{n\to\infty} P\Big(\bigcup_{k\geq n} A_k\Big) = \lim_{n\to\infty} P\Big(\bigcup_{m=n}^{\infty}\bigcup_{k=n}^{m} A_k\Big) \\ &= \lim_{n\to\infty} P\Big(\lim_{m\to\infty}\bigcup_{k=n}^{m} A_k\Big) = \lim_{n\to\infty}\lim_{m\to\infty} P\Big(\bigcup_{k=n}^{m} A_k\Big) \\ &= \lim_{n\to\infty}\lim_{m\to\infty}\Big[1 - P\Big(\Big(\bigcup_{k=n}^{m} A_k\Big)^c\Big)\Big] \\ &= \lim_{n\to\infty}\lim_{m\to\infty}\Big[1 - P\Big(\bigcap_{k=n}^{m} A_k^c\Big)\Big] \\ &= \lim_{n\to\infty}\lim_{m\to\infty}\Big[1 - \prod_{k=n}^{m} P(A_k^c)\Big] \quad \text{by Result 6.4.1} \\ &= \lim_{n\to\infty}\lim_{m\to\infty}\Big[1 - \prod_{k=n}^{m}(1 - P(A_k))\Big] \\ &\geq \lim_{n\to\infty}\lim_{m\to\infty}\Big[1 - \prod_{k=n}^{m} e^{-P(A_k)}\Big] \quad \text{by Result 6.4.2} \\ &= \lim_{n\to\infty}\lim_{m\to\infty}\Big[1 - \exp\Big(-\sum_{k=n}^{m} P(A_k)\Big)\Big]. \end{aligned}$$

Hence,

$$P(\limsup A_n) \geq 1 - \exp\left[-\lim_{n\to\infty}\lim_{m\to\infty}\sum_{k=n}^{m} P(A_k)\right]$$
$$= 1 - \exp\left[-\lim_{n\to\infty} g_n\right],$$

where $g_n = \lim_{m\to\infty}\sum_{k=n}^{m} P(A_k)$. Now it is given that $\sum_{n=1}^{\infty} P(A_n) = \infty$, hence for each $n \geq 1$, $g_n = \infty$, which further implies that $\lim_{n\to\infty} g_n = \infty$. Hence, $P(\limsup A_n) = 1$. □

The first part of Borel zero-one law, is known as the first Borel-Cantelli lemma and the second part of Borel zero-one law is known as the second Borel-Cantelli lemma (Athreya and Lahiri [3]). Borel zero-one law is useful in Chapter 9, in the proof of Kolmogorov's strong law of large numbers.

Borel zero-one law is stated in different words in the following corollaries.

Corollary 6.4.1. *For a sequence $\{A_n, n \geq 1\}$ of independent events,*

$$P(\limsup A_n) = 1 \iff \sum_{n=1}^{\infty} P(A_n) = \infty.$$

Proof. Suppose $\sum_{n=1}^{\infty} P(A_n) = \infty$. Then $P(\limsup A_n) = 1$ follows from part (ii) of Borel zero-one law. Now suppose $P(\limsup A_n) = 1$. We have to prove that $\sum_{n=1}^{\infty} P(A_n) = \infty$. We assume the contrary that $\sum_{n=1}^{\infty} P(A_n) < \infty$. But then by the Borel-Cantelli lemma $P(\limsup A_n) = 0$ which is a contradiction to the given setup that $P(\limsup A_n) = 1$. Thus, if $P(\limsup A_n) = 1$, then $\sum_{n=1}^{\infty} P(A_n) = \infty$. □

Corollary 6.4.2. *For a sequence $\{A_n, n \geq 1\}$ of independent events,*

$$P(\limsup A_n) = 0 \iff \sum_{n=1}^{\infty} P(A_n) < \infty.$$

Proof. Suppose $\sum_{n=1}^{\infty} P(A_n) < \infty$. Then $P(\limsup A_n) = 0$ follows from part (i) of Borel zero-one law. Now suppose $P(\limsup A_n) = 0$. We have to prove that $\sum_{n=1}^{\infty} P(A_n) < \infty$. We assume the contrary that $\sum_{n=1}^{\infty} P(A_n) = \infty$. But then by the Borel zero-one law $P(\limsup A_n) = 1$ which is a contradiction to the given setup that $P(\limsup A_n) = 0$. Thus, if $P(\limsup A_n) = 0$, then $\sum_{n=1}^{\infty} P(A_n) < \infty$. □

Corollary 6.4.3. *Suppose $\{A_n, n \geq 1\}$ is a sequence of independent events such that $A_n \to A$ as $n \to \infty$. Then (i) $P(A) = 0$ or 1. (ii) A is independent of itself.*

Proof. Note that

$$A = \lim_{n\to\infty} A_n = \limsup_{n\to\infty} A_n \Rightarrow P(A) = P(\limsup A_n).$$

Further, $\sum_{n=1}^{\infty} P(A_n)$ is either convergent or divergent. Hence, by the Borel zero-one law,

$$\sum_{n=1}^{\infty} P(A_n) < \infty \Rightarrow P(A) = 0 \quad \& \quad \sum_{n=1}^{\infty} P(A_n) = \infty \Rightarrow P(A) = 1.$$

(ii) Since $P(A) = 0$ or 1 A is independent of itself, as proved in Theorem 5.2.1. □

Corollary 6.3.1 states that

$$\forall \;\; k \geq 1, \quad \sum_{n=1}^{\infty} P[|X_n - X| > 1/k] < \infty \quad \Rightarrow \quad X_n \stackrel{a.s.}{\to} X.$$

Thus, as stated in Remark 6.3.1, if the sequence $\{X_n, n \geq 1\}$ converges completely to a random variable X, then $X_n \stackrel{a.s.}{\to} X$. In the following theorem, we prove that the converse of this result is true under the additional condition that $\{X_n, n \geq 1\}$ is a sequence of independent random variables. Thus, for a sequence of independent random variables complete convergence and the almost sure convergence are equivalent and we get a necessary and sufficient condition for the almost sure convergence.

Theorem 6.4.3. *Suppose $\{X_n, n \geq 1\}$ is a sequence of independent random variables. Then*

$$X_n \stackrel{a.s.}{\to} 0 \quad \Longleftrightarrow \quad \sum_{n=1}^{\infty} P[|X_n| > 1/k] < \infty \quad \forall \;\; k \geq 1$$

$$\textit{or equivalently,} \;\; X_n \stackrel{a.s.}{\to} 0 \quad \Longleftrightarrow \quad \sum_{n=1}^{\infty} P[|X_n| > \epsilon] < \infty \;\; \forall \;\; \epsilon \geq 0.$$

Proof. In Corollary 6.3.1, we have already proved that

$$\sum_{n=1}^{\infty} P[|X_n| > 1/k] < \infty \quad \forall \;\; k \geq 1 \quad \Rightarrow \quad X_n \stackrel{a.s.}{\to} 0.$$

To prove the converse, observe that $\{X_n, n \geq 1\}$ is a sequence of independent random variables implies $\{\sigma(X_n), n \geq 1\}$ is a sequence of independent sigma fields, that is, for every $A_n \in \sigma(X_n)$, $\{A_n, n \geq 1\}$ is a sequence of independent events. In particular, suppose $A_n = [|X_n| > 1/k]$. Then by the Borel zero-one law,

$$P[\limsup[|X_n| > 1/k]] = 0 \quad \Rightarrow \quad \sum_{n=1}^{\infty} P[|X_n| > 1/k] < \infty \, .$$

As in Section 3, suppose an event D_k is defined as

$$D_k = \bigcap_{n \geq 1} \bigcup_{m \geq n} \{\omega | |X_m(\omega)| \geq 1/k\} = \limsup[|X_n| \geq 1/k].$$

Now from the definition of almost sure convergence, we have

$$\begin{aligned} X_n \xrightarrow{a.s.} 0 &\iff P[X_n \nrightarrow 0] = 0 \\ &\iff P\Big[\bigcup_{k\geq 1}\bigcap_{n\geq 1}\bigcup_{m\geq n}[|X_m| \geq 1/k]\Big] = P\Big[\bigcup_{k\geq 1} D_k\Big] = 0 \\ &\iff \forall\ k \geq 1\ \ P(D_k) = 0 \\ &\iff \forall\ k \geq 1\ \ P[\limsup[|X_n| \geq 1/k]] = 0. \end{aligned}$$

Thus,

$$X_n \xrightarrow{a.s.} 0 \Rightarrow \forall\, k \geq 1,\ P[\limsup[|X_n| \geq 1/k]] = 0 \Rightarrow \sum_{n=1}^{\infty} P[|X_n| > 1/k] < \infty.$$

The result in terms of ϵ follows by noting that for every $\epsilon > 0$, there exists a positive integer k such that $1/k < \epsilon$. □

Remark 6.4.2. From Theorem 6.4.3, we note that for a sequence $\{X_n, n \geq 1\}$ of independent random variables, if $\sum_{n=1}^{\infty} P[|X_n| > \epsilon] = \infty\ \ \forall\ \ \epsilon \geq 0$ then $X_n \overset{a.s.}{\nrightarrow} 0$, since $X_n \xrightarrow{a.s.} 0 \ \Rightarrow\ \sum_{n=1}^{\infty} P[|X_n| > \epsilon] < \infty$.

The next example illustrates Remark 6.4.2. It also shows that convergence in probability does not imply almost sure convergence.

Example 6.4.5. Suppose $\{X_n, n \geq 1\}$ is a sequence of independent random variables such that $P[X_n = 1] = 1/n$ and $P[X_n = 0] = 1 - 1/n$. Suppose $X \equiv 0$. Observe that $\forall\ \epsilon > 0$,

$$P[|X_n| > \epsilon] = P[X_n = 1] = 1/n \to 0 \ \Rightarrow\ X_n \xrightarrow{P} 0.$$

$$\text{However, } \sum_{n=1}^{\infty} P[|X_n| > \epsilon] = \sum_{n=1}^{\infty} 1/n = \infty \Rightarrow X_n \overset{a.s.}{\nrightarrow} 0,$$

by Theorem 6.4.3. □

Following example also illustrates Remark 6.4.2 and various modes of convergence.

Example 6.4.6. Suppose in an experiment consisting of independent repeated tosses of a coin, an event A_n occurs if the n-th toss results in a head. Suppose $P(A_n) = p \in (0,1)$. A random variable X_n is defined as $X_n = I_{A_n}$. Thus possible values of X_n are 0 and 1 with probabilities $1-p$ and p respectively. Note that $\forall\ \omega \in \Omega$

$$\liminf_{n\to\infty} X_n(\omega) = 0 \ \&\ \limsup_{n\to\infty} X_n(\omega) = 1 \ \Rightarrow\ \lim_{n\to\infty} X_n(\omega) \text{ does not exist.}$$

Thus X_n does not converge pointwise. Further, note that in view of independence of tosses, $\{A_n, n \geq 1\}$ is a sequence of independent events and

$\{X_n, n \geq 1\}$ is a sequence of independent random variables. If the coin is fair, $P(A_n) = 1/2$. By Borel zero-one law,

$$\begin{aligned} P(A_n) &= 1/2 \ \forall \ n \ \Rightarrow \ \sum_{n=1}^{\infty} P(A_n) = \infty \ \Rightarrow \ P(\limsup A_n) = 1 \\ &\Rightarrow \ P[X_n = 1 \ \text{ infinitely often}] = 1 \\ P(A_n^c) &= 1/2 \ \forall \ n \ \Rightarrow \ \sum_{n=1}^{\infty} P(A_n^c) = \infty \ \Rightarrow \ P(\limsup A_n^c) = 1 \\ &\Rightarrow \ P[X_n = 0 \ \text{ infinitely often}] = 1. \end{aligned}$$

Thus, with probability 1, the sequence X_n oscillates between 0 and 1. Suppose now that the coin is not fair and $P(A_n) = p \in (0,1)$. In this setup also $\sum_{n=1}^{\infty} P(A_n) = \infty$ and hence $P(\limsup A_n) = 1$. Similarly, $\sum_{n=1}^{\infty} P(A_n^c) = \infty$ and hence $P(\limsup A_n^c) = 1$. It is to be noted that however small p may be, probability that head will occur infinitely often is 1 and probability that tail will occur infinitely often is 1. Observe that $\forall \ \epsilon > 0$,

$$\begin{aligned} P[|X_n| > \epsilon] &= P[X_n = 1] = P(A_n) = p \nrightarrow 0 \quad \Rightarrow \quad X_n \overset{P}{\nrightarrow} 0 \\ P[|X_n - 1| > \epsilon] &= P[X_n = 0] = P(A_n^c) = 1 - p \nrightarrow 0 \quad \Rightarrow \quad X_n \overset{P}{\nrightarrow} 1 \\ \sum_{n \geq 1} P[|X_n| > \epsilon] &= \infty \quad \Rightarrow \quad X_n \overset{a.s}{\nrightarrow} 0 \\ \sum_{n \geq 1} P[|X_n - 1| > \epsilon] &= \infty \quad \Rightarrow \quad X_n \overset{a.s}{\nrightarrow} 1 \\ E(|X_n|^r) &= p \nrightarrow 0 \quad \Rightarrow \quad X_n \overset{r}{\nrightarrow} 0 \\ E(|X_n - 1|^r) &= 1 - p \nrightarrow 0 \quad \Rightarrow \quad X_n \overset{r}{\nrightarrow} 1. \end{aligned}$$

Note that the distribution function $F_n(\cdot)$ of X_n is given by,

$$F_n(x) = \begin{cases} 0, & \text{if} \quad x < 0 \\ 1 - p, & \text{if} \quad 0 \leq x < 1 \\ 1, & \text{if} \quad x \geq 1. \end{cases}$$

Observe that $F_n(\cdot)$ is the same $\forall \ n \geq 1$ and hence $F_n(x) \to F(x) \ \forall \ x \in \mathbb{R}$, where F is the same as F_n. Thus, $X_n \xrightarrow{L} X$, where the distribution function of X is F. Thus, X_n does not converge pointwise, completely, almost surely, in probability or in r-th mean to either 0 or 1, but converges in law, which is the weakest mode of convergence. It is to be noted that if $P(A_n) = c/n^2, n \geq 1$, where $c = (\sum_{n \geq 1} 1/n^2)^{-1}$, then X_n converges completely, almost surely, in probability and in r-th mean to 0. □

Remark 6.4.3. In Section 3, we have shown that $X_n \xrightarrow{a.s.} X$ implies that

$$P[\limsup[|X_n - X| \geq 1/k]] = P[|X_n - X| \geq 1/k, \ \text{ infinitely often}] = 0.$$

Thus, almost sure convergence is concerned with $\limsup A_n$ where A_n is defined as $A_n = [|X_n - X| \geq 1/k]$. In Section 2.3, we have noted that for point-wise convergence, $\limsup X_n$, $\liminf X_n$ and $\lim X_n$, whenever it exists, are tail events. Almost sure convergence is a point-wise convergence on a set with probability 1. Borel zero-one law states that for a sequence of independent events $P(\limsup A_n) = 0$ or 1, where $\limsup A_n$ is a tail event. Thus, Borel zero-one law and its corollaries have relation with Kolmogorov zero-one law, proved in Section 5.4. However, Borel zero-one law gives a precise criterion for $P(\limsup A_n)$ to be 0 or 1 and its proof is comparatively more simple.

Following examples support the comments in Remark 6.4.3.

Example 6.4.7. Suppose $\{A_n, n \geq 1\}$ is a sequence of independent events. Then $\{I_{A_n}, n \geq 1\}$ is a sequence of independent random variables. Hence, $\limsup I_{A_n}$, $\liminf I_{A_n}$ and $\lim I_{A_n}$, if it exists, are tail measurable random variables and hence by Kolmogorov zero-one law, are degenerate random variables. It is known that,

$$\limsup I_{A_n} = I_{\limsup A_n}, \quad \liminf I_{A_n} = I_{\liminf A_n} \quad \& \quad \lim I_{A_n} = I_{\lim A_n}.$$

Hence, $I_{\limsup A_n}$ is either 0 almost surely or 1 almost surely, which implies that $P(\limsup A_n)$ is either 0 or 1. Similarly, $P(\liminf A_n)$ is 0 or 1 and $P(\lim A_n)$ is 0 or 1, if $\lim A_n$ exists. Thus, the probabilities of tail events are either 0 or 1. Borel zero-one law gives a criterion for $P(\limsup A_n)$ to be 0 or 1. □

Example 6.4.8. Suppose $\{A_n, n \geq 1\}$ is a sequence of events. From Borel-Cantelli lemma,

$$\begin{aligned}\sum_{n=1}^{\infty} P(A_n) < \infty \quad &\Rightarrow \quad P(\limsup A_n) = 0 \\ &\Rightarrow \quad P(\liminf A_n) = 0, \quad \text{as} \quad \liminf A_n \subset \limsup A_n.\end{aligned}$$

Similarly,

$$\begin{aligned}\sum_{n=1}^{\infty} P(A_n^c) < \infty \quad &\Rightarrow \quad P(\limsup A_n^c) = 0 \\ &\Rightarrow \quad P((\liminf A_n)^c) = 0, \quad \Rightarrow \quad P(\liminf A_n) = 1.\end{aligned}$$

However, both $\sum_{n=1}^{\infty} P(A_n)$ and $\sum_{n=1}^{\infty} P(A_n^c)$ cannot be convergent, since if both are convergent then their sum which is a series of constant 1 will be convergent. Thus, if $\sum_{n=1}^{\infty} P(A_n) < \infty$ then $\sum_{n=1}^{\infty} P(A_n^c) = \infty$. If $\{A_n, n \geq 1\}$ is a sequence of independent events, then $\{A_n^c, n \geq 1\}$ is also a sequence of independent events and hence $P(\limsup A_n^c) = 1 \Rightarrow P(\liminf A_n) = 0$. It is to be noted that $\limsup A_n$ and $\liminf A_n$ are tail events and their probabilities are 0 or 1, supporting Kolmogorov zero-one law. In particular, suppose

$P(A_n) = 1/n^2$, then $\sum_{n=1}^{\infty} P(A_n) < \infty$ and hence $P(\limsup A_n) = 0$. However, $\sum_{n=1}^{\infty} P(A_n^c) = \infty$. Suppose $\{A_n, n \geq 1\}$ is a sequence of independent events with $P(A_n) = 1/2$. Then both $\sum_{n=1}^{\infty} P(A_n)$ and $\sum_{n=1}^{\infty} P(A_n^c)$ are divergent. Hence, by Borel zero-one law

$$P(\limsup A_n) = 1 \;\; \& \;\; P(\limsup A_n^c) = 1 \;\; \Rightarrow \;\; P(\liminf A_n) = 0$$

and again the probabilities of tail events are 0 or 1. It is to be noted here that $P(\limsup A_n) = 1$ & $P(\liminf A_n) = 0$ and hence $\limsup A_n \neq \liminf A_n$ and hence $\lim A_n$ does not exist. □

The next example illustrates an application of the Borel-Cantelli lemma in asymptotic inference.

Example 6.4.9. Suppose $X \sim N(\theta, 1)$ distribution, where $\theta \in \Theta = I$, a set of integers. It can be shown that (Deshmukh and Kulkarni [11]), the maximum likelihood estimator $\hat{\theta}_n$ of θ is given by,

$$\hat{\theta}_n = k \quad \text{if} \quad \overline{X}_n \in [k - 1/2, k + 1/2), \quad \text{where} \quad k \in I.$$

$\hat{\theta}_n$ can also be expressed as $\hat{\theta}_n = [\overline{X}_n + 1/2]$, integer part of $\overline{X}_n + 1/2$. Using the sufficient condition (i) of almost sure convergence as given in Theorem 6.4.1, we examine whether $\hat{\theta}_n \xrightarrow{a.s.} \theta$. We first find the probability mass function of $\hat{\theta}_n$ as follows. Suppose $\Phi(\cdot)$ denotes the distribution function of the standard normal distribution. For $k \in I$,

$$\begin{aligned} P_\theta[\hat{\theta}_n = \theta] &= P_\theta[\theta - 1/2 \leq \overline{X}_n < \theta + 1/2] \\ &= \Phi(\sqrt{n}(\theta + 1/2 - \theta)) - \Phi(\sqrt{n}(\theta - 1/2 - \theta)) \\ &= \Phi(\sqrt{n}/2) - \Phi(-\sqrt{n}/2) \\ &= 1 - 2\Phi(-\sqrt{n}/2) \;\; \forall \;\; \theta \in I. \end{aligned}$$

It is to be noted that the probability mass function of $\hat{\theta}_n$ is the same for all $\theta \in I$. To examine the convergence of the series $\sum_{n \geq 1} P_\theta[|\hat{\theta}_n - \theta| > \epsilon]$, we approximate the expression of $P_\theta[\hat{\theta}_n = \theta]$ as follows. We use the result that for $x > 0$ and sufficiently large, $\Phi(-x) = \phi(-x)/x$, where $\phi(\cdot)$ denotes the probability density function of the standard normal distribution. With such a relation we have,

$$\begin{aligned} P_\theta[\hat{\theta}_n \neq \theta] &= 2\Phi(-\sqrt{n}/2) = 2\phi\left(\sqrt{n}/2\right)(2/\sqrt{n}) = (4/\sqrt{2\pi})(e^{-n/8}/\sqrt{n}) \\ &\Rightarrow \sum_{n \geq 1} P_\theta[\hat{\theta}_n \neq \theta] = (4/\sqrt{2\pi}) \sum_{n \geq 1} (e^{-n/8}/\sqrt{n}) < \infty \\ &\Rightarrow \sum_{n \geq 1} P_\theta[|\hat{\theta}_n - \theta| > \epsilon] \leq \sum_{n \geq 1} P_\theta[\hat{\theta}_n \neq \theta] < \infty, \;\; \forall \;\; \epsilon > 0. \end{aligned}$$

Convergence of the series follows from the ratio test. Thus, $\hat{\theta}_n \to \theta$ completely and hence by the sufficient condition of almost sure convergence as given in Theorem 6.4.1, $\hat{\theta}_n \stackrel{a.s.}{\to} \theta$.
Suppose an event A_n is defined as $A_n = \{\omega | \hat{\theta}_n(\omega) \neq \theta\}$, then $\sum_{n\geq 1} P(A_n) < \infty$. Now by the Borel-Cantelli lemma,

$$\sum_{n\geq 1} P(A_n) < \infty \;\;\Rightarrow\;\; P(\limsup A_n) = 0 \;\;\Longleftrightarrow\;\; P(\liminf A_n^c) = 1$$

$$\Rightarrow\;\; P_\theta\{\omega | \hat{\theta}_n(\omega) = \theta, \;\; \forall \;\; n \geq n_0(\omega)\} = 1. \qquad \square$$

In Example 6.4.9, we have noted that for large n, $\hat{\theta}_n = \theta$ almost surely. In the next example using Code 6.4.1, we find $\hat{\theta}_n$ for various values of n and examine whether it remains the same after a stage. We also find n_0 for which the series $\sum_{n\geq 1} P_\theta[\hat{\theta}_n \neq \theta]$ stabilizes.

Example 6.4.10. Suppose $X \sim N(\theta, 1)$ distribution, where $\theta \in \Theta = I$. In Example 6.4.9, we have noted that $\hat{\theta}_n = [\overline{X}_n + 1/2]$. Using the following code, we find $\hat{\theta}_n$ for various values of n by simulating random samples from $N(\theta, 1)$. We take $\theta = -2$.

Code 6.4.1.
```
th=-2; mle=p=c(); n=c(10,30,50,70,90,110,130,150)
for(i in 1:length(n))
{
x=rnorm(n[i],th,1)
mle[i]=floor(mean(x)+.5)
p[i]=2*pnorm(-.5*n[i]^(.5))
}
d=round(data.frame(n,mle,p),4);d
# Using the following code, we compute the series
# and note when it stabilizes.
p=function(n)
{
a=c()
for(k in 1:n){a[k]=2*pnorm(-.5*k^(.5))}
p=sum(a);return(p)
}
N=seq(50,150,20); s=c()
for(j in N){s[j]=p(j)}
s1=s[N];d=round(data.frame(N,s1),4);d
```

The output from data frame d is organized in Table 6.2. From Table 6.2, we note that for all n, even for $n = 10$, $\hat{\theta}_n = \theta = -2$. The third row displays $P_\theta[\hat{\theta}_n \neq \theta]$ and we observe that for $n > 50$ it is almost 0.

TABLE 6.2
MLE of θ for $N(\theta, 1)$ Distribution, $\theta \in I$

n	10	30	50	70	90	110	130	150
$\hat{\theta}_n$	-2	-2	-2	-2	-2	-2	-2	-2
$P_\theta[\hat{\theta}_n \neq \theta]$	0.1138	0.0062	0.0004	0	0	0	0	0

TABLE 6.3
$\sum_{n=1}^{N} P_\theta[\hat{\theta}_n \neq \theta]$

N	50	70	90	110	130	150
$\sum_{n=1}^{N} P_\theta[\hat{\theta}_n \neq \theta]$	3.5796	3.5823	3.5825	3.5825	3.5825	3.5825

Table 6.3 displays the values of the convergent series $\sum_{n\geq 1} P_\theta[\hat{\theta}_n \neq \theta]$. Note that it stabilizes at 3.5825 for $n = 90$. Such a computation reveals that the estimator $\hat{\theta}_n$ is a good estimator, in the sense that its value will be very close to the value of θ which generated the sample. □

Following is a quick recap of the results discussed in this chapter.

Summary

1. Almost sure convergence: $X_n \xrightarrow{a.s.} X$, if as $n \to \infty$, $X_n(\omega) \to X(\omega)$ $\forall\ \omega \in N^c$, such that $P(N) = 0$.
2. Convergence in probability: $X_n \xrightarrow{P} X$, if $\forall\ \epsilon > 0$, $P[|X_n - X| < \epsilon] \to 1$ as $n \to \infty$.
3. Convergence in law: $X_n \xrightarrow{L} X$, if $F_n(x) = P[X_n \leq x] \to P[X \leq x] = F(x)$, $\forall\ x \in C(F_X)$, where $C(F_X)$ is a set of points of continuity of the distribution function of X.
4. Convergence in r-th mean: $X_n \xrightarrow{r} X$ if $E(|X_n - X|^r) \to 0$ as $n \to \infty$, $r > 0$, where $\forall\ n \geq 1, X_n\ \&\ X \in L_r$ space. If $r = 2$ then the convergence is referred to as convergence in quadratic mean.
5. For almost sure convergence, a sequence $\{X_n, n \geq 1\}$ of random variables and X are defined on the same probability space.
6. For almost sure convergence, convergence and Cauchy convergence are equivalent.
7. $X_n \xrightarrow{a.s.} X$ if and only if $\lim_{n\to\infty} P\Big[\bigcup_{m\geq n}[|X_m - X| > 1/k]\Big] = 0\ \ \forall\ k \geq 1$.
8. $X_n \xrightarrow{a.s.} X$ if (i) $\sum_{n=1}^{\infty} P[|X_n - X| \geq 1/k] < \infty\ \ \forall\ \ k \geq 1$ or (ii) if $\sum_{n=1}^{\infty} E(X_n - X)^r < \infty$ for some $r > 0$, where $\forall\ \ n \geq 1$, $X_n\ \&\ \ X \in \mathbb{L}_r$.

9. In almost sure convergence, the limit random variable is almost surely unique.

10. $X_n \stackrel{a.s.}{\to} X \quad \Rightarrow \quad X_n \stackrel{P}{\to} X$.

11. $X_n \stackrel{P}{\to} X \quad \Longleftrightarrow \quad X_n \stackrel{a.s.}{\to} X$ if the sequence $\{X_n, n \geq 1\}$ is monotone.

12. Almost sure convergence is closed under arithmetic operations.

13. Almost sure convergence is closed under continuous transformation.

14. Borel-Cantelli lemma: Suppose $\{A_n, n \geq 1\}$ is a sequence of events defined on $(\Omega, \mathbb{A}, P)$. If $\sum_{n=1}^{\infty} P(A_n) < \infty$ then $P(\limsup A_n) = 0$.

15. Borel zero-one law: Suppose $\{A_n, n \geq 1\}$ is a sequence of independent events. (i) If $\sum_{n=1}^{\infty} P(A_n) < \infty$, then $P(\limsup A_n) = 0$.
(ii) If $\sum_{n=1}^{\infty} P(A_n) = \infty$, then $P(\limsup A_n) = 1$.

16. For a sequence $\{A_n, n \geq 1\}$ of independent events, $P(\limsup A_n) = 1$, if and only if $\sum_{n=1}^{\infty} P(A_n) = \infty$ and $P(\limsup A_n) = 0$, if and only if $\sum_{n=1}^{\infty} P(A_n) < \infty$.

17. Suppose $\{A_n, n \geq 1\}$ is a convergent sequence of independent events such that $A_n \to A$ as $n \to \infty$. Then $P(A) = 0$ or 1 according as $\sum_{n=1}^{\infty} P(A_n) < \infty$ or $\sum_{n=1}^{\infty} P(A_n) = \infty$.

18. Suppose $\{X_n, n \geq 1\}$ is a sequence of independent random variables. Then

$$X_n \stackrel{a.s.}{\to} 0 \quad \Longleftrightarrow \quad \sum_{n=1}^{\infty} P[|X_n| > 1/k] < \infty \quad \forall \ k \geq 1.$$

6.5 Conceptual Exercises

6.5.1 Suppose $\{X_n, n \geq 1\}$ is a sequence of random variables such that $P[X_n = \pm n] = 1/2$. Examine if $X_n \stackrel{a.s.}{\to} 0$.

6.5.2 Suppose $\{X_n, n \geq 1\}$ is a sequence of random variables defined on the probability space $(\Omega, \mathbb{A}, P)$ where $\Omega = [0, 1]$, $\mathbb{A}$ is a sigma field of subsets of $[0, 1]$ and $P((a, b)) = b - a$ for $(a, b) \in \mathbb{A}$. Suppose $\{X_n\}$ is defined as

$$X_n(\omega) = \begin{cases} 1, & \text{if } 0 \leq \omega < (n+1)/2n \\ 0, & \text{otherwise.} \end{cases}$$

X is a random variable defined as $X(\omega) = 1$ if $0 \leq \omega < 1/2$ and 0 otherwise. Examine whether $X_n \stackrel{a.s.}{\to} X$.

6.5.3 Suppose $\{X_n, n \geq 1\}$ is a sequence of continuous random variables with support $[-1/n, 1/n]$ and with probability density function given by,

$$f_{X_n}(x) = \begin{cases} n/2, & \text{if } x \in [-1/n, 1/n] \\ 0, & \text{if } x \notin [-1/n, 1/n]. \end{cases}$$

Examine whether the sequence $\{X_n, n \geq 1\}$ converges almost surely.

6.5.4 Suppose $X_n \sim G(n, n)$. Examine if $X_n \xrightarrow{a.s.} 1$.

6.5.5 Suppose $X_n \xrightarrow{a.s.} X$. Show that $\liminf X_n$ and $\limsup X_n$ are equivalent random variables.

6.5.6 Suppose $(\Omega, \mathbb{A}, P)$ is a probability space, where $\Omega = [0, 1]$, $\mathbb{A}$ is a sigma field of subsets of Ω and P is a Lebesgue measure on $(\Omega, \mathbb{A})$. Suppose $A_n = [1/n, 1]$ and $X_n = I_{A_n}, n \geq 1$. Examine whether X_n converges point-wise, almost surely and in probability to 1. Examine whether $\{X_n, n \geq 1\}$ is a monotone sequence of random variables.

6.5.7 Suppose $\{X_n, n \geq 1\}$ is a sequence of random variables. If $X_n \xrightarrow{a.s.} 0$, show that $\overline{X}_n \xrightarrow{a.s.} 0$.

6.5.8 Suppose $(\Omega, \mathbb{A}, P)$ is a probability space where $\Omega = \{1, 2, \cdots, \}$ and probability measure P assigns probability $1/(k(k+1))$ to a set $\{k\}, k = 1, 2, \cdots$.
(i) Define $\mathbb{A}$ appropriately so that sets $\{k\}$, $k = 1, 2, \cdots$, are in $\mathbb{A}$. If $A_n = \{n, n+1, \cdots\}$, show that it is in $\mathbb{A}$. (ii) Find $P(A_n)$ and $\lim_{n\to\infty} P(A_n)$. (iii) Show that $\{A_n, n \geq 1\}$ is a convergent sequence and find its limit A and $P(A)$. (iv) Verify whether converse of the Borel-Cantelli lemma is true for this sequence of sets. If not, why? Justify your answer.

6.5.9 If $\{A_n, n \geq 1\}$ is a sequence of independent events converging to A, prove that A is independent of itself.

6.5.10 Prove or disprove: If $\sum_{n\geq 1} P(A_n) < \infty$ then $P(\liminf A_n) = 0$.

6.5.11 Suppose $(\Omega, \mathbb{A}, P)$ is a probability space and $\{A_n, n \geq 1\}$ is a sequence of independent events from $\mathbb{A}$ such that $P(A_n) < 1 \;\forall\; n \geq 1$. Show that
(i) $P(\bigcup_{i=1}^n A_i) = 1 - \prod_{i=1}^n (1 - P(A_i))$,
(ii) $P(\bigcup_{i=1}^n A_i) \geq 1 - \exp(-\sum_{i=1}^n P(A_i))$,
(iii) $\sum_{n\geq 1} P(A_n) = \infty \Rightarrow P(\bigcup_{n\geq 1} A_n) = 1$ and
(iv) $P(\limsup A_n) = 1 \iff P(\bigcup_{n\geq 1} A_n) = 1$.

6.5.12 Suppose $(\Omega, \mathbb{A}, P)$ is a probability space and $\{A_n, n \geq 1\}$ is a sequence of independent events from $\mathbb{A}$. If $\sum_{n\geq 1} P(A \cap A_n) = \infty$ for any $A \in \mathbb{A}$ such that $P(A) > 0$, then show that $P(\limsup A_n) = 1$.

6.6 Computational Exercises

6.6.1 Suppose a probability space $(\Omega, \mathbb{A}, P)$ is defined as $\Omega = [0, \infty)$ and $\mathbb{A}$ is a sigma field of subsets of Ω. Suppose a sequence $\{X_n, n \geq 1\}$ of $\mathbb{A}$ measurable random variables and a $\mathbb{A}$ measurable random variable X are defined as follows.

$$X_n(\omega) = \exp(-n\omega), \ n \geq 1 \ \ \& \ \ X(\omega) = 0 \quad \forall \ \ \omega \in \Omega.$$

(i) Examine graphically if $X_n \stackrel{a.s.}{\to} X$, if $P(\{0\}) = 0$ and $P(\{0\}) > 0$.

6.6.2 Suppose $X \sim N(\theta, 1)$ distribution, where $\theta \in \Theta = [2, 7]$. Find the maximum likelihood estimator $\hat{\theta}_n$ for various values of n and examine whether it remains the same after a stage. Also find n_0 for which the series $\sum_{n \geq 1} P_\theta[\hat{\theta}_n \neq \theta]$ stabilizes.

6.7 Multiple Choice Questions

Note: In each of the questions, multiple options may be correct. Unless specified otherwise, identify which of the statement(s) is/are correct. Answers are given in Chapter 11, after the solutions of conceptual exercises of Chapter 6.

6.7.1 $X_n \stackrel{a.s.}{\to} X$ implies

(a) $P[\limsup[|X_n - X| \geq 1/k]] = 0$
(b) $P[\limsup[|X_n - X| \geq 1/k]] = 1$
(c) $P[\liminf[|X_n - X| \geq 1/k]] = 0$
(d) $P[\liminf[|X_n - X| \geq 1/k]] = 1$

6.7.2 $X_n \stackrel{a.s.}{\to} X$ implies

(a) $P[\liminf[|X_n - X| < 1/k]] = 1$
(b) $P[\liminf[|X_n - X| < 1/k]] = 0$
(c) $P[\limsup[|X_n - X| < 1/k]] = 1$
(d) $P[\limsup[|X_n - X| < 1/k]] = 0$

6.7.3 $X_n \stackrel{a.s.}{\to} X$ if

(a) $\sum_{n=1}^{\infty} P[|X_n - X| \geq 1/k] = \infty \ \forall \ k \geq 1$
(b) $\sum_{n=1}^{\infty} P[|X_n - X| \geq 1/k] < \infty \ \forall \ k \geq 1$
(c) $\sum_{n=1}^{\infty} P[|X_n - X| \leq 1/k] < \infty \ \forall \ k \geq 1$
(d) $\sum_{n=1}^{\infty} P[|X_n - X| \leq 1/k] = \infty \ \forall \ k \geq 1$

6.7.4 Suppose for all $n \geq 1$, X_n and X belong to $\mathbb{L}_r$, then a sufficient condition for $X_n \xrightarrow{a.s.} X$ is $\sum_{n=1}^{\infty} E(X_n - X)^r < \infty$. Following are two statements. (I) The condition should be satisfied for all $r > 0$. (II) It is enough if the condition is satisfied for some $r > 0$.

(a) Both (I) and (II) are true
(b) Both (I) and (II) are false
(c) (I) is true but (II) is false
(d) (I) is false but (II) is true

6.7.5 Suppose $\{A_n, n \geq 1\}$ is a sequence of events on a probability space $(\Omega, \mathbb{A}, P)$. Then

(a) $P(\limsup A_n) = 0 \;\Rightarrow\; \sum_{n\geq 1} P(A_n) < \infty$
(b) $P(\limsup A_n) = 1 \;\Rightarrow\; \sum_{n\geq 1} P(A_n) = \infty$
(c) $\sum_{n\geq 1} P(A_n) < \infty \;\Rightarrow\; P(\limsup A_n) = 0$
(d) $\sum_{n\geq 1} P(A_n) = \infty \;\Rightarrow\; P(\limsup A_n) = 1$

6.7.6 Suppose $\{A_n, n \geq 1\}$ is a sequence of events on a probability space $(\Omega, \mathbb{A}, P)$. Then

(a) $P(\limsup A_n) = 0 \;\Rightarrow\; \sum_{n\geq 1} P(A_n) < \infty$
(b) $P(\limsup A_n) = 1 \;\Rightarrow\; \sum_{n\geq 1} P(A_n) = \infty$
(c) $\sum_{n\geq 1} P(A_n) < \infty \;\Rightarrow\; P(\liminf A_n) = 0$
(d) $\sum_{n\geq 1} P(A_n) = \infty \;\Rightarrow\; P(\liminf A_n) = 1$

6.7.7 Suppose $\{A_n, n \geq 1\}$ is a sequence of independent events on a probability space $(\Omega, \mathbb{A}, P)$. Then

(a) $P(\limsup A_n) = 0 \;\Rightarrow\; \sum_{n\geq 1} P(A_n) < \infty$
(b) $P(\limsup A_n) = 1 \;\Rightarrow\; \sum_{n\geq 1} P(A_n) = \infty$
(c) $\sum_{n\geq 1} P(A_n) < \infty \;\Rightarrow\; P(\limsup A_n) = 0$
(d) $\sum_{n\geq 1} P(A_n) = \infty \;\Rightarrow\; P(\limsup A_n) = 1$

6.7.8 Suppose $\{A_n, n \geq 1\}$ is a sequence of independent events on a probability space $(\Omega, \mathbb{A}, P)$. Then

(a) $P(\limsup A_n) = 0 \;\Rightarrow\; \sum_{n\geq 1} P(A_n) < \infty$
(b) $P(\liminf A_n) = 1 \;\Rightarrow\; \sum_{n\geq 1} P(A_n) = \infty$
(c) $\sum_{n\geq 1} P(A_n) < \infty \;\Rightarrow\; P(\liminf A_n) = 0$
(d) $\sum_{n\geq 1} P(A_n) = \infty \;\Rightarrow\; P(\limsup A_n) = 1$

7

Convergence in Probability, in Law and in *r*-th Mean

7.1 Introduction

Suppose $\{X_n, n \geq 1\}$ is a sequence of random variables defined on a probability space $(\Omega, \mathbb{A}, P)$. In Chapter 6, we defined different modes of convergence and discussed almost sure convergence in detail and studied Borel zero-one law. The present chapter is devoted to the remaining three modes of convergence, namely, convergence in probability, convergence in law and convergence in r-th mean. In all these modes of convergence, except convergence in law, it is assumed that a sequence $\{X_n, n \geq 1\}$ of random variables and the limit random variable X are defined on the same probability space $(\Omega, \mathbb{A}, P)$.

In large sample inference, the weak consistency of an estimator is defined in terms of convergence in probability. A sequence of estimators $\{T_n, n \geq 1\}$ of a parameter θ, is said to be consistent for θ if $T_n \xrightarrow{P_\theta} \theta, \ \forall \ \theta \in \Theta$ as $n \to \infty$. The weak law of large numbers, which we discuss in Chapter 9, is also concerned with convergence in probability. In the next section, we study some of its properties and illustrate with conceptual as well as numerical examples using R.

7.2 Convergence in Probability

We have already defined convergence in probability in Section 6.2. However, for ready reference, it is given below.

Definition 7.2.1. *Convergence in probability:* $X_n \xrightarrow{P} X$, *if*

$$\forall \ \epsilon > 0, \quad \lim_{n \to \infty} P[|X_n - X| < \epsilon] = 1 \quad \iff \quad \lim_{n \to \infty} P[|X_n - X| \geq \epsilon] = 0,$$

or equivalently, given

$$\delta > 0 \ \& \ k \geq 1, \ \exists \ n_0(\delta, k) \ \textit{such that} \ \forall \ n \geq n_0(\delta, k), \ P[|X_n - X| > 1/k] < \delta.$$

DOI: 10.1201/9781032619057-7

We begin with some examples to illustrate the concept of convergence in probability.

Example 7.2.1. If $\{A_n, n \geq 1\}$ is a sequence of events such that $P(A_n) \to 0$, then for any real random variable X and for all $\epsilon > 0$,

$$P[|XI_{A_n}| > \epsilon] \leq P[I_{A_n} = 1] = P(A_n) \to 0 \quad \Rightarrow \quad XI_{A_n} \xrightarrow{P} 0. \qquad \square$$

Example 7.2.2. Suppose as in Example 6.3.1, the probability space $(\Omega, \mathbb{A}, P)$ is defined as follows. $\Omega = [0, 1]$, $\mathbb{A}$ is sigma field of subsets of Ω and for $A = (a, b) \subseteq [0, 1]$, $P(A) = b - a$. Suppose a random variable X and a sequence of random variables $\{X_n, n \geq 1\}$ are defined as $X_n(\omega) = e^\omega + \omega^n$, $n \geq 1$ and $X(\omega) = e^\omega$. In Example 6.3.1, we have noted that the sequence of random variables $\{X_n, n \geq 1\}$ does not converge point-wise to X but $X_n \xrightarrow{a.s.} X$ with $\{1\}$ as a P-null set. In Section 6.3, it is proved that almost sure convergence implies convergence in probability. Thus, $X_n \xrightarrow{P} X$. We now examine it using the definition of convergence in probability. Observe that

$$P[|X_n - X| < \epsilon] = P[|\omega^n| < \epsilon] = P[\omega^n < \epsilon] \quad = \quad \begin{cases} 1, & \text{if } \epsilon > 1 \\ \epsilon^{1/n} \to 1, & \text{if } \epsilon \leq 1. \end{cases}$$

Hence, $X_n \xrightarrow{P} X$ as $n \to \infty$. $\square$

In the following example, we compute $P[|X_n - X| < \epsilon] = \epsilon^{1/n}$ for some values of $\epsilon \leq 1$ and some values of n and examine its nature as ϵ changes and as n increases.

Example 7.2.3. In Example 7.2.2, we have discussed convergence in probability and has shown that $P[|X_n - X| < \epsilon] = \epsilon^{1/n}$. We use the following code to compute these values for some values of ϵ.

Code 7.2.1. Values of ϵ are specified in a variable `epvec` and values of n are specified in a variable `nvec`. The values $\epsilon^{1/n}$ are stored in a matrix `probmat`.

```
prob=function(epsilon,n)
{
 pr=epsilon^(1/n)
 return(pr)
}
epvec=c(0.01,0.05,0.1,0.5); epl=length(epvec)
nvec=c(100,200,400,800,1000)
nn=length(nvec); probmat=matrix(nrow=nn,ncol=epl)
for(i in 1:nn)
{
 for(j in 1:epl)
```

```
{
 n=nvec[i]
 ep=epvec[j]
 probmat[i,j]=prob(ep,n)
}
}
P=round(probmat,4);P
plot(nvec,probmat[,1],type="o",lwd=2,xlab="n",ylab=
"Probability",
ylim=c(.95,1),pch=20)
for(i in 2:epl)
{
 lines(nvec,probmat[,i],type="o",lwd=2,lty=i,col=i,pch=20)
}
legend("bottomright",legend=c(epvec[1],epvec[2],epvec[3],
epvec[4]),
lty=1:epl,col=1:epl,lwd=2,title=expression(paste(epsilon)))
```

Table 7.1 displays the values of $P[|X_n - X| < \epsilon]$ for four values of ϵ and five values of n.

TABLE 7.1
$P[|X_n - X| < \epsilon]$ for Given n and ϵ

n \ ϵ	0.01	0.05	0.1	0.5
100	0.9550	0.9705	0.9772	0.9931
200	0.9772	0.9851	0.9886	0.9965
400	0.9886	0.9925	0.9943	0.9983
800	0.9943	0.9963	0.9971	0.9991
1000	0.9954	0.9970	0.9977	0.9993

From Table 7.1, we observe that for $\epsilon = 0.5$, $P[|X_n - X| < \epsilon]$ is 0.9931 for $n = 100$ while for $\epsilon = 0.01$, $P[|X_n - X| < \epsilon]$ is 0.9954 for $n = 1000$. Thus, the rate of convergence depends on ϵ, as is expected. As n increases, the probabilities increase to 1, supporting the conclusion that $X_n \xrightarrow{P} X$ as $n \to \infty$. Figure 7.1 also conveys these features. It is to be noted that even for $n = 200$, $P[|X_n - X| < \epsilon]$ is close to 99%, except for $\epsilon = 0.01$. □

Example 7.2.4. Suppose $\{X_1, X_2, \cdots, X_n\}$ are independent and identically distributed random variables each having Cauchy $C(\theta, 1)$ distribution, with location parameter $\theta \in \mathbb{R}$ and scale parameter 1. The common characteristic function $\phi(t)$ of $Y_i - \theta$ is given by, $\phi(t) = \exp\{-|t|\}$. Hence the characteristic function $\phi_n(t)$ of $\overline{Y}_n = \overline{X}_n - \theta$ is given by, $\phi_n(t) = \exp\{n(-|t/n|)\} =$

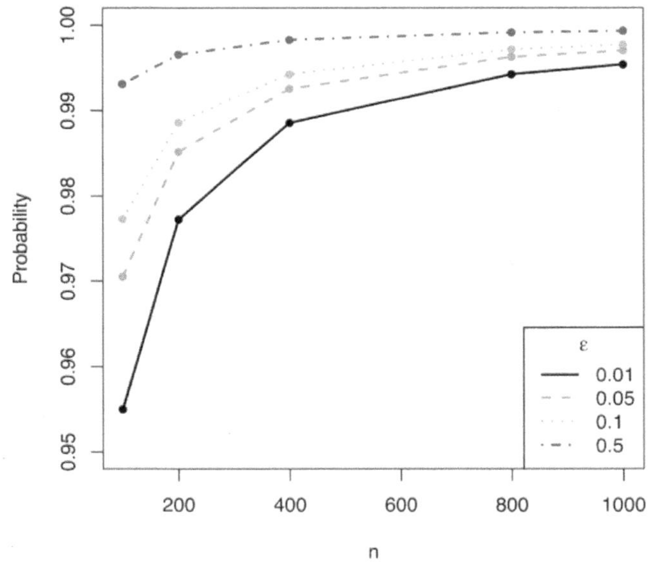

FIGURE 7.1
Convergence in Probability of $X_n = e^\omega + \omega^n$ to $X = e^\omega$

$\exp\{-|t|\}$. Thus, $\overline{X}_n = \overline{Y}_n + \theta$ follows the Cauchy distribution with location parameter θ and scale parameter 1. Hence for any $\epsilon > 0$,

$$\begin{aligned} P[|\overline{X}_n - \theta| < \epsilon] &= \int_{\theta-\epsilon}^{\theta+\epsilon} \frac{1}{\pi}\frac{1}{1+(x-\theta)^2}dx = \frac{1}{\pi}\left[\tan^{-1}(x-\theta)\right]_{\theta-\epsilon}^{\theta+\epsilon} \\ &= 2\tan^{-1}(\epsilon)/\pi \ . \end{aligned}$$

It is free from n and hence does not converge to 1 as $n \to \infty$. Hence, $\overline{X}_n \overset{P}{\nrightarrow} \theta$. □

Example 7.2.5. Suppose $\{X_1, X_2, \cdots, X_n\}$ are independent and identically distributed random variables each having uniform $U(0,1)$ distribution. Suppose $\{X_{(1)}, X_{(2)}, \cdots, X_{(n)}\}$ is the corresponding order statistics. If $X \sim U(0,1)$, then its distribution function $F_X(x)$ is given by,

$$F_X(x) = \begin{cases} 0, & \text{if} \quad x < 0 \\ x, & \text{if} \quad 0 \le x < 1 \\ 1, & \text{if} \quad x \ge 1. \end{cases}$$

Note that, being Borel functions of $\{X_1, X_2, \cdots, X_n\}$, $Y_n = X_{(1)}$ and $Z_n = X_{(n)}$ are also random variables. We now discuss the convergence in probability of Y_n and Z_n. The distribution functions $F_{Y_n}(x)$ of Y_n and $F_{Z_n}(x)$ of Z_n are given by,

$$F_{Y_n}(x) = \begin{cases} 0, & \text{if} \quad x < 0 \\ 1-(1-x)^n, & \text{if} \quad 0 \le x < 1 \\ 1, & \text{if} \quad x \ge 1 \end{cases} \& \ F_{Z_n}(x) = \begin{cases} 0, & \text{if} \quad x < 0 \\ x^n, & \text{if} \quad 0 \le x < 1 \\ 1, & \text{if} \quad x \ge 1. \end{cases}$$

Observe that,

$$\begin{aligned}
\text{if } \epsilon > 1, P[|X_{(1)}| < \epsilon] &= P[-\epsilon < X_{(1)} < \epsilon] = P[0 < X_{(1)} < \epsilon] = 1 \\
\& \text{ if } \epsilon \leq 1, \ P[|X_{(1)}| < \epsilon] &= P[0 < X_{(1)} < \epsilon] = F_{X_{(1)}}(\epsilon) \\
&= 1 - (1-\epsilon)^n \to 1 \ \Rightarrow \ X_{(1)} \xrightarrow{P} 0.
\end{aligned}$$

$$\begin{aligned}
\text{Further if } \epsilon > 1, \ P[|X_{(n)} - 1| < \epsilon] &= P[1 - \epsilon < X_{(n)} < 1 + \epsilon] \\
&= P[1 - \epsilon < X_{(n)} < 1] = 1. \\
\text{If } \epsilon \leq 1, \ P[|X_{(n)} - 1| < \epsilon] &= P[1 - \epsilon < X_{(n)} < 1] \\
&= F_{X_{(n)}}(1) - F_{X_{(n)}}(1-\epsilon) \\
&= 1 - (1-\epsilon)^n \to 1 \ \Rightarrow \ X_{(n)} \xrightarrow{P} 1.
\end{aligned}$$

Thus, if $X \sim U(0,1)$, then $X_{(1)} \xrightarrow{P} 0$ and $X_{(n)} \xrightarrow{P} 1$. We now examine whether $X_{(1)} \xrightarrow{P} a$ for any a which is close to 0. Suppose we take $a = 0.01$. Observe that for $\epsilon < 0.01$,

$$\begin{aligned}
P[|X_{(1)} - 0.01| < \epsilon] &= P[0.01 - \epsilon < X_{(1)} < 0.01 + \epsilon] \\
&= 1 - (1 - 0.01 - \epsilon)^n - 1 + (1 - 0.01 + \epsilon)^n \\
&= (0.99 + \epsilon)^n - (0.99 - \epsilon)^n \to 0 \ \Rightarrow \ X_{(1)} \stackrel{P}{\nrightarrow} 0.01,
\end{aligned}$$

since there are some values of ϵ for which $P[|X_{(1)} - 0.01| < \epsilon] \nrightarrow 1$. By the definition of convergence in probability, $X_n \xrightarrow{P} X$ if $P[|X_n - X| < \epsilon] \to 1$ for all $\epsilon > 0$. On similar lines it can be shown that $X_{(1)} \stackrel{P}{\nrightarrow} a \neq 0$.

Similarly, we examine whether $X_{(n)} \xrightarrow{P} b$ for any b which is close to 1. Suppose we take $b = 0.99$. Observe that for $\epsilon < 0.01$,

$$\begin{aligned}
P[|X_{(n)} - 0.99| < \epsilon] &= P[0.99 - \epsilon < X_{(n)} < 0.99 + \epsilon] \\
&= (0.99 + \epsilon)^n - (0.99 - \epsilon)^n \to 0 \ \Rightarrow \ X_{(n)} \stackrel{P}{\nrightarrow} 0.99.
\end{aligned}$$

Using similar arguments, it can be shown that $X_{(n)} \stackrel{P}{\nrightarrow} b \neq 1$. □

Remark 7.2.1. As in Example 7.2.3, we can compute $P[|X_{(1)}| < \epsilon]$ and $P[|X_{(n)} - 1| < \epsilon]$ from their expressions and verify numerically that these probabilities approach 1 as n increases. In Example 7.2.16, we verify by simulation that $X_{(1)} \xrightarrow{P} 0$ and $X_{(n)} \xrightarrow{P} 1$.

In Example 7.2.5, we noted that $X_{(n)} \xrightarrow{P} 1$ but $X_{(n)} \stackrel{P}{\nrightarrow} 0.99$. Similarly, $X_{(1)} \xrightarrow{P} 0$ but $X_{(1)} \stackrel{P}{\nrightarrow} 0.01$. It indicates that the limit random variable in convergence in probability may be unique. We prove in the following theorem that, as in the almost sure convergence, the limit random variable in convergence in probability is almost surely unique. In asymptotic inference, this result is useful to conclude that a given estimator cannot be consistent for two different parametric functions.

Theorem 7.2.1. *If $X_n \xrightarrow{P} X$ and $X_n \xrightarrow{P} Y$, then X and Y are equivalent random variables, that is, $X = Y$ a.s.*

Proof. From the definition of convergence in probability, $X_n \xrightarrow{P} X$ implies that given $\delta_1 > 0$ and $k \geq 1$, $\exists\ n_0(\delta_1, k)$ such that $P[|X_n - X| > 1/(2k)] < \delta_1$, for all $n \geq n_0(\delta_1, k)$. Similarly, $X_n \xrightarrow{P} Y$ implies that given $\delta_2 > 0$ and $k \geq 1$, $\exists\ n_1(\delta_2, k)$ such that $\forall\ n \geq n_1(\delta_2, k)$, $P[|X_n - Y| > 1/(2k)] < \delta_2$. Suppose $n_2(k) = \max\{n_0(\delta_1, k), n_1(\delta_2, k)\}$. To prove that $X = Y$ a.s. we prove that $P[X \neq Y] = 0$. Observe that,

$$
\begin{aligned}
P[X \neq Y] &= P\Big[\bigcup_{k \geq 1}[|X - Y| > 1/k]\Big] \\
\text{Further, } 1/k < |X - Y| &= |X - X_n + X_n - Y| < |X_n - X| + |X_n - Y| \\
\Rightarrow\ [|X - Y| > 1/k] &\subseteq [|X_n - X| > (1/2k)] \cup [|X_n - Y| > (1/2k)] \\
\Rightarrow\ P[|X - Y| > 1/k] &\leq P[|X_n - X| > (1/2k)] + P[|X_n - Y| > (1/2k)] \\
&< \delta_1 + \delta_2 = \delta, \text{ say}, \ \forall\ n \geq n_2(k).
\end{aligned}
$$

It is true for all $\delta > 0$, hence $P[|X - Y| > 1/k] = 0\ \ \forall\ \ k \geq 1$ for large n. It then follows that,

$$
P[X \neq Y] = P\Big[\bigcup_{k \geq 1}[|X - Y| > 1/k]\Big] \leq \sum_{k \geq 1} P[|X - Y| > 1/k] = 0
$$

for large n, which proves that $X = Y$ a.s. □

We have already proved that the limit random variable in almost sure convergence is almost surely unique. The same result follows from the above theorem. It is proved in the following corollary.

Corollary 7.2.1. *If $X_n \xrightarrow{a.s.} X$ and $X_n \xrightarrow{a.s.} Y$, then $X = Y$ a.s.*

Proof. If $X_n \xrightarrow{a.s.} X$ and $X_n \xrightarrow{a.s.} Y$ then $X_n \xrightarrow{P} X$ and $X_n \xrightarrow{P} Y$ and hence $X = Y$ a.s. □

In Section 6.3, it is shown that almost sure convergence implies convergence in probability. We now examine more links of convergence in probability with the other modes of convergence.

Example 7.2.6. Suppose $\{X_n, n \geq 1\}$ is a sequence of random variables such that $X_n \sim N(1/n, 1/n)$ distribution. Then

$$
E(X_n - 0)^2 = E(X_n^2) = 1/n + 1/n^2 \to 0 \ \text{ as } \ n \to \infty \ \ \Rightarrow \ \ X_n \xrightarrow{q.m.} 0.
$$

Now by Chebyshev's inequality, for $\epsilon > 0$

$$
P[|X_n| > \epsilon] \leq E(X_n^2)/\epsilon^2 = (1/\epsilon^2)\left(1/n + 1/n^2\right) \to 0 \text{ as } n \to \infty \Rightarrow X_n \xrightarrow{P} 0. \ \square
$$

The above example shows that convergence in quadratic mean implies convergence in probability. It is true in general and can be proved using the Markov inequality, proved in Chapter 4. Moreover, it provides a sufficient condition for convergence in probability, as is clear from the following theorem.

Theorem 7.2.2. *Convergence in r-th mean implies convergence in probability.*

Proof. By the Markov inequality, for all $\epsilon > 0$ and for some $r > 0$,

$$\begin{aligned} P[|X_n - X| > \epsilon] &\leq E|X_n - X|^r/\epsilon^r \to 0 \text{ since } X_n \xrightarrow{r} X \\ \Rightarrow P[|X_n - X| > \epsilon] &\to 0 \Rightarrow X_n \xrightarrow{P} X. \end{aligned}$$

Thus, $X_n \xrightarrow{r} X \Rightarrow X_n \xrightarrow{P} X$. □

Following examples illustrate this implication.

Theorem 7.2.2 is very useful in verifying the consistency of an estimator, as illustrated in the following example.

Example 7.2.7. Suppose X is a random variable with mean μ and finite variance σ^2. Suppose $\{X_1, X_2, \cdots, X_n\}$ is a random sample of size n from the distribution of X, that is, $\{X_1, X_2, \cdots, X_n\}$ are independent and identically distributed random variables, each having the same distribution function F. Suppose $\overline{X}_n$ denotes the sample mean based on this random sample. Then

$$E(\overline{X}_n - \mu)^2 = Var(\overline{X}_n) = \sigma^2/n \to 0 \Rightarrow \overline{X}_n \xrightarrow{q.m.} \mu \Rightarrow \overline{X}_n \xrightarrow{P} \mu.$$

In particular suppose X has Poisson $P(\theta)$ distribution, then $\overline{X}_n \xrightarrow{q.m.} \theta$ which implies $\overline{X}_n \xrightarrow{P} \theta$. Suppose if possible, $\overline{X}_n \xrightarrow{P} g(\theta)$ where g is any real valued function. However, as proved in Theorem 7.2.1, the limit random variable in convergence in probability is almost surely unique. Hence, $g(\theta) = \theta$, that is, if $\overline{X}_n$ is a consistent estimator of θ, then it cannot be a consistent estimator for any other parametric function of θ. □

In the following example, we show that the r-th order sample raw moment converges in probability to the corresponding population moment under certain conditions.

Example 7.2.8. Suppose X is a random variable with the r-th order raw moment $\mu'_r = E(X^r)$. Suppose $\mu'_{2r} < \infty$. Hence by Lemma 4.5.1, all the moments of lower order are finite. Suppose $m'_r = \sum_{i=1}^{n} X_i^r/n$ denotes the r-th order sample raw moment based on a random sample of size n from the distribution of X. If $Y_i = X_i^r$, $i = 1, 2, \cdots, n$, then $\mu'_r = E(X^r) = E(Y_1)$, $Var(Y_1) = \mu'_{2r} - (\mu'_r)^2$ is finite. Hence, as shown in Example 7.2.7, $m'_r = \overline{Y}_n \xrightarrow{P} \mu'_r$. Thus, the r-th order sample raw moment converges in probability to the corresponding population moment. □

In Chapter 9, we prove that the requirement $\mu'_{2r} < \infty$ can be replaced by $\mu'_r < \infty$.

Example 7.2.9. Suppose $\{X_n, n \geq 1\}$ is a sequence of random variables such that $X_n \sim N(\theta, \sigma_n^2)$ distribution. Then $E(X_n - \theta)^2 = \sigma_n^2$. Suppose $\sigma_n^2 \to 0$, then $X_n \stackrel{q.m.}{\to} \theta$ and hence $X_n \stackrel{P}{\to} \theta$. Now suppose $X_n \stackrel{P}{\to} \theta$, then $\forall \;\; \epsilon > 0$,

$$\begin{aligned} P[|X_n - \theta| < \epsilon] &= P\left[|X_n - \theta|/\sigma_n < \epsilon/\sigma_n\right] \\ &= \Phi\left(\epsilon/\sigma_n\right) - \Phi\left(-\epsilon/\sigma_n\right) = 2\Phi\left(\epsilon/\sigma_n\right) - 1 \to 1 \\ &\Rightarrow \Phi\left(\epsilon/\sigma_n\right) \to 1 \Rightarrow \sigma_n \to 0 \Rightarrow X_n \stackrel{q.m.}{\to} \theta. \end{aligned}$$

Thus, in this example, the two modes of convergence are equivalent. □

However, the converse of Theorem 7.2.2 is not true. Convergence in probability does not imply convergence of sequence of moments or convergence in r-th mean, as is clear from the following example.

Example 7.2.10. Suppose $\{X_n, n \geq 1\}$ is a sequence of random variables, and the probability mass function of X_n is given by,

$$X_n = \begin{cases} 1, & \text{with probability} \;\; 1 - p_n \\ n, & \text{with probability} \;\; p_n \,. \end{cases}$$

From the given probability distribution of X_n, it is clear that for any $\epsilon > 0$

$$P[|X_n - 1| < \epsilon] \geq P[X_n = 1] = 1 - p_n \to 1 \;\; \text{provided} \;\; p_n \to 0\,.$$

Suppose $p_n = a/n^\delta, \; \delta > 0, \; a > 0$ is such that $0 < p_n < 1, \;\; \forall \;\; n \geq 1$. Thus, $p_n \to 0 \;\;\; \forall \;\; \delta > 0$ and $X_n \stackrel{P}{\to} X \equiv 1$. From the distribution of X_n we have,

$$\begin{aligned} &E(X_n) = (1 - p_n) + np_n = 1 + (n-1)p_n, \\ &E(X_n^2) = (1 - p_n) + n^2 p_n = 1 + (n^2 - 1)p_n \;\;\Rightarrow\;\; E(X_n - 1)^2 = (n-1)^2 p_n. \end{aligned}$$

With $p_n = a/n^\delta, \; \delta > 0, \; a > 0$, observe that,

$$E(X_n) \to \begin{cases} 1 + a, & \text{if} \;\; \delta = 1 \\ 1, & \text{if} \;\; \delta > 1 \\ \infty, & \text{if} \;\; \delta < 1 \end{cases} \quad \& \quad E(X_n^2) \to \begin{cases} 1 + a, & \text{if} \;\; \delta = 2 \\ 1, & \text{if} \;\; \delta > 2 \\ \infty, & \text{if} \;\; \delta < 2 \end{cases}$$

$$\text{Further,} \;\; E(X_n - 1)^2 = (n-1)^2 p_n \quad \to \quad \begin{cases} a, & \text{if} \;\; \delta = 2 \\ 0, & \text{if} \;\; \delta > 2 \\ \infty, & \text{if} \;\; \delta < 2\,. \end{cases}$$

It is to be noted that for $\delta > 2$, $X_n \stackrel{P}{\to} 1$, $E(X_n) \to 1$, $E(X_n^2) \to 1$ and $E(X_n - 1)^2 \to 0$. However, for $1 < \delta \leq 2$, $X_n \stackrel{P}{\to} X \equiv 1$ and $E(X_n) \to 1$, but $E(X_n^2) \nrightarrow 1$ and $E(X_n - 1)^2 \nrightarrow 0$. Thus, $X_n \stackrel{P}{\to} 1$ but $X_n \stackrel{q.m.}{\nrightarrow} 1$. □

This example clearly shows the difference between the convergence in probability of a sequence $\{X_n, n \geq 1\}$ of random variables and the limiting behaviour of sequence of moments. Observe that $X_n \xrightarrow{P} C$, states that for large n, X_n is very likely to be close to C. However, it says nothing about the location of the remaining small probability mass assigned to points that are not close to C and can strongly affect the mean and other moments. This distinction depends on the assumption that the distribution of X_n has a positive probability mass for arbitrarily large or small values. Thus, the two statements - (a) $X_n \xrightarrow{P} C$ and (b) $E(X_n) \to C$ are clearly distinct. Statement (a) is concerned with the bulk of the probability mass and is independent of the position of the small remainder. However, statement (b) heavily depends on the behaviour of X_n in the extreme tails. The situation changes if the X_n's are uniformly bounded.

Definition 7.2.2. *A sequence $\{X_n, n \geq 1\}$ of random variables is said to be uniformly bounded if there exists $M > 0$ such that $P[|X_n| < M] = 1 \ \forall \ n \geq 1$.*

Theorem 7.2.3. *Suppose a sequence $\{X_n, n \geq 1\}$ of random variables is uniformly bounded. Then*

$$X_n \xrightarrow{P} C \ \Rightarrow \ E(X_n) \to C, \ E(X_n - C)^2 \to 0 \ \text{and} \ E(|X_n - C|^r) \to 0, \ r > 0.$$

Proof. The proof follows from the basic inequality given by,

$$\frac{E(g(X)) - g(a)}{a.s.sup(g(X))} \leq P[|X| \geq a] \leq \frac{E(g(X))}{g(a)}, \qquad a \geq 0,$$

where g is a non-negative, even and non-decreasing function on $\mathbb{R}^+$. Now, X_n is uniformly bounded, hence there exists $M > 0$ such that $P[|X_n| < M] = 1$ for all n. Thus,

$$\begin{aligned}
&C \geq 0 \ \& \ -M < X_n < M \ \Rightarrow \ -M - C < X_n - C < M - C < M + C \\
&\quad \Rightarrow |X_n - C| < M + C \\
&C < 0 \ \& \ -M < X_n < M \ \Rightarrow \ -M + C < -M - C < X_n - C < M - C \\
&\quad \Rightarrow \ |X_n - C| < M - C.
\end{aligned}$$

Thus if X_n is uniformly bounded, then $|X_n - C|$ is also uniformly bounded. Suppose a function g is defined as $g(x) = |x|, \ x \in \mathbb{R}$. Then for any $\epsilon > 0$,

$$(E(|X_n - C|) - g(\epsilon))/M_1 \leq P[|X_n - C| \geq \epsilon] \to 0 \ \Rightarrow \ E(|X_n - C|) \to 0,$$

where M_1 is almost sure supremum of $|X_n - C|$ and it is finite. Now,

$$E(X_n - C) \leq E(|X_n - C|) \to 0 \ \Rightarrow \ E(X_n) \to C .$$

To prove that $E(X_n - C)^2 \to 0$, suppose $g(x) = x^2, \ x \in \mathbb{R}$. Now for any $\epsilon > 0$,

$$(E(X_n - C)^2 - \epsilon^2)/M_2 \leq P[|X_n - C| \geq \epsilon] \to 0 \ \Rightarrow \ E(X_n - C)^2 \to 0,$$

where M_2 is almost sure supremum of $|X_n - C|^2$. Thus, $X_n \xrightarrow{P} C \Rightarrow X_n \xrightarrow{q.m.} C$. It further shows that $E(X_n^2) \to C^2$. With $g(x) = |x|^r$, $x \in \mathbb{R}$, for any $\epsilon > 0$,

$$(E(|X_n - C|^r) - \epsilon^r)/M_r \leq P[|X_n - C| \geq \epsilon] \to 0 \Rightarrow E(|X_n - C|^r) \to 0,$$

where M_r is almost sure supremum of $|X_n - C|^r$. Thus, $X_n \xrightarrow{P} C$ implies $X_n \xrightarrow{r} C$. □

In the following examples, we illustrate Theorem 7.2.3 and verify by simulation.

Example 7.2.11. In Example 7.2.5, it is shown that if $X \sim U(0,1)$, then $X_{(n)} \xrightarrow{P} Y_1 \equiv 1$ and $X_{(1)} \xrightarrow{P} Y_0 \equiv 0$. In Example 7.2.16, we verify these results by simulation using R. Using the distribution of $X_{(1)}$ and $X_{(n)}$ we can find their moments. Thus,

$$E(X_{(n)}) = n/(n+1) \to 1 = E(Y_1) \ \ \& \ \ E(X_{(1)}) = 1/(n+1) \to 0 = E(Y_0).$$

Further, $|X_{(n)}| = X_{(n)} \leq 1$ and $|X_{(1)}| = X_{(1)} \leq 1$. Hence, $X_{(n)}$ and $X_{(1)}$ both are uniformly bounded random variables. Thus by Theorem, 7.2.3

$$\begin{aligned} \lim_{n\to\infty} E(X_{(n)}) &= 1 \ \ \& \ \ \lim_{n\to\infty} E(X_{(1)}) = 0. \\ \text{Now, } E(X_{(n)}^2) &= n/(n+2) \ \ \& \ \ E(X_{(1)}^2) = 2/(n+1)(n+2) \\ &\Rightarrow \lim_{n\to\infty} E((X_{(n)} - 1)^2) = 0 \ \ \& \ \ \lim_{n\to\infty} E(X_{(1)}^2) = 0. \end{aligned}$$

□

We verify these results by simulation using Code 7.2.2 in the following example.

Example 7.2.12. Suppose $\{X_1, X_2, \cdots, X_n\}$ are independent and identically distributed random variables each having uniform $U(0,1)$ distribution. We take $n = 100$ to 1100 with an increment of 300 and simulate 300 observations for each n. From 300 simulated values of $X_{(n)}$ and $X_{(1)}$, we find mean M_n and m_n of these values as estimates of $E(X_{(n)})$ and $E(X_{(1)})$ respectively. We also find V_n and W_n as mean of $(X_{(n)} - 1)^2$ and $(X_{(1)})^2$ as estimates of $E((X_{(n)} - 1)^2)$ and $E((X_{(1)} - 0)^2)$ respectively. We use the following code.

Code 7.2.2.
```
N=seq(100,1100,200); nsim=300; umax=umin=m1=m2=
v1=v2=c()
for(j in 1:length(N))
{
 for(m in 1:nsim)
{
 set.seed(m)
 x=runif(N[j],0,1)
 umax[m]=max(x)
 umin[m]=min(x)
```

TABLE 7.2
Verification of Theorem 7.2.3 for $U(0,1)$ Distribution

n	M_n	m_n	V_n	W_n
100	0.9896	0.0111	0.0002	0.0002
300	0.9964	0.0038	0	0
500	0.9980	0.0023	0	0
700	0.9985	0.0015	0	0
900	0.9988	0.0011	0	0
1100	0.9991	0.0009	0	0

```
}
 m1[j]=mean(umax)
 m2[j]=mean(umin)
 v1[j]=mean((umax-1)^2)
 v2[j]=mean(umin^2)
}
d=round(data.frame(N,m1,m2,v1,v2),4);d
```

We examine whether $M_n \to 1$, $m_n \to 0$ and both V_n and W_n converge to 0 as n increases. The output in data frame d is organized in Table 7.2.

From Table 7.2, we note that $M_n \to 1$ and $m_n \to 0$ as n increases. Further and both V_n and W_n are almost 0 for all n. Thus, the results proved in Theorem 7.2.3 are verified by simulation. □

As in almost sure convergence, it is of interest to examine whether $X_n \xrightarrow{P} X$ implies that $g(X_n) \xrightarrow{P} g(X)$, for some reasonably behaved function g. The following two theorems address this issue. In the first theorem, we take $X \equiv C$, a degenerate limit random variable; in the second, it is a non-degenerate random variable.

Theorem 7.2.4. *If $X_n \xrightarrow{P} C$, then $g(X_n) \xrightarrow{P} g(C)$, provided g is a continuous function on $\mathbb{R}$.*

Proof. Continuity of the function g implies that g is a Borel function, hence $g(X_n)$ is a random variable for each $n \geq 1$ and $\{g(X_n), n \geq 1\}$ is a sequence of random variables defined on the same probability space $(\Omega, \mathbb{A}, P)$. Since g is continuous on $\mathbb{R}$, g is continuous at each $a \in \mathbb{R}$. Hence

$$\forall \ \epsilon > 0, \ \exists \ \delta(a,\epsilon) > 0, \ \text{ such that } \ |x-a| < \delta(a,\epsilon) \Rightarrow |g(x) - g(a)| < \epsilon.$$

Hence, $[|X_n - C| < \delta] \subseteq [|g(X_n) - g(C)| < \epsilon]$, where δ depends on C and ϵ. As a consequence,

$$P[|g(X_n) - g(C)| < \epsilon] \ \geq \ P[|X_n - C| < \delta] \ \to 1 \ \text{ since } \ X_n \xrightarrow{P} C.$$

Hence, $\forall \ \epsilon > 0, \ P[|g(X_n) - g(C)| < \epsilon] \to 1$ and $g(X_n) \xrightarrow{P} g(C)$. □

In the above proof, δ depends on ϵ and C. Suppose $X_n \xrightarrow{P} X$ where X is a non-degenerate random variable. If we use arguments similar to those in Theorem 7.2.4, then continuity of g at $X(\omega)$ implies that for all $\epsilon > 0$, $\exists\ \delta > 0$ which depends on ϵ and $X(\omega)$ and hence on $\omega \in \Omega$. Hence using continuity of g, we cannot conclude that $[|X_n(\omega) - X(\omega)| < \delta(\epsilon, \omega)]$ is included in $[|g(X_n)(\omega) - g(X)(\omega)| < \epsilon]$. This inclusion is valid if g is uniformly continuous, that is, when δ does not depend on ω. A function g is said to be uniformly continuous on S, if $\forall\ \epsilon > 0$, $\exists\ \delta(\epsilon) > 0$ such that $x_1, x_2 \in S$ with $|x_1 - x_2| < \delta$ implies that $|g(x_1) - g(x_2)| < \epsilon$. Thus, if $X_n \xrightarrow{P} X$, then $g(X_n) \xrightarrow{P} g(X)$, provided g is uniformly continuous. In the following theorem, we prove that it is not necessary to impose this additional condition of uniform continuity on g. In the proof, we use the fact that a continuous function on a compact set is uniformly continuous, and secondly, X being a real random variable is bounded in probability.

Theorem 7.2.5. *$X_n \xrightarrow{P} X \;\Rightarrow\; g(X_n) \xrightarrow{P} g(X)$, provided g is a continuous function.*

Proof. Continuity of the function g implies that g is a Borel function, hence $g(X)$ is a random variable and $\{g(X_n), n \geq 1\}$ is a sequence of random variables defined on the same probability space $(\Omega, \mathbb{A}, P)$. It is given that X is a real random variable and hence by Lemma 4.4.1, it is bounded in probability, that is, for any $\epsilon > 0$, there exists a constant K such that, $P[|X| \leq K] \geq 1 - \epsilon$. Further, $X_n \xrightarrow{P} X$ implies that given $\epsilon_1 > 0$ & $\delta > 0$, $\exists\ n_0(\epsilon_1, \delta)$ such that $P[|X_n - X| \leq \delta] \geq 1 - \epsilon_1 \;\; \forall\ n \geq n_0$. Observe that,

$$\begin{aligned} |X_n - X| \leq \delta \;\&\; |X| \leq K &\Rightarrow X - \delta \leq X_n \leq X + \delta \;\&\; -K \leq X \leq K \\ &\Rightarrow -K - \delta \leq X - \delta \leq X_n \leq X + \delta \leq K + \delta \\ &\Rightarrow |X_n - X| < \delta \;\&\; X_n, X \in [-K - \delta, K + \delta]\,. \end{aligned}$$

Now g is a continuous function. Hence it is uniformly continuous on $[-K - \delta, K + \delta]$. Hence, when $X_n, X \in [-K - \delta, K + \delta]$, given $\epsilon_2 > 0\ \exists\ \delta_1(\epsilon_2) > 0$ such that $|X_n - X| \leq \delta_1(\epsilon_2) \;\Rightarrow\; |g(X_n) - g(X)| < \epsilon_2$. Suppose $\delta_2 = \min\{\delta, \delta_1\}$. Thus we have,

$$|X_n - X| \leq \delta_2 \;\&\; |X| \leq K \;\Rightarrow\; |g(X_n) - g(X)| < \epsilon_2\,.$$

Now observe that $\forall\ n \geq n_0$,

$$\begin{aligned} 1 - \epsilon_1 &\leq P[|X_n - X| \leq \delta_2] \\ &= P[|X_n - X| \leq \delta_2,\ |X| \leq K] + P[|X_n - X| \leq \delta_2,\ |X| > K] \\ &\leq P[|X_n - X| \leq \delta_2,\ |X| \leq K] + P[|X| > K] \\ &\leq P[|X_n - X| \leq \delta_2,\ |X| \leq K] + \epsilon\,. \end{aligned}$$

Hence, $P[|X_n - X| \leq \delta_2,\ |X| \leq K] \geq 1 - \epsilon_1 - \epsilon$. Further $\forall\ n \geq n_0$,

$$P[|g(X_n) - g(X)| < \epsilon_2] \geq P[|X_n - X| \leq \delta_2,\ |X| \leq K] \geq 1 - \epsilon_1 - \epsilon$$

and hence $g(X_n) \xrightarrow{P} g(X)$. □

This result is stated as convergence in probability is invariant under continuous transformation. Such a property of convergence in probability is intuitively appealing. If with high probability X_n is close to X, then by continuity, $g(X_n)$ will be close to $g(X)$ with high probability. In the setup of inference, the above theorem states that the consistency of an estimator is invariant under continuous transformation, and this result is heavily used to verify the consistency of the estimators for parametric functions. For example, in Example 7.2.7 it is shown that the sample mean $\overline{X}_n$ based on a random sample of size n from Poisson $P(\theta)$ distribution is consistent for θ. Since $g(x) = \exp(-x)$ is a continuous function, it then follows that $\exp(-\overline{X}_n)$ is a consistent estimator of $\exp(-\theta) = P[X = 0]$, where $X \sim P(\theta)$.

In Theorem 7.2.5, it is proved that if g is a continuous function, then $X_n \xrightarrow{P} X$ implies $g(X_n) \xrightarrow{P} g(X)$. If g is not a continuous function, then we cannot conclude that $g(X_n) \overset{P}{\nrightarrow} g(X)$. One has to use the definition to arrive at a conclusion. In some cases, $g(X_n) \xrightarrow{P} g(X)$ and in some cases, it does not. In the next example, we note that $X_n \xrightarrow{P} X$, g is not continuous and $g(X_n) \overset{P}{\nrightarrow} g(X)$ (Stoyanov [23]). We discuss one example in the next section, where g is not continuous, but still $g(X_n) \xrightarrow{P} g(X)$, under the condition that the probability measure assigned to the point of discontinuity is 0.

Example 7.2.13. Suppose $\{X_n, n \geq 1\}$ is a sequence of random variables such that for $n \geq 2$, $P[X_n = 1] = 1/n$ and $P[X_n = 1/n] = 1 - 1/n$. Note that

$$E(X_n^2) = 1/n + 1/n^2 - 1/n^3 \to 0 \quad \Rightarrow \quad X_n \xrightarrow{q.m.} X \equiv 0 \quad \Rightarrow \quad X_n \xrightarrow{P} X \equiv 0.$$

Suppose a function $g : \mathbb{R} \to \mathbb{R}$ is defined as

$$g(x) = \begin{cases} 0, & \text{if} \quad x \leq 0 \\ 1, & \text{if} \quad x > 0. \end{cases}$$

Note that g is not a continuous function and 0 is a point of discontinuity. Further, for each n, $X_n > 0$, hence $g(X_n) = 1$. However, $X \equiv 0$ which implies $g(X) \equiv 0$. Hence $|g(X_n) - g(X)| \equiv 1$. Thus, for $\epsilon < 1, P[|g(X_n) - g(X)| < \epsilon] = 0$. Thus, $g(X_n) \overset{P}{\nrightarrow} g(X)$. □

The following example gives an application of Theorem 7.2.5.

Example 7.2.14. Suppose X is a random variable with an absolutely continuous distribution, with support $[a, b]$ and distribution function F. Suppose $\{X_1, X_2, \cdots, X_n\}$ are independent and identically distributed random variables, each having the distribution same as that of X. It is known that if the distribution of X is absolutely continuous with distribution function $F(\cdot)$, then by the probability integral transformation, $U = F(X) \sim U(0, 1)$ distribution. As a consequence, $(F(X_{(1)}), F(X_{(n)}))'$ are distributed as $(U_{(1)}, U_{(n)})'$, minimum and maximum order statistics corresponding to a random sample of size n from $U(0,1)$. In Example 7.2.5, it is shown that if

$$U \sim U(0,1), \quad \text{then} \quad U_{(1)} \xrightarrow{P} 0 \quad \& \quad U_{(n)} \xrightarrow{P} 1 \,.$$

Suppose $F^{-1}(\cdot)$ exists. It is continuous, as F is continuous, in view of the inverse function theorem (Apostol [1]). Hence, by the invariance property of convergence in probability under continuous transformation, it follows that

$$U_{(1)} = F(X_{(1)}) \xrightarrow{P} 0 \quad \Rightarrow \quad F^{-1}(F(X_{(1)})) = X_{(1)} \xrightarrow{P} F^{-1}(0) = a$$

$$\& \quad U_{(n)} = F(X_{(n)}) \xrightarrow{P} 1 \quad \Rightarrow \quad F^{-1}(F(X_{(n)})) = X_{(n)} \xrightarrow{P} F^{-1}(1) = b \,,$$

for all a and b. □

In Example 7.2.8, we have shown that the r-th order sample raw moment converges in probability to the corresponding population moment. Using Theorem 7.2.5, we now prove that sample quantiles converge in probability to the corresponding population quantiles under certain conditions.

Suppose X is an absolutely continuous random variable with distribution function $F(x)$. Suppose a_p is such that

$$P[X \leq a_p] \geq p \quad \& \quad P[X \geq a_p] \geq 1 - p.$$

Then a_p is known as the p-th population quantile or fractile. There may be multiple values of a_p unless the distribution function $F(x)$ is strictly monotone. We assume that the distribution function $F(x)$ is strictly monotone. Hence p-th population quantile a_p is a unique solution of $F(a_p) = p, \; 0 < p < 1$. We assume that the solution of the equation $F(a_p) = p$ exists. As an illustration, suppose X follows an exponential distribution with scale parameter θ. Its distribution function $F(x, \theta)$ is given by, $F(x, \theta) = 1 - \exp(-\theta x)$ for $x > 0$. Hence the solution of the equation $F(a_p(\theta), \theta) = p$ is given by, $a_p(\theta) = -\log(1-p)/\theta$, $0 < p < 1$ and it is the p-th population quantile of the distribution of X.

To define the corresponding sample quantile, suppose $\{X_1, X_2, \cdots, X_n\}$ is a random sample from the distribution of X with distribution function $F(x)$. Suppose $\{X_{(1)}, X_{(2)}, \cdots, X_{(n)}\}$ is the corresponding order statistic. Then $X_{(r_n)}$ is defined as a p-th sample quantile, where $r_n = [np]+1, \; 0 < p < 1$. In the following theorem, we prove that the p-th sample quantile converges in probability to the p-th population quantile.

Theorem 7.2.6. *Suppose X is an absolutely continuous random variable with distribution function $F(x)$ and $\{X_1, X_2, \cdots, X_n\}$ are independent and identically distributed random variables with the distribution function F. Suppose F is strictly increasing and p-th population quantile a_p is a unique solution of $F(a_p) = p, \; 0 < p < 1$. Then the p-th sample quantile $X_{(r_n)}$ converges in probability to the p-th population quantile a_p, where $r_n = [np] + 1$.*

Proof. Since the distribution of X is absolutely continuous with distribution function $F(x)$, by the probability integral transformation $U = F(X) \sim$

$U(0,1)$ distribution. As a consequence, $\{F(X_{(1)}), F(X_{(2)}), \cdots, F(X_{(n)})\}$ are distributed as an order statistic corresponding to a random sample of size n from uniform $U(0,1)$ distribution. Thus, the distribution of $U_{(r_n)} = F(X_{(r_n)})$ is the same as that of the r_n-th order statistic from uniform $U(0,1)$ distribution. The probability density function of $U_{(r_n)}$ is given by,

$$g_{r_n}(u) = \frac{n!}{(r_n-1)!(n-r_n)!} u^{r_n-1}(1-u)^{n-r_n}, \quad 0 < u < 1.$$

It then follows from the definition of a beta function that

$$\begin{aligned} E(U_{(r_n)}) &= \frac{r_n}{n+1} \ \& \ E(U^2_{(r_n)}) = \frac{r_n(r_n+1)}{(n+1)(n+2)} \\ \Rightarrow \ E(U_{(r_n)} - p)^2 &= E(U^2_{(r_n)}) - 2pE(U_{(r_n)}) + p^2 \\ &= \frac{r_n(r_n+1)}{(n+1)(n+2)} - 2p\frac{r_n}{n+1} + p^2. \end{aligned}$$

To find the limit of $E(U_{(r_n)} - p)^2$, observe that

$$\begin{aligned} r_n = [np] + 1 \ &\Rightarrow \ np < r_n \le np + 1 \\ &\Rightarrow \ \frac{np}{n+1} < \frac{r_n}{n+1} \le \frac{np+1}{n+1} \\ &\Rightarrow \ \lim_{n\to\infty} \frac{np}{n+1} \le \lim_{n\to\infty} \frac{r_n}{n+1} \le \lim_{n\to\infty} \frac{np+1}{n+1} \\ &\Rightarrow \ p \le \lim_{n\to\infty} r_n/(n+1) \le p \\ &\Rightarrow \ \lim_{n\to\infty} r_n/(n+1) = p \\ \Rightarrow \ \lim_{n\to\infty} \frac{r_n+1}{n+2} &= \lim_{n\to\infty} \frac{r_n}{n+1}\frac{n+1}{n+2} + \frac{1}{n+2} = p. \end{aligned}$$

Hence,

$$\begin{aligned} \lim_{n\to\infty} E(U_{(r_n)} - p)^2 &= \lim_{n\to\infty} \left\{ \frac{r_n(r_n+1)}{(n+1)(n+2)} - 2p\frac{r_n}{n+1} + p^2 \right\} = 0 \\ &\Rightarrow \ U_{(r_n)} \xrightarrow{q.m.} p \ \Rightarrow \ U_{(r_n)} \xrightarrow{P} p \ \Rightarrow \ F(X_{(r_n)}) \xrightarrow{P} p \\ &\Rightarrow \ F^{-1}(F(X_{(r_n)})) \xrightarrow{P} F^{-1}(p) \ \Rightarrow \ X_{(r_n)} \xrightarrow{P} a_p. \end{aligned}$$

Thus, the p-th sample quantile $X_{(r_n)}$ converges in probability to the p-th population quantile a_p. □

This result is helpful in finding a consistent estimator for the parameter based on a sample quantile. The procedure is illustrated in the following example.

Example 7.2.15. (i) For normal $N(\theta, 1)$ and Cauchy $C(\theta, 1)$ distributions, θ is a population median. Hence in both the cases, by Theorem 7.2.6, the

sample median $X_{([n/2]+1)} \xrightarrow{P} \theta$.
(ii) For a uniform $U(0,\theta)$ distribution, $\theta/2$ is a population median. Hence, $X_{([n/2]+1)} \xrightarrow{P} \theta/2$. Using invariance of convergence in probability under continuous transformation, $2X_{([n/2]+1)} \xrightarrow{P} \theta$.
(iii) More generally, for a normal $N(\theta,1)$ distribution, the p-th quantile is $a_p(\theta) = \theta + \Phi^{-1}(p)$, hence $X_{([np]+1)} - \Phi^{-1}(p) \xrightarrow{P} \theta$, $0 < p < 1$.
(iv) For a uniform $U(0,\theta)$ distribution, the p-th quantile is given by, $a_p(\theta) = \theta p$, hence $X_{([np]+1)}/p \xrightarrow{P} \theta$, $0 < p < 1$.
(v) For a Cauchy $C(\theta,1)$ distribution, the p-th quantile is given by, $a_p(\theta) = \theta + \tan(\pi(p-1/2))$, hence $X_{([np]+1)} - \tan(\pi(p-1/2)) \xrightarrow{P} \theta$, $0 < p < 1$.
In illustrations (iii), (iv) and (v), we use the result that convergence in probability is invariant under continuous transformation. In the inference setting, we thus have an uncountable family of consistent estimators for θ as p varies over $(0,1)$ for normal $N(\theta,1)$, uniform $U(0,\theta)$ and Cauchy $C(\theta,1)$ distributions. We have to choose p appropriately to get a better estimator. Note that a given estimator cannot be consistent for more than one parametric function, but there can be more than one consistent estimators for a given parametric function. □

In Example 7.2.8, it is shown that r-th order sample raw moment converges in probability to the r-th order population moment when $2r$-th order population moment is finite. Theorem 7.2.6 proves that p-th sample quantile converges in probability to the p-th population quantile. We verify these results when $\{X_1, X_2, \cdots, X_n\}$ are independent and identically distributed random variables, each having uniform $U(0,1)$ distribution. Thus, we examine by simulation whether the sample mean converges in probability to 1/2 and the sample median converges in probability to 1/2. In Example 7.2.5, we have shown that $X_{(n)} \xrightarrow{P} 1$ and $X_{(1)} \xrightarrow{P} 0$. We can verify these results by computing $P[|X_n - X| < \epsilon]$ for some ϵ and some n, since we have the explicit expressions for $P[|X_n - X| < \epsilon]$. However, if we do not have the explicit expression for $P[|X_n - X| < \epsilon]$, we cannot adopt such a procedure for verification. For example, suppose $Y_n = \overline{X}_n$, where $\{X_1, X_2, \cdots, X_n\}$ are independent and identically distributed random variables each having uniform $U(0,1)$ distribution. Then we cannot find $P[|Y_n - 1/2| < \epsilon]$ explicitly to verify whether $Y_n \xrightarrow{P} 1/2$. Hence, we now discuss another approach to verify convergence in probability by simulation, using R software. It is useful even if we do not have the explicit expression for $P[|X_n - X| < \epsilon]$. According to the definition of convergence in probability,

$$\forall\ \epsilon > 0,\ \ P[|X_n - X| < \epsilon] \ \to \ 1 \ \text{as} \ n \to \infty \ \Rightarrow \ X_n \xrightarrow{P} X.$$

To verify it by simulation, we generate m observations from the distributions of X_n and X and find the relative frequency rf_n as an estimate of the probability $p_n(\epsilon) = P[|X_n - X| < \epsilon]$. It is defined as $rf_n = \left(\sum_{i=1}^m I_{[|X_n - X| < \epsilon]}\right)/m$. We then examine whether rf_n approaches 1 as n increases for various values of ϵ. The stepwise procedure is as follows.

1. Fix a value of n and ϵ.
2. Generate m observations from the distribution of X_n and X.
3. Find the estimate rf_n of the probability $p_n(\epsilon)$.
4. Increase n and repeat the steps 2 and 3 and examine whether $rf_n \to 1$ as n increases.
5. Repeat steps 2 to 4 for various values of ϵ.

In the following example, we illustrate the procedure for the results derived in Example 7.2.5 and for the sample mean and the sample median.

Example 7.2.16. In Example 7.2.5, it is shown that if $X \sim U(0,1)$, then $X_{(n)} \xrightarrow{P} 1$ and $X_{(1)} \xrightarrow{P} 0$. Further, suppose $Y_n = \overline{X}_n$ and M_n is a sample median based on a random sample $\{X_1, X_2, \cdots, X_n\}$. From Example 7.2.8, it follows that $Y_n \xrightarrow{P} 1/2$ and from Theorem 7.2.6, we know that $M_n \xrightarrow{P} 1/2$. We generate samples from the uniform distribution and obtain estimates of $P[|X_{(n)} - 1| < \epsilon]$, $P[|X_{(1)}| < \epsilon]$, $P[|Y_n - 1/2| < \epsilon]$ and $P[|M_n - 1/2| < \epsilon]$. Using the stepwise procedure outlined above, we examine their behaviour as n increases to verify the convergence in probability. We use the following code for these computations.

Code 7.2.3.
```
eps=c(0.01,0.02,0.03); nsim=500; Max=Min=me=med=c()
N=seq(100,700,100)
p1=p2=p3=p4=matrix(nrow=length(eps),ncol=length(N))
for(i in 1:length(eps))
{
 for(j in 1:length(N))
{
 for(m in 1:nsim)
{
 set.seed(m)
 x=runif(N[j],0,1)
 Max[m]=max(x)
 Min[m]=min(x)
 me[m]=mean(x)
 med[m]=median(x)
}
 p1[i,j]=length(which(abs(Max-1)<eps[i]))/nsim
 p2[i,j]=length(which(abs(Min)<eps[i]))/nsim
```

```
 p3[i,j]=length(which(abs(me-.5)<eps[i]))/nsim
 p4[i,j]=length(which(abs(med-.5)<eps[i]))/nsim
}
}
p1;p2;p3;p4; p=list(p1,p2,p3,p4)
var=c("Maximum","Minimum","Sample Mean","Sample Median")
par(mfrow=c(2,2))
name=paste(var,sep=" ")
for(i in 1:4)
{
 plot(N,p[[i]][1,],type="o",xlab="n",ylab="r_n",main=name[i],
 lty=1,col="dark blue",ylim=range(.2,1.02),pch=20,lwd=1)
 lines(N,p[[i]][2,],lty=2,col="purple",type="o",pch=20,lwd=1)
 lines(N,p[[i]][3,],lty=3,col="blue",type="o",pch=20,lwd=1)
 legend("bottomright",legend=c(eps[1],eps[2],eps[3]),
 lty=c(1,2,3),col=c("dark blue","purple","blue"),
 title=expression(paste(epsilon)))
}
```

The output is organized in Table 7.3.

Corresponding to three values of ϵ and $n = 100, 200, \cdots, 700$, the second, third and fourth rows display the estimate of $P[|X_{(n)} - 1| < \epsilon]$, the next three display estimate of $P[|X_{(1)}| < \epsilon]$, the next three display the estimate of $P[|\overline{X}_n - 1/2| < \epsilon]$ and the last three display the estimate of $P[|M_n - 1/2| < \epsilon]$. Figure 7.2 displays the nature of rf_n as n increases for all four sequences.

TABLE 7.3
U(0,1) Distribution: Convergence in Probability

ϵ		100	200	300	400	500	600	700
0.01		0.624	0.856	0.946	0.978	0.998	1.000	1.000
0.02	$X_{(n)}$	0.870	0.990	0.998	1.000	1.000	1.000	1.000
0.03		0.936	0.994	0.998	1.000	1.000	1.000	1.000
0.01		0.576	0.840	0.928	0.970	0.992	0.998	1.000
0.02	$X_{(1)}$	0.836	0.982	1.000	1.000	1.000	1.000	1.000
0.03		0.942	0.998	1.000	1.000	1.000	1.000	1.000
0.01		0.260	0.370	0.474	0.506	0.566	0.618	0.666
0.02	$\overline{X}_n$	0.528	0.706	0.778	0.844	0.878	0.914	0.944
0.03		0.694	0.858	0.910	0.956	0.976	0.986	0.996
0.01		0.142	0.214	0.296	0.320	0.342	0.378	0.376
0.02	M_n	0.282	0.430	0.550	0.614	0.646	0.698	0.710
0.03		0.432	0.600	0.734	0.780	0.834	0.902	0.892

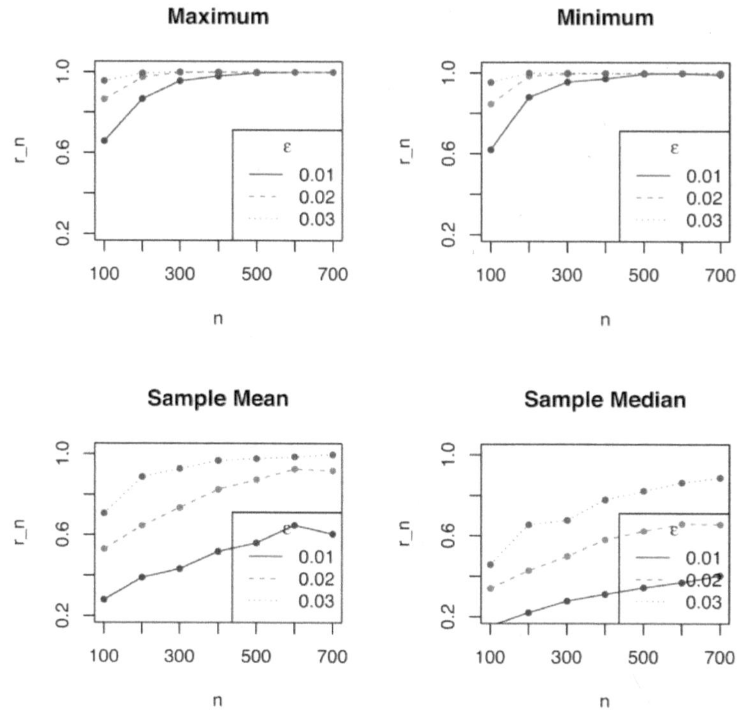

FIGURE 7.2
U(0,1) Distribution: Convergence in Probability of $X_{(n)}$, $X_{(1)}$, $\overline{X}_n$ and Sample Median

From Table 7.3 and Figure 7.2, we note that as n increases, the estimate of the probability $p_n(\epsilon)$ converges to 1. The rate depends on ϵ as is expected. The convergence is faster for larger ϵ. Further, we observe that rate of convergence to 1 of rf_n for sample mean and sample median is slow as compared to that of $X_{(n)}$ and $X_{(1)}$. This feature is clearly revealed in Figure 7.2. In general, the rate of convergence depends on the probability structure of the sequence $\{X_n, n \geq 1\}$. Thus, by simulation, we have verified that if $X \sim U(0,1)$, then $X_{(1)} \xrightarrow{P} 0$, $X_{(n)} \xrightarrow{P} 1$, the sample mean and the sample median converge in probability to $1/2$, which is the mean and median of the uniform $U(0,1)$ distribution. □

In Example 7.2.15, we have noted that for the normal distribution sample median converges in probability to the population median. Likewise, from Example 7.2.8, it follows that the sample mean also converges in probability to the population mean. In the following example, we verify these results by simulation for the standard normal distribution and compare the rate of convergence.

Example 7.2.17. Suppose $X_1, X_2, \cdots, X_n$ are independent and identically distributed random variables, each following the standard normal distribution.

Then from Example 7.2.8, $\overline{X}_n \xrightarrow{P} 0$ and from Example 7.2.15, $M_n \xrightarrow{P} 0$, where M_n denotes the sample median. We simulate the samples from the standard normal distribution for $n = 100, 300, 500, 700, 900$ and find the relative frequency rf_n. We then examine whether it converges to 1. We further examine the rate of its convergence to 1 for the sample mean and the sample median. We use the following code.

Code 7.2.4. `eps=c(0.05,0.08,0.1); nsim=300; me=med=c()`

```
N=seq(100,900,200)
p1=p2=matrix(nrow=length(eps),ncol=length(N))
for(i in 1:length(eps))
{
 for(j in 1:length(N))
{
 for(m in 1:nsim)
{
 set.seed(m)
 x=rnorm(N[j],0,1)
 me[m]=mean(x)
 med[m]=median(x)
}
 p1[i,j]=length(which(abs(me)<eps[i]))/nsim
 p2[i,j]=length(which(abs(med)<eps[i]))/nsim
}
}
p1=round(p1,4);p1;p2=round(p2,4); p2;
par(mfrow=c(3,1))
namevec=paste("epsilon",eps,sep="=")
for(i in 1:length(eps))
{
 plot(N-10,p1[i,],type="h",xlab="n",ylab="r_n",main=namevec[i],
 lty=1,col="dark blue",ylim=range(.2,1.02),pch=20,lwd=2,
 xaxt="n",xlim=c(50,1200))
 axis(1,at=N)
 points(N-10,p1[i,],pch=16)
 lines(N+10,p2[i,],lty=2,col="purple",type="h",pch=20,lwd=2)
 points(N+10,p2[i,],pch=16)
 legend("bottomright",legend=c("Mean","Median"),
 lty=c(1,2),col=c("dark blue","purple"),lwd=2)
}
```

TABLE 7.4
Values of rf_n for Sample Mean and Sample Median

ϵ		100	300	500	700	900
0.05	Sample Mean	0.3633	0.6133	0.7200	0.8133	0.8600
0.05	Sample Median	0.3467	0.5333	0.6533	0.7167	0.7833
0.08	Sample Mean	0.5733	0.8533	0.9267	0.9733	0.9800
0.08	Sample Median	0.5400	0.7633	0.8400	0.9200	0.9433
0.10	Sample Mean	0.7067	0.9267	0.9800	0.9933	1.0000
0.10	Sample Median	0.6400	0.8633	0.9300	0.9800	0.9867

Values of r_n for $n = 100, 300, 500, 700, 900$ and $\epsilon = 0.05, 0.08, 0.1$ are presented in Table 7.4. From the table, we note that values of rf_n are higher for the sample mean than that for the sample median for all values of n and ϵ.

Figure 7.3 displays the vertical lines corresponding to rf_n for the sample mean and the sample median adjacent to each other for $n = 100, 300, 500, 700, 900$ and $\epsilon = 0.05, 0.08, 0.1$. From Figure 7.3, we note that the height of vertical lines, representing the value of rf_n for the sample mean, is higher than that for the sample median, for all values of n and ϵ. Further, the difference is small for $\epsilon = 0.1$ and $n = 700$, and $n = 900$. Thus, from Table 7.4 and Figure 7.3, we conclude that the sample mean converges in probability to 0 faster than the sample median. □

We have introduced the concept of a sequence of random variables being bounded in probability in Section 4.4. In Example 7.2.10, $\{X_n, n \geq 1\}$ is bounded in probability if $p_n \to 0$ but not if $p_n \to p > 0$. In Lemma 4.4.1, it is proved that a real random variable is always bounded in probability or is finite almost surely. Using this lemma, we now prove that if $X_n \xrightarrow{P} X$, then the sequence $\{X_n, n \geq 1\}$ is bounded in probability.

Lemma 7.2.1. *If $X_n \xrightarrow{P} X$, then the sequence $\{X_n, n \geq 1\}$ is bounded in probability.*

Proof. By definition, $X_n \xrightarrow{P} X$ implies that $\forall\ \epsilon > 0\ \&\ \delta_1 > 0$, $\exists\ n_0 \in \mathbb{N}$ such that $\forall\ n \geq n_0$, $P[|X_n - X| > \epsilon] < \delta_1$. Further, X is a real random variable and hence by Lemma 4.4.1, it is bounded in probability, that is,

$$\forall\ \delta_2 > 0 \quad \exists\ M > 0\ , \quad \text{such that} \quad P[|X| > M] < \delta_2.$$

Now for $K > M$ we have,

$$\begin{aligned} P[|X_n| > K] &= P[|X_n| > K, |X| \leq M] + P[|X_n| > K, |X| > M] \\ &\leq P[|X_n - X| > K - M] + P[|X| > M] \\ &\leq \delta_1 + \delta_2 = \delta, \quad \text{say, with} \quad \epsilon = K - M, \quad \forall\ n \geq n_0. \end{aligned}$$

Thus, the sequence $\{X_n, n \geq 1\}$ is bounded in probability. □

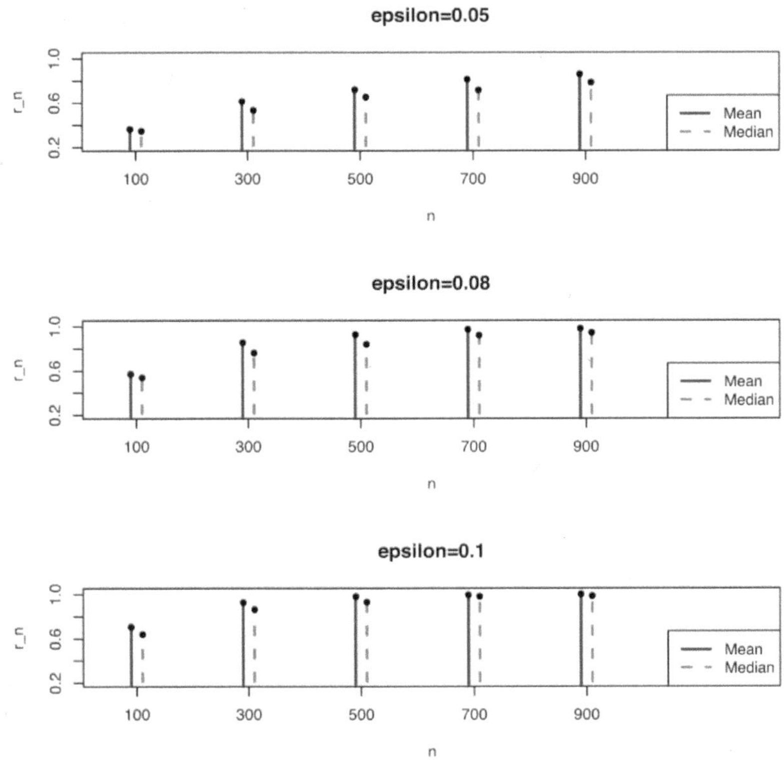

FIGURE 7.3
Rate of Convergence of Sample Mean and Sample Median

Lemma 7.2.1 is useful in many examples and derivations in subsequent sections and chapters.

Theorem 7.2.7. *Suppose $\{Y_n, n \geq 1\}$ is a sequence of random variables such that $Y_n \xrightarrow{P} 0$. (i) If X is a real random variable then $XY_n \xrightarrow{P} 0$. (ii) Suppose $\{X_n, n \geq 1\}$ is a sequence of random variables such that $X_n \xrightarrow{P} X$. Then $X_nY_n \xrightarrow{P} 0$.*

Proof. (i) If $X \equiv 0$, then the result is trivially true. Hence we assume $X \neq 0$. For any $\epsilon > 0$ and $M > 0$, observe that

$$\begin{aligned} P[|XY_n| > \epsilon] &= P[|XY_n| > \epsilon, |X| > M] + P[|XY_n| > \epsilon, |X| \leq M] \\ &\leq P[|X| > M] + P[|Y_n| > \epsilon/M] \to 0, \end{aligned}$$

using the fact that X is bounded in probability for the first term and $Y_n \xrightarrow{P} 0$ for the second term. Thus, $XY_n \xrightarrow{P} 0$.
(ii) By Theorem 7.2.1, if $X_n \xrightarrow{P} X$, then the sequence $\{X_n, n \geq 1\}$ is bounded

in probability. Further, $Y_n \xrightarrow{P} 0$ and hence using similar arguments as in (i), it follows that $X_nY_n \xrightarrow{P} 0$. □

As in almost sure convergence, convergence in probability is also closed under arithmetic operations. It is proved in the following theorem, using Theorem 7.2.7.

Theorem 7.2.8. *Suppose $\{X_n, n \geq 1\}$ and $\{Y_n, n \geq 1\}$ are sequences of random variables defined on the same probability space $(\Omega, \mathbb{A}, P)$. If $X_n \xrightarrow{P} X$ and $Y_n \xrightarrow{P} Y$, then (i) $aX_n \xrightarrow{P} aX$. (ii) $X_n \pm Y_n \xrightarrow{P} X \pm Y$. (iii) $X_nY_n \xrightarrow{P} XY$. (iv) $X_n/Y_n \xrightarrow{P} X/Y$, provided X_n/Y_n and X/Y are defined.*

Proof. (i) $X_n \xrightarrow{P} X$ implies $aX_n \xrightarrow{P} aX$ is trivially true if $a = 0$. Suppose $a \neq 0$, then observe that,

$$P[|aX_n - aX| > \epsilon] = P[|X_n - X| > \epsilon/|a|] = P[|X_n - X| > \epsilon^*] \to 0, \text{ as } n \to \infty,$$

where $\epsilon^* = \epsilon/|a|$.
(ii) $X_n \xrightarrow{P} X$ and $Y_n \xrightarrow{P} Y$ implies that,

$$\forall\ \epsilon > 0,\ \ P[|X_n - X| > \epsilon] \to 0 \text{ and } P[|Y_n - Y| > \epsilon] \to 0.$$

Observe that for $\epsilon > 0$,

$$\epsilon < |(X_n + Y_n) - (X + Y)| = |(X_n - X) - (Y_n - Y)| < |X_n - X| + |Y_n - Y|.$$

It is clear that if the sum of two non-negative numbers exceeds ϵ, then at least one of the two has to be greater than $\epsilon/2$. Thus as $n \to \infty$,

$$\begin{aligned} P[|(X_n + Y_n) - (X + Y)| > \epsilon] &\leq P[|X_n - X| > \epsilon/2] + P[|Y_n - Y| > \epsilon/2] \\ &\to 0 \Rightarrow X_n + Y_n \xrightarrow{P} X + Y. \end{aligned}$$

Another approach is as follows. If

$$\begin{aligned} E_n &= [|X_n - X| < \epsilon/2] \ \ \&\ \ F_n = [|Y_n - Y| < \epsilon/2] \\ &\Rightarrow P(E_n) \to 1 \ \&\ P(F_n) \to 1 \ \Rightarrow\ P(E_n \cap F_n) \to 1. \end{aligned}$$

If both $|X_n - X|$ and $|Y_n - Y|$ are $< \epsilon/2$ then $|(X_n + Y_n) - (X + Y)| < \epsilon]$. Hence $n \to \infty$,

$$\begin{aligned} P[|(X_n + Y_n) - (X + Y)| < \epsilon] &\geq P[|X_n - X| < \epsilon/2, |Y_n - Y| < \epsilon/2] \\ \to 1 &\Rightarrow X_n + Y_n \xrightarrow{P} X + Y. \end{aligned}$$

Using (i) if $Y_n \xrightarrow{P} Y$ then $-Y_n \xrightarrow{P} -Y$ hence

$$X_n - Y_n = X_n + (-Y_n) \xrightarrow{P} X - Y\ .$$

(iii) Observe that for $\epsilon > 0$,

$$\begin{aligned}\epsilon &< |X_nY_n - XY| = |X_nY_n - X_nY + X_nY - XY)| \\ &\le |X_n(Y_n - Y)| + |Y(X_n - X)| \\ &\Rightarrow P[|X_nY_n - XY| > \epsilon] \le P[|X_n(Y_n - Y)| > \epsilon/2] + P[|Y(X_n - X)| > \epsilon/2].\end{aligned}$$

Note that $Y_n - Y \xrightarrow{P} 0$ and by Lemma 7.2.1, $\{X_n, n \ge 1\}$ is bounded in probability. Hence by the result (ii) of Theorem 7.2.7, $X_n(Y_n - Y) \xrightarrow{P} 0$. Further, $X_n - X \xrightarrow{P} 0$ and Y is bounded in probability. Hence by the result (i) of Theorem 7.2.7, $Y(X_n - X) \xrightarrow{P} 0$. As a consequence,

$$\begin{aligned}P[|X_nY_n - XY| > \epsilon] &\le P[|X_n(Y_n - Y)| > \epsilon/2] + P[|Y(X_n - X)| > \epsilon/2] \\ &\to 0 \;\; \Rightarrow X_nY_n \xrightarrow{P} XY.\end{aligned}$$

(iv) To prove that $X_n/Y_n \xrightarrow{P} X/Y$, it is enough to prove that $1/Y_n \xrightarrow{P} 1/Y$ and use result (iii). Suppose a function g is defined as $g(x) = 1/x, \ x \ne 0$. Then g is a continuous function. Hence

$$Y_n \xrightarrow{P} Y \;\; \Rightarrow \;\; g(Y_n) = 1/Y_n \xrightarrow{P} g(Y) = 1/Y \ .$$

Hence, together with result (iii) we get $X_n/Y_n \xrightarrow{P} X/Y$. □

In Example 7.2.8, it is shown that the r-th order sample raw moment converges in probability to the corresponding population raw moment. In the next example, using Theorem 7.2.8, we show that the r-th order sample central moment converges in probability to the corresponding population central moment.

Example 7.2.18. Suppose X is a random variable with the r-th order central moment $\mu_r = E((X - E(X))^r)$ and $\mu'_{2r} < \infty$. Suppose $m_r = \sum_{i=1}^n (X_i - m'_1)^r/n$ denotes the r-th order sample central moment based on a random sample of size n from the distribution of X. By the binomial theorem, we have,

$$\begin{aligned}\mu_r &= E(X - E(X))^r = \sum_{j=0}^{r} \binom{r}{j} E(X^j)(-1)^{r-j}(E(X))^{r-j} \\ &= \sum_{j=0}^{r} \binom{r}{j} \mu'_j(-1)^{r-j}(\mu'_1)^{r-j}.\end{aligned}$$

On similar lines, m_r can be expressed as,

$$\begin{aligned}m_r &= \Big(\sum_{i=1}^{n}(X_i - m'_1)^r\Big)\Big/n = \Big(\sum_{i=1}^{n}\sum_{j=0}^{r}\binom{r}{j}X_i^j(-1)^{r-j}(m'_1)^{r-j}\Big)\Big/n \\ &= \sum_{j=0}^{r}\binom{r}{j} m'_j(-1)^{r-j}(m'_1)^{r-j}.\end{aligned}$$

In Example 7.2.8, it is shown that $m'_j \xrightarrow{P} \mu'_j$. In Theorem 7.2.8, it is proved that convergence in probability is closed under all arithmetic operations. Hence, $m_r \xrightarrow{P} \mu_r$. Thus, the sample central moment m_r converges in probability to the population central moment μ_r. □

In the setup of inference, Theorem 7.2.8 implies that a class of consistent estimators is closed under all arithmetic operations. The following example illustrates this result.

Example 7.2.19. Suppose the probability density function of an absolutely continuous random variable X is given by $f(x,\theta) = \theta x^{\theta-1}$, $0 < x < 1, \theta > 0$. Then $E(X) = \theta/(\theta+1)$ and $Var(X) = \theta/((\theta+2)(\theta+1)^2)$. Suppose $\overline{X}_n$ denotes the sample mean based on a random sample of size n from the distribution of X. Then, by Chebyshev's inequality, $\overline{X}_n \xrightarrow{q.m.} \theta/(\theta+1)$ and hence $\overline{X}_n \xrightarrow{P} \theta/(\theta+1)$. Now using the fact that convergence in probability is closed under the arithmetic operations, we have

$$\begin{aligned} \overline{X}_n \xrightarrow{P_\theta} \frac{\theta}{\theta+1} & \Rightarrow \frac{1}{\overline{X}_n} \xrightarrow{P_\theta} \frac{\theta+1}{\theta} = \frac{1}{\theta} + 1 \Rightarrow \frac{1}{\overline{X}_n} - 1 \xrightarrow{P_\theta} \frac{1}{\theta} \\ & \Rightarrow \frac{\overline{X}_n}{1-\overline{X}_n} \xrightarrow{P_\theta} \theta. \end{aligned}$$

Thus, $\overline{X}_n/(1-\overline{X}_n)$ is a consistent estimator of θ. □

We have a following corollary to Theorem 7.2.8, which is useful in many examples and derivations.

Corollary 7.2.2. *Suppose $\{X_n, n \geq 1\}$ and $\{Y_n, n \geq 1\}$ are sequences of random variables defined on the same probability space $(\Omega, \mathbb{A}, P)$. If $X_n - Y_n \xrightarrow{P} 0$ and $Y_n \xrightarrow{P} X$, then $X_n \xrightarrow{P} X$.*

Proof. Observe that $X_n = (X_n - Y_n) + Y_n \xrightarrow{P} X$ by Theorem 7.2.8. □

The following example illustrates Corollary 7.2.2.

Example 7.2.20. In Example 7.2.5, it is shown that if $X \sim U(0,1)$, then $X_{(n)} \xrightarrow{P} 1$. We now examine whether $X_{(n-1)} \xrightarrow{P} 1$. Observe that for $\epsilon > 0$

$$\begin{aligned} P[|X_{(n-1)} - X_{(n)}| < \epsilon] & \geq P[X_{(n-1)} = X_{(n)}] = P[X_{(n-1)} \geq X_n] \\ & = E(P[X_{(n-1)} \geq X_n | X_n]) = E(1 - X_n^{n-1}) \\ & = \int_0^1 (1-u^{n-1})\, du = 1 - 1/n \ \to \ 1 \\ & \Rightarrow X_{(n-1)} - X_{(n)} \xrightarrow{P} 0 \\ & \Rightarrow X_{(n-1)} \xrightarrow{P} 1 \ \text{ since } \ X_{(n)} \xrightarrow{P} 1. \end{aligned}$$ □

We use Corollary 7.2.2 in the following theorem.

Theorem 7.2.9. *(i) Suppose $\{X_n, n \geq 1\}$ is a sequence of random variables defined on $(\Omega, \mathbb{A}, P)$ such that $X_n \xrightarrow{P} X$ and $\{A_n, n \geq 1\}$ is a sequence of sets in $\mathbb{A}$. (i) If $A_n \downarrow \emptyset$, then $XI_{A_n} \xrightarrow{P} 0$ and $X_n I_{A_n} \xrightarrow{P} 0$. (ii) If $A_n \uparrow \Omega$, then $XI_{A_n} \xrightarrow{P} X$ and $X_n I_{A_n} \xrightarrow{P} X$.*

Proof. It is proved in Theorem 2.3.9 that $A_n \to A \Rightarrow I_{A_n} \to I_A$ point-wise, hence almost surely and hence in probability.
(i) If $A_n \downarrow \emptyset$, then $I_{A_n} \xrightarrow{P} I_\emptyset \equiv 0$. Further, X being a real random variable, X is bounded in probability. Hence by Theorem 7.2.7, $XI_{A_n} \xrightarrow{P} 0$. Note that $X_n \xrightarrow{P} X$ implies that X_n is bounded in probability and hence $X_n I_{A_n} \xrightarrow{P} 0$.
(ii) If $A_n \uparrow \Omega$, then $I_{A_n} \xrightarrow{P} I_\Omega \equiv 1 \Rightarrow (I_{A_n} - 1) \xrightarrow{P} 0$. By Theorem 7.2.7, $X(I_{A_n} - 1) \xrightarrow{P} 0$ which implies that $XI_{A_n} \xrightarrow{P} X$, by Corollary 7.2.2. Further, $I_{A_n} \to I_A$ point-wise and hence in probability. Thus, I_{A_n} is bounded in probability. It is given that $X_n \xrightarrow{P} X$. Consequently, $(X_n - X)I_{A_n} \xrightarrow{P} 0$, but $XI_{A_n} \xrightarrow{P} X$ and hence $X_n I_{A_n} \xrightarrow{P} X$. □

The next example demonstrates various approaches to show that $X_n \xrightarrow{P} X$.

Example 7.2.21. Suppose $\{X_n, n \geq 1\}$ is a sequence of random variables defined as $X_n = (1 + 1/n)X$, where X is a discrete random variable with support $\{0, 1\}$ and $P[X = 0] = 2/3$. We show that $X_n \xrightarrow{P} X$ by various approaches.
(i) From the distribution of X it is clear that $E(X^r) = 1/3 \;\; \forall \;\; r \geq 1$. Thus for any $\epsilon > 0$, as $n \to \infty$,

$$P[|X_n - X| < \epsilon] = P\left[|X/n| < \epsilon\right] \geq 1 - E(X/n)^2/\epsilon^2 = 1 - 1/3n^2\epsilon^2 \to 1.$$

Hence, $X_n \xrightarrow{P} X$.
(ii) Observe that $X_n - X = (1/n)X$. X is a real random variable and hence is bounded in probability, $1/n \to 0$ and hence by Theorem 7.2.7, $(1/n)X \xrightarrow{P} 0$ and hence $X_n \xrightarrow{P} X$.
(iii) Further for any $r > 0$, $E(|X_n - X|^r) = (1/n^r)E(|X|^r) = 1/(3n^r) \to 0$, hence $X_n \xrightarrow{r} X$ and hence $X_n \xrightarrow{P} X$.
(iv) For any

$$\omega \in \Omega, \;\; X_n(\omega) = (1 + 1/n)X(\omega) \to X(\omega),$$

implying that $X_n \to X$ point-wise, hence almost surely and hence in probability.
(v) Further $\forall \;\; r > 1$,

$$\sum_{n=1}^{\infty} E(|X_n - X|^r) = \sum_{n=1}^{\infty} 1/(3n^r) < \infty \;\; \Rightarrow \;\; X_n \xrightarrow{a.s.} X \;\; \Rightarrow \;\; X_n \xrightarrow{P} X. \;\; □$$

We now define the Cauchy criterion of convergence in probability. It is also known as mutual convergence in probability.

Definition 7.2.3. *A sequence $\{X_n, n \geq 1\}$ of random variables is said to be Cauchy (fundamental) in probability if $\forall \ \epsilon > 0$, $P[|X_n - X_m| > \epsilon] \to 0$ as $m, n \to \infty$.*

Convergence in probability implies mutual convergence in probability, as shown in the following theorem.

Theorem 7.2.10. *If $X_n \stackrel{P}{\to} X$ as $n \to \infty$ then $X_n - X_m \stackrel{P}{\to} 0$ as $m, n \to \infty$.*

Proof. Since X is a real random variable, it is finite almost surely, that is, on N^c where $P(N) = 0$. Thus for all $\omega \in N^c$ and for any $\epsilon > 0$, we have

$$\begin{aligned} [|X_n - X_m| > \epsilon] &\subseteq [|X_n - X| > \epsilon/2] \cup [|X_m - X| > \epsilon/2] \\ \Rightarrow \quad P[|X_n - X_m| > \epsilon] &\leq P[|X_n - X| > \epsilon/2] + P[|X_m - X| > \epsilon/2] \to 0 \\ \Rightarrow \quad X_n - X_m &\stackrel{P}{\to} 0 \text{ as } m, n \to \infty. \end{aligned}$$

□

To prove that Cauchy convergence in probability implies convergence in probability, we need to prove the following theorem. It is also helpful in proving Lebesgue dominated probability convergence theorem in Chapter 8.

Theorem 7.2.11. *Suppose $\{X_n, n \geq 1\}$ is a sequence of random variables such that $X_n \stackrel{P}{\to} X$. Then there exists a sub-sequence of $\{X_n, n \geq 1\}$ which converges to X almost surely and hence in probability.*

Proof. By definition

$$\begin{aligned} X_n \stackrel{P}{\to} X &\Rightarrow \forall \ \epsilon > 0, \ \lim_{n \to \infty} P[|X_n - X| > \epsilon] = 0 \\ &\iff \forall \ k \geq 1, \ \exists \ n(k) \text{ such that } P[|X_n - X| > 1/2^k] < 1/2^k, \end{aligned}$$

$\forall \ n \geq n(k)$. We define $n_1 = n(1)$, $n_2 = \max\{n_1 + 1, n(2)\}, \cdots$, $n_k = \max\{n_{k-1} + 1, n(k)\}$. Thus, $\{n_k, k \geq 1\}$ is an increasing sequence and $n_k \to \infty$ as $k \to \infty$. Thus, $\{X_{n_k}, k \geq 1\}$ is a sub-sequence of the sequence $\{X_n, n \geq 1\}$. Now,

$$\sum_{n=1}^{\infty} P[|X_{n_k} - X| > 1/2^k] < \sum_{n=1}^{\infty} 1/2^k < \infty \quad \Rightarrow \quad X_{n_k} \stackrel{a.s}{\to} X$$

by Corollary 6.3.1. Since almost sure convergence implies convergence in probability, $X_{n_k} \stackrel{P}{\to} X$ □

Using Theorem 7.2.11, we now prove that Cauchy convergence in probability implies convergence in probability.

Theorem 7.2.12. *A sequence $\{X_n, n \geq 1\}$ of random variables is Cauchy in probability implies that $X_n \xrightarrow{P} X$, for some X.*

Proof. A sequence $\{X_n, n \geq 1\}$ is Cauchy in probability implies that $X_n - X_m \xrightarrow{P} 0$ as $m, n \to \infty$. Hence, $\forall\ k \geq 1$, $\exists\ n(k)$ such that $\forall\ n, m \geq n(k)$, $P[|X_n - X_m| \geq 1/2^k] < 1/2^k$. Consequently, as shown in Theorem 7.2.11, there exists a sub-sequence $\{X_{n_k} - X_{m_k}, k \geq 1\}$ of the sequence $\{X_n - X_m, n \geq 1\}$ such that $X_{n_k} - X_{m_k} \xrightarrow{a.s} 0$. In Theorem 6.3.6, it is proved that almost sure convergence and almost sure Cauchy convergence are equivalent. Hence, $X_{n_k} \xrightarrow{a.s} X$ for some X. Almost sure convergence implies convergence in probability and hence $X_{n_k} \xrightarrow{P} X$. Thus, $\forall\ \ \epsilon > 0$,

$$P[|X_n - X| > \epsilon] \leq P[|X_n - X_{n_k}| > \epsilon/2] + P[|X_{n_k} - X| > \epsilon/2] \to 0\ ,$$

as $n,\ k \to \infty$, where $P[|X_n - X_{n_k}| > \epsilon/2] \to 0$ in view of Cauchy convergence in probability and $P[|X_{n_k} - X| > \epsilon/2] \to 0$ as $X_{n_k} \xrightarrow{P} X$. Thus, $X_n \xrightarrow{P} X$, for some X. □

The following theorem gives a necessary and sufficient condition for convergence in probability, using the basic inequality discussed in Chapter 4.

Theorem 7.2.13. $X_n \xrightarrow{P} 0$ *if and only if* $E\left(|X_n|/(1 + |X_n|)\right) \to 0$ *as* $n \to \infty$.

Proof. Suppose a function g is defined as $g(x) = |x|/(1 + |x|),\ x \in \mathbb{R}$. It is clear that g is a non-negative and even function. Further,

$$|x_1| \leq |x_2| \Rightarrow 1 + \frac{1}{|x_1|} \geq 1 + \frac{1}{|x_2|} \Rightarrow \frac{|x_1|}{1 + |x_1|} \leq \frac{|x_2|}{1 + |x_2|}.$$

Thus, $g(x) = |x|/(1+|x|)$ is a non-decreasing function. By definition, $g(x) \leq 1$, thus supremum of $g(x)$ is 1 and almost sure supremum of $g(X)$ is 1. With such a choice of a function g, we replace X by X_n and a by ϵ in the basic inequality to get,

$$E\left(\frac{|X_n|}{1 + |X_n|}\right) - \frac{\epsilon}{1 + \epsilon} \ \leq\ P[|X_n| \geq \epsilon] \ \leq\ \left(E\left(\frac{|X_n|}{1 + |X_n|}\right)\right) \Big/ \left(\frac{\epsilon}{1 + \epsilon}\right).$$

Suppose $E\left(|X_n|/(1 + |X_n|)\right) \to 0$ as $n \to \infty$. Then by the right hand side of the above inequality, we get $P[|X_n| \geq \epsilon] \to 0$ as $n \to \infty$, $\forall\ \epsilon > 0$ and hence $X_n \xrightarrow{P} 0$. Conversely, assume that $X_n \xrightarrow{P} 0$, that is, $P[|X_n| \geq \epsilon] \to 0$ as $n \to \infty\ \ \forall\ \epsilon > 0$, hence by the left hand side of the above inequality, we get $E\left(|X_n|/(1 + |X_n|)\right) \to 0$. □

The next section is concerned with an important mode of convergence, convergence in law.

7.3 Convergence in Law

Convergence in law is the most frequently used mode of convergence in asymptotic inference, in large sample interval estimation and to find the asymptotic null distribution of a test statistic. The important theorem in this mode of convergence is the central limit theorem, which we discuss in detail in Chapter 10. In the present section, we study some aspects listed at the end of Section 6.2 for this mode of convergence and illustrate by examples. We state below its definition, which is already given in Section 6.2.

Definition 7.3.1. *Convergence in law: Suppose $\{X_n, n \geq 1\}$ is a sequence of random variables defined on a probability space $(\Omega, \mathbb{A}, P)$. Suppose $F_n(\cdot)$ is a distribution function of X_n and $F_X(\cdot)$ is a distribution function of X. A sequence $\{X_n, n \geq 1\}$ is said to converge in law to a random variable X, not necessarily defined on the same probability space, if*

$$F_n(x) = P[X_n \leq x] \to P[X \leq x] = F_X(x), \quad \forall \ x \in C(F_X),$$

where $C(F_X)$ is a set of points of continuity of the distribution function F_X.

We illustrate below the convergence in law with some examples.

Example 7.3.1. Suppose $\{X_n, n \geq 1\}$ is a sequence of random variables with $P[X_n = \pm 1/n] = 1/2$. Then note that the distribution function F_n of X_n, given by,

$$F_{X_n}(x) = \begin{cases} 0, & \text{if} \quad x < -1/n \\ 1/2, & \text{if} \quad -1/n \leq x < 1/n \\ 1, & \text{if} \quad x \geq 1/n \end{cases} \quad \to \ F(x) = \begin{cases} 0, & \text{if} \quad x < 0 \\ 1, & \text{if} \quad x > 0. \end{cases}$$

If we define $X \equiv 0$, then 0 is the only point of discontinuity of a distribution function F. Thus, $F_{X_n}(x) \to F_X(x)$ for all $x \in C(F_X)$. Hence, we conclude that $X_n \xrightarrow{L} X$. □

Example 7.3.2. Suppose $\{X_n, n \geq 1\}$ is a sequence of random variables such that X_n has exponential distribution with mean $\theta_n = 1 + 1/n$. For $x < 0$, $F_n(x) = 0 \ \forall \ n \geq 1$ and hence converges to 0 as $n \to \infty$. For $x \geq 0$,

$$F_n(x) = P[X_n \leq x] = 1 - \exp\{-x/\theta_n\} \quad \to \quad 1 - \exp\{-x\} = F_X(x) \ ,$$

where F is a distribution function of a random variable following the exponential distribution with mean 1. Thus, $F_n(x) \to F_X(x) \ \forall \ x \in \mathbb{R} = C(F_X)$ and hence $X_n \xrightarrow{L} X$ where X has exponential distribution with mean 1. □

Example 7.3.3. Suppose $X_1, X_2, \cdots, X_n$ are independent and identically distributed random variables each having uniform $U(0,\theta)$ distribution. Then as shown in Example 7.2.5, distribution function of $X_{(n)}$ given by,

$$F_{X_{(n)}}(x) = [F_X(x)]^n \quad = \quad \begin{cases} 0, & \text{if} \quad x < 0 \\ (x/\theta)^n, & \text{if} \quad 0 \le x < \theta \\ 1, & \text{if} \quad x \ge \theta. \end{cases}$$

Using similar arguments as in Example 7.2.5, it can be shown that $X_{(n)} \xrightarrow{P} \theta$. Suppose $Y_n = n(\theta - X_{(n)})$. We derive its distribution function $G_{Y_n}(y)$ for $y \in \mathbb{R}$ from the distribution function of $X_{(n)}$. Since $X_{(n)} \le \theta$, $Y_n \ge 0$, hence for $y < 0$, $G_{Y_n}(y) = 0$. Suppose $y \ge 0$, then

$$G_{Y_n}(y) = P_\theta[n(\theta - X_{(n)}) \le y] = P_\theta[X_{(n)} \ge \theta - y/n] = 1 - F_{X_{(n)}}(\theta - y/n).$$

Hence,

$$G_{Y_n}(y) \quad = \quad \begin{cases} 1 - 0, & \text{if} \quad \theta - y/n < 0 \\ & \text{that is} \quad y \ge n\theta \\ 1 - \left(\frac{\theta - y/n}{\theta}\right)^n = 1 - (1 - \frac{y}{n\theta})^n, & \text{if} \quad 0 < y \le n\theta \\ 1 - 1, & \text{if} \quad y \le 0. \end{cases}$$

As $n \to \infty$,

$$G_{Y_n}(y) = G_n(y) \quad \to \quad F_Y(y) = \begin{cases} 0, & \text{if} \quad y < 0 \\ 1 - e^{-(y/\theta)}, & \text{if} \quad y \ge 0. \end{cases}$$

However, $F_Y(y)$ is a distribution function of a random variable Y, which has an exponential distribution with location parameter 0 and scale parameter $1/\theta$. Thus,

$$Y_n = n(\theta - X_{(n)}) \xrightarrow{L} Y \quad \Longleftrightarrow \quad Y_n/\theta = U_n \xrightarrow{L} U$$

where U which has exponential distribution with location parameter 0 and scale parameter 1. □

In the following example, using R, we examine the convergence of $G_n(\cdot)$ to $F_Y(\cdot)$ as proved in Example 7.3.3.

Example 7.3.4. Suppose $\{X_1, X_2, \cdots, X_n\}$ are independent and identically distributed random variables, each having uniform $U(0,\theta)$ distribution. Then it is shown in Example 7.3.3 that $Y_n \xrightarrow{L} Y$, where Y has an exponential distribution with location parameter 0 and scale parameter $1/\theta$. We use the following code to examine the convergence of $G_n(\cdot)$ to $F_Y(\cdot)$. We take $n = 40, 60, 80, 100$ and $\theta = 0.2$.

Code 7.3.1.
```
th=0.2; nvec=c(40,60,80,100);l=length(nvec)
u=0:20; x=1-exp(-u/th); y=z=list(); par(mfrow=c(2,2))
no=paste("n=",nvec,sep=" ")
for(j in 1:l)
{
 y[[j]]=seq(0,nvec[j]*th,length=100)
 z[[j]]=1-(1-y[[j]]/(nvec[j]*th))^(nvec[j])
 plot(y[[j]],z[[j]],type="o", col="green",main=no[j],pch=19,
 ylab=expression(paste(Gn(y))),xlab=expression(paste(y)),lwd=1)
 lines(u,x,type="o",col="dark green",pch=20)
 legend("bottomright",legend=c("Gn(y)","F(y)"),lwd=c(1,1),
 pch=c(19,20),col=c("green","dark green"))
}
```

Figure 7.4 displays graphs of $G_n(y)$ for four values of n, and the graph of $F_Y(y)$ is imposed on it. We note that as n increases, the curves of two distribution functions are closer to each other. For $n = 100$, these are very close to each other. □

We now discuss how to verify convergence in law by simulation using R software and illustrate for Example 7.3.3. We generate m observations from the given distribution of X_n and obtain a histogram of simulated values of X_n or some function of X_n, with relative frequencies on the Y-axis. For Example 7.3.3, we obtain the histogram of U_n. From the visual inspection, we may guess the limiting distribution. We then impose a curve of the corresponding probability law on it and can judge the closeness. We can justify the claim analytically using the chi-square goodness of fit test or the Komogorov-Smirnov test.

For example, if the histogram is bell-shaped, we impose the curve of the probability density function of the normal distribution on it. We observe whether the two are close to each other as the sample size increases. We may also draw a box plot and QQ plot to assess the normality of X_n for large n. In addition, we draw a graph of the empirical distribution function of simulated values of X_n and impose the graph of the distribution function of the normal distribution on it. The visual impression produced by these four types of plots can be judged analytically by the Shapiro-Wilk test. We adopt such a procedure in Example 7.3.25.

The stepwise procedure to verify the convergence in law is as follows.

(i) Generate m observations from the distribution of X_n.

(ii) Corresponding to m values of X_n, draw a histogram with relative frequencies on the Y-axis and impose a curve of the probability density function of a suitable distribution on it. Further, carry out the appropriate goodness of fit test.

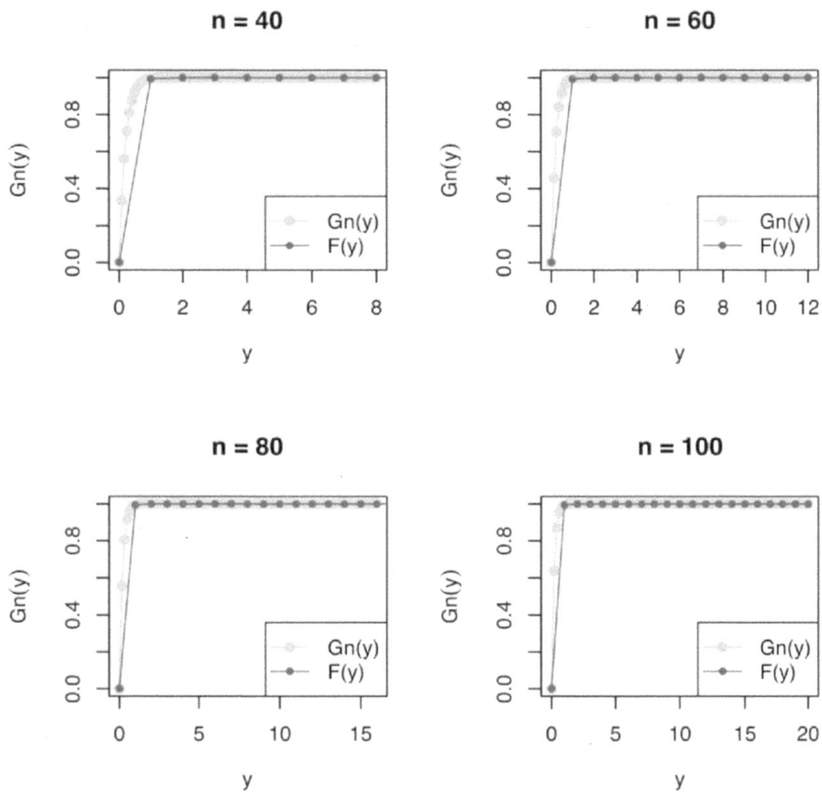

FIGURE 7.4
Convergence in law of Y_n

(iii) Increase the sample size n and repeat the steps (i) and (ii).

We illustrate the above procedure for Example 7.3.3, using Code 7.3.2.

Example 7.3.5. Suppose $\{X_1, X_2, \cdots, X_n\}$ are independent and identically distributed random variables, each having uniform $U(0, \theta)$ distribution. Suppose $U_n = n(\theta - X_{(n)})/\theta$, then it is shown in Example 7.3.3 that $U_n \xrightarrow{L} U$, where U has exponential distribution with location parameter 0 and scale parameter 1. In Example 7.3.4, we have noted that the two distribution functions are close to each other for $n = 100$. Hence, we verify this result by simulation for $n = 100$ and $\theta = 0.2$ using the following code.

Code 7.3.2.
```
n=100; th=0.2; m=500; t=c()
for(i in 1:m)
{
 set.seed(i)
 x=runif(n,0,th)
```

```
 t[i]=max(x)
}
tn=n*(th-t)/th; me=mean(tn);me; v=var(tn)*(m-1)/m;v
sd=v^.5; sd; r=seq(0,8,.1); y=dexp(r,rate=1)
par(mfrow=c(1,1))
hist(tn,freq=FALSE,main="Histogram of U_n",ylim=c(0,0.6),
xlab = expression(paste("U"[n])),col="light blue")
lines(r,y,"o",pch=20,col="dark blue",lty=1)
legend("topright",legend="pdf of exponential distribution
with mean 1",lty=1,pch=20,col="dark blue",cex=.9)
O=hist(tn,plot=FALSE)$counts; sum(O)
bk=hist(tn,plot=FALSE)$breaks; bk
M=max(bk); u=seq(0,M,.1); y1=dexp(u, rate=1)
e=exp(1); ep=c()
for(i in 1:(length(bk)-1))
{
 ep[i]=e^(-bk[i])-e^(-bk[i+1])
}
a=1-sum(ep); a; ep1=c(ep, a); ef=sum(O)*ep1; O1=c(O,0)
d=data.frame(O1,round(ef,2)); d
ts=sum((O1-ef)^2/ef);ts
df=length(ef)-1; df; b=qchisq(.95,df); b
p1=1-pchisq(ts,df); p1; chisq.test(O1,p=ep1)
# pooling of frequencies less than 5 is ignored
```

From the output, we note that sample mean 1.0371 and sample standard deviation 1.0284 of the simulated values of U_n are close to each other, supporting the result that for large n the distribution of U_n can be approximated by the exponential distribution. Further, the histogram indicates that the limiting distribution is skew. We have imposed the curve of the probability density function of the exponential distribution with mean 1. We note that there is a close agreement. We carry out the goodness of fit test to test the null hypothesis that for large n, the distribution of U_n is an exponential distribution with a mean one. We use Karl Pearson's chi-square test statistic T given by $T = \sum_{i=1}^{k}(o_i - e_i)^2/e_i$, where k denotes the number of classes in which m observations on U_n are classified, o_i and e_i denote the observed frequency and expected frequency respectively, expected under the exponential distribution with mean 1, for the i-th class. Under the null setup, T has χ^2_{k-1} distribution. For the simulated data, $T = 8.3287$, with 8 degrees of freedom, the cut-off is 15.5073, and the p-value is 0.4020. The built-in function `chisq.test(O1,p=ep1)` gives the same results. From the cut-off point,

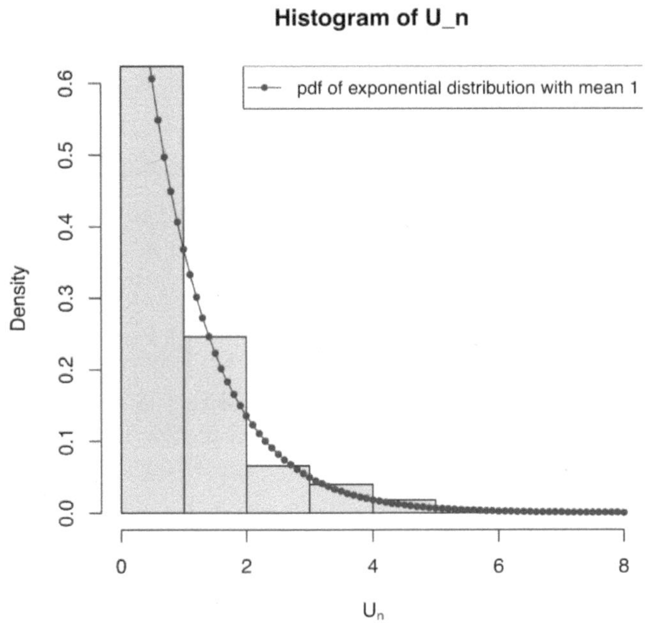

FIGURE 7.5
$U(0,\theta)$ Distribution: Convergence in Law of $n(\theta - X_{(n)})/\theta$

p-value, results of the built-in function and Figure 7.5, we note that for large n, the distribution of $U_n = n(\theta - X_{(n)})/\theta$ can be approximated by that of U, where U has an exponential distribution with location parameter 0 and scale parameter 1. □

It is a well-known result that a binomial distribution under certain conditions converges to a Poisson distribution. We prove it below by showing that the probability mass function of a binomial distribution converges to that of a Poisson distribution. Convergence of the corresponding distribution functions then follows by Scheffe's lemma, refer to p.228 of Gut [13].

Example 7.3.6. Suppose $X_n \sim B(n, p_n)$ distribution. Suppose as $n \to \infty$, $p_n \to 0$ such that $np_n \to \lambda$ where λ is a positive constant. Under these conditions, we examine whether the probability mass function of a binomial distribution tends to that of a Poisson distribution. Suppose $q_n = 1 - p_n$, then

$$\begin{aligned}
\lim_{n\to\infty} n \log q_n &= \lim_{n\to\infty} \{-n(p_n + p_n^2/2 + p_n^3/3 + \cdots)\} \\
&= \lim_{n\to\infty} \{-np_n - (np_n)(p_n/2) - (np_n)(p_n^2/3) - \cdots\} \\
&= -\lambda \text{ as } np_n \to \lambda \ \& \ p_n \to 0 \\
\Rightarrow \quad \lim_{n\to\infty} q_n^n &= e^{-\lambda}.
\end{aligned}$$

With this limit, observe that,

$$\begin{aligned} P[X_n = k] &= \binom{n}{k} p_n^k q_n^{n-k} = \frac{n(n-1)\cdots(n-k+1)}{k(k-1)\cdots 1} \frac{p_n^k}{q_n^k} q_n^n \\ &= \frac{(1-1/n)(1-2/n)\cdots(1-(k-1)/n)}{k!} \frac{(np_n)^k}{(1-p_n)^k} q_n^n \\ &\to (1/k!)\lambda^k e^{-\lambda} = P[X = k], \end{aligned}$$

as $n \to \infty$ and $p_n \to 0$ such that $np_n \to \lambda$. Note that $X \sim P(\lambda)$. □

In the following example, we verify the results of Example 7.3.6 by drawing the graphs of the probability mass function of binomial $B(n, p_n)$ distribution for different values of n and comparing these with the graph of Poisson distribution with parameter λ. It gives some idea of how large n is needed and the rate of convergence of p_n to 0.

Example 7.3.7. Suppose $X_n \sim B(n, p_n)$. We have proved in the above example that binomial distribution can be approximated by the Poisson $P(\lambda)$ distribution as $n \to \infty$, $p_n \to 0$ such that $np_n \to \lambda$. We use the following code for verification.

Code 7.3.3. We take $n = 10, 50, 100, 500$, $\lambda = 2$ and $p_n = \lambda/(n+1)$. We plot the probability mass function of $B(n, p_n)$ and $P(\lambda)$ and their distribution functions to observe for which n these are close. Figure 7.6 displays these graphs.

```
lambda=2; nvec=c(10,50,100,500)
k=1;pvec=lambda/(nvec+k); xvec=0:10
PMFPois=dpois(xvec,lambda)
PMFBin=matrix(ncol=length(xvec),nrow=length(nvec))
for(i in 1:length(nvec))
{
 n=nvec[i]
 p=pvec[i]
 PMFBin[i,]=dbinom(xvec,size=n,prob=p)
}
PMF=rbind(PMFBin,PMFPois); colnames(PMF)=xvec
rownames(PMF)=c(nvec,"Poisson"); PMF
par(mfrow=c(1,2))
M=max(PMFPois,PMFBin)
plot(xvec+0.05*(length(nvec)+2),PMFPois,col=1,type="h",
ylim=c(0,M+0.1),xaxt="n",xlab="X",ylab="PMF",
xlim=c(-0.5,max(xvec)+0.5))
for(i in 1:length(nvec))
```

```
{
 lines(xvec+0.05*i,PMFBin[i,],type="h",col=i+1)
}
axis(1,at=xvec+0.25,labels=xvec)
labvec=paste("n=", nvec, sep="")
labvec=c("Poisson",labvec)
legend("topright",col=1:(length(nvec)+1),legend=labvec,
lty=1,xjust=1)
CDF=t(apply(PMF,1,cumsum))
View(CDF)
plot(xvec,CDF[length(nvec)+1,],col=1,type="s",ylim=c(0,1.1),
xlab="X",ylab="CDF",yaxs="i",lwd=2)
for(i in 1:length(nvec))
{
 lines(xvec,CDF[i,],type="s",col=i+1)
}
legend("bottomright",col=1:(length(nvec)+1),legend=labvec,
lty=1,xjust=1)
```

In the code for both the distributions, we compute probabilities for $x = 0$ to $x = 10$ since for $x > 10$, these are almost 0. From Figure 7.6, we observe that for $n = 100$ and 500, the probability mass function, as well as the distribution function of the two distributions, are close to each other. From the output, we note that

$$P[X_n = 0] = q^n = 0.1344, 0.1353, 0.1353, 0.1353 \quad \text{for} \quad n = 10, 50, 100, 500$$

respectively and $P[X = 0] = \exp(-\lambda) = 0.1353$, where $X \sim P(\lambda)$. In the left panel of the graph, displaying the probability mass function, the five vertical lines at $x = 0$ are almost identical and in the right panel of the graph, displaying the distribution function, the five horizontal lines from $x = 0$ to $x = 1$ overlap each other. Thus, the result $\lim_{n\to\infty} q_n^n = e^{-\lambda}$ proved in Example 7.3.6, is verified numerically. We note that the limit is attained at $n = 100$. A minute observation of the graph in the left panel shows that the height of vertical lines corresponding to the probability mass function of the binomial distribution is not monotonically increasing or decreasing, it increases initially and then decreases. Such a nature results from the fact that the total probability has to add to 1. □

We now investigate the links between convergence in law and other modes of convergence. In the following theorem, we prove that convergence in probability implies convergence in law. We have already proved that almost sure convergence and convergence in r-th mean imply convergence in probability.

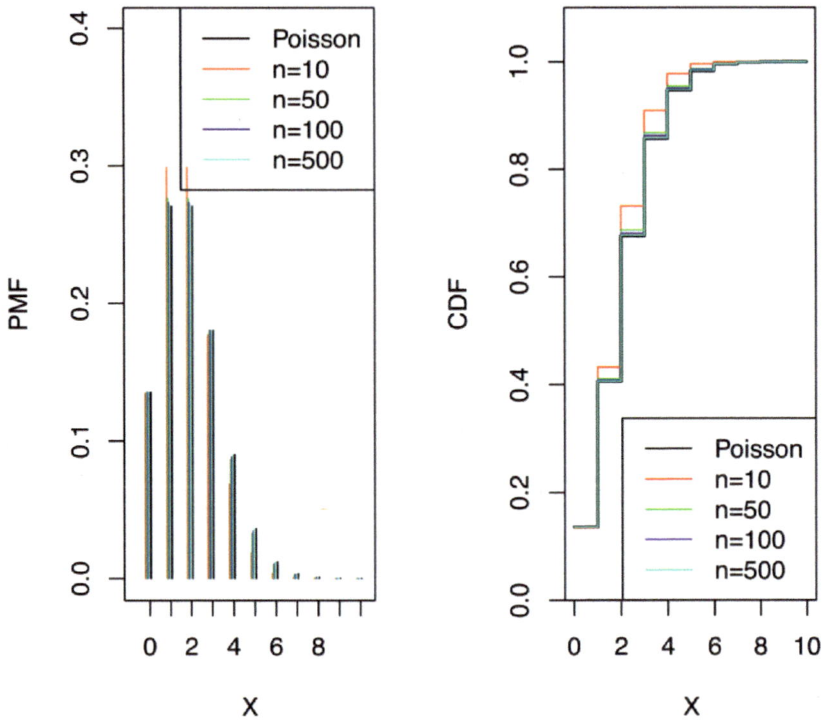

FIGURE 7.6
Poisson Distribution Approximation to Binomial Distribution

Thus, both the almost sure convergence and convergence in r-th mean imply convergence in law. Consequently, convergence in law is the weakest form of convergence discussed in Chapters 6 and 7.

Theorem 7.3.1. *Suppose* $\{X_n, n \geq 1\}$ *is a sequence of random variables defined on the probability space* $(\Omega, \mathbb{A}, P)$*. Then* $X_n \stackrel{P}{\to} X \;\Rightarrow\; X_n \stackrel{L}{\to} X$.

Proof. Suppose $F_n(\cdot)$ and $F_X(\cdot)$ are distribution functions of X_n and X respectively. $X_n \stackrel{L}{\to} X$ if $F_n(x) \to F_X(x)\ \forall\ x \in C(F_X)$, provided $\lim_{n\to\infty} F_n(x)$ exists. We first check whether $\lim_{n\to\infty} F_n(x)$ exists, that is, whether $\limsup_{n\to\infty} F_n(x) = \liminf_{n\to\infty} F_n(x)$, for all $x \in C(F_X)$. Note that $\liminf_{n\to\infty} F_n(x) \leq \limsup_{n\to\infty} F_n(x)$ always. Thus it is enough to prove that

$$F_X(x) \leq \liminf_{n\to\infty} F_n(x) \quad \& \quad \limsup_{n\to\infty} F_n(x) \leq F_X(x)\ \ \forall\ \ x \in C(F_X)\ .$$

Observe that for any $\epsilon > 0$,

$$\begin{aligned} X_n \ \stackrel{P}{\to}\ X \ \Rightarrow\ & \lim_{n\to\infty} P[|X_n - X| > \epsilon] = 0 \\ \Rightarrow\ & \limsup_{n\to\infty} P[|X_n - X| > \epsilon] = \liminf_{n\to\infty} P[|X_n - X| > \epsilon] = 0. \quad (7.1) \end{aligned}$$

Note that for any real numbers x and x_1 such that $x_1 < x$,

$$\begin{aligned}
P[X \leq x_1] &= P[X \leq x_1, X_n \leq x] + P[X \leq x_1, X_n > x] \\
&\leq P[X_n \leq x] + P[|X_n - X| > x - x_1] \\
\Rightarrow \quad F_X(x_1) &\leq F_n(x) + P[|X_n - X| > x - x_1] \\
\Rightarrow \liminf_{n\to\infty} F_X(x_1) &\leq \liminf_{n\to\infty} F_n(x) + \liminf_{n\to\infty} P[|X_n - X| > x - x_1] \\
\Rightarrow F_X(x_1) &\leq \liminf_{n\to\infty} F_n(x), \quad \text{by Equation (7.1).}
\end{aligned}$$

Similarly, for any real numbers x and x_2 such that $x < x_2$,

$$\begin{aligned}
P[X_n \leq x] &= P[X_n \leq x, X \leq x_2] + P[X_n \leq x, X > x_2] \\
&\leq P[X \leq x_2] + P[|X_n - X| > x_2 - x] \\
\Rightarrow F_n(x) &\leq F_X(x_2) + P[|X_n - X| > x_2 - x] \\
\Rightarrow \limsup_{n\to\infty} F_n(x) &\leq \limsup_{n\to\infty} F_X(x_2) + \limsup_{n\Rightarrow\infty} P[|X_n - X| > x_2 - x] \\
\Rightarrow \limsup_{n\to\infty} F_n(x) &\leq F_X(x_2), \quad \text{by Equation (7.1).}
\end{aligned}$$

Consequently,

$$F_X(x_1) \leq \liminf_{n\to\infty} F_n(x) \leq \limsup_{n\to\infty} F_n(x) \leq F_X(x_2). \tag{7.2}$$

Suppose $x_1 = x - h_1$ and $x_2 = x + h_2$, $h_1, \ h_2 > 0$ and suppose x is a point of continuity so that F is right continuous as well as left continuous at x. Hence,

$$\begin{aligned}
\text{as } \ h_1 \to 0, \quad F_X(x_1) &= F_X(x - h_1) \to F_X(x) \\
\text{and as } \quad h_2 \to 0, \quad F_X(x_2) &= F_X(x + h_2) \to F_X(x).
\end{aligned}$$

If $h_1, \ h_2 \to 0$ in Equation (7.2), then

$$F_X(x) \leq \liminf_{n\to\infty} F_n(x) \leq \limsup_{n\to\infty} F_n(x) \leq F_X(x).$$

Thus, $\forall \ x \in C(F_X)$, $\lim_{n\to\infty} F_n(x)$ exists and is given by $F_X(x)$ and hence, $X_n \xrightarrow{L} X$. □

The next example illustrates Theorem 7.3.1.

Example 7.3.8. Suppose $\{X_1, X_2, \cdots, X_n\}$ are independent and identically distributed random variables each having uniform $U(0,1)$ distribution. In Example 7.2.5, it is shown that $X_{(1)} \xrightarrow{P} 0$ & $X_{(n)} \xrightarrow{P} 1$. Hence by Theorem 7.3.1, $X_{(1)} \xrightarrow{L} 0$ & $X_{(n)} \xrightarrow{L} 1$. We now prove the convergence in law using the definition. From Example 7.2.5 we have,

$$F_{X_{(1)}}(x) = \begin{cases} 0, & \text{if} \quad x < 0 \\ 1 - (1-x)^n, & \text{if} \quad 0 \leq x < 1 \\ 1, & \text{if} \quad x \geq 1 \end{cases} \quad \& F_{X_{(n)}}(x) = \begin{cases} 0, & \text{if} \quad x < 0 \\ x^n, & \text{if} \quad 0 \leq x < 1 \\ 1, & \text{if} \quad x \geq 1. \end{cases}$$

Observe that,

$$F_{X_{(1)}}(x) \to \begin{cases} 0, & \text{if} \quad x < 0 \\ 1, & \text{if} \quad x \geq 0 \end{cases} \quad \& \quad F_{X_{(n)}}(x) \to \begin{cases} 0, & \text{if} \quad x < 1 \\ 1, & \text{if} \quad x \geq 1. \end{cases}$$

Hence, $X_{(1)} \xrightarrow{L} 0$ & $X_{(n)} \xrightarrow{L} 1$. □

In Theorem 7.3.1, we have proved that $X_n \xrightarrow{P} X \Rightarrow X_n \xrightarrow{L} X$. Thus, $X_n \overset{L}{\nrightarrow} X \Rightarrow X_n \overset{P}{\nrightarrow} X$. The next example illustrates this assertion.

Example 7.3.9. Suppose $\{X_n, n \geq 1\}$ is a sequence of random variables such that $X_n \sim U(-n, n)$ distribution. The distribution function of X_n is given by,

$$F_n(x) = \begin{cases} 0, & \text{if} \quad x < -n \\ (x+n)/2n, & \text{if} \quad -n \leq x < n \\ 1, & \text{if} \quad x \geq n. \end{cases}$$

Then $\forall\ x \in \mathbb{R},\ F_n(x) \to 1/2$. Thus, $F_n(x)$ does not converge to a distribution function and $X_n \overset{L}{\nrightarrow} X$, say. Now observe that,

$$\begin{aligned} \text{for} \quad \epsilon < n, \quad P[|X_n| < \epsilon] &= P[-\epsilon < X_n < \epsilon] = (\epsilon + n)/2n - (-\epsilon + n)/2n \\ &= \epsilon/n \to 0 \quad \text{as} \quad n \to \infty \\ \Rightarrow X_n &\overset{P}{\nrightarrow} 0 \Rightarrow X_n \overset{r}{\nrightarrow} 0 \ \textit{and} \ X_n \overset{a.s.}{\nrightarrow} 0. \end{aligned}$$

Thus, in this example, $\{X_n, n \geq 1\}$ does not converge almost surely, in probability, in r-th mean and in law. □

The converse Theorem 7.3.1 is not true in general. We illustrate this assertion in the following two examples.

Example 7.3.10. Suppose an unbiased die is rolled many times. A random variable X_n is defined as follows.

$$X_n = \begin{cases} 1, & \text{if} \quad n\text{-th toss results in an even number} \\ 0, & \text{if} \quad n\text{-th toss results in an odd number.} \end{cases}$$

The probability distribution of X_n is given by, $P[X_n = 1] = P[X_n = 0] = 1/2$, since the die is unbiased. Hence its distribution function is given by,

$$F_n(x) = \begin{cases} 0, & \text{if} \quad x \leq 0 \\ 1/2, & \text{if} \quad 0 \leq x < 1 \\ 1, & \text{if} \quad x \geq 1. \end{cases}$$

It is the same for all n, and hence its limit is the same distribution function as specified above. Thus, $X_n \xrightarrow{L} X$, where the probability distribution of X is the same as that of X_n. With $Y = 1 - X$, the probability distribution of Y is given by, $P[Y = 1] = P[Y = 0] = 1/2$ and its distribution function is the same as that of X_n and X. Hence, $X_n \xrightarrow{L} Y$ also. Since their distribution functions are the same, X and Y are identically distributed random variables. However, observe that if $X(\omega) = 0$ then $Y(\omega) = 1$ and if $X(\omega) = 1$ then $Y(\omega) = 0$. Thus, $P[X = Y] = 0$ and X and Y are not equivalent random variables. Now using the result that for each $n \geq 1$, X_n and X are identically distributed and $Y(\omega) = 1 - X(\omega)$, we have

$$P[|X_n - Y| > \epsilon] = P[|X - (1 - X)| > \epsilon] = P[|2X - 1| > \epsilon] = 1 \text{ if } \epsilon < 1$$

and hence $X_n \overset{P}{\nrightarrow} Y$. Thus, $X_n \xrightarrow{L} Y$ but $X_n \overset{P}{\nrightarrow} Y$. □

Example 7.3.11. Suppose $\{X_n, n \geq 1\}$ is a sequence of random variables such that $X_n = -X, \ \forall \ n \geq 1$ where $X \sim N(0,1)$. Further, $-X \sim N(0,1)$. Thus,
$X_n \overset{d}{=} X \ \Rightarrow \ X_n \xrightarrow{L} X \sim N(0,1)$. However, for any $\epsilon > 0$,

$$P[|X_n - X| > \epsilon] = P[|X| > \epsilon/2] \nrightarrow 0, \quad \text{as } n \to \infty.$$

Hence, $X_n \overset{P}{\nrightarrow} X$. Thus, $X_n \xrightarrow{L} X$ but $X_n \overset{P}{\nrightarrow} X$. □

As noted in the above two examples, the converse of Theorem 7.3.1 is not true. However, it is true if the limit random variable is a degenerate random variable. We prove it in the following theorem.

Theorem 7.3.2. *Suppose $\{X_n, n \geq 1\}$ is a sequence of random variables defined on the probability space $(\Omega, \mathbb{A}, P)$. Then $X_n \xrightarrow{L} C \iff X_n \xrightarrow{P} C$, where C is a constant.*

Proof. By Theorem 7.3.1, $X_n \xrightarrow{P} C \Rightarrow X_n \xrightarrow{L} C$. Suppose $X_n \xrightarrow{L} C$, that is, $F_n(x) \to F_C(x), \ \forall \ x \in C(F_C)$, where $F_C(x)$ is given by,

$$F_C(x) = \begin{cases} 0, & \text{if } x < C \\ 1, & \text{if } x \geq C. \end{cases}$$

Hence, the set $C(F_C)$, of points of continuity of F_C is given by $\mathbb{R} - \{C\}$. To check if $X_n \xrightarrow{P} C$, consider for $\epsilon > 0$,

$$\begin{aligned} P[|X_n - C| \leq \epsilon] &= P[C - \epsilon \leq X_n \leq C + \epsilon] = F_n(C + \epsilon) - F_n(C - \epsilon-) \\ &= F_n(C + \epsilon) - F_n(C - \epsilon) \quad \to 1 \text{ as } n \to \infty \end{aligned}$$

and hence, $X_n \xrightarrow{P} C$. □

The following examples illustrate Theorem 7.3.2.

Example 7.3.12. Suppose $\{X_1, X_2, \cdots, X_n\}$ are independent and identically distributed random variables each having uniform $U(0,1)$ distribution. In Example 7.2.5, it is shown that $X_{(1)} \xrightarrow{P} 0$ & $X_{(n)} \xrightarrow{P} 1$. In Example 7.3.8, it is shown that $X_{(1)} \xrightarrow{L} 0$ & $X_{(n)} \xrightarrow{L} 1$. Thus as proved in Theorem 7.3.2, if the limit random variable is degenerate, then convergence in probability and law are equivalent. □

Example 7.3.13. Suppose $X_n \sim N(0, 1/n)$ distribution. Then its distribution function $F_n(x)$ is given by,

$$F_n(x) = P[X_n \leq x] = P[\sqrt{n}X_n \leq \sqrt{n}x] = \Phi(\sqrt{n}x).$$

$$\text{Hence,} \quad F_n(x) \quad \to \quad \begin{cases} 0, & \text{if} \quad x < 0 \\ 1/2, & \text{if} \quad x = 0 \\ 1, & \text{if} \quad x > 0. \end{cases}$$

Note that the distribution function $F_X(x)$ of $X \equiv 0$ is given by,

$$F_X(x) \quad = \quad \begin{cases} 0, & \text{if} \quad x < 0 \\ 1, & \text{if} \quad x \geq 0, \end{cases}$$

Thus, as $n \to \infty$,

$$\begin{aligned} F_n(x) \quad &\to \quad F_X(x) \ \forall \ x \in C(F_X) = \mathbb{R} - \{0\} \ \Rightarrow \ X_n \xrightarrow{L} X \equiv 0 \\ &\Rightarrow \quad X_n \xrightarrow{P} X \equiv 0, \quad \text{by Theorem 7.3.2.} \end{aligned}$$

We now verify whether $X_n \xrightarrow{P} X \equiv 0$ using the definition. Observe that,

$$P[|X_n| > \epsilon] \ \leq \ E(X_n^2)/\epsilon^2 = 1/n\epsilon^2 \to 0 \ \forall \ \epsilon > 0 \ \Rightarrow \ X_n \xrightarrow{P} 0.$$

Note that the limit law is degenerate and hence the sequence convergence in law as well as in probability. □

Remark 7.3.1. The above example illustrates that a sequence of absolutely continuous random variables converges in law to a discrete random variable.

Using Theorem 7.3.1, we prove an interesting result in the following theorem, which is useful in many theorems and examples. Result (i) is heavily used to show that under certain regularity conditions, with probability approaching 1, the likelihood equation has a consistent solution, refer to Deshmukh and Kulkarni [11].

Theorem 7.3.3. *(i) If $X_n \xrightarrow{P} C < 0$, then $P[X_n \leq 0] \to 1$ as $n \to \infty$.*
(ii) If $X_n \xrightarrow{P} C > 0$, then $P[X_n > 0] \to 1$ as $n \to \infty$.

Proof. From Theorem 7.3.1, $X_n \xrightarrow{P} C \ \Rightarrow \ X_n \xrightarrow{L} C$ and hence $F_{X_n}(x)$ converges to the distribution function of $X \equiv C$ at all real numbers except C, it being a point of discontinuity. Thus,

$$F_{X_n}(x) = P[X_n \leq x] \quad \to \quad \begin{cases} 0, & \text{if} \quad x < C \\ 1, & \text{if} \quad x > C\,. \end{cases}$$

(i) Observe that if $C < 0$, $P[X_n \leq 0] \to 1$.
(ii) Further, if $C > 0$, $P[X_n \leq 0] \to 0 \iff P[X_n > 0] \to 1$. □

In Theorem 7.2.5, it is proved that convergence in probability is invariant under continuous transforms. Following example shows that $X_n \xrightarrow{P} C$ and $g(X_n) \xrightarrow{P} g(C)$, even if g is not continuous. It illustrates the application of Theorem 7.3.3.

Example 7.3.14. Suppose $X_n \xrightarrow{P} C$, $C \in \mathbb{R}$. Suppose a function $g : \mathbb{R} \to \mathbb{R}$ is defined as

$$g(x) \quad = \quad \begin{cases} -1, & \text{if} \quad x < 1 \\ 1, & \text{if} \quad x \geq 1. \end{cases}$$

It is clear that g is not a continuous function, 1 being a point of discontinuity, and hence we cannot use Theorem 7.2.5 to claim $g(X_n) \xrightarrow{P} g(C)$. We verify whether $g(X_n) \xrightarrow{P} g(C)$ from the first principles. We assume that $P[X_n = 1] = 0$. Suppose $C \geq 1$, then $g(C) = 1$. Now,

$$P[|g(X_n) - g(C)| < \epsilon] = P[|g(X_n) - 1| < \epsilon] = P[1 - \epsilon < g(X_n) < 1 + \epsilon] = 1,$$

if $\epsilon > 2$, as possible values of $g(X_n)$ are -1 and 1.

$$\text{For} \quad 0 < \epsilon \leq 2, \; P[1 - \epsilon < g(X_n) < 1 + \epsilon] = P[g(X_n) = 1] = P[X_n \geq 1] \,.$$

Suppose now $C < 1$, then $g(C) = -1$. Observe that for $\epsilon > 2$,

$$P[|g(X_n) - g(C)| < \epsilon] = P[|g(X_n) - (-1)| < \epsilon] = P[-1 - \epsilon < g(X_n) < -1 + \epsilon] = 1 \,.$$

$$\text{For} \quad 0 < \epsilon \leq 2, \; P[-1 - \epsilon < g(X_n) < -1 + \epsilon] = P[g(X_n) = -1] = P[X_n < 1] \,.$$

Now, we use Theorem 7.3.3 to show that $P[X_n \geq 1]$ and $P[X_n < 1]$ converge to 1. Suppose $C < 1$, then

$$\begin{aligned} X_n \xrightarrow{P} C \quad & \Rightarrow \quad X_n - 1 \xrightarrow{P} C - 1 < 0 \\ & \Rightarrow \quad P[X_n - 1 \leq 0] \to 1 \quad \text{as} \quad n \to \infty \\ & \Rightarrow \quad P[X_n \leq 1] \to 1 \quad \text{as} \quad n \to \infty. \end{aligned}$$

Now with the assumption that $P[X_n = 1] = 0$, $[X_n \leq 1] = [X_n < 1] \to 1$. Thus, if $C < 1$ then $P[X_n < 1] \to 1$. Suppose $C \geq 1$, then

$$\begin{aligned} X_n \xrightarrow{P} C \quad & \Rightarrow \quad 1 - X_n \xrightarrow{P} 1 - C \leq 0 \\ & \Rightarrow \quad P[1 - X_n \leq 0] \to 1 \quad \text{as} \quad n \to \infty \\ & \Rightarrow \quad P[X_n \geq 1] \to 1 \quad \text{as} \quad n \to \infty. \end{aligned}$$

Thus, if $C \geq 1$ then $P[X_n \geq 1] \to 1$. Hence we claim that $g(X_n) \xrightarrow{P} g(C)$ for all $C \in \mathbb{R}$. Observe that for

$$\epsilon = 2 \;\; \& \;\; C = 8, \quad P[|g(X_n) - C| < \epsilon] = P[6 < g(X_n) < 10] = 0$$

and hence $g(X_n)$ does not converge in probability to C, which has to be true in view of the fact that the limit random variable in convergence in probability is almost surely unique. □

Remark 7.3.2. It is to be noted that in the above example, the probability measure assigned to the set of points of discontinuity is 0.

In Example 7.3.10, it is shown that $X_n \xrightarrow{L} X$ and $X_n \xrightarrow{L} Y = 1 - X$, where X and Y are identically distributed. It is also noted that,

$$\begin{aligned}\{\omega | X(\omega) = 1\} &= \{\omega | Y(\omega) = 0\} \;\; \& \;\; \{\omega | X(\omega) = 0\} = \{\omega | Y(\omega) = 1\} \\ &\Rightarrow P[X = Y] = 0.\end{aligned}$$

Thus, X and Y are not equivalent random variables, but they are identically distributed random variables. From this example, we note that, as in almost sure convergence and convergence in probability, the limit random variables in convergence in law are not equivalent random variables. However, they are identically distributed random variables. It is true in general and is proved in the following theorem.

Theorem 7.3.4. *If $X_n \xrightarrow{L} X$ and $X_n \xrightarrow{L} Y$, then X and Y are identically distributed random variables.*

Proof. By definition of convergence in law,

$$\begin{aligned} X_n \xrightarrow{L} X &\Rightarrow F_n(x) \to F_X(x) \;\; \forall \;\; x \in C(F_X) \\ \& \;\; X_n \xrightarrow{L} Y &\Rightarrow F_n(x) \to F_Y(x) \;\; \forall \;\; x \in C(F_Y),\end{aligned}$$

where $C(F_X)$ and $C(F_Y)$ are sets of points of continuity of $F_X(x)$ and $F_Y(x)$ respectively. By Theorem 3.2.4, $C(F_X) \cap C(F_Y) \neq \emptyset$. Now,

$$\forall \;\; x \in C(F_X) \cap C(F_Y), \;\; F_n(x) \to F_X(x) \;\; \text{and} \;\; F_n(x) \to F_Y(x).$$

Note that $\{F_n(x), n \geq 1\}$ is a sequence of real numbers converging to $F_X(x)$ and $F_Y(x)$. Since the limit of a convergent sequence of real numbers is unique,

$$F_X(x) = F_Y(x), \qquad \forall \;\; x \in C(F_X) \cap C(F_Y).$$

Now suppose $x \in (C(F_X) \cap C(F_Y))^c$. Again as in Theorem 3.2.4, if $x \in (C(F_X) \cap C(F_Y))^c$, then there exists a sequence $\{x + r_n, n \geq 1\}$ of points in $C(F_X) \cap C(F_Y)$ such that $x + r_n \to x$, as $r_n \downarrow 0$. Further, using the right continuity of distribution functions, we have for $x \in (C(F_X) \cap C(F_Y))^c$,

$$\begin{aligned} F_X(x) &= \lim_{h \downarrow 0} F_X(x + h) = \lim_{r_n \downarrow 0} F_X(x + r_n) \\ &= \lim_{r_n \downarrow 0} F_Y(x + r_n) = \lim_{h \downarrow 0} F_Y(x + h) = F_Y(x).\end{aligned}$$

Thus, $F_X(x) = F_Y(x) \;\; \forall \;\; x \in \mathbb{R}$ and hence X and Y are identically distributed random variables. □

The following theorem states that if the difference $X_n - Y_n$ converges to zero in probability, then both $\{X_n, n \geq 1\}$ and $\{Y_n, n \geq 1\}$ have the same limit law. It is an important result since it is heavily used in asymptotic inference theory to decide the asymptotic distribution of a suitably normalized estimator. The proof is similar to that of Theorem 7.3.1.

Theorem 7.3.5. *Suppose $\{X_n, n \geq 1\}$ and $\{Y_n, n \geq 1\}$ are sequences of random variables defined on the same probability space $(\Omega, \mathbb{A}, P)$. If $X_n - Y_n \xrightarrow{P} 0$ and $Y_n \xrightarrow{L} X$, then $X_n \xrightarrow{L} X$.*

Proof. It is given that $X_n - Y_n \xrightarrow{P} 0$, hence $\forall \ \epsilon > 0$

$$\begin{aligned} \lim_{n\to\infty} P[|X_n - Y_n| > \epsilon] &= 0 \\ \Rightarrow \liminf_{n\to\infty} P[|X_n - Y_n| > \epsilon] &= \limsup_{n\to\infty} P[|X_n - Y_n| > \epsilon] = 0. \quad (7.3) \end{aligned}$$

To prove that $X_n \xrightarrow{L} X$, we show that $F_{X_n}(x) \to F_X(x) \ \forall \ x \in C(F_X)$, provided $\lim_{n\to\infty} F_{X_n}(x)$ exists, that is, $\limsup_{n\to\infty} F_{X_n}(x) = \liminf_{n\to\infty} F_{X_n}(x)$, for all $x \in C(F_X)$. Further, $\liminf_{n\to\infty} F_{X_n}(x) \leq \limsup_{n\to\infty} F_{X_n}(x)$ is always true. Thus as in Theorem 7.3.1, it is enough to prove that

$$F_X(x) \leq \liminf_{n\to\infty} F_{X_n}(x) \quad \& \quad \limsup_{n\to\infty} F_{X_n}(x) \leq F_X(x) \quad \forall \ x \in C(F_X)\,.$$

$$\begin{aligned} \text{Now,} \quad Y_n \xrightarrow{L} X \quad &\Rightarrow \quad F_{Y_n}(x) \to F_X(x), \quad \forall \ x \in C(F_X) \\ &\Rightarrow \quad \liminf_{n\to\infty} F_{Y_n}(x) = \limsup_{n\to\infty} F_{Y_n}(x) = F_X(x) \quad \forall \ x \in C(F_X). \end{aligned}$$

These results suggest that we have to find some inequalities between the distribution functions of X_n and Y_n. Observe that for any real numbers x and x_1 such that $x_1 < x$,

$$\begin{aligned} P[Y_n \leq x_1] &= P[Y_n \leq x_1, X_n \leq x] + P[Y_n \leq x_1, X_n > x] \\ &\leq P[X_n \leq x] + P[|X_n - Y_n| > x - x_1] \\ \Rightarrow F_{Y_n}(x_1) &\leq F_{X_n}(x) + P[|X_n - Y_n| > x - x_1] \\ \Rightarrow \liminf_{n\to\infty} F_{Y_n}(x_1) &\leq \liminf_{n\to\infty} F_{X_n}(x) + \liminf_{n\to\infty} P[|X_n - Y_n| > x - x_1] \\ \Rightarrow \liminf_{n\to\infty} F_{Y_n}(x_1) &\leq \liminf_{n\to\infty} F_{X_n}(x) \text{ by Equation (7.3).} \end{aligned}$$

Now, for any real numbers x and x_2 such that $x < x_2$,

$$\begin{aligned} P[X_n \leq x] &= P[X_n \leq x, Y_n \leq x_2] + P[X_n \leq x, Y_n > x_2] \\ &\leq P[Y_n \leq x_2] + P[|X_n - Y_n| > x_2 - x] \\ \Rightarrow F_{X_n}(x) &\leq F_{Y_n}(x_2) + P[|X_n - Y_n| > x_2 - x] \\ \Rightarrow \limsup_{n\to\infty} F_{X_n}(x) &\leq \limsup_{n\to\infty} F_{Y_n}(x_2) + \limsup_{n\to\infty} P[|X_n - Y_n| > x_2 - x] \\ \Rightarrow \limsup_{n\to\infty} F_{X_n}(x) &\leq \limsup_{n\to\infty} F_{Y_n}(x_2) \text{ by Equation (7.3).} \end{aligned}$$

Thus we have,

$$\liminf_{n\to\infty} F_{Y_n}(x_1) \le \liminf_{n\to\infty} F_{X_n}(x) \le \limsup_{n\to\infty} F_{X_n}(x) \le \limsup_{n\to\infty} F_{Y_n}(x_2)\ . \quad (7.4)$$

Suppose x_1 and x_2 are points of continuity of F_X. Hence,

$$\begin{aligned} \liminf_{n\to\infty} F_{Y_n}(x_1) &= \lim_{n\to\infty} F_{Y_n}(x_1) = F_X(x_1) \\ \& \ \limsup_{n\to\infty} F_{Y_n}(x_2) &= \lim_{n\to\infty} F_{Y_n}(x_2) = F_X(x_2)\ . \end{aligned}$$

Consequently, Equation (7.4) reduces to

$$F_X(x_1) \le \liminf_{n\to\infty} F_{X_n}(x) \ \le \ \limsup_{n\to\infty} F_{X_n}(x) \ \le \ F_X(x_2). \quad (7.5)$$

Suppose $x_1 = x - h_1$ and $x_2 = x + h_2$, $h_1,\ h_2 > 0$ and x is a point of continuity of F_X. Hence,

$$\begin{aligned} h_1 \to 0 \quad &\Rightarrow \quad F_X(x_1) = F_X(x - h_1) \to F_X(x) \\ \& \ h_2 \to 0 \quad &\Rightarrow \quad F_X(x_2) = F_X(x + h_2) \to F_X(x). \end{aligned}$$

Suppose $h_1,\ h_2 \to 0$ in Equation (7.5), then

$$F_X(x) \le \liminf_{n\to\infty} F_{X_n}(x) \ \le \ \limsup_{n\to\infty} F_{X_n}(x) \ \le \ F_X(x)\ .$$

Thus, $\forall\ x \in C(F_X)$, $\lim_{n\to\infty} F_{X_n}(x)$ exists and is given by $F_X(x)$ and hence $X_n \xrightarrow{L} X$. □

The following example illustrates the result obtained in Theorem 7.3.5 and how to identify the limiting distribution. The limit law in this example comes out to be a distribution function which is neither discrete nor continuous. We decompose it according to the Jordan decomposition theorem, as discussed in Chapter 3.

Example 7.3.15. Suppose $X \sim N(\theta, 1)$ distribution and $\theta \in \Theta = [a, b]$. It can be shown that (Deshmukh and Kulkarni [11]) the maximum likelihood estimator $\hat{\theta}_n$ of θ, based on a random sample of size n from the distribution of X is given by,

$$\hat{\theta}_n = \begin{cases} a, & \text{if} \quad \overline{X}_n < a \\ \overline{X}_n, & \text{if} \quad \overline{X}_n \in [a, b] \\ b, & \text{if} \quad \overline{X}_n > b. \end{cases}$$

We first examine whether $\hat{\theta}_n \xrightarrow{P} \theta$ and investigate the limit law of $U_n = \sqrt{n}(\hat{\theta}_n - \theta)$. Since $X \sim N(\theta, 1)$, $\overline{X}_n \sim N(\theta, 1/n)$ distribution for each n and hence $\sqrt{n}(\overline{X}_n - \theta) \sim N(0, 1)$ distribution for each n. Hence for all $\epsilon > 0$ and for all $\theta \in \Theta$,

$$P_\theta[|\overline{X}_n - \theta| < \epsilon] = \Phi(\epsilon\sqrt{n}) - \Phi(-\epsilon\sqrt{n}) \ \to \ 1 \ \Rightarrow \ \overline{X}_n \xrightarrow{P} \theta \ \Rightarrow \ \overline{X}_n \xrightarrow{L} \theta.$$

Now for $\epsilon > 0$ and $\theta \in (a, b)$,

$$\begin{aligned} P_\theta[|\hat{\theta}_n - \overline{X}_n| < \epsilon] &\geq P_\theta[\hat{\theta}_n = \overline{X}_n] = P_\theta[a \leq \overline{X}_n \leq b] \\ &= \Phi(\sqrt{n}(b-\theta)) - \Phi(\sqrt{n}(a-\theta)) \to 1 \\ &\Rightarrow \hat{\theta}_n - \overline{X}_n \xrightarrow{P} 0 \\ &\Rightarrow \hat{\theta}_n \xrightarrow{L} \theta \Rightarrow \hat{\theta}_n \xrightarrow{P} \theta, \end{aligned}$$

by Theorem 7.3.5 and Theorem 7.3.2. Now we examine convergence in probability at the boundary points a and b. Since $\hat{\theta}_n \geq a$, for $\theta = a$ we have,

$$P_a[|\hat{\theta}_n - a| > \epsilon] = \begin{cases} P_a[\hat{\theta}_n > a + \epsilon] = 0\ , & \text{if } \epsilon > b - a \\ P_a[\overline{X}_n > a + \epsilon] = 1 - \Phi(\sqrt{n}\epsilon) \to 0, & \text{if } \epsilon \leq b - a. \end{cases}$$

Thus, $\hat{\theta}_n \xrightarrow{P_a} a$. Further, $\hat{\theta}_n \leq b$ implies that the event $[|\hat{\theta}_n - b| > \epsilon]$ is the same as $[b - \hat{\theta}_n > \epsilon]$. Hence for the boundary point b,

$$P_b[|\hat{\theta}_n - b| > \epsilon] = \begin{cases} P_b[\hat{\theta}_n < b - \epsilon] = 0\ , & \text{if } \epsilon > b - a \\ P_b[\overline{X}_n < b - \epsilon] = \Phi(-\sqrt{n}\epsilon) \to 0, & \text{if } \epsilon \leq b - a. \end{cases}$$

Hence, $\hat{\theta}_n \xrightarrow{P_b} b$. Thus, we have shown that $\forall\ \theta \in [a, b]$, $\hat{\theta}_n \xrightarrow{P} \theta$. We now investigate the limit law of $U_n = \sqrt{n}(\hat{\theta}_n - \theta)$. If $X \sim N(\theta, 1)$ distribution then $Y_n = \sqrt{n}(\overline{X}_n - \theta)$ has the standard normal distribution for all $\theta \in [a, b]$ and for all n. Thus, $Y_n \xrightarrow{L} Z$, where $Z \sim N(0, 1)$. Now
$U_n - Y_n = \sqrt{n}(\hat{\theta}_n - \theta) - \sqrt{n}(\overline{X}_n - \theta) = \sqrt{n}(\hat{\theta}_n - \overline{X}_n)$ and $\forall\ \epsilon > 0$,

$$\begin{aligned} P[|U_n - Y_n| < \epsilon] &= P[\sqrt{n}|\hat{\theta}_n - \overline{X}_n| < \epsilon] \\ &\geq P[\hat{\theta}_n = \overline{X}_n] = P[a \leq \overline{X}_n \leq b] \\ &= \Phi\left(\sqrt{n}(b-\theta)\right) - \Phi\left(\sqrt{n}(a-\theta)\right) \\ &\to 1 - 0 = 1 \text{ as } n \to \infty, \ \forall\ \theta \in (a, b). \end{aligned}$$

Thus, $U_n - Y_n \xrightarrow{P} 0$. Hence, the limit law of U_n and Y_n is the same $\forall\, \theta \in (a, b)$. But for all $\theta \in (a, b)$ $Y_n \xrightarrow{L} Z$ and hence by Theorem 7.3.5, $U_n \xrightarrow{L} Z \sim N(0, 1)$. To find the limit law of U_n at $\theta = a$, we study the limit of $P_a[\sqrt{n}(\hat{\theta}_n - a) \leq x]$ for all $x \in \mathbb{R}$. Since $\hat{\theta}_n \geq a$, for $x < 0$, $P_a[\sqrt{n}(\hat{\theta}_n - a) \leq x] = 0$. Suppose $x = 0$. Then

$$\begin{aligned} P_a[\sqrt{n}(\hat{\theta}_n - a) \leq 0] &= P_a[\sqrt{n}(\hat{\theta}_n - a) = 0] = P_a[\hat{\theta}_n = a] = P_a[\overline{X}_n < a] \\ &= P_a[\sqrt{n}(\overline{X}_n - a) < 0] = \Phi(0) = 1/2. \end{aligned}$$

From the expression of $\hat{\theta}_n$, we have $\hat{\theta}_n \leq b \Rightarrow \sqrt{n}(\hat{\theta}_n - a) \leq \sqrt{n}(b - a)$. Suppose $0 < x < \sqrt{n}(b - a)$. Then

$$\begin{aligned} P_a[\sqrt{n}(\hat{\theta}_n - a) \leq x] &= P_a[\sqrt{n}(\hat{\theta}_n - a) \leq 0] + P_a[0 < \sqrt{n}(\hat{\theta}_n - a) \leq x] \\ &= 1/2 + P_a[0 < \sqrt{n}(\overline{X}_n - a) \leq x] \\ &= 1/2 + \Phi(x) - 1/2 = \Phi(x). \end{aligned}$$

If $x \geq \sqrt{n}(b-a)$, then $P_a[\sqrt{n}(\hat{\theta}_n - a) \leq x] = 1$. Thus,

$$P_a[\sqrt{n}(\hat{\theta}_n - a) \leq x] = \begin{cases} 0, & \text{if} \quad x < 0 \\ 1/2 = \Phi(x), & \text{if} \quad x = 0 \\ \Phi(x), & \text{if} \quad 0 < x < \sqrt{n}(b-a) \\ 1, & \text{if} \quad x \geq \sqrt{n}(b-a) \ . \end{cases}$$

$$\text{Hence,} \quad P_a[\sqrt{n}(\hat{\theta}_n - a) \leq x] \to \begin{cases} 0, & \text{if} \quad x < 0 \\ \Phi(x), & \text{if} \quad x \geq 0 \ . \end{cases}$$

Thus for $\theta = a$, the limit law of U_n is not normal, and 0 is a point of discontinuity. We now show that it is a mixture of discrete and continuous distributions, and the continuous distribution is related to the standard normal distribution. Suppose W_1 is a random variable with a distribution degenerate at 0. Then its distribution function is given by,

$$F_{W_1}(x) = \begin{cases} 0, & \text{if} \quad x < 0 \\ 1, & \text{if} \quad x \geq 0 \ . \end{cases}$$

Suppose a random variable W_2 is defined as $W_2 = |Z|$ where $Z \sim N(0,1)$. Then $P[W_2 \leq x] = 0$ if $x < 0$. Suppose $x \geq 0$, then

$$P[W_2 \leq x] = P[|Z| \leq x] = P[-x \leq Z \leq x] = \Phi(x) - \Phi(-x) = 2\Phi(x) - 1.$$

Thus, the distribution function of W_2 is given by,

$$F_{W_2}(x) = \begin{cases} 0, & \text{if} \quad x < 0 \\ 2\Phi(x) - 1, & \text{if} \quad x \geq 0 \ . \end{cases}$$

It is easy to verify that

$$P_a[\sqrt{n}(\hat{\theta}_n - a) \leq x] \to (1/2)F_{W_1}(x) + (1/2)F_{W_2}(x).$$

To find the limit law of U_n at $\theta = b$, we adopt the similar procedure as that for $\theta = a$ and study the limit of $P_b[\sqrt{n}(\hat{\theta}_n - b) \leq x]$ for all $x \in \mathbb{R}$. By definition $\hat{\theta}_n \geq a$, so that $\sqrt{n}(\hat{\theta}_n - b) \geq \sqrt{n}(a-b) < 0$. Hence, for $x < \sqrt{n}(a-b)$, $P_b[\sqrt{n}(\hat{\theta}_n - b) \leq x] = 0$. Suppose $\sqrt{n}(a-b) \leq x < 0$, then

$$P_b[\sqrt{n}(\hat{\theta}_n - b) \leq x] = P_b[\sqrt{n}(\overline{X}_n - b) \leq x] = \Phi(x).$$

Suppose $x = 0$. Then $P_b[\sqrt{n}(\hat{\theta}_n - b) \leq 0] = P_b[\hat{\theta}_n \leq b] = 1$. Suppose $x > 0$. By definition, $\hat{\theta}_n \leq b$ implying that $\sqrt{n}(\hat{\theta}_n - b) \leq 0$, hence $P_b[\sqrt{n}(\hat{\theta}_n - b) \leq x] = 1, \ \forall \ x > 0$. Thus,

$$P_b[\sqrt{n}(\hat{\theta}_n - b) \leq x] = \begin{cases} 0, & \text{if} \quad x < \sqrt{n}(a-b) \\ \Phi(x), & \text{if} \quad \sqrt{n}(a-b) \leq x < 0 \\ 1, & \text{if} \quad x \geq 0 \ . \end{cases}$$

Consequently, as $n \to \infty$,

$$P_b[\sqrt{n}(\hat{\theta}_n - b) \leq x] \quad \to \quad \begin{cases} \Phi(x), & \text{if} \quad x < 0 \\ 1, & \text{if} \quad x \geq 0 \,. \end{cases}$$

Note that as in the case of $\theta = a$, the limit law of U_n is not normal at $\theta = b$, and 0 is a point of discontinuity. Further, the limit law is a mixture of discrete and continuous distributions, as shown below. Suppose a random variable W_3 is defined as $W_3 = -|Z|$ where $Z \sim N(0,1)$. Since $W_3 \leq 0$, $P[W_3 \leq x] = 1$ for all $x \geq 0$. Suppose $x < 0$, then

$$P[W_3 \leq x] = P[|Z| \geq -x] = 1 - P[x \leq Z \leq -x] = 1 - \Phi(-x) + \Phi(x) = 2\Phi(x).$$

Thus, the distribution function of W_3 is given by,

$$F_{W_3}(x) \quad = \quad \begin{cases} 2\Phi(x), & \text{if} \quad x < 0 \\ 1, & \text{if} \quad x \geq 0 \,. \end{cases}$$

It then follows that $P_b[\sqrt{n}(\hat{\theta}_n - b) \leq x] \to (1/2)F_{W_1}(x) + (1/2)F_{W_3}(x)$. □

In Section 2, we have proved that convergence in probability is closed under continuous transformation. Similarly, convergence in law is also closed under continuous transformation. This result is known as a continuous mapping theorem. We state it below. For proof, refer to Billingsley [5].

Theorem 7.3.6. *Continuous mapping theorem: If $X_n \xrightarrow{L} X$ and g is a continuous function, then $g(X_n) \xrightarrow{L} g(X)$.*

Following example presents some illustrations of the continuous mapping theorem.

Example 7.3.16. Suppose $X_n \xrightarrow{L} X$. With different choices for a continuous function g, we have the following results. (i) $-X_n \xrightarrow{L} -X$. (ii) $|X_n| \xrightarrow{L} |X|$. (iii) Suppose $g(x) = a_0 + a_1 x + a_2 x^2$. Then $a_0 + a_1 X_n + a_2 X n^2 \xrightarrow{L} a_0 + a_1 X + a_2 X^2$. (iv) Suppose X has exponential distribution with location parameter 0 and scale parameter 1. Then the distribution of $U = \exp(-X)$ is uniform $U(0,1)$. Hence, $\exp(-X_n) \xrightarrow{L} \exp(-X) = U \sim U(0,1)$ distribution. (v) If $X \sim N(0,1)$ distribution, then $X_n^2 \xrightarrow{L} X^2 \sim \chi_1^2$.

It is proved in Lemma 7.2.1 that if the sequence $\{X_n, n \geq 1\}$ converges in probability, then it is bounded in probability. A similar result is true if the sequence $\{X_n, n \geq 1\}$ converges in law. It is proved in the following lemma.

Lemma 7.3.1. *If the sequence $\{X_n, n \geq 1\}$ converges in law, then it is bounded in probability.*

Proof. Suppose $X_n \xrightarrow{L} X$, that is, $\forall \ x \in C(F_X)$, $F_n(x) \to F_X(x)$. To prove that $\{X_n, n \geq 1\}$ is bounded in probability, given $\epsilon > 0$ we have to find M

and n_0 such that $P[|X_n| \leq M] > 1-\epsilon$ for all $n \geq n_0$. Since the set of points of continuities of a distribution function is uncountable, any distribution function can have arbitrarily large continuity points. Suppose $M_1 > 0$ and $-M_2 < 0$ are points of continuity of F_X such that $F_X(M_1) > 1-\epsilon/4$ and $F_X(-M_2) < \epsilon/4$. Since M_1 and $-M_2$ are points of continuity of F_X, $F_n(M_1) \to F_X(M_1)$ and $F_n(-M_2) \to F_X(-M_2)$. Thus, for $\epsilon > 0$, there exists n_0 such that $\forall\ n \geq n_0$,

$$\begin{aligned} |F_n(M_1) - F_X(M_1)| &< \epsilon/4 \ \ \& \ \ |F_n(-M_2) - F_X(-M_2)| < \epsilon/4 \\ &\Rightarrow F_n(M_1) > F_X(M_1) - \epsilon/4 > 1 - \epsilon/2 \\ &\quad\ \& \ F_n(-M_2) < F_X(-M_2) + \epsilon/4 < \epsilon/2 \\ &\Rightarrow P[-M_2 < X_n \leq M_1] = F_n(M_1) - F_n(-M_2) > 1 - \epsilon \\ &\Rightarrow P[|X_n| \leq M] \geq P[-M_2 < X_n \leq M_1] > 1 - \epsilon, \end{aligned}$$

where $M = \max\{|M_1|, |M_2|\}+1$. Thus, the sequence $\{X_n, n \geq 1\}$ is bounded in probability. □

Theorem 7.3.7. *If $X_n \xrightarrow{L} X$ and if $Y_n \xrightarrow{P} 0$, then $X_nY_n \xrightarrow{P} 0$ and $X_nY_n \xrightarrow{L} 0$.*

Proof. Since $X_n \xrightarrow{L} X$, the sequence $\{X_n, n \geq 1\}$ is bounded in probability. Hence, by Theorem 7.2.7, we claim that $X_nY_n \xrightarrow{P} 0$. The result can be proved using the definition of convergence in probability as follows. For any $\epsilon > 0$, note that $P[|X_nY_n| > \epsilon] = 0$ for all those ω for which $Y_n(\omega) = 0$. Further, for any $\epsilon > 0$ and $\delta > 0$ and for all ω such that $Y_n(\omega) \neq 0$, observe that

$$\begin{aligned} P[|X_nY_n| > \epsilon] &= P[|X_nY_n| > \epsilon, |Y_n| \leq \delta] + P[|X_nY_n| > \epsilon, |Y_n| > \delta] \\ &\leq P\left[|X_n| > \epsilon/\delta\right] + P[|Y_n| > \delta] \to 0, \end{aligned}$$

using the result that the sequence $\{X_n, n \geq 1\}$ is bounded in probability for the first term, and $Y_n \xrightarrow{P} 0$ for the second term. Thus, $X_nY_n \xrightarrow{P} 0$ and hence $X_nY_n \xrightarrow{L} 0$. □

As in almost sure convergence and convergence in probability, convergence in law is closed under arithmetic operations, under a restrictive setup, in the sense that one of the two limit laws must be degenerate. Of course, if the limit law is degenerate, convergence in probability and convergence in law are equivalent, as proved in Theorem 7.3.2. The theorem proving the closure of convergence in law under arithmetic operations is known as Slutsky's theorem. It is also referred to as Cramer's theorem (Gut [13]). It is heavily used in the studentization procedure to construct a large sample confidence interval and also in deriving the large sample null distribution of some test statistics. We prove it below.

Theorem 7.3.8. *Slutsky's theorem: Suppose $\{X_n, n \geq 1\}$ and $\{Y_n, n \geq 1\}$ are sequences of random variables defined on the same probability space $(\Omega, \mathbb{A}, P)$.*

If $X_n \xrightarrow{L} X$ and $Y_n \xrightarrow{L} C$, where C is a degenerate random variable, then (i) $X_n + Y_n \xrightarrow{L} X + C$, (ii) $X_n - Y_n \xrightarrow{L} X - C$, (iii) $X_nY_n \xrightarrow{L} XC$ and (iv) $X_n/Y_n \xrightarrow{L} X/C$, provided X_n/Y_n and X/C are defined.

Proof. (i) Note that,

$$Y_n \xrightarrow{L} C \;\Rightarrow\; Y_n \xrightarrow{P} C \;\Rightarrow\; (X_n + Y_n) - (X_n + C) = Y_n - C \xrightarrow{P} 0.$$

Hence by Theorem 7.3.5, limit law of (X_n+Y_n) is the same as that of (X_n+C). We find the limit law of $(X_n + C)$ as follows. Observe that

$$\begin{aligned} F_{X_n+C}(x) &= P[X_n + C \le x] = P[X_n \le x - C] \\ &= F_n(x - C) \to F_X(x - C) = P[X \le x - C] = F_{X+C}(x), \end{aligned}$$

provided $x - C$ is a point of continuity of F_X. Now, $X_n + C \xrightarrow{L} X + C$ if x is a point of continuity of F_{X+C}. To establish it, suppose $x - C$ is a point of continuity of F_X. F_X being a distribution function, is always right continuous. Hence, $x - C$ is a point of continuity of F_X implies that,

$$\begin{aligned} \lim_{h\to 0} F_X(x - C - h) &= F_X(x - C) \\ \iff \lim_{h\to 0} P[X \le x - C - h] &= P[X \le x - C] \\ \iff \lim_{h\to 0} P[X + C \le x - h] &= P[X + C \le x] \\ \iff \lim_{h\to 0} F_{X+C}(x - h) &= F_{X+C}(x). \end{aligned}$$

Thus, x is a point of continuity of F_{X+C}. Thus, $X_n + C \xrightarrow{L} X + C$ and hence, $X_n + Y_n \xrightarrow{L} X + C$.
(ii) Observe that $Y_n \xrightarrow{P} C \;\Rightarrow\; -Y_n \xrightarrow{P} -C$. Hence by (i) $X_n - Y_n \xrightarrow{L} X - C$.
(iii) If $C = 0$, then the result follows from Theorem 7.3.7. Hence we now assume that $C \neq 0$. To prove (iii), we adopt an approach different than in (i). We make two cases depending on $C > 0$ and $C < 0$.
Case(a) $C > 0$: Suppose $x \in \mathbb{R}$ is such that $x/C \in C(F_X)$. Hence,

$$\begin{aligned} \lim_{h\to 0} F_X(x/C - h) &= F_X(x/C) \\ \iff \lim_{h\to 0} P[X \le x/C - h] &= P[X \le x/C] \\ \iff \lim_{h\to 0} P[XC \le x - Ch] &= P[XC \le x] \\ \iff \lim_{h\to 0} F_{XC}(x - Ch) &= F_{XC}(x). \end{aligned}$$

Thus, x is a point of continuity of F_{XC}. Suppose $\epsilon > 0$ is such that $x/(C - \epsilon)$ and $x/(C + \epsilon)$ are in $C(F_X)$. It is posiible since $C(F_X)$ is dense in $\mathbb{R}$. For large n we assume $Y_n > 0$, it is valid since by Theorem 7.3.3, $P[Y_n > 0] \to 1$. Observe that

$$|Y_n - C| \le \epsilon \;\Rightarrow\; 1/(C + \epsilon) \le 1/Y_n \le 1/(C - \epsilon). \tag{7.6}$$

Suppose $x > 0$. Note that

$$\begin{aligned}
F_{X_nY_n}(x) &= P[X_nY_n \leq x] = P[X_nY_n \leq x, |Y_n - C| \leq \epsilon] \\
&+ P[X_nY_n \leq x, |Y_n - C| > \epsilon] \\
&\leq P[X_n \leq x/Y_n, |Y_n - C| \leq \epsilon] + P[|Y_n - C| > \epsilon] \\
&\leq P[X_n \leq x/(C - \epsilon)] + P[|Y_n - C| > \epsilon] \\
\Rightarrow \limsup F_{X_nY_n}(x) &\leq \limsup P[X_n \leq x/(C - \epsilon)] + \limsup P[|Y_n - C| > \epsilon] \\
\Rightarrow \limsup F_{X_nY_n}(x) &\leq \limsup F_{X_n}(x/(C - \epsilon)). \qquad (7.7)
\end{aligned}$$

Further observe that,

$$\begin{aligned}
F_{X_n}(x/(C + \epsilon)) &= P[X_n \leq x/(C + \epsilon)] \\
&= P[X_n \leq x/(C + \epsilon), |Y_n - C| \leq \epsilon] \\
&+ P[X_n \leq x/(C + \epsilon), |Y_n - C| > \epsilon] \\
&\leq P[X_n \leq x/Y_n] + P[|Y_n - C| > \epsilon] \\
\Rightarrow \liminf F_{X_n}(x/(C + \epsilon)) &\leq \liminf P[X_nY_n \leq x] + \liminf P[|Y_n - C| > \epsilon] \\
\Rightarrow \liminf F_{X_n}(x/(C + \epsilon)) &\leq \liminf F_{X_nY_n}(x). \qquad (7.8)
\end{aligned}$$

From Equation (7.7) and Equation (7.8), we have

$$\begin{aligned}
\liminf F_{X_n}(x/(C + \epsilon)) &\leq \liminf P[X_nY_n \leq x] \leq \limsup F_{X_nY_n}(x) \\
&\leq \limsup P[X_n \leq x/(C - \epsilon)] \\
\Rightarrow \lim F_{X_n}(x/(C + \epsilon)) &\leq \liminf P[X_nY_n \leq x] \leq \limsup F_{X_nY_n}(x) \\
&\leq \lim P[X_n \leq x/(C - \epsilon)] \\
\Rightarrow F_{X_nY_n}(x) &\to F_X(x/C) = F_{XC}(x),
\end{aligned}$$

since ϵ is arbitrary. If $x \leq 0$, then the entire derivation of Case (a) goes through with $C - \epsilon$ interchanged with $C + \epsilon$.

Case(b) $C < 0$: Suppose $x \in \mathbb{R}$ is such that $x/C \in C(F_X)$. Hence, $P[X \leq x/C] = P[X < x/C]$. We select h such that as $h \downarrow 0$, $x/C - h \in C(F_X)$, it is possible since $C(F_X)$ is dense in $\mathbb{R}$. Thus,

$$\begin{aligned}
\lim_{h\to 0} F_X(x/C - h) &= F_X(x/C) \\
\Longleftrightarrow \lim_{h\to 0} P[X \leq x/C - h] &= P[X \leq x/C] \\
\Longleftrightarrow \lim_{h\to 0} P[X < x/C - h] &= P[X < x/C] \\
\Longleftrightarrow \lim_{h\to 0} P[XC > x - Ch] &= P[XC > x] \\
\Longleftrightarrow \lim_{h\to 0} 1 - F_{XC}(x - Ch) &= 1 - F_{XC}(x) \\
\Longleftrightarrow \lim_{h\to 0} F_{XC}(x - Ch) &= F_{XC}(x)
\end{aligned}$$

Thus, x is a point of continuity of F_{XC}. Suppose $\epsilon > 0$ is such that $x/(C - \epsilon)$ and $x/(C + \epsilon)$ are in $C(F_X)$. For large n we assume $Y_n \leq 0$, it is valid since

Theorem 7.3.3, $P[Y_n \le 0] \to 1$. By Equation (7.6), $1/(C+\epsilon) < 1/Y_n < 1/(C-\epsilon)$. Suppose $x > 0$. Note that

$$\begin{aligned} F_{X_nY_n}(x) &= P[X_nY_n \le x] = P[X_nY_n \le x, |Y_n - C| < \epsilon] \\ &\quad + P[X_nY_n \le x, |Y_n - C| \ge \epsilon] \\ &\le P[X_n \ge x/Y_n, |Y_n - C| < \epsilon] + P[|Y_n - C| \ge \epsilon] \\ &\le P[X_n > x/(C+\epsilon)] + P[|Y_n - C| \ge \epsilon] \end{aligned}$$

$$\Rightarrow \ \limsup F_{X_nY_n}(x) \le \limsup P[X_n > x/(C+\epsilon)] + \limsup P[|Y_n - C| > \epsilon]$$

$$\Rightarrow \ \limsup F_{X_nY_n}(x) \le \limsup P[X_n > x/(C+\epsilon)]. \tag{7.9}$$

Further,

$$\begin{aligned} P[X_n > x/(C-\epsilon)] &= P[X_n > x/(C-\epsilon), |Y_n - C| < \epsilon] \\ &\quad + P[X_n > x/(C-\epsilon), |Y_n - C| \ge \epsilon] \\ &\le P[X_n > x/Y_n] + P[|Y_n - C| \ge \epsilon] \end{aligned}$$

$$\Rightarrow \ \liminf P[X_n > x/(C-\epsilon)] \le \liminf P[X_nY_n < x] + \liminf P[|Y_n - C| \ge \epsilon]$$

$$\Rightarrow \ \liminf P[X_n > x/(C-\epsilon)] \le \liminf P[X_nY_n \le x]$$

$$\Rightarrow \ \liminf P[X_n > x/(C-\epsilon)] \le \liminf F_{X_nY_n}(x). \tag{7.10}$$

In the second last step, note that $P[X_nY_n < x] \le P[X_nY_n \le x]$. Since $x/(C-\epsilon) \in C(F_X)$, observe that

$$\begin{aligned} F_{X_n}(x/(C-\epsilon)) &\to F_X(x/(C-\epsilon)) \\ \Rightarrow \ P[X_n \le x/(C-\epsilon)] &\to P[X \le x/(C-\epsilon)] \\ \Rightarrow \ P[X_n > x/(C-\epsilon)] &\to P[X > x/(C-\epsilon)] = P[X \ge x/(C-\epsilon)]. \end{aligned} \tag{7.11}$$

Similarly, $x/(C+\epsilon) \in C(F_X)$ implies that

$$P[X_n > x/(C+\epsilon)] \to P[X > x/(C+\epsilon)] = P[X \ge x/(C+\epsilon)]. \tag{7.12}$$

From Equations (7.9), (7.10), (7.11) and (7.12), we have

$$\begin{aligned} \liminf P[X_n > x/(C-\epsilon)] &\le \liminf F_{X_nY_n}(x) \le \limsup F_{X_nY_n}(x) \\ &\le \limsup P[X_n > x/(C+\epsilon)] \\ \Rightarrow \ P[X \ge x/(C-\epsilon)] &\le \liminf P[X_nY_n \le x] \le \limsup F_{X_nY_n}(x) \\ &\le P[X \ge x/(C+\epsilon)] \\ \Rightarrow \ \lim P[X_nY_n \le x] &= P[X \ge x/C] \text{ as } \epsilon \text{ is arbitrary} \\ \Rightarrow \ \lim P[X_nY_n \le x] &= P[XC \le x] = F_{XC}(x) \text{ as } C < 0. \end{aligned}$$

If $x \le 0$, then the entire derivation of Case(b) goes through with $C - \epsilon$ interchanged with $C+\epsilon$. Thus, it is proved that if $X_n \xrightarrow{L} X$ and $Y_n \xrightarrow{L} C$, then $X_nY_n \xrightarrow{L} XC$ for all $C \in \mathbb{R}$.

(iv) Note that $Y_n \xrightarrow{P} C \Rightarrow 1/Y_n \xrightarrow{P} 1/C$. Hence by (iii) $X_n/Y_n \xrightarrow{L} X/C$, provided X_n/Y_n and X/C are defined. □

Following corollaries follow immediately from Slutsky's theorem.

Corollary 7.3.1. *If $X_n \xrightarrow{L} X$, then $a+bX_n \xrightarrow{L} a+bX$. The result also follows from the continuous mapping theorem. As a particular case, with $a=0$ and $b=-1$, $-X_n \xrightarrow{L} -X$.*

Corollary 7.3.2. *If $X_n \xrightarrow{L} X$ and U_n and V_n converge in probability to constants a and b respectively, then $U_n + V_nX_n \xrightarrow{L} a+bX$.*

The following example illustrates Slutsky's theorem, in which we show that Student's t_n distribution converges in law to the standard normal distribution as the degrees of freedom $n \to \infty$.

Example 7.3.17. Suppose a random variable Y_n is defined as $Y_n = U_n/\sqrt{V_n/n}$ where $U_n \sim N(0,1)$ distribution, $V_n \sim \chi^2_n$ distribution and U_n and V_n are independent random variables. Then $Y_n \sim t_n$ distribution. Since $U_n \sim N(0,1)$ distribution for each n, $U_n \xrightarrow{L} Z \sim N(0,1)$. Further, $V_n \sim \chi^2_n$ distribution hence $E(V_n) = n$ and $Var(V_n) = 2n$. By Chebyshev's inequality, as $n \to \infty$,

$$P\left[|V_n/n - 1| > \epsilon\right] \le \left(E\left(V_n/n - 1\right)^2\right)/\epsilon^2 = Var(V_n)/n^2\epsilon^2 = 2/n\epsilon^2 \to 0.$$

Thus, $V_n/n \xrightarrow{P} 1$. Hence, by the invariance of convergence in probability under continuous transformation, $\sqrt{V_n/n} \xrightarrow{P} 1$. Further, by Slutsky's theorem $Y_n = U_n/\sqrt{V_n/n} \xrightarrow{L} Z \sim N(0,1)$. Thus, Student's t_n distribution converges in law to the standard normal distribution as degrees of freedom $n \to \infty$. □

Theorem 7.3.7 is heavily used in large sample inference to show that the remainder term converges to 0, as is clear from the proof of the following result, which is the well-known delta method in asymptotic inference. We prove it below in inference terminology. We also use Slutsky's theorem. We first prove a lemma.

Lemma 7.3.2. *If $U_n = \sqrt{n}(X_n - \mu) \xrightarrow{L} X$, then $X_n \xrightarrow{P} \mu$.*

Proof. It is given that $U_n \xrightarrow{L} X$ and hence by Lemma 7.3.1, U_n is bounded in probability. Further, $1/\sqrt{n} \to 0$ and hence by Theorem 7.3.7,

$$(X_n - \mu) = (1/\sqrt{n})(\sqrt{n}(X_n - \mu)) = (1/\sqrt{n})U_n \xrightarrow{P} 0 \;\Rightarrow\; X_n \xrightarrow{P} \mu.$$

□

As an application of this lemma, if an estimator T_n of a parameter μ, is such that $\sqrt{n}(T_n - \mu) \xrightarrow{L} X$, then it is consistent for μ.

Theorem 7.3.9. *Delta method: Suppose T_n is an estimator of μ such that $\sqrt{n}(T_n - \mu) \xrightarrow{L} X \sim N(0,\sigma^2)$ distribution. Suppose g is a differentiable function with $g'(\mu) \neq 0$, then $\sqrt{n}(g(T_n) - g(\mu)) \xrightarrow{L} Y \sim N(0,(g'(\mu))^2\sigma^2)$ distribution.*

Proof. It is given that $U_n = \sqrt{n}(T_n - \mu) \xrightarrow{L} X \sim N(0, \sigma^2)$ distribution. Since g is a differentiable function with $g'(\mu) \neq 0$, by the Taylor series expansion,

$$g(T_n) = g(\mu) + (T_n - \mu)g'(\mu) + R_n, \quad \text{where} \quad |R_n| \leq M|T_n - \mu|^{1+\delta}, \ \ \delta > 0 \ .$$

Thus,

$$\sqrt{n}(g(T_n) - g(\mu)) = \sqrt{n}(T_n - \mu)g'(\mu) + \sqrt{n}R_n \ \xrightarrow{L} \ Y \sim N(0, (g'(\mu))^2\sigma^2),$$

provided $\sqrt{n}R_n \xrightarrow{P} 0$. Now, $|\sqrt{n}R_n| \leq M\sqrt{n}|T_n - \mu||T_n - \mu|^{\delta}$. As shown in Lemma 7.3.2, $(T_n - \mu) \xrightarrow{P} 0$, hence $|T_n - \mu|^{\delta} \xrightarrow{P} 0$, since convergence in probability is invariant under continuous transformation. Further, $\sqrt{n}(T_n - \mu) \xrightarrow{L} X \Rightarrow \sqrt{n}|T_n - \mu| \xrightarrow{L} |X|$, by continuous mapping theorem and hence $\sqrt{n}|T_n - \mu|$ is bounded in probability. As a consequence,

$$\sqrt{n}|T_n - \mu||T_n - \mu|^{\delta} \xrightarrow{P} 0 \ \Rightarrow \ \sqrt{n}R_n \xrightarrow{P} 0 \ \Rightarrow \ \sqrt{n}(g(T_n) - g(\mu)) \xrightarrow{L} Y \ ,$$

by Slutsky's theorem, where Y follows normal $N(0, (g'(\mu))^2\sigma^2)$ distribution. □

Remark 7.3.3. In the delta method, it is assumed that g is differentiable and hence it is continuous. Thus, by the continuous mapping theorem $g(\sqrt{n}(T_n - \mu)) \xrightarrow{L} g(X)$. It is known that if X is normally distributed, then only linear functions of X will have a normal distribution. Observe that $g(\sqrt{n}(T_n - \mu)) = \sqrt{n}(g(T_n) - g(\mu))$ if g is a linear function. However, the delta method is valid for any differentiable function g such that $g'(\mu) \neq 0$ and for any such function, $g(X)$ has a normal distribution. Hence the result, which says that $\sqrt{n}(g(T_n) - g(\mu))$ also converges to a normal distribution, may seem surprising. However, from the proof of the delta method, we note that $\sqrt{n}(g(T_n) - g(\mu))$ is expressed as a linear function of $\sqrt{n}(T_n - \mu)$ with an additional term which converges to zero in probability, and hence we get that $\sqrt{n}(g(T_n) - g(\mu))$ converges to a normal distribution, even if g is not a linear function.

The delta method provides a basis for deriving variance stabilizing transformations, that is, transformations leading to an asymptotic variance free from the parameter. Such transformations are helpful in finding large sample confidence intervals for the parameter of interest. It is illustrated in the following example.

Example 7.3.18. Suppose $\{X_1, X_2, \cdots, X_n\}$ is a random sample from normal $N(\theta, \theta^2)$ distribution, $\theta > 0$. Then $T_n = \sqrt{n}(\overline{X}_n - \theta) \sim N(0, \theta^2)$ distribution. We now find a differentiable function g such that $(g'(\theta))^2 \neq 0$ and $Var(g(\overline{X}_n))$ is free from θ, that is $(g'(\theta))^2\theta^2 = c^2$, that is, $g(\theta) = \int \frac{c \, d\theta}{\theta} = c \log \theta$. Hence,

$$Q_n = \sqrt{n}(\log \overline{X}_n - \log \theta) \ \xrightarrow{L} \ Z \sim N(0, 1) \ \ \Rightarrow Q_n \ \text{ is a pivotal quantity.}$$

Thus, we find $a_{(1-\alpha/2)}$ such that $P[-a_{(1-\alpha/2)} < Q_n < a_{(1-\alpha/2)}] = 1-\alpha$, the given confidence coefficient. Now, $-a_{(1-\alpha/2)} < Q_n < a_{(1-\alpha/2)}$ is equivalent to $\log \overline{X}_n - a_{(1-\alpha/2)}/\sqrt{n} < \log\theta < \log \overline{X}_n + a_{(1-\alpha/2)}/\sqrt{n}$. Thus, the asymptotic confidence interval for θ with confidence coefficient $(1-\alpha)$ is given by,

$$\left(\exp\left(\log \overline{X}_n - a_{(1-\alpha/2)}/\sqrt{n}\right), \quad \exp\left(\log \overline{X}_n + a_{(1-\alpha/2)}/\sqrt{n}\right)\right) . \quad \square$$

Remark 7.3.4. It is to be noted that $\overline{X}_n \sim N(\theta, \theta^2/n)$ distribution for each n but $\log \overline{X}_n \sim N(\log\theta, 1/n)$ distribution for large n.

One of the most important tools to establish convergence in law of a suitably scaled and centred average is the central limit theorem (CLT). We discuss in detail various versions of CLT in Chapter 10. We state below the simplest version of CLT, which is used to solve some examples.

Theorem 7.3.10. *Lindeberg-Levy CLT: Suppose $\{X_n, n \geq 1\}$ is a sequence of independent and identically distributed random variables with mean μ and positive, finite variance σ^2, then*

$$Y_n = \Big(\sum_{i=1}^{n} X_i - n\mu\Big)/(\sqrt{n}\sigma) = (S_n - n\mu)\sqrt{n}\sigma = \sqrt{n}(\overline{X}_n - \mu)/\sigma \xrightarrow{L} Z \sim N(0,1).$$

The following example illustrates how CLT and Slutsky's theorem are useful for constructing a large sample confidence interval.

Example 7.3.19. Suppose $\{X_1, X_2, \cdots, X_n\}$ is a random sample from Poisson distribution with mean θ. Then $Var(X) = \theta < \infty$ and $Var(\overline{X}_n) = \theta/n \to 0$. Hence

$$\overline{X}_n \xrightarrow{P} \theta \ \ \& \ \text{by the CLT} \quad Q_n = \sqrt{n/\theta}(\overline{X}_n - \theta) \xrightarrow{L} Z \sim N(0,1).$$

Thus, Q_n is a pivotal quantity, for large n. However, it is not useful to construct a confidence interval for θ. We define $\tilde{Q}_n$ as

$$\tilde{Q}_n = \frac{\sqrt{n}}{\sqrt{\overline{X}_n}}(\overline{X}_n - \theta) = \left\{\frac{\sqrt{\theta}}{\sqrt{\overline{X}_n}}\right\}\left\{\frac{\sqrt{n}}{\sqrt{\theta}}(\overline{X}_n - \theta)\right\} \xrightarrow{L} 1 \times Z \sim N(0,1)$$

by CLT and Slutsky's theorem. Thus, $\tilde{Q}_n$ can be taken as a pivotal quantity for large n, which is suitable to find the asymptotic confidence interval for θ. Hence, given the confidence coefficient $(1-\alpha)$, we find $a_{(1-\alpha/2)}$ such that $P[-a_{(1-\alpha/2)} < \tilde{Q}_n < a_{(1-\alpha/2)}] = 1-\alpha$. Now, $-a_{(1-\alpha/2)} < \tilde{Q}_n < a_{(1-\alpha/2)}$ is equivalent to $\overline{X}_n - \sqrt{\overline{X}_n/n}\ a_{(1-\alpha/2)} < \theta < \overline{X}_n + \sqrt{\overline{X}_n/n}\ a_{(1-\alpha/2)}$. Thus, asymptotic confidence interval for θ with confidence coefficient $(1-\alpha)$ is given by,

$$\left(\overline{X}_n - \sqrt{\overline{X}_n/n}\ a_{(1-\alpha/2)}, \ \overline{X}_n + \sqrt{\overline{X}_n/n}\ a_{(1-\alpha/2)}\right),$$

which can also be written as

$$\left(\overline{X}_n - s.e.(\overline{X}_n)a_{(1-\alpha/2)}, \ \ \overline{X}_n + s.e.(\overline{X}_n)a_{(1-\alpha/2)} \right),$$

where $\sqrt{\overline{X}_n/n}$ is the standard error of $\overline{X}_n$. □

Generating functions, such as characteristic function, moment generating function or probability generating function for discrete integer-valued random variables, are always used to derive the distribution of sums of independent random variables. One essential feature underlying these derivations is the uniqueness theorem which states that if the generating functions of two random variables are the same, then their distributions are the same. In particular, if the moment generating function or probability generating function of two random variables X and Y are the same on the domain, then X and Y are identically distributed random variables. In Chapter 4, we have proved the uniqueness theorem for characteristic functions. This idea can be extended to decide the limit distribution in convergence in law, that is, it is possible to identify the limit law if one identifies the limit of a sequence of generating functions corresponding to a sequence of random variables. We state some theorems which are useful to find the limit law of X_n using generating functions. In Chapter 4, we noted that the domain of a characteristic function is $\mathbb{R}$. Hence, we work with a characteristic function rather than the corresponding moment generating function. For a discrete integer-valued random variable, it is better to deal with a probability generating function defined on $[-1, 1]$. We state below some theorems which convey a connection between convergence in law and convergence of a sequence of characteristic functions.

Theorem 7.3.11. *Suppose $\{X_n, n \geq 1\}$ is a sequence of random variables with $\phi_n(t)$, $t \in \mathbb{R}$, being a characteristic function of $X_n, n \geq 1$. Suppose X is a random variable with $\phi(t)$, $t \in \mathbb{R}$ as its characteristic function. Then*

$$\phi_n(t) \to \phi(t) \ \ \forall \ \ t \in \mathbb{R} \quad \Longleftrightarrow \quad X_n \stackrel{L}{\to} X.$$

For proof, we refer to Gut [13].

Suppose $\{X_n, n \geq 1\}$ and $\{Y_n, n \geq 1\}$ are sequences of random variables. If $X_n \stackrel{L}{\to} X$ and $Y_n \stackrel{L}{\to} C$, then Slutsky's theorem states that $X_n + Y_n \stackrel{L}{\to} X + C$. If $Y_n \stackrel{L}{\to} Y$, then the result $X_n + Y_n \stackrel{L}{\to} X + Y$ is not always true. It is true if $\forall \ n \geq 1$, X_n and Y_n are independent. Theorem 7.3.11 is useful to prove this result.

Theorem 7.3.12. *Suppose $X_n \stackrel{L}{\to} X$, $Y_n \stackrel{L}{\to} Y$ and $\forall \ n \geq 1$, X_n and Y_n are independent. Suppose X and Y are also independent. Then $X_n + Y_n \stackrel{L}{\to} X + Y$.*

Proof. From Theorem 7.3.11

$$\begin{aligned} X_n \stackrel{L}{\to} X \ \ \& \ \ Y_n \stackrel{L}{\to} Y \quad & \Rightarrow \quad \phi_{X_n}(t) \to \phi_X(t) \ \ \& \ \ \phi_{Y_n}(t) \to \phi_Y(t) \\ & \Rightarrow \quad \phi_{X_n}(t)\phi_{Y_n}(t) \to \phi_X(t)\phi_Y(t) \\ & \Rightarrow \quad \phi_{X_n+Y_n}(t) \ \to \ \phi_{X+Y}(t). \end{aligned}$$

The last step follows in view of the independence of X_n and Y_n and of X and Y. Note that $\phi_{X+Y}(t)$ is a characteristic function of $X+Y$. Hence again by Theorem 7.3.11, $X_n + Y_n \xrightarrow{L} X + Y$. □

The following example illustrates that Theorem 7.3.12 is not true if we drop the independence condition (Stoyanov [23]).

Example 7.3.20. Suppose $\{X_n, n \geq 1\}$ and $\{Y_n, n \geq 1\}$ are sequences of random variables. Suppose $X_n \xrightarrow{L} X$ and $Y_n \xrightarrow{L} Y$ where $X \sim N(0,1)$ and $Y \sim N(0,1)$. If X_n and Y_n are independent $\forall\ n \geq 1$, then $X_n + Y_n \sim N(0,2)$ distribution. Moreover, in this case, the distribution of $(X_n, Y_n)'$ converges to the bivariate normal distribution with zero mean vector and dispersion matrix diag$(1,1)$. Suppose now $\{X_n, n \geq 1\}$ is a sequence of random variables, such that $X_n \xrightarrow{L} X \sim N(0,1)$. If $Y_n = X_n\ \ \forall\ \ n \geq 1$, then $Y_n \xrightarrow{L} X \sim N(0,1)$. However, X_n and Y_n are not independent. Further, $X_n + Y_n = 2X_n$ and $2X_n \xrightarrow{L} 2X \sim N(0,4)$ distribution and not $N(0,2)$ distribution. □

The theorem similar to Theorem 7.3.12 is stated below. For proof, refer to Grimmett and Stirzaker [12].

Theorem 7.3.13. *Suppose $X_n \xrightarrow{L} X$, $Y_n \xrightarrow{L} Y$ and the joint distribution $(X_n, Y_n)'$ converges to that of $(X,Y)'$. Suppose X and Y are independent. Then $X_n + Y_n \xrightarrow{L} X + Y$.*

Suppose $\phi_n(t)$ is a characteristic function of $X_n, n \geq 1$. If one can show that a sequence $\{\phi_n(t), n \geq 1\}$ of characteristic functions converges to some function, we can conclude that a sequence $\{X_n, n \geq 1\}$ converges in law, under one condition on the limit of $\phi_n(t)$. A statement of the theorem is given below. It is referred to as a continuity theorem for characteristic functions and is heavily used in identifying the limit law of a sequence of random variables.

Theorem 7.3.14. *Continuity theorem for characteristic functions: Suppose $\{X_n, n \geq 1\}$ is a sequence of random variables with $\phi_n(t)$, $t \in \mathbb{R}$ being a characteristic function of $X_n, n \geq 1$. Suppose $\phi_n(t) \to \phi(t)\ \ \forall\ \ t \in \mathbb{R}$ as $n \to \infty$, where $\phi(t)$ is continuous at $t = 0$. Then there exists a random variable X such that $X_n \xrightarrow{L} X$ and $\phi(t)$ is a characteristic function of X.*

For proof, we refer to Gut [13]. The following example shows how the continuity of $\phi(t)$ at $t = 0$ is crucial.

Example 7.3.21. Suppose $\{X_n, n \geq 1\}$ is a sequence of random variables such that X_n has uniform $U(-n,n)$ distribution. Then the characteristic function $\phi_{X_n}(t)$ of X_n is given by

$$\begin{aligned}\phi_{X_n}(t) &= \frac{1}{2n}\int_{-n}^{n} e^{itx}\,dx = \frac{1}{2n}2\int_{0}^{n} \cos(tx)\,dx \\ &= 1 \text{ if } t = 0 \ \ \& \ \ \sin(nt)/nt \text{ if } t \neq 0.\end{aligned}$$

If $t \neq 0$ then $\lim_{n\to\infty} \phi_{X_n}(t) = \lim_{n\to\infty} \sin(nt)/nt = 0$ using the result that $|\sin(nt)| \leq 1$. Thus,

$$\phi_{X_n}(t) \quad \to \quad \begin{cases} 1, & \text{if} \quad t = 0 \\ 0, & \text{if} \quad t \neq 0. \end{cases}$$

It is to be noted that $\lim_{n\to\infty} \phi_{X_n}(t)$ is not continuous at $t = 0$ and Theorem 7.3.14 is not applicable. In Example 7.3.9, we have shown that X_n does not converge in law by finding the limit of the distribution function of X_n. □

We discuss some examples below to illustrate Theorem 7.3.14 and the uniqueness theorem of a characteristic function, proved in Chapter 4. Continuity and uniqueness theorems for characteristic functions are also useful to establish a weak law of large numbers and the central limit theorem, which we discuss in Chapters 9 and 10 respectively.

Example 7.3.22. Suppose $\{X_n, n \geq 1\}$ is a sequence of random variables such that characteristic function $\phi_{X_n}(t) \to \exp(itC)$ for some $C \in \mathbb{R}$ and $\forall \ t \in \mathbb{R}$. It is to be noted that $\exp(itC)$ is continuous at $t = 0$, and it is a characteristic function of a random variable degenerate at C. Hence by Theorem 7.3.14, $X_n \xrightarrow{L} X$ and by the uniqueness theorem for characteristic functions $X \equiv C$. Since the limit random variable is degenerate, convergence in law implies convergence in probability, thus $X_n \xrightarrow{P} C$. □

The following simple example illustrates both Theorem 7.3.11 and Theorem 7.3.14.

Example 7.3.23. Suppose X_n has normal $N(\theta_n, \sigma_n^2)$ distribution and $\theta_n \to \theta$ and $\sigma_n^2 \to \sigma^2$. Then observe that,

$$F_n(x) = P[X_n \leq x] = P\left[\frac{X_n - \theta_n}{\sigma_n} \leq \frac{x - \theta_n}{\sigma_n}\right] = \Phi\left(\frac{x - \theta_n}{\sigma_n}\right) \to \Phi\left(\frac{x - \theta}{\sigma}\right),$$

$\forall \ x \in \mathbb{R} = C(F_X)$. Thus, $X_n \xrightarrow{L} X \sim N(\theta, \sigma^2)$. Thus, the large sample distribution of X_n is $N(\theta, \sigma^2)$. Alternatively, we can prove the same result using the continuity theorem for characteristic functions. Since $X_n \sim N(\theta_n, \sigma_n^2)$ distribution, its characteristic function $\phi_{X_n}(t)$ is given by, $\phi_{X_n}(t) = \exp(it\theta_n - t^2\sigma_n^2/2)$. Note that as $\theta_n \to \theta$ and $\sigma_n^2 \to \sigma^2$,

$$\phi_{X_n}(t) = \exp(it\theta_n - t^2\sigma_n^2/2) \quad \to \quad \exp(it\theta - t^2\sigma^2/2) = \phi_X(t), \ \forall \, t \in \mathbb{R}.$$

Further, $\phi_X(t)$ is continuous at $t = 0$ and it is a characteristic function of $X \sim N(\theta, \sigma^2)$. Hence by the continuity theorem and the uniqueness theorem for characteristic functions, we conclude that $X_n \xrightarrow{L} X \sim N(\theta, \sigma^2)$. Conversely, suppose now $X_n \xrightarrow{L} X \sim N(\theta, \sigma^2)$, where $X_n \sim N(\theta_n, \sigma_n^2)$. We examine whether $\theta_n \to \theta$ and $\sigma_n^2 \to \sigma^2$. It is known that for any complex number $z = a + ib$,

$$e^z = e^a e^{ib} = e^a(\cos b + i \sin b) \quad \Rightarrow \quad |e^z| = (e^{2a}\cos^2 b + e^{2a}\sin^2 b)^{1/2} = e^a.$$

Thus,

$$\begin{aligned} X_n \xrightarrow{L} X \quad & \Rightarrow \quad \phi_{X_n}(t) \to \phi_X(t) \ \ \forall \ \ t \in \mathbb{R} \quad \text{by Theorem 7.3.11} \\ & \Rightarrow \quad \left|e^{it\theta_n - t^2\sigma_n^2/2}\right| \to \left|e^{it\theta - t^2\sigma^2/2}\right| \ \ \forall \ \ t \in \mathbb{R} \\ & \Rightarrow \quad e^{-t^2\sigma_n^2/2} \to e^{-t^2\sigma^2/2} \ \Rightarrow \ \sigma_n^2 \to \sigma^2 \ \ \text{as } n \to \infty \\ \text{Further,} \ \ \lim_{n\to\infty} e^{it\theta_n} \quad & = \quad \lim_{n\to\infty} \frac{\phi_{X_n}(t)}{|\phi_{X_n}(t)|} = \frac{\phi_X(t)}{|\phi_X(t)|} = e^{it\theta} \ \Rightarrow \ \theta_n \to \theta, \end{aligned}$$

as $n \to \infty$. □

Example 7.3.24. Suppose $X_n \sim G(n, n)$ distribution with probability density function, mean and variance given by,

$$f(x, n) = \frac{n^n}{\Gamma(n)} \exp(-nx)x^{n-1}, \ x > 0, \ n \in \mathbb{N}, \ E(X_n) = 1 \ \& \ Var(X_n) = 1/n.$$

Its characteristic function is given by $\phi_{X_n}(t) = (1 - it/n)^{-n}$.

$$\text{As } \ n \to \infty, \ \ \forall \ \ t \in \mathbb{R}, \ \phi_{X_n}(t) = (1 - it/n)^{-n} \to \exp(it) = \phi(t), \ \ \text{say.}$$

$\phi(t)$ is continuous at $t = 0$ and it is the characteristic function of $X \equiv 1$. By Theorem 7.3.14, $X_n \xrightarrow{L} 1$ and since the limit random variable is degenerate, $X_n \xrightarrow{P} 1$. Further,

$$E(X_n - 1)^2 = Var(X_n) = 1/n \to 0 \ \Rightarrow \ X_n \xrightarrow{q.m.} 1 \ \Rightarrow \ X_n \xrightarrow{P} 1 \ \Rightarrow \ X_n \xrightarrow{L} 1.$$

Suppose now we centre and scale X_n and define $Y_n = \sqrt{n}(X_n - 1)$ and study its limit distribution. Suppose $\phi_n(\cdot)$ denotes the characteristic function of Y_n. Then $\forall \ \ t \in \mathbb{R}$

$$\begin{aligned} \phi_n(t) = E(e^{itY_n}) \quad & = \quad E(e^{it\sqrt{n}(X_n - 1)}) = e^{-it\sqrt{n}} E(e^{it\sqrt{n}X_n}) \\ & = \quad e^{-it\sqrt{n}}(1 - it\sqrt{n}/n)^{-n} \\ \Rightarrow \quad \log \phi_n(t) \quad & = \quad -it\sqrt{n} - n\log(1 - it/\sqrt{n}) \\ & = \quad -it\sqrt{n} + n(it/\sqrt{n} - t^2/2n - it^3/3n^{3/2} + \cdots) \\ & = \quad -t^2/2 + \delta_n \ \ \to \ \ -t^2/2 = \phi(t), \ \ \text{say} \end{aligned}$$

since $\delta_n \to 0$ as $n \to \infty$. Thus, the sequence $\{\phi_n(t), n \geq 1\}$ of characteristic functions converges to $\phi(t) = \exp(-t^2/2)$, which is continuous at $t = 0$. Hence, by the continuity theorem for characteristic functions $Y_n \xrightarrow{L} Z$, where $\phi(t) = \exp(-t^2/2)$ is a characteristic function of Z. It is known that $\phi(t) = \exp(-t^2/2)$ is a characteristic function of the standard normal distribution. Hence by the uniqueness theorem of a characteristic function, $Z \sim N(0, 1)$ distribution. □

In the following example, we verify the results shown in Example 7.3.24 by simulation.

Example 7.3.25. Suppose $X_n \sim G(n,n)$. Using Code 7.3.4, by simulation, we verify convergence in probability and quadratic mean of $\{X_n, n \geq 1\}$ to 1, and convergence to the standard normal distribution of $Y_n = \sqrt{n}(X_n - 1)$. We verify that $X_n \xrightarrow{L} 1$ in the next example, as we need large values of n to arrive at the desired result. To verify that $X_n \xrightarrow{P} 1$, as in Example 7.2.16, we obtain the estimate of the probability $p_n(\epsilon)$ as $rf_n = \sum_{i=1}^{m} I_{[|X_i-1|<\epsilon]}/m$, where X_i is the realized observation from the given gamma distribution. We examine whether it approaches 1 as n increases. To examine if $X_n \xrightarrow{q.m} 1$, we compute $q_n = (1/m)\sum_{i=1}^{m}(X_i - 1)^2$ and observe whether it approaches 0 as n increases. To examine $Y_n = \sqrt{n}(X_n - 1) \xrightarrow{L} Z \sim N(0,1)$, based on m realized values of Y_n, we plot the histogram with relative frequencies on Y-axis and impose on it the curve of the probability density function of the standard normal distribution. We then observe how close the two are. To support the visual impression, we carry out the Shapiro-Wilk test of normality and decide based on p-values. To judge the normality, we also draw a box plot, a QQ plot with a QQ line imposed on it and a graph of the empirical distribution function of simulated values of Y_n and impose the graph of the distribution function of the standard normal distribution on it. We adopt this approach for $n = 1800$. We use the following code.

Code 7.3.4.
```
eps=0.05; p=me=v=q=p=pval=c(); M=500
s=c(300,600,900,1200,1500,1800)# vector of values of n
y=matrix(nrow=M,ncol=length(s))
for(i in 1:length(s))
{
 n=s[i]
 set.seed(i)
 x=rgamma(M,shape=n,scale=1/n )
 y[,i]= (n)^0.5*(x-1)
 me[i]=mean(x)
 v[i]=var(x)*(M-1)/M
 q[i]=mean(x-1)^2
 p[i]=length(which(abs(x-1)<eps))/M
 pval[i]=shapiro.test(y[,i])$p.value
}
d=round(data.frame(s,p,me,v,q,pval),4); d
r=seq(-4,4,.1); u=dnorm(r)
par(mfrow=c(3,2))
no=paste("n =",s,sep=" ")
for(j in 1:length(s))
{
```

```
 hist(y[,j],freq=FALSE,main=no[j],ylim=c(0,0.5),col="light blue",
 xlim=c(-4,4),xlab="")
 lines(r,u,"o",pch=20,col="dark blue")
}
par(mfrow=c(2,2))
hist(y[,6],freq=FALSE,main="Histogram of Y_1800",ylim=c(0,0.5),
col="light blue",xlim=c(-4,4),xlab="")
lines(r,u,"o",pch=20,col="dark blue")
boxplot(y[,6],main="Box Plot of Y_1800")
qqnorm(y[,6]); qqline(y[,6])
plot(ecdf(y[,6]),main="Empirical df of Y_1800",col="light blue",
xlab=expression(paste("Y"[n])),ylab=expression(paste("F"[n]
(t))),lty=1)
lines(r,pnorm(r),col="dark blue",lty=2)
legend("bottomright",legend=c("ecdf","df"),
col=c("light blue","dark blue"),lty=c(1,2))
```

From Table 7.5, we note that as n increases, $rf_n \to 1$, $q_n \to 0$. Thus, $X_n \stackrel{P}{\to} 1$ and $X_n \stackrel{q.m.}{\to} 1$. We also note that the mean and the variance of the simulated values approach 1 and 0, respectively. The p-values in the last column support the theoretical derivation that for large n, the distribution of Y_n can be approximated by the standard normal distribution.

From Figure 7.7, we note that the curve of the probability density function, imposed on the histogram, also supports the approximation. The four plots in Figure 7.8 drawn for $n = 1800$ convey the same findings. □

In the next example, using Code 7.3.5, we verify by simulation that $X_n \stackrel{L}{\to} 1$ where $X_n \sim G(n, n)$.

Example 7.3.26. Suppose $X_n \sim G(n, n)$. To verify that $X_n \stackrel{L}{\to} 1$, as in Example 7.2.16, we examine whether as n increases, the distribution function $F_n(x)$ of X_n approaches to the distribution function F of $X \equiv 1$ for all

TABLE 7.5
Gamma Distribution: Convergence in Probability and in Quadratic Mean

n	rf_n	Mean	Variance	q_n	p-val
300	0.590	1.0004	0.0034	0	0.3625
600	0.754	1.0033	0.0017	0	0.2577
900	0.878	1.0002	0.0011	0	0.3160
1200	0.924	0.9998	0.0009	0	0.3043
1500	0.952	0.9995	0.0006	0	0.7862
1800	0.970	0.9994	0.0005	0	0.7911

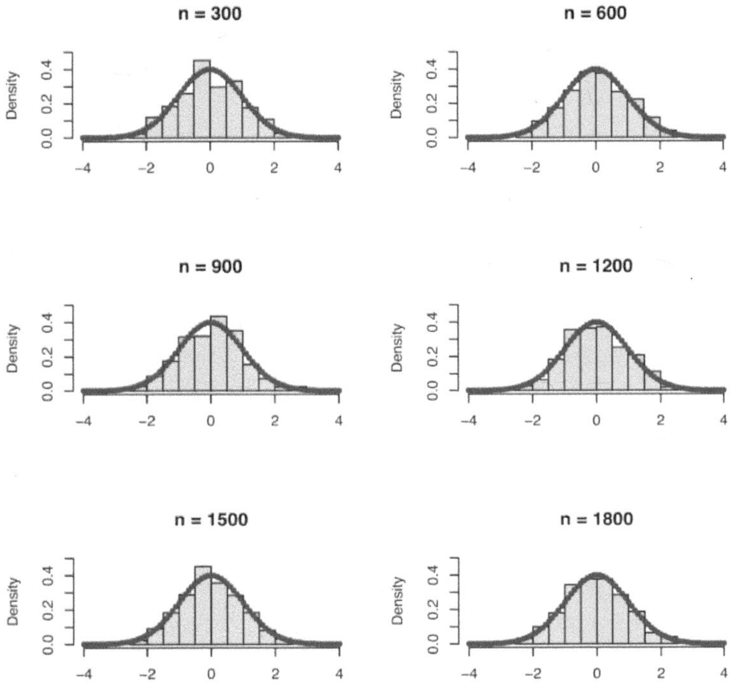

FIGURE 7.7
Convergence in Law of $Y_n = \sqrt{n}(X_n - 1)$: Histograms

$x \in \mathbb{R} - \{1\}$. Based on the simulated observations from X_n, we can find the empirical distribution function $G_n(x)$ and check whether, as n increases, it also approaches to that of $X \equiv 1$. In Chapter 9, using strong law of large numbers, we prove that $G_n(x) \stackrel{a.s.}{\to} F(x)$ for all $x \in \mathbb{R} - \{1\}$. Glivenko-Cantelli theorem (Gut [13]) states that the convergence is uniform in x. Using the following code, we verify whether $F_n(x)$ and $G_n(x)$ approach $F(x)$.

Code 7.3.5.
```
M=500; s=c(500,1500,2500,3500)# vector of values of n
x=matrix(nrow=M,ncol=length(s))
z=seq(.9,1,.001);length(z)
z1=seq(1,1.2,.001);length(z1); z2=c(z,z1)
w=c(rep(0,length(z)),rep(1,length(z1)))
w1=seq(.92,1.1,.001)
w2=matrix(nrow=length(w1),ncol=length(s))
par(mfrow=c(2,2))
for(i in 1:length(s))
{
 n=s[i]
 no=paste("n =",s,sep=" ")
```

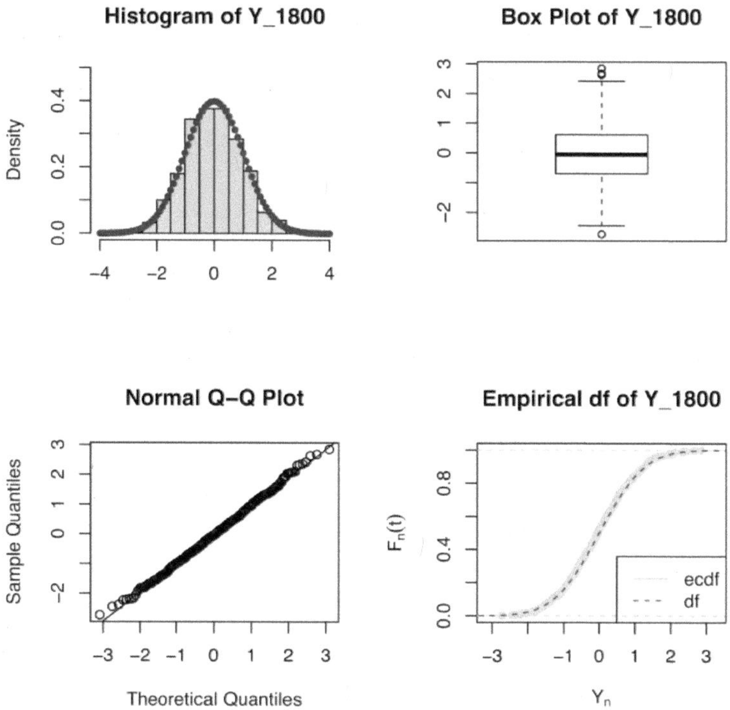

FIGURE 7.8
Convergence in Law of $Y_n = \sqrt{n}(X_n - 1)$

```
  set.seed(i)
  x[,i]=rgamma(M,shape=n,scale=1/n )
  plot(ecdf(x[,i]),main=s[i],col="blue",lwd=2)
  w2[,i]=pgamma(w1,shape=n,scale=1/n )
  lines(w1,w2[,i],main=s[i],col="red",lwd=1)
  lines(z2,w,col="dark blue")
}
legend("bottomright",legend=c("ecdf","df"),col=c("blue","red"),
lwd=c(2,1))
```

From Figure 7.9, we note that the curves of F_n and G_n are close to each other for all the four values of n. However, not very close to the curve of F. It may still need a larger n. It may be in view of the approximation of a continuous distribution function by a discrete one. This example also illustrates that a sequence of absolutely continuous random variables converges in law to a discrete random variable. □

We now state the analogue of Theorem 7.3.11 for a sequence probability generating functions of discrete integer-valued random variables. For proof, we

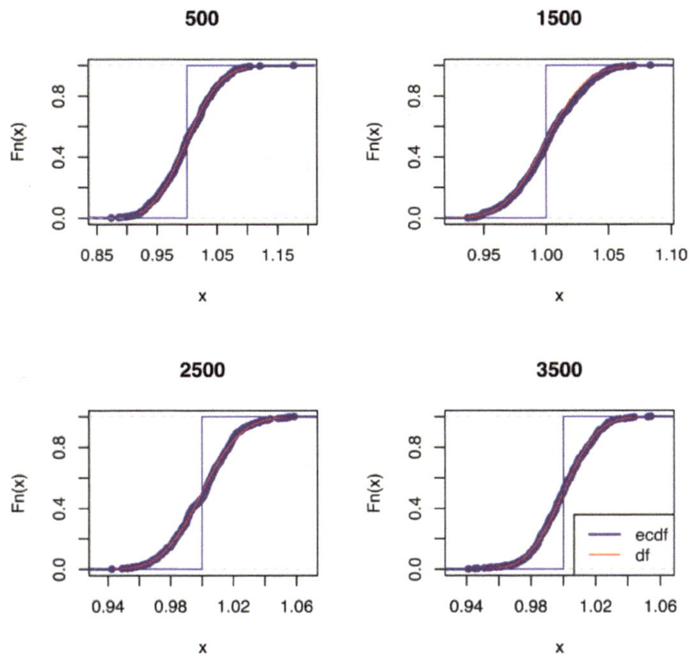

FIGURE 7.9
$X_n \sim G(n,n)$: Convergence in Law to 1

refer to Gut [13]. As the characteristic function determines the distribution uniquely, the probability generating function also determines the distribution uniquely.

Theorem 7.3.15. *Continuity theorem for probability generating functions: Suppose $\{X_n, n \geq 1\}$ is a sequence of discrete integer valued random variables with $P_n(t)$ as the probability generating function of X_n, $t \in [-1,1]$. Suppose probability generating function of X is $P(t)$, $t \in [-1,1]$. If as $n \to \infty$, $P_n(t) \to P(t)\ \forall\ t \in [-1,1]$ then $X_n \stackrel{L}{\to} X$.*

We have shown in Example 7.3.6 that under certain conditions, binomial distribution converges to Poisson distribution by showing that the probability mass function of a binomial distribution converges to that of a Poisson distribution. In the following example, we discuss the same result using Theorem 7.3.15.

Example 7.3.27. Suppose $X_n \sim B(n, \lambda/n)$ distribution. Its probability generating function is given by $P_n(t) = (1 - \lambda/n + \lambda t/n)^n$. Then as $n \to \infty$, $\forall\ t \in [-1,1]$,

$$P_n(t) = \left(1 - \frac{\lambda}{n} + \frac{\lambda t}{n}\right)^n = \left(1 + \frac{\lambda(t-1)}{n}\right)^n \ \to\ e^{\lambda(t-1)} = P(t), \text{ say.}$$

Hence by Theorem 7.3.15, binomial distribution converges to some distribution. However, $P(t)$ is the probability generating function of $X \sim P(\lambda)$ distribution. Hence from the uniqueness theorem of probability generating functions, we conclude that the limiting distribution is a Poisson distribution. More generally, suppose $X_n \sim B(n, p_n)$ distribution, where $n \to \infty$ in such a way that $p_n \to 0$ and $np_n \to \lambda$. Then as $n \to \infty$, $\forall\ t \in [-1, 1]$,,

$$\begin{aligned} P_n(t) &= (1 - p_n + p_n t)^n = ((t-1)p_n + 1)^n + 1 + n(t-1)p_n \\ &+ \frac{n(n-1)}{2!}(t-1)^2 p_n^2 + \frac{n(n-1)(n-2)}{3!}(t-1)^3 p_n^3 + \cdots \\ &\to 1 + (t-1)\lambda + \frac{(t-1)^2\lambda^2}{2!} + \frac{(t-1)^3\lambda^3}{3!} + \cdots = e^{\lambda(t-1)} = P(t), \end{aligned}$$

where $P(t)$ is the probability generating function of $X \sim P(\lambda)$ distribution. Hence, binomial $B(n, p_n)$ distribution converges to Poisson distribution as $n \to \infty$ in such a way that $p_n \to 0$ and $np_n \to \lambda$. □

Remark 7.3.5. While verifying that binomial distribution converges to Poisson distribution, under certain conditions, we can also use Theorem 7.3.14. Thus, if $X_n \sim B(n, \lambda/n)$ distribution. The limit of its characteristic function is given by

$$\phi_n(t) = \left(1 - \lambda/n + \lambda e^{it}/n\right)^n = \left(1 + \lambda(e^{it} - 1)/n\right)^n \to e^{\lambda(e^{it}-1)} = \phi(t),$$

say. It is continuous at $t = 0$ and hence $X_n \xrightarrow{L} X$. Note that $\phi(t)$ is a characteristic function of $P(\lambda)$ distribution. Hence by the uniqueness theorem of characteristic functions, $X \sim P(\lambda)$ distribution.

We now briefly discuss the weak convergence of a sequence of distribution functions and its relation with convergence in law. We begin with an example.

Example 7.3.28. Suppose $\{X_n, n \geq 1\}$ is a sequence of random variables such that $X_n \sim N(0, \sigma_n^2)$ distribution. The distribution function of X_n is given by,

$$\begin{aligned} F_n(x) &= P[X_n \leq x] = \Phi(x/\sigma_n) \ \to \ \Phi(0) \\ &= 1/2 \ \text{if} \sigma_n \to \infty \ \text{ as } n \to \ \infty, \ \forall\, x \in \mathbb{R}. \end{aligned}$$

Thus, $F_n(x)$ does not converge to a distribution function. □

In the above example and in Example 7.3.9, a sequence of distribution functions does not converge to a proper distribution function. Convergence of this type is known as weak convergence or convergence in a wide sense or vague convergence of a sequence of distribution functions. The definition of weak convergence is as given below.

Definition 7.3.2. *Weak convergence: A sequence of distribution functions $\{F_n(x), n \geq 1\}$ is said to converge weakly to a non-decreasing and right-continuous function F if $F_n(x) \to F(x)$, $\forall\ x \in C(F)$ as $n \to \infty$, where $C(F)$ is a set of points of continuity of F.*

In this definition, F is not necessarily a distribution function of a random variable. Weak convergence of $\{F_n(x), n \geq 1\}$ is denoted by $F_n \xrightarrow{w} F$. It can be proved that set $D(F)$ of points of discontinuity of F is at most countable, which in turn implies that set $C(F)$ of points of continuity is dense in $\mathbb{R}$. If $\lim_{x\to\infty} F(x) = 1$ and $\lim_{x\to-\infty} F(x) = 0$, then we say that F_n converges completely to F. In particular, if F_n is a distribution function of a random variable $X_n, n \geq 1$ and F is a distribution function of a random variable X, and if $F_n \xrightarrow{w} F$ then $X_n \xrightarrow{L} X$. Thus, the distinction between weak convergence and convergence in law is that the limit random variable in the former case may not be a proper or real random variable but may be an extended valued random variable. In weak convergence, the limit distribution is referred to as a pseudo distribution or sub-probability distribution and the limit random variable is referred to as an improper random variable. Another subtle difference between the two is that in convergence in law, the probability measure P in $F_n(x) = P[X_n \leq x]$ is fixed, and we get a sequence of distribution functions corresponding to the sequence of the random variables. In weak convergence, F_n need not be a distribution function of a random variable and $F_n((a,b]) = P_n((a,b])$ and thus we have a sequence of measures $\{P_n, n \geq 1\}$ and P_n need not be a probability measure.

It is to be noted that convergence in law implies weak convergence. Thus, convergence in law is a stronger mode of convergence than weak convergence.

The following section presents some results related to convergence in r-th mean.

7.4 Convergence in r-th Mean

As defined in Section 6.1, $X_n \xrightarrow{r} X$ if $E(|X_n - X|^r) \to 0,\ r > 0$. In almost sure convergence and in convergence in probability, it is proved that the limit random variable is almost surely unique. The same result holds for convergence in r-th mean, which follows from the C_r inequality proved in Section 4.5.

Theorem 7.4.1. *If $X_n \xrightarrow{r} X$ and $X_n \xrightarrow{r} Y$, then X and Y are equivalent random variables, that is, $X = Y$ a.s.*

Proof. By C_r inequality,

$$\begin{aligned} E(|X-Y|^r) &= E(|X - X_n + X_n - Y|^r) \\ &\leq C_r[E(|X_n - X|^r) + E(|X_n - Y|^r)] \to 0, \quad \text{as } n \to \infty. \end{aligned}$$

Thus, $E(|X-Y|^r)=0$. Suppose $Z=|X-Y|^{r/2}$. Observe that

$$\begin{aligned} E(|X-Y|^r)=0 &\Rightarrow E(|X-Y|^{r/2})^2=0 \Rightarrow E(Z^2)=0 \\ &\Rightarrow Var(Z)\le E(Z^2)=0 \Rightarrow Var(Z)=0 \\ Var(Z)=0 \ \& \ E(Z^2)=0 &\Rightarrow E(Z)=0. \end{aligned}$$

Thus, Z is a degenerate random variable, degenerate at $E(Z)=0$. Hence, $P[Z=0]=1$, that is, $X=Y$ a.s. Alternatively, using Theorem 7.2.2,

$$X_n \xrightarrow{r} X \Rightarrow X_n \xrightarrow{P} X \ \& \ X_n \xrightarrow{r} Y \Rightarrow X_n \xrightarrow{P} Y \Rightarrow X=Y \ a.s.$$

by Theorem 7.2.1. □

In Section 2, we have noted that convergence in r-th mean implies convergence in probability. The converse is not true in general, as shown in Example 7.2.10. It is also clear from the following example. However, the converse is true if the random variables X_n's are almost surely bounded. We discuss it in the next theorem.

Example 7.4.1. Suppose a probability mass function of a random variable X_n is given by,

$$P[X_n=1/n]=1-1/n^2=1-P[X_n=n].$$

From the given distribution of X_n, it is clear that,

$$P[|X_n|<\epsilon] = \begin{cases} 1-1/n^2, & \text{if } \epsilon<n \\ 1, & \text{if } \epsilon\ge n \end{cases}$$

and it converges to 1 as $n\to\infty$, for every $\epsilon>0$. Hence, $X_n \xrightarrow{P} X\equiv 0$ and hence in law. Now,

$$E(|X_n-X|)=E(|X_n|)=E(X_n)=2/n-1/n^3 \to 0$$

but

$$E(|X_n-X|)^2=E(|X_n|)^2=E(X_n^2)=1+1/n^2-1/n^4\to 1.$$

Thus, $E(X_n)\to 0$ but $E(X_n^2)\to 1$ and hence $X_n \overset{q.m.}{\nrightarrow} 0$. □

In Theorem 7.2.3 it is proved that for a sequence $\{X_n, n\ge 1\}$ of uniformly bounded random variables $X_n \xrightarrow{P} C$ implies $E(X_n-C)^2\to 0$. The following theorem proves a similar result when X_n's are almost surely bounded, and the limit random variable is non-degenerate.

Theorem 7.4.2. *Suppose $\{X_n, n\ge 1\}$ is a sequence of random variables defined on the probability space $(\Omega, \mathbb{A}, P)$. If X_n's are almost surely bounded, then $X_n \xrightarrow{P} X \Rightarrow X_n \xrightarrow{r} X$.*

Proof. It is given that $|X_n(\omega)| \leq M \quad \forall \quad \omega \in N^c$, where $P(N) = 0$. Since $|X_n| \leq M$, its distribution function $F_n(x)$ is given by

$$F_n(x) \quad = \quad \begin{cases} 0, & \text{if} \quad x \leq -M \\ h_n(x) & \text{if} \quad -M \leq x < M \\ 1, & \text{if} \quad x \geq M. \end{cases}$$

Since $X_n \xrightarrow{P} X \Rightarrow X_n \xrightarrow{L} X$, $F_n(x) \to F_X(x) \quad \forall \quad x \in C(F_X)$ and $F_X(x)$ is given by,

$$F_X(x) \quad = \quad \begin{cases} 0, & \text{if} \quad x \leq -M \\ \lim_{n\to\infty} h_n(x) & \text{if} \quad -M \leq x < M \\ 1, & \text{if} \quad x \geq M, \end{cases}$$

where $\lim_{n\to\infty} h_n(x)$ exists as $F_n(x) \to F_X(x) \quad \forall \quad x \in C(F_X)$. From the form of $F_X(x)$ it is clear that

$$|X| \leq M \quad \Rightarrow \quad |X_n - X| \leq |X_n| + |X| = 2M \quad \Rightarrow \quad |X_n - X|^r \leq (2M)^r = S,$$

say. Now from the basic inequality we have for all $\epsilon > 0$,

$$(E(|X_n - X|^r) - \epsilon^r)/S \quad \leq \quad P[|X_n - X| > \epsilon] \to 0 \quad \Rightarrow \quad X_n \xrightarrow{r} X.$$

□

The following theorem gives a necessary and sufficient condition for convergence in quadratic mean of a sequence of random variables to a constant.

Theorem 7.4.3. *Suppose $\{X_n, n \geq 1\}$ is a sequence of random variables defined on the probability space $(\Omega, \mathbb{A}, P)$. $X_n \xrightarrow{q.m.} b$ if and only if $E(X_n) \to b$ and $Var(X_n) \to 0$, where b is a constant.*

Proof. Suppose $E(X_n) \to b$ and $Var(X_n) \to 0$. Observe that

$$\begin{aligned} E(|X_n - b|^2) &= E(|X_n - E(X_n) + E(X_n) - b|^2) \\ &= E(X_n - E(X_n))^2 + (E(X_n) - b)^2 \\ &= Var(X_n) + (E(X_n) - b)^2 \to 0 \Rightarrow X_n \xrightarrow{q.m.} b. \end{aligned}$$

Now assume that $X_n \xrightarrow{q.m.} b$. Note that

$$\begin{aligned} X_n \xrightarrow{q.m.} b &\Rightarrow E(|X_n - b|^2) \to 0 \Rightarrow Var(X_n) + (E(X_n) - b)^2 \to 0 \\ &\Rightarrow E(X_n) \to b \;\; \& \;\; Var(X_n) \to 0, \end{aligned}$$

since the sum of two non-negative numbers converges to zero, each must converge to zero. □

Theorem 7.4.3 is useful to verify the consistency of an estimator T_n of a parameter θ, provided its mean squared error exists. If the bias of T_n and $Var(T_n)$ both converge to zero, that is, if the mean squared error converges to zero, then T_n is a consistent estimator of θ. For example, suppose $\overline{X}_n$ is a sample mean based on a random sample of size n from the distribution of X with mean μ and variance σ^2. Then $E(\overline{X}_n) = \mu$ and $Var(\overline{X}_n) = \sigma^2/n$. Thus $\overline{X}_n$ is an unbiased estimator of μ, and its variance converges to 0 as $n \to \infty$. Hence sample mean $\overline{X}_n$ is always a consistent estimator of the population mean μ.

In Example 7.4.1 we have noted that, $E(X_n) \to 0$ but $Var(X_n) \nrightarrow 0$ and $X_n \overset{q.m.}{\nrightarrow} 0$. The following example also demonstrates that both the conditions in Theorem 7.4.3 are not satisfied, we do not get convergence in quadratic mean. It also shows that convergence in probability does not imply convergence in r-th mean.

Example 7.4.2. Suppose a probability mass function of a random variable X_n is given by,

$$P[X_n = 0] = 1 - 1/n = 1 - P[X_n = n].$$

From the given distribution of X_n, we have,

$$P[|X_n| < \epsilon] = \begin{cases} 1 - 1/n, & \text{if } \epsilon < n \\ 1, & \text{if } \epsilon \ge n \end{cases}$$

and it converges to 1 as $n \to \infty$, for every $\epsilon > 0$. Hence, $X_n \overset{P}{\to} X \equiv 0$. Now $\forall\ n \ge 1$,

$$\begin{aligned} E(|X_n - X|) &= E(|X_n|) = E(X_n) = n.1/n = 1 \to 1 \Rightarrow X_n \overset{1}{\nrightarrow} 0 \\ \&\ E(|X_n - X|)^2 &= E(|X_n|)^2 = E(X_n^2) = n^2.1/n = n \to \infty \\ &\Rightarrow X_n \overset{q.m.}{\nrightarrow} 0. \end{aligned}$$

□

It is interesting to note that the convergence of moments differs from convergence in r-th mean. Convergence in r-th mean implies convergence of r-th order moments, but the converse is not true. We discuss these issues below.

Example 7.4.3. Suppose $\{X_n, n \ge 1\}$ is a sequence of random variables defined on $(\Omega, \mathbb{A}, P)$ as $X_n = I_A,\ A \in \mathbb{A}$. Suppose X on $(\Omega, \mathbb{A}, P)$ is defined as $X = I_{A^c}$. Suppose $P(A) = 1/2$, then $\forall\ r \ge 1$ and $\forall\ n \ge 1$,

$$\begin{aligned} E(|X|^r) &= 1/2\ \&\ E(|X_n|^r) = 1/2 \Rightarrow E(|X_n|^r) \to E(|X|^r) \\ \text{Now, } X_n - X &= 1 \text{ if } \omega \in A\ \&\ X_n - X = -1 \text{ if } \omega \in A^c \\ \Rightarrow E(|X_n - X|^r) &= 1/2 + 1/2 = 1\ \forall\ n, r \ge 1 \Rightarrow X_n \overset{r}{\nrightarrow} X. \end{aligned}$$

Thus, in general, the convergence of moments does not imply convergence in r-th mean. □

In the following theorem, we prove that convergence in r-th mean implies convergence of r-th order moments.

Theorem 7.4.4. *Suppose $\{X_n, n \geq 1\}$ is a sequence of random variables defined on a probability space $(\Omega, \mathbb{A}, P)$. Then $X_n \xrightarrow{r} X \Rightarrow E(|X_n|^r) \to E(|X|^r)$, $r > 0$.*

Proof. By C_r inequality,

$$E(|X_n|^r) = E(|X_n - X + X|^r) \leq C_r\{E(|X_n - X|^r) + E(|X|^r)\}.$$

Suppose $r \leq 1$, then $C_r = 1$ and hence, $E(|X_n|^r) - E(|X|^r) \leq E(|X_n - X|^r)$. Interchanging X_n and X we get, $E(|X|^r) - E(|X_n|^r) \leq E(|X - X_n|^r)$. Hence,

$$\left|E(|X_n|^r) - E(|X|^r)\right| \leq E(|X_n - X|^r) \to 0 \quad \Rightarrow E(|X_n|^r) \to E(|X|^r).$$

Suppose $r > 1$, then by Minkowski's inequality

$$(E|X_n|^r)^{1/r} = (E|X_n - X + X|^r)^{1/r} \leq \{(E|X_n - X|^r)^{1/r} + (E|X|^r)^{1/r}\}.$$

Hence, $(E|X_n|^r)^{1/r} - (E|X|^r)^{1/r} \leq (E|X_n - X|^r)^{1/r}$. Interchanging X_n and X we get, $(E|X|^r)^{1/r} - (E|X_n|^r)^{1/r} \leq (E|X - X_n|^r)^{1/r}$. Hence,

$$\left|(E|X_n|^r)^{1/r} - (E|X|^r)^{1/r}\right| \leq (E|X_n - X|^r)^{1/r} \to 0 \quad \Rightarrow E(|X_n|^r) \to E(|X|^r).$$

□

In Chapter 4 we have noted that if $X \in L_r$ then $X \in L_s, \ \forall \ s < r$. On similar lines, if the sequence $\{X_n, n \geq 1\}$ converges in r-th mean, then it also converges in s-th mean for all $s < r$. We prove it in the following theorem using Jensen's inequality.

Theorem 7.4.5. *Suppose $\{X_n, n \geq 1\}$ is a sequence of random variables defined on a probability space $(\Omega, \mathbb{A}, P)$. Then*

$$X_n \xrightarrow{r} X \Rightarrow X_n \xrightarrow{s} X, \ \ \forall \ \ s < r \ .$$

Proof. From the moment inequality studied in Lemma 4.5.1, we know that if $X \in L_r$ then $X \in L_s$ for all $s < r$. Observe that

$$E(|X_n - X|^s) = E((|X_n - X|^r))^{s/r} = E(g(|X_n - X|^r)), \text{where} g(x) = x^{s/r}, \ x > 0.$$

Now for $x > 0$, if $g(x) = x^u$, then $g'(x) = ux^{u-1}$ and $g''(x) = u(u-1)x^{u-2} < 0$ for $u < 1$. A necessary and sufficient condition for a differentiable function to be concave is that its second derivative is negative. Thus, $g(x) = x^u, \ x > 0$ is a concave function for $u < 1$ and hence by Jensen's inequality $E(g(X)) \leq g(E(X))$. Now $s < r \ \Rightarrow \ s/r < 1$. Hence,

$$\begin{aligned} E(|X_n - X|^s) &= E((|X_n - X|^r))^{s/r} = E(g(|X_n - X|^r)) \\ &\leq g(E(|X_n - X|^r)) = (E(|X_n - X|^r))^{s/r} \to 0 \ \text{ as } \ n \to \infty \\ &\Rightarrow X_n \xrightarrow{s} X, \ \ \forall \ \ s < r. \end{aligned}$$

□

However, the converse of the Theorem 7.4.5 is not true. It is illustrated in the following example.

Example 7.4.4. Suppose $\{X_n, n \geq 1\}$ is a sequence of random variables such that

$$\begin{aligned} P[X_n = n] &= n^{-(r+s)/2} = 1 - P[X_n = 0], \quad n \geq 1 \;\&\; 0 < s < r \\ \text{Observe that } E(|X_n|^s) &= n^s n^{-(r+s)/2} = n^{(s-r)/2} \to 0 \text{ since } s < r \\ E(|X_n|^r) &= n^r n^{-(r+s)/2} = n^{(r-s)/2} \to \infty \text{ since } s < r. \end{aligned}$$

Thus if $s < r$, $X_n \stackrel{s}{\to} 0$, however $X_n \stackrel{r}{\nrightarrow} 0$. □

In general, almost sure convergence neither implies nor is implied by the convergence in r-th mean. The next two examples illustrate that almost sure convergence does not imply convergence in quadratic mean.

Example 7.4.5. Suppose $\{X_n, n \geq 1\}$ is a sequence of random variables defined on a probability space $(\Omega, \mathbb{A}, P)$ where $\Omega = [0, 1]$, $\mathbb{A}$ is a sigma field of subsets of Ω and P is the Lebesgue measure on $(\Omega, \mathbb{A})$. The random variable X_n is defined as,

$$X_n(\omega) = \begin{cases} 2^n, & \text{if } \omega \in (0, 1/n) \\ 0 & \text{if } \omega \notin (0, 1/n). \end{cases}$$

It is to be noted that $\forall\ \omega \in \Omega$, $\exists\ n_0(\omega)$ such that $\omega > 1/n\ \forall\ n \geq n_0$. Hence,
$\forall\ \omega \in \Omega$, $\&\ \forall\ n \geq n_0$, $X_n(\omega) = 0$ which implies that $\forall\ \omega \in \Omega$, $X_n(\omega) \to 0$ and hence $X_n \stackrel{a.s.}{\to} 0$. However, $E(X_n^2) = 2^{2n}/n \to \infty$ as $n \to \infty$. Thus, $X_n \stackrel{q.m.}{\nrightarrow} 0$. □

Example 7.4.6. Suppose $\{X_n, n \geq 1\}$ is a sequence of random variables such that $P[X_n = 0] = 1 - 1/n^2$ and $P[X_n = e^n] = 1/n^2$. Note that for $\epsilon > 0$, $P[|X_n| > \epsilon] = 1/n^2 \to 0$ and hence $X_n \stackrel{P}{\to} 0$. Further,

$$\sum_{n\geq 1} P[|X_n| > \epsilon] = \sum_{n\geq 1} 1/n^2 < \infty \quad \Rightarrow \quad X_n \stackrel{a.s.}{\to} 0.$$

However, $E(|X_n|^r) = e^{nr}/n^2 \to \infty$ as $n \to \infty$. Hence, $X_n \stackrel{r}{\nrightarrow} 0$. □

The following example illustrates that convergence in r-th mean does not imply almost sure convergence.

Example 7.4.7. Suppose $\{X_n, n \geq 1\}$ is a sequence of independent random variables such that $P[X_n = 1] = 1/n\ \&\ P[X_n = 0] = 1 - 1/n$, $n \geq 1$. It is clear that $E(X_n^r) = 1/n \to 0$ as $n \to \infty$. Hence, $X_n \stackrel{r}{\to} 0\ \forall\ r > 0$. It is to be noted that $\{X_n(\omega), n \geq 1\}$ is a sequence of 0's and 1's and hence $X_n(\omega) \to 0$ if and only if $X_n(\omega) = 0$ for sufficiently large n. Thus we compute

$P[X_n(\omega) = 0 \ \forall \ n \geq n_0]$ where n_0 is a large positive integer and examine its behaviour.

$$\begin{aligned}
P[X_n(\omega) = 0 \ \forall \ n \geq n_0] &= P[\bigcap_{n=n_0}^{\infty} [X_n = 0]] = P[\lim_{N\to\infty} \bigcap_{n=n_0}^{N} [X_n = 0]] \\
&= \lim_{N\to\infty} P[\bigcap_{n=n_0}^{N} [X_n = 0]] \\
&= \lim_{N\to\infty} \prod_{n=n_0}^{N} P[X_n = 0] \ \text{ in view of independence} \\
&= \lim_{N\to\infty} \prod_{n=n_0}^{N} (1 - 1/n) \\
&= \lim_{N\to\infty} \frac{n_0 - 1}{n_0} \frac{n_0}{n_0 + 1} \frac{n_0 + 1}{n_0 + 2} \cdots \frac{N-2}{N-1} \frac{N-1}{N} \\
&= \lim_{N\to\infty} (n_0 - 1)/N = 0
\end{aligned}$$

Thus, $X_n \overset{a.s.}{\nrightarrow} 0$. Alternatively, note that

$$\sum_{n\geq 1} P[|X_n| > \epsilon] = \sum_{n\geq 1} (1/n) = \infty \quad \Rightarrow \quad X_n \overset{a.s.}{\nrightarrow} 0,$$

in view of Theorem 6.4.3. □

Remark 7.4.1. In the above example, it is to be noted that for any $r > 0$ $E(|X_n|^r) = 1/n$ and hence $\sum_{n=1}^{\infty} E(|X_n|^r)$ is a divergent series. Similarly, for any $\epsilon > 0$, $P[|X_n| > \epsilon] = 1/n$ and hence $\sum_{n=1}^{\infty} P[|X_n| > \epsilon]$ is also a divergent series. Hence, we cannot use the sufficient conditions for the almost sure convergence as proved in Section 6.3.

In the following example, the sequence neither converges in r-th mean nor almost surely.

Example 7.4.8. Suppose $\{X_n, n \geq 1\}$ is a sequence of random variables such that X_n has uniform $U(-n, n)$ distribution. In Example 7.3.9, we have shown that X_n does not converge in law by finding the limit of the distribution function of X_n. Hence, $X_n \overset{P}{\nrightarrow} 0$ and hence $X_n \overset{a.s.}{\nrightarrow} 0$. Now,

$$E(|X_n|^2) = E(X_n^2) = Var(X_n) + (E(X_n))^2 = n^2/3 \to \infty \quad \Rightarrow \quad X_n \overset{q.m.}{\nrightarrow} X. \square$$

In the following example, the sequence converges in both the modes, r-th mean and almost surely.

Example 7.4.9. Suppose $\{X_n, n \geq 1\}$ is a sequence of random variables with probability mass function given by, $P[X_n = \pm 1/n] = 1/2$. It is clear that

$E|X_n|^r = 1/n^r \to 0$ hence $X_n \xrightarrow{r} X \equiv 0$ and $X_n \xrightarrow{P} X \equiv 0$ Further for all $n \geq 1, E(X_n) = 0$, hence $E(X_n) \to E(X) = 0$. Thus, we have convergence in $\mathbb{L}_r$ and also the convergence of a sequence of means. In Example 6.3.4, we have shown that $X_n \xrightarrow{a.s.} 0$. □

The following example illustrates that almost sure convergence does not imply convergence of moments.

Example 7.4.10. Suppose $\{X_n, n \geq 1\}$ is a sequence of random variables defined on a probability space $(\Omega, \mathbb{A}, P)$ where $\Omega = (0,1)$, $\mathbb{A}$ is a sigma field of subsets of Ω and P is the Lebesgue measure on $(\Omega, \mathbb{A})$. The random variable X_n is defined as,

$$X_n(\omega) = \begin{cases} 2^{n/r}, & \text{if} \quad \omega \in (0, 1/n), \quad r > 0 \\ 0, & \text{if} \quad \omega \notin (0, 1/n). \end{cases}$$

It is to be noted that $\forall \ \omega \in \Omega, \ \exists \ n_0(\omega)$ such that $\omega > 1/n \ \forall \ n \geq n_0$. Hence,
$\forall \ \omega \in \Omega, \ \& \ \forall \ n \geq n_0, \ X_n(\omega) = 0 \ \Rightarrow \ \forall \ \omega \in \Omega, \ X_n(\omega) \to X = 0$ and hence $X_n \xrightarrow{a.s.} 0$. However, $E(|X_n|^r) = (2^{n/r})^r(1/n) = 2^n/n \nrightarrow 0 = E(|X|^r)$ as $n \to \infty$. Thus, $E(|X_n|^r) \nrightarrow E(|X|^r)$. □

The following example illustrates convergence in all four modes of convergence.

Example 7.4.11. Suppose $\{X_n, n \geq 1\}$ is a sequence of random variables such that $X_n \sim N(1/n, 1/n)$ distribution. Then as $n \to \infty$,

$$E(X_n - 0)^2 = E(X_n^2) = 1/n + 1/n^2 \to 0 \Rightarrow X_n \xrightarrow{q.m.} 0 \Rightarrow X_n \xrightarrow{P} 0 \Rightarrow X_n \xrightarrow{L} 0.$$

To examine whether $X_n \xrightarrow{a.s.} 0$, we examine whether for some $r > 0$, $\sum_{n \geq 1} E(|X_n - X|^r)$ is finite. Note that as in Example 6.4.3, $r = 2$ will not be helpful. Suppose we take $r = 4$. To find $E(|X_n|^4) = E(X_n^4)$, note that $Y_n = \sqrt{n}(X_n - 1/n) \sim N(0,1)$ and hence $E(Y_n^3) = 0$ which implies $E(X_n^3) = 4/n^3$. Further, $E(Y_n^4) = 3$ implies that

$$\begin{aligned} 3 &= n^2 E(X_n^4 - 4X_n^3/n + 6X_n^2/n^2 - 4X_n/n^3 + 1/n^4) \\ \Rightarrow \ E(X_n^4) &= 13/n^4 - 6/n^3 + 3/n^2 = (3n(n-2) + 13)/n^4 > 0 \\ \Rightarrow \ \sum_{n \geq 1} E(|X_n|^4) &< \infty \ \Rightarrow \ X_n \xrightarrow{a.s.} 0. \end{aligned}$$

In this example $X_n \xrightarrow{q.m.} 0$ and $X_n \xrightarrow{a.s.} 0$. However, it may not always be true, as noted in some examples above. □

The following example illustrates the implications among various modes of convergence.

Example 7.4.12. Suppose $\{X_n, n \geq 1\}$ is a sequence of independent random variables such that

$$P[X_n = 0] = 1 - 1/n^\alpha \quad \& \quad P[X_n = n] = 1/n^\alpha, \quad n \geq 1, \quad \alpha > 0 .$$

We examine whether $\{X_n, n \geq 1\}$ converges in probability, in law, almost surely and in r-th mean. For any

$$\epsilon > 0, \;\; P[|X_n| > \epsilon] = 1/n^\alpha \to 0 \;\; \forall \;\; \alpha > 0 \;\; \Rightarrow \;\; X_n \xrightarrow{P} 0 \;\; \Rightarrow \;\; X_n \xrightarrow{L} 0.$$

To examine almost sure convergence, observe that by Theorem 6.4.3,

$$\forall \;\; \alpha > 1, \;\; \sum_{n=1}^{\infty} P[|X_n| > \epsilon] < \infty \;\; \Rightarrow \;\; X_n \xrightarrow{a.s.} 0 .$$

It is given that $\{X_n, n \geq 1\}$ is a sequence of independent random variables. Hence, by Theorem 6.4.3,

$$\forall \;\; \alpha \leq 1 \;\; \sum_{n=1}^{\infty} P[|X_n| > \epsilon] = \infty \;\; \Rightarrow \;\; X_n \stackrel{a.s.}{\nrightarrow} 0 .$$

It is to be noted that $X_n \xrightarrow{P} 0 \;\; \forall \;\; \alpha > 0$ but $X_n \stackrel{a.s.}{\nrightarrow} 0$ for $\alpha \leq 1$. Now to examine convergence in r-th mean observe that,

$$E(|X_n|^r) = 1/n^{\alpha - r} \quad \to \quad \begin{cases} 0, & \text{if} \quad r < \alpha \\ 1, & \text{if} \quad r = \alpha \\ \infty, & \text{if} \quad r > \alpha. \end{cases}$$

Hence, $X_n \xrightarrow{r} 0$ if $r < \alpha$. In particular if $\alpha = 1$ then $X_n \xrightarrow{P} 0$, but $X_n \stackrel{a.s.}{\nrightarrow} 0$ and $X_n \stackrel{q.m.}{\nrightarrow} 0$. □

In the following example, the sequence converges in quadratic mean, probability, law and almost surely.

Example 7.4.13. Suppose the joint probability density function of (X_n, Y) for $n \geq 1$ is given by, $f(x, y) = ne^{-y}, \; 0 < y < x < y + 1/n$. We examine various modes of convergence for the sequence $\{X_n, n \geq 1\}$. To examine whether $X_n \xrightarrow{q.m.} Y$, we compute $E(X_n - Y)^2$ from the given joint probability distribution as follows. With $u = x - y$,

$$\begin{aligned} E(X_n - Y)^2 &= n \int_0^\infty \Big(\int_y^{y+1/n} e^{-y}(x-y)^2 \, dx \Big) \, dy = n \int_0^\infty e^{-y} \Big(\int_0^{1/n} u^2 \, du \Big) \, dy \\ &= n \int_0^\infty e^{-y}(1/3n^3) \, dy \;\; = 1/3n^2. \end{aligned}$$

$E(X_n - Y)^2$ can also be computed as $E(X_n - Y)^2 = E(E(X_n - Y)^2|Y)$. Thus, we first find the conditional distribution of X_n given Y. The marginal distribution of Y is given by,

$$f(y) = \int_y^{y+1/n} ne^{-y}\, dx = ne^{-y}(1/n) = e^{-y}, \quad y \geq 0.$$

Thus, Y follows exponential distribution with $E(Y) = 1$ & $E(Y^2) = 2$. The conditional distribution of X_n given Y is given by, $f(x|y) = n, \quad y < x < y + 1/n$. Thus, $X_n|Y \sim U(Y, Y + 1/n)$. Hence, $E(X_n|Y) = (2Y + 1/n)/2 = Y + 1/2n$ and

$$\begin{aligned}
E(X_n^2|Y) &= (n/3)[(Y+1/n)^3 - Y^3] = Y^2 + Y/n + 1/3n^2 \\
\Rightarrow E((X_n - Y)^2|Y) &= Y^2 + Y/n + 1/3n^2 - 2Y(Y + 1/2n) + Y^2 \\
\Rightarrow E(X_n - Y)^2 &= E(E(X_n - Y)^2|Y) = 1/3n^2 \to 0 \text{ as } n \to \infty \\
\Rightarrow X_n &\xrightarrow{q.m.} Y \Rightarrow X_n \xrightarrow{P} Y \Rightarrow X_n \xrightarrow{L} Y \\
\text{Further } \sum_{n\geq 1} E(X_n - Y)^2 &= \sum_{n\geq 1} 1/3n^2 < \infty \Rightarrow X_n \xrightarrow{a.s} Y.
\end{aligned}$$

Observe that $E(X_n) = 1 + 1/2n \to 1 = E(Y) = 1$ and $E(X_n^2) = E(E(X_n^2|Y)) = 1/3n^2 + 2 + 1/n \to 2 = E(Y^2) = 2$. In general,

$$\begin{aligned}
E(X_n^r) &= E(E(X_n^r|Y)) = E\Big(n\int_Y^{Y+1/n} x^r\, dx\Big) \\
&= E\Big((n/(r+1))[(Y + 1/n)^{r+1} - Y^{r+1}]\Big) \\
&= E\Big(\frac{n}{r+1}\Big[Y^{r+1} + \frac{(r+1)}{n}Y^r + \frac{(r+1)r}{2n^2}Y^{r-1} + \cdots + \frac{1}{n^r} - Y^{r+1}\Big]\Big) \\
&= E\Big(Y^r + \frac{(r+1)r}{2n}Y^{r-1} + \cdots + \frac{1}{n^{r-1}}\Big) \\
&\to E(Y^r) = r! < \infty \ \forall \text{ finite } r.
\end{aligned}$$

Thus, $E(X_n^r) \to E(Y^r)$ for $r \geq 1$. We can also show that $X_n \xrightarrow{P} Y$ by using the definition of convergence in probability. Thus, using the conditional distribution of X_n given Y, for $\epsilon > 0$ we find $P[|X_n - Y| < \epsilon]$ as follows.

$$\begin{aligned}
P[|X_n - Y| < \epsilon] &= E(P[|X_n - Y| < \epsilon|Y]) = E(P[Y - \epsilon < X_n < Y + \epsilon|Y]) \\
&= 1 \text{ if } \epsilon > 1/n, \text{ since } Y + \epsilon > Y + 1/n \ \& \ Y - \epsilon < Y \\
&= E(P[Y < X_n < Y + \epsilon|Y]) = n\epsilon \leq 1, \text{ if } \epsilon \leq 1/n.
\end{aligned}$$

As $n \to \infty$, for $\epsilon > 0$, $P[|X_n - Y| < \epsilon] \to 1$ and hence $X_n \xrightarrow{P} Y$. □

In the following example, we verify the results established in Example 7.4.13 by simulation using R.

Example 7.4.14. Suppose the joint probability density function of (X_n, Y) for $n \geq 1$ is given by, $f(x, y) = ne^{-y}$, $0 < y < x < y + 1/n$. We have noted that the marginal distribution of Y is exponential with mean 1, while the conditional distribution of X_n given Y is uniform $U(Y, Y + 1/n)$. To obtain a realization from (X_n, Y), we first obtain a realization from the distribution of Y and then, corresponding to the realized value of $Y = y$, obtain a realization from the conditional distribution of X_n given $Y = y$. On the basis of m observations on (X_n, Y), for some values of n, we obtain the estimate rf_n, estimate $q_n = (1/m)\sum_{i=1}^{m}(X_{ni} - Y_i)^2$ of $E(X_n - Y)^2$ and estimate $m'_{rn} = (1/m)\sum_{i=1}^{m}(X_{ni})^r$ of $E(X_n^r)$ for $r = 1, 2, 3, 4$. We examine whether as n increases, $rf_n \to 1$, $q_n \to 0$ and $m'_{rn} \to E(Y^r)$, to decide whether $X_n \xrightarrow{P} Y$, $X_n \xrightarrow{q.m.} Y$ and $E(|X_n|^r) \to E(|Y|^r)$ respectively. Since Y follows the exponential distribution with mean 1, $E(Y^r) = r!$, $r \geq 1$. We use the following code.

Code 7.4.1. `eps=0.01; me=m2=m3=m4=q=p=c(); M=400`

```
s=c(50,100,150,200,250,300,350,400)# vector of values of n
y=matrix(nrow=M,ncol=length(s))
z=matrix(nrow=M,ncol=length(s))
for(i in 1:length(s))
{
 n=s[i]
 set.seed(i)
 y[,i]=rexp(M,rate=1)
 set.seed(i+10)
 z[,i]=runif(M,y[,i],y[,i]+1/n )
 me[i]=mean(z[,i])
 m2[i]=mean(z[,i]^2)
 m3[i]=mean(z[,i]^3)
 m4[i]=mean(z[,i]^4)
 q[i]=mean(z[,i]-y[,i])^2
 p[i]=length(which(abs(z[,i]-y[,i])<eps))/M
}
d=round(data.frame(s,p,me,m2,m3,m4,q),4); d
```

The output in data frame `d` is organized in Table 7.6. From Table 7.6, we note that as n increases, $rf_n \to 1$, $q_n \to 0$ and $m'_{rn} \to E(Y^r), r = 1, 2, 3, 4$. Thus, $X_n \xrightarrow{P} Y$, $X_n \xrightarrow{q.m.} Y$ and $E(|X_n|^r \to E|Y|^r$, $r = 1, 2, 3, 4$. It is to be noted that for $\epsilon = 0.01$ and $n = 50$, $\epsilon < 1/n$, hence as we have derived above, $P[|X_n - Y| < \epsilon] = n\epsilon = 0.5$. From the simulated data, we get it as 0.5225, close to 0.5. Since we have shown that $X_n \xrightarrow{P} Y$, we conclude that $X_n \xrightarrow{L} Y$.

TABLE 7.6
Convergence in Probability, in Quadratic Mean and Convergence of Moments

n	rf_n	$E(\|X_n\|)$	$E(\|X_n\|^2)$	$E(\|X_n\|^3)$	$E(\|X_n\|^4)$	$E(\|X_n - Y\|^2)$
50	0.5225	0.9829	1.7705	4.6202	15.7312	0.0001
100	1.0000	1.0437	2.1255	5.9997	20.0075	0
150	1.0000	0.9426	1.6497	3.8887	10.8662	0
200	1.0000	1.0791	2.4553	8.6685	39.8851	0
250	1.0000	1.0777	2.2721	7.4608	34.2127	0
300	1.0000	0.9182	1.6727	4.4615	15.6058	0
350	1.0000	0.9794	1.8759	5.2096	18.0000	0
400	1.0000	0.9863	1.9713	5.9920	23.8426	0

In the next example, we now prove it using the definition and verify it by simulation. □

Example 7.4.15. Suppose the joint probability density function of (X_n, Y) is given by, $f(x, y) = ne^{-y}$, $0 < y < x < y + 1/n$, $n \geq 1$. We have already noted that the marginal distribution of Y is exponential with mean 1. To examine whether $X_n \xrightarrow{L} Y$, we find the distribution function of X_n and study its limiting behaviour. Now, the marginal distribution of X_n is given by,

$$\begin{aligned} f_{X_n}(x) &= \int_{x-1/n}^{x} ne^{-y}\,dy = n(e^{1/n} - 1)e^{-x}, \quad 1/n < x < \infty \\ &= e^{-x}(e^{1/n} - 1)/(1/n) \to e^{-x}\log_e e = e^{-x} \quad \text{as} \quad n \to \infty. \end{aligned}$$

Thus, the probability density function of X_n converges to that of the exponential distribution with mean 1. Hence, by Scheffe's lemma (refer to p.227 of Gut [13]) $X_n \xrightarrow{L} Y$. It can also be shown by the definition of convergence in law. Observe that the distribution function $F_n(x)$ of X_n for $x > 0$ is given by,

$$\begin{aligned} F_n(x) &= n(e^{1/n} - 1)\int_{1/n}^{x} e^{-u}\,du = n(e^{1/n} - 1)(e^{-1/n} - e^{-x}) \\ &= (e^{-1/n} - e^{-x})(e^{1/n} - 1)/(1/n) \;\to 1 - e^{-x} = F(x), \quad x > 0 \\ &\Rightarrow \quad X_n \xrightarrow{L} Y. \end{aligned}$$

Since we have the explicit expressions for $F_n(x)$ and $F(x)$, we draw the curves of F and F_n for $n = 10, 50, 100, 200$ and examine for which n these are close. We also draw the box plots of values of $|F_n(x) - F(x)|$ and compute $T_n = \sup_x |F_n(x) - F(x)|$, for these values of n, to judge the closeness of $F_n(x)$ with $F(x)$. We use the following code.

Code 7.4.2.
```
nvec=c(10,50,100,200); l=length(nvec)
a=seq(0,7,length=200); F=pexp(a)
```

```
FnMat=DiffMat=matrix(nrow=length(a),ncol=l)
par(mfrow=c(1,1))
plot(a,F,type="l",col="dark blue",ylim=c(0,1),xlab="x",
ylab="Distribution Function",lty=1,lwd=2)
for(j in 1:l)
{
  n=nvec[j]
  cm=(exp(1/n)-1)/(1/n)
  FnMat[,j]=cm*(exp(-1/n)-exp(-a))
  DiffMat[,j]=FnMat[,j]-F
  lines(a,FnMat[,j],lty=j,col=j+1,type="l")
}
nlab=paste("n=",nvec,sep="")
legend("bottomright",legend=c("exp(1)",nlab),
col=c("dark blue",2:(l+1)),lty=1:(l+1))
colnames(DiffMat)=nlab
boxplot(abs(DiffMat),col="light blue")
diff=apply(abs(DiffMat),2,max)
lines(diff,col=2,type="b",pch=16,lwd=2)
legend("topright","sup |Fn - F|", col=2, lwd=2,pch=16)
DisTab=round(cbind(nvec,diff),4)
colnames(DisTab)=c("n","sup|Fn-F|")
rownames(DisTab)=c(); DisTab
```

Figure 7.10 displays the distribution functions $F(x)$ and $F_n(x)$. From the graph, we note that except for $n = 10$, curves of F_n overlap on the curve of F, supporting the convergence of distribution function F_n to F. Table 7.7 presents the values of T_n for $n = 10, 50, 100, 200$.

We again note that except for $n = 10$, values of T_n are very close to 0. Figure 7.11 displays box plots of values of $|F_n(x) - F(x)|$ and the values of T_n. These indicate that except for $n = 10$, values of $|F_n(x) - F(x)|$ are very small. Further, the values of T_n decrease as n increases.

Thus, Figure 7.10, Figure 7.11 and Table 7.7 support the theoretical result that $F_n(x) \to F(x)$ and the convergence is achieved at $n = 200$. □

In the next example, we verify $X_n \xrightarrow{L} Y$ by simulation.

TABLE 7.7
Values of T_n

n	10	50	100	200
T_n	0.1001	0.0200	0.0100	0.0050

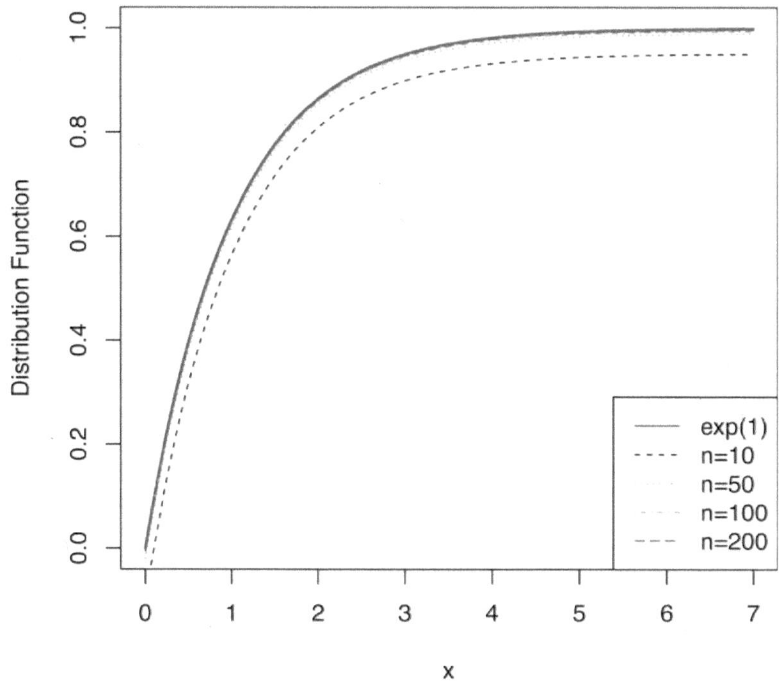

FIGURE 7.10
Distribution Functions F_n and F

Example 7.4.16. Suppose the joint probability density function of (X_n, Y), is given by, $f(x, y) = ne^{-y}$, $0 < y < x < y + 1/n$, $n \geq 1$. In Example 7.4.15, we have already noted that $X_n \xrightarrow{L} Y$, where Y follows the exponential distribution with mean 1. To verify $X_n \xrightarrow{L} Y$, by simulation, we plot histograms based on simulated data for $n = 50, 100, 150, 200$ and on the histograms, impose the graph of the probability density function of the exponential distribution with mean 1. We also carry out the test for goodness of fit. We use the following code to draw these plots and perform the test procedure.

Code 7.4.3.
```
M=300; s=c(50,100,150,200)# vector of values of n
y=matrix(nrow=M,ncol=length(s))
z=matrix(nrow=M,ncol=length(s))
w1=seq(.1,20,.1);length(w1)
w2=matrix(nrow=length(w1),ncol=length(s))
par(mfrow=c(2,2))
for(i in 1:length(s))
{
 n=s[i]
 set.seed(i+20)
```

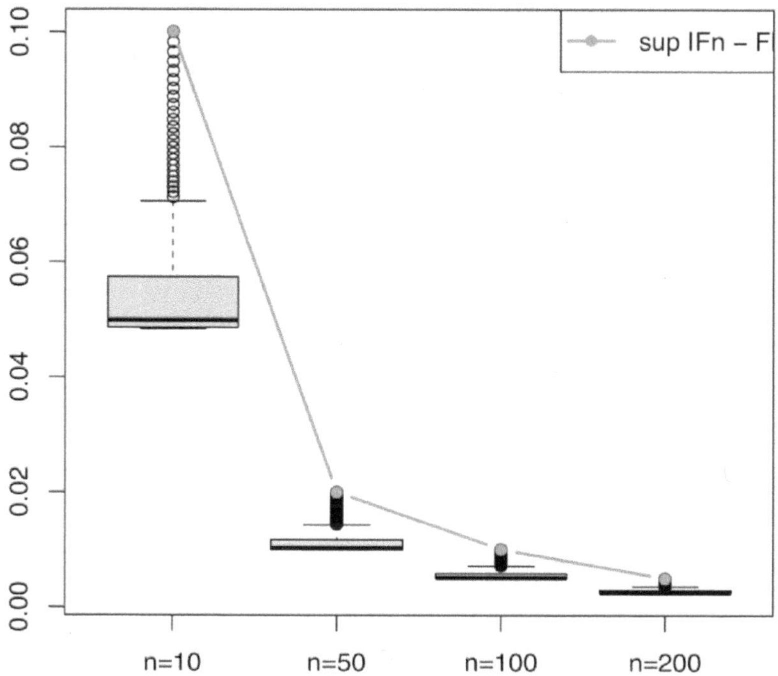

FIGURE 7.11
Box Plots of $|F_n(x) - F(x)|$ and Values of T_n

```
 y[,i]=rexp(M,rate=1)
 set.seed(i+5)
 z[,i]=runif(M,y[,i],y[,i]+1/n )
 no=paste("n =",s,sep=" ")
 hist((z[,i]),freq=FALSE,main=no[i],col="light blue",
 xlab=expression(paste(X_n)))
 w2[,i]=dexp(w1,rate=1 )
 lines(w1,w2[,i],main=s[i],col="dark blue",lwd=1,type="o",
 pch=20)
}
O=hist(z[,4],plot=FALSE)$counts; sum(O)
bk=hist(z[,4],plot=FALSE)$breaks; bk
m=max(bk); e=exp(1); ep=c()
for(i in 1:(length(bk)-1))
{
 ep[i]=e^(-bk[i])-e^(-bk[i+1])
}
a=1-sum(ep); a; ep1=c(ep, a); ef=sum(O)*ep1; O1=c(O,0)
```

```
d=data.frame(O1,round(ef,2)); d
ts=sum((O1-ef)^2/ef);ts
df=length(ef)-1; df; b=qchisq(.95,df); b
p1=1-pchisq(ts,df); p1; chisq.test(O1,p=ep1)
# pooling of frequencies less than 5 is ignored
```

From Figure 7.12, we note that for all n, there is a close agreement between the two curves, which supports the convergence in law of X_n to Y. We carry out the goodness of fit test for $n = 200$. The value of the test statistic is 10.8026, the 95% cut-off of χ^2_{13} is 22.3620 and the p-value comes out to be 0.6274. Thus, from the graph and the goodness of fit test, we claim that $X_n \xrightarrow{L} Y$. □

As in the mode of almost sure convergence or convergence in probability, we have Cauchy convergence in r-th mean as defined below.

Definition 7.4.1. *A sequence $\{X_n, n \geq 1\}$ is said to be Cauchy in r-th mean if $E(|X_n - X_m|^r) \to 0$ as , n & $m \to \infty$.*

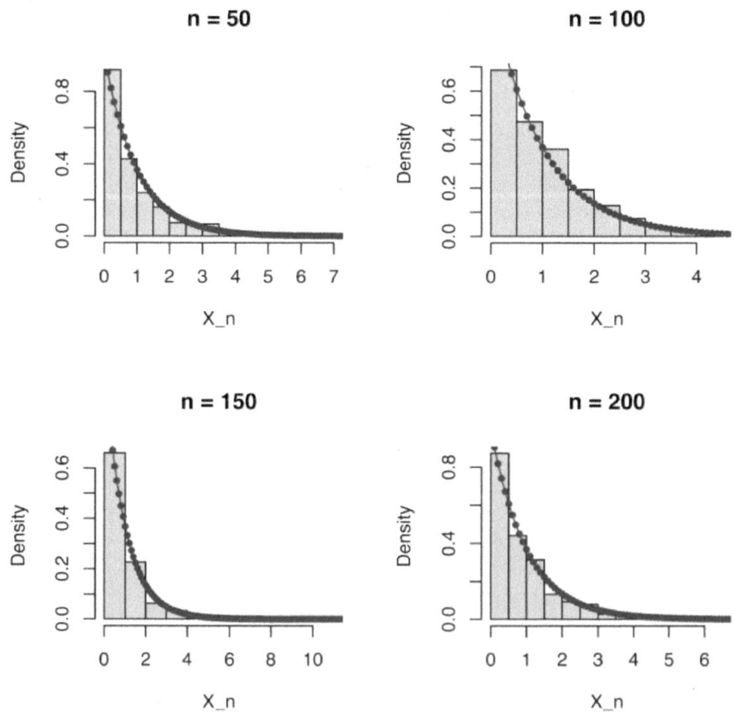

FIGURE 7.12
Convergence in Law of X_n to Y

The advantage of the Cauchy criterion of convergence is that it permits checking for convergence in r-th mean without identifying the limit of a sequence. We will prove the equivalence of Cauchy convergence in r-th mean and convergence in r-th mean in Chapter 8 after discussing Fatou's lemma.

Following is a quick recap of the results discussed in this chapter.

Summary

1. Convergence in probability: $X_n \xrightarrow{P} X$, if $P[|X_n - X| < \epsilon] \to 1$ as $n \to \infty, \ \forall \ \epsilon > 0$.

2. Convergence in law: If $F_n(x) = P[X_n \leq x] \to P[X \leq x] = F(x)$, $\forall \ x \in C(F_X)$, then $X_n \xrightarrow{L} X$, where $C(F_X)$ is a set of points of continuity of the distribution function of X.

3. Convergence in r-th mean: $X_n \xrightarrow{r} X$ if $E(|X_n - X|^r) \to 0$ as $n \to \infty$, $r > 0$, where $\forall \ n \geq 1, X_n \ \& \ X \in L_r$ space. If $r = 2$, then the convergence is referred to as convergence in quadratic mean.

4. For all the modes of convergence, except convergence in law, a sequence $\{X_n, n \geq 1\}$ of random variables and X are defined on the same probability space.

5. For all the modes of convergence, except convergence in law, convergence and Cauchy convergence are equivalent.

6. In convergence in probability and convergence in r-th mean, the limit random variable is almost surely unique. However, in convergence in law, limit random variables are identically distributed.

7. $X_n \xrightarrow{a.s.} X \ \Rightarrow \ X_n \xrightarrow{P} X \ \Rightarrow \ X_n \xrightarrow{L} X$.

8. $X_n \xrightarrow{P} X \iff X_n \xrightarrow{L} X$ if and only if X is a degenerate random variable.

9. $X_n \xrightarrow{r} X \ \Rightarrow \ X_n \xrightarrow{P} X$.

10. $X_n \xrightarrow{P} X \ \Rightarrow \ X_n \xrightarrow{r} X$, if X_n's are almost surely bounded.

11. If $X_n \xrightarrow{P} X$, then there exists a sub-sequence of $\{X_n, n \geq 1\}$ which converges to X almost surely.

12. Convergence in probability does not imply convergence of moments or convergence in r-th mean.

13. Convergence in r-th mean implies convergence of r-th order moments, but the converse is not true.

14. In general, almost sure convergence neither implies nor is implied by the convergence in r-th mean.

15. $X_n \xrightarrow{r} X \Rightarrow X_n \xrightarrow{s} X, \;\; \forall \; s < r$.

16. $X_n \overset{q.m.}{\rightarrow} b$ if and only if $E(X_n) \rightarrow b$ and $Var(X_n) \rightarrow 0$, where b is a constant.

17. If a sequence $\{X_n, n \geq 1\}$ of random variables is uniformly bounded, then $X_n \xrightarrow{P} C$ implies $E(X_n) \rightarrow C$ and $E(X_n - C)^2 \rightarrow 0$.

18. Convergence in probability is closed under arithmetic operations.

19. Convergence in law and convergence in probability are closed under continuous transformation.

20. If $X_n \xrightarrow{P} X$ or if $X_n \xrightarrow{L} X$, then the sequence $\{X_n, n \geq 1\}$ is bounded in probability.

21. If $Y_n \xrightarrow{P} 0$ and if X is a real random variable then $XY_n \xrightarrow{P} 0$.

22. If $X_n \xrightarrow{P} C < 0$, then $P[X_n \leq 0] \rightarrow 1$ as $n \rightarrow \infty$. If $X_n \xrightarrow{P} C > 0$, then $P[X_n > 0] \rightarrow 1$ as $n \rightarrow \infty$.

23. If $X_n - Y_n \xrightarrow{P} 0$ and $Y_n \xrightarrow{L} X$, then $X_n \xrightarrow{L} X$.

24. If the sequence $\{X_n, n \geq 1\}$ is bounded in probability and if $Y_n \xrightarrow{P} 0$, then $X_nY_n \xrightarrow{P} 0$.

25. Slutsky's theorem: If $X_n \xrightarrow{L} X$ and $Y_n \xrightarrow{L} C$, then (i) $X_n \pm Y_n \xrightarrow{L} X \pm C$. (ii) $X_nY_n \xrightarrow{L} XC$. (iii) $X_n/Y_n \xrightarrow{L} X/C$, provided X_n/Y_n and X/C are defined.

26. Suppose $\{X_n, n \geq 1\}$ is a sequence of random variables with $\phi_n(t),\; t \in \mathbb{R}$ being a characteristic function of $X_n, n \geq 1$. Suppose X is a random variable with $\phi(t),\; t \in \mathbb{R}$ as its characteristic function. Then

$$\phi_n(t) \rightarrow \phi(t) \;\; \forall \;\; t \in \mathbb{R} \quad \Longleftrightarrow \quad X_n \xrightarrow{L} X \; .$$

27. Suppose $\{X_n, n \geq 1\}$ is a sequence of random variables with $\phi_n(t),\; t \in \mathbb{R}$ being a characteristic function of $X_n, n \geq 1$. Suppose $\phi_n(t) \rightarrow \phi(t) \;\; \forall \;\; t \in \mathbb{R}$ as $n \rightarrow \infty$, where $\phi(t)$ is continuous at $t = 0$. Then there exists a random variable X with characteristic function $\phi(t)$ such that $X_n \xrightarrow{L} X$.

7.5 Conceptual Exercises

7.5.1 Suppose $\{X_n, n \geq 1\}$ is a sequence of random variables such that $P[X_n = \pm n] = 1/2$. Using the definitions, examine if $X_n \xrightarrow{L} X$, $X_n \xrightarrow{P} 0$ and $X_n \xrightarrow{r} 0$. Verify the known implications among these modes of convergence. Examine if $X_n \xrightarrow{a.s.} X$.

7.5.2 Suppose $X_n \equiv 1/n$ and $Y_n \equiv -1/n$. Using the definitions examine whether X_n and Y_n converge in law, in probability, r-th mean and almost surely.

7.5.3 Suppose$\{X_n, n \geq 1\}$ is a sequence of random variables such that $P[X_n = 0] = 1 - 1/n^2$ & $P[X_n = e^n] = 1/n^2$. Using the definitions, examine if $X_n \xrightarrow{L} 0$, $X_n \xrightarrow{P} 0$ and $X_n \xrightarrow{r} 0$. Using a sufficient condition, examine if $X_n \xrightarrow{a.s.} 0$. Hence, verify the known implications among these modes of convergence.

7.5.4 Suppose a probability mass function of a random variable X_n is given by,

$$P[X_n = \sqrt{n}] = 1/n = 1 - P[X_n = 0].$$

Examine whether $X_n \xrightarrow{P} 0$, $X_n \xrightarrow{L} 0$ and $X_n \xrightarrow{q.m.} 0$.

7.5.5 Suppose a probability mass function of a random variable X_n is given by,

$$P[X_n = 0] = 1/n^2 = 1 - P[X_n = n].$$

Examine whether $X_n \xrightarrow{P} 0$, $X_n \xrightarrow{L} 0$, $X_n \xrightarrow{a.s.} 0$ and $X_n \xrightarrow{q.m.} 0$.

7.5.6 Suppose $\{X_n, n \geq 1\}$ is a sequence of discrete random variables with probability mass function given by,

$$P[X_n = x] = \begin{cases} 1/n, & \text{if } x = n \\ 1 - 1/n, & \text{if } x = 0. \end{cases}$$

Discuss the limiting behaviour of $\{X_n, n \geq 1\}$ in probability, in law and in quadratic mean.

7.5.7 Suppose $\{X_n, n \geq 1\}$ is a sequence of random variables defined as $X_n = X + Y_n$ where X is a random variable and $\{Y_n, n \geq 1\}$ is a sequence of random variables such that $E(Y_n) = 1/n$ and $Var(Y_n) = \sigma^2/n$ where $\sigma > 0$ is a constant. Verify whether $X_n \xrightarrow{P} X$.

7.5.8 Suppose $\{X_n, n \geq 1\}$ is a sequence of random variables defined on the probability space $(\Omega, \mathbb{A}, P)$ where $\Omega = [0, 1]$, $\mathbb{A}$ is a sigma field of subsets of $[0, 1]$ and $P((a, b)) = b - a$ for $(a, b) \in \mathbb{A}$. Suppose $\{X_n\}$ is defined as

$$X_n(\omega) = \begin{cases} 1, & \text{if } 0 \leq \omega < (n+1)/2n \\ 0, & \text{otherwise.} \end{cases}$$

X is a random variable defined as $X(\omega) = 1$ if $0 \leq \omega < 1/2$ and 0 otherwise. Examine whether $X_n \stackrel{a.s.}{\to} X$, $X_n \stackrel{P}{\to} X$, $X_n \stackrel{L}{\to} X$ and $X_n \stackrel{q.m.}{\to} X$.

7.5.9 Suppose $\{X_n, n \geq 1\}$ is a sequence of discrete random variables with probability mass function given by,

$$\begin{aligned} P[X_n = -(n+4)] &= 1/(n+4), \quad P[X_n = -1] = 1 - 4/(n+4) \\ \& \;\; P[X_n = n+4] &= 3/(n+4). \end{aligned}$$

Discuss the limiting behaviour of $\{X_n\}$ in probability, in law and in quadratic mean to $X \equiv -1$ and $X \equiv 0$. Examine whether $E(X_n)$ converges.

7.5.10 Suppose $\{X_n, n \geq 1\}$ is a sequence of discrete random variables with probability mass function given by, $P[X_n = \pm 1/n] = 1/2$. Discuss the limiting behaviour of $\{X_n\}$ in probability, in law, almost surely and in quadratic mean to X. Examine whether $E(X_n) \to E(X)$.

7.5.11 Suppose $\{X_n, n \geq 1\}$ is a sequence of continuous random variables with support $[-1/n, 1/n]$ and with the probability density function given by,

$$f_{X_n}(x) = \begin{cases} n/2, & \text{if } \; x \in [-1/n, 1/n] \\ 0, & \text{if } \; x \notin [-1/n, 1/n]. \end{cases}$$

Examine whether the sequence $\{X_n, n \geq 1\}$ converges in quadratic mean, in probability, in law and almost surely.

7.5.12 Suppose $(\Omega, \mathbb{A}, P)$ is a probability space, where $\Omega = [0, 1]$, $\mathbb{A}$ is a sigma field of subsets of Ω and $P(A_n) = 1/n$ for $A_n = [0, 1/n)$. Suppose $X_n = e^n I_{A_n}, n \geq 1$. Show that $X_n \stackrel{P}{\to} 0$, but $X_n \stackrel{r}{\nrightarrow} 0$.

7.5.13 Suppose $\{X_1, X_2, \cdots, X_n\}$ are independent and identically distributed random variables with the probability density function $f(x) = \alpha x^{-\alpha-1}$ for $x > 1, \alpha > 0$. Suppose $Y_n = n^{-1/\alpha} X_{(n)}$. Examine whether Y_n converges in law.

7.5.14 Suppose $\{X_n, n \geq 1\}$ is a sequence of independent and identically distributed random variables, such that $X_n = 1$ or 0 with probability 1/2 each. Suppose $Y_n = 1 - X_n$. Show that $X_n \stackrel{L}{\to} X$ and $Y_n \stackrel{L}{\to} Y$, where X and Y are identically distributed random variables, with values 1 and 0 with equal probabilities 1/2. Examine whether $X_n + Y_n \stackrel{L}{\to} X + Y$. Examine whether X_n and Y_n are independent random variables.

7.5.15 Suppose $\{X_n, n \geq 1\}$ is a sequence of random variables such that X_n follows $N(0, \sigma^2/n)$. Suppose a function $g : \mathbb{R} \to \mathbb{R}$ is defined as

$$g(x) = \begin{cases} 0, & \text{if } \; x \leq 0 \\ 1, & \text{if } \; x > 0. \end{cases}$$

Show that $X_n \stackrel{P}{\to} X \equiv 0$, but $g(X_n) \stackrel{P}{\nrightarrow} g(X)$.

7.5.16 Suppose $X_n \sim G(n, n)$. Examine if $X_n \xrightarrow{r} 1$ for $r > 1$ and $X_n \xrightarrow{a.s.} 1$.

7.5.17 Suppose $\{X_n, n \geq 1\}$ is a sequence of random variables such that X_n has Cauchy distribution with probability density function given by $f(x, n) = (n/\pi)(1/(1 + n^2x^2))$, $x \in \mathbb{R}$. Examine whether $X_n \xrightarrow{P} 0$, $X_n \xrightarrow{r} 0$, $r \geq 1$ and $X_n \xrightarrow{L} 0$.

7.5.18 Suppose $\{X_n, n \geq 1\}$ is a sequence of random variables such that X_n has Cauchy distribution with probability density function given by $f(x, n) = (n/\pi)(1/(n^2 + x^2))$, $x \in \mathbb{R}$. Examine whether $X_n \xrightarrow{P} 0$, $X_n \xrightarrow{r} 0$, $r \geq 1$, $X_n \xrightarrow{L} 0$ and $X_n \xrightarrow{a.s.} 0$.

7.5.19 Suppose X is a real random variable defined on a probability space on $(\Omega, \mathbb{A}, P)$ and $\{A_n, n \geq 1\}$ is a sequence of sets in $\mathbb{A}$ such that $P(A_n) \to 0$. Then show that $XI_{A_n} \xrightarrow{P} 0$.

7.5.20 Show that equivalent random variables are identically distributed.

7.5.21 Suppose $X_n = \min\{Y_1, Y_2, \cdots, Y_n\}$, where $\{Y_n, n \geq 1\}$ is a sequence of independent and identically distributed random variables, each having an exponential distribution with location parameter θ and scale parameter 1. (i) Show that $X_n \xrightarrow{P} \theta$ and $X_n \xrightarrow{a.s.} \theta$. (ii) Suppose $U_n = n(X_n - \theta)$, then show that $U_n \xrightarrow{L} U$, where U has exponential distribution with location parameter 0 and scale parameter 1. (iii) From the limit law of U_n, find the limit law of $V_n = \sqrt{n}(X_n - \theta)$. (iv) From the limit law of U_n, find the limit law of $\exp(-U_n)$.

7.5.22 Suppose $X_n \xrightarrow{L} X$ and $Y_n \xrightarrow{L} C$, where $C \neq 0$ is a degenerate random variable. Using the definition of convergence in law, show that (i) $X_nC \xrightarrow{L} XC$ and (ii) $X_n/C \xrightarrow{L} X/C$, provided these are defined. Hence show that $X_nY_n \xrightarrow{L} XC$, and $X_n/Y_n \xrightarrow{L} X/C$, provided these are defined.

7.5.23 Suppose $X_n \sim \chi^2_n$, $n \geq 1$ and $Y_n = (X_n - n)/\sqrt{2n}$. Examine whether (i) $X_n/n \xrightarrow{P} 1$, $X_n/n \xrightarrow{q.m.} 1$ and (ii) $Y_n \xrightarrow{L} Z \sim N(0, 1)$. What is the distribution of Y_n^2 as $n \to \infty$?

7.5.24 Suppose $X_n \sim P(\lambda_n)$ and $Y_n = (X_n - \lambda_n)/\sqrt{\lambda_n}$. Show that $Y_n \xrightarrow{L} Z$, which follows $N(0, 1)$ distribution as $\lambda_n \to \infty$. What is the distribution of Y_n^2 as $\lambda_n \to \infty$?

7.5.25 Suppose $\{X_n, n \geq 1\}$ is a sequence of independent random variables each having uniform $U(0, 1)$ distribution. Suppose $X_{(1)} = \min\{X_1, X_2, \cdots, X_n\}$ and $X_{(n)} = \max\{X_1, X_2, \cdots, X_n\}$. Show that (i) $nX_{(1)} \xrightarrow{L} X$, where X has exponential distribution with mean 1, (ii)

$1-(X_{(1)}+X_{(n)}) \xrightarrow{P} 0$ and

(iii) $n(1-X_{(n)})X_{(1)} \xrightarrow{P} 0$.

7.5.26 Suppose $\{X_n, n \geq 1\}$ is a sequence of independent and identically distributed random variables, each having $U(0,1)$ distribution. Find the limiting distribution of $n(1-X_{(n-1)})$.

7.5.27 Suppose $\{X_1, X_2, \cdots, X_n\}$ is a random sample from a normal $N(\theta, 1)$ distribution and $\theta \in [0, \infty)$. The maximum likelihood estimator $\hat{\theta}_n$ of θ is given by,

$$\hat{\theta}_n = \begin{cases} \overline{X}_n, & \text{if } \overline{X}_n \geq 0 \\ 0, & \text{if } \overline{X}_n < 0, \end{cases}$$

Examine whether $\hat{\theta}_n \xrightarrow{P} \theta$. Find the limit law of $Y_n = \sqrt{n}(\hat{\theta}_n - \theta)$. Identify the limiting distribution at $\theta = 0$.

7.5.28 Suppose $X_n \xrightarrow{P} X$. Show that $|X_n| \xrightarrow{L} |X|$.

7.5.29 Suppose $X_n \xrightarrow{L} X$ where X is a continuous random variable. Using the definition of convergence in law, show that $|X_n| \xrightarrow{L} |X|$.

7.5.30 Suppose $(\Omega, \mathbb{A}, P)$ is a probability space, where $\Omega = [0,1]$, $\mathbb{A}$ is a sigma field of subsets of Ω and $P(A_n) = 1/n$ for $A_n = [0, 1/n)$. Suppose the random variable X and X_n, $n \geq 1$ are defined on this space as follows.

$$X(\omega) = \begin{cases} 0, & \text{if } 0 \leq \omega \leq 1/2 \\ 1, & \text{if } 1/2 < \omega \leq 1 \end{cases} \quad \& \quad X_n(\omega) = \begin{cases} 1, & \text{if } 0 \leq \omega \leq 1/2 \\ 0, & \text{if } 1/2 < \omega \leq 1. \end{cases}$$

Examine whether $X_n \xrightarrow{L} X$ and $X_n \xrightarrow{P} X$.

7.5.31 Suppose $\{X_n, n \geq 1\}$ is a sequence of independent and identically distributed random variables, each having the exponential distribution with scale parameter 1. Examine whether $X_{(n)} - \log n \xrightarrow{L} X$, where the distribution function of X is $F_X(x) = \exp(-e^{-x})$, $\quad -\infty < x < \infty$.

7.5.32 Suppose m_n is a median of a random variable X_n, $n \geq 1$. Show that if $X_n \xrightarrow{L} X$, then any limit point of m_n is a median of X.

7.5.33 If $\{X_n, n \geq 1\}$ is Cauchy in r-th mean, show that it is Cauchy in probability.

7.5.34 If $X_n \xrightarrow{r} X$ and $Y_n \xrightarrow{r} Y$, examine whether (i) $X_n \pm Y_n \xrightarrow{r} X \pm Y$ and (ii) $X_n Y_n \xrightarrow{r} XY$.

7.5.35 If $X_n \xrightarrow{r} X$ and $Y_n \xrightarrow{r} Y$, examine whether $E(X_n Y_n) \to E(XY)$.

7.5.36 Suppose $\{X_n, n \geq 1\}$ is a sequence of independent random variables such that $\sum_{i=1}^{n}(X_i - E(X_i))/s_n \xrightarrow{L} Z \sim N(0,1)$, where $s_n^2 = \sum_{i=1}^{n} Var(X_i)$. Show that $\sum_{i=1}^{n}(X_i - E(X_i))/n \xrightarrow{P} 0$ if and only if $s_n/n \to 0$ as $n \to \infty$.

7.5.37 Suppose $\{X_1, X_2, \cdots, X_n\}$ is a random sample from a normal $N(\theta, 1)$ distribution, where $\theta \in \{0, 1\}$. The maximum likelihood estimator $\hat{\theta}_n$ of θ is given by,

$$\hat{\theta}_n = \begin{cases} 1, & \text{if } \overline{X}_n > \frac{1}{2} \\ 0, & \text{if } \overline{X}_n \leq \frac{1}{2}. \end{cases}$$

Examine whether (i) $\hat{\theta}_n \xrightarrow{P} \theta$. (ii) Suppose $Y_n = \sqrt{n}(\hat{\theta}_n - \theta)$. Examine the convergence in law of Y_n. (iii) Examine whether $\hat{\theta}_n \xrightarrow{a.s.} \theta$. (iv) Examine whether for large n, $\hat{\theta}_n = \theta$ almost surely.

7.6 Computational Exercises

7.6.1 Suppose $\{X_1, X_2, \cdots, X_n\}$ is a random sample from a normal $N(\theta, 1)$ distribution, where $\theta \in \{0, 1\}$. The maximum likelihood estimator $\hat{\theta}_n$ of θ is given by,

$$\hat{\theta}_n = \begin{cases} 1, & \text{if } \overline{X}_n > \frac{1}{2} \\ 0, & \text{if } \overline{X}_n \leq \frac{1}{2}. \end{cases}$$

Examine by simulation whether (i) $\hat{\theta}_n \xrightarrow{P} \theta$. (ii) Suppose $Y_n = \sqrt{n}(\hat{\theta}_n - \theta)$. Examine the convergence in law of Y_n by simulation.

7.6.2 Suppose $X_n \sim B(n, p)$. In Section 3, it is shown that $X_n \xrightarrow{L} X$, as $p \to 0$ & $n \to \infty$ such that $np \to \lambda$, where $X \sim P(\lambda)$. Verify the result by Karl Pearson's goodness of fit test based on the simulated data.

7.6.3 Suppose $X_n = \min\{Y_1, Y_2, \cdots, Y_n\}$, where $\{Y_n, n \geq 1\}$ is a sequence of independent and identically distributed random variables, each having an exponential distribution with location parameter θ and scale parameter 1. Examine by simulation whether (i) $X_n \xrightarrow{P} \theta$ and (ii) $U_n \xrightarrow{L} U$, where U has exponential distribution with location parameter 0 and scale parameter 1 and $U_n = n(X_n - \theta)$.

7.6.4 Suppose $X_n, n \geq 1$ be a sequence of random variables such that $P[X_n = 0] = 1 - 1/n^2$ & $P[X_n = e^n] = 1/n^2$. In Exercise 3, we have examined if $X_n \xrightarrow{L} 0$, $X_n \xrightarrow{P} 0$ and $X_n \xrightarrow{r} 0$. Verify these modes of convergence via simulations.

7.6.5 Suppose $\{X_n, n \geq 1\}$ is a sequence of independent and identically distributed random variables, each having an exponential distribution with location parameter 0 and scale parameter 1. Examine by simulation whether $X_{(n)} - \log n \stackrel{L}{\to} X$ where the distribution function of X is $F_X(x) = \exp(-e^{-x}), \quad -\infty < x < \infty$.

7.6.6 Suppose $X_n \sim \chi^2_n, n \geq 1$ and $Y_n = (X_n - n)/\sqrt{2n}$. Verify the following results by simulation. (i) $X_n/n \stackrel{P}{\to} 1$, (ii)$X_n/n \stackrel{q.m.}{\to} 1$ and (iii) $Y_n \stackrel{L}{\to} Z \sim N(0,1)$ distribution. Use Karl Pearson's goodness of fit test for (iii).

7.6.7 Suppose $X_n \sim P(\lambda_n)$ and $Y_n = (X_n - \lambda_n)/\sqrt{\lambda_n}$. Verify the result $Y_n \stackrel{L}{\to} Z \sim N(0,1)$ as $\lambda_n \to \infty$, by simulation, graphically and by Karl Pearson's goodness of fit test.

7.6.8 Suppose $\{X_1, X_2, \cdots, X_n\}$ are independent and identically distributed random variables with the probability density function $f(x) = \alpha x^{-\alpha-1}$ for $x > 1, \alpha > 0$. Suppose $Y_n = n^{-1/\alpha} X_{(n)}$. In Exercise 7.5.13, it is shown that Y_n converges in law. Verify it by simulation. Use Karl Pearson's goodness of fit test to verify convergence in law. Also verify by simulation that Y_n converges in r-th mean and in probability.

7.7 Multiple Choice Questions

Note: In each question, multiple options may be correct. Unless specified otherwise, identify which of the statement(s) is/are correct. Answers are given in Chapter 11, after the solutions of conceptual exercises of Chapter 7.

7.7.1 Among the various modes of convergence of a sequence of random variables,

(a) almost sure convergence implies convergence in probability
(b) convergence in probability always implies almost sure convergence
(c) almost sure convergence implies convergence in quadratic mean
(d) convergence in quadratic mean implies convergence in law

7.7.2 Among the various modes of convergence of a sequence of random variables,

(a) almost sure convergence implies convergence in probability
(b) convergence in probability implies almost sure convergence for monotone sequences
(c) almost sure convergence implies convergence in quadratic mean
(d) convergence in quadratic mean implies almost sure convergence

7.7.3 Suppose a probability space $(\Omega, \mathbb{A}, P)$ is defined as follows. $\Omega = [0,1]$, $\mathbb{A}$ is a sigma field of subsets of Ω and P is a Lebesgue measure. Suppose a sequence $\{X_n, n \geq 1\}$ of random variables and a random variable X are defined as $X_n(\omega) = e^{\omega} + \omega^n$, $n \geq 1$ and $X(\omega) = e^{\omega}$ on $(\Omega, \mathbb{A}, P)$. Then

(a) $X_n \stackrel{a.s.}{\to} X$
(b) $X_n \stackrel{P}{\to} X$
(c) $X_n \stackrel{L}{\to} X$
(d) $X_n \to X$ pointwise.

7.7.4 Suppose $\{X_n, n \geq 1\}$ is a sequence of independent random variables with
$E(X_n) = 0$ and $V(X_n) = \sigma_n^2$. In which of the following cases, $S_n/n^2 \stackrel{P}{\to} 0$, where $S_n = \sum_{k=1}^{n} X_k$?

(a) $\sigma_n^2 = 1$
(b) $\sigma_n^2 = n$
(c) $\sigma_n^2 = n^2$
(d) $\sigma_n^2 = n^3$

7.7.5 In which of the following cases, $X = Y$ almost surely?

(a) $X_n \stackrel{P}{\to} X, X_n \stackrel{P}{\to} Y$
(b) $X_n \stackrel{a.s.}{\to} X, X_n \stackrel{a.s.}{\to} Y$
(c) $X_n \stackrel{L}{\to} X, X_n \stackrel{L}{\to} Y$
(d) $X_n \stackrel{q.m.}{\to} X, X_n \stackrel{q.m.}{\to} Y$

7.7.6 In which of the following cases, $P[X \neq Y] > 0$?

(a) $X_n \stackrel{P}{\to} X, X_n \stackrel{P}{\to} Y$
(b) $X_n \stackrel{a.s.}{\to} X, X_n \stackrel{a.s.}{\to} Y$
(c) $X_n \stackrel{L}{\to} X, X_n \stackrel{L}{\to} Y$
(d) $X_n \stackrel{q.m.}{\to} X, X_n \stackrel{q.m.}{\to} Y$

7.7.7 In which of the following cases, X and Y are identically distributed?

(a) $X_n \stackrel{P}{\to} X, X_n \stackrel{P}{\to} Y$
(b) $X_n \stackrel{a.s.}{\to} X, X_n \stackrel{a.s.}{\to} Y$
(c) $X_n \stackrel{L}{\to} X, X_n \stackrel{L}{\to} Y$
(d) $X_n \stackrel{q.m.}{\to} X, X_n \stackrel{q.m.}{\to} Y$

7.7.8 Which of the following types of convergence is/are always invariant under arithmetic operations?

(a) Almost sure convergence
(b) Point-wise convergence
(c) Convergence in probability
(d) Convergence in distribution

7.7.9 Which of the following types of convergence is/are invariant under continuous transformation?

(a) Almost sure convergence
(b) Point-wise convergence
(c) Convergence in probability
(d) Convergence in distribution

7.7.10 Suppose $(\Omega, \mathbb{A}, P)$ is a probability space, where $\Omega = [0,1]$, $\mathbb{A}$ is a sigma field of subsets of Ω and $P(A_n) = 1/n$ for $A_n = [0, 1/n)$. Suppose $X_n = I_{A_n}, n \geq 1$. Then

(a) $X_n \stackrel{a.s.}{\to} 0$
(b) $X_n \stackrel{P}{\to} 0$
(c) $X_n \stackrel{L}{\to} 0$
(d) $X_n \stackrel{r}{\to} 0$

7.7.11 Suppose $\{X_n, n \geq 1\}$ is a sequence of random variables such that $X_n \sim N(1/n, 1/n)$ distribution. Then

(a) $X_n \stackrel{q.m.}{\to} 0$
(b) $X_n \stackrel{P}{\to} 0$
(c) $X_n \stackrel{L}{\to} 0$
(d) $\sqrt{n}(X_n - 1/n) \stackrel{L}{\to} Z \sim N(0,1)$

7.7.12 Suppose $\{X_n, n \geq 1\}$ is a sequence of random variables such that $X_n \sim U(-n, n)$ distribution. Then

(a) $X_n \stackrel{L}{\nrightarrow} 0$
(b) $X_n \stackrel{P}{\nrightarrow} 0$
(c) $X_n \stackrel{a.s}{\nrightarrow} 0$
(d) $X_n \stackrel{q.m.}{\nrightarrow} 0$

7.7.13 Suppose $X_n \stackrel{L}{\to} X$ and $X_n \stackrel{L}{\to} Y$. The following are three statements. (I) X and Y are identically distributed random variables. (II) X and Y are identical random variables. (III) X and Y are equivalent random variables. Then which of the following is/are true?

(a) Only I is true
(b) Both I and II are true
(c) Both II and III are true
(d) Both I and III are true

7.7.14 Following are three statements. Suppose $X_n - Y_n \stackrel{P}{\to} 0$ and $Y_n \stackrel{L}{\to} X$. Then (I) $X_n \stackrel{P}{\to} X$, (II) $X_n \stackrel{L}{\to} X$ and (III) the characteristic function of X_n converges to that of X on $\mathbb{R}$. Then which of the following is/are true?

(a) Only I is true
(b) Both I and II are true
(c) Both II and III are true
(d) Only II is true

7.7.15 Suppose $\{X_n, n \geq 1\}$ is a sequence of random variables such that

$$P[X_n = 0] = 1 - 1/n^3 \quad \& \quad P[X_n = n] = 1/n^3, \quad n \geq 1 .$$

Following are four statements. (I)$X_n \xrightarrow{P} 0$. (II) $X_n \xrightarrow{L} 0$. (III) $X_n \xrightarrow{a.s.} 0$. (IV) $X_n \overset{r}{\nrightarrow} 0$ for $r = 3, 4$. Then which of the following is/are true?

(a) Only I and III are true
(b) Only I and II are true
(c) Only I, II and III are true
(d) All four are true

7.7.16 If $U_n = \sqrt{n}(X_n - \mu) \xrightarrow{L} X$, then which of the following is always true?

(a) $X_n \xrightarrow{L} \mu$
(b) $X_n \xrightarrow{a.s.} \mu$
(c) $X_n \xrightarrow{P} \mu$
(d) $X_n \xrightarrow{r} \mu$

7.7.17 Following are two statements. Suppose $X_n \xrightarrow{L} X$ and $Y_n \xrightarrow{L} Y$. Then (I) $X_n \pm Y_n \xrightarrow{L} X \pm Y$ always, (II) $X_n \pm Y_n \xrightarrow{L} X \pm Y$ if X_n is independent of Y_n for all n and if X is independent of Y. Which of the following is true?

(a) Both (I) and (II) are false
(b) Both (I) and (II) are true
(c) (I) is true but (II) is false
(d) (I) is false but (II) is true

7.7.18 If $X_n \xrightarrow{L} X$ and $Y_n \xrightarrow{L} C$, where C is a constant. Then

(a) $X_n + Y_n \xrightarrow{L} X + C$
(b) $X_n Y_n \xrightarrow{L} XC$
(c) $X_n/Y_n \xrightarrow{L} X/C$, provided X_n/Y_n and X/C are defined
(d) $X_n - Y_n \xrightarrow{L} X - C$

7.7.19 Suppose $X_n \sim G(n, n)$ with probability density function $f(x, n) = \frac{n^n}{\Gamma(n)} \exp(-nx)x^{n-1}, \ x > 0, \ n \in I^+$. Then

(a) $X_n \xrightarrow{L} 1$
(b) $X_n \xrightarrow{P} 1$
(c) $X_n \xrightarrow{q.m.} 1$

(d) $E(X_n) \to 1$

7.7.20 $X_n \overset{q.m.}{\to} b$ if and only if

(a) $E(X_n) \to b$, where b is a constant
(b) $E(X_n) \to b$ and $Var(X_n) \to b$
(c) $E(X_n) \to b$ and $Var(X_n) \to 0$
(d) $E(X_n) \to 0$ and $Var(X_n) \to b$

7.7.21 Suppose X_n and $X \in \mathbb{L}_r$. Then

(a) $E(|X_n - X|^r) \to 0 \;\Rightarrow\; E(|X_n|^r) \to E(|X|^r)$
(b) $E(|X_n - X|^r) \to 0 \;\Rightarrow\; X_n \overset{P}{\to} X$
(c) $E(|X_n|^r) \to E(|X|^r) \;\Rightarrow\; E(|X_n - X|^r) \to 0$
(d) $E(|X_n - X|^r) \to 0 \;\Rightarrow\; X_n \overset{L}{\to} X$

7.7.22 If $X_n \overset{a.s.}{\to} X$, then which of the following is/are always true?

(a) $e^{X_n} \overset{a.s.}{\to} e^X$
(b) $e^{X_n} \overset{P}{\to} e^X$
(c) $e^{X_n} \overset{L}{\to} e^X$
(d) $e^{X_n} \overset{q.m.}{\to} e^X$

7.7.23 suppose $X_n \overset{a.s.}{\to} X$, Y_n and Y are defined as $Y_n = X_n^3 + 2X_n^2 + 3X_n + 2$ and $Y = X^3 + 2X^2 + 3X + 2$. The following are three statements. (I) $Y_n \overset{a.s.}{\to} Y$ (II) $Y_n \overset{P}{\to} Y$ (III) $Y_n \overset{L}{\to} Y$. Which of the following is the correct option?

(a) Only I is true
(b) Only I and II are true
(c) Only II and III are true
(d) All three are true

7.7.24 If $X_n \overset{a.s.}{\to} X$, then which of the following statements is/are not always correct?

(a) $e^{X_n} \overset{a.s.}{\to} e^X$
(b) $e^{X_n} \overset{P}{\to} e^X$
(c) $e^{X_n} \overset{L}{\to} e^X$
(d) $e^{X_n} \overset{q.m.}{\to} e^X$

7.7.25 If $X_n \overset{r}{\to} X$, then which of the following statements is/are always correct?

(a) $|X_n| \overset{P}{\to} |X|$
(b) $|X_n| \overset{L}{\to} |X|$
(c) $|X_n| \overset{a.s.}{\to} |X|$
(d) $E(|X_n|^r) \to E(|X|^r), r > 0$ as $n \to \infty$

7.7.26 If $X_n \xrightarrow{P} X$, then which of the following statements is/are always true?

(a) $E(|X_n|^r) \to E(|X|^r), r > 0$ as $n \to \infty$
(b) $E(|X_n - X|^r) \to 0, r > 0$ as $n \to \infty$
(c) $E(|X_n - X|^r) \to 0, r > 0$ as $n \to \infty$, if X_n's are almost surely bounded.
(d) $E(X_n) \to E(X)$ as $n \to \infty$

7.7.27 Suppose X is a real random variable on a probability space $(\Omega, \mathbb{A}, P)$ and
$\{A_n, n \geq 1\}$ is a sequence of sets in $\mathbb{A}$. Then

(a) $A_n \downarrow \emptyset \Rightarrow XI_{A_n} \xrightarrow{P} 0$
(b) $A_n \downarrow \emptyset \Rightarrow XI_{A_n} \xrightarrow{P} X$
(c) $A_n \downarrow \emptyset \Rightarrow XI_{A_n} \xrightarrow{P} 1$
(d) $A_n \downarrow \emptyset \Rightarrow I_{A_n} \xrightarrow{L} 0$

7.7.28 Suppose X is a real random variable on a probability space $(\Omega, \mathbb{A}, P)$ and
$\{A_n, n \geq 1\}$ is a sequence of events in $\mathbb{A}$. Then

(a) $A_n \uparrow \Omega \Rightarrow XI_{A_n} \xrightarrow{P} 0$
(b) $A_n \uparrow \Omega \Rightarrow XI_{A_n} \xrightarrow{P} X$
(c) $A_n \uparrow \Omega \Rightarrow XI_{A_n} \xrightarrow{P} 1$
(d) $A_n \uparrow \Omega \Rightarrow I_{A_n} \xrightarrow{L} 1$

7.7.29 Suppose $\{X_n, n \geq 1\}$ is a sequence of random variables defined as $X_n = (1 + 1/n)X$, where X is a discrete random variable with support $\{0, 1\}$ and $P[X = 0] = 2/3$. Then

(a) $X_n \xrightarrow{P} X$
(b) $X_n \xrightarrow{r} X$
(c) $X_n \xrightarrow{a.s.} X$
(d) $X_n \xrightarrow{L} X$

7.7.30 Suppose $\{X_n, n \geq 1\}$ be a sequence of random variables such that X_n has uniform $U(-n, n)$ distribution. Then

(a) X_n does not converge in law
(b) X_n converges to 0 in probability
(c) X_n does not converge in quadratic mean
(d) X_n converges to 0 almost surely

7.7.31 Following are two statements: (I) $X_n \xrightarrow{L} C \Rightarrow X_n \xrightarrow{P} C$, where C is a constant. (II) $X_n \xrightarrow{P} C \Rightarrow X_n \xrightarrow{L} C$, where C is a constant. Then which of the following statements is true?

(a) Both (I) and (II) are false

(b) Both (I) and (II) are true
(c) (I) is true but (II) is false
(d) (I) is false but (II) is true

7.7.32 Suppose X is a non-degenerate random variable. Following are two statements: (I) $X_n \xrightarrow{L} X \Rightarrow X_n \xrightarrow{P} X$. (II) $X_n \xrightarrow{P} X \Rightarrow X_n \xrightarrow{L} X$. Then which of the following statements is true?

(a) Both (I) and (II) are false
(b) Both (I) and (II) are true
(c) (I) is true but (II) is false
(d) (I) is false but (II) is true

7.7.33 Suppose X_n has normal $N(0, 1/n)$ distribution. Then

(a) $X_n \xrightarrow{L} 0$
(b) $X_n \xrightarrow{P} 0$
(c) $X_n \xrightarrow{q.m.} 0$
(d) $X_n \xrightarrow{a.s.} 0$

7.7.34 Suppose X_n has normal $N(0, 1/n)$ distribution. Then

(a) $X_n \xrightarrow{L} 0$
(b) $X_n \xrightarrow{P} 0$
(c) $X_n \xrightarrow{q.m.} 0$
(d) $\sqrt{n}X_n \xrightarrow{L} Z \sim N(0, 1)$

7.7.35 Suppose $\sqrt{n}(X_n - \mu) \xrightarrow{L} X \sim N(0, \sigma^2)$ distribution. Suppose g is a differentiable function with $g'(\mu) \neq 0$ and $U_n = \sqrt{n}(g(X_n) - g(\mu))$. Following are four statements. (I)$U_n \xrightarrow{L} U \sim N(0, g(\sigma^2))$ distribution. (II) $U_n \xrightarrow{L} U \sim N(0, g'(\mu)\sigma^2)$ distribution. (III) $U_n \xrightarrow{L} U \sim N(0, (g'(\mu))^2\sigma^2)$ distribution. (IV) The given information is not sufficient to find the limit law of U_n. Then which of the following statements is true?

(a) Only (I) is true
(b) Only (II) is true
(c) Only (III) is true
(d) Only (IV) is true

7.7.36 Suppose $(\Omega, \mathbb{A}, P)$ is a probability space, where $\Omega = [0, 1]$, $\mathbb{A}$ is a sigma field of subsets of Ω and $P(A_n) = 1/n$ for $A_n = [0, 1/n)$. Suppose the random variable X and X_n, $n \geq 1$ are defined as on this space as follows.

$$X(\omega) = \begin{cases} 0, & \text{if} \quad 0 \leq \omega \leq 1/2 \\ 1, & \text{if} \quad 1/2 < \omega \leq 1 \end{cases} \quad \& \quad X_n(\omega) = \begin{cases} 1, & \text{if} \quad 0 \leq \omega \leq 1/2 \\ 0, & \text{if} \quad 1/2 < \omega \leq 1. \end{cases}$$

Following are two statements. (I) $X_n \xrightarrow{L} X$. (II) $X_n \xrightarrow{P} X$. Then which of the following statements is true?

(a) Both (I) and (II) are false
(b) Both (I) and (II) are true
(c) (I) is true but (II) is false
(d) (I) is false but (II) is true

7.7.37 Suppose $\{X_n, n \geq 1\}$ is Cauchy in r-th mean. Following are two statements. (I) $\{X_n, n \geq 1\}$ is Cauchy in probability. (II) $\{X_n, n \geq 1\}$ converges in law. Then which of the following statements is true?

(a) Both (I) and (II) are false
(b) Both (I) and (II) are true
(c) (I) is true but (II) is false
(d) (I) is false but (II) is true

7.7.38 Suppose $\{X_n, n \geq 1\}$ is Cauchy in r-th mean. Following are two statements. (I) $\{X_n, n \geq 1\}$ is Cauchy in probability. (II) $\{X_n, n \geq 1\}$ converges in probability. Then which of the following statements is true?

(a) Both (I) and (II) are false
(b) Both (I) and (II) are true
(c) (I) is true but (II) is false
(d) (I) is false but (II) is true

7.7.39 Suppose $\{X_n, n \geq 1\}$ is Cauchy in r-th mean. Following are two statements. (I) $\{X_n, n \geq 1\}$ is Cauchy in probability. (II) $\{X_n, n \geq 1\}$ converges almost surely. Then which of the following statements is true?

(a) Both (I) and (II) are false
(b) Both (I) and (II) are true
(c) (I) is true but (II) is false
(d) (I) is false but (II) is true

7.7.40 Suppose $\{X_n, n \geq 1\}$ is a sequence random variables such that $P[X_n = 1] = 1/\log n$ and $P[X_n = 0] = 1 - 1/\log n$. Following are two statements. (I) $X_n \xrightarrow{P} 0$ (II) $X_n \xrightarrow{r} 0$. Then which of the following statements is true?

(a) Both (I) and (II) are false
(b) Both (I) and (II) are true
(c) (I) is true but (II) is false
(d) (I) is false but (II) is true

7.7.41 Suppose $\{X_n, n \geq 1\}$ is a sequence random variables such that $P[X_n = 1] = 1/\log n$ and $P[X_n = 0] = 1 - 1/\log n$. The following are three statements. (I) $X_n \xrightarrow{P} 0$. (II). $X_n \xrightarrow{r} 0$. (III) $X_n \xrightarrow{L} 0$. Then which of the following statements is/are true?

(a) Only (I) is true
(b) Only (II) is true
(c) Only (I) and (II) are true

(d) All three are true

7.7.42 Suppose $\{X_n, n \geq 1\}$ is a sequence random variables such that $P[X_n = 0] = 1/n^2$ and $P[X_n = n] = 1 - 1/n^2$. Following are four statements. (I) $X_n \xrightarrow{L} 0$. (II) $X_n \xrightarrow{r} 0$. (III) $X_n \xrightarrow{P} 0$. (IV) $X_n \xrightarrow{a.s} 0$. Then which of the following statements is true?

(a) Only (I) is true
(b) Only (II) is true
(c) Only (I) and (II) are true
(d) None of the statements is true

7.7.43 Suppose $\{X_n, n \geq 1\}$ is a sequence random variables such that $P[X_n = 0] = 1 - 1/n^3$ and $P[X_n = n] = 1/n^3$. Following are four statements. (I) $X_n \xrightarrow{P} 0$. (II) $X_n \xrightarrow{q.m.} 0$. (III) $X_n \xrightarrow{L} 0$. (IV) $X_n \xrightarrow{a.s.} 0$. Then which of the following statements is true?

(a) Only (I) is true
(b) Only (II) is true
(c) Only (I), (II) and (III) are true
(d) All four are true

7.7.44 Suppose $\{X_n, n \geq 1\}$ is a sequence of random variables such that X_n follows Cauchy distribution with probability density function $f(x, n) = (n/\pi)(1/(n^2 + x^2))$, $x \in \mathbb{R}$. Following are three statements. (I) $X_n \xrightarrow{L} 0$. (II) $X_n \xrightarrow{P} 0$. (III) $X_n \xrightarrow{a.s} 0$. Then which of the following statements is true?

(a) Only (I) is false
(b) Only (II) is false
(c) Only (I) and (II) are false
(d) All the statements are false

7.7.45 Suppose $\{X_n, n \geq 1\}$ is a sequence random variables such that $P[X_n = \pm n] = 1/(2n^\alpha)$ and $P[X_n = 0] = 1 - 1/n^\alpha$, $0 < \alpha \leq 2$. Following are two statements. (I) $X_n \xrightarrow{P} 0$ (II) $X_n \xrightarrow{q.m.} 0$. Then which of the following statements is true?

(a) Both (I) and (II) are false
(b) Both (I) and (II) are true
(c) (I) is true but (II) is false
(d) (I) is false but (II) is true

7.7.46 Suppose $\{X_n, n \geq 1\}$ is a sequence of independent and identically distributed random variables, each following uniform $U(2, 6)$ distribution. Following are four statements. (I) $X_{(n)} \xrightarrow{P} 6$ (II) $X_{(1)} \xrightarrow{P} 2$. (III) $X_{(1)} \xrightarrow{L} 2$.

(IV) $M_n \xrightarrow{P} 4$, where M_n is the median of $\{X_1, X_2, \cdots, X_n\}$. Then which of the following statements is true?

(a) Only (I) and (II) are true
(b) Only (I) and (IV) are true
(c) Only (I), (II) and (III) are true
(d) All four are true

7.7.47 Suppose $\{X_n, n \geq 1\}$ is a sequence of random variables with $\phi_n(t)$, $t \in \mathbb{R}$, being a characteristic function of $X_n, n \geq 1$. Suppose X is a random variable with $\phi(t)$, $t \in \mathbb{R}$ as its characteristic function. Then which of the following statements is/are true?

(a) $X_n \xrightarrow{P} X \Rightarrow \phi_n(t) \to \phi(t) \;\forall\; t \in \mathbb{R}$
(b) $X_n \xrightarrow{q.m.} X \Rightarrow \phi_n(t) \to \phi(t) \;\forall\; t \in \mathbb{R}$
(c) $X_n \xrightarrow{a.s.} X \Rightarrow \phi_n(t) \to \phi(t) \;\forall\; t \in \mathbb{R}$
(d) $X_n \xrightarrow{L} X \Rightarrow \phi_n(t) \to \phi(t) \;\forall\; t \in \mathbb{R}$

8

Convergence of a Sequence of Expectations

8.1 Introduction

In Chapters 6 and 7, we studied various modes of convergence of a sequence of random variables. It is of interest to examine which of these modes imply convergence of a sequence of expectations of random variables. We have discussed some examples and theorems, where we have the convergence of a sequence of expectations corresponding to two modes of convergence. These are listed below.

(i) In Example 7.2.10, we have noted that convergence in probability does not in general imply convergence of the corresponding moments. But if the random variables are uniformly bounded, then it is proved in Theorem 7.2.3 that

$$X_n \xrightarrow{P} C \Rightarrow E(X_n) \to C \;\;\&\;\; X_n \xrightarrow{r} C, \;\; C \in \mathbb{R}.$$

(ii) Theorem 7.4.4 proves that $X_n \xrightarrow{r} X \Rightarrow E(|X_n|^r) \to E(|X|^r)$ for $r > 0$.

(iii) In Theorem 7.4.2, it is proved that if X_n's are almost surely bounded, then

$$X_n \xrightarrow{P} X \Rightarrow X_n \xrightarrow{r} X \Rightarrow E(|X_n|^r) \to E(|X|^r), \;\; r > 0.$$

In the present chapter, we discuss the following three results in which we have convergence of sequence of expectations of random variables, corresponding to pointwise convergence, almost sure convergence and convergence in probability.

(i) If $\{X_n, n \geq 1\}$ is a non-decreasing sequence of non-negative random variables such that $X_n \to X$ on Ω then $E(X_n) \to E(X)$.

(ii) If $|X_n| \leq Y$ such that $E(Y) < \infty$ and if $X_n \xrightarrow{a.s.} X$ then $E(X_n) \to E(X)$.

(iii) If $|X_n| \leq Y$ a.s. such that $E(Y) < \infty$ and if $X_n \xrightarrow{P} X$ then $E(X_n) \to E(X)$.

DOI: 10.1201/9781032619057-8

The first result is known as the monotone convergence theorem. We prove in the next section that the result is true for a monotone sequence converging to X on Ω. In literature there is some confusion in labeling the results (ii) and (iii). Many authors label result (ii) as Lebesgue dominated convergence theorem or dominated convergence theorem, for example refer to Athreya and Lahiri [3], Billingsley [5], Gut [13] and Shao [21]. In result (iii) almost sure convergence from (ii) is replaced by convergence in probability. This result is also referred to as the dominated convergence theorem, for example refer to Bhat [4] and Loeve [16]. To avoid the confusion, we label result (ii) as the Lebesgue dominated almost sure convergence theorem and result (iii) as the Lebesgue dominated probability convergence theorem.

We begin with the monotone convergence theorem in the next section and using it prove the Lebesgue dominated almost sure convergence theorem. The Lebesgue dominated probability convergence theorem is discussed in Section 3.

8.2 Monotone Convergence Theorem

In Chapter 4 we have defined the expectation of a random variable as an integral with respect to a probability measure. The expectation of a non-negative random variable X is defined as $E(X) = \lim_{n\to\infty} E(X_n)$ where $\{X_n, n \geq 1\}$ is a non-decreasing sequence of non-negative simple random variables such that $X_n \to X$ on Ω. The definition of expectation of a non-negative random variable X can be expressed as $E(X) = E(\lim_{n\to\infty} X_n) = \lim_{n\to\infty} E(X_n)$. It is to be noted that here limit and expectation are interchanged. It is of interest to see in which other cases we can interchange limit and expectation. In this chapter we study three theorems, which specify the conditions under which we can interchange limit and expectation.

Following lemma is heavily used in the proof of the monotone convergence theorem.

Lemma 8.2.1. *Suppose $U_1, U_2, \cdots, U_k$ are simple random variables defined on a probability space $(\Omega, \mathbb{A}, P)$. Then $W_k = \max\{U_1, U_2, \cdots, U_k\}$ is also a simple random variables defined on the probability space $(\Omega, \mathbb{A}, P)$.*

Proof. Suppose $U_1 = \sum_{i=1}^{m} a_i I_{A_i}$ and $U_2 = \sum_{j=1}^{n} b_j I_{B_j}$ where $\{A_1, A_2, \cdots, A_m\}$ and $\{B_1, B_2, \cdots, B_n\}$ form measurable partitions of Ω and $\{a_1, a_2, \cdots, a_m\}$ and $\{b_1, b_2, \cdots, b_n\}$ are real numbers. Now for

$$\omega \in C_{ij} = A_i \cap B_j, W_2(\omega) = \max\{U_1(\omega), U_2(\omega)\} = \max\{a_i, b_j\} = c_{ij}, \quad \text{say.}$$

Further, when $\{A_1, A_2, \cdots, A_m\}$ and $\{B_1, B_2, \cdots, B_n\}$ are measurable partitions of Ω, as shown in Theorem 2.4.1, $\{C_{ij}, i = 1, 2, \cdots, m,\ j = 1, 2, \cdots, n\}$ is also a measurable partition of Ω. Thus W_2 can be expressed as

$W_2 = \sum_{i=1}^{m}\sum_{j=1}^{n} c_{ij} I_{C_{ij}}$ and hence W_2 is a simple random variable. Thus, the lemma is true for $k = 2$. Suppose $k = 3$, then

$$W_3 = \max\{U_1, U_2, U_3\} = \max\{\max\{U_1, U_2\}, U_3\} = \max\{W_2, U_3\}$$

is again a simple random variable. Continuing in this manner, it follows that W_k is a simple random variable. □

In Theorem 2.4.3 we have proved that, if X is a non-negative random variable then there exists a non-decreasing sequence $\{X_n, n \geq 1\}$ of non-negative simple random variables, such that $X_n \to X$ on Ω. Expectation $E(X)$ of X is then defined as $\lim_{n\to\infty} E(X_n)$. Thus, if $X_n \to X$ on Ω then $E(X_n) \to E(X)$. Monotone convergence theorem, proved below states that $X_n \to X$ on Ω implies $E(X_n) \to E(X)$, even if $\{X_n, n \geq 1\}$ is not a sequence of simple random variables but it has to be a monotone sequence. In the proof, we use Theorem 2.4.3 and the definition of expectation of a non-negative random variable as given above. We prove it in four steps as follows: $\{X_n, n \geq 1\}$ is such that (i) $0 \leq X_n \uparrow X$, (ii) $Y \leq X_n \uparrow X$ with $E(Y) < \infty$, (iii) $0 \geq X_n \downarrow X$ and (iv) $Y \geq X_n \downarrow X$ with $E(Y) < \infty$. The first case is traditionally known as the monotone convergence theorem and the other cases are given as the corollaries to the monotone convergence theorem. We begin with case (i).

Theorem 8.2.1. *Monotone convergence theorem for non-decreasing sequence of non-negative random variables: Suppose $\{X_n, n \geq 1\}$ is a non-decreasing sequence of non-negative random variables defined on a probability space $(\Omega, \mathbb{A}, P)$. If $X_n \uparrow X$ on Ω, then $E(X_n) \uparrow E(X)$.*

Proof. From the properties of expectation, $\forall\ n \geq 1$

$$X_n \geq 0 \Rightarrow E(X_n) \geq 0 \quad \& \quad X_n \leq X_{n+1} \Rightarrow E(X_n) \leq E(X_{n+1}).$$

Thus, $\{E(X_n), n \geq 1\}$ is a non-decreasing sequence of non-negative real numbers and hence it has a limit, which may be finite or infinite. Now $X_k \geq 0$, hence by Theorem 2.4.3, corresponding to each X_k there exists a non-decreasing sequence $\{X_{kn}, n \geq 1\}$ of non-negative simple random variables, such that $X_{kn} \to X_k$ on Ω as $n \to \infty$. Thus, we have

$$\begin{aligned}
&0 \leq X_{11} \leq X_{12} \leq X_{13} \leq \cdots \to X_1\\
&0 \leq X_{21} \leq X_{22} \leq X_{23} \leq \cdots \to X_2\\
&0 \leq X_{31} \leq X_{32} \leq X_{33} \leq \cdots \to X_3\\
&\vdots \qquad\qquad\qquad\qquad\qquad\qquad \vdots\\
&0 \leq X_{n1} \leq X_{n2} \leq X_{n3} \leq \cdots \to X_n.
\end{aligned} \tag{8.1}$$

We define

$$\begin{aligned} Y_1 &= \max\{X_{k1}, k \leq 1\} = X_{11} \\ Y_2 &= \max\{X_{k2}, k \leq 2\} = \max\{X_{12}, X_{22}\} = \max\{X_{11}, X_{12}, X_{21}, X_{22}\} \\ &= \max_{i\leq 2, j\leq 2}\{X_{ij}\}, \quad \text{since } \ X_{11} \leq X_{12} \ \ \& \ \ X_{21} \leq X_{22}. \end{aligned}$$

Continuing in this manner we define

$$Y_n = \max\{X_{kn}, k \leq n\} = \max_{i\leq n, j\leq n}\{X_{ij}\},$$

which follows from the array presented in (8.1). In Lemma 8.2.1 we have proved that the maximum of simple random variables is again a simple random variable, hence Y_n is a simple random variable. Further, $0 \leq Y_n \leq Y_{n+1} \ \forall \ n \geq 1$. Thus, $\{Y_n, n \geq 1\}$ is a non-decreasing sequence of non-negative simple random variables. To find its limit, from the definitions of Y_n and X_{kn} we note that,

$$X_{kn} \leq Y_n \leq X_n \text{ on } \Omega \ \ \Rightarrow \ E(X_{kn}) \leq E(Y_n) \leq E(X_n). \tag{8.2}$$

Allowing $n \to \infty$ in (8.2) we get,

$$X_k \leq \lim_{n\to\infty} Y_n \leq X \text{ on } \Omega \ \& \ \lim_{n\to\infty} E(X_{kn}) = E(X_k) \leq \lim_{n\to\infty} E(Y_n) \leq \lim_{n\to\infty} E(X_n), \tag{8.3}$$

as $\{X_{kn}, n \geq 1\}$ is a sequence of simple random variables converging to X_k. Allowing $k \to \infty$ in (8.3) we get,

$$X \leq \lim_{n\to\infty} Y_n \leq X \text{ on } \Omega \ \ \& \ \ \lim_{k\to\infty} E(X_k) \leq \lim_{n\to\infty} E(Y_n) \leq \lim_{n\to\infty} E(X_n). \tag{8.4}$$

Equation (8.4) further implies that,

$$\lim_{n\to\infty} Y_n = X \ \text{ on } \Omega \ \ \& \ \ \lim_{n\to\infty} E(Y_n) = \lim_{n\to\infty} E(X_n). \tag{8.5}$$

Thus, $\{Y_n, n \geq 1\}$ is a non-decreasing sequence of non-negative simple random variables converging to $X \geq 0$ on Ω. Hence, by the definition of expectation of a non-negative random variable,

$$\begin{aligned} E(X) &= \lim_{n\to\infty} E(Y_n) = \lim_{n\to\infty} E(X_n) \quad \text{by Equation (8.5)} \\ \iff \quad E(\lim_{n\to\infty} X_n) &= \lim_{n\to\infty} E(X_n). \end{aligned}$$

□

Following are some corollaries to Theorem 8.2.1.

Corollary 8.2.1. *Suppose $Y \leq X_n \uparrow X$ on Ω and Y is an integrable random variable. Then $E(Y) \leq E(X_n) \uparrow E(X)$.*

Proof. Suppose $Z_n = X_n - Y$ then $0 \leq Z_n \uparrow X - Y$ on Ω. Hence by Theorem 8.2.1, $E(Z_n) = E(X_n) - E(Y) \uparrow E(X) - E(Y)$. It is given that Y is an integrable random variable, hence $E(Y) < \infty$. Thus, adding $E(Y)$ on both sides we get $E(X_n) \uparrow E(X)$. □

Remark 8.2.1. Corollary 8.2.1 conveys that X_n's need not be non-negative but the sequence $\{X_n, n \geq 1\}$ must be non-decreasing and bounded from below by an integrable random variable.

Corollary 8.2.2. *Suppose* $0 \geq X_n \downarrow X$ *on* Ω, *then* $E(X_n) \downarrow E(X)$.

Proof. Since $0 \geq X_n \downarrow X$, $0 \leq -X_n \uparrow -X$ on Ω. Hence by Theorem 8.2.1, $E(-X_n) = -E(X_n) \uparrow E(-X) = -E(X)$. Thus, $E(X_n) \downarrow E(X)$. □

Corollary 8.2.3. *Suppose* $Y \geq X_n \downarrow X$ *on* Ω *and* Y *is an integrable random variable. Then* $E(X_n) \downarrow E(X)$.

Proof. Suppose $Z_n = X_n - Y$ then $0 \geq Z_n \downarrow X - Y$ on Ω. Hence by Corollary 8.2.2, $E(Z_n) = E(X_n) - E(Y) \downarrow E(X) - E(Y)$, which implies that $E(X_n) \downarrow E(X)$ since $E(Y) < \infty$. □

Remark 8.2.2. Corollary 8.2.3 conveys that X_n's need not be non-positive but the sequence $\{X_n, n \geq 1\}$ must be non-increasing and bounded from above by an integrable random variable.

Remark 8.2.3. Theorem 8.2.1 and Corollaries 8.2.1, 8.2.2 and 8.2.3 convey that if $\{X_n, n \geq 1\}$ is non-decreasing and bounded from below or non-increasing and bounded from above, by an integrable random variable, then $\lim E(X_n) = E(\lim X_n)$. Hence, we may state the monotone convergence theorem as follows. The proof follows by combining the proofs of Theorem 8.2.1 and Corollaries 8.2.1, 8.2.2 and 8.2.3.

Theorem 8.2.2. *Monotone convergence theorem: Suppose* $\{X_n, n \geq 1\}$ *is a non-decreasing sequence bounded from below by an integrable random variable or a non-increasing sequence bounded from above by an integrable random variable. If* $X_n \to X$ *on* Ω, *then* $\lim_{n\to\infty} E(X_n) = E(\lim_{n\to\infty} X_n)$.

In Chapter 4, we have discussed the linearity property of the expectation, which states that expectation of a finite sum of random variables is the sum of their expectations. In many applications we would like to have such a property for a countably infinite sum of random variables. The following two corollaries convey that it is possible, using the monotone convergence theorem. Thus, if we have a monotone sequence of random variables, then summation and expectation can be interchanged. In other words, "E" operation is sigma additive on the family of such random variables.

Corollary 8.2.4. *Suppose* $\{X_n, n \geq 1\}$ *is a sequence of non-negative random variables then* $E(\sum_{n\geq 1} X_n) = \sum_{n\geq 1} E(X_n)$.

Proof. Suppose $Z_n = \sum_{i=1}^n X_i$. Then $0 \le Z_n \uparrow \sum_{i\ge1} X_i = X$, say on Ω. It is to be noted that $\lim Z_n = \sum_{i\ge1} X_i$ is well defined, it may be infinite for some ω. Hence by the monotone convergence theorem, $\lim E(Z_n) = E(\lim Z_n)$. Now

$$\begin{aligned} \lim_{n\to\infty} E(Z_n) &= \lim_{n\to\infty} E\Big(\sum_{i=1}^n X_i\Big) = \lim_{n\to\infty} \sum_{i=1}^n E(X_i) = \sum_{i\ge1} E(X_i) \\ \& \quad E(\lim_{n\to\infty} Z_n) &= E\Big(\sum_{i\ge1} X_i\Big) \\ \text{Hence,} \quad \lim_{n\to\infty} E(Z_n) &= E(\lim_{n\to\infty} Z_n) \;\Rightarrow\; E\Big(\sum_{n\ge1} X_n\Big) = \sum_{n\ge1} E(X_n). \end{aligned}$$

□

Corollary 8.2.5. *Suppose $\{X_n, n \ge 1\}$ is a sequence of non-positive random variables then $E(\sum_{n\ge1} X_n) = \sum_{n\ge1} E(X_n)$.*

Proof. Suppose $Y_n = -X_n$. Then $\{Y_n, n \ge 1\}$ is a sequence of non-negative random variables. Hence by Corollary 8.2.4,

$$\begin{aligned} E\Big(\sum_{n\ge1} Y_n\Big) = \sum_{n\ge1} E(Y_n) \quad &\iff \quad E\Big(\sum_{n\ge1}(-X_n)\Big) = \sum_{n\ge1} E(-X_n) \\ &\iff \quad E\Big(\sum_{n\ge1} X_n\Big) = \sum_{n\ge1} E(X_n). \end{aligned}$$

□

Example 8.2.1. Suppose $(\Omega, \mathbb{A}, P)$ is a probability space, where $\Omega = [0,1]$, $\mathbb{A}$ is a sigma field of subsets of Ω, $A_n = [0, 1/n)$ and $X_n = I_{A_n}, n \ge 1$. In Example 6.3.7, we have noted that $\{A_n, n \ge 1\}$ is a decreasing sequence of sets and hence is convergent with $\lim_{n\to\infty} A_n = \bigcap_{n\ge1} A_n = \emptyset$. By Theorem 2.3.9 $I_{A_n} \downarrow I_\emptyset$ on $\Omega \iff X_n \downarrow X \equiv 0$ on Ω. Note that by Corollary 8.2.3,

$$1 \ge I_{A_n} \downarrow I_\emptyset \text{ on } \Omega \;\Rightarrow\; E(I_{A_n}) \downarrow E(I_\emptyset) \Rightarrow P(A_n) \downarrow 0,$$

which also follows from the continuity theorem for a probability measure. □

In the following theorem, we prove an interesting result which states that a random variable X is integrable if and only if $\sum_{n\ge1} P[|X| \ge n] < \infty$. In Theorem 4.3.5, it is proved that for a discrete random variable X with support as the set W of whole numbers $E(X) = \sum_{n\ge1} P[X \ge n]$. We use Theorem 4.3.5 in the proof of the theorem. We also need the following lemma in the proof. The proof is based on the Corollary 8.2.4.

Lemma 8.2.2. *Suppose $(\Omega, \mathbb{A}, P)$ is a probability space and $\{A_n, n \ge 1\}$ is a measurable partition of Ω, then for $\mathbb{A}$ measurable random variable X*

$$E(X) = \int_\Omega X\, dP = \sum_{n\ge1} \int_{A_n} X\, dP.$$

Proof. Case (i): Suppose $X \geq 0$. By the definition of expectation,

$$\begin{aligned}
E(X) &= \int_\Omega X\, dP = \int_\Omega X\, (I_{\sum_{n\geq 1} A_n}) dP = \int_\Omega \sum_{n\geq 1} X I_{A_n} dP = E\Big(\sum_{n\geq 1} X I_{A_n}\Big) \\
&= \sum_{n\geq 1} E(X I_{A_n}) \text{ by Corollary 8.2.4} \\
&= \sum_{n\geq 1} \int_\Omega X I_{A_n} dP = \sum_{n\geq 1} \int_{A_n} X dP.
\end{aligned}$$

(ii) Suppose $X \leq 0 \Rightarrow Y = -X \geq 0$. Hence by case(i)

$$\begin{aligned}
E(Y) = \sum_{n\geq 1} \int_{A_n} Y dP \quad &\Longleftrightarrow \quad E(-X) = \sum_{n\geq 1} \int_{A_n} (-X) dP \\
&\Longleftrightarrow \quad E(X) = \sum_{n\geq 1} \int_{A_n} X dP.
\end{aligned}$$

□

Theorem 8.2.3. *For any random variable X defined on $(\Omega, \mathbb{A}, P)$*

$$\sum_{n\geq 1} P[|X| \geq n] \;\leq\; E(|X|) \;\leq\; 1 + \sum_{n\geq 1} P[|X| \geq n]$$

and X is integrable random variable if and only if $\sum_{n\geq 1} P[|X| \geq n] < \infty$.

Proof. Since X is $\mathbb{A}$ measurable, $|X|$ is also $\mathbb{A}$ measurable random variable. Suppose an event A_n is defined as $A_n = [n \leq |X| < n+1] = |X|^{-1}(S_n) \in \mathbb{A}$, where $S_n = [n, n+1), n \geq 0$. For $n \geq 0$, $S_n = [n, n+1)$ are disjoint sets and hence by Result (v) of Lemma 2.2.1, it follows that $A_n = |X|^{-1}(S_n), n \geq 0$ are disjoint events. Further by Result (iii) of Lemma 2.2.1,

$$\begin{aligned}
\bigcup_{n\geq 0} S_n = \mathbb{R}^+ \quad &\Rightarrow \quad |X|^{-1}\Big(\bigcup_{n\geq 0} S_n\Big) = |X|^{-1}(\mathbb{R}^+) \Rightarrow \bigcup_{n\geq 0} |X|^{-1}(S_n) = |X|^{-1}(\mathbb{R}^+) \\
&\Rightarrow \quad \bigcup_{n\geq 0} A_n = \sum_{n\geq 0} A_n = \Omega.
\end{aligned}$$

Thus, $\{A_n, n \geq 0\}$ is a measurable partition of Ω. Hence, by Lemma 8.2.2, $E(|X|) = \sum_{n\geq 0} \int_{A_n} |X|\, dP$. Observe that on A_n

$$\begin{aligned}
n \;&\leq\; |X| \;<\; n+1 \\
\Rightarrow \int_{A_n} n\, dP \;&\leq\; \int_{A_n} |X|\, dP \;<\; \int_{A_n} (n+1)\, dP \\
\Rightarrow nP(A_n) \;&\leq\; \int_{A_n} |X|\, dP \;<\; (n+1)P(A_n).
\end{aligned}$$

Hence,

$$\begin{aligned}
\sum_{n\geq 0} nP(A_n) &\leq \sum_{n\geq 0}\int_{A_n} |X|\, dP < \sum_{n\geq 0}(n+1)P(A_n) \\
\Rightarrow \sum_{n\geq 0} nP(A_n) &\leq E(|X|) < \sum_{n\geq 0} P(A_n) + \sum_{n\geq 0} nP(A_n) \\
\Rightarrow \sum_{n\geq 1} nP(A_n) &\leq E(|X|) < P\Big(\sum_{n\geq 0} A_n\Big) + \sum_{n\geq 1} nP(A_n) \\
\Rightarrow \sum_{n\geq 1} nP(A_n) &\leq E(|X|) < 1 + \sum_{n\geq 1} nP(A_n). \qquad (8.6)
\end{aligned}$$

To prove the theorem, we need to prove that $\sum_{n\geq 1} nP(A_n) = \sum_{n\geq 1} P[|X| \geq n]$. To prove it, suppose an elementary random variable Y is defined as $Y = \sum_{n\geq 0} nI_{A_n} = \sum_{n\geq 1} nI_{A_n}$. Then by Corollary 8.2.4

$$\begin{aligned}
E(Y) &= E\Big(\sum_{n\geq 1} nI_{A_n}\Big) = \sum_{n\geq 1} E(nI_{A_n}) \\
&= \sum_{n\geq 1} nP(A_n) = \sum_{n\geq 1} nP[Y = n] = \sum_{n\geq 1} P[Y \geq n],
\end{aligned}$$

by Theorem 4.3.5, since Y is a discrete random variable with support as the set W of whole numbers. Now, $Y \geq n$ if $\omega \in A_k$ for any one $k \geq n$. Hence,

$$\begin{aligned}
P[Y \geq n] &= P\Big(\sum_{k=n}^{\infty} A_k\Big) = P\Big(\sum_{k=n}^{\infty}[k \leq |X| < k+1]\Big) = P[|X| \geq n] \\
\Rightarrow E(Y) &= \sum_{n\geq 1} nP(A_n) = \sum_{n\geq 1} P[Y \geq n] = \sum_{n\geq 1} P[|X| \geq n]. \qquad (8.7)
\end{aligned}$$

Now using Equation (8.7) in Equation (8.6), we get

$$\begin{aligned}
\sum_{n\geq 1} nP(A_n) &\leq E(|X|) < 1 + \sum_{n\geq 1} nP(A_n) \\
\Longleftrightarrow \sum_{n\geq 1} P[|X| \geq n] &\leq E(|X|) \leq 1 + \sum_{n\geq 1} P[|X| \geq n].
\end{aligned}$$

From the above inequality it follows that X is integrable if and only if $\sum_{n\geq 1} P[|X| \geq n] < \infty$. □

We use this result in the next chapter in Khintchine's weak laws of large numbers and Kolmogorov's strong laws of large numbers.

Following lemma, known as Fatou's lemma, is based on the monotone convergence theorem and is heavily used in the proof of the Lebesgue dominated almost sure convergence theorem. It is also known as Fatou-Lebesgue theorem, (Loeve [16]).

Lemma 8.2.3. *Fatou's lemma: Suppose $\{X_n, n \geq 1\}$ is a sequence of random variables defined on a probability space $(\Omega, \mathbb{A}, P)$. If Y and Z are integrable random variables, then*

(i) $Y \leq X_n \;\Rightarrow\; E(\liminf X_n) \leq \liminf E(X_n)$.

(ii) $Z \geq X_n \;\Rightarrow\; E(\limsup X_n) \geq \limsup E(X_n)$.

(iii) $Y \leq X_n \leq Z \;\&\; X_n \stackrel{a.s.}{\to} X \;\Rightarrow\; \lim_{n\to\infty} E(X_n) = E(X)$.

Proof.

(i) Note that on Ω

$$\begin{aligned} Y \leq X_n &\Rightarrow \text{the sequence } \{X_n, n \geq 1\} \text{ is bounded from below} \\ &\Rightarrow \liminf X_n < \infty \end{aligned}$$

Similarly $Y \leq X_n \;\Rightarrow\; E(Y) \leq E(X_n) \;\Rightarrow\; \liminf E(X_n) < \infty$.

Suppose $Y_n = \inf_{k\geq n} X_k$, then $\{Y_n, n \geq 1\}$ is a non-decreasing sequence of random variables and

$$\lim_{n\to\infty} Y_n = \lim_{n\to\infty}(\inf_{k\geq n} X_k) = \sup_{n\geq 1}(\inf_{k\geq n} X_k) = \liminf X_n.$$

Case (a): Suppose $Y_n \geq 0$. Then by the monotone convergence theorem,

$$0 \leq Y_n \uparrow \liminf X_n \;\Rightarrow\; E(Y_n) \uparrow E(\liminf X_n).$$

Observe that,

$$\begin{aligned} X_n \geq Y_n &\Rightarrow E(X_n) \geq E(Y_n) \\ &\Rightarrow \liminf E(X_n) \geq \liminf E(Y_n) = \lim_{n\to\infty} E(Y_n) = E(\liminf X_n) \\ &\Rightarrow E(\liminf X_n) \leq \liminf E(X_n). \end{aligned}$$

Case(b): Suppose Y_n's are arbitrary. If $Z_n = X_n - Y$, then $Z_n \geq 0$. Suppose $U_n = \inf_{k\geq n} Z_k$, then $\{U_n, n \geq 1\}$ is a non-decreasing sequence of non-negative random variables. Hence,

$$\lim_{n\to\infty} U_n = \lim_{n\to\infty}(\inf_{k\geq n} Z_k) = \sup_{n\geq 1}(\inf_{k\geq n} Z_k) = \liminf Z_n.$$

By the monotone convergence theorem,

$$0 \leq U_n \uparrow \liminf Z_n \;\Rightarrow\; E(U_n) \uparrow E(\liminf Z_n).$$

Observe that,

$$\begin{aligned} Z_n \geq U_n &\Rightarrow E(Z_n) \geq E(U_n) \\ &\Rightarrow \liminf E(Z_n) \geq \liminf E(U_n) = \lim_{n\to\infty} E(U_n) = E(\liminf Z_n) \\ &\Rightarrow E(\liminf Z_n) \leq \liminf E(Z_n) \\ &\iff E(\liminf(X_n - Y)) \leq \liminf E(X_n - Y) \\ &\iff E(\liminf X_n) \leq \liminf E(X_n) \quad \text{since} \quad E(Y) < \infty. \end{aligned}$$

(ii) It is given that on Ω,

$$\begin{aligned} X_n \leq Z &\Rightarrow \text{the sequence } \{X_n, n \geq 1\} \text{ is bounded from above} \\ &\Rightarrow \limsup X_n < \infty. \end{aligned}$$

Further, $X_n \leq Z \Rightarrow E(X_n) \leq E(Z) \Rightarrow \limsup E(X_n) < \infty.$

Now $X_n \leq Z \Rightarrow V_n = Z - X_n \geq 0$, hence by part (i) we get,

$$\begin{aligned} E(\liminf V_n) &\leq \liminf E(V_n) \\ \iff E(\liminf(Z - X_n)) &\leq \liminf E(Z - X_n) \\ \iff E(Z) - E(\limsup X_n) &\leq E(Z) - \limsup E(X_n) \\ \iff E(\limsup X_n) &\geq \limsup E(X_n) \quad \text{as} \quad E(Z) < \infty. \end{aligned}$$

(iii) Note that

$$\begin{aligned} X_n \overset{a.s.}{\to} X &\Rightarrow \liminf X_n = \limsup X_n = X \text{ a.s.} \\ &\Rightarrow E(\liminf X_n) = E(\limsup X_n) = E(X). \end{aligned}$$

It is given that, $Y \leq X_n \leq Z$ hence, $E(Y) \leq E(X_n) \leq E(Z)$. Thus, $\liminf E(X_n)$ and $\limsup E(X_n)$ both are finite. Using part (i) and part (ii) we get,

$$\begin{aligned} E(\liminf X_n) &\leq \liminf E(X_n) \leq \limsup E(X_n) \leq E(\limsup X_n) \\ \iff E(X) &\leq \liminf E(X_n) \leq \limsup E(X_n) \leq E(X) \\ &\Rightarrow \liminf E(X_n) = \limsup E(X_n) = E(X) \\ &\Rightarrow \lim_{n\to\infty} E(X_n) = E(X). \end{aligned}$$

□

Following theorem is an immediate consequence of Fatou's Lemma. As mentioned in Section 1, we label it as the Lebesgue dominated almost sure convergence theorem.

Theorem 8.2.4. *Lebesgue dominated almost sure convergence theorem: Suppose $\{X_n, n \geq 1\}$ is a sequence of random variables defined on a probability space $(\Omega, \mathbb{A}, P)$. If Y is an integrable random variable such that $|X_n| \leq Y$ a.s. and if $X_n \overset{a.s.}{\to} X$, then $\lim_{n\to\infty} E(X_n) = E(X)$.*

Proof. Note that $|X_n| \leq Y$ a.s. $\Rightarrow -Y \leq X_n \leq Y$ a.s. where Y is an integrable random variable. Further, $X_n \overset{a.s.}{\to} X$ and hence by (iii) of Fatou's lemma, we conclude that $\lim_{n\to\infty} E(X_n) = E(X)$. □

In particular if Y is degenerate at constant C, that is, if the random variables X_n's are uniformly bounded, the Lebesgue dominated almost sure convergence theorem is called as the Lebesgue bounded convergence theorem, refer to Billingsley [5].

Example 8.2.2. Suppose $\{A_n, n \geq 1\}$ is a sequence of events in $\mathbb{A}$ and suppose $X_n = I_{A_n}$. Then $|I_{A_n}| \leq Y \equiv 1$. By Fatou's Lemma and Lemma 2.3.2,

$$\begin{aligned} E(\liminf X_n) \leq \liminf E(X_n) &\Rightarrow E(\liminf I_{A_n}) \leq \liminf E(I_{A_n}) \\ \Rightarrow E(I_{\liminf A_n}) \leq \liminf E(I_{A_n}) &\Rightarrow P(\liminf A_n) \leq \liminf P(A_n) \ . \end{aligned}$$

Similarly,

$$\begin{aligned} \limsup E(X_n) \leq E(\limsup X_n) &\Rightarrow \limsup E(I_{A_n}) \leq E(\limsup I_{A_n}) \\ \Rightarrow \limsup E(I_{A_n}) \leq E(I_{\limsup A_n}) &\Rightarrow \limsup P(A_n) \leq P(\limsup A_n) \ . \end{aligned}$$

Combining two results we have

$$P(\liminf A_n) \leq \liminf P(A_n) \leq \limsup P(A_n) \leq P(\limsup A_n) \ .$$

The same result is proved in Section 1.5. Further, it is proved in Theorem 2.3.9 that $A_n \to A \Rightarrow I_{A_n} \to I_A$ pointwise and hence almost surely. Moreover, $|I_{A_n}| \leq Y \equiv 1$. Hence by Theorem 8.2.4 we get,

$$\lim_{n\to\infty} E(I_{A_n}) = \lim_{n\to\infty} P(A_n) = E(I_A) = P(A),$$

which is the continuity theorem for a probability measure. In particular, if $A_n \to \emptyset \ \Rightarrow \ P(A_n) \to 0$.

We have defined convergence in quadratic mean and Cauchy convergence in quadratic mean in Section 7.4. We now prove their equivalence using Fatou's lemma. This result is known as $\mathbb{L}_2$ completeness theorem.

Theorem 8.2.5. $\mathbb{L}_2$ *completeness theorem: Suppose a sequence* $\{X_n, n \geq 1\}$ *of random variables and* X *are defined on the same probability space* $(\Omega, \mathbb{A}, P)$ *and* $X, \ X_n \in \mathbb{L}_2$ *for all* $n \geq 1$. *Then* $X_n \stackrel{q.m.}{\to} X$ *as* $n \to \infty$ *if and only if* $X_n - X_m \stackrel{q.m.}{\to} 0$ *as* $m, \ n \to \infty$.

Proof. Suppose $X_n \stackrel{q.m.}{\to} X$. Then by C_r inequality, with $r = 2$

$$\begin{aligned} E(|X_n - X_m|^2) &\leq 2[E(|X_n - X|^2) + E(|X_m - X|^2)] \to 0 \ \text{ as } \ m, \ n \to \infty \\ &\Rightarrow X_n - X_m \stackrel{q.m.}{\to} 0. \end{aligned}$$

Conversely, suppose $X_n - X_m \stackrel{q.m.}{\to} 0$. Now as $m, \ n \to \infty$,

$$X_n - X_m \stackrel{q.m.}{\to} 0 \ \Rightarrow \ X_n - X_m \stackrel{P}{\to} 0 \ \Rightarrow \ X_n \stackrel{P}{\to} X \ \text{ for some } \ X$$

since convergence in quadratic mean implies convergence in probability and Cauchy convergence in probability implies convergence in probability, as proved in Section 7.2. Further, by Theorem 7.2.11, there exists a subsequence $\{X_{n_k}, k \geq 1\}$ of $\{X_n, n \geq 1\}$ such that $X_{n_k} \stackrel{a.s.}{\to} X$ as $k \to \infty$. Hence for fixed

m, $X_m - X_{n_k} \overset{a.s.}{\to} X_m - X$ as $k \to \infty$ and $(X_m - X_{n_k})^2 \overset{a.s.}{\to} (X_m - X)^2$ as $k \to \infty$. Thus,

$$E((X_m - X)^2) = E(\lim_{k\to\infty} (X_m - X_{n_k})^2) = E(\liminf_{k\to\infty}(X_m - X_{n_k})^2). \quad (8.8)$$

Now $0 \le (X_m - X_{n_k})^2$. As a consequence, by the first part of Fatou's lemma,

$$E(\liminf_{k\to\infty}(X_m - X_{n_k})^2) \le \liminf_{k\to\infty} E(X_m - X_{n_k})^2 = \lim_{k\to\infty} E(X_m - X_{n_k})^2 = 0, \quad (8.9)$$

since $X_m - X_{n_k} \overset{q.m.}{\to} 0$ as $m,\ k \to \infty$. Using result in (8.9) in Equation (8.8), we get

$$E(|X_m - X|^2) \quad \to \quad 0 \text{ as } m \to \infty \Rightarrow X_m \overset{q.m.}{\to} X.$$

□

In Example 7.4.10, it is shown that almost sure convergence does not imply convergence of moments. However, using Fatou's lemma we can find a certain relation between $E(|X_n|^r)$ and $E(|X|^r)$ when $X_n \overset{a.s.}{\to} X$, as shown in the following theorem.

Theorem 8.2.6. *If* $X_n \overset{a.s.}{\to} X$ *then* $\forall \ r > 0$, *(i)* $E(|X|^r) \le \liminf_{n\to\infty} E(|X_n|^r)$ *and (ii)* $E(|X|^r) \le \limsup_{n\to\infty} E(|X_n|^r)$.

Proof. $X_n \overset{a.s.}{\to} X \Rightarrow |X_n|^r(\omega) \overset{a.s.}{\to} |X|^r(\omega) \ \forall\, \omega \in N^c$ where $P(N) = 0$.
(i) By Fatou's lemma,

$$\begin{aligned}
0 \le |X_n|^r \quad &\Rightarrow \quad \int_\Omega \liminf_{n\to\infty} |X_n|^r dP \le \liminf_{n\to\infty} \int_\Omega |X_n|^r dP \\
\Rightarrow E(|X|^r) \quad &= \quad \int_\Omega |X|^r dP = \int_{N^c} \lim_{n\to\infty} |X_n|^r dP = \int_{N^c} \liminf_{n\to\infty} |X_n|^r dP \\
&\le \quad \liminf_{n\to\infty} \int_{N^c} |X_n|^r dP = \liminf_{n\to\infty} \int_\Omega |X_n|^r dP = \liminf_{n\to\infty} E(|X_n|^r),
\end{aligned}$$

the last step follows since $P(N) = 0$.
(ii) It follows immediately from (i), since

$$E(|X|^r) \ \le \ \liminf_{n\to\infty} E(|X_n|^r) \ \le \ \limsup_{n\to\infty} E(|X_n|^r).$$

□

In the next section, we prove a theorem, labeled as the Lebesgue dominated probability convergence theorem. In this theorem we prove the result of the Lebesgue dominated almost sure convergence theorem, under a weaker condition, when the almost sure convergence is replaced by the convergence in probability.

8.3 Lebesgue Dominated Probability Convergence Theorem

In Theorem 7.2.3, it is proved that if $X_n \xrightarrow{P} C$ and if X_n's are uniformly bounded, that is $|X_n| < M$, $\forall\ n \geq 1$, then $\lim_{n\to\infty} E(X_n) = E(\lim_{n\to\infty} X_n)$. In the Lebesgue dominated probability convergence theorem, M is replaced by an integrable random variable Y and limit random variable in convergence in probability is a non-degenerate random variable.

In the proof, we heavily use the result that every subsequence of a convergent sequence of real numbers is convergent and has the same limit. We need one more result proved in Theorem 7.2.11, which states that if $X_n \xrightarrow{P} X$, then there exists a subsequence of $\{X_n, n \geq 1\}$ which converges to X almost surely. We also use the Lebesgue dominated almost sure convergence theorem. Since the almost sure convergence is replaced by a weaker condition, the proof is rather involved as compared to that of the Lebesgue dominated almost sure convergence theorem.

Theorem 8.3.1. *Lebesgue dominated probability convergence theorem: Suppose $\{X_n, n \geq 1\}$ is a sequence of random variables defined on $(\Omega, \mathbb{A}, P)$. If Y is an integrable random variables such that $|X_n| \leq Y$ a.s. and if $X_n \xrightarrow{P} X$, then $\lim_{n\to\infty} E(X_n) = E(X)$.*

Proof. It is given that $|X_n| \leq Y$ a.s. $\Rightarrow$ $E(|X_n|) \leq E(Y)$, that is, $-E(Y) \leq E(X_n) \leq E(Y)$. Thus the sequence $\{E(X_n), n \geq 1\}$ is bounded from above and below and hence $\liminf E(X_n)$ and $\limsup E(X_n)$ both are finite. We want to show that both are the same and the common value is $E(X)$.
Case (i) Suppose $X \equiv 0$. By definition of convergence in probability,

$$X_n \xrightarrow{P} X \equiv 0 \quad \iff \quad \forall\ \epsilon > 0, \quad p_{\epsilon n} = P[|X_n| > \epsilon] \to 0 \quad \text{as } n \to \infty.$$

Thus, $\{p_{\epsilon n}, n \geq 1\}$ is a sequence of real numbers in $[0, 1]$ converging to 0, hence every subsequence $\{p_{\epsilon n_k}, k \geq 1\}$ of $\{p_{\epsilon n}, n \geq 1\}$ is convergent and has the same limit 0 for all $\epsilon > 0$. Thus, $X_n \xrightarrow{P} 0$ as $n \to \infty$ implies $X_{n_k} \xrightarrow{P} 0$ as $k \to \infty$ for any subsequence. Suppose we divide all subsequences of $\{X_n, n \geq 1\}$ in two classes $\mathbb{C}_1$ and $\mathbb{C}_2$. $\mathbb{C}_1$ consists of all subsequences $\{X_{n'}, n' \geq 1\}$ such that $E(X_{n'}) \to \liminf E(X_n)$ and $\mathbb{C}_2$ consists of all subsequences $\{X_{n''}, n'' \geq 1\}$ such that $E(X_{n''}) \to \limsup E(X_n)$. It is to be noted that, for every subsequence from $\mathbb{C}_1$, the sequence $\{E(X_{n'}), n' \geq 1\}$ is a convergent sequence and its every subsequence has the same limit as $\liminf E(X_n)$. To identify the value of the limit, consider a subsequence $\{X_{n'_k}, n'_k \geq 1\}$ of $\{X_{n'}, n' \geq 1\}$, such that $X_{n'} \xrightarrow{P} 0$ and $X_{n'_k} \xrightarrow{a.s.} 0$, which exists in view of Theorem 7.2.11.

Now,
$$\begin{aligned} |X_n| \leq Y \text{ a.s.} \quad &\Rightarrow \quad |X_{n'_k}| \leq Y \text{ a.s.} \\ &\Rightarrow \quad \lim_{n\to\infty} E(X_{n'_k}) = 0 \text{ by Theorem 8.2.4} \\ &\Rightarrow \quad \lim_{n\to\infty} E(X_{n'}) = \liminf E(X_n) = 0, \end{aligned}$$

as a subsequence $\{E(X_{n'_k}), n'_k \geq 1\}$ of $\{E(X_{n'}), n' \geq 1\}$ has limit 0 and it implies that $\liminf E(X_n) = 0$. Using similar arguments, it can be shown that $\limsup E(X_n) = 0$. Hence, $\lim_{n\to\infty} E(X_n)$ exists and is 0. Since $X \equiv 0$, $E(X) = 0$. Thus, $\lim_{n\to\infty} E(X_n) = E(X)$.
Case (ii) Suppose $X \neq 0$. If $Z_n = (X_n - X)$ then $X_n \xrightarrow{P} X \Rightarrow Z_n = (X_n - X) \xrightarrow{P} 0$. If Z_n is bounded by an integrable random variable, then by case (i) we get the required result. To prove it, observe that,

$$\begin{aligned} |X_n| \leq Y \text{ a.s.} \quad &\Rightarrow \quad |X_n(\omega)| \leq Y(\omega) \ \forall \ \omega \in N_1^c \ \& \ P(N_1) = 0 \\ &\Rightarrow \quad Z_n(\omega) = |X_n(\omega) - X(\omega)| \leq |X_n(\omega)| + |X(\omega)| \\ &\leq \quad Y(\omega) + |X(\omega)| \ \forall \ \omega \in N_1^c. \end{aligned}$$

To examine whether X is bounded by an integrable random variable, we again use Theorem 7.2.11. Thus,

$$\begin{aligned} X_n \xrightarrow{P} X &\Rightarrow \exists \text{ a subsequence } \{X_{n_k}, k \geq 1\} \text{ such that } X_{n_k} \xrightarrow{a.s.} X \\ &\Rightarrow |X(\omega)| = \lim_{n_k\to\infty} |X_{n_k}(\omega)| \ \forall \ \omega \in N_2^c \ \& \ P(N_2) = 0 \\ &\Rightarrow |X(\omega)| = \lim_{n_k\to\infty} |X_{n_k}(\omega)| \leq Y(\omega) \ \forall \ \omega \in N^c = N_1^c \cap N_2^c \\ &\Rightarrow |Z_n(\omega)| = |X_n(\omega) - X(\omega)| \leq Y(\omega) + Y(\omega) = 2Y(\omega) \ \forall \ \omega \in N^c. \end{aligned}$$

Note that $P(N) = 0$. Consequently, $\{Z_n, n \geq 1\}$ is a sequence of random variables such that $|Z_n| \leq 2Y$ a.s. where $2Y$ is an integrable random variable. Further, $Z_n \xrightarrow{P} 0$. Hence by case (i),

$$\lim_{n\to\infty} E(Z_n) = 0 \quad \Rightarrow \quad \lim_{n\to\infty} E(X_n) = E(X).$$

□

It is to be noted that in both the dominated convergence theorems, X_n is dominated by an integrable random variable Y and hence the theorems are labeled as the dominated convergence theorems.

Remark 8.3.1. In the proof of the Lebesgue dominated probability convergence theorem, we have considered subsequences at two levels and used the result about almost sure convergence at the second level. It is necessary to proceed on these lines in view of the following problems. We are given that $X_n \xrightarrow{P} 0$, hence by Theorem 7.2.11, there exists a subsequence $\{X_{n'}, n' \geq 1\}$ such that $X_{n'} \xrightarrow{a.s.} 0$. By Fatou's lemma, then $E(X_{n'}) \to 0$. However, from this we cannot conclude that sequence of expectations of every subsequence converges to 0. This is the reason we have to consider two levels of subsequences.

Corollary 8.3.1. *Suppose X is a random variable on $(\Omega, \mathbb{A}, P)$ such that $E(|X|) < \infty$. If $\{A_n, n \geq 1\}$ is a sequence of sets in $\mathbb{A}$ such that $A_n \downarrow \emptyset$, then $\int_{A_n} X\ dP \to 0$. If $A_n \uparrow \Omega$, then $\int_{A_n} X\ dP \to E(X)$.*

Proof. It is proved in Corollary 7.2.9 that if X is a random variable defined on $(\Omega, \mathbb{A}, P)$ and $\{A_n, n \geq 1\}$ is a sequence of sets in $\mathbb{A}$, then $A_n \downarrow \emptyset$ implies $XI_{A_n} \xrightarrow{P} 0$ and $A_n \uparrow \Omega$ implies that $XI_{A_n} \xrightarrow{P} X$. Further, $|XI_{A_n}| \leq |X|$ and it is given that $|X|$ is integrable. Hence by the Lebesgue dominated probability convergence theorem,

$$\begin{aligned} A_n \downarrow \emptyset \quad &\Rightarrow \quad \int_{A_n} X\ dP = \int_{\Omega} I_{A_n} X\ dP = E(XI_{A_n}) \to 0 \\ \&\ A_n \uparrow \Omega \quad &\Rightarrow \quad \int_{A_n} X\ dP = \int_{\Omega} I_{A_n} X\ dP = E(XI_{A_n}) \to E(X)\ . \end{aligned}$$

□

Dominated convergence theorems are heavily used in all branches of statistics, to justify interchange of summation and integration or derivatives and integration or limits and integration. In Section 4.4, in the proof of the inversion theorem for a characteristic function, we have used the dominated convergence theorem to interchange limit and integration. We use the Lebesgue dominated probability convergence theorem in Chapter 9 in the proof of Khintchine's weak law of large numbers and Kolmogorov's strong law of large numbers, and also in Lindeberg-Levy central limit theorem in Chapter 10 .

In the next example, we verify the Lebesgue dominated probability convergence theorem and Theorem 7.2.3 by simulation.

Example 8.3.1. Suppose $X \sim N(\theta, 1)$ distribution and $\theta \in \Theta = [a, b]$. The maximum likelihood estimator $\hat{\theta}_n$ of θ, based on a random sample of size n from the distribution of X, is given by,

$$\hat{\theta}_n = \begin{cases} a, & \text{if} \quad \overline{X}_n < a \\ \overline{X}_n, & \text{if} \quad \overline{X}_n \in [a, b] \\ b, & \text{if} \quad \overline{X}_n > b. \end{cases}$$

In Example 7.3.15, we have shown that $\hat{\theta}_n \xrightarrow{P} \theta$. Now $\hat{\theta}_n \in [a, b]$, with $a < b$, we have $|\hat{\theta}_n| \leq \max\{|a|, |b|\}$. Hence, by the Lebesgue dominated probability convergence theorem and Theorem 7.2.3 $E(\hat{\theta}_n) \to \theta$. Further, by Theorem 7.2.3, $E(\hat{\theta}_n - \theta)^2 \to 0$. We verify these results using the following code.

Code 8.3.1. We take $a = 2, b = 4$ and generate a random sample of size n from normal $N(\theta, 1)$ with $\theta = 2.5$ and $n = 100$ to $n = 2500$ with increment of 300. On the basis of sample we find $\hat{\theta}_n$. On the basis of 400 simulated values of $\hat{\theta}_n$ we compute rf_n, an estimate of $P[|\hat{\theta}_n - \theta| < \epsilon]$, m_n, an estimate of $E(\hat{\theta}_n)$ and v_n, an estimate of $E(\hat{\theta}_n - \theta)^2$. We examine whether, $rf_n \to 1, m_n \to \theta = 2.5$ and $v_n \to 0$ as n increases.

```
N=seq(100,2500,300); nsim=400; th=2.5; a=2; b=4
me=mle=u=v=p=c(); ep=0.05
for(j in 1:length(N))
{
 for(m in 1:nsim)
{
 set.seed(m)
 x=rnorm(N[j],th,1)
 if(mean(x)>b)
{
 mle[m]=b
 if(mean(x)<a)
 mle[m]=a
}
 else
{
 mle[m]=mean(x)
}
}
 p[j]=length(which(abs(mle-th)<ep))/nsim
 u[j]=mean(mle)
 v[j]=mean((mle-th)^2)
}
d=round(data.frame(N,p,u,v),4); d
```

The output is organized in Table 8.1.

TABLE 8.1
Verification of LDCT for $N(\theta, 1)$ Distribution, $\theta \in [a, b]$

n	r_n	m_n	v_n
100	0.3775	2.5034	0.0094
400	0.6975	2.5017	0.0023
700	0.8125	2.5017	0.0014
1000	0.8825	2.5004	0.0011
1300	0.8975	2.5010	0.0008
1600	0.9525	2.5006	0.0007
1900	0.9825	2.5008	0.0005
2200	0.9850	2.5011	0.0004
2500	0.9950	2.5010	0.0004

From the table we note that $rf_n \to 1, m_n \to \theta = 2.5$ and $v_n \to 0$ as n increases. Thus, we have verified by simulation that if $\hat{\theta}_n \xrightarrow{P} \theta$, $E(\hat{\theta}_n) \to \theta$, as proved in the Lebesgue dominated probability convergence theorem and $E(\hat{\theta}_n - \theta)^2 \to 0$, as proved in Theorem 7.2.3. Using similar code, it can be verified that the results remain true for any $\theta \in [a, b]$, even at boundary points a and b. □

Following is a quick recap of the results discussed in this chapter.

Summary

1. Suppose $\{X_n, n \geq 1\}$ is a non-decreasing sequence of non-negative random variables defined on a probability space $(\Omega, \mathbb{A}, P)$. If $X_n \uparrow X$ on Ω, then $E(X_n) \uparrow E(X)$. If $Y \leq X_n \uparrow X$ on Ω and Y is an integrable random variable, then $E(Y) \leq E(X_n) \uparrow E(X)$.

2. Suppose $0 \geq X_n \downarrow X$ on Ω, then $E(X_n) \downarrow E(X)$. If $Y \geq X_n \downarrow X$ on Ω and Y is an integrable random variable, then $E(X_n) \downarrow E(X)$.

3. Suppose $\{X_n, n \geq 1\}$ is a non-decreasing sequence bounded from below by an integrable random variable or a non-increasing sequence bounded from above an integrable random variable. If $X_n \to X$ on Ω, then $\lim_{n\to\infty} E(X_n) = E(\lim_{n\to\infty} X_n)$.

4. Suppose $\{X_n, n \geq 1\}$ is a sequence of random variables then $E(\sum_{n\geq 1} X_n) = \sum_{n\geq 1} E(X_n)$.

5. Fatou's lemma: Suppose $\{X_n, n \geq 1\}$ is a sequence of random variables defined on a probability space $(\Omega, \mathbb{A}, P)$.
(i) If Y is an integrable random variable such that $Y \leq X_n$, then $E(\liminf X_n) \leq \liminf E(X_n)$.
(ii) If Z is an integrable random variable such that $X_n \leq Z$, then $\limsup E(X_n) \leq E(\limsup X_n)$.
(iii) If $Y$and Z are integrable random variables such that $Y \leq X_n \leq Z$, and if $X_n \xrightarrow{a.s.} X$, then $\lim_{n\to\infty} E(X_n) = E(X)$.

6. Lebesgue dominated almost sure convergence theorem: Suppose $\{X_n, n \geq 1\}$ is a sequence of random variables defined on a probability space $(\Omega, \mathbb{A}, P)$. If Y is an integrable random variables such that $|X_n| \leq Y$ and if $X_n \xrightarrow{a.s.} X$, then $\lim_{n\to\infty} E(X_n) = E(X)$.

7. Lebesgue dominated probability convergence theorem: Suppose $\{X_n, n \geq 1\}$ is a sequence of random variables defined on a probability space $(\Omega, \mathbb{A}, P)$. If Y is an integrable random variables such that $|X_n| \leq Y$ a.s. and if $X_n \xrightarrow{P} X$, then $\lim_{n\to\infty} E(X_n) = E(X)$.

8. Suppose X is a random variable on $(\Omega, \mathbb{A}, P)$ such that $E(|X|) < \infty$. If $\{A_n, n \geq 1\}$ is a sequence of sets in $\mathbb{A}$ such that $A_n \downarrow \emptyset$, then $\int_{A_n} X\, dP \to 0$. If $A_n \uparrow \Omega$, then $\int_{A_n} X\, dP \to E(X)$.

8.4 Conceptual Exercises

8.4.1 Suppose a sequence $\{(X_n, Y), n \geq 1\}$ of random variables is such that the conditional distribution of X_n given Y is uniform $U(Y - 1/n, Y + 1/n)$,
$n \geq 1$ and the marginal distribution of Y is exponential with location parameter 1 and scale parameter 1. Examine whether the sequence $\{X_n, n \geq 1\}$ converges in quadratic mean, probability, law and almost surely to Y. Examine whether a sequence $\{E(X_n), n \geq 1\}$ converges to $E(Y)$. Check whether the conditions of the Lebesgue dominated probability convergence theorem are satisfied.

8.4.2 Suppose $\{X_n, n \geq 1\}$ is a sequence of random variables such that X_n follows exponential distribution with mean θ^n, $0 < \theta < 1$. Show that $E(\sum_{n\geq 1} X_n) = \theta/(1-\theta)$. State the results that you may use.

8.4.3 Suppose $\{X_n, n \geq 1\}$ is a sequence of identically distributed random variables such that $E(|X_1|) < \infty$. Suppose $Y_n = (1/n)\max_{1\leq i\leq n} |X_i|$. Show that $\lim_{n\to\infty} E(Y_n) = 0$.

8.4.4 Show that X is an integrable random variable if and only if $\sum_{n\geq 1} P[|X| \geq cn] < \infty$, for some $c > 0$. In particular, deduce that if the above series converges for some $c > 0$ then it converges for all $c > 0$.

8.4.5 Suppose X is a non-negative random variable. Show that (i) $\lim_{n\to\infty} nE(I_{[X>n]}/X) = 0$ and (ii) $\lim_{n\to\infty}(1/n)E(I_{[X>1/n]}/X) = 0$.

8.4.6 Suppose $\{X_n, n \geq 1\}$ is a sequence of random variables defined on $(\Omega, \mathbb{A}, P)$ and Y is an integrable random variables such that $|X_n| \leq Y$ a.s. Prove that if $X_n \xrightarrow{L} 0$, then $\lim_{n\to\infty} E(X_n) = 0$.

8.5 Computational Exercises

8.5.1 Suppose $X \sim N(\theta, 1)$ distribution and $\theta \in \Theta = \{0, 1\}$. Verify the Lebesgue dominated probability convergence theorem and Theorem 7.2.3 by simulation.

8.6 Multiple Choice Questions

Note: In each question, multiple options may be correct. Unless specified otherwise, identify which of the statement(s) is/are correct. Answers are given in Chapter 11, after the solutions of conceptual exercises of Chapter 8.

8.6.1 $E(X_n) \to E(X)$ if

(a) $\{X_n, n \geq 1\}$ is a non-decreasing sequence of non-negative random variables converging to X on Ω
(b) $\{X_n, n \geq 1\}$ is a non-increasing sequence of negative random variables converging to X on Ω
(c) $\{X_n, n \geq 1\}$ is a non-decreasing sequence of random variables converging to X almost surely and the sequence is bounded below by an integrable random variable
(d) $\{X_n, n \geq 1\}$ is a non-increasing sequence of random variables converging to X almost surely and the sequence is bounded above by an integrable random variable

8.6.2 $E(X_n) \to E(X)$ always if

(a) $\{X_n, n \geq 1\}$ is a non-decreasing sequence of non-negative random variables converging to X on Ω
(b) $\{X_n, n \geq 1\}$ is a non-increasing sequence of negative random variables converging to X almost surely
(c) $\{X_n, n \geq 1\}$ is a non-decreasing sequence of random variables converging to X almost surely and the sequence is bounded below by an integrable random variable
(d) $\{X_n, n \geq 1\}$ is a non-increasing sequence of random variables converging to X almost surely and the sequence is bounded above by an integrable random variable

8.6.3 If Y is an integrable random variable such that $Y \leq X_n$, then

(a) $E(\liminf X_n) \geq \liminf E(X_n)$
(b) $E(\limsup X_n) \leq \limsup E(X_n)$
(c) $E(\liminf X_n) \leq \limsup E(X_n)$
(d) $E(\limsup X_n) \leq \liminf E(X_n)$

8.6.4 If Z is an integrable random variable such that $X_n \leq Z$, then

(a) $\limsup E(X_n) \geq E(\limsup X_n)$
(b) $\liminf E(X_n) \leq E(\limsup X_n)$
(c) $\limsup E(X_n) \leq E(\limsup X_n)$
(d) $\limsup E(X_n) \geq E(\liminf X_n)$

8.6.5 If Y is an integrable random variables such that $|X_n| \leq Y$, then $\lim_{n\to\infty} E(X_n) = E(X)$ always, if

(a) $X_n \overset{a.s.}{\to} X$
(b) $X_n \overset{P}{\to} X$
(c) $X_n \overset{q.m.}{\to} X$
(d) $X_n \overset{L}{\to} X$

8.6.6 If Y is an integrable random variables such that $|X_n| \leq Y$, then $\lim_{n\to\infty} E(X_n) = E(X)$ always, if (I) $X_n \stackrel{a.s.}{\to} X$. (II) $X_n \stackrel{P}{\to} X$. (III) $X_n \stackrel{q.m.}{\to} X$. (IV) $X_n \stackrel{L}{\to} X$. Which of the following is a correct option?

(a) Only (I) is true
(b) Only (II) is true
(c) Both (II) and (IV) are true
(d) (I), (II) and (III) are true

8.6.7 If Y is an integrable random variables such that $|X_n| \leq Y$, then $\lim_{n\to\infty} E(X_n) = 0$, if

(a) $X_n \stackrel{a.s.}{\to} 0$
(b) $X_n \stackrel{P}{\to} 0$
(c) $X_n \stackrel{q.m.}{\to} 0$
(d) $X_n \stackrel{L}{\to} 0$

9

Laws of Large Numbers

9.1 Introduction

In probability theory we study limit theorems of two types: (i) strong limit theorems which deal with the almost sure convergence of a sequence of random variables and (ii) weak limit theorems which deal with convergence in probability of a sequence of random variables as well as convergence of a sequence of distribution functions. We have discussed in detail these modes of convergence in Chapters 6 and 7. In the present chapter, we study almost sure convergence and convergence in probability of a sequence $\{S_n, n \geq 1\}$ derived from a sequence $\{X_n, n \geq 1\}$ by defining $S_n = \sum_{i=1}^{n} X_i$. S_n is known as n-th partial sum. The laws of large numbers are concerned with the limiting behaviour of $\{S_n, n \geq 1\}$. These state conditions under which the sequence $\{S_n, n \geq 1\}$ of partial sums converges almost surely, in probability and in distribution, with suitable norming. The weak law of large numbers (WLLN) refers to convergence in probability, whereas the strong law of large numbers (SLLN) refers to the almost sure convergence. Central limit theorem (CLT) is concerned with the convergence in distribution of the appropriately normed sequence $\{S_n, n \geq 1\}$. We define these below.

Definition 9.1.1. *Suppose $\{X_n, n \geq 1\}$ is a sequence of random variables and $S_n = \sum_{i=1}^{n} X_i$. Suppose there exists two sequences $\{a_n, n \geq 1\}$ of real numbers and $\{b_n, n \geq 1\}$ of positive real numbers such that $b_n \to \infty$. A sequence $\{X_n, n \geq 1\}$ of random variables is said to obey a*

(i) SLLN, if $(S_n - a_n)/b_n \xrightarrow{a.s.} 0$ as $n \to \infty$.

(ii) WLLN, if $(S_n - a_n)/b_n \xrightarrow{P} 0$ as $n \to \infty$.

(iii) CLT, if $(S_n - a_n)/b_n \xrightarrow{L} Z$ as $n \to \infty$ where $Z \sim N(0,1)$.

In the next section, in some examples, we note that we can have more than one sequences $\{a_n, n \geq 1\}$ and $\{b_n, n \geq 1\}$ for which the WLLN holds. In Chapter 10, we note that in many cases for $b_n^2 = Var(S_n)$, CLT is valid. When a sequence $\{X_n, n \geq 1\}$ satisfies WLLN, it is said to be stable in probability, when it satisfies SLLN, it is said to be almost surely stable and when it satisfies CLT, it is said to be stable in law.

DOI: 10.1201/9781032619057-9

Laws of large numbers have important applications in establishing asymptotic properties, such as consistency and asymptotic normality of estimators. From WLLN one can conclude that the mean of the large sample of observations is close to the population mean with very high chance. Depending on the conditions, there are various versions of laws of large numbers and CLT. In the next two sections, we discuss WLLN and SLLN. Various types of CLT are studied in the next chapter.

9.2 Weak Law of Large Numbers

WLLN is considered as a great achievement in probability in view of its number of applications. As defined in Section 1, it deals with convergence in probability of the sequence $\{S_n, n \geq 1\}$ of partial sums. We begin with some examples.

Example 9.2.1. Suppose $\{X_n, n \geq 1\}$ is a sequence of independent random variables with $P[X_n = 2^n] = P[X_n = -2^n] = 1/2$. It is clear that

$$E(X_n) = 0 \,\&\, Var(X_n) = 4^n \Rightarrow Var(S_n) = \sum_{i=1}^{n} Var(X_i) = \sum_{i=1}^{n} 4^i = \frac{4(4^n - 1)}{3}.$$

We now examine whether $\{X_n, n \geq 1\}$ follows WLLN. $E(X_n) = 0$ suggests $a_n = 0$. To find $b_n > 0$ and tending to ∞, observe that

$$Var(S_n/b_n) = 4(4^n - 1)/3b_n^2 = 4((2^n)^2 - 1)/3b_n^2 \;\to 0 \text{ if } b_n = (2+\delta)^n, \;\delta > 0.$$

Thus by Chebyshev's inequality, for any $\epsilon > 0$

$$P\left[|(S_n - E(S_n))/b_n| > \epsilon\right] \;\leq\; (1/\epsilon^2)Var(S_n/b_n) \;\to 0, \text{ with } b_n = (2+\delta)^n.$$

Thus, $\{X_n, n \geq 1\}$ follows WLLN with $a_n = 0$ and $b_n = (2+\delta)^n$, $\delta > 0$. Note that corresponding to different values of δ, we have an uncountable family of sequences $\{b_n, n \geq 1\}$ with which WLLN holds. Now we investigate whether we have some more sequences $\{a_n, n \geq 1\}$, apart from $a_n \equiv 0$ which satisfy WLLN. Note that $E(S_n) = 0$ and hence $E(S_n^2) = Var(S_n)$. Observe that by Chebyshev's inequality,

$$\begin{aligned} P\left[|(S_n - a_n)/b_n| > \epsilon\right] &\leq (1/\epsilon^2)E((S_n - a_n)/b_n)^2 \\ &= (1/\epsilon^2 b_n^2)(E(S_n^2) - 2a_n E(S_n) + a_n^2) \\ &= (1/\epsilon^2 b_n^2)(4(4^n - 1)/3 + a_n^2) \;\to\; 0 \text{ if } a_n^2/b_n^2 \to 0. \end{aligned}$$

For $a_n^2/b_n^2 \to 0$, we may select $a_n^2 = n$ or n^2 or 2^n, in general any sequence $\{a_n, n \geq 1\}$, in which $a_n^2 \to \infty$ at a slower rate than b_n^2 tending to infinity. □

Remark 9.2.1. Example 9.2.1 coveys that, in general we have more than one sequences $\{a_n, n \geq 1\}$ and $\{b_n, n \geq 1\}$ with which the WLLN holds.

Example 9.2.2. Suppose $\{X_n, n \geq 1\}$ is a sequence of independent random variables with

$$P[X_n = 2^n] = P[X_n = -2^n] = 2^{-(2n+1)} \;\; \& \;\; P[X_n = 0] = 1 - 2^{-2n}.$$

Note that $E(X_n) = 0$, $Var(X_n) = 1$ and hence $Var(S_n/b_n) = n/b_n^2$. This expression of $Var(S_n/b_n)$ suggests that if we select $b_n^2 = n^{1+\delta}, \delta > 0$, then $b_n > 0$, $b_n \to \infty$ and $Var(S_n/b_n) \to 0$. Hence by Chebyshev's inequality, it follows that $(S_n - a_n)/b_n \xrightarrow{P} 0$, with $a_n = 0$ and with $b_n^2 = n^{1+\delta}$. Thus, the sequence $\{X_n, n \geq 1\}$ follows WLLN. With different values of δ, we have many sequences $\{b_n, n \geq 1\}$ for which the WLLN holds. As in Example 9.2.1, can find more than one sequences $\{a_n, n \geq 1\}$ with which the WLLN holds.□

In the following theorems and corollaries, we investigate under which conditions, the sequence $\{X_n, n \geq 1\}$ follows WLLN.

Theorem 9.2.1. *Chebyshev's WLLN: Suppose $\{X_n, n \geq 1\}$ is a sequence of uncorrelated random variables with uniformly bounded variances, then the sequence follows WLLN with $a_n = E(S_n)$ and $b_n = n$.*

Proof. It is given that $\{X_n, n \geq 1\}$ is a sequence of uncorrelated random variables with uniformly bounded variances. Hence,

$$\begin{aligned}
&Cov(X_n, X_m) = 0 \;\; \forall \;\; n \neq m \;\; \& \;\; Var(X_n) \leq M, \;\; \forall \;\; n \geq 1 \\
\Rightarrow \;\; & Var(S_n - E(S_n)) = Var(S_n) = \sum_{i=1}^{n} Var(X_i) \;\; \leq \;\; nM \\
\Rightarrow \;\; & Var\,(S_n - E(S_n))/n\,) = Var(S_n - E(S_n))/n^2 \;\; \leq \;\; M/n \\
\Rightarrow \;\; & P\left[\frac{|S_n - E(S_n)|}{n} > \epsilon\right] \leq E\left(\frac{S_n - E(S_n)}{n\epsilon}\right)^2 \text{ by Chebyshev's inequality} \\
& \leq M/n\epsilon^2 \to 0, \;\; \text{as} \;\; n \to \infty \;\; \forall \;\; \epsilon > 0 \\
\Rightarrow \;\; & (S_n - E(S_n))/n \xrightarrow{P} 0.
\end{aligned}$$

Thus, $\{X_n, n \geq 1\}$ follows WLLN with $a_n = E(S_n)$ and $b_n = n$. □

Corollary 9.2.1. *Suppose $\{X_n, n \geq 1\}$ is a sequence of independent random variables with uniformly bounded variances, then the sequence follows WLLN with $a_n = E(S_n)$ and $b_n = n$.*

Proof. Since$\{X_n, n \geq 1\}$ is a sequence of independent random variables, it follows that $\{X_n, n \geq 1\}$ is a sequence of uncorrelated random variables and hence the result follows from Theorem 9.2.1. □

Corollary 9.2.2. *Suppose $\{X_n, n \geq 1\}$ is a sequence of independent and identically distributed random variables with mean μ and finite variance σ^2, then the sequence follows WLLN with $a_n = E(S_n) = n\mu$ and $b_n = n$.*

Proof. Since $\{X_n, n \geq 1\}$ is a sequence of independent and identically distributed random variables with mean μ and finite variance σ^2, $E(S_n) = n\mu$ and $Var(S_n) = n\sigma^2$. Hence,

$$Var(S_n - E(S_n))/n = Var(S_n)/n^2 = \sigma^2/n \to 0 \text{ as } n \to \infty.$$

The proof now follows by Chebyshev's inequality as in Theorem 9.2.1. □

Corollary 9.2.3. *Suppose $\{X_n, n \geq 1\}$ is a sequence of independent and identically distributed random variables each having Bernoulli $B(1,p)$ distribution. Then $P_n = \overline{X}_n \xrightarrow{P} p$.*

Proof. Since $\{X_n, n \geq 1\}$ is a sequence of independent and identically distributed random variables each having Bernoulli $B(1,p)$ distribution,

$$\begin{aligned} E(X_n) = p \;\&\; Var(X_n) = p(1-p) &\Rightarrow E(S_n) = np \;\&\; Var(S_n) = np(1-p) \\ \Rightarrow (S_n - np)/n \xrightarrow{P} 0 &\iff S_n/n = \overline{X}_n = P_n \xrightarrow{P} p, \end{aligned}$$

by Corollary 9.2.2, □

Corollary 9.2.3 states that in a sequence of independent Bernoulli trials, sample proportion P_n of successes, which is the same as the relative frequency of success, stabilizes at p. In the setup of large sample inference, it states that sample proportion P_n is a consistent estimator of probability p of success. Corollary 9.2.3 is known as a Bernoulli WLLN.

In two examples, in Theorem 9.2.1 and in the corollaries to the theorem, we heavily used Chebyshev's inequality, to prove that the sequence satisfies the WLLN. Hence the underlying assumption is that random variables of the sequence are in $\mathbb{L}_2$. In the following theorem, known as Khintchine's WLLN, we note that the condition of finiteness of variance can be relaxed. As a consequence, the proof to establish WLLN requires some more techniques. The proof is based on a truncation technique. We briefly describe it below. A random variable X is said to be truncated at $c, 0 < c < \infty$ if X^* is defined as a function of X as follows.

$$X^* = \begin{cases} X, & \text{if } |X| \leq c \\ 0, & \text{if } |X| > c. \end{cases}$$

Thus with this technique, the values of X such that $|X| > c$ are replaced by 0 and hence an arbitrary random variable X is replaced by an almost surely bounded random variable X^*. As a consequence, all moments of X^* exist and are finite. Thus,

$$E(X^*) = \int_{\Omega} X^* dP = \int_{[|X| \leq c]} X^* dP + \int_{[|X|>c]} X^* dP = \int_{[|X| \leq c]} X dP$$
$$\leq c \int_{\Omega} dP = c$$
$$\& \ \ Var(X^*) \leq E(X^*)^2 = \int_{[|X| \leq c]} X^2 dP \ \leq \ c^2 \int_{[|X| \leq c]} dP \leq c^2 \int_{\Omega} dP = c^2.$$

Further, $P[X \neq X^*] = P[|X| > c]$. It can be made arbitrarily small by selecting c large, since X is bounded in probability, being a real random variable.

As a pre-requisite in the proofs of Khintchine's WLLN and Kolmogorov's SLLN in Section 3, we need certain results. We discuss these below.

Definition 9.2.1. *Suppose $\{X_n, n \geq 1\}$ and $\{Y_n, n \geq 1\}$ are sequences of random variables defined on the same probability space. The two sequences are said to be equivalent sequences if $\sum_{n \geq 1} P[X_n \neq Y_n] < \infty$.*

The limiting behaviour of the two sequences is the same as elaborated in the following theorem.

If $\{Z_n, n \geq 1\}$ is a sequence of random variables, then we say that $\sum_{i=1}^{\infty} Z_i$ is defined if $\lim_{n \to \infty} \sum_{i=1}^{n} Z_i$ exists in some sense and is finite almost surely.

Theorem 9.2.2. *Suppose $\{X_n, n \geq 1\}$ and $\{Y_n, n \geq 1\}$ are equivalent sequences of random variables defined on the same probability space. Then*

(i) $\sum_{n \geq 1}(X_n - Y_n)$ *converges a.s.*

(ii) $\sum_{i=1}^{n}(X_i - Y_i)/b_n \overset{a.s.}{\to} 0$ *as* $n \to \infty$, *where* $\{b_n, n \geq 1\}$ *is a sequence of positive real numbers such that* $b_n \to \infty$ *as* $n \to \infty$.

(iii) $\sum_{i=1}^{n} Y_i/b_n \overset{a.s.}{\to} Y \ \ \Rightarrow \ \ \sum_{i=1}^{n} X_i/b_n \overset{a.s.}{\to} Y$ *as* $n \to \infty$.

(iv) $\sum_{i=1}^{n} Y_i/b_n \overset{P}{\to} Y \ \ \Rightarrow \ \ \sum_{i=1}^{n} X_i/b_n \overset{P}{\to} Y$ *as* $n \to \infty$.

Proof.

(i) Suppose $\{X_n, n \geq 1\}$ and $\{Y_n, n \geq 1\}$ are equivalent sequences of random variables. Then by the Borel-Cantelli lemma,

$$\sum_{n \geq 1} P[X_n \neq Y_n] < \infty$$
$$\Rightarrow \ P[\limsup[X_n \neq Y_n]] = 0 \iff P[\liminf[X_n = Y_n]] = 1.$$

Thus, there exists a p-null set N and an integer $n_0(\omega)$, $\omega \in N^c$, such that $X_n(\omega) = Y_n(\omega) \ \forall \ n \geq n_0 \ \ \& \ \ \forall \ \omega \in N^c$. Thus,

$$\forall \, \omega \in N^c, \sum_{n=1}^{\infty}(X_n - Y_n) = \sum_{n=1}^{n_0}(X_n - Y_n) < \infty \Rightarrow \sum_{n \geq 1}(X_n - Y_n) \text{ converges a.s.}$$

(ii) In view of (i), $\sum_{i=1}^{n}(X_i - Y_i)$ converges to a finite number almost surely and $b_n \to \infty$ as $n \to \infty$, hence we get (ii).

(iii) From the result in (ii)

$$(1/b_n)\sum_{i=1}^{n} X_i = (1/b_n)\sum_{i=1}^{n}(X_i - Y_i) + (1/b_n)\sum_{i=1}^{n} Y_i \overset{a.s.}{\to} 0 + Y = Y.$$

(iv) From the result in (ii),

$$(1/b_n)\sum_{i=1}^{n}(X_i - Y_i) \overset{a.s.}{\to} 0 \Rightarrow (1/b_n)\sum_{i=1}^{n}(X_i - Y_i) \overset{P}{\to} 0$$

$$\Rightarrow (1/b_n)\sum_{i=1}^{n} X_i = (1/b_n)\sum_{i=1}^{n}(X_i - Y_i) + (1/b_n)\sum_{i=1}^{n} Y_i \overset{P}{\to} 0 + Y = Y.$$

□

In Theorem 8.2.3, we have proved that for any random variable X defined on $(\Omega, \mathbb{A}, P)$,

$$\sum_{n\geq 1} P[|X| \geq n] \leq E(|X|) \leq 1 + \sum_{n\geq 1} P[|X| \geq n].$$

Thus, a random variable X is integrable if and only if $\sum_{n\geq 1} P[|X| \geq n] < \infty$.
In the proof of Khintchine's WLLN, we use Toeplitz' lemma. It is stated below, Loeve [16].

Lemma 9.2.1. *Suppose $\{x_n, n \geq 1\}$ is a sequence of real numbers such that $x_n \to x$ as $n \to \infty$. Then as $n \to \infty$,*

$$b_n = \sum_{k=1}^{n} a_k \to \infty \Rightarrow (1/b_n)\sum_{k=1}^{n} x_k a_k \to x.$$

As a particular case, suppose $a_k = 1 \ \forall \ k = 1, 2, \cdots, n$. Then $b_n = n \to \infty$ as $n \to \infty$ and $(1/n)\sum_{k=1}^{n} x_k = \overline{x}_n \to x$, whenever $x_n \to x$.
In the proof of Khintchine's WLLN in the next theorem, we use truncation technique and all these results.

Theorem 9.2.3. *Khintchine's WLLN: Suppose $\{X_n, n \geq 1\}$ is a sequence of independent and identically distributed random variables with finite mean μ, then the sequence follows WLLN with $a_n = E(S_n) = n\mu$ and $b_n = n$, that is, $(S_n - n\mu)/n = \overline{X}_n - \mu \overset{P}{\to} 0$.*

Proof. Using truncation technique, for every $n \geq 1$, a random variable X_n^* is defined as follows.

$$X_n^* = \begin{cases} X_n, & \text{if } |X_n| < n \\ 0, & \text{if } |X_n| \geq n. \end{cases}$$

Note that being Borel functions of X_n, $\{X_n^*, n \geq 1\}$ is also a sequence of independent random variables. Suppose $S_n^* = \sum_{k=1}^{n} X_k^*$ and an event E_n is defined as $E_n = [|X_n| \geq n]$, $n \geq 0$ with $E_0 = \Omega$. Thus, $X_n^* = X_n$ on E_n^c and $X_n^* \neq X_n$ on E_n. By Theorem 8.2.3,

$$\begin{aligned}
\sum_{n\geq 1} P[X_n \neq X_n^*] &= \sum_{n\geq 1} P(E_n) = \sum_{n\geq 1} P[|X_n| \geq n] \leq E(|X_n|) < \infty \\
&\Rightarrow \{X_n, n \geq 1\} \ \& \ \{X_n^*, n \geq 1\} \text{ are equivalent sequences} \\
&\Rightarrow (S_n - S_n^*)/n \overset{a.s.}{\to} 0 \text{ by result (ii) of Theorem 9.2.2} \\
&\Rightarrow (S_n - S_n^*)/n \overset{P}{\to} 0. \qquad (9.1)
\end{aligned}$$

Thus, by result (iii) and (iv) of Theorem 9.2.2, the limiting behaviour of S_n/n is the same as that of S_n^*/n. We now study the limiting behaviour of S_n^*/n. Suppose $X_n, n \geq 1$ are distributed as X with finite mean μ. Observe that for each $n \geq 1$,

$$E(X_n^*) = \int_{\Omega} X_n^* dP = \int_{E_n} X_n^* dP + \int_{E_n^c} X_n^* dP = \int_{E_n^c} X_n dP = \int_{E_n^c} X dP = E(XI_{E_n^c}),$$

since X_n is distributed as X. Further, observe that $|XI_{E_n^c}| \leq |X|$ and $E(|X|) < \infty$. From the definition of E_n^c, note that for $n_1 < n_2$,

$$\begin{aligned}
|X_n(\omega)| &< n_1 \Rightarrow |X_n(\omega)| < n_2 \Rightarrow E_{n_1}^c \subset E_{n_2}^c \\
&\Rightarrow \{E_n^c, n \geq 1\} \uparrow \bigcup_{n\geq 1} E_n^c = \Omega \text{ as } n \to \infty \\
&\Rightarrow I_{E_n^c} \overset{a.s.}{\to} I_\Omega \Rightarrow XI_{E_n^c} \overset{a.s.}{\to} XI_\Omega = X \text{ as } n \to \infty \\
&\Rightarrow E(XI_{E_n^c}) \to E(X) \text{ by dominated a.s. convergence theorem} \\
&\Rightarrow E(X_n^*) \to \mu \quad \text{as } n \to \infty \\
&\Rightarrow E(S_n^*/n) = \frac{1}{n}\sum_{k=1}^{n} E(X_k^*) \to \mu \quad \text{by Toeplitz' lemma.} \qquad (9.2)
\end{aligned}$$

Now we study the nature of $Var(S_n^*/n)$. Suppose $\{a_n, n \geq 1\}$ is a sequence positive integers such that $0 < a_n < n$ and $a_n \to \infty$ in such a way that $a_n/n \to 0$ as $n \to \infty$. Then $Var(S_n^*)$ can be expressed as

$$\begin{aligned}
Var(S_n^*) &= \sum_{i=1}^{n} Var(X_i^*) \leq \sum_{i=1}^{n} E(X_i^*)^2 \leq \sum_{i=1}^{n} \int_{[|X|\leq i]} X^2 \, dP \\
&= \sum_{i=1}^{a_n} \int_{[|X|\leq i]} X^2 \, dP + \sum_{i=a_n+1}^{n} \int_{[|X|\leq i]} X^2 \, dP \\
&= \sum_{i=1}^{a_n} \int_{[|X|\leq i]} X^2 \, dP + \sum_{i=a_n+1}^{n} \int_{[|X|\leq a_n]} X^2 \, dP + \sum_{i=a_n+1}^{n} \int_{[a_n<|X|\leq i]} X^2 \, dP.
\end{aligned}$$

Further note that

$$
\begin{aligned}
Var(S_n^*) &\leq \sum_{i=1}^{a_n} a_n \int_{[|X|\leq a_n]} |X|\, dP + \sum_{i=a_n+1}^{n} a_n \int_{[|X|\leq a_n]} |X|\, dP \\
&+ \sum_{i=a_n+1}^{n} n \int_{[a_n<|X|\leq n]} |X|\, dP \\
&= na_n \int_{[|X|\leq a_n]} |X|\, dP + n(n-a_n) \int_{[a_n<|X|\leq n]} |X|\, dP \\
&\leq na_n \int_{[|X|\in \mathbb{R}^+]} |X|\, dP + n^2 \int_{[|X|>a_n]} |X|\, dP \quad \text{as } n-a_n<n.
\end{aligned}
$$

It is given that $E(X) < \infty$, hence $\int_{[|X|\in\mathbb{R}^+]} |X|\, dP = E(|X|) = M < \infty$. Observe that $I_{[|X|>a_n]}|X| < |X|$ and $|X|$ is integrable. Further, $a_n \to \infty$ as $n \to \infty$, hence $I_{[|X|>a_n]} \to I_{\emptyset} \equiv 0$ a.s. Hence $|X|I_{[|X|>a_n]} \to 0$ a.s. Then by the Lebesgue dominated almost sure convergence theorem, $\int_{[|X|>a_n]} |X|\, dP = E(I_{[|X|>a_n]}|X|) \to 0$. As a consequence,

$$
\begin{aligned}
Var(S_n^*/n) &\leq (a_n/n)M + \int_{[|X|>a_n]} |X|\, dP \;\to\; 0 \\
&\Rightarrow E\left((S_n^* - E(S_n^*))/n\right)^2 \;\to\; 0 \\
&\Rightarrow (S_n^* - E(S_n^*))/n \overset{q.m.}{\to} 0 \Rightarrow (S_n^* - E(S_n^*))/n \overset{P}{\to} 0. \quad (9.3)
\end{aligned}
$$

Using results in Equation (9.1), Equation (9.2) and Equation (9.3), we have

$$
\begin{aligned}
S_n/n - \mu &= (S_n - S_n^* + S_n^*)/n - \mu + E(S_n^*)/n - E(S_n^*)/n \\
&= (S_n - S_n^*)/n + (S_n^* - E(S_n^*))/n + E(S_n^*)/n - \mu \overset{P}{\to} 0 \\
&\Rightarrow S_n/n \overset{P}{\to} \mu \iff \overline{X}_n - \mu \overset{P}{\to} 0
\end{aligned}
$$

and the theorem is proved. □

There is one more approach to prove Khintchine's WLLN which does not use truncation technique but uses some results related to a characteristic function. We prove some of these below. These are also useful to prove a central limit theorem in the next chapter. We refer to Gut [13] for these results.

Lemma 9.2.2. *Suppose $u_1, u_2, ..., u_m$ and $w_1, w_2, ..., w_m$ are complex numbers where $|u_i| \leq 1, |w_i| \leq 1,\ i = 1, 2, \cdots, m$. Then,*

$$
\left|\prod_{i=1}^{m} u_i - \prod_{i=1}^{m} w_i\right| \leq \sum_{i=1}^{m} |u_i - w_i|.
$$

Proof. The proof is by induction. Result is obviously true for $m = 1$. For $m = 2$ observe that,

$$\begin{aligned}\left|\prod_{i=1}^{2} u_i - \prod_{i=1}^{2} w_i\right| &= |(u_2 - w_2)u_1 + w_2(u_1 - w_1)| \\ &\leq |(u_2 - w_2)||u_1| + |w_2||(u_1 - w_1)| \\ &\leq \sum_{i=1}^{2} |u_i - w_i|, \quad \text{since } |u_1| \leq 1, |w_2| \leq 1.\end{aligned}$$

We assume the result is true for $m = k$. Now

$$\prod_{i=1}^{k+1} u_i - \prod_{i=1}^{k+1} w_i = \left[(u_{k+1} - w_{k+1})\prod_{i=1}^{k} u_i\right] + \left[w_{k+1}\left(\prod_{i=1}^{k} u_i - \prod_{i=1}^{k} w_i\right)\right]$$

$$\begin{aligned}\Rightarrow \left|\prod_{i=1}^{k+1} u_i - \prod_{i=1}^{k+1} w_i\right| &\leq |u_{k+1} - w_{k+1}| + \left|\prod_{i=1}^{k} u_i - \prod_{i=1}^{k} w_i\right| \\ &\leq |u_{k+1} - w_{k+1}| + \sum_{i=1}^{k} |u_i - w_i|, \text{by induction hypothesis} \\ &= \sum_{i=1}^{k+1} |u_i - w_i|,\end{aligned}$$

The second step follows since $|\prod_{i=1}^{k} u_i|$ and $|w_{k+1}| \leq 1$. □

We state a lemma below, for proof refer to Lemma A 1.2, p. 556 of Gut [13].

Lemma 9.2.3. *For all real numbers x and $i = \sqrt{-1}$,*

$$\left|e^{ix} - \sum_{j=0}^{k} (ix)^j/j!\right| \leq \min\left\{2|x|^k/k!, \ |x|^{k+1}/(k+1)!\right\}.$$

We use Lemma 9.2.3 in the following theorem, which is useful in proving Khintchine's WLLN and Lindeberg-Levy CLT in the next chapter.

Theorem 9.2.4. *If $E(|X|^k) < \infty$ and if ϕ is a characteristic function of X, then $\forall\, t \in \mathbb{R}$,*

$$\left|\phi(t) - \sum_{j=0}^{k} \frac{(it)^j}{j!} E(X^j)\right| \leq E\left(\min\left\{\frac{2|tX|^k}{k!}, \ \frac{|tX|^{k+1}}{(k+1)!}\right\}\right).$$

Proof. From the definition of a characteristic function we have

$$\begin{aligned}\Big|\phi(t) - \sum_{j=0}^{k}\frac{(it)^j}{j!}E(X^j)\Big| &= \Big|E(e^{itX}) - E\Big(\sum_{j=0}^{k}\frac{(it)^j}{j!}X^j\Big)\Big| \\ &\leq E\Big(\Big|e^{itX} - \sum_{j=0}^{k}(itX)^j/j!\Big|\Big) \\ &\leq E\Big(\min\Big\{\frac{2|tX|^k}{k!},\ \frac{|tX|^{k+1}}{(k+1)!}\Big\}\Big), \text{ by Lemma 9.2.3.}\end{aligned}$$

□

In the following theorem, we give another proof of Khintchine's WLLN which is based on these results about a characteristic function.

Theorem 9.2.5. *Khintchine's WLLN: Suppose $\{X_n, n \geq 1\}$ is a sequence of independent and identically distributed random variables with finite mean μ, then $(S_n - n\mu)/n = \overline{X}_n - \mu \xrightarrow{P} 0$.*

Proof. The proof is in two steps: (i) $\mu = 0$ and (ii) $\mu \neq 0$.
Case (i): Suppose $\mu = 0$. By Theorem 7.3.2, $S_n/n \xrightarrow{P} 0 \iff S_n/n \xrightarrow{L} 0$. Hence we show that $S_n/n \xrightarrow{L} 0$ by proving that the characteristic function of S_n/n converges to the characteristic function $\phi_0(t) \equiv 1$ of $X \equiv 0$. Suppose $\phi(\cdot)$ denote the common characteristic function of X_i, $i = 1, 2, \cdots, n$. Then by Theorem 5.3.5, the characteristic function of S_n/n is $\phi^n(t/n)$. With $k = 1$ and $\mu = 0$ in Theorem 9.2.4 we have,

$$\begin{aligned}|\phi(t/n) - (1 + itE(X))| = |\phi(t/n) - 1| &\leq E\left(\min\{2|tX|/n,\ |tX|^2/2n^2\}\right) \\ &= (1/n)E\left(\min\{2|tX|,\ |tX|^2/2n\}\right) \\ \Rightarrow n|\phi(t/n) - 1| \leq E(W_n), \text{where } W_n &= \min\left\{2|tX|, \frac{|tX|^2}{2n}\right\}. \quad (9.4)\end{aligned}$$

Further, note that $W_n \leq 2|tX|\ \forall\ n$ and $E(|X|) < \infty$. Hence, W_n is dominated by an integrable random variable. Observe that for large n, $W_n = |tX|^2/2n$. Now X is bounded in probability, being a real random variable which implies that $W_n \xrightarrow{P} 0$ as $n \to \infty$. Hence, by the Lebesgue dominated probability convergence theorem, $\lim_{n\to\infty} E(W_n) = E(\lim_{n\to\infty} W_n) = 0$. Observe that,

$$\begin{aligned}|\phi_{S_n/n}(t) - 1| &= |\phi^n(t/n) - 1| = |\phi^n(t/n) - 1^n| \\ &\leq \sum_{k=1}^{n}|\phi(t/n) - 1|, \quad \text{by Lemma 9.2.2} \\ &= n|\phi(t/n) - 1| \ \leq\ E(W_n) \ \text{ by (9.4)} \\ \Rightarrow \lim_{n\to\infty}|\phi_{S_n/n}(t) - 1| &\leq \lim_{n\to\infty} E(W_n) = E\left(\lim_{n\to\infty} W_n\right) = 0.\end{aligned}$$

Thus, $\forall\ t \in \mathbb{R},\ \phi_{S_n/n}(t) \to 1 \equiv \phi_0(t)$ and $\phi_0(t) = 1$ is continuous at $t = 0$. Hence by the continuity theorem and the uniqueness theorem for characteristic functions, we conclude that

$$S_n/n \xrightarrow{L} 0 \iff S_n/n \xrightarrow{P} 0.$$

Case-(ii): When $\mu \neq 0$, we define a random variables $Y_n = X_n - \mu$ so that $\{Y_n, n \geq 1\}$ is a sequence of independent and identically distributed random variables with mean zero. Hence from case (i),

$$(Y_1 + Y_2 + \cdots + Y_n)/n \xrightarrow{P} 0 \iff S_n/n \xrightarrow{P} \mu.$$

□

Remark 9.2.2. We have already proved in Corollary 9.2.2 that a sequence $\{X_n, n \geq 1\}$ of independent and identically distributed random variables with mean μ and finite variance σ^2 follows WLLN with $a_n = E(S_n) = n\mu$ and $b_n = n$. The same result follows from the Theorem 9.2.5 since, $Var(X_1) < \infty$ implies $E(X_1) < \infty$.

In Example 7.2.8, we have shown that the r-th order sample raw moment converges in probability to the corresponding population moment, provided $\mu'_{2r} < \infty$. In the following example, we show that this result is true even if the condition $\mu'_{2r} < \infty$ is relaxed to $\mu'_r < \infty$. It follows from Khintchine's WLLN.

Example 9.2.3. Suppose $\{X_n, n \geq 1\}$ is a sequence of independent and identically distributed random variables with finite r-th order raw moment $\mu'_r = E(X_1^r)$, $r \geq 1$. Being Borel functions, $\{X_n^r, n \geq 1\}$ is also a sequence of independent and identically distributed random variables with finite mean μ'_r. Hence, by Khintchine's WLLN, the r-th order sample raw moment $m'_r = \sum_{i=1}^n X_i^r/n \xrightarrow{P} \mu'_r$. Further using binomial theorem, r-th order central moment μ_r can be expressed as a polynomial function of r-th order raw moments for $r \leq 1$ as follows.

$$\mu_r = \binom{r}{0}\mu'_r + \binom{r}{1}\mu'_{r-1}\mu'_1 + \binom{r}{2}\mu'_{r-2}(\mu'_1)^2 + \cdots + \binom{r}{r}(\mu'_1)^r,\ r \geq 1.$$

Similarly, r-th order sample central moment $m_r = \sum_{i=1}^n (X_i - \overline{X}_n)^r/n$ can be written as

$$m_r = \binom{r}{0}m'_r + \binom{r}{1}m'_{r-1}m'_1 + \binom{r}{2}m'_{r-2}(m'_1)^2 + \cdots + \binom{r}{r}(m'_1)^r,\ r \geq 1.$$

In view of the result that the convergence in probability is closed under all arithmetic operations, $m_r \xrightarrow{P} \mu_r$. □

From this example, it thus follows that raw moments of a random sample are consistent for the corresponding population raw moments and further by invariance property of convergence in probability under arithmetic operations, sample central moments are also consistent for the corresponding population central moments. In particular, the sample mean and the sample variance are consistent for population mean and population variance respectively.

Khintchine's WLLN is a foundation of large sample inference procedures. Suppose an estimator based on a random sample $\{X_1, X_2, \cdots, X_n\}$ is in the form of average $T_n = \sum_{i=1}^{n} g(X_i)/n$, say, where g is a Borel function. Then Khintchine's WLLN conveys that $T_n \xrightarrow{P} E(g(X))$, provided it is finite. If $E(g(X)) = h(\theta)$ and h^{-1} exists and is continuous then by the invariance property of convergence in probability under continuous transformation, $h^{-1}(T_n)$ will be a consistent estimator of θ.

In the following example using `R`, we examine convergence in probability of some raw and central moments to the corresponding population moments for some distributions.

Example 9.2.4. In this example we verify that sample mean converges in probability to the mean of the distribution of X. For comparison of the rate of convergence we consider three distributions, exponential, geometric and Poisson. For uniformity, we set the parameters of the distributions so that mean of each is 1. As discussed in Section 7.2, to verify the convergence in probability, we simulate m random samples, each of size n from the distributions and find the estimate rf_n. We examine whether $rf_n \to 1$ as $n \to \infty$. In this example for illustration, we have fixed $\epsilon = 0.05$, $m = 500$ and n takes values from 500 to 3000 with an increment of 500.

Code 9.2.1. We use the following code to compute rf_n for three distributions.

```
eps=0.05; nsim=500; N=seq(500,3000,500)
meex=meg=mep=p1=p2=p3=c()
for(j in 1:length(N))
{
 for(m in 1:nsim)
{
 set.seed(m)
 x=rexp(N[j],1)
 y=rgeom(N[j],.5)
 z=rpois(N[j],1)
 meex[m]=mean(x)
 meg[m]=mean(y)
 mep[m]=mean(z)
}
 p1[j]=length(which(abs(meex-1)<eps))/nsim
 p2[j]=length(which(abs(meg-1)<eps))/nsim
 p3[j]=length(which(abs(mep-1)<eps))/nsim
}
d=data.frame(N,p1,p2,p3);d
```

TABLE 9.1
Verification of WLLN: Exponential, Geometric and Poisson Distributions

n	500	1000	1500	2000	2500	3000
Exponential	0.738	0.884	0.946	0.978	0.982	0.990
Geometric	0.586	0.766	0.818	0.896	0.914	0.948
Poisson	0.720	0.904	0.952	0.966	0.992	0.986

The output is organized in Table 9.1.

From Table 9.1, we note that $rf_n \to 1$ as n increases. The rate is comparatively slow for the geometric distribution. □

In the next example, we verify WLLN by examining that the sample mean, the sample variance and the sample second raw moment converge in probability to respective population characteristics.

Example 9.2.5. We simulate m random samples, each of size n from the binomial $B(6, 1/3)$ distribution and find the estimate rf_n. We examine whether $rf_n \to 1$ as $n \to \infty$. We have fixed $\epsilon = 0.05$, $m = 500$ and n takes values from 1000 to 6000 with an increment of 1000.

Code 9.2.2. We use the following code to compute rf_n for three characteristics, the sample mean, the sample variance and the sample second raw moment.

```
eps=0.05; nsim=1000; N=seq(1000,6000,1000)
me=me2=p1=p2=p3=c()
for(j in 1:length(N))
{
 for(m in 1:nsim)
{
 set.seed(m)
 x=rbinom(N[j],6,1/3)
 me[m]=mean(x)
 me2[m]=mean(x^2)
}
 v=me2-(me)^2
 p1[j]=length(which(abs(me-2)<eps))/nsim
 p2[j]=length(which(abs(me2-16/3)<eps))/nsim
 p3[j]=length(which(abs(v-4/3)<eps))/nsim
}
d=data.frame(N,p1,p2,p3);d
```

TABLE 9.2
Verification of WLLN: Binomial Distribution

n	1000	2000	3000	4000	5000	6000
Sample Mean	0.822	0.936	0.974	0.994	0.996	0.998
m_2'	0.232	0.317	0.387	0.440	0.494	0.520
Sample variance	0.627	0.805	0.871	0.917	0.948	0.967

The output is organized in Table 9.2.

From Table 9.2, we observe that $rf_n \to 1$ as n increases for all the sequences. However, rf_n converges faster to 1 for the sample mean, the rate is comparatively slow for sample variance and very slow for the sample second raw moment. □

The next section is devoted to SLLN.

9.3 Strong Law of Large Numbers

SLLN is similar to WLLN, the word "strong" basically replaces convergence in probability by the almost sure convergence as is clear from Definition 9.1.1. However, almost sure convergence being stronger than the convergence in probability, the techniques needed to establish SLLN are slightly involved than those for WLLN. In the Bernoulli WLLN as given in Corollary 9.2.3, we have noted that if $\{X_n, n \geq 1\}$ is a sequence of independent and identically distributed random variables each having Bernoulli $B(1,p)$ distribution, then the sample proportion $P_n = \overline{X}_n \xrightarrow{P} p$. In the following theorem, known as Bernoulli SLLN, we prove that the convergence in probability in Bernoulli WLLN can be extended to the almost sure convergence. To prove it we use the sufficient condition for almost sure convergence as given in (i) of Corollary 6.3.1.

Theorem 9.3.1. *Bernoulli SLLN: Suppose $\{X_n, n \geq 1\}$ is a sequence of independent and identically distributed random variables each having Bernoulli $B(1,p)$ distribution. Then*

$$(S_n - np)/n \xrightarrow{a.s.} 0 \quad \iff \quad P_n = \overline{X}_n \xrightarrow{a.s.} p.$$

Proof. Since $X_i \sim B(1,p)$ for each $i \geq 1$, we have

$$\begin{aligned} S_n \sim B(n,p) \;&\Rightarrow\; E(S_n) = np, \;\; Var(S_n) = np(1-p) \\ \& \;\; E((S_n - np)/n)^2 \;&=\; p(1-p)/n \\ &\Rightarrow\; P\left[|(S_n - np)/n| > \epsilon\right] \leq p(1-p)/n\epsilon^2 \;\; \forall \;\; \epsilon > 0, \end{aligned}$$

by Chebyshev's inequality. With the same arguments, replacing n by n^2 we have $\forall\ \epsilon > 0$,

$$\begin{aligned} P\left[|(S_{n^2} - n^2p)/n^2| > \epsilon\right] &\leq p(1-p)/n^2\epsilon^2 \\ \Rightarrow \sum_{n\geq 1} P\left[|(S_{n^2} - n^2p)/n^2| > \epsilon\right] &\leq \sum_{n\geq 1} p(1-p)/n^2\epsilon^2 < \infty \\ \Rightarrow\ S_{n^2}/n^2 &\stackrel{a.s.}{\to} p. \end{aligned}$$

Suppose k_n is an integer and $k_n \to \infty$ as $n \to \infty$, then using the same arguments we have, $S_{k_n^2}/k_n^2 \stackrel{a.s.}{\to} p$. We now prove that $|S_n/n - S_{k_n^2}/k_n^2| \to 0$ pointwise and hence almost surely. As a consequence, $S_{k_n^2}/k_n^2 \stackrel{a.s.}{\to} p$ implies that $S_n/n \stackrel{a.s.}{\to} p$. It is to be noted that $\forall\ n \geq 1\ \exists\ k_n$ such that $k_n^2 \leq n < (k_n+1)^2$. Now

$$k_n^2 \leq n < (k_n+1)^2 \ \Rightarrow\ 0 \leq n - k_n^2 < 2k_n + 1 \ \Rightarrow\ 0 \leq n - k_n^2 \leq 2k_n,$$

$n - k_n^2$ being an integer. Further as $n \to \infty$, $k_n \to \infty$. Since $X_i = 0$ or 1, for $n > k_n^2$ and $\forall\ \omega \in \Omega$ observe that,

$$\begin{aligned} \left|\frac{S_n}{n} - \frac{S_{k_n^2}}{k_n^2}\right| &= \left|\sum_{i=1}^{n} X_i/n - \sum_{i=1}^{k_n^2} X_i/k_n^2\right| = \left|\left(\frac{1}{n} - \frac{1}{k_n^2}\right)\sum_{i=1}^{k_n^2} X_i + \frac{1}{n}\sum_{i=k_n^2+1}^{n} X_i\right| \\ &\leq ((n-k_n^2)/nk_n^2)\sum_{i=1}^{k_n^2} X_i + (1/n)\sum_{i=k_n^2+1}^{n} X_i \\ &\leq ((n-k_n^2)k_n^2/nk_n^2)) + (n-k_n^2)/n = 2(n-k_n^2)/n \ \text{ as } \ X_i = 0, 1 \\ &\leq 4k_n/n \ \text{ as } \ n - k_n^2 \leq 2k_n \\ &\leq 4k_n/k_n^2 \ \text{ as } \ n > k_n^2 \\ &= 4/k_n \ \to 0 \ \text{ as } \ n \to \infty. \end{aligned}$$

Thus, $|S_n/n - S_{k_n^2}/k_n^2| \to 0$ pointwise and hence almost surely. As a consequence we have,

$$S_n/n - p \ = \ (S_n/n - S_{k_n^2}/k_n^2) + (S_{k_n^2}/k_n^2 - p) \ \stackrel{a.s.}{\to}\ p.$$

□

Remark 9.3.1. Another simpler proof for Bernoulli SLLN also uses the sufficient condition for almost sure convergence as given in (ii) of Corollary 6.3.1 and as illustrated in Example 6.3.8. Observe that $S_n \sim B(n,p)$, hence

$$\begin{aligned} E(S_n - np)^4 &= 3(npq)^2 + npq(1-6pq) \\ \Rightarrow\ E\left((S_n - np)/n\right)^4 &= (3(npq)^2 + npq(1-6pq))/n^4 \\ &= 3(pq)^2/n^2 + pq(1-6pq)/n^3 \\ \Rightarrow\ \sum_{n=1}^{\infty} E\left((S_n - np)/n\right)^4 &= \sum_{n=1}^{\infty}\left[3(pq)^2/n^2 + pq(1-6pq)/n^3\right] < \infty \\ \Rightarrow\ S_n/n &\stackrel{a.s.}{\to} p. \end{aligned}$$

Using Bernoulli SLLN we show in the following example that the empirical distribution function converges almost surely to the distribution function from which the sample is generated.

Example 9.3.1. Suppose $\{X_1, X_2, \cdots, X_n\}$ are independent and identically distributed random variables each having the distribution function $F(x), x \in \mathbb{R}$. For each fixed $x \in \mathbb{R}$, the empirical distribution function $F_n(x)$, also known as the sample distribution function, corresponding to $\{X_1, X_2, \cdots, X_n\}$ is defined as

$$F_n(x) = (\text{number of } \ X_i \leq x)/n \ = \sum_{i=1}^{n} Y_i(x)/n, \quad \text{where}$$

$$Y_i(x) \quad = \quad \begin{cases} 1, & \text{if} \quad X_i \leq x \\ 0, & \text{if} \quad X_i > x. \end{cases}$$

for $i = 1, 2, \cdots, n$. Thus, $Y_i(x)$ is a Borel function of X_i and hence $\{Y_i(x), i = 1, 2, \cdots, n\}$ are also independent and identically distributed random variables, each having Bernoulli distribution with $E(Y_i(x)) = F(x) < \infty$. Hence by the Bernoulli WLLN, $F_n(x) = \sum_{i=1}^{n} Y_i(x)/n \xrightarrow{P} E(Y_i) = F(x)$. Further,

$$nF_n(x) \sim B(n, F(x)) \ \Rightarrow \ Var(F_n(x)) = F(x)(1 - F(x))/n \ \leq 1/4n.$$

Thus, by Chebyshev's inequality it can be shown that the convergence in probability is uniform in x. By the Bernoulli SLLN, $F_n(x) = \sum_{i=1}^{n} Y_i(x)/n \xrightarrow{a.s.} F(x)$. It can further be proved that the almost sure convergence is also uniform in x, that is, $P[\sup_{x \in \mathbb{R}} |F_n(x) - F(x)| \to 0] = 1$, Loeve [16]. This result is a well known Glivenko Cantelli theorem. It states that the sample distribution function converges almost surely uniformly to the distribution function from which we have drawn the random sample. In other words, for large n, the sample distribution function is a good approximation for $F(x)$. It is a very important result and forms the basis of non-parametric inference. Hence, it is also referred to as the central statistical theorem, Loeve [16]. □

Remark 9.3.2. In Example 7.3.26 using Code 7.3.5, we have verified the convergence in law of $\{X_n, n \geq 1\}$ where $X_n \sim G(n, n)$, by showing that as n increases, the distribution function $F_n(x)$ of X_n approaches to the distribution function F of $X \equiv 1$ for all $x \in \mathbb{R} - \{1\}$. Based on simulated observations from X_n, we have obtained the empirical distribution function $G_n(x)$ and has verified that as n increases, it also approaches to that of $X \equiv 1$, see Figure 7.9.

We now proceed to the most important result, it is Kolmogorov's SLLN. It states that for a sequence of independent and identically distributed random variables with finite mean μ, sample mean converges almost surely to μ. As in

Khintchine's WLLN, we use truncation technique and related results as given in Theorem 9.2.2 and Lemma 8.2.3. The proof of Kolmogorov's SLLN is as outlined in Breiman [6]. We need one more result, known as Kolmogorov's proposition, in the proof of Kolmogorov's SLLN. We state it below.

Theorem 9.3.2. *Kolmogorov's proposition: If $\{X_n, n \geq 1\}$ is a sequence of independent random variables such that $\sum_{n\geq 1} Var(X_n)/n^2 < \infty$, then* $(S_n - E(S_n))/n \overset{a.s.}{\to} 0$.

The proof is based on Kolmogorov's inequality concerning the maximum of the partial sums. For the proof of Kolmogorov's inequality and Kolmogorov's proposition we refer to Loeve [16].

Using Theorem 9.2.2, Lemma 8.2.3 and Kolmogorov's proposition, we prove Kolmogorov's SLLN in the following theorem. The proof proceeds on the same lines as that of Khintchine's WLLN, with the only difference that to claim $(S_n^* - E(S_n^*))/n \overset{a.s.}{\to} 0$ we have to use Kolmogorov's proposition. The initial part of the proof is the same as that of Khintchine's WLLN.

Theorem 9.3.3. *Kolmogorov's SLLN: Suppose $\{X_n, n \geq 1\}$ is a sequence of independent and identically distributed random variables with finite mean μ, then the sequence follows SLLN with $a_n = E(S_n) = n\mu$ and $b_n = n$, that is, $(S_n - E(S_n))/n = \overline{X}_n - \mu \overset{a.s.}{\to} 0$. Conversely if, $\overline{X}_n \overset{a.s.}{\to} C$, then X_n's are integrable random variables and $C = \mu$ is the common expectation.*

Proof. (I) Suppose $\{X_n, n \geq 1\}$ is a sequence of independent and identically distributed random variables each having distribution same as that of X, with finite mean μ. As in Khintchine's WLLN, for every $n \geq 1$, we define a random variable X_n^* as follows.

$$X_n^* \quad = \quad \begin{cases} X_n, & \text{if} \quad |X_n| < n \\ 0, & \text{if} \quad |X_n| \geq n. \end{cases}$$

Note that being Borel functions of X_n, $\{X_n^*, n \geq 1\}$ is also a sequence of independent random variables. Suppose $S_n^* = \sum_{k=1}^{n} X_k^*$ and an event E_n is defined as $E_n = [|X_n| \geq n]$, $n \geq 0$ with $E_0 = \Omega$. Thus, $X_n^* = X_n$ on E_n^c and $X_n^* \neq X_n$ on E_n. By Theorem 8.2.3,

$$\begin{aligned} \sum_{n\geq 1} P[X_n \neq X_n^*] &= \sum_{n\geq 1} P(E_n) = \sum_{n\geq 1} P[|X_n| \geq n] \;\leq\; E(|X_n|) < \infty \\ &\Rightarrow \{X_n, n \geq 1\} \;\&\; \{X_n^*, n \geq 1\} \;\; \text{are equivalent sequences} \\ &\Rightarrow (S_n - S_n^*)/n \overset{a.s.}{\to} 0, \qquad (9.5) \end{aligned}$$

the last step follows by result (ii) of Theorem 9.2.2. Thus, limiting behaviour of S_n/n is the same as that of S_n^*/n. We now study the limiting behaviour of S_n^*/n. Observe that for each $n \geq 1$,

$$E(X_n^*) = \int_\Omega X_n^* dP = \int_{E_n} X_n^* dP + \int_{E_n^c} X_n^* dP = \int_{E_n^c} X_n dP = \int_{E_n^c} X dP = E(X I_{E_n^c}),$$

since X_n is distributed as X. Further, observe that $|XI_{E_n^c}| \leq |X|$ and $E(|X|) < \infty$. From the definition of E_n^c, note that for $n_1 < n_2$,

$$
\begin{aligned}
|X_n(\omega)| &< n_1 \Rightarrow |X_n(\omega)| < n_2 \Rightarrow E_{n_1}^c \subset E_{n_2}^c \\
&\Rightarrow \{E_n^c, n \geq 1\} \uparrow \bigcup_{n \geq 1} E_n^c = \Omega \ \text{ as } n \to \infty \\
&\Rightarrow I_{E_n^c} \overset{a.s.}{\to} I_\Omega \Rightarrow XI_{E_n^c} \overset{a.s.}{\to} XI_\Omega = X \text{ as } n \to \infty \\
&\Rightarrow E(XI_{E_n^c}) \to E(X) \text{ by dominated a.s. convergence theorem} \\
&\Rightarrow E(X_n^*) \to \mu \quad \text{as } n \to \infty \\
&\Rightarrow E(S_n^*/n) = (1/n)\sum_{k=1}^{n} E(X_k^*) \to \mu \ \text{ by Toeplitz' lemma.} \quad (9.6)
\end{aligned}
$$

To decide the nature of $\sum_{n \geq 1} Var(X_n^*)/n^2$, we define an event A_k as $A_k = [k \leq |X_n| < k+1]$, $k \geq 0$. Further as proved in Lemma 8.2.3, these events are mutually exclusive. Then the event E_n^c can be expressed as,

$$
\begin{aligned}
E_n^c &= [|X_n| < n] = \bigcup_{k=1}^{n} [k-1 \leq |X_n| < k] = \sum_{k=1}^{n} A_{k-1}, \ n \geq 1 \\
\Rightarrow Var(X_n^*) &\leq E((X_n^*)^2) = \int_{E_n^c} X_n^2 dP = \int_{\sum_{k=1}^{n} A_{k-1}} X_n^2 dP \\
&= \sum_{k=1}^{n} \int_{A_{k-1}} X_n^2 dP \ \leq \ \sum_{k=1}^{n} k^2 \int_{A_{k-1}} dP \\
&= \sum_{k=1}^{n} k^2 P(A_{k-1}), \text{ since on } A_{k-1} \ X_n^2 \leq k^2.
\end{aligned}
$$

Since X_n's are distributed as X,

$$P(A_{k-1}) = P[k \leq |X_n| < k+1] = P[k \leq |X| < k+1].$$

Hence, $\displaystyle\sum_{n \geq 1} \frac{Var(X_n^*)}{n^2} \leq \sum_{n \geq 1} \sum_{k=1}^{n} (k^2/n^2) P(A_{k-1}) = \sum_{k \geq 1} k^2 P(A_{k-1}) \sum_{n=k}^{\infty} (1/n^2).$

To find a bound on $\sum_{n=k}^{\infty}(1/n^2)$, suppose $B_n = (n-1, n]$, $n \geq 1$ so that $\sum_{n=k+1}^{\infty} B_n = (k, \infty) = B$, say. Note that

$$
\begin{aligned}
\int_k^\infty \frac{1}{x^2}\, dx &= \int_B \frac{1}{x^2}\, dx = \sum_{n=k+1}^{\infty} \int_{B_n} \frac{1}{x^2}\, dx \ \geq \ \sum_{n=k+1}^{\infty} \frac{1}{n^2} \int_{B_n} dx \\
&= \sum_{n=k+1}^{\infty} \frac{1}{n^2} \int_{n-1}^{n} dx = \sum_{n=k+1}^{\infty} \frac{1}{n^2}
\end{aligned}
$$

$$\text{Hence, } \sum_{n=k}^{\infty} \frac{1}{n^2} = \frac{1}{k^2} + \sum_{n=k+1}^{\infty} \frac{1}{n^2} \le \frac{1}{k^2} + \int_k^{\infty} \frac{1}{x^2}\,dx$$

$$= \frac{1}{k^2} + \frac{1}{k} \le 2/k \quad \text{as } k^2 \ge k \iff 1/k^2 \le 1/k$$

$$\begin{aligned}
\text{Hence, } \sum_{n\ge1} Var(X_n^*)/n^2 &\le \sum_{k\ge1} k^2 P(A_{k-1}) \sum_{n=k}^{\infty}(1/n^2) \le \sum_{k\ge1} k^2 P(A_{k-1})(2/k) \\
&= 2\sum_{k\ge1} kP(A_{k-1}) = 2\sum_{k\ge0}(k+1)P(A_k) \\
&= 2(1+\sum_{n\ge1} P(E_n)) \quad \text{as in Theorem 8.2.3} \\
&= 2(1+\sum_{n\ge1} P[|X_n| \ge n]) = 2(1+\sum_{n\ge1} P[|X| \ge n]) \\
&\le 2(1+E(X)) < \infty.
\end{aligned}$$

The second last step follows by Theorem 8.2.3. Thus, $\{X_n^*, n \ge 1\}$ is a sequence of independent random variables such that $\sum_{n\ge1} Var(X_n^*)/n^2 < \infty$. Hence, by Kolmogorov's proposition

$$(S_n^* - E(S_n^*))/n \stackrel{a.s.}{\to} 0 \tag{9.7}$$

Using the results in Equation (9.5), Equation (9.6) and Equation (9.7), we have

$$\begin{aligned}
S_n/n - \mu &= (S_n - S_n^* + S_n^*)/n - \mu + E(S_n^*)/n - E(S_n^*)/n \\
&= (S_n - S_n^*)/n + (S_n^* - E(S_n^*))/n + E(S_n^*)/n - \mu \stackrel{a.s}{\to} 0 \\
\Rightarrow & \ S_n/n \stackrel{a.s.}{\to} \mu.
\end{aligned}$$

(II) Conversely, suppose $S_n/n \stackrel{a.s.}{\to} C$. To prove that X_n's are integrable random variables, we use the fact that $\{X_n, n \ge 1\}$ are independent random variables and hence the Borel zero-one law is applicable. Observe that,

$$\begin{aligned}
X_n/n &= (S_n - S_{n-1})/n = S_n/n - ((n-1)/n)(S_{n-1}/(n-1)) \stackrel{a.s.}{\to} 0 \\
&\Rightarrow P\left[\limsup[|X_n/n| > \epsilon]\right] = 0 \ \forall \ \epsilon > 0 \\
&\Rightarrow P[\limsup[|X_n| > n]] = 0 \ \text{ with } \ \epsilon = 1 \\
&\Rightarrow \sum_{n\ge1} P[|X_n| > n] < \infty \text{ by Borel zero-one law} \\
&\Rightarrow E(X_n) < \infty \ \text{ by Theorem 8.2.3,}
\end{aligned}$$

that is, X_n's are integrable random variables. Thus, $\{X_n, n \ge 1\}$ is a sequence of independent and identically distributed random variables with finite mean $E(X)$. Hence by part (I), $S_n/n \stackrel{a.s.}{\to} E(X)$. Since the limit random variable in almost sure convergence is almost surely unique, $C = E(X)$. □

SLLN is useful to establish strong consistency of estimators which are averages. From SLLN one can conclude that the r-th raw moment based on a random sample is a strongly consistent estimator for the corresponding population moment. Using the property that almost sure convergence is closed under arithmetic operations, it follows that r-th central moment based on a random sample is a strongly consistent estimator for the corresponding population moment.

Example 9.3.2. Suppose $\{X_1, X_2, \cdots, X_n\}$ are independent and identically distributed random variables with mean μ and variance $\sigma^2 < \infty$. By SLLN,

$$(1/n)\sum_{i=1}^{n} X_i \stackrel{a.s.}{\to} \mu \ \& \ (1/n)\sum_{i=1}^{n} X_i^2 \stackrel{a.s.}{\to} \sigma^2 + \mu^2$$

$$\Rightarrow \sum_{i=1}^{n} X_i \Big/ \sum_{i=1}^{n} X_i^2 \stackrel{a.s.}{\to} \mu/(\sigma^2 + \mu^2). \qquad \square$$

We summarize below the results discussed in this chapter.

Summary

1. A sequence $\{X_n, n \geq 1\}$ of random variables is said to obey a SLLN, if there exists two sequences $\{a_n, n \geq 1\}$ of real numbers and $\{b_n, n \geq 1\}$ of positive real numbers tending to ∞ such that, $(S_n - a_n)/b_n \stackrel{a.s.}{\to} 0$ and a WLLN, if $(S_n - a_n)/b_n \stackrel{P}{\to} 0$.

2. Chebyshev's WLLN: Suppose $\{X_n, n \geq 1\}$ is a sequence of uncorrelated random variables with uniformly bounded variances, then the sequence follows WLLN with $a_n = E(S_n)$ and $b_n = n$.

3. Suppose $\{X_n, n \geq 1\}$ is a sequence of independent random variables with uniformly bounded variances, then the sequence follows WLLN with $a_n = E(S_n)$ and $b_n = n$.

4. Suppose $\{X_n, n \geq 1\}$ is a sequence of independent and identically distributed random variables with mean μ and finite variance σ^2, then the sequence follows WLLN with $a_n = E(S_n) = n\mu$ and $b_n = n$.

5. Khintchine's WLLN: Suppose $\{X_n, n \geq 1\}$ is a sequence of independent and identically distributed random variables with finite mean μ, then the sequence follows WLLN with $a_n = E(S_n) = n\mu$ and $b_n = n$, that is, $(S_n - E(S_n))/n = \overline{X}_n - \mu \stackrel{P}{\to} 0$.

6. Bernoulli SLLN: Suppose $\{X_n, n \geq 1\}$ is a sequence of independent and identically distributed random variables each having Bernoulli $B(1, p)$ distribution. Then $(S_n - np)/n \stackrel{a.s.}{\to} 0 \iff P_n = \overline{X}_n \stackrel{a.s.}{\to} p$.

7. Kolmogorov's SLLN: Suppose $\{X_n, n \geq 1\}$ is a sequence of independent and identically distributed random variables with finite mean μ, then the sequence follows SLLN with $a_n = E(S_n) = n\mu$ and $b_n = n$, that is, $(S_n - E(S_n))/n = \overline{X}_n - \mu \stackrel{a.s.}{\rightarrow} 0$. Conversely if, $\overline{X}_n \stackrel{a.s.}{\rightarrow} C$, then X_n's are integrable random variables and $C = \mu$ is the common expectation.

9.4 Conceptual Exercises

9.4.1 Suppose $\{X_n, n \geq 1\}$ is a sequence of independent random variables such that $P[X_n = 2^n] = 1/2^n,\ n \geq 1$. Examine whether the sequence $\{X_n, n \geq 1\}$ obeys the WLLN. Find more that one sequences $\{a_n, n \geq 1\}$ and $\{b_n, n \geq 1\}$ with which WLLN holds.

9.4.2 Suppose $\{X_n, n \geq 1\}$ is a sequence of independent random variables with $E(X_n) = 0$ and $V(X_n) = n^2$. Examine whether the sequence $\{X_n, n \geq 1\}$ obeys the WLLN.

9.4.3 Suppose $\{X_n, n \geq 1\}$ is a sequence of independent random variables with $E(X_n) = 0$ and $V(X_n) = n$. Examine whether the sequence $\{X_n, n \geq 1\}$ obeys the WLLN.

9.4.4 Suppose $\{X_n, n \geq 1\}$ is a sequence of independent random variables with $P[X_n = 2^n] = P[X_n = -2^n] = 2^{-2(n+1)}$ and $P[X_n = 0] = 1 - 2^{-2n-1}$. Examine whether the sequence $\{X_n, n \geq 1\}$ obeys the WLLN.

9.4.5 Suppose $\{X_n, n \geq 1\}$ is a sequence of independent random variables with $P[X_n = n] = P[X_n = -n] = 1/(2n^\lambda)$ and $P[X_n = 0] = 1 - 1/n^\lambda,\ \lambda > 0$. Examine whether the sequence $\{X_n, n \geq 1\}$ obeys the WLLN.

9.4.6 Suppose $\{X_n, n \geq 1\}$ is a sequence of independent random variables with $P[X_n = n^\alpha] = P[X_n = -n^\alpha] = 1/2n^2$ and $P[X_n = 0] = 1 - 1/n^2$, where $\alpha < 3/2$. Examine whether the sequence $\{X_n, n \geq 1\}$ obeys the WLLN.

9.4.7 Suppose $\{X_n, n \geq 1\}$ is a sequence of independent random variables with $P[X_n = n/\log n] = P[X_n = -n/\log n] = \log n/2n$ and $P[X_n = 0] = 1 - \log n/n$. Examine whether the sequence $\{X_n, n \geq 1\}$ obeys the WLLN.

9.4.8 Suppose $\{X_n, n \geq 1\}$ is a sequence of independent and identically distributed random variables with the characteristic function

$$\phi(t) = \begin{cases} 1 - |t|, & \text{if } |t| \leq 1 \\ 0, & \text{if } |t| > 1. \end{cases}$$

Show that $S_n/n \xrightarrow{L} X$, where X has Cauchy distribution with location parameter 0 and scale parameter 1 and hence WLLN does not hold.

9.4.9 Suppose $\{X_n, n \geq 1\}$ is a sequence of independent random variables such that $P[X_n = \sqrt{n}] = P[X_n = -\sqrt{n}] = 1/2$. (i) Show that the characteristic function $\phi_n(t)$ of S_n/n is $\prod_{k=1}^{n} \cos\left(t\sqrt{k}/n\right)$ and converges to $\exp(-t^2/4)$ as $n \to \infty$. (ii) Hence, show that $S_n/n \stackrel{P}{\nrightarrow} 0$. (iii) Find a sequence $\{a_n, n \geq 1\}$ and $\{b_n, n \geq 1\}$ such that $(S_n - a_n)/b_n \xrightarrow{P} 0$ as $n \to \infty$.

9.4.10 Suppose $\{X_n, n \geq 1\}$ is a sequence of random variables such that

$$E(X_n) = 0, \quad Var(X_n) = \sigma^2 \;\;\&\;\; Cov(X_n, X_m) = \sigma^2\alpha^{|m-n|}, \; |\alpha| < 1.$$

Verify whether WLLN holds.

9.4.11 Suppose $\{X_n, n \geq 1\}$ is a sequence of independent and identically distributed random variables with $Var(X_1) < \infty$. Show that $(2/n(n+1)) \sum_{i=1}^{n} iX_i \xrightarrow{P} E(X_1)$.

9.5 Computational Exercises

9.5.1 Repeat Example 9.2.4 for any two discrete and continuous distributions, other than those in Example 9.2.4.

9.5.2 Repeat Example 9.2.5 when random samples, each of size n are simulated from the Poisson $Poi(1.2)$ distribution.

9.6 Multiple Choice Questions

Note: In each question, multiple options may be correct. Unless specified otherwise, identify which of the statement(s) is/are correct. Answers are given in Chapter 11, after the solutions of conceptual exercises of Chapter 9.

9.6.1 A sequence $\{X_n, n \geq 1\}$ of random variables is said to obey a SLLN, if there exists two sequences $\{a_n, n \geq 1\}$ of real numbers and $\{b_n, n \geq 1\}$ of positive real numbers, with $b_n \to \infty$, such that $n \to \infty$,

(a) $(S_n - a_n)/b_n \xrightarrow{P} 0$ where $S_n = \sum_{i=1}^{n} X_i$

(b) $(S_n - a_n)/b_n \xrightarrow{a.s.} 0$
(c) $(S_n - a_n)/b_n \xrightarrow{L} 0$
(d) $(S_n - a_n)/b_n \xrightarrow{q.m.} 0$

9.6.2 A sequence $\{X_n, n \geq 1\}$ of random variables is said to obey a WLLN, if there exists two sequences $\{a_n, n \geq 1\}$ of real numbers and $\{b_n, n \geq 1\}$ of positive real numbers, with $b_n \to \infty$, such that $n \to \infty$,

(a) $(S_n - a_n)/b_n \xrightarrow{L} 0$ where $S_n = \sum_{i=1}^{n} X_i$
(b) $(S_n - a_n)/b_n \xrightarrow{a.s.} 0$
(c) $(S_n - a_n)/b_n \xrightarrow{P} 0$
(d) $(S_n - a_n)/b_n \xrightarrow{q.m.} 0$

9.6.3 A sequence $\{X_n, n \geq 1\}$ of random variables satisfies a WLLN, if as $n \to \infty$

(a) $(S_n - a_n)/b_n \xrightarrow{L} 0$
(b) $(S_n - a_n)/b_n \xrightarrow{a.s.} 0$
(c) $(S_n - a_n)/b_n \xrightarrow{P} 0$
(d) $(S_n - a_n)/b_n \xrightarrow{q.m.} 0$
where $S_n = \sum_{i=1}^{n} X_i$, $\{a_n, n \geq 1\}$ is a sequence of real numbers and $\{b_n, n \geq 1\}$ is a sequence of positive real numbers with $b_n \to \infty$ as $n \to \infty$.

9.6.4 Suppose $\{X_n, n \geq 1\}$ is a sequence of independent random variables. Kolmogorov's SLLN is applicable, if

(a) $X_n \sim$ Poisson(2)
(b) $X_n \sim$ Exp(0.1)
(c) $X_n \sim t_1$
(d) $X_n \sim$ Beta(1, 1)

9.6.5 Suppose $\{X_n, n \geq 1\}$ is a sequence of independent random variables. Khintchine's WLLN is not applicable if

(a) $X_n \sim$ Poisson(2)
(b) $X_n \sim$ Exp(0.1)
(c) $X_n \sim t_1$
(d) $X_n \sim$ Beta(1, 1)

9.6.6 Suppose $\{X_n, n \geq 1\}$ is a sequence of independent random variables. Khintchine's WLLN is applicable, if

(a) $X_n \sim$ Geometric(0.2)
(b) $X_n \sim \chi^2_5$
(c) $X_n \sim t_1$
(d) $X_n \sim$ Beta(1, 1)

9.6.7 Suppose $\{X_n, n \geq 1\}$ is a sequence of independent and identically distributed random variables with finite mean μ and $S_n = \sum_{k=1}^{n} X_k$. Following are two statements:
(I) $S_n/n \xrightarrow{P} \mu$. (II) $S_n/n \xrightarrow{a.s.} \mu$. Then

(a) Both (I) and (II) are false
(b) Both (I) and (II) are true
(c) (I) is true but (II) is false
(d) (I) is false but (II) is true

9.6.8 Suppose $\{X_1, X_2, \cdots, X_n\}$ are independent and identically distributed random variables, each having the distribution function $F(x), x \in \mathbb{R}$. Suppose $F_n(x)$ is the empirical distribution function. Then for each fixed $x \in \mathbb{R}$,

(a) $F_n(x) \xrightarrow{a.s.} F(x)$, uniformly in x
(b) $F_n(x) \xrightarrow{a.s.} F(x)$, for each fixed x
(c) $F_n(x) \xrightarrow{P} F(x)$, for each fixed x
(d) $\sqrt{n}(F_n(x) - F(x))$ converges in law, for each fixed x

9.6.9 Glivenko-Cantelli theorem states that

(a) $F_n(x) \xrightarrow{a.s.} F(x)$, uniformly in x
(b) $F_n(x) \xrightarrow{a.s.} F(x)$, for each fixed x
(c) $F_n(x) \xrightarrow{P} F(x)$, for each fixed x
(d) $\sqrt{n}(F_n(x) - F(x))$ converges in law, for each fixed x

10

Central Limit Theorem

10.1 Introduction

The sample mean is the most frequently used statistic in statistical estimation and in testing of hypotheses and its large sample behaviour is quite important in these inference procedures. The WLLN and SLLN state that the distribution of the sample mean piles up near the population mean, however, this information is not enough to approximate the probability statements about the sample mean. Using WLLN and SLLN, one can claim that the distribution of the sample mean is degenerate at the population mean, but it is not useful to obtain confidence intervals. With suitable normalization, its limiting non-degenerate distribution is obtained in one of the most fundamental theorems in statistics and it is the central limit theorem (CLT).

CLT is the most useful result in probability theory. It states that the mean of a large number of random variables, no matter what their distribution is, with the only condition on the variance of the random variables, is approximately normally distributed. Empirical findings in applied sciences, dating back to the seventeenth century, showed that the averages of laboratory measurements on various physical quantities tend to have a bell shaped distribution. The CLT provides a theoretical justification for this observation. The importance of CLT is that it is one of the first theorems that allowed respectability to probability theory, which was initially thought of as only related to gambling and card playing.

CLT plays a central role in probability theory and in asymptotic inference and hence is named as the central limit theorem by Polya in 1920. A remarkable feature of the result is that it is distribution-free, that is, there is no assumption on the distribution of the random variables, it could be uniform or exponential or Poisson or chi-square. In the setup of independent and identically distributed random variables, the only condition is finite positive second moment. In view of this distribution-free feature, it is possible to derive from it statistical inference procedures which are asymptotically valid without specific distributional assumptions.

CLT is not a single theorem, but it embodies a variety of results concerned with a sum of a large number of random variables which, suitably normalized, has a limit distribution as a normal distribution. Lindeberg-Levy CLT is the simplest version of CLT, a generalization of the De Moivre-Laplace theorem.

DOI: 10.1201/9781032619057-10

The gain in generality from the De Moivre-Laplace to the Lindeberg-Levy CLT is enormous. The De Moivre-Laplace theorem states that if X_n has binomial $B(n,p)$ distribution, then

$$\sqrt{n}(X_n/n - p) \xrightarrow{L} X \sim N(0, p(1-p)) \text{ distribution.}$$

In Lindeberg-Levy CLT, starting from virtually no assumptions about the distribution of random variables in the sum, apart from the finiteness of variance and variance being positive, we end up with normality. The condition that the variance is positive excludes the case that the random variables have degenerate distribution while finiteness of variance excludes the distributions with infinite variance, such as Cauchy distribution. Lindeberg-Feller CLT and Lyapounov's CLT are further extensions of Lindeberg-Levy CLT, when the random variables in the sum need not be identically distributed. Random sum CLT is concerned with the setup when the number of summands is a discrete integer valued random variable. Multivariate CLT deals with the sequence of random vectors. CLT is also established in the setup of Markov dependence and martingales.

The usefulness of the CLT is greatly increased by Slutsky's theorem and by the delta method which extends the limit result to smooth functions of random variables having asymptotically normal distribution.

The next section deals with the simplest but the most heavily applied version of CLT. It is the Lindeberg-Levy CLT, which is stated in Chapter 7.

10.2 Lindeberg-Levy CLT

The proof of Lindeberg-Levy CLT is based on Theorem 7.3.14, that is, the continuity theorem for characteristic function and one more result proved in Theorem 5.3.5, which states that a characteristic function of sum of n independent and identically distributed random variables is the n-th power of the common characteristic function of the summands. We give two proofs of the theorem, one in Theorem 10.2.1, which is rather approximate and another in Theorem 10.2.2, which is rigorous.

Theorem 10.2.1. *Lindeberg-Levy CLT: Suppose $\{X_n, n \geq 1\}$ is a sequence of independent and identically distributed random variables with mean μ and positive, finite variance σ^2. Suppose $S_n = \sum_{i=1}^n X_i$. Then*

$$Y_n = (S_n - n\mu)/\sqrt{n}\sigma = \sqrt{n}(\overline{X}_n - \mu)/\sigma \quad \xrightarrow{L} \quad Z \sim N(0,1).$$

Proof. Suppose $\phi_n(t)$ and $\phi(t)$ denote the characteristic functions of Y_n and of X_1 respectively for $t \in \mathbb{R}$. Observe that,

$$\begin{aligned}
\phi_n(t) &= E(\exp(itY_n)) = E\left(\exp it\left((S_n - n\mu)/\sqrt{n}\sigma\right)\right) \\
&= \exp(-it\sqrt{n}\mu/\sigma)E(\exp(itS_n/\sqrt{n}\sigma)) = \exp(-it\sqrt{n}\mu/\sigma)\phi_{S_n}(t/\sqrt{n}\sigma) \\
&= \exp(-it\sqrt{n}\mu/\sigma)(\phi(t/\sqrt{n}\sigma))^n.
\end{aligned}$$

Now taking logarithm of both sides, using series expansion of e^x and $\log(1+x)$, which is valid for complex x as well, we get,

$$\begin{aligned}\log\phi_n(t) &= -it\sqrt{n}\mu/\sigma + n\log(\phi(t/\sqrt{n}\sigma)) \\ &= -it\sqrt{n}\mu/\sigma + n\log(E(\exp(itX/\sqrt{n}\sigma))) \\ &= -it\sqrt{n}\mu/\sigma + n\log\left[E\left\{1+\frac{itX}{\sqrt{n}\sigma}-\frac{t^2X^2}{2n\sigma^2}+\cdots\right\}\right] \\ &= -it\sqrt{n}\mu/\sigma + n\log\left[1+\frac{it\mu}{\sqrt{n}\sigma}-\frac{t^2E(X^2)}{2n\sigma^2}+\cdots\right]\end{aligned}$$

Suppose $u_n = it\mu/\sqrt{n}\sigma - t^2E(X^2)/2n\sigma^2 + \cdots$. Then $\log(1+u_n) = u_n - u_n^2/2 + u_n^3/3 - \cdots$. Substituting for u_n and simplifying we get,

$$\begin{aligned}\log\phi_n(t) &= -it\sqrt{n}\mu/\sigma + it\sqrt{n}\mu/\sigma - t^2E(X^2)/2\sigma^2 + t^2\mu^2/2\sigma^2 + \delta_n \\ &= -(t^2/2\sigma^2)(E(X^2)-\mu^2) + \delta_n = -t^2/2 + \delta_n,\end{aligned}$$

where δ_n consists of all terms which tend to 0 as n tends to ∞. Thus, $\phi_n(t) \to \exp(-t^2/2)$ as $n \to \infty$. Observe that $\exp(-t^2/2)$ is continuous at $t = 0$ and it is the characteristic function of $Z \sim N(0,1)$. Hence, by Theorem 7.3.14 and the uniqueness theorem for characteristic function, $Y_n \xrightarrow{L} Z \sim N(0,1)$. □

In the above proof, the term δ_n consists of moments of order larger than two. We do not have any assumption about these higher order moments. To avoid these issues, we give below another more rigorous and elegant proof of the same theorem. Similar approach is adopted in the Lindeberg-Feller CLT in the next section, when random variables in the sequence are independent but not identically distributed. The proof is based on Lemma 9.2.2, which states that if $u_1, u_2, ..., u_m$ and $w_1, w_2, ..., w_m$ are complex numbers with $|u_i| \leq 1$ and $|w_i| \leq 1,\ i = 1, 2, \cdots, m$. Then,

$$\left|\prod_{i=1}^m u_i - \prod_{i=1}^m w_i\right| \leq \sum_{i=1}^m |u_i - w_i|.$$

We also need Theorem 9.2.4, which states that if $E(|X|^k) < \infty$ and if ϕ is a characteristic function of X, then $\forall\ t \in \mathbb{R}$,

$$\left|\phi(t) - \sum_{j=0}^k (it)^j E(X^j)/j!\right| \leq E\Big(\min\Big\{2|tX|^k/k!,\ \ |tX|^{k+1}/(k+1)!\Big\}\Big).$$

Thus with $k = 2$,

$$\left|\phi(t) - 1 - itE(X) + t^2E(X^2)/2\right| \leq E\Big(\min\Big\{|tX|^2,\ \ |tX|^3/6\Big\}\Big).$$

Theorem 10.2.2. *Lindeberg-Levy CLT: Suppose $\{X_n, n \geq 1\}$ is a sequence of independent and identically distributed random variables with mean μ and positive, finite variance σ^2. Suppose $S_n = \sum_{i=1}^n X_i$. Then*

$$Y_n = (S_n - n\mu)/\sqrt{n}\sigma = \sqrt{n}(\overline{X}_n - \mu)/\sigma \quad \xrightarrow{L} \quad Z \sim N(0,1).$$

Proof. Without loss of generality, we assume that $\mu = 0$ and $\sigma = 1$. Otherwise, we define $W_n = (X_n - \mu)/\sigma$ and work with the sequence $\{W_n, n \geq 1\}$ which is a sequence of independent and identically distributed random variables with mean 0 and variance 1. Suppose $\phi(\cdot)$ is the common characteristic function of X_n's, then by Theorem 5.3.5, the characteristic function of $S_n/\sqrt{n}$ is given by $\phi^n(t/\sqrt{n})$. It is known that $\phi_Z(t) = \exp(-t^2/2)$ is the characteristic function of $Z \sim N(0,1)$ distribution, which can be expressed as

$$\phi_Z(t) = e^{-t^2/2} = \left\{e^{-t^2/2n}\right\}^n = \phi_Z^n(t/\sqrt{n}).$$

Now observe that,

$$\begin{aligned}
\left|\phi_{S_n/\sqrt{n}}(t) - \phi_Z(t)\right| &= \left|\phi^n(t/\sqrt{n}) - e^{-t^2/2}\right| = \left|\phi^n(t/\sqrt{n}) - \phi_Z^n(t/\sqrt{n})\right| \\
&= \left|\prod_{k=1}^n \phi(t/\sqrt{n}) - \prod_{k=1}^n \phi_Z(t/\sqrt{n})\right| \\
&\leq \sum_{k=1}^n |\phi(t/\sqrt{n}) - \phi_Z(t/\sqrt{n})| \quad \text{by Lemma 9.2.2} \\
&= n \times |\phi(t/\sqrt{n}) - \phi_Z(t/\sqrt{n})| \\
&= n \times |\phi(t/\sqrt{n}) - (1 - t^2/2n) + (1 - t^2/2n) - \phi_Z(t/\sqrt{n})| \\
&\leq n \times |\phi(t/\sqrt{n}) - (1 - t^2/2n)| \\
&\quad + n \times |\phi_Z(t/\sqrt{n}) - (1 - t^2/2n)|. \qquad (10.1)
\end{aligned}$$

Since $E(X_1^2)$ & $E(Z^2) < \infty$, with $k = 2$ in Theorem 9.2.4 we have,

$$\begin{aligned}
|\phi(t) - (1 - t^2/2)| &\leq E\left(\min\left\{|tX_1|^2,\ |tX_1|^3/6\right\}\right) \\
\& \ |\phi_Z(t) - (1 - t^2/2)| &\leq E\left(\min\left\{|tZ|^2,\ |tZ|^3/6\right\}\right).
\end{aligned}$$

Consequently,

$$\begin{aligned}
|\phi(t/\sqrt{n}) - (1 - t^2/2n)| &\leq E\left(\min\left\{|tX_1|^2/n,\ |tX_1|^3/6n^{3/2}\right\}\right) \\
\& \ |\phi_Z(t/\sqrt{n}) - (1 - t^2/2n)| &\leq E\left(\min\left\{|tZ|^2/n,\ |tZ|^3/6n^{3/2}\right\}\right).
\end{aligned}$$

With these substitutions, inequality in (10.1) reduces as follows.

$$\begin{aligned}
\left|\phi_{S_n/\sqrt{n}}(t) - \phi_Z(t)\right| &\leq n \times E\left(\min\left\{|tX_1|^2/n,\ |tX_1|^3/6n^{3/2}\right\}\right) \\
&+ n \times E\left(\min\left\{|tZ|^2/n,\ |tZ|^3/6n^{3/2}\right\}\right) \\
&= E(U_n) + E(V_n), \qquad (10.2)
\end{aligned}$$

where $U_n = \min\left\{|tX_1|^2,\ \ |tX_1|^3/6n^{1/2}\right\}$ and $V_n = \min\left\{|tZ|^2,\ |tZ|^3/6n^{1/2}\right\}$. It is to be noted that

$$\forall\ n \geq 1,\ \ U_n \leq |tX_1|^2\quad \&\quad E(X_1^2) < \infty,\quad V_n \leq |tZ|^2\quad \&\quad E(Z^2) < \infty.$$

Thus, both U_n and V_n are dominated by integrable random variables. Further, for large n, $U_n = |tX_1|^3/6n^{1/2}$ and $V_n = |tZ|^3/6n^{1/2}$. Note that X_1 and Z are bounded in probability, being real random variables and hence $n \to \infty$

$$\begin{aligned} U_n &= |tX_1|^3/6n^{1/2} \xrightarrow{P} 0\ \ \&\ \ V_n = |tZ|^3/6n^{1/2} \xrightarrow{P} 0 \\ &\Rightarrow E(U_n) \to 0\ \ \&\ \ E(V_n) \to 0, \end{aligned}$$

using the Lebesgue dominated probability convergence theorem. Taking limits as $n \to \infty$ on both sides of the inequality in (10.2) we get,

$$\lim_{n\to\infty} \left|\phi_{S_n/\sqrt{n}}(t) - \phi_Z(t)\right| = 0 \ \Rightarrow\ \lim_{n\to\infty} \phi_{S_n/\sqrt{n}}(t) = \phi_Z(t) = e^{-t^2/2}.$$

Thus, the sequence $\{\phi_{S_n/\sqrt{n}}(t), n \geq 1\}$ of characteristic functions of $S_n/\sqrt{n}$ converges to $\exp(-t^2/2)$, which is continuous at zero. Hence by the continuity theorem for characteristic functions, the corresponding sequence of distribution functions converges in law to a distribution function with characteristic function $\exp(-t^2/2)$. But $\exp(-t^2/2)$ is the characteristic function of $Z \sim N(0,1)$ and hence by the uniqueness theorem of characteristic functions $S_n/\sqrt{n} \xrightarrow{L} Z \sim N(0,1)$. □

Remark 10.2.1. When $\{X_n, n \geq 1\}$ is a sequence of independent and identically distributed random variables, each having Bernoulli $B(1,p)$ distribution, then the Lindeberg-Levy theorem reduces to Demoivre-Lapalce CLT.

For all standard discrete and continuous distributions with mean μ and positive and finite variance σ^2, we can claim that the large sample distribution of the sample mean is normal, using the Lindeberg-Levy CLT. If the distribution function of $\sqrt{n}(\overline{X}_n - \mu)/\sigma$ is denoted by $F_n(x)$, then the CLT states that $F_n(x) \to \Phi(x)$ as $n \to \infty$. However, it provides no indication of the speed of the convergence of $F_n(x)$ to $\Phi(x)$ and hence of the error to be expected when approximating $F_n(x)$ by $\Phi(x)$. Thus, although the CLT gives a useful general approximation, there is no way of knowing how good the approximation is. In fact, goodness of the approximation is a function of the original distribution and so must be checked case by case. In Example 10.2.1 and in Example 10.2.2, we illustrate how the approximation can be judged by simulation.

The approximation we discussed above is the first order approximation. Such approximations are somewhat crude and can be improved. However, the resulting second or higher order approximations are more complicated and require more knowledge of the underlying distribution. The CLT is strengthened in two directions: (i) The Berry-Essen theorem provides a bound for the error of the normal approximation and (ii) a better approximation can be obtained through the first terms of Edgeworth expansion. We will not elaborate on these issues but state below two results related to these. The following result gives the bound for the error.

Theorem 10.2.3. *Berry-Esseen theorem: Suppose $\{X_n, n \geq 1\}$ is a sequence of independent and identically distributed random variables with mean μ, variance σ^2 and finite third moment, then there exists a constant C, independent of common distribution F of X_n, such that for all x,*

$$|F_n(x) - \Phi(x)| \leq (C/\sqrt{n})(E|X_1 - \mu|^3/\sigma^3).$$

The important feature of the theorem is that the constant C is independent of F. Further, the bound given by the theorem is not a limit statement but is an exact statement valid for any F, n and x. The Edgeworth expansion stated below conveys that the order of the error bound in Berry-Esseen theorem is $1/\sqrt{n}$

Theorem 10.2.4. *Suppose $\{X_n, n \geq 1\}$ is a sequence of independent and identically distributed random variables with mean μ, variance σ^2 and finite third central moment μ_3, then,*

$$F_n(x) = \Phi(x) + \mu_3(1 - x^2)\phi(x)/(6\sigma^3\sqrt{n}) + o(1/\sqrt{n}).$$

The second term on the right hand side of the above identity is known as the first Edgeworth correction.

We have noted that it is difficult to assess the approximations theoretically, since the accuracy of these approximations is not distribution-free, it very much depends both on the underlying distribution as well as on the values of parameters. We now proceed to discuss in the following examples, how the approximations can be judged by simulation using `R`.

The following example is concerned with three distributions, exponential, geometric and Poisson, which are skew distributions. Using `R`, we verify the CLT for these distributions and compare the approximations.

Example 10.2.1. Suppose X follows either exponential or geometric or Poisson distribution with mean μ and variance σ^2. Suppose $\{X_1, X_2, \cdots, X_n\}$ is a random sample from the distribution of X. By the CLT, $Y_n = \sqrt{n}(\overline{X}_n - \mu)/\sigma \xrightarrow{L} Z \sim N(0, 1)$. We fix the parameters of these distributions so that $\mu = 1.4$ is the same for all. Thus, for exponential distribution $\mu = \sigma = 1.4$, for Poisson distribution $\mu = \sigma^2 = 1.4$ and for geometric distribution success probability $p = 1/(1 + \mu) = 0.41667$ and $\sigma^2 = \mu(1 + \mu) = 0.5833$. We examine the values of n for which normality of Y_n is acceptable based on 300 simulations, using the following code.

Code 10.2.1.
```
s=c(30,60,90,120); mu=1.4; nsim=300; p1=p2=p3=c()
y=z=t=matrix(nrow=nsim,ncol=length(s))
for(j in 1:length(s))
```

```
{
n=s[j]
for(i in 1:nsim)
{
set.seed(i)
x=rexp(n,rate=1/mu)
y[i,j]= n^(.5)*(mean(x)-mu)/mu
set.seed(i)
w=rpois(n,mu)
z[i,j]= n^(.5)*(mean(w)-mu)/(mu^(.5))
set.seed(i)
v=rgeom(n,1/(1+mu))
t[i,j]= n^(.5)*(mean(v)-mu)/((mu*(1+mu))^(.5))
}
p1[j]=shapiro.test(y[,j])$p.value
p2[j]=shapiro.test(z[,j])$p.value
p3[j]=shapiro.test(t[,j])$p.value
}
d=round(data.frame(s,p1,p2,p3),4); d
r=seq(-4,4,.03); u=dnorm(r)
par(mfrow= c(3,1))
hist(y[,3],freq=FALSE,main="Exponential Distribution",
col="light blue",xlim=c(-4,4),xlab="")
lines(r,u,"o",pch=20,col="dark blue")
hist(z[,3],freq=FALSE,main="Poisson Distribution",
col="light blue",ylim=range(0,.4),xlim=c(-4,4),xlab="")
lines(r,u,"o",pch=20,col="dark blue")
hist(t[,3],freq=FALSE,main="Geometric Distribution",
col="light blue",ylim=range(0,.4),xlim=c(-4,4),xlab="")
lines(r,u,"o",pch=20,col="dark blue")
```

The p-values of Shapiro-Wilk test are reported in Table 10.1.

TABLE 10.1
p-values: Exponential, Poisson and Geometric Distributions

n	Exponential	Poisson	Geometric
30	0.0049	0.0453	0.0011
60	0.4664	0.0106	0.0007
90	0.2870	0.0895	0.0217
120	0.2122	0.2324	0.9326

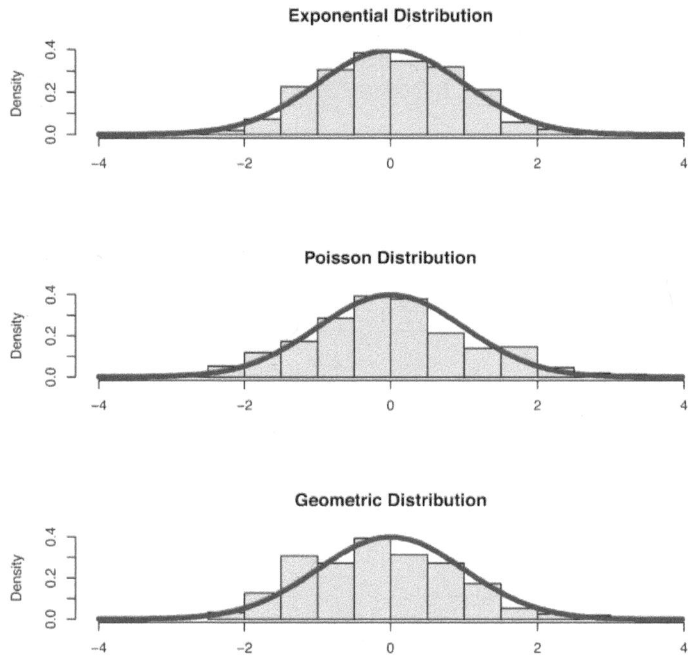

FIGURE 10.1
CLT: Exponential, Poisson and Geometric Distributions

From these p-values, we note that the values of n for which normality of Y_n is acceptable are different for three distributions. For exponential distribution for $n = 60$ normality of Y_n is acceptable, while for Poisson, it is 90 and for geometric, it is 120. These values are corresponding to fixed value of μ and fixed number of simulations. These may change as μ and number of simulations change.

Figure 10.1 displays the histogram of Y_n values with the curve of the standard normal distribution imposed on it for $n = 90$. For exponential and Poisson distribution the curve of the standard normal distribution closely approximates the histogram, but not for the geometric distribution. These results are consistent with the p-values reported in Table 10.1. □

The next example is concerned with a binomial distribution $B(m,p)$ with mean $\mu = mp$ and variance $\sigma^2 = mp(1-p) < \infty$. When $p = 1/2$ it is a symmetric distribution. We simulate random samples from a binomial distribution $B(m,p)$ distribution for two different values of p and examine the limit law of the sample mean and whether it depends on the values of p. In Example 10.2.1, we have noted that normal approximation depends on the underlying distribution. In the next example, we note that for the same distribution, it may depend on the values of the parameters.

Example 10.2.2. Suppose $X \sim B(m,p)$ and $\{X_1, X_2, \cdots, X_n\}$ is a random sample from the distribution of X. Then as $n \to \infty$ by the CLT,

$Y_n = \sqrt{n}(\overline{X}_n - mp)/\sqrt{mp(1-p)} \xrightarrow{L} Z \sim N(0,1)$ distribution. We consider $m = 10$, two values of p as 0.5 and 0.1 and examine the values of n where normality of Y_n is acceptable based on 1000 simulations, using the following code.

Code 10.2.2. `nsim=1000; m=10; s=c(20,50,100) # sample sizes`

```
p1=p2=c(); y=z=matrix(nrow=nsim,ncol=length(s))
for(j in 1:length(s))
{
 n=s[j]
 for(i in 1:nsim)
{
 set.seed(i)
 x=rbinom(n,size=m,prob=.5)
 y[i,j]=n^(.5)*(mean(x)-m*.5)/((m*.5*(1-.5))^(.5))
 set.seed(i)
 w=rbinom(n,size=m, prob=.1)
 z[i,j]=n^(.5)*(mean(w)-m*.1)/((m*.1*(1-.1))^(.5))
}
 p1[j]=shapiro.test(y[,j])$p.value
 p2[j]=shapiro.test(z[,j])$p.value
}
d=round(data.frame(s,p1,p2),4); d
set.seed(2); u=rnorm(nsim)
par(mfrow=c(2,1))
plot(density(y[,1]),main="B(10,0.5)",lwd=2)
lines(density(y[,2]),col=2,lty=2,lwd=2)
lines(density(y[,3]),col=3,lty=3,lwd=2)
lines(density(u),col=4,lty=4,lwd=2)
abline(v=0)
legend("topright",legend=c("n=20","n=50","n=100","N(0,1)"),
col=1:4,lty=1:4)
plot(density(z[,1]),main="B(10,0.1)",ylim=range(0,.42),lwd=2)
lines(density(z[,2]),col=2,lty=2,lwd=2)
lines(density(z[,3]),col=3,lty=3,lwd=2)
lines(density(u),col=4,lty=4,lwd=2)
abline(v=0)
legend("topright",legend=c("n=20","n=50","n=100","N(0,1)"),
col=1:4,lty=1:4)
```

TABLE 10.2
p-values: CLT for Binomial Distribution

n	$B(10, 0.5)$	$B(10, 0.1)$
20	0.0349	0.0000
50	0.2449	0.0009
100	0.2288	0.0148

The p-values of the Shapiro-Wilk test are presented in Table 10.2.

From Table 10.2, we observe that when $p = 1/2$, that is, when the binomial distribution is symmetric, based on p-values of Shapiro-wilk test, normality of Y_n is acceptable even for $n = 50$. On the other hand, normality of Y_n is not acceptable even for $n = 100$ when $p = 0.1$ and when the distribution is skew.

Figure 10.2 presents the same scenario. When $p = 0.5$, curves of Y_n for $n = 50, 80$ and curve of the probability density function of the standard normal distribution are in close agreement as compared to that for $p = 0.1$, as expected in view of symmetry for $p = 0.5$. This example thus illustrates that the approximation by the CLT depends not only on the distribution but also on the parameters of the distribution. □

The applicability of the CLT is greatly extended by combining it with Slutsky's theorem, in large sample inference to construct confidence intervals and to decide the asymptotic null distribution of a test statistic. It is illustrated in Example 7.3.17, in Example 7.3.19 in Chapter 7 and in the following example. In this example we show that the sample variance and the unbiased estimator of the population variance, both converge in probability to population variance and with suitable scaling their large sample distributions are normal. Many results such as, (i) if $X_n \xrightarrow{P} X$ or $X_n \xrightarrow{L} X$, then $\{X_n, n \geq 1\}$ is bounded in probability, (ii) Slutsky's theorem and (iii) CLT, are illustrated in this example.

Example 10.2.3. Suppose $\{X_1, X_2, \cdots, X_n\}$ are independent and identically distributed random variables with mean 0, variance σ^2 and finite fourth order moment. We show that the asymptotic distribution of the sample variance $S_n^2 = \sum_{i=1}^n X_i^2/n - \overline{X}_n^2$, with suitable normalization, is normal. Observe that by WLLN,

$$\overline{X}_n = \frac{1}{n}\sum_{i=1}^n X_i \xrightarrow{P} E(X_1) = 0 \ \& \ \frac{1}{n}\sum_{i=1}^n X_i^2 \xrightarrow{P} E(X_1^2) = \sigma^2 \Rightarrow S_n^2 \xrightarrow{P} \sigma^2$$

Further,

$$\sqrt{n}(S_n^2 - \sigma^2) = \sqrt{n}\Big(\frac{1}{n}\sum_{i=1}^n X_i^2 - \overline{X}_n^2 - \sigma^2\Big) = \frac{1}{\sqrt{n}}\Big(\sum_{i=1}^n X_i^2 - n\sigma^2\Big) - (\sqrt{n}\overline{X}_n)\overline{X}_n.$$

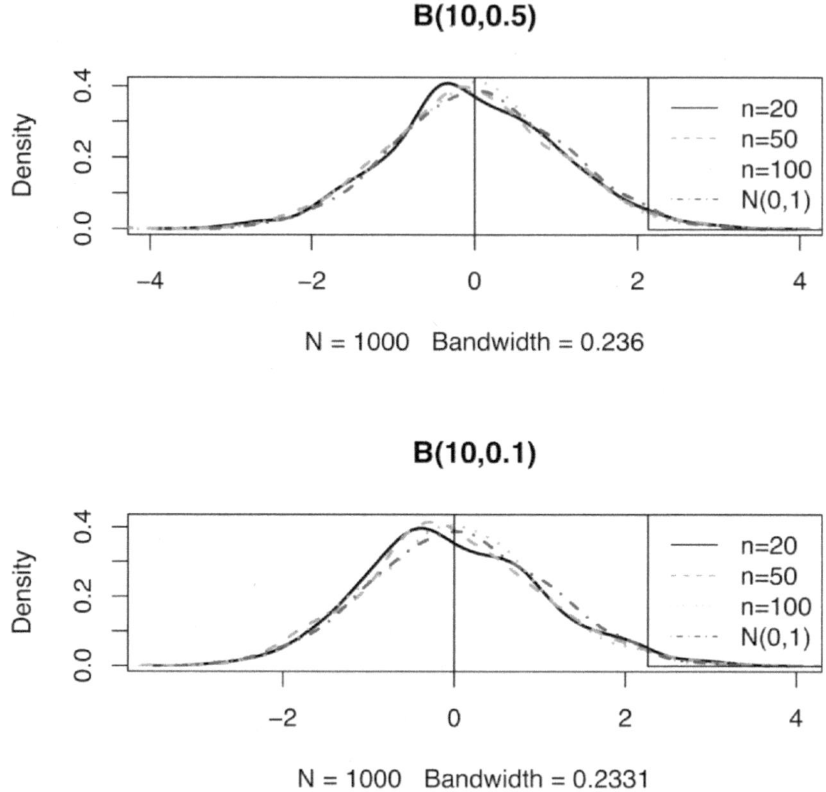

FIGURE 10.2
Verification of CLT for Binomial Distribution

It is to be noted that by the CLT, $\sqrt{n}\overline{X}_n \xrightarrow{L} Z_1 \sim N(0, \sigma^2)$ and hence $\sqrt{n}\overline{X}_n$ is bounded in probability. Further, $\overline{X}_n \xrightarrow{P} 0$. Hence $(\sqrt{n}\overline{X}_n)\overline{X}_n \xrightarrow{P} 0$. Now, $\{X_1, X_2, \cdots, X_n\}$ are independent and identically distributed random variables implies that $\{X_1^2, X_2^2, \cdots, X_n^2\}$ are also independent and identically distributed random variables with mean σ^2 and variance $V = E(X_1^4) - (E(X_1^2))^2 < \infty$. Hence, by the CLT $(\sum_{i=1}^{n} X_i^2 - n\sigma^2)/\sqrt{n} \xrightarrow{L} Z_2 \sim N(0, V)$. Thus by Slutsky's theorem,

$$\sqrt{n}(S_n^2 - \sigma^2) = \frac{1}{\sqrt{n}}\Big(\sum_{i=1}^{n} X_i^2 - n\sigma^2\Big) - (\sqrt{n}\overline{X}_n)\overline{X}_n \xrightarrow{L} Z_2 \sim N(0, V).$$

The unbiased estimator of σ^2 is given by

$$T_n = \frac{1}{n-1}\sum_{i=1}^{n}(X_i - \overline{X}_n)^2 = \frac{n}{n-1}S_n^2 = a_n S_n^2, \quad \text{where } a_n = \frac{n}{n-1} \to 1.$$

Further, $S_n^2 \xrightarrow{P} \sigma^2$, hence $T_n \xrightarrow{P} \sigma^2$. To find the asymptotic distribution of T_n consider,

$$\begin{aligned}\sqrt{n}(T_n - \sigma^2) - \sqrt{n}(S_n^2 - \sigma^2) &= \sqrt{n}(a_n S_n^2 - \sigma^2) - \sqrt{n}(S_n^2 - \sigma^2)\\ &= \sqrt{n}S_n^2(a_n - 1) = \frac{\sqrt{n}}{n-1}S_n^2 \xrightarrow{P} 0,\end{aligned}$$

as $S_n^2 \xrightarrow{P} \sigma^2$ and hence is bounded in probability. But $\sqrt{n}(S_n^2 - \sigma^2) \xrightarrow{L} Z_2 \sim N(0, V)$ and hence $\sqrt{n}(T_n - \sigma^2) \xrightarrow{L} Z_2 \sim N(0, V)$. These results remain valid even if $E(X) \neq 0$, since

$$S_n^2 = \frac{1}{n}\sum_{i=1}^{n}(X_i - \overline{X}_n)^2 = \frac{1}{n}\sum_{i=1}^{n}((X_i - E(X) - (\overline{X}_n - E(X)))^2 = \frac{1}{n}\sum_{i=1}^{n}(Y_i - \overline{Y}_n)^2,$$

where $Y_i = X_i - E(X)$, $i = 1, 2, \cdots, n$ and $E(Y_i) = 0$, $i = 1, 2, \cdots, n$. We then proceed exactly on similar lines as above to find the limiting distribution of S_n^2 and of T_n. □

Remark 10.2.2. In the setup of large sample inference, the results of Example 10.2.3 are interpreted as follows. If X is a random variable with variance σ^2 and finite fourth order moment, then based on a random sample $\{X_1, X_2, \cdots, X_n\}$ from the distribution of X, both S_n^2 and T_n are consistent and asymptotically normal estimators of σ^2 with approximate variance $(E(X^4) - (E(X^2))^2)/n$.

Example 10.2.4. Suppose $\{X_1, X_2, \cdots, X_n\}$ are independent and identically distributed random variables with mean μ and variance σ^2. On the basis of these n observations, suppose we wish to test a null hypothesis $H_0 : \mu = \mu_0$ against the alternative $H_1 : \mu \neq \mu_0$, when σ^2 is unknown. Suppose a test statistic T_n is proposed as,

$$T_n = \frac{\overline{X}_n - \mu_0}{\text{s.e.of } \overline{X}_n} = \frac{\overline{X}_n - \mu_0}{S_n/\sqrt{n}}, \quad \text{where } S_n^2 = \sum_{i=1}^{n}(X_i - \overline{X}_n)^2/n$$

and s.e.of $\overline{X}_n$ is the standard error of $\overline{X}_n$. To obtain the p-value, we need to know the null distribution of test statistic T_n. As a particular case, suppose $X_i \sim N(\mu, \sigma^2)$ distribution. Then under H_0, $\overline{X}_n$ has normal $N(\mu_0, \sigma^2/n)$ distribution, nS_n^2/σ^2 has chi-square distribution with $n-1$ degrees of freedom and $\overline{X}_n$ and S_n^2 are independent. Suppose the test statistic T_n is expressed as

$$\begin{aligned}T_n &= \frac{\overline{X}_n - \mu_0}{S_n/\sqrt{n}} = \frac{U_n}{V_n} \quad \text{where } U_n = \frac{\sqrt{n}(\overline{X}_n - \mu_0)}{\sigma}\\ \& \ V_n &= \sqrt{\frac{\sum_{i=1}^{n}(X_i - \overline{X}_n)^2}{\sigma^2(n-1)}\frac{n-1}{n}}.\end{aligned}$$

Now, $U_n \sim N(0,1)$ & $\displaystyle\sum_{i=1}^{n}(X_i - \overline{X}_n)^2/\sigma^2 \sim \chi^2_{n-1} \;\Rightarrow\; \sqrt{\frac{n}{n-1}}T_n \sim t_{n-1}$ distribution for each finite n. If the common distribution of X_i's is not normal, then it is in general difficult to obtain the null distribution of T_n for finite n. However, using CLT and Slutsky's theorem we can obtain its asymptotic null distribution, assuming $0 < \sigma^2 < \infty$. Observe that,

$$
\begin{aligned}
U_n &= \frac{\sqrt{n}(\overline{X}_n - \mu_0)}{\sigma} \xrightarrow{L} Z \sim N(0,1), \quad \frac{\sum_{i=1}^{n}(X_i - \overline{X}_n)^2}{\sigma^2(n-1)} \xrightarrow{P} 1 \\
(n-1)/n &\to 1 \quad \Rightarrow \quad V_n \xrightarrow{P} 1 \quad \Rightarrow \quad T_n = U_n/V_n \xrightarrow{L} Z \sim N(0,1),
\end{aligned}
$$

by Slutsky's theorem. Thus, the asymptotic null distribution of T_n is $N(0,1)$. □

Example 10.2.5. This example illustrates how Lindeberg-Levy CLT is useful to evaluate $\lim_{n\to\infty}\sum_{i=0}^{[n\lambda]}\frac{e^{-n\lambda}(n\lambda)^i}{i!}$. It is clear that the summands are the probabilities $P[X = i], i = 0, 1, \cdots, [n\lambda]$ where $X \sim P(n\lambda)$ distribution. Suppose $\{X_n, n \geq 1\}$ is a sequence of independent and identically distributed random variables each having Poisson $P(\lambda)$ distribution. Then $X \stackrel{d}{=} \sum_{i=1}^{n} X_i$. Since $Var(X_n) = \lambda$ is positive and finite, by Lindeberg-Levy CLT

$$
\begin{aligned}
Y_n &= \sum_{i=1}^{n}(X_i - \lambda)/\sqrt{n\lambda} \xrightarrow{L} Z \sim N(0,1) \\
\Rightarrow \lim_{n\to\infty}\sum_{i=0}^{[n\lambda]}\frac{e^{-n\lambda}(n\lambda)^i}{i!} &= \lim_{n\to\infty} P[X \leq n\lambda] = \lim_{n\to\infty} P\left[\sum_{i=1}^{n} X_i \leq n\lambda\right] \\
&= \lim_{n\to\infty} P\left[\sum_{i=1}^{n}(X_i - \lambda)/\sqrt{n\lambda} \leq 0\right] \\
&= \lim_{n\to\infty} F_{Y_n}(0) = P[Z \leq 0] = \Phi(0) = 1/2 \,. \quad \square
\end{aligned}
$$

Example 10.2.6. Similar to Example 10.2.5, this example illustrates how to evaluate

$$
\lim_{n\to\infty}\int_0^{n+t\sqrt{2n}} \frac{(1/2)^{n/2}}{\Gamma(n/2)} x^{n/2-1} e^{-x/2}\, dx
$$

using Lindeberg-Levy CLT. From the integrand, we note that it is a probability density function of $G(1/2, n/2)$ distribution which is the same as the chi-square distribution with n degrees of freedom. Thus, we have to find the limit as $n \to \infty$ of $P[Y_n \leq n + t\sqrt{2n}]$ where $Y_n \sim \chi^2_n$. Note that $Y_n \stackrel{d}{=} \sum_{i=1}^{n} X_i$, where $\{X_1, X_2, \cdots, X_n\}$ are independent and identically distributed random variables each having χ^2_1 distribution, with mean 1 and positive finite variance

2. Thus by Lindeberg-Levy CLT,

$$
\begin{aligned}
U_n &= \Big(\sum_{i=1}^{n} X_i - n\Big)/\sqrt{2n} = (Y_n - n)/\sqrt{2n} \xrightarrow{L} Z \\
\Rightarrow \lim_{n\to\infty} P[Y \le n + t\sqrt{2n}] &= \lim_{n\to\infty} P\left[(Y_n - n)/\sqrt{2n} \le t\right] \\
&= \lim_{n\to\infty} F_{U_n}(t) = P[Z \le t] = \frac{1}{\sqrt{2\pi}} \int_{-\infty}^{t} e^{-u^2/2} du,
\end{aligned}
$$

where $Z \sim N(0,1)$. In particular when $t = 0$,

$$\lim_{n\to\infty} \int_0^n \frac{(1/2)^{n/2}}{\Gamma(n/2)} x^{n/2-1} e^{-x/2}\, dx = 1/2. \qquad \square$$

Example 10.2.7. Suppose $\{X_n, n \ge 1\}$ is a sequence of independent and identically distributed random variables with mean μ and positive, finite variance σ^2. By Lindeberg-Levy CLT,
$Y_n = (\sum_{i=1}^n X_i - n\mu)/\sqrt{n}\sigma \xrightarrow{L} Z \sim N(0,1)$. Hence, $\forall\ x \in \mathbb{R}$

$$
\begin{aligned}
\lim_{n\to\infty} P[Y_n \le x] &= P[Z \le x] \\
\Rightarrow \lim_{n\to\infty} P[Y_n \le 0] &= P[Z \le 0] = 1/2, \text{with } \ x = 0 \\
\Rightarrow \lim_{n\to\infty} P[\overline{X}_n \le \mu] &= 1/2 \iff \lim_{n\to\infty} P[\overline{X}_n > \mu] = 1/2.
\end{aligned}
$$

Thus, in the long run, the probability that sample mean is larger than the population mean is 1/2. It is to be noted that for large n, distribution $\overline{X}_n$ is approximated by the normal distribution $N(\mu, \sigma^2)$ and μ is the median of the limiting distribution and hence the result seems reasonable. $\square$

In the following example, we verify the result of Example 10.2.7 by simulation for three distributions.

Example 10.2.8. We consider three distributions with the same mean $\mu = 2$, exponential, Poisson and geometric with support as a set of whole numbers. We generate 1000 samples of size n and find the proportion of times in 1000 samples $\overline{X}_n \le \mu$. As n goes on increasing, we examine whether the proportion approaches 1/2. Following is the code for these computations.

Code 10.2.3.
```
mu=2; nsim=1000; N=seq(50,250,50);
me=mp=mg=pe=pp=pg=c()
for(j in 1:length(N))
{
 n = N[j]
for(m in 1:nsim)
{
```

```
 set.seed(m)
 x=rexp(n,rate=1/mu)
 y=rpois(n,mu)
 z=rgeom(n,1/(1+mu))
 me[m]=mean(x)
 mp[m]=mean(y)
 mg[m]=mean(z)
}
 pe[j]=length(which(me<=mu))/nsim
 pp[j]=length(which(mp<=mu))/nsim
 pg[j]=length(which(mg<=mu))/nsim
}
d=round(data.frame(N,pe,pp,pg),4); d
```

The output is organized in Table 10.3.

From Table 10.3, we note that as n increases, $P[\overline{X}_n \leq \mu]$ is close to 1/2 for the three distributions. □

Example 10.2.9. Suppose $\{X_n, n \geq 1\}$ is a sequence of independent and identically distributed random variables with mean finite μ and as in Example 10.2.7, we wish to find $\lim_{n\to\infty} P[\overline{X}_n \leq \mu]$. It is to be noted that nothing is specified about variance, so as in Example 10.2.7 we cannot use Lindeberg-Levy CLT to find the limit. Since it is given that mean is finite, we can use Khintchine's WLLN and conclude that $\overline{X}_n \xrightarrow{P} \mu$. Convergence in probability implies convergence in law, hence $\overline{X}_n \xrightarrow{L} \mu$. Thus, $\forall \;\; x \in \mathbb{R}$

$$P[\overline{X}_n \leq x] \quad \to \quad \begin{cases} 0, & \text{if} \quad x < \mu \\ 1, & \text{if} \quad x > \mu, \end{cases} \tag{10.3}$$

μ being the only point of discontinuity of a distribution function of a random variable which is degenerate at μ. However, we cannot find $\lim\limits_{n\to\infty} P[\overline{X}_n \leq \mu]$ on the basis of given information. □

TABLE 10.3
$P[\overline{X}_n \leq \mu] \;\to\; 1/2$: Exponential, Poisson and Geometric Distributions

n	Exponential	Poisson	Geometric
50	0.508	0.529	0.534
100	0.494	0.525	0.520
150	0.486	0.508	0.506
200	0.503	0.532	0.515
250	0.505	0.509	0.512

Example 10.2.10. In this example we consider a distribution whose mean is finite but variance is not finite. Suppose X is a random variable with probability density function given by $f(x, \theta) = 2\theta^2/x^3,\ x \geq \theta$. Then $E(X) = 2\theta$ and the integral corresponding to $E(X^2)$ is divergent hence variance is infinite. As discussed in Example 10.2.9, we cannot use the CLT to find $\lim_{n\to\infty} P[\overline{X}_n \leq \mu]$. Hence, we find it by simulation using R. We generate m random samples from this distribution, each of size n for $n = 500$ to $n = 7500$ with an increment of 1000. We then find the proportion of times $\overline{X}_n \leq x$ in m samples for $x < \mu, x = \mu$ and $x > \mu$. To generate a random sample from the given distribution we use probability integral transformation. The distribution function of X for $x \geq \theta$ is given by, $F(x, \theta) = 1 - (\theta/x)^2$. We use the following code to generate the samples and find the proportions.

Code 10.2.4.
```
th=1.5; N=seq(500,7500,1000); nsim=500
mean=p1=p2=p3=c()
for(j in 1:length(N))
{
for(m in 1:nsim)
{
 set.seed(m)
 u=runif(N[j],0,1)
 x=th/((1-u)^(1/2))
 mean[m]=mean(x)
}
 p1[j]=length(which(mean<=2*th-1))/nsim
 p2[j]=length(which(mean<=2*th))/nsim
 p3[j]=length(which(mean<=2*th+1))/nsim
}
d=round(data.frame(N,p1,p2,p3),4);d
```

The output is organized in Table 10.4. From Table 10.4, we note that $P[\overline{X}_n \leq x] \to 0$ if $x < \mu$, $P[\overline{X}_n \leq x] \to 1$ if $x > \mu$, as shown in Equation 10.3. Further, $P[\overline{X}_n \leq \mu]$ approaches 0.5, it is 0.548 for sample size $n = 7500$. We may require still large n for it to be close to 0.5, the reason may be the infinite variance of X. □

In Lindeberg-Levy theorem, n is deterministic. The CLT continues to hold even if n is random with some additional condition. It is known as a random sum CLT. We state it below. For proof we refer to Billingsley [5] or page 346 of Gut [13].

Theorem 10.2.5. *Random sum CLT: Suppose $\{X_n, n \geq 1\}$ is a sequence of independent and identically distributed random variables with mean* 0 *and*

TABLE 10.4
$P[\overline{X}_n \leq x] : x < \mu, x = \mu, x > \mu$

n	$P[\overline{X}_n \leq 2\theta - 1]$	$P[\overline{X}_n \leq 2\theta]$	$P[\overline{X}_n \leq 2\theta + 1]$
500	0	0.628	0.994
1500	0	0.596	1.000
2500	0	0.552	1.000
3500	0	0.544	1.000
4500	0	0.570	1.000
5500	0	0.556	1.000
6500	0	0.558	1.000
7500	0	0.548	1.000

variance 1. *Suppose* $\{N(t), t \geq 0\}$ *is a family of positive integer valued random variables where* $N(t) \xrightarrow{a.s.} \infty$ *such that* $N(t)/t \xrightarrow{P} c$, *where* $c \in (0, \infty)$ *is a constant. Then as* $t \to \infty$,

$$Y_{N(t)} = \sum_{i=1}^{N(t)} X_i/\sqrt{N(t)} \xrightarrow{L} Z \sim N(0,1).$$

Random sum CLT has applications in a variety of fields, such as in risk theory in general insurance, in a compound Poison process, in a renewal process and in branching processes. Following example illustrates an application of random sum CLT in a compound Poison process. If $\{N(t), t \geq 0\}$ is a Poisson process, with rate λ, then for each fixed t, $N(t) \sim P(\lambda t)$ distribution. It can be shown that $N(t)/t \xrightarrow{P} \lambda$. If $\{X_n, n \geq 1\}$ is a sequence of independent and identically distributed random variables with mean μ and variance σ^2, then by random sum CLT,

$$Y_{N(t)} = \frac{\sum_{i=1}^{N(t)} (X_i - \mu)/\sigma}{\sqrt{N(t)}} = \frac{(X(t) - E(X(t))}{\sqrt{Var(X(t))}} \xrightarrow{L} Z \sim N(0,1),$$

where $X(t) = \sum_{i=1}^{N(t)} X_i$. Following example illustrates the application of the random sum CLT, (Sivaprasad and Deshmukh [22]).

Example 10.2.11. Customers arrive at the ATM according to a Poisson process with rate 15 per hour. The amount of money withdrawn on each transaction is a random variable with mean Rs 5000 and standard deviation 1000. The machine is in use for 24 hours. The total daily withdrawal is then modelled as a compound Poisson process as $X(t) = \sum_{i=1}^{N(t)} X_i$, where X_i denotes the amount withdrawn by the i-th individual. Thus, expected total daily withdrawal and its variance are given by,

$$\begin{aligned} E(X(24)) &= 15 \times 24 \times 5000 = 1800000 \\ \text{and } Var(X(24)) &= 15 \times 24 \times (1000^2 + 5000^2) = (9.36) \times 10^9. \end{aligned}$$

The standard deviation of $X(24)$ is 96747.09. By the random sum CLT, distribution of $X(24)$ can be approximated by the normal distribution with mean 18×10^5 and standard deviation 96747.09. We can then compute the approximate probability that the total daily withdrawal is less than 1900000 as follows.

$$\begin{aligned} P[X(24)) \leq 1900000] &= P\left[Z \leq (1900000 - 18 \times 10^5)/96747.09\right] \\ &= P[Z \leq 1.033623] = 0.8493. \end{aligned}$$

Using similar arguments we get that

$$P[X(24)) \leq 2000000] = 0.9806 \quad \& \quad P[X(24)) \leq 2100000] = 0.99904.$$

Such an analysis is helpful for the management to decide on how much cash to be deposited daily at the ATM. □

Lindeberg-Levy CLT is extended to k dimensional random vectors, using Cramer-Wold device as stated below. Suppose $\underline{X}_n$ and $\underline{X}$ are k dimensional random vectors. Then Cramer-Wold device states that

$$\underline{X}_n \xrightarrow{L} \underline{X} \quad \text{if and only if} \quad \underline{L}'\underline{X}_n \xrightarrow{L} \underline{L}'\underline{X} \ \ \forall \ \ \underline{L} \in \mathbb{R}^k, \ \ \underline{L} \neq \underline{0} \, .$$

Cramer-Wold device allows higher-dimensional problems to reduce to the one-dimensional case.

Theorem 10.2.6. *Multivariate CLT: Suppose $\underline{X}$ is a k dimensional random vector with mean vector $E(\underline{X}) = \underline{\mu}$ and dispersion matrix Σ, which is positive definite. Suppose $\{\underline{X}_1, \underline{X}_2, \cdots, \underline{X}_n\}$ are independent and identically distributed random vectors each distributed as $\underline{X}$. Suppose $\overline{\underline{X}_n}$ denotes the sample mean vector, then as $n \to \infty$*

$$\sqrt{n}(\overline{\underline{X}_n} - \underline{\mu}) \xrightarrow{L} \underline{U} \sim N_k(\underline{0}, \Sigma) \quad \textit{distribution.}$$

Proof. Suppose a random variable Y_i is defined as $Y_i = \underline{L}'\underline{X}_i$, $i = 1, 2, \cdots, n$, where $\underline{L} \neq \underline{0}$ is a k dimensional vector of real numbers. Then $\{Y_1, Y_2, \cdots, Y_n\}$ are independent and identically distributed random variables with common mean $\theta = \underline{L}'\underline{\mu}$ and variance $\sigma^2 = \underline{L}'\Sigma\underline{L}$. Since Σ is positive definite and $\underline{L} \neq 0$, σ^2 is positive. We assume it to be finite. Then by the Lindeberg-Levy CLT,

$$\sqrt{n}(\overline{Y}_n - \theta) \xrightarrow{L} Z_1 \sim N(0, \sigma^2) \quad \Longleftrightarrow \quad \underline{L}'(\sqrt{n}(\overline{\underline{X}_n} - \underline{\mu})) \xrightarrow{L} Z_1 \sim N(0, \underline{L}'\Sigma\underline{L})$$

Hence, by Cramer-Wold device $\sqrt{n}(\overline{\underline{X}_n} - \underline{\mu})) \xrightarrow{L} \underline{U} \sim N_k(\underline{0}, \Sigma)$ distribution. □

Multivariate CLT is heavily used in multivariate analysis and asymptotic inference when the distribution of a random variable is indexed by a vector parameter. We illustrate below one application of multivariate CLT in the asymptotic inference setup, for more illustrations in the asymptotic inference setup refer to (Deshmukh and Kulkarni [11]). Suppose X is a random variable or a random vector defined on a probability space $(\Omega, \mathbb{A}, P)$ where the probability measure P is indexed by a vector parameter $\underline{\theta} \in \Theta \subset \mathbb{R}^k$. Suppose $\underline{\theta} = (\theta_1, \theta_2, \cdots, \theta_k)'$. Given a random sample $\{X_1, X_2, \cdots, X_n\}$ of size n from the distribution of X, suppose $\underline{T}_n = (T_{1n}, T_{2n}, \cdots, T_{kn})'$ is an estimator of $\underline{\theta}$. If it is in the form of an average, then Multivariate CLT is useful to verify whether the asymptotic distribution of $\underline{T}_n$, with suitable centering and scaling is normal. In the following example, it is shown that the large sample distribution of a vector of the first k sample raw moments, with suitable centering and scaling is multivariate normal.

Example 10.2.12. Suppose $\{X_1, X_2, \cdots, X_n\}$ are independent and identically distributed random variables with $\mu'_{2k} < \infty$. Suppose $\underline{T}_n = (m'_1, m'_2, \cdots, m'_k)'$ is a random vector, where $m'_r = \sum_{i=1}^n X_i^r/n$, $r = 1, 2, \cdots, k$. Suppose $\underline{\mu} = (\mu'_1, \mu'_2, \cdots, \mu'_k)'$. From the WLLN, it follows that $m'_r \xrightarrow{P} \mu'_r$, $r = 1, 2, \cdots, k$. We now examine whether $\sqrt{n}(\underline{T}_n - \underline{\mu}) \xrightarrow{L} \underline{Z} \sim N_k(\underline{0}, \Sigma)$. Suppose a random vector $\underline{Z}$ is defined as $\underline{Z} = (X, X^2, \cdots, X^k)$. Then $E(\underline{Z}) = \underline{\mu}$ and the dispersion matrix of $\underline{Z}$ is $\Sigma = [\sigma_{ij}]$, where $\sigma_{ij} = Cov(X^i, X^j)$, $i, j = 1, 2, \cdots, k$. To examine whether Σ is positive definite, suppose a random variable Y is defined as $Y = \sum_{i=1}^k a_i X^i$. Without loss of generality we assume that X is a non-degenerate random variable which implies that Y is also a non-degenerate random variable and hence $Var(Y) > 0$. Observe that for any non-zero vector $\underline{a} = (a_1, a_2, \cdots, a_k)'$ of real numbers

$$0 < Var(Y) = Var\left(\sum_{i=1}^k a_i X^i\right) = \sum_{i=1}^k \sum_{j=1}^k a_i a_j Cov(X^i, X^j) = \underline{a}'\Sigma\underline{a},$$

which proves that Σ is a positive definite matrix. Thus $\underline{Z}$ is a random vector with mean vector $\underline{\mu}$ and positive definite dispersion matrix Σ. Now $\{X_1, X_2, \cdots, X_n\}$ are independent and identically distributed random variables implies that $\{\underline{Z}_1, \underline{Z}_2, \cdots, \underline{Z}_n\}$ are independent and identically distributed random vectors, each following the distribution same as that of $\underline{Z}$. Suppose $\overline{\underline{Z}}_n = \sum_{i=1}^n \underline{Z}_i/n$ denotes the sample mean vector. Then by the multivariate CLT,

$$\sqrt{n}(\overline{\underline{Z}}_n - \underline{\mu}) \xrightarrow{L} \underline{Z} \sim N_k(0, \Sigma) \quad \Rightarrow \quad \sqrt{n}(\underline{T}_n - \underline{\mu}) \xrightarrow{L} \underline{Z} \sim N_k(\underline{0}, \Sigma)$$

distribution as $n \to \infty$ since $\overline{\underline{Z}}_n = T_n$. □

In Theorem 7.3.9, we have proved that if $\sqrt{n}(X_n - \mu) \xrightarrow{L} X \sim N(0, \sigma^2)$ distribution and g is a differentiable function with $g'(\mu) \neq 0$, then $\sqrt{n}(g(X_n) - g(\mu)) \xrightarrow{L} Y \sim N(0, (g'(\mu))^2\sigma^2)$ distribution. The result is true for a random vector $\underline{X}_n$. We state it below, it is also known as a delta method.

Theorem 10.2.7. *Suppose $\{\underline{X}_n, n \geq 1\}$ is a sequence of random vectors such that $\sqrt{n}(\underline{X}_n - \underline{\mu}) \xrightarrow{L} \underline{Z} \sim N_k(\underline{0}, \Sigma)$, where Σ is a positive definite dispersion matrix. Suppose $g : \mathbb{R}^k \to \mathbb{R}$ is a totally differentiable function. Then*

$$\sqrt{n}(g(\underline{X}_n) - g(\underline{\mu})) \xrightarrow{L} \underline{U} \sim N(0, v), \quad as \ \ n \to \infty,$$

where $v = \Delta'\Sigma\Delta$ *and* Δ *is a gradient vector of order* $k \times 1$ *of* g, *with i-th component given by* $\frac{\partial g}{\partial x_i}$.

Using multivariate CLT and Theorem 10.2.7, in the next example we show that the properly scaled sample correlation coefficient, based on the random sample from a bivariate normal distribution, converges to the standard normal distribution. The famous arctan transformation of Fisher, popularly known as Fisher's Z transformation, for the correlation parameter in the bivariate normal distribution is obtained from this result using the variance stabilization technique. For details refer to Deshmukh and Kulkarni [11]. This example illustrates number of results proved in Chapter 7.

Example 10.2.13. Suppose $(X, Y)'$ has a bivariate normal distribution with zero mean vector and dispersion matrix $\Sigma = [\sigma_{ij}]$, where $\sigma_{11} = \sigma_{22} = 1$ and $\sigma_{12} = \sigma_{21} = \rho \in (-1, 1)$. Suppose $\{(X_i, Y_i)',\ i = 1, 2, \cdots, n\}$ are independent and identically distributed random vectors each having the distribution same as that of $(X, Y)'$. The correlation coefficient R_n is defined as

$$\begin{aligned} R_n &= \frac{S^2_{XY}}{S_X S_Y} \quad \text{where } S^2_{XY} = \frac{1}{n}\sum_{i=1}^{n}(X_i - \overline{X}_n)(Y_i - \overline{Y}_n), \\ S^2_X &= \frac{1}{n}\sum_{i=1}^{n}(X_i - \overline{X}_n)^2 \quad \& \quad S^2_Y = \frac{1}{n}\sum_{i=1}^{n}(Y_i - \overline{Y}_n)^2. \end{aligned}$$

As shown in Example 10.2.3, $S^2_X \xrightarrow{P} 1$ and $S^2_Y \xrightarrow{P} 1$. Similarly by the WLLN, $(1/n)\sum_{i=1}^{n} X_i Y_i \xrightarrow{P} E(XY)$ and hence $S^2_{XY} \xrightarrow{P} E(XY) = Cov(X, Y)$. In view of the result that convergence in probability is closed under arithmetic operations, it follows that $R_n \xrightarrow{P} \rho$. We now examine whether suitably scaled R_n converges in law to the standard normal distribution. Note that $X \sim N(0, 1)$, $Y \sim N(0, 1)$ and the conditional distribution of X given $Y = y$ is also normal $N(\rho y, 1 - \rho^2)$. Suppose a random vector $\underline{U}$ is defined as $\underline{U} = (X^2, Y^2, XY)'$. Then $\underline{\mu} = E(\underline{U}) = (1, 1, \rho)'$. To find the dispersion matrix V of $\underline{U}$, note that

$$Var(X^2) = Var(Y^2) = 2 \,\&\, Var(XY) = E(X^2Y^2) - (E(XY))^2 = E(X^2Y^2) - \rho^2.$$

Using the conditional distribution of X given $Y = y$, we find $E(X^2Y^2)$ and the remaining elements of V as follows.

$$\begin{aligned} E(X^2Y^2) &= E(E(X^2Y^2)|Y) = E(Y^2E(X^2|Y)) \\ &= E(Y^2(Var(X|Y) + (E(X|Y))^2)) \\ &= E(Y^2(1 - \rho^2 + \rho^2Y^2)) \\ &= 1 - \rho^2 + \rho^2E(Y^4) = 1 - \rho^2 + 3\rho^2 = 1 + 2\rho^2. \end{aligned}$$

Hence,

$$\begin{aligned}
Var(XY) &= 1+\rho^2 \;\;\&\;\; Cov(X^2,Y^2) = 1+2\rho^2-1 = 2\rho^2 \\
Cov(X^2,XY) &= E(X^3Y)-E(X^2)E(XY) = E(E(X^3Y|Y))-\rho \\
&= E(YE(X^3|Y))-\rho \\
E((X-\rho Y)^3|Y) &= 0 \text{ since } X|Y=y \sim N(\rho y, 1-\rho^2) \\
\Rightarrow \;\; E(X^3|Y) &= 3\rho Y E(X^2|Y) - 3\rho^2 Y^2 E(X|Y) + \rho^3 Y^3 \\
&= 3\rho(1-\rho^2)Y + \rho^3 Y^3 \\
\Rightarrow \;\; E(X^3Y) &= E(YE(X^3|Y)) = 3\rho(1-\rho^2)E(Y^2) + \rho^3 E(Y^4) = 3\rho \\
\Rightarrow \;\; Cov(X^2,XY) &= 2\rho \text{ similarly } Cov(Y^2,XY) = 2\rho.
\end{aligned}$$

Hence the dispersion matrix V of $\underline{U}$ is given by,

$$V = \begin{pmatrix} 2 & 2\rho^2 & 2\rho \\ 2\rho^2 & 2 & 2\rho \\ 2\rho & 2\rho & 1+\rho^2 \end{pmatrix}.$$

From the principal minors of V, it follows that it is a positive definite matrix with determinant $4(1-\rho^2)^2 \neq 0$. Note that $\{(X_i, Y_i)', \; i = 1, 2, \cdots, n\}$ are independent and identically distributed random vectors implies that $\{\underline{U}_1, \underline{U}_2, \cdots, \underline{U}_n\}$ are also independent and identically distributed random vectors, each having the distribution same as that of $\underline{U}$. Hence by the multivariate CLT,

$$\sqrt{n}(\overline{\underline{U}_n} - \underline{\mu}) \xrightarrow{L} W \sim N_3(\underline{0}, V) \text{ where } \overline{\underline{U}_n}$$
$$= \Big(\sum_{i=1}^{n} X_i^2/n, \sum_{i=1}^{n} Y_i^2/n, \sum_{i=1}^{n} X_iY_i/n\Big)'.$$

It is to be noted that

$$\sqrt{n}((S_X^2, S_Y^2, S_{XY})' - \underline{\mu}) = \sqrt{n}(\overline{\underline{U}_n} - \underline{\mu}) - \sqrt{n}(\overline{X}_n^2, \; \overline{Y}_n^2, \; \overline{X}_n\overline{Y}_n)'.$$

Now,

$$\begin{aligned}
X \sim N(0,1) \;\; &\Rightarrow \;\; \sqrt{n}\,\overline{X}_n \sim N(0,1) \;\&\; \overline{X}_n \xrightarrow{P} 0 \text{ by WLLN} \\
Y \sim N(0,1) \;\; &\Rightarrow \;\; \sqrt{n}\,\overline{Y}_n \sim N(0,1) \;\&\; \overline{Y}_n \xrightarrow{P} 0 \text{ by WLLN} \\
&\Rightarrow \;\; (\sqrt{n}\,\overline{X}_n)\overline{X}_n \xrightarrow{P} 0 \text{ as } \sqrt{n}\overline{X}_n \text{ is bounded in probability} \\
&\Rightarrow \;\; (\sqrt{n}\,\overline{Y}_n)\overline{Y}_n \xrightarrow{P} 0 \text{ as } \sqrt{n}\overline{Y}_n \text{ is bounded in probability} \\
&\Rightarrow \;\; (\sqrt{n}\,\overline{X}_n)\overline{Y}_n \xrightarrow{P} 0 \text{ as } \sqrt{n}\overline{X}_n \text{ is bounded in probability} \\
&\Rightarrow \;\; \sqrt{n}((S_X^2, S_Y^2, S_{XY})' - \underline{\mu}) - \sqrt{n}(\overline{\underline{U}_n} - \underline{\mu}) \xrightarrow{P} \underline{0} \\
&\Rightarrow \;\; \sqrt{n}((S_X^2, S_Y^2, S_{XY})' - \underline{\mu}) \xrightarrow{L} W \sim N_3(\underline{0}, V),
\end{aligned}$$

since $\sqrt{n}(\underline{\overline{U}}_n-\underline{\mu}) \xrightarrow{L} W$. To examine the limit law of the correlation coefficient R_n, observe that

$$R_n = \frac{S_{XY}^2}{S_X S_Y} = g(S_X^2, S_Y^2, S_{XY}) \text{ where } g(x_1, x_2, x_3) = x_3/(x_1x_2)^{1/2},$$

a function from $\mathbb{R}^3 - \{(0,0,0)\} \to (-1,1)$. Hence, the vector of partial derivatives of g is given by,

$$\begin{aligned}\frac{\partial}{\partial x_1}g(x_1,x_2,x_3) &= \frac{-x_3}{2x_1^{3/2}x_2^{1/2}}, \quad \frac{\partial}{\partial x_2}g(x_1,x_2,x_3) = \frac{-x_3}{2x_1^{1/2}x_2^{3/2}} \\ \frac{\partial}{\partial x_3}g(x_1,x_2,x_3) &= \frac{1}{x_1^{1/2}x_2^{1/2}}.\end{aligned}$$

By Theorem 10.2.7,

$$\sqrt{n}(g(S_X^2, S_Y^2, S_{XY}) - g(\underline{\mu})) = \sqrt{n}(R_n - \rho) \xrightarrow{L} W_1 \sim N(0, \sigma^2) \text{ where } \sigma^2 = \Delta' V \Delta$$

and $\Delta' = (-\rho/2, -\rho/2, 1)$. Hence $\sigma^2 = (1-\rho^2)^2$. □

The next section is concerned with the CLT for a sequence of independent, but not necessarily identically distributed random variables.

10.3 CLT for Independent Random Variables

The classical Lindeberg-Levy CLT is related to a sequence of independent and identically distributed random variables. We now relax the assumption that random variables in the sequence are identically distributed and retain the assumption of independence. Thus, suppose $\{X_n, n \geq 1\}$ is a sequence of independent random variables defined on $(\Omega, \mathbb{A}, P)$ such that $E(X_k) = \mu_k$, $Var(X_k) = \sigma_k^2 < \infty$ and F_k is a distribution function of X_k, $k \geq 1$. Suppose $S_n = \sum_{k=1}^n (X_k - \mu_k)$ and $s_n^2 = Var(S_n) = \sum_{k=1}^n \sigma_k^2$. It is assumed that $\sigma_k^2 > 0$ for at least one $k = 1, 2, \cdots, n$, which implies that $s_n^2 > 0$. In such a setup, we have two CLTs, one is Lyapounov's CLT and the other is Lindeberg-Feller CLT. These state that S_n/s_n has approximately normal distribution for large n, under certain conditions, such as Lyapounov's condition, Lindeberg's condition and Feller's condition (Shao [21]). We state below these conditions.

(i) Lyapounov's condition: A sequence $\{X_n, n \geq 1\}$ satisfies Lyapounov's condition if for some $\delta > 0$, $\sum_{k=1}^n E(|X_k - \mu_k|^{2+\delta})/s_n^{2+\delta} \to 0$ as $n \to \infty$.

(ii) Lindeberg's condition: Suppose for $k = 1, 2, \cdots, n$, a subset A_{nk} of $\mathbb{R}$ is defined as $A_{nk} = \{x_k | x_k - \mu_k| > \epsilon s_n\}$. A sequence $\{X_n, n \geq 1\}$ satisfies Lindeberg's condition if $\forall\ \epsilon > 0$, as $n \to \infty$,

$$\begin{aligned} g_n(\epsilon) &= \frac{1}{s_n^2}\sum_{k=1}^{n}\int_{A_{nk}}(x_k-\mu_k)^2 dF_k(x_k) \\ &= \sum_{k=1}^{n} E\left(((X_k-\mu_k)/s_n)^2 I_{A_{nk}}\right) \to 0, \end{aligned}$$

where expectation is a Riemann-Stieltjes integral with respect to F_k. It can also be written as

$$g_n(\epsilon) = \sum_{k=1}^{n} E\left(\left(\frac{X_k-\mu_k}{s_n}\right)^2 I_{B_{nk}}\right) = \sum_{k=1}^{n}\int_{\Omega}\left(\frac{X_k-\mu_k}{s_n}\right)^2 I_{B_{nk}} dP \to 0,$$

where expectation is an integral with respect to probability measure P and $B_{nk} = \{\omega | X_k(\omega) - \mu_k| > \epsilon s_n\}$. Note that $B_{nk} = X_k^{-1}(A_{nk})$.

(iii) Feller's condition: A sequence $\{X_n, n \geq 1\}$ satisfies Feller's condition if $\max_{k\leq n} \sigma_k^2/s_n^2 \to 0$ as $n \to \infty$.

Feller's condition is also stated as, $\max_{k\leq n} P(B_{nk}) \to 0$ as $n \to \infty$ for any $\epsilon > 0$, refer to p. 348 of Athreya and Lahiri [3]. It follows from Chebyshev's inequality. Observe that as $n \to \infty$,

$$P(B_{nk}) = P[|X_k - \mu_k| > \epsilon s_n] \leq \sigma_k^2/\epsilon^2 s_n^2 \Rightarrow \max_{k\leq n} P(B_{nk}) \leq \max_{k\leq n} \sigma_k^2/\epsilon^2 s_n^2 \to 0.$$

We now analyze Lindeberg's condition. Suppose the sequence $\{X_n, n \geq 1\}$ satisfies Lindeberg's condition. Note that on A_{nk}, $1 < (x-\mu_k)^2/\epsilon^2 s_n^2$, $k = 1, 2, \cdots, n$. Hence,

$$\begin{aligned} P\Big[\max_{1\leq k\leq n}|X_k-\mu_k| > \epsilon s_n\Big] &= P\Big(\bigcup_{k=1}^{n} B_{nk}\Big) \leq \sum_{k=1}^{n} P(B_{nk}) = \sum_{k=1}^{n}\int_{A_{nk}} 1 dF_k(x) \\ &< \sum_{k=1}^{n}\int_{A_{nk}}\frac{(x-\mu_k)^2}{\epsilon^2 s_n^2} dF_k(x) = g_n(\epsilon)/\epsilon^2 \to 0 . \end{aligned}$$

Thus, if the sequence $\{X_n, n \geq 1\}$ satisfies the Lindeberg's condition, then

$$\forall\ \epsilon > 0,\ P\Big[\max_{1\leq k\leq n}|X_k-\mu_k| > \epsilon s_n\Big] \to 0$$

and we say that the deviations $|X_n - \mu_n|$'s are uniformly asymptotically negligible.

We now discuss the implications among the three conditions stated above.

Lemma 10.3.1. *Feller's condition implies that $s_n^2 \to \infty$ as $n \to \infty$.*

Proof. It is assumed that $Var(X_k) = \sigma_k^2 > 0$ for at least one $k = 1, 2, \cdots, n$. We assume that $\exists\ m$ such that $\sigma_m^2 > 0$. Hence,

$$\forall\ n > m,\ \sigma_m^2 > 0 \Rightarrow \frac{\sigma_m^2}{s_n^2} \leq \max_{k\leq n}\frac{\sigma_k^2}{s_n^2} \to 0 \Rightarrow s_n^2 \to \infty \text{ as } n \to \infty.$$

□

In the next lemma, we prove that Lyapounov's condition implies Lindeberg's condition.

Lemma 10.3.2. *Lyapounov's condition implies Lindeberg's condition.*

Proof. Suppose the sequence $\{X_n, n \geq 1\}$ satisfies Lyapounov's condition. Observe that on A_{nk}, $|x_k - \mu_k|/\epsilon s_n > 1$. Hence $\forall\ \epsilon > 0$ and $\delta > 0$,

$$\begin{aligned} g_n(\epsilon) &= \frac{1}{s_n^2}\sum_{k=1}^{n}\int_{A_{nk}}(x_k-\mu_k)^2 dF_k(x_k) \\ &\leq \frac{1}{s_n^2}\sum_{k=1}^{n}\int_{A_{nk}}(x_k-\mu_k)^2\left[\frac{|x_k-\mu_k|}{\epsilon s_n}\right]^{\delta} dF_k(x_k) \\ &\leq \frac{1}{\epsilon^\delta s_n^{2+\delta}}\sum_{k=1}^{n}\int_{\mathbb{R}}|x_k-\mu_k|^{2+\delta}dF_k(x_k) = \frac{1}{\epsilon^\delta s_n^{2+\delta}}\sum_{k=1}^{n}E(|X_k-\mu_k|^{2+\delta}) \\ &\to 0, \end{aligned}$$

if Lyapounov's condition is satisfied. Thus, Lyapounov's condition implies Lindeberg's condition. □

In view of Lemma 10.3.2, we first prove the Lindeberg-Feller CLT. Lyapounov's CLT then follows from the Lindeberg-Feller CLT. It is proved in the first part of the Lindeberg-Feller CLT that Lindeberg's condition implies Feller's condition. Thus, Lyapounov's condition implies Lindeberg's condition and Lindeberg's condition implies Feller's condition.

We prove below a lemma involving a trignometric function, which is used in the proof of Lindeberg-Feller CLT.

Lemma 10.3.3. $\cos(x) - 1 + x^2/2 \geq 0\ \ \forall\ x \in \mathbb{R}$.

Proof. Suppose $h(x) = \cos(x) - 1 + x^2/2$. Observe that $h(-x) = h(x)$, thus h is an even function and $h(0) = 0$. Suppose $x \geq 0$. Note that

$$\begin{aligned} &h'(x) = -\sin(x) + x\ \ \&\ \ h'(0) = 0,\ \ h''(x) = -\cos(x) + 1 \geq 0\ \forall\ x \in \mathbb{R} \\ &\quad\Rightarrow h'(x)\ \text{ is a non-decreasing function } \Rightarrow\ \text{ for } x \geq 0,\ h'(x) \geq h'(0) = 0 \\ &\quad\Rightarrow h(x)\ \text{ is a non-decreasing function } \Rightarrow\ \text{ for }\ x \geq 0,\ \ h(x) \geq h(0) = 0 \\ &\quad\Rightarrow \text{For }\ x < 0\ \ h(x) = h(-x) \geq 0\ \Rightarrow\ h(x) \geq 0,\ \ \forall\ x \in \mathbb{R}. \end{aligned}$$

□

Theorem 10.3.1. *Lindeberg-Feller CLT: Suppose* $\{X_n, n \geq 1\}$ *is a sequence of independent random variables such that* $E(X_k) = \mu_k,\ \ Var(X_k) = \sigma_k^2 > 0$ *for at least one* k *and* F_k *is a distribution function of* X_k, $k \geq 1$. *Suppose* $A_{nk} = \{x_k | |x_k - \mu_k| > \epsilon s_n\},\ \ k = 1, 2, \cdots, n$ *and* $s_n^2 = \sum_{k=1}^{n}\sigma_k^2$.

$$\textit{If}\ \ \forall\ \ \epsilon > 0,\ \ g_n(\epsilon) = \frac{1}{s_n^2}\sum_{k=1}^{n}\int_{A_{nk}}(x_k-\mu_k)^2 dF_k(x) \ \to\ 0 \qquad (10.4)$$

$$\textit{then,}\ \ \sum_{k=1}^{n}(X_k-\mu_k)/s_n \ \xrightarrow{L}\ \ Z \sim N(0,1)\ \ \&\ \ \max_{k\leq n}\sigma_k/s_n \to 0. \qquad (10.5)$$

Conversely, if both the results in Equation (10.5) *are satisfied, then Equation* (10.4) *is satisfied, that is,* $\forall\ \epsilon > 0,\ \ g_n(\epsilon) \to 0.$

Proof. Part (i): We first prove that Equation (10.4) implies Equation (10.5). For notational convenience, for $\forall\ n \geq 1$ and $k = 1, 2 \cdots, n$, we define

$$X_{nk} = \frac{X_k - \mu_k}{s_n},\ S_n = \sum_{k=1}^{n} X_{nk}\ \&\ A_{nk} = \{x_{nk} | x_{nk}| > \epsilon\},\ \text{with } x_{nk} = X_{nk}(\omega)$$

$$\Rightarrow E(X_{nk}) = 0,\ \ Var(X_{nk}) = \sigma_k^2/s_n^2 = \sigma_{nk}^2,\ \ \text{say}$$

$$\Rightarrow E(S_n) = 0\ \ \&\ \ Var(S_n) = \sum_{k=1}^{n} \sigma_{nk}^2 = 1.$$

In view of the above notation we prove that $\forall\ \epsilon > 0$,

$$g_n(\epsilon) = \sum_{k=1}^{n} \int_{A_{nk}} x_{nk}^2 dF_k(x)\ \to\ 0\ \Rightarrow\ S_n \xrightarrow{L} Z\ \sim\ N(0,1)\ \&\ \max_{k \leq n} \sigma_{nk} \to 0.$$

We begin by showing that $g_n(\epsilon) \to 0\ \Rightarrow\ \max_{k \leq n} \sigma_{nk} \to 0$. Note that for all $k = 1, 2, \cdots, n$,

$$\begin{aligned}
\sigma_{nk}^2 &= E(X_{nk}^2) = \int_{[|x_{nk}| \leq \epsilon]} x_{nk}^2 dF_k(x) + \int_{[|x_{nk}| > \epsilon]} x_{nk}^2 dF_k(x) \\
&\leq \epsilon^2 P[|X_{nk}| \leq \epsilon] + \sum_{k=1}^{n} \int_{A_{nk}} x_{nk}^2 dF_k(x) \leq \epsilon^2 + g_n(\epsilon) \\
\Rightarrow\ \max_{1 \leq k \leq n} \sigma_{nk}^2 &\leq \epsilon^2 + g_n(\epsilon)\ \to\ 0\ \text{ if }\ g_n(\epsilon) \to 0\ \text{ as }\ n \to \infty.
\end{aligned}$$

Thus, Lindeberg's condition implies Feller's condition. We now prove that Lindberg's condition implies $S_n \xrightarrow{L} Z \sim N(0,1)$. Suppose

$$\Psi_{nk}(t) = E(e^{itX_{nk}})\ \ \&\ \ \Psi_n(t) = E(e^{itS_n}) = \prod_{k=1}^{n} \Psi_{nk}(t),\ t \in \mathbb{R}$$

are characteristic functions of X_{nk} and S_n respectively. Observe that,

$$S_n \xrightarrow{L} Z\ \sim\ N(0,1)\qquad \Longleftrightarrow \qquad \Psi_n(t)\ \to\ e^{-t^2/2}\ \ \forall\ \ t \in \mathbb{R},$$

in view of the continuity and uniqueness theorems for characteristic functions. Since $\sum_{k=1}^{n} \sigma_{nk}^2 = 1$, note that

$$e^{-t^2/2} = \prod_{k=1}^{n} e^{-(t^2\sigma_{nk}^2)/2} = \prod_{k=1}^{n} \phi_{nk}(t),\ \ \text{where}\ \ \phi_{nk}(t) = e^{-(t^2\sigma_{nk}^2)/2}$$

is a characteristic function of $Z_{nk} \sim N(0, \sigma_{nk}^2)$ distribution and for each fixed n, $Z_{nk}, k = 1, 2, \cdots, n$ are independent random variables. Now, by Lemma 9.2.2

$$\begin{aligned}
\left|\Psi_n(t) - e^{-t^2/2}\right| &= \left|\prod_{k=1}^{n} \Psi_{nk}(t) - \prod_{k=1}^{n} \phi_{nk}(t)\right| \leq \sum_{k=1}^{n} |\Psi_{nk}(t) - \phi_{nk}(t)| \\
&= \sum_{k=1}^{n} \left|\Psi_{nk}(t) - (1 - t^2\sigma_{nk}^2/2) + (1 - t^2\sigma_{nk}^2/2) - \phi_{nk}(t)\right| \\
&\leq \sum_{k=1}^{n} \left|\Psi_{nk}(t) - (1 - t^2\sigma_{nk}^2/2)\right| + \sum_{k=1}^{n} \left|\phi_{nk}(t) - (1 - t^2\sigma_{nk}^2/2)\right| \\
&\leq \sum_{k=1}^{n} E\left(\min\left\{|tX_{nk}|^2,\ |tX_{nk}|^3/6\right\}\right) \\
&+ \sum_{k=1}^{n} E\left(\min\left\{|tZ_{nk}|^2,\ |tZ_{nk}|^3/6\right\}\right), \qquad (10.6)
\end{aligned}$$

with $k = 2$ in Theorem 9.2.4. It is to be noted that $S_n \xrightarrow{L} Z \sim N(0,1)$, if both the terms on the right hand side of (10.6) converge to 0 as $n \to \infty$. We now examine whether the first term converges to 0. Suppose $U_{nk} = \min\left\{|tX_{nk}|^2,\ |tX_{nk}|^3/6\right\}$ and $V_{nk} = \min\left\{|tZ_{nk}|^2,\ |tZ_{nk}|^3/6\right\}$. Observe that for small ϵ,

$$\begin{aligned}
|X_{nk}| &\leq \epsilon \Rightarrow U_{nk} = |tX_{nk}|^3/6 < |tX_{nk}|^3 \\
\& \ |X_{nk}| &> \epsilon \Rightarrow U_{nk} = |tX_{nk}|^2.
\end{aligned}$$

Hence,

$$\begin{aligned}
E(U_{nk}) &= \int_{\Omega} U_{nk} dP = \int_{[|X_{nk}| \leq \epsilon]} U_{nk} dP + \int_{[|X_{nk}| > \epsilon]} U_{nk} dP \\
&\leq |t|^3 \epsilon \int_{[|X_{nk}| \leq \epsilon]} |X_{nk}|^2 dP + t^2 \int_{[|X_{nk}| > \epsilon]} |X_{nk}|^2 dP \\
&\leq |t|^3 \epsilon \int_{\Omega} |X_{nk}|^2 dP + t^2 \int_{[|X_{nk}| > \epsilon]} |X_{nk}|^2 dP \\
&= |t|^3 \epsilon\, \sigma_{nk}^2 + t^2 \int_{[|X_{nk}| > \epsilon]} |X_{nk}|^2 dP \\
\Rightarrow \sum_{k=1}^{n} E(U_{nk}) &\leq \epsilon |t|^3 \sum_{k=1}^{n} \sigma_{nk}^2 + t^2 \sum_{k=1}^{n} \int_{[|X_{nk}| > \epsilon]} |X_{nk}|^2 dP \\
&= \epsilon\, |t|^3 + t^2\, g_n(\epsilon).
\end{aligned}$$

Now taking limits as $n \to \infty$ on both sides of the above inequality and using Lindeberg condition we have,

$$\lim_{n\to\infty} \sum_{k=1}^{n} E(U_{nk}) \ \leq \ \epsilon |t|^3 = 0,$$

since ϵ is arbitrary. The proof of convergence to 0 of the second term in (10.6) follows when Z_{nk} are substituted for X_{nk} in the above steps. Hence,

$$|\Psi_n(t) - e^{-t^2/2}| \to 0 \ \forall \ t \in \mathbb{R} \ \Rightarrow \ \sum_{k=1}^{n} (X_k - \mu_k)/s_n \xrightarrow{L} Z \sim N(0,1).$$

Part (ii): We now proceed to prove that if both the results in Equation (10.5) are satisfied then, $\forall \ \epsilon > 0, \ g_n(\epsilon) \to 0$. With $k = 1$ in Theorem 9.2.4 and using the result that $\max_{1\leq k\leq n} \sigma_{nk}^2 \to 0$ as $n \to \infty$, we have

$$\begin{aligned} |1 - \Psi_{nk}(t)| &\leq E\left(\min\left\{2|tX_{nk}|, \ |tX_{nk}|^2/2\right\}\right) \\ &\leq E\left(|tX_{nk}|^2/2\right) = t^2\sigma_{nk}^2/2 \\ \Rightarrow \max_{1\leq k\leq n} |1 - \Psi_{nk}(t)| &\leq (t^2/2) \max_{1\leq k\leq n} \sigma_{nk}^2 \to 0 \\ \Rightarrow \lim_{n\to\infty} |1 - \Psi_{nk}(t)| &= 0 \ \text{uniformly in} \ \ k. \qquad (10.7) \end{aligned}$$

$$\begin{aligned} \text{Hence,} \ \sum_{k=1}^{n} |1 - \Psi_{nk}(t)|^2 &\leq \max_{1\leq k\leq n} |1 - \Psi_{nk}(t)| \sum_{k=1}^{n} |1 - \Psi_{nk}(t)| \\ &\leq (t^2/2) \max_{1\leq k\leq n} \sigma_{nk}^2 \sum_{k=1}^{n} t^2\sigma_{nk}^2/2 \\ &\leq (t^4/4) \max_{1\leq k\leq n} \sigma_{nk}^2, \qquad (10.8) \end{aligned}$$

since $\sum_{k=1}^{n} \sigma_{nk}^2 = 1$. We have already noted in Chapter 7 that for any complex number $z = a + ib$,

$$\begin{aligned} e^z &= e^a e^{ib} = e^a(\cos b + i \sin b) \Rightarrow |e^z| = (e^{2a}\cos^2 b + e^{2a}\sin^2 b)^{1/2} = e^a \\ \Rightarrow z &= \Psi_{nk}(t) - 1 = E(\cos tX_{nk}) - 1 + iE(\sin tX_{nk}) \\ &\Rightarrow |e^z| = e^{E(\cos tX_{nk})-1} \ \leq \ e^{-1}e = 1. \end{aligned}$$

In Equation (10.7), it is proved that $|1 - \Psi_{nk}(t)| \to 0$ as $n \to \infty$ uniformly in k, hence for large n, one can assume that $|z| \leq 1/2$. Hence using the result from p. 555 of Gut [13], we have $|e^z - 1 - z| \leq |z|^2$. Now, $|\Psi_{nk}(t)| \leq$

1 & $|e^{\Psi_{nk}(t)-1}| \leq 1$. Hence by Lemma 9.2.2

$$\begin{aligned}
\left|\prod_{k=1}^{n} e^{\Psi_{nk}(t)-1} - \Psi_n(t)\right| &= \left|\prod_{k=1}^{n} e^{\Psi_{nk}(t)-1} - \prod_{k=1}^{n} \Psi_{nk}(t)\right| \\
&\leq \sum_{k=1}^{n} \left|e^{\Psi_{nk}(t)-1} - \Psi_{nk}(t)\right| \\
&= \sum_{k=1}^{n} \left|e^{\Psi_{nk}(t)-1} - (\Psi_{nk}(t) - 1) - 1\right| \\
&\leq \sum_{k=1}^{n} |\Psi_{nk}(t) - 1|^2 \quad \text{using } |e^z - 1 - z| \leq |z|^2 \\
&\leq (t^4/4) \max_{1\leq k\leq n} \sigma_{nk}^2 \qquad \text{by (10.8)} \\
&\to 0\,, \qquad (10.9)
\end{aligned}$$

since it is given that $\max_{1\leq k\leq n} \sigma_{nk}^2 \to 0$. We further know that

$$S_n \xrightarrow{L} Z \sim N(0,1) \iff \Psi_n(t) - e^{-t^2/2} \to 0 \ \forall \ t \in \mathbb{R}. \qquad (10.10)$$

Hence,

$$\begin{aligned}
\left|\prod_{k=1}^{n} e^{\Psi_{nk}(t)-1} - e^{-t^2/2}\right| &= \left|\prod_{k=1}^{n} e^{\Psi_{nk}(t)-1} - \Psi_n(t) + \Psi_n(t) - e^{-t^2/2}\right| \\
&\leq \left|\prod_{k=1}^{n} e^{\Psi_{nk}(t)-1} - \Psi_n(t)\right| + \left|\Psi_n(t) - e^{-t^2/2}\right| \\
&\to 0 \ \text{ by (10.9) and (10.10)} \\
\Rightarrow \prod_{k=1}^{n} e^{\Psi_{nk}(t)-1} &\to e^{-t^2/2} = e^{-(t^2/2)\sum_{k=1}^{n}\sigma_{nk}^2} \ \text{ as } \sum_{k=1}^{n} \sigma_{nk}^2 = 1 \\
&\Rightarrow \prod_{k=1}^{n} e^{\Psi_{nk}(t)-1} e^{(t^2/2)\sum_{k=1}^{n}\sigma_{nk}^2} \to 1.
\end{aligned}$$

Thus,

$$\begin{aligned}
\exp\left\{\sum_{k=1}^{n} \left(\Psi_{nk}(t) - 1 + (t^2/2)\sigma_{nk}^2\right)\right\} &\to 1 \\
\Rightarrow \sum_{k=1}^{n} \left(\Psi_{nk}(t) - 1 + (t^2/2)\sigma_{nk}^2\right) &\to 0.
\end{aligned}$$

It is to be noted that the limit of a sequence of complex numbers is 0 and hence the real part must converge to 0. Thus,

$$\begin{aligned}\sum_{k=1}^{n}(E(\cos(tX_{nk}))-1+(t^2/2)\sigma_{nk}^2) &= \sum_{k=1}^{n}(E(\cos(tX_{nk}))-1+(t^2/2)E(X_{nk}^2))\\ &= \sum_{k=1}^{n}\int_{\Omega}(\cos(tX_{nk})-1+(t^2/2)X_{nk}^2)\,dP\\ &\to 0 \ \text{ as } \ n\to\infty.\end{aligned}$$

Now by Lemma 10.3.3, the integrand $(\cos(x)-1+x^2/2)\geq 0\ \forall\ x\in\mathbb{R}$. Thus, if integral over Ω converges to 0, integral over any subset of Ω must converge to 0. We take a subset to be $[|X_{nk}|>\epsilon]$. Suppose $u_n=\sum_{k=1}^{n}E\left((1-\cos(tX_{nk}))I_{[|X_{nk}|>\epsilon]}\right)$. Now, $|\cos(x)-1|\leq 2,\ \ \forall\ x\in\mathbb{R}$. Hence,

$$\begin{aligned}|u_n| &\leq \sum_{k=1}^{n}E\left(|\cos(tX_{nk})-1|I_{[|X_{nk}|>\epsilon]}\right) \leq \sum_{k=1}^{n}2P[|X_{nk}|>\epsilon]\\ &\leq 2\sum_{k=1}^{n}\sigma_{nk}^2/\epsilon^2 = 2/\epsilon^2 \Rightarrow \ \limsup_{n\to\infty}u_n\leq 2/\epsilon^2.\end{aligned}$$

In the second step we use Chebyshev's inequality. Now for $\epsilon>0$,

$$\begin{aligned}\sum_{k=1}^{n}E(\cos(tX_{nk})-1+(t^2/2)X_{nk}^2) &\to 0\\ \Rightarrow \sum_{k=1}^{n}E\left((\cos(tX_{nk})-1+(t^2/2)X_{nk}^2)I_{[|X_{nk}|>\epsilon]}\right) &\to 0\\ \Rightarrow \sum_{k=1}^{n}(t^2/2)E\left(X_{nk}^2)I_{[|X_{nk}|>\epsilon]}\right)-u_n &\to 0\\ \Rightarrow (t^2/2)g_n(\epsilon)-u_n &\to 0\\ \Rightarrow g_n(\epsilon)-(2/t^2)u_n &\to 0,\end{aligned}$$

Hence,
$$\begin{aligned}\limsup_{n\to\infty}g_n(\epsilon) &= \limsup_{n\to\infty}\Big(g_n(\epsilon)-(2/t^2)u_n+(2/t^2)u_n\Big)\\ &\leq \limsup_{n\to\infty}\Big(g_n(\epsilon)-(2/t^2)u_n\Big)+\limsup_{n\to\infty}(2/t^2)u_n\\ &\leq \lim_{n\to\infty}\Big(g_n(\epsilon)-(2/t^2)u_n\Big)+(2/t^2)(2/\epsilon^2)=0+4/t^2\epsilon^2,\end{aligned}$$

which can be made arbitrarily small by selecting t sufficiently large for each $\epsilon>0$. Further $g_n(\epsilon)\geq 0$, hence we conclude that $\forall\ \epsilon>0,\ g_n(\epsilon)\to 0$ as $n\to\infty$ and the theorem is proved. □

Remark 10.3.1. In the first part of the proof, it is proved that Lindeberg's condition implies Feller's condition. From the second part of the theorem, we note that if Feller's condition is assumed, then Lindeberg's condition is not only sufficient but also necessary for the normality of S_n. The first part of the proof is established by Lindeberg and the second part is proved by Feller. Hence, it is known as a Lindeberg-Feller CLT. The second part of the theorem is also referred to as Feller's theorem (Athreya and Lahiri [3]).

We now show that Lindeberg-Levy CLT follows from Lindeberg-Feller CLT, as is expected.

Theorem 10.3.2. *Lindeberg-Feller CLT implies Lindeberg-Levy CLT.*

Proof. Suppose $\{X_n, n \geq 1\}$ is a sequence of independent and identically distributed random variables such that $E(X_k) = \mu$, $Var(X_k) = \sigma^2$, $0 < \sigma^2 < \infty$ and F is the common distribution function. Thus, $s_n^2 = n\sigma^2$. We now verify whether the Lindeberg condition is satisfied. Observe that $\forall\ \epsilon > 0$,

$$\begin{aligned} g_n(\epsilon) &= \frac{1}{s_n^2}\sum_{k=1}^{n}\int_{A_{nk}}(x_k-\mu_k)^2 dF_k(x_k) \\ &= \frac{1}{n\sigma^2}\sum_{k=1}^{n}\int_{[|x-\mu|>\epsilon\sigma\sqrt{n}]}(x-\mu)^2 dF(x) = \frac{1}{\sigma^2}\int_{[|x-\mu|>\epsilon\sigma\sqrt{n}]}(x-\mu)^2 dF(x). \end{aligned}$$

It is to be noted that $P\left[|X-\mu|/\sigma > \epsilon\sqrt{n}\right] \to 0$ in view of the fact that the real random variable $(X-\mu)/\sigma$ is bounded in probability. Further, $(X-\mu)^2$ is an integrable random variable. Hence, $\int_{[|x-\mu|>\epsilon\sigma\sqrt{n}]}(x-\mu)^2 dF(x) \to 0$, by Corollary 8.3.1. Thus, $\forall\ \epsilon > 0,\ g_n(\epsilon) \to 0$ and the Lindeberg's condition is satisfied by the sequence $\{X_n, n \geq 1\}$ of independent and identically distributed random variables. Hence the sequence $\{X_n, n \geq 1\}$ satisfies the Lindeberg-Feller CLT. Thus,

$$\sum_{k=1}^{n}(X_k-\mu_k)/s_n = \sum_{k=1}^{n}(X_k-\mu)/\sqrt{n}\sigma \xrightarrow{L} Z \sim N(0,1).$$

Thus the sequence $\{X_n, n \geq 1\}$ satisfies the Lindeberg-Levy CLT. □

We now prove Lyapounov's CLT in the following theorem.

Theorem 10.3.3. *Lyapounov's CLT: Suppose $\{X_n, n \geq 1\}$ is a sequence of independent random variables with mean μ_k and $Var(X_k) = \sigma_k^2 > 0$. Suppose $S_n = \sum_{k=1}^{n}(X_k-\mu_k)$ and $s_n^2 = Var(S_n) < \infty$. In addition, suppose $E(|X_k-\mu_k|^{2+\delta}) > 0$ for some $\delta > 0$. Then for some $\delta > 0$,*

$$\sum_{k=1}^{n}E(|X_k-\mu_k|^{2+\delta})/s_n^{2+\delta} \to 0 \ \Rightarrow\ S_n/s_n \xrightarrow{L} Z \sim N(0,1).$$

Proof. In Lemma 10.3.2 it is proved that Lyapounov's condition implies Lindeberg's condition. Hence, Lyapounov's CLT follows from the Lindeberg-Feller CLT and hence $S_n/s_n \xrightarrow{L} Z \sim N(0,1)$ distribution. □

Remark 10.3.2. It is in general difficult to verify Lindeberg's condition. As noted above a sufficient condition for Lindeberg's condition is Lyapounov's condition, which is somewhat easier to verify. However, one criticism about Lyapounov's condition is that it involves moments of higher order than the moments in the formulation of the problem.

Corollary 10.3.1. *Suppose $\{X_n, n \geq 1\}$ is a sequence of independent random variables with mean 0 and X_n's are uniformly bounded. Suppose $S_n = \sum_{i=1}^n X_i$ and $s_n^2 = Var(S_n)$. If $s_n \to \infty$, then $S_n/s_n \xrightarrow{L} Z \sim N(0,1)$.*

Proof. Observe that

$$\begin{aligned}
|X_k| \leq C < \infty \quad &\Rightarrow \quad E(|X_k|^{2+\delta}) \leq C^\delta E(|X_k^2|) = C^\delta \sigma_k^2 \\
&\Rightarrow \quad \frac{\sum_{k=1}^n E(|X_k|^{2+\delta})}{s_n^{2+\delta}} \leq \frac{C^\delta}{s_n^\delta} \to 0, \quad \text{as } s_n \to \infty \\
&\Rightarrow \quad S_n/s_n \xrightarrow{L} Z \sim N(0,1),
\end{aligned}$$

by Lyapounov's CLT. □

Example 10.3.1. Suppose $\{X_n, n \geq 1\}$ is a sequence of independent random variables, where X_k follows Bernoulli $B(1, p_k)$ distribution, $0 < p_k < 1$. Then X_k's are uniformly bounded by 1. If $s_n^2 = \sum_{k=1}^n p_k(1-p_k) \to \infty$, then by the above corollary, $(S_n - E(S_n))/s_n \xrightarrow{L} Z \sim N(0,1)$. □

In Lyapounov's theorem, it is assumed that Lyapounov's condition holds for some $\delta > 0$. It is proved below that if Lyapounov's condition holds for δ_2, then it also holds with δ_1 where $0 < \delta_1 < \delta_2$. In applications, Lyapounov's condition is verified first for $\delta = 1$ or 2 and if it fails, then verified for $\delta < 1$. In Example 10.3.2 and in Example 10.3.3, the condition is verified for $\delta = 1$, while in Example 10.3.10 it is verified for $\delta = 2$.

Lemma 10.3.4. *Lyapounov's condition with $\delta_2 > 0$ is satisfied implies that Lyapounov's condition is satisfied for every δ_1, such that $0 < \delta_1 < \delta_2$, that is,*

$$\lim_{n\to\infty} \sum_{k=1}^n E(|X_{nk}|^{2+\delta_2}) = 0 \quad \Rightarrow \quad \lim_{n\to\infty} \sum_{k=1}^n E\left(|X_{nk}|^{2+\delta_1}\right) = 0, \quad \forall \ \delta_1 \in (0, \delta_2),$$

where $X_{nk} = (X_k - \mu_k)/s_n$, $k = 1, 2 \cdots, n$.

Proof. Suppose the set $\{x_{nk}||x_{nk}| \le \epsilon\}$ is denoted by $[|x_{nk}| \le \epsilon]$. For $\epsilon > 0$ observe that,

$$\begin{aligned}
E(|X_{nk}|^{2+\delta_1}) &= \int\limits_{[|x_{nk}|\le\epsilon]} |x_{nk}|^{2+\delta_1} dF_{nk}(x) + \int\limits_{[|x_{nk}>\epsilon]} |x_{nk}|^{2+\delta_1} dF_{nk}(x) \\
&\le \epsilon^{\delta_1} \int\limits_{[|x_{nk}|\le\epsilon]} |x_{nk}|^2 dF_{nk}(x) + \int\limits_{[|x_{nk}|>\epsilon]} \frac{|x_{nk}|^{2+\delta_2}}{|x_{nk}|^{\delta_2-\delta_1}} dF_{nk}(x) \\
&\le \epsilon^{\delta_1} \int\limits_{\mathbb{R}} |x_{nk}|^2 dF_{nk}(x) + \int\limits_{[|x_{nk}|>\epsilon]} \frac{|x_{nk}|^{2+\delta_2}}{\epsilon^{\delta_2-\delta_1}} dF_{nk}(x) \\
&\le \epsilon^{\delta_1} E(|X_{nk}|^2) + \frac{1}{\epsilon^{\delta_2-\delta_1}} \int\limits_{\mathbb{R}} |x_{nk}|^{2+\delta_2} dF_{nk}(x) \\
&= \epsilon^{\delta_1}\sigma_{nk}^2 + E(|X_{nk}|^{2+\delta_2})/\epsilon^{\delta_2-\delta_1}.
\end{aligned}$$

$$\begin{aligned}
\text{Hence, } \sum_{k=1}^{n} E(|X_{nk}|^{2+\delta_1}) &\le \epsilon^{\delta_1} \sum_{k=1}^{n} \sigma_{nk}^2 + \sum_{k=1}^{n} E|X_{nk}|^{2+\delta_2}/\epsilon^{\delta_2-\delta_1} \\
&= \epsilon^{\delta_1} + \sum_{k=1}^{n} E|X_{nk}|^{2+\delta_2}/\epsilon^{\delta_2-\delta_1} \\
\Rightarrow \limsup_{n\to\infty} \sum_{k=1}^{n} E(|X_{nk}|^{2+\delta_1}) &\le \epsilon^{\delta_1} + \limsup_{n\to\infty} \sum_{k=1}^{n} E(|X_{nk}|^{2+\delta_2})/\epsilon^{\delta_2-\delta_1}
\end{aligned}$$

Now, the second term on the right hand side is zero. By taking ϵ to be sufficiently small it follows that $\limsup_{n\to\infty} \sum\limits_{k=1}^{n} E(|X_{nk}|^{2+\delta_1}) = 0$ and the result follows. □

Following examples illustrate the verification of the three conditions and also illustrate Lindeberg-Feller CLT and Lyapounov's CLT.

Example 10.3.2. Suppose $\{X_n, n \ge 1\}$ is a sequence of independent random variables where $X_n \sim B(1, p_n)$ distribution, $0 < p_n < 1$. Then $E(X_n) = p_n$ and $Var(X_n) = p_n q_n$, where $q_n = 1 - p_n$. Suppose $S_n = \sum_{k=1}^{n}(X_k - p_k)$, then $s_n^2 = Var(S_n) = \sum_{k=1}^{n} p_k q_k$. Note that $s_n^2 \le n/4$. Suppose $s_n^2 \to \infty$. We now show that all the three conditions are satisfied. If $X_k \sim B(1, p_k)$, then using the fact that $p_k^2 + q_k^2 \le 2$, we have

$$\begin{aligned}
E(|X_k - E(X_k)|^3) &= p_k^3 q_k + p_k q_k^3 = p_k q_k(p_k^2 + q_k^2) \le 2p_k q_k \\
\Rightarrow \sum_{k=1}^{n} E(|X_k - \mu_k|^3)/s_n^3 &\le 2\sum_{k=1}^{n} p_k q_k/s_n^3 = 2s_n^2/s_n^3 \to 0.
\end{aligned}$$

Thus, Lyapounov's condition is satisfied with $\delta = 1$. It is shown in Example 10.3.1 that Lyapounov's CLT is satisfied as the random variables X_k are uniformly bounded by 1. It is proved in Lemma 10.3.2 that

$$g_n(\epsilon) \leq \frac{1}{\epsilon^\delta s_n^{2+\delta}} \sum_{k=1}^{n} E(|X_k - \mu_k|^{2+\delta}).$$

Hence, with $\delta = 1$, it follows that $g_n(\epsilon) \leq 2/s_n \to 0$. Thus, Lindeberg's condition is also satisfied. We can also verify it as follows. Note that for each $k = 1, 2, \cdots, n$, the deviations $|X_k - p_k| \in (0, 1)$. Given $\epsilon > 0$, we can select n_0 such that $\epsilon > 1/n_0$. Since $s_n \to \infty$, corresponding to this n_0, $\exists \ n_1 \in \mathbb{N}$ such that $s_n > n_0 \ \ \forall \ n \geq n_1$. Suppose $B_{nk} = [|X_k - p_k| > \epsilon s_n]$. Note that $\forall \ n \geq n_1$,

$$1/s_n < 1/n_0 < \epsilon \ \Rightarrow \ \epsilon s_n > 1 \ \Rightarrow \ B_{nk} = \emptyset \ \Rightarrow \ I_{B_{nk}} = 0.$$

Hence,

$$\begin{aligned} g_n(\epsilon) &= \frac{1}{s_n^2} \sum_{k=1}^{n} E((X_k - p_k)^2 I_{B_{nk}}) = \frac{1}{s_n^2} \sum_{k=1}^{n_1 - 1} E((X_k - p_k)^2 I_{B_{nk}}) \\ &+ \frac{1}{s_n^2} \sum_{k=n_1}^{n} E((X_k - p_k)^2 I_{B_{nk}}) = \frac{1}{s_n^2} \sum_{k=1}^{n_1 - 1} E((X_k - p_k)^2 I_{B_{nk}}) \ \to \ 0, \end{aligned}$$

as $n \to \infty$, since the sum is finite and $s_n \to \infty$ as $n \to \infty$. Thus, it is verified that Lindeberg's condition is satisfied. Observe that $\max_{k \leq n} p_k q_k / s_n \leq 1/4s_n \to 0$ and hence Feller's condition is satisfied. Thus if $s_n^2 \to \infty$, then the sequence $\{X_n, n \geq 1\}$ satisfies CLT.

Suppose $s_n^2 \to c < \infty$, then we can choose ϵ, for example $\epsilon = 1/2s_n$, such that $P[|X_k - p_k| > \epsilon s_n] = P[|X_k - p_k| \geq 1/2] > 0, \ \ \forall \ k \geq 1$. Then $g_n(\epsilon)$ does not converge to 0 for all $\epsilon > 0$. Thus, $g_n(\epsilon) \to 0$ only if $s_n^2 \to \infty$. Hence, for a sequence $\{X_n, n \geq 1\}$ of independent random variables where $X_n \sim B(1, p_n)$ distribution, the necessary and sufficient condition for CLT to hold is $s_n^2 = \sum_{k=1}^{n} p_k q_k \to \infty$ as $n \to \infty$. If $p_n = p, \ \ \forall \ n \geq 1$, then $s_n^2 \to \infty$. Thus, for a sequence $\{X_n, n \geq 1\}$ of independent and identically distributed random variables where $X_n \sim B(1, p)$ distribution, CLT always holds. It also follows from Lindeberg-Levy CLT. □

Example 10.3.3. Suppose $\{X_n, n \geq 1\}$ is a sequence of independent random variables such that $X_n \sim U(-n, n)$ distribution. We examine whether CLT is satisfied by $\{X_n, n \geq 1\}$. Observe that

$$\begin{aligned} X_k \sim U(-k, k) \ &\Rightarrow \ E(X_k) = \mu_k = 0 \\ Var(X_k) = k^2/3 \ &\Rightarrow \ s_n^2 = \sum_{k=1}^{n} k^2/3 = n(n+1)(2n+1)/18 \to \infty. \end{aligned}$$

Further,

$$|X_k| \sim U(0,k) \;\Rightarrow\; E(|X_k^3|) = k^3/4 \;\Rightarrow\; \sum_{k=1}^{n} E(|X_k^3|) = n^2(n+1)^2/4$$
$$\Rightarrow\; \sum_{k=1}^{n} E(|X_k - \mu_k|^3)/s_n^3 = \sum_{k=1}^{n} E(|X_k|^3)/s_n^3 \to 0.$$

Thus, Lyapounov's condition is satisfied for $\delta = 1$ and hence $S_n/s_n \xrightarrow{L} Z \sim N(0,1)$ distribution. Thus, CLT holds for the sequence $\{X_n, n \geq 1\}$. Since the sequence satisfies Lyapounov's condition, by Lemma 10.3.2 Lindeberg's condition is also satisfied, which further implies that Feller's condition is satisfied. We verify these below. To verify Lindeberg's condition, we have to examine whether $\forall\ \epsilon > 0,\ g_n(\epsilon) \to 0$ as $n \to \infty$. Observe that $n/s_n \to 0$, which implies that given $\epsilon > 0,\ \exists\ n_0$, such that $n < \epsilon s_n$ for all $n \geq n_0$. Further, $|X_k| > \epsilon s_n > n$ for all $n \geq n_0$ and $k = 1, 2, \cdots, n$. Hence, in view of the fact that $P[|X_n| > n] = 0$, $P(B_{nk}) = P[|X_k| > \epsilon s_n] = 0$ for all $n \geq n_0$. Hence as $n \to \infty,\ \forall\ \epsilon > 0$

$$\begin{aligned}
\int_{A_{nk}} x_k^2 dF_k(x_k) &\to 0 \\
&\Rightarrow (1/n)\sum_{k=1}^{n} \int_{A_{nk}} x_k^2 dF_k(x_k) \to 0 \text{ by Toeplitz' lemma} \\
\Rightarrow g_n(\epsilon) &= \frac{1}{s_n}\frac{n}{s_n}\Big(\frac{1}{n}\sum_{k=1}^{n} \int_{A_{nk}} x_k^2 dF_k(x_k)\Big) \to 0.
\end{aligned}$$

It is easy to check that

$$\max_{1 \leq k \leq n} Var(X_k)/s_n^2 = (n^2/3)(18/n(n+1)(2n+1) \to 0 \text{ as } n \to \infty$$

and Feller's condition is satisfied. □

In the next example, we verify by simulation the normal approximation of S_n/s_n shown in Example 10.3.3.

Example 10.3.4. Suppose $\{X_n, n \geq 1\}$ is a sequence of independent random variables such that $X_n \sim U(-n,n)$ distribution. In Example 10.3.3 we have examined that the sequence $\{X_n, n \geq 1\}$ satisfies CLT and hence $S_n/s_n \xrightarrow{L} Z \sim N(0,1)$ distribution. We verify this result by simulation using the following code and find the values of n for which normal approximation is acceptable.

Code 10.3.1.
```
s=c(50,100,150,200);nsim=300; x=p=p1=v=c()
y=matrix(nrow=nsim,ncol=length(s))
```

```
for(j in 1:length(s))
{
 n=s[j]
for(i in 1:nsim)
{
 set.seed(i)
for(k in 1:n)
{
 x[k]=runif(1,-k,k)
}
 v[j]=(n*(n+1)*(2*n+1)/18)^(.5)
 y[i,j]=sum(x)/v[j]
}
 p1[j]=shapiro.test(y[,j])$p.value
}
d=round(data.frame(s,p1,v),4); d
```

The output is organized in Table 10.5.

TABLE 10.5
Verification of Lindeberg-Feller CLT: $X_k \sim U(-k,k)$

n	p-value	s_n
50	0.8354	119.6174
100	0.2749	335.8323
150	0.9828	615.4335
200	0.5867	946.3438

From p-values of the Shapiro-Wilk test we note that normal approximation of S_n/s_n is acceptable even for $n = 50$. We also note that s_n increases as n increases. □

In the next example, we note that Lyapounov's condition and Lindeberg's condition are not satisfied, but Feller's condition is satisfied.

Example 10.3.5. Suppose $\{X_n, n \geq 1\}$ is a sequence of independent random variables such that $P[X_n = \pm n] = 1/2n^2$ and $P[X_n = 0] = 1 - 1/n^2$. Hence,

$$\begin{aligned} E(X_n) &= 0, \; Var(X_n) = 1 \;\; \Rightarrow \;\; s_n^2 = \sum_{k=1}^{n} Var(X_k) = n \\ E(|X_k|^{2+\delta}) &= k^\delta \;\; \Rightarrow \;\; \sum_{k=1}^{n} E(|X_k|^{2+\delta}) \approx \int_0^n x^\delta \, dx = \frac{n^{\delta+1}}{\delta+1}. \end{aligned}$$

Hence,

$$\frac{1}{s_n^{2+\delta}}\sum_{k=1}^{n}E(|X_k|^{2+\delta})=\frac{n^{\delta+1}}{(\delta+1)n^{\delta/2+1}}=\frac{n^{\delta/2}}{\delta+1}\nrightarrow 0,$$

since $\delta > 0$. Thus, Lyapounov's condition is not satisfied. Observe that for $\epsilon > 0$, with $A_{nk} = \{x_k||x_k| > \epsilon s_n\}$,

$$\begin{aligned}\int_{A_{nk}} x_k^2 dF_k(x_k) &> \epsilon^2 s_n^2 P[|X_k| > \epsilon n] = \epsilon^2 s_n^2 P[|X_k| \neq 0] = \epsilon^2 s_n^2 (1/k^2)\\ \Rightarrow\ g_n(\epsilon) &= (1/s_n^2)\sum_{k=1}^{n}\int_{A_{nk}} x_k^2 dF_k(x_k) > \epsilon^2\sum_{k=1}^{n}(1/k^2)\\ &= \epsilon^2 A, \quad \text{where } A=\sum_{k=1}^{n}(1/k^2)<\infty\\ \Rightarrow\ g_n(\epsilon) &\nrightarrow 0.\end{aligned}$$

Thus, Lindeberg's condition is also not satisfied. However, note that

$$\max_{1\le k\le n} Var(X_k)/s_n^2 = 1/n \to 0,$$

that is, Feller's condition is satisfied. □

Example 10.3.6. Suppose $\{Y_n, n \geq 1\}$ and $\{Z_n, n \geq 1\}$ are independent sequences of independent random variables such that $\{Y_n, n \geq 1\}$ is a sequence of independent and identically distributed random variables with $E(Y_n) = 0$ & $Var(Y_n) = 1$. $\{Z_n, n \geq 1\}$ is a sequence of independent random variables such that $P[Z_n = n] = P[Z_n = -n] = 1/2n^2$ and $P[Z_n = 0] = 1 - 1/n^2$. Thus, $E(Z_n) = 0$ & $Var(Z_n) = 1$. If X_n is defined as $X_n = Y_n + Z_n$, then $\{X_n, n \geq 1\}$ is a sequence of independent random variables and

$$E(X_n)=0,\quad Var(X_n)=2\ \&\ s_n^2=\sum_{k=1}^{n}Var(X_k)=2n.$$

Suppose $U_n = \sum_{k=1}^n Y_k, V_n = \sum_{k=1}^n Z_k$ and $S_n = \sum_{k=1}^n X_k$. $\{Y_n, n \geq 1\}$ is a sequence of independent and identically distributed random variables with positive finite variance 1, hence $U_n/\sqrt{n} \xrightarrow{L} Z \sim N(0,1)$ distribution by the Lindeberg-Levy CLT. Further $\forall\ \epsilon > 0$,

$$\sum_{n=1}^{\infty}P[|Z_n|>\epsilon]=\sum_{n=1}^{\infty}1/n^2<\infty \Rightarrow Z_n \xrightarrow{a.s.} 0, \text{ as } n\to\infty, \text{ by Theorem 6.4.1.}$$

Hence, $\exists\ N$ with $P(N) = 0$ & n_0 such that $Z_n(\omega) = 0\ \forall\ n \geq n_0$ and $\forall\ \omega \in N^c$. As a consequence, $\forall\ \omega \in N^c$,

$$\begin{aligned}\sum_{k=1}^{\infty}Z_k(\omega) &= \sum_{k=1}^{n_0-1}Z_k(\omega)+\sum_{k=n_0}^{\infty}Z_k(\omega)=\sum_{k=1}^{n_0-1}Z_k(\omega)=M, \text{ say}\\ &\Rightarrow \sum_{k=1}^{\infty}Z_k(\omega)=M<\infty \Rightarrow \sum_{k=1}^{\infty}Z_k \xrightarrow{a.s.} M.\end{aligned}$$

Hence,

$$\begin{aligned}\sum_{k=1}^{n} Z_k/\sqrt{n} &= V_n/\sqrt{n} \xrightarrow{a.s.} 0 \\ &\Rightarrow S_n/\sqrt{n} = U_n/\sqrt{n} + V_n/\sqrt{n} \xrightarrow{L} Z \sim N(0,1) \\ &\Rightarrow S_n/\sqrt{2n} = S_n/s_n \xrightarrow{L} Z_1 \sim N(0,1/2).\end{aligned}$$

The second last step follows from Slutsky's theorem. Further note that, $\max_{1\leq k\leq n} Var(X_k)/s_n^2 = 2/2n \to 0$, which implies that Feller's condition is satisfied. Thus, S_n/s_n has normal distribution for large n and Feller's condition is satisfied and hence by the Lindeberg-Feller CLT, Lindeberg's condition must be satisfied. We examine it below. Observe that $|X_k| = |Y_k|$ with probability $1-1/k^2$, $|X_k| = |Y_k+k|$ with probability $1/2k^2$ and $|X_k| = |Y_k-k|$ with probability $1/2k^2$. Hence for any distribution of Y_n with mean 0 and variance 1, large k and $\epsilon = 1/\sqrt{n}$,

$$\begin{aligned}P[|X_k| > \epsilon\sqrt{2n}] &= P[|X_k| > \sqrt{2}] > 0 \Rightarrow \int_{A_{nk}} x_k^2 dF_k(x_k) > 0 \\ \Rightarrow g_n(\epsilon) &= \frac{1}{s_n^2}\sum_{k=1}^{n}\int_{A_{nk}} x_k^2 dF_k(x_k) \\ &> \frac{1}{2n}\epsilon^2 2n \int_{A_{nn}} 1dF_n(x_n) > \epsilon^2 P(A_{nn}) > 0.\end{aligned}$$

Thus, $\lim_{n\to\infty} g_n(\epsilon) > 0$ and hence Lindeberg's condition is not satisfied. This result may appear contradictory to the second part of the Lindeberg-Feller theorem. Note that the Lindeberg-Feller theorem requires the asymptotic normal distribution of S_n/s_n to be the standard normal. In this case it is not $N(0,1)$ but $N(0,1/2)$. Thus, there is no contradiction to the second part of the Lindeberg-Feller theorem. □

Following example illustrates that neither Lindeberg's condition nor Feller's condition is necessary for normality of S_n. It also illustrates that if one the two results in (10.5) is not satisfied, then Lindeberg's condition is not satisfied.

Example 10.3.7. Suppose $\{X_n, n \geq 1\}$ is a sequence of independent random variables where $X_n \sim N(0, \theta^n)$ distribution, $\theta > 1$. Suppose $S_n = \sum_{k=1}^{n} X_k$, then $Var(S_n)$ is given by $s_n^2 = \sum_{k=1}^{n} \theta^k = \theta(\theta^n - 1)/(\theta - 1)$. Thus, for each $n \geq 1$, $S_n \sim N(0, s_n^2)$ distribution. Hence, $S_n/s_n \xrightarrow{L} Z \sim N(0,1)$ distribution and we do not need any CLT to claim normality of S_n/s_n for large n. Observe that

$$\max_{1\leq k\leq n}\frac{Var(X_k)}{s_n^2} = \max_{1\leq k\leq n}\frac{\theta^k}{s_n^2} = \frac{\theta^n(\theta-1)}{\theta(\theta^n-1)} = \frac{\theta-1}{\theta}\frac{1}{(1-\theta^{-n})} \to \frac{\theta-1}{\theta}.$$

Thus, Feller's condition is not satisfied. As a consequence Lindeberg's condition is also not be satisfied. We verify it below. Observe that $k = 1, 2, \cdots, n$ and $\forall \ \epsilon > 0$,

$$\begin{aligned} P[|X_k| > \epsilon s_n] &= P[|X_k|/\sigma_k > \epsilon s_n/\sigma_k] \\ &= [1 - \Phi(\epsilon s_n/\sigma_k) + \Phi(-\epsilon s_n/\sigma_k)] > 0, \end{aligned}$$

since $X_k/\sigma_k \sim N(0,1)$ distribution, where $\sigma_k = \theta^{k/2}$. Hence, taking only the last term from $g_n(\epsilon)$, we have

$$\begin{aligned} g_n(\epsilon) &= \frac{1}{s_n^2}\sum_{k=1}^{n}\int_{A_{nk}} x_k^2 dF_k(x_k) > \int_{A_{nn}} \frac{x_n^2}{s_n^2} dF_n(x_n) \\ &> \epsilon^2 \int_{A_{nn}} dF_n(x_n) = \epsilon^2 P[|X_n| > \epsilon s_n] \\ &= \epsilon^2 P[|X_n|/\sigma_n > \epsilon s_n/\sigma_n] \\ &= \epsilon^2 [1 - \Phi(\epsilon s_n/\sigma_n) + \Phi(-\epsilon s_n/\sigma_n)] \\ &\to \epsilon^2 \left[1 - \Phi(\sqrt{\theta_1}\epsilon) + \Phi(-\sqrt{\theta_1}\epsilon)\right] \neq 0 \ \text{ where } \ \theta_1 = \theta/(\theta - 1). \end{aligned}$$

Thus, Lindeberg's condition is not satisfied and hence Lyapounov's condition is not satisfied. □

Remark 10.3.3. In the statement of Lindeberg-Feller CLT, we have noted that Equation (10.4) implies Equation (10.5), which consists of two results. Hence, if at least one the two results in (10.5) is not satisfied, then Equation (10.4) is not satisfied. Example 10.3.7 supports this statement as in this example $S_n/s_n \xrightarrow{L} Z \sim N(0,1)$ distribution, but the second result is not satisfied, since $\max_{k \leq n} \sigma_k^2/s_n^2 \to (\theta - 1)/\theta$. Thus, one the two results in (10.5) is not satisfied and hence (10.4) is not satisfied. It is to be noted that $S_n/s_n \xrightarrow{L} Z \sim N(0,1)$ distribution, in view of the fact that X_n for each n has normal distribution. Observe that in this example σ_n^2 and s_n^2 both increase to ∞ very fast, at the same rate of θ^n. As a consequence, σ_n^2/s_n^2 converges to a constant $(\theta - 1)/\theta$.

In the following example, none of the three conditions are satisfied, but still asymptotic distribution of S_n/s_n is normal.

Example 10.3.8. Suppose $\{X_n, n \geq 1\}$ is a sequence of independent random variables where $X_n \sim N(0, 2^{-n})$ distribution. Then $S_n = \sum_{k=1}^n X_k$ follows $N(0, s_n^2)$ distribution where $s_n^2 = Var(S_n) = \sum_{k=1}^n 2^{-k} = (1 - 2^{-n}) \ \forall \ n \geq 1$. Hence $S_n/s_n \xrightarrow{L} Z \sim N(0,1)$ distribution. Observe that $s_n^2 \uparrow 1$ and

$$\max_{1 \leq k \leq n} Var(X_k)/s_n^2 = \max_{1 \leq k \leq n} 2^{-k}/s_n^2 = (1/2)/(1 - 2^{-n}) \ \to \ 1/2.$$

Thus, Feller's condition is not satisfied which implies that Lindeberg's condition is not satisfied, which further implies that Lyapounov's condition is not satisfied. We verify these below. We now examine whether Lindeberg's condition is satisfied. Since $X_k \sim N(0, 2^{-k})$ distribution, $P[|X_k| > \epsilon s_n] > 0$ for all $k = 1, 2, \cdots, n$. Hence,

$$\begin{aligned} g_n(\epsilon) &= \frac{1}{s_n^2}\sum_{k=1}^{n}\int_{A_{nk}} x_k^2 dF_k(x_k) > \int_{A_{n1}} \frac{x_1^2}{s_n^2} dF_n(x_1) \\ &> \epsilon^2 \int_{A_{n1}} dF_n(x_1) = \epsilon^2 P[|X_1| > \epsilon s_n] > 0. \end{aligned}$$

Note that in this case $s_n^2 \to 1$. Thus, Lindeberg's condition is not satisfied. We show below that Lyapounov's condition is not satisfied for $\delta = 2$. Note that,

$$\begin{aligned} E(X_k^4) = 3 \times 2^{-2k} \quad &\Rightarrow \quad \sum_{k=1}^{n} E(X_k^4) = 1 - 4^{-n} \\ &\Rightarrow \quad \frac{1}{s_n^4}\sum_{k=1}^{n} E(|X_k|^4) = \frac{(1-4^{-n})}{(1-2^{-n})^2} \to 1. \end{aligned}$$

Thus, Lyapounov's condition is not satisfied for $\delta = 2$. □

We have proved at the beginning of this section that if Feller's condition is satisfied, then $s_n^2 \to \infty$. In the above example, $s_n^2 \to 1$ and none of the three conditions is satisfied, still $S_n/s_n \xrightarrow{L} Z \sim N(0,1)$ distribution. It is to be noted that normality of S_n/s_n results from that fact that X_n has normal distribution for each n.

It is proved that if $s_n^2 \uparrow s^2 < \infty$, then each X_k in the sum S_n must have a normal distribution, refer to p.340 of Gut [13]. As a consequence S_n will have a normal distribution and there is nothing to prove. Thus, in order to obtain non-trivial results, it is necessary that $s_n^2 \to \infty$.

As in Example 10.3.8, in the next example also, none of the conditions are satisfied, under certain setup.

In a birth process and in a birth-death process, we come across with a sequence of independent but not identically distributed random variables. Suppose X_i denotes the sojourn time in state $i \geq 1$ of a birth process with birth rates $\{\lambda_i, i \geq 1\}$. Then $\{X_i, i \geq 1\}$ is a sequence of independent random variables and X_i follows exponential distribution with scale parameter λ_i, $i \geq 1$. The distribution of $S_n = \sum_{i=1}^{n} X_i$ is known as a hypoexponential distribution. In the next two examples, we examine whether $\{X_i, i \geq 1\}$ satisfies CLT for two different sets of birth rates.

Example 10.3.9. Suppose $\{X_n, n \geq 1\}$ is a sequence of independent random variables such that X_n follows exponential distribution with rate parameter $n\lambda$. Observe that from the moment generating function $(1 - t/n\lambda)^{-1}$ of X_n,

$$\begin{aligned} E(X_k) = \mu_k = 1/k\lambda, \quad Var(X_k) = 1/(k\lambda)^2 &\Rightarrow s_n^2 = (1/\lambda^2)\sum_{k=1}^{n}(1/k^2) \\ E(X_k^4) = 4!/(k\lambda)^4 &\Rightarrow \mu_4 = E(|X_k - \mu_k|^4) = 9/(k\lambda)^4. \end{aligned}$$

It is to be noted that $\sum_{k=1}^{\infty}(1/k^2)$ is a convergent series. Thus, $s_n^2 \to s < \infty$, where s is a positive real number. Hence,

$$\max_{1\le k\le n} Var(X_k)/s_n^2 = 1/(\lambda^2 s_n^2) \to 1/(\lambda^2 s) \neq 0$$

which implies that Feller's condition is not satisfied. In Lemma 10.3.1, we have proved that Feller's condition implies that $s_n^2 \to \infty$. Thus, $s_n^2 \to s < \infty$ implies that Feller's condition is not satisfied. By Equation (10.5), it further implies that Lindeberg's condition will not be satisfied and hence Lyapounov's condition is also not satisfied. Observe that, $\sum_{k=1}^{\infty}(1/k^4)$ is a convergent series. Hence,

$$\sum_{k=1}^{n} E(|X_k - \mu_k|^4)/s_n^4 = 9\sum_{k=1}^{n}(1/k^4)/(\sum_{k=1}^{n}(1/k^2))^2 \nrightarrow 0,$$

being a ratio of two non-negative real numbers. Thus, Lyapounov's condition is not satisfied for $\delta = 2$ and hence will not be satisfied for any $\delta > 2$. We have thus shown that for a sequence $\{X_n, n \ge 1\}$, none of the conditions are satisfied. As a consequence, $S_n/s_n \overset{L}{\nrightarrow} Z \sim N(0,1)$. Using the following code we show that when $\lambda = 0.05$, even for $n = 1000$ or $n = 2000$, normality of S_n/s_n is not acceptable, based on the p-values of Shapiro-Wilk test.

Code 10.3.2.
```
la=.05; s=c(1000,2000);nsim=500; x=p=p1=v=mu=a=c()
y=matrix(nrow=nsim,ncol=length(s))
for(j in 1:length(s))
{
 n=s[j]
for(i in 1:nsim)
{
 set.seed(i)
for(k in 1:n)
{
 x[k]=rexp(1,rate=k*la)
 mu[k]=1/k*la
 v[k]=1/(k*la)^2
}
 a[j]=sum(v)^(.5)
 y[i,j]=sum(x-mu)/a[j]
}
 p1[j]=shapiro.test(y[,j])$p.value
}
d=round(data.frame(s,p1,a),4); d
```

TABLE 10.6
Hypoexponential Distribution: p-values

n	p-value	s_n
1000	0	25.6432
2000	0	25.6471

The output is organized in Table 10.6. From Table 10.6, we note that for $n = 1000$ and $n = 2000$, p-value of the Shapiro-Wilk test is 0. Further, s_n stabilizes at 25.64. □

In the next example, we take $\lambda_i = \lambda/\sqrt{i}$ and show that all the three conditions are satisfied, thus the scenario of Example 10.3.9 changes as birth rates change.

Example 10.3.10. Suppose $\{X_n, n \geq 1\}$ is a sequence of independent random variables such that X_n follows exponential distribution with rate parameter $\lambda/\sqrt{n}$. Then,

$$\begin{aligned} E(X_k) &= \mu_k = \sqrt{k}/\lambda, \;\; Var(X_k) = k/\lambda^2 \;\Rightarrow\; s_n^2 = \sum_{k=1}^{n} k/\lambda^2 \\ &= n(n+1)/2\lambda^2 \to \infty. \end{aligned}$$

Further observe that,

$$\begin{aligned} E(X_k^4) = 4!k^2/\lambda^4 \;\; &\Rightarrow \;\; \mu_4 = E(|X_k - \mu_k|^4) = 9k^2/\lambda^4 \\ &\Rightarrow \;\; \sum_{k=1}^{n} E(|X_k - \mu_k|^4)/s_n^4 = 36\sum_{k=1}^{n} k^2/(n(n+1))^2 \\ &= \;\; 6(2n+1)/n(n+1) \;\; \to \;\; 0. \end{aligned}$$

Thus, Lyapounov's condition is satisfied with $\delta = 2$ and hence for all $\delta < 2$. Hence by Lyapounov's CLT, $S_n/s_n \xrightarrow{L} Z \sim N(0,1)$ distribution. Since the sequence satisfies Lyapounov's condition, Lindeberg's condition is also satisfied, which further implies that Feller's condition is satisfied. We verify Feller's condition below. It is easy to see that

$$\max_{1\leq k\leq n} Var(X_k)/s_n^2 = 2n/n(n+1) \to 0 \;\text{ as }\; n \to \infty$$

and Feller's condition is satisfied. Using the following code, we examine normality of S_n/s_n for four values of n, based on the p-values of Shapiro-Wilk test.

Code 10.3.3. `la=.05; s=c(100,200,300,400);nsim=200;`

```
x=p=p1=v=mu=a=c()
y=matrix(nrow=nsim,ncol=length(s))
for(j in 1:length(s))
```

```
{
 n=s[j]
for(i in 1:nsim)
{
 set.seed(i)
for(k in 1:n)
{
 x[k]=rexp(1,rate=la/k^(.5))
 mu[k]=k^(.5)/la
 v[k]=k/la^2
}
 a[j]=sum(v)^(.5)
 y[i,j]=sum(x-mu)/a[j]
}
 p1[j]=shapiro.test(y[,j])$p.value
}
d=round(data.frame(s,p1,a),4); d
```

TABLE 10.7
Lindeberg-Feller CLT: Hypoexponential Distribution

n	p-value	s_n
100	0.1502	1421.267
200	0.8754	2835.489
300	0.6770	4249.706
400	0.4665	5663.921

The output is organized in Table 10.7. From p-values of the Shapiro-Wilk test we note that normal approximation of S_n/s_n is acceptable. The values of s_n increase as n increases, as expected. □

We summarize below the results discussed in this chapter.

Summary

1. Lindeberg-Levy CLT: Suppose $\{X_n, n \geq 1\}$ is a sequence of independent and identically distributed random variables with mean μ and positive, finite variance σ^2, then $\sqrt{n}(\overline{X}_n - \mu)/\sigma \ \stackrel{L}{\to} \ Z \sim N(0,1)$.

2. Random sum CLT: Suppose $\{X_n, n \geq 1\}$ is a sequence of independent and identically distributed random variables with mean 0 and variance

1. Suppose $\{N(t), t \geq 0\}$ is a family of positive integer valued random variables where $N(t) \xrightarrow{a.s.} \infty$ such that $N(t)/t \xrightarrow{P} c$, where $c \in (0, \infty)$ is a constant. Then as $t \to \infty$, $\sum_{i=1}^{N(t)} X_i/\sqrt{N(t)} \xrightarrow{L} Z \sim N(0,1)$.

3. Multivariate CLT: Suppose $\underline{X}$ is a k dimensional random vector with mean vector $E(\underline{X}) = \underline{\mu}$ and dispersion matrix Σ, which is positive definite. Suppose $\{\underline{X}_1, \underline{X}_2, \cdots, \underline{X}_n\}$ are independent and identically distributed random vectors each distributed as $\underline{X}$. Suppose $\overline{\underline{X}}_n$ denotes the sample mean vector, then $\sqrt{n}(\overline{\underline{X}}_n - \underline{\mu}) \xrightarrow{L} \underline{U} \sim N_k(\underline{0}, \Sigma)$ distribution as $n \to \infty$.

4. Suppose $\{X_n, n \geq 1\}$ is a sequence of independent random variables such that $E(X_k) = \mu_k$, $Var(X_k) = \sigma_k^2 < \infty$ and F_k is a distribution function of $X_k, k = 1, 2, \cdots, n$. Suppose $S_n = \sum_{k=1}^n (X_k - \mu_k)$ and $s_n^2 = Var(S_n) = \sum_{k=1}^n \sigma_k^2$. Then Lyapounov's condition, Lindeberg's condition and Feller's condition are as follows.

   (i) Lyapounov's condition: $\sum_{k=1}^n E(|X_k - \mu_k|^{2+\delta})/s_n^{2+\delta} \to 0$ as $n \to \infty$, for some $\delta > 0$.

   (ii) Lindeberg's condition: With $A_{nk} = \{x_k | x_k - \mu_k| > \epsilon s_n\}$, $\forall\ \epsilon > 0$,

$$g_n(\epsilon) = \frac{1}{s_n^2} \sum_{k=1}^n \int_{A_{nk}} (x_k - \mu_k)^2 dF_k(x_k) \to 0 \text{ as } n \to \infty.$$

   (iii) Feller's condition: $\max_{k \leq n} \sigma_k^2/s_n^2 \to 0$ as $n \to \infty$.

5. Lyapounov's condition implies Lindeberg's condition and Lindeberg's condition implies Feller's condition

6. Lindeberg-Feller CLT: Suppose $\{X_n, n \geq 1\}$ is a sequence of independent random variables such that $E(X_k) = \mu_k$, $Var(X_k) = \sigma_k^2$ and F_k is a distribution function of X_k. Suppose $S_n = \sum_{k=1}^n (X_k - \mu_k)$ and $s_n^2 = Var(S_n) = \sum_{k=1}^n \sigma_k^2$. If $\forall\ \epsilon > 0$, $g_n(\epsilon) \to 0$, then $S_n/s_n \xrightarrow{L} Z \sim N(0,1)$ and $\max_{k \leq n} \sigma_k/s_n \to 0$. Conversely if $S_n/s_n \xrightarrow{L} Z \sim N(0,1)$ and Feller's condition is satisfied, then Lindeberg's condition is satisfied.

7. Lyapounov's CLT: Suppose $\{X_n, n \geq 1\}$ is a sequence of independent random variables with mean μ_k and $Var(X_k) = \sigma_k^2 > 0$. Suppose $S_n = \sum_{k=1}^n (X_k - \mu_k)$ and $s_n^2 = Var(S_n) = \sum_{k=1}^n \sigma_k^2$. In addition, suppose $E(|X_k - \mu_k|^{2+\delta}) > 0$ for some $\delta > 0$. Then $S_n/s_n \xrightarrow{L} Z \sim N(0,1)$ if

$$\sum_{k=1}^n E(|X_k - \mu_k|^{2+\delta})/s_n^{2+\delta} \to 0 \text{ for some } \delta > 0.$$

10.4 Conceptual Exercises

10.4.1 Suppose $\{X_n, n \geq 1\}$ is a sequence of independent and identically distributed random variables, each having $P(\lambda)$ distribution. Suppose $S_n = \sum_{i=1}^{n} X_i$. Find $\lim_{n\to\infty} P[S_n \leq n\lambda + \sqrt{n\lambda}]$.

10.4.2 Suppose $\{X_n, n \geq 1\}$ is a sequence of independent and identically distributed random variables, each having Pareto-type distribution, with probability density function $f(x) = c/(|x|^3(\log|X|)^2)$ for $|x| > 2$ and 0 otherwise, where c is a normalizing constant. Examine whether the sequence satisfies CLT. Examine whether one can verify Lyapounov's condition.

10.4.3 Prove that $\lim_{n\to\infty} e^{-n} \sum_{i=0}^{n} n^i/i! = 1/2$.

10.4.4 Evaluate $\lim_{n\to\infty} \sum_{i=0}^{[nq/p]} \binom{i+k-1}{i} p^k q^{i-k}$, where $0 < p < 1$ and $q = 1 - p$.

10.4.5 Examine whether $(X_n - E(X_n))/\sqrt{V(X_n)} \xrightarrow{L} Z \sim N(0,1)$ as $n \to \infty$, if $X_n \sim G(\alpha, n)$, where n is a natural number.

10.4.6 Evaluate $\lim_{n\to\infty} \int_0^{n/\alpha + 2\sqrt{n}/\alpha} \frac{\alpha^n}{\Gamma(n)} x^{n-1} e^{-\alpha x}\, dx$.

10.4.7 Suppose $\{X_n, n \geq 1\}$ is a sequence of independent and identically distributed random variables each having standard normal distribution. Suppose $Y_n = \sum_{i=1}^{n} X_i^2$. Show that $(Y_n - n)/\sqrt{(2n)} \xrightarrow{L} Z \sim N(0,1)$. What is the distribution of $(Y_n - n)^2/(2n)$ for large n? Justify.

10.4.8 Suppose $X \sim P(\lambda)$, $\lambda > 0$. Prove that for large λ, distribution of $(X - \lambda)^2/\lambda$ can be approximated by the chi-square distribution.

10.4.9 Suppose $\{X_n, n \geq 1\}$ is a sequence of independent and identically distributed random variables, each following $U(0, \theta)$ distribution, $\theta > 0$. Suppose $S_n = \left(\prod_{i=1}^{n} X_i\right)^{1/n}$. Find the non-degenerate limiting distribution of $\sqrt{n}(S_n - \theta e^{-1})$.

10.4.10 Suppose $\{X_n, n \geq 1\}$ is a sequence of independent and identically distributed positive random variables. Suppose $S_n = \left(\prod_{i=1}^{n} X_i\right)^{1/n}$. Find the non-degenerate limiting distribution of suitably normalized S_n.

10.4.11 Suppose $\{X_n, n \geq 1\}$ is a sequence of independent and identically distributed random variables, each following a Poisson $P(\theta)$ distribution. Find the non-degenerate limiting distribution of suitably normalized sample variance.

10.4.12 Suppose $X_n \sim B(n,p)$, $0 < p < 1$, $n \geq 1$ and $Y_n = \log(X_n/n)$ if $X_n \neq 0$ and $Y_n = 1$ if $X_n = 0$. Show that $\sqrt{n}(Y_n - \log p) \xrightarrow{L} Z_1 \sim N(0, q/p)$, where $q = 1 - p$.

10.4.13 Suppose $\{X_n, n \geq 1\}$ is a sequence of independent random variables with $P[X_n = \pm n] = n^{-\lambda}/2$ and $P[X_n = 0] = 1 - n^{-\lambda}$, $\lambda > 0$. Show that Liapounov's condition for the central limit theorem holds for this sequence for some λ.

10.4.14 Suppose $\{X_n, n \geq 1\}$ is a sequence of independent random variables with $P[X_n = \pm n^{\lambda}] = 1/2$. Examine whether Liapounov's condition, Lindeberg's condition and Feller's condition are satisfied. Examine if CLT holds for some values of λ.

10.4.15 Suppose $\{X_n, n \geq 1\}$ is a sequence of independent random variables such that for $\alpha \geq 1/2$,

$$P[X_n = \pm n^{\alpha}] = n^{1-2\alpha}/2 \;\;\&\;\; P[X_n = 0] = 1 - n^{1-2\alpha}, \; n \geq 1.$$

Show that $\forall \;\alpha \in [1/2, 1)$, $S_n/s_n \xrightarrow{L} Z \sim N(0,1)$.

10.4.16 Suppose $\{X_n, n \geq 1\}$ is a sequence of independent random variables such that $X_k \sim U(0,k)$ distribution. Suppose $Y_k = X_k/n^{1/2}$, $k = 1, 2, \cdots, n$, $n \geq 1$. Show that $\{Y_k, k \geq 1\}$ satisfies Liapounov's, Lindeberg's and Feller's conditions and hence CLT is valid for $S_n = \sum_{k=1}^{n}(Y_k - E(Y_k))$.

10.5 Computational Exercises

10.5.1 Verify conceptual exercise 10.4.3 by simulation.

10.5.2 Suppose $X_n \sim B(n,p)$ $0 < p < 1$, $n \geq 1$ and $Y_n = \log(X_n/n)$ if $X_n = 1$ and $Y_n = 1$ if $X_n = 0$. By simulation verify that $\sqrt{n}(Y_n - \log p) \xrightarrow{L} Z_1 \sim N(0, q/p)$, where $q = 1 - p$.

10.5.3 Suppose $\{X_n, n \geq 1\}$ is a sequence of independent and identically distributed random variables, each following $U(0,\theta)$ distribution, $\theta > 0$. Suppose $S_n = \left(\prod_{i=1}^{n} X_i\right)^{1/n}$. By simulation verify that the non-degenerate limiting distribution of $\sqrt{n}(S_n - \theta e^{-1})$ is normal.

10.5.4 Suppose $X \sim P(\lambda)$, $\lambda > 0$. Verify by simulation that for large λ, distribution of $(X - \lambda)^2/\lambda$ can be approximated by the chi-square distribution.

10.6 Multiple Choice Questions

Note: In each question, multiple options may be correct. Unless specified otherwise, identify which of the statement(s) is/are correct. Answers are given in Chapter 11, after the solutions of conceptual exercises of Chapter 10.

10.6.1 A sequence $\{X_n, n \geq 1\}$ of random variables is said to obey a CLT, if there exists two sequences $\{a_n, n \geq 1\}$ of real numbers and $\{b_n, n \geq 1\}$ of positive real numbers, with $b_n \to \infty$, such that $n \to \infty$,

(a) $\frac{S_n - a_n}{b_n} \stackrel{a.s.}{\to} Z$ where $Z \sim N(0,1)$ and $S_n = \sum_{i=1}^{n} X_i$

(b) $\frac{S_n - a_n}{b_n} \stackrel{P}{\to} Z$

(c) $\frac{S_n - a_n}{b_n} \stackrel{q.m.}{\to} Z$

(d) $\frac{S_n - a_n}{b_n} \stackrel{L}{\to} Z$

10.6.2 A sequence $\{X_n, n \geq 1\}$ of random variables satisfies a CLT, if

(a) $\frac{S_n - a_n}{b_n} \stackrel{a.s.}{\to} Z$ where $Z \sim N(0,1)$

(b) $\frac{S_n - a_n}{b_n} \stackrel{P}{\to} Z$

(c) $\frac{S_n - a_n}{b_n} \stackrel{q.m.}{\to} Z$

(d) $\frac{S_n - a_n}{b_n} \stackrel{L}{\to} Z$, where $S_n = \sum_{i=1}^{n} X_i$, $\{a_n, n \geq 1\}$ is a sequence of real numbers and $\{b_n, n \geq 1\}$ is a sequence of positive real numbers with $b_n \to \infty$ as $n \to \infty$

10.6.3 Suppose $\{X_n, n \geq 1\}$ is a sequence of independent random variables and $S_n = \sum_{k=1}^{n} X_k$, $s_n^2 = \sum_{k=1}^{n} V(X_k)$. Then $(S_n - E(S_n)/s_n \stackrel{L}{\to} Z \sim N(0,1)$, if

(a) $X_n \sim$ Binomial$(5, 0.3)$

(b) $X_n \sim$ t(2)

(c) $X_n \sim$ Beta$(1,1)$

(d) $X_n \sim G(\alpha, n)$, with scale parameter α and shape parameter n.

10.6.4 Suppose $\{X_n, n \geq 1\}$ is a sequence of independent and identically distributed random variables and $S_n = \sum_{k=1}^{n} X_k$. $s_n^2 = \sum_{k=1}^{n} V(X_k)$. Then $(S_n - E(S_n)/s_n \stackrel{L}{\to} Z \sim N(0,1)$ if
(I) $X_n \sim$ Binomial$(5, 0.3)$. (II) $X_n \sim$ t(2). (III) $X_n \sim$ Beta$(1,1)$.
(IV) $X_n \sim G(\alpha, n)$, with scale parameter α and shape parameter n.
Which of the following is correct?

(a) (I) and (II)

(b) (I) and (III)

(c) (II) and (III)

(d) (I) and (IV).

10.6.5 $(X - E(X))/\sqrt{V(X)} \xrightarrow{L} Z \sim N(0,1)$ as $n \to \infty$, if

(a) $X \sim$ Poisson(n)
(b) $X \sim \chi^2_n$
(c) $X \sim$ Gamma(α, n)
(d) $X \sim N(\mu, \sigma^2/n)$

10.6.6 Suppose $\{X_n, n \geq 1\}$ is a sequence of independent and identically distributed random variables with mean 0 and variance 1. Which of the following is/are correct?

(a) $\sqrt{n}\overline{X}_n \xrightarrow{L} U$, where U has normal distribution
(b) $\sqrt{n}\overline{X}_n^2 \xrightarrow{L} U$, where U has chi-square distribution
(c) $n\overline{X}_n^2 \xrightarrow{L} U$, where U has chi-square distribution
(d) $\sqrt{n}\overline{X}_n^2 \xrightarrow{P} U$, where distribution of U is degenerate at 0

10.6.7 Suppose $\{X_n, n \geq 1\}$ is a sequence of independent and identically distributed random variables with mean 0 and variance 1. Which of the following statements is not correct?

(a) $\sqrt{n}\overline{X}_n \xrightarrow{L} U$, where U has normal distribution
(b) $\sqrt{n}\overline{X}_n^2 \xrightarrow{L} U$, where U has chi-square distribution
(c) $n\overline{X}_n^2 \xrightarrow{L} U$, where U has chi-square distribution
(d) $\sqrt{n}\overline{X}_n^2 \xrightarrow{P} U$, where distribution of U is degenerate at 0

10.6.8 Suppose $\{X_n, n \geq 1\}$ is a sequence of independent and identically distributed random variables with mean 0 and variance 1. Which of the following is/are correct?

(a) $\sqrt{n}\overline{X}_n^2 \xrightarrow{L} U$, where U has normal distribution
(b) $\sqrt{n}\overline{X}_n^2 \xrightarrow{L} U$, where U has chi-square distribution
(c) $\sqrt{n}\overline{X}_n^2 \xrightarrow{L} U$, where distribution of U is degenerate at 0
(d) $\sqrt{n}\overline{X}_n^2 \xrightarrow{P} U$, where distribution of U is degenerate at 0

10.6.9 Suppose $\{X_n, n \geq 1\}$ is a sequence of independent and identically distributed random variables with mean 0 and finite moment of order 4. Suppose m_1', m_2' denote the sample raw moments of order 1 and 2 respectively and μ_2' denote the population raw moment of order 2. Which of the following is/are correct?

(a) $\sqrt{n}(m_2' - \mu_2') \xrightarrow{L} U$, where U has normal distribution
(b) $\sqrt{n}(m_2' - \mu_2')m_1' \xrightarrow{L} U$, where U has normal distribution
(c) $\sqrt{n}(m_2' - \mu_2')\mu_1' \xrightarrow{L} U$, where distribution of U is degenerate at 0
(d) $m_2'm_1' \xrightarrow{P} U$, where distribution of U is degenerate at 0

10.6.10 Suppose $\{X_n, n \geq 1\}$ is a sequence of independent random variables with $E(X_k) = 0$, $Var(X_k) = \sigma_k^2 < \infty$, $k = 1, 2, \cdots, n$. Suppose $S_n = \sum_{k=1}^n X_k$, $s_n^2 = Var(S_n) = \sum_{k=1}^n \sigma_k^2$. Then Lyapounov's condition for $S_n/s_n \xrightarrow{L} Z \sim N(0,1)$ is

(a) $\sum_{k=1}^n E(|X_k|^{2+\delta})/s_n^{2+\delta} \to \infty \ \forall \ \delta > 0$ as $n \to \infty$
(b) $\sum_{k=1}^n E(|X_k|^{2+\delta})/s_n^{2+\delta} \to 0$ for some $\delta > 0$ as $n \to \infty$
(c) $\sum_{k=1}^n E(|X_k|^{2})/s_n^{2+\delta} \to 0$ for some $\delta > 0$ as $n \to \infty$
(d) $\sum_{k=1}^n E(|X_k|^{2+\delta})/s_n^{2} \to 0$ for some $\delta > 0$ as $n \to \infty$

10.6.11 Suppose $\{X_n, n \geq 1\}$ is a sequence of independent random variables with $E(X_k) = 0$, $Var(X_k) = \sigma_k^2 < \infty$, $k = 1, 2, \cdots, n$. Suppose $S_n = \sum_{k=1}^n X_k$, $s_n^2 = \sum_{k=1}^n \sigma_k^2$ and $B_{nk} = [|X_k| > \epsilon s_n]$, $k = 1, 2, \cdots, n$, $\epsilon > 0$. Then Lindeberg's condition for $S_n/s_n \xrightarrow{L} Z \sim N(0,1)$ is

(a) $(1/s_n^2)\sum_{k=1}^n E\left(X_k^2 I_{B_{nk}}\right) \to 1$ as $n \to \infty$, $\forall \ \epsilon > 0$
(b) $(1/s_n^2)\sum_{k=1}^n E\left(X_k^2 I_{B_{nk}}\right) \to 0$ as $n \to \infty$, for some $\epsilon > 0$
(c) $(1/s_n^2)\sum_{k=1}^n E\left(X_k^2 I_{B_{nk}}\right) \to 0$ as $n \to \infty$, $\forall \ \epsilon > 0$
(d) $(1/s_n^2)\sum_{k=1}^n E\left(|X_k| I_{B_{nk}}\right) \to 0$ as $n \to \infty$, $\forall \ \epsilon > 0$

10.6.12 Suppose $\{X_n, n \geq 1\}$ is a sequence of independent random variables such that $E(X_k) = 0$, $Var(X_k) = \sigma_k^2 < \infty$, $k = 1, 2, \cdots, n$. Suppose $S_n = \sum_{k=1}^n X_k$ and $s_n^2 = Var(S_n) = \sum_{k=1}^n \sigma_k^2$, $k = 1, 2, \cdots, n$. Then Feller's condition is

(a) $\max_{k \leq n} \sigma_k^2/s_n^2 \to 0$ as $n \to \infty$
(b) $\max_{k \leq n} \sigma_k^2/s_n^2 \to 1$ as $n \to \infty$
(c) $\max_{k \leq n} \sigma_k^2/s_n \to 0$ as $n \to \infty$
(d) $\min_{k \leq n} \sigma_k^2/s_n^2 \to 0$ as $n \to \infty$

10.6.13 Suppose $\{X_n, n \geq 1\}$ is a sequence of independent random variables such that $E(X_k) = 0$, $Var(X_k) = \sigma_k^2 < \infty$, $k = 1, 2, \cdots, n$. Following are two statements related to conditions for asymptotic normality of $S_n = \sum_{k=1}^n X_k$. (I) Lyapounov's condition implies Lindeberg's condition. (II) Lindeberg's condition implies Feller's condition.

(a) Both (I) and (II) are false
(b) Both (I) and (II) are true
(c) (I) is true but (II) is false
(d) (I) is false but (II) is true

10.6.14 Suppose $\{X_n, n \geq 1\}$ is a sequence of independent random variables such that $E(X_k) = 0$, $Var(X_k) = \sigma_k^2 < \infty$, $k = 1, 2, \cdots, n$. Following are two statements related to conditions for asymptotic normality of $S_n = \sum_{k=1}^n X_k$. (I) Lyapounov's condition implies Feller's condition. (II) Lindeberg's condition implies Lyapounov's condition.

(a) Both (I) and (II) are false

(b) Both (I) and (II) are true
(c) (I) is true but (II) is false
(d) (I) is false but (II) is true

10.6.15 Suppose $\{X_n, n \geq 1\}$ is a sequence of independent random variables such that $E(X_k) = 0$, $Var(X_k) = \sigma_k^2 < \infty$, $k = 1, 2, \cdots, n$. Following are two statements related to conditions for asymptotic normality of $S_n = \sum_{k=1}^{n} X_k$. (I) If Feller's condition is not satisfied then Lyapounov's condition is also not satisfied. (II) Lyapounov's condition is not satisfied then Lindeberg's is also not satisfied.

(a) Both (I) and (II) are false
(b) Both (I) and (II) are true
(c) (I) is true but (II) is false
(d) (I) is false but (II) is true

10.6.16 Suppose $\{X_n, n \geq 1\}$ is a sequence of independent random variables such that $E(X_k) = 0$, $Var(X_k) = \sigma_k^2 < \infty$, $k = 1, 2, \cdots, n$. Following are two statements related to conditions for asymptotic normality of $S_n = \sum_{k=1}^{n} X_k$. (I) If Feller's condition is satisfied then $Var(S_n) \to 0$. (II) If Lindeberg's condition is satisfied then $Var(S_n) \to 0$.

(a) Both (I) and (II) are false
(b) Both (I) and (II) are true
(c) (I) is true but (II) is false
(d) (I) is false but (II) is true

10.6.17 Suppose $\{X_n, n \geq 1\}$ is a sequence of independent random variables such that $E(X_k) = 0$, $Var(X_k) = \sigma_k^2 < \infty$, $k = 1, 2, \cdots, n$ and $s_n^2 = Var(S_n)$, where $S_n = \sum_{k=1}^{n} X_k$. Following are two statements related to conditions for asymptotic normality of S_n. $P[\max_{1 \leq k \leq n} |X_k| > \epsilon s_n] \to 0 \;\; \forall \;\; \epsilon > 0$ as $n \to \infty$ (I) if Lyapounov's condition is satisfied (II) if Lindeberg's condition is satisfied .

(a) Both (I) and (II) are false
(b) Both (I) and (II) are true
(c) (I) is true but (II) is false
(d) (I) is false but (II) is true

11

Solutions to Conceptual Exercises

11.1 Chapter 1

1.6.1 Verify whether the following sequences of the sets are monotone.
(i) $\{(a - \frac{1}{n}, b + \frac{1}{n})\}, n = 1, 2, \ldots$. (ii) $\{(a + \frac{1}{n}, b - \frac{1}{n})\}, n = 1, 2, \ldots$.
(iii) $\{(a + \frac{1}{n}, b + \frac{1}{n})\}, n = 1, 2, \ldots$. (iv) Suppose $\{A_n\}$ is a sequence of the sets defined as $A_n = \{(x, y)|0 \leq x \leq n, 0 \leq y \leq 1/n\}$.

Solution: Sequence in (i) is decreasing while in (ii) is increasing. (iii) and (iv) are not monotone.

1.6.2 For the following sequences $\{A_n, n \geq 1\}$ of sets, examine whether it has a limit. If yes, find it.
(i) $A_n = (a - 1/n, a + 1/n)$, (ii)$A_n = [a - 1/n, a + 1/n]$,
(iii) $A_n = (a - 1/n, a + 1/n]$, (iv) $A_n = [a - 1/n, a + 1/n)$,
(v) $A_n = (a - 1/n, a]$, (vi) $A_n = (a, b + 1/n)$, (vii) $A_n = (a, b + 1/n]$,
(viii) $A_n = [a, b + 1/n)$, (ix) $A_n = [a, b + 1/n]$, (x) $A_n = (a, b - 1/n)$,
(xi) $A_n = (a, b - 1/n]$, (xii) $A_n = [a, b - 1/n)$, (xiii) $A_n = [a, b - 1/n]$,
(xiv) $A_n = (a - 1/n, b + 1/n)$, (xv) $A_n = [a - 1/n, b + 1/n]$,
(xvi) $A_n = (a - 1/n, b + 1/n]$, (xvii) $A_n = [a - 1/n, b + 1/n)$.

Solution: For the first four sequences, limit exists and is $\{a\}$. (v) $\{a\}$, (vi),(vii) $(a, b]$,(viii) $[a, b]$, (ix) $[a, b]$, (x),(xi) (a, b), (xii),(xiii) $[a, b)$. For the last four sequences limit exists and is $[a, b]$.

1.6.3 For the following sequences $\{A_n, n \geq 1\}$ of sets examine whether it has a limit. If yes, find it.

$$(i)\ A_n = \begin{cases} (0, 1/n), & \text{if } n \text{ is odd} \\ (-1/n, 1/n), & \text{if } n \text{ is even.} \end{cases}$$

$$(ii)\ A_n = \begin{cases} (0, 1/n), & \text{if } n \text{ is odd} \\ (1 - 1/n, 1), & \text{if } n \text{ is even.} \end{cases}$$

$$(iii)\ A_n = \begin{cases} (0, 1 - 1/n), & \text{if } n \text{ is odd} \\ (1/n, 1), & \text{if } n \text{ is even.} \end{cases}$$

DOI: 10.1201/9781032619057-11

$$(iv)\ A_n = \begin{cases} \left(2-\frac{1}{n}, 2+\frac{1}{n}\right), & \text{if } n \text{ is odd} \\ \emptyset, & \text{if } n \text{ is even.} \end{cases}$$

$$(v)\ A_n = \begin{cases} \left(2-\frac{1}{n}, 2\right), & \text{if } n \text{ is odd} \\ \emptyset, & \text{if } n \text{ is even.} \end{cases}$$

(vi) $A_n = \{0, 1/n, 2/n, \cdots, 1-1/n, 1\}$.
(vii) A_n : set of rationals in $\left(1-\frac{1}{n+1}, 1+\frac{1}{n}\right]$,
(viii) $A_{2n} = \left(0, \frac{1}{2n}\right), A_{2n+1} = \left[-1, \frac{1}{2n+1}\right]$, $A_1 = [-1, 1]$,
(ix) A_n: set of rationals in $(1-\frac{1}{n}, 1+\frac{1}{n})$.

Solution: (i) By the definition, $\limsup A_n = \bigcap_{n\geq 1}\bigcup_{k\geq n} A_k = \bigcap_{n\geq 1} C_n$ where $C_n = \bigcup_{k\geq n} A_k$. Hence, for $n = 2m$,
$C_{2m} = (-1/(2m), 1/(2m)) \cup (0, 1/(2m+1)) \cup (-1/(2m+2), 1/(2m+2)\cdots$ which is $(-1/(2m), 1/(2m))$. For $n = 2m-1$,
$C_{2m-1} = \bigcup_{k\geq 2m-1} A_k = (0, 1/(2m-1)) \cup (-1/(2m), 1/(2m))\cdots$ which is $(-1/2m, 1/(2m-1))$. Hence, $\limsup A_n = \bigcap_{n\geq 1} C_n = (-1/2, 1) \cap (-1/4, 1/4) \cap (-1/6, 1/5] \cdots = \{0\}$ Now, $\liminf A_n = \bigcup_{n\geq 1}\bigcap_{k\geq n} A_k = \bigcup_{n\geq 1} B_n$ where $B_n = \bigcap_{k\geq n} A_k$. For $n = 2m$,
$B_{2m} = (-1/(2m), 1/(2m)) \cap (0, 1/(2m+1)) \cap (-1/(2m+2), 1/(2m+2)\cdots$, which is $\emptyset$. For $n = 2m-1$,
$B_{2m-1} = \bigcap_{k\geq 2m-1} A_k = (0, 1/(2m-1)) \cap (-1/(2m), 1/(2m))\cdots = \emptyset$.
Hence, $\liminf A_n = \emptyset$ which implies that $\lim A_n$ does not exist.
(ii) We proceed as in (i) and note that

$$\begin{aligned} C_{2m} &= \bigcup_{k\geq 2m} A_k = (1-1/2m, 1) \cup (0.1/(2m+1)\cdots = (0,1) \\ C_{2m-1} &= \bigcup_{k\geq 2m-1} A_k = (0, 1/(2m-1)) \cup (1-1/(2m), 1)\cdots = (0,1). \end{aligned}$$

Hence, $\limsup A_n = (0, 1)$. Similarly,

$$\begin{aligned} B_{2m} &= \bigcap_{k\geq 2m} A_k = (1-1/2m, 1) \cap (0.1/(2m+1)\cdots = \emptyset \\ B_{2m-1} &= \bigcap_{k\geq 2m-1} A_k = (0, 1/(2m-1)) \cap (1-1/(2m), 1)\cdots = \emptyset. \end{aligned}$$

Hence, $\liminf A_n = \emptyset$ and $\lim A_n$ does not exist.
(iii) Observe that $A_1 = \emptyset, A_2 = (1/2, 1), A_3 = (0, 2/3), A_4 = (1/4, 1), A_5 = (0, 4/5), A_6 = (1/6, 1)\cdots$ Hence, $C_n = (0, 1)\ \forall\ n \geq 1$ which implies that $\limsup A_n = (0, 1)$. Now note that $B_1 = \emptyset$, $B_2 = (1/2, 2/3), B_3 = (1/4, 2/3), B_4 = (1/4, 4/5), B_5 = (1/6, 4/5)$. Hence, $\liminf A_n = (0, 1)$ which implies that $\lim A_n$ exists and is $(0, 1)$.

(iv) By definition, $\liminf A_n = \bigcup_{n\geq 1}\bigcap_{k\geq n} A_k = \bigcup_{n\geq 1}\emptyset = \emptyset$. Now, $\limsup A_n = \bigcap_{n\geq 1}\bigcup_{k\geq n} A_k$. Suppose n is odd. Then

$$\begin{aligned}\bigcup_{k\geq n} A_k &= \left(2-\frac{1}{n},2+\frac{1}{n}\right)\cup\emptyset\cup\left(2-\frac{1}{n+2},2+\frac{1}{n+2}\right)\cdots\\ &= \left(2-\frac{1}{n},2+\frac{1}{n}\right).\end{aligned}$$

Hence, $\limsup A_n = \bigcap_{n\geq 1}\left(2-\frac{1}{n},2+\frac{1}{n}\right) = [2,2] = \{2\}$. Thus, $\lim A_n$ does not exist.
(v) With $C_n = \bigcup_{k\geq n} A_k$, note that

$$\begin{aligned}C_1 &= (1,2)\cup\emptyset\cup(5/3,2)\emptyset\cup(9/5,2)\cdots = (1,2)\\ C_2 &= \emptyset\cup(5/3,2)\emptyset\cup(9/5,2)\cdots = (5/3,2)\\ C_3 &= (5/3,2)\emptyset\cup(9/5,2)\cdots = (5/3,2) = C_2.\end{aligned}$$

Thus, we get $C_{2m} = C_{2m+1} = (2-1/(2m+1),2)\ \forall\ m\geq 1$. Note that it is decreasing sequence with limit $\cap C_n = \emptyset$. Hence $\limsup A_n = \emptyset$. Since for odd values of n, $A_n = \emptyset$, $B_n = \bigcap_{k\geq n} A_k = \emptyset$. Hence $\liminf A_n = \emptyset$. Thus, limit of the sequence exists and it is $\emptyset$.
(vi) It is given that $A_n = \{0, 1/n, 2/n, \cdots, 1-1/n, 1\}$. Note that any rational number in $[0,1]$ can be written as p/q where p and q are integers such that $0\leq p\leq q\leq 1$. Thus, $p/q\in A_q,\ q\in\mathbb{N}$. To find $\limsup A_n = \bigcap_{n\geq 1} C_n$, observe that

$$C_1 = A_1\cup A_2\cup A_3\cup\cdots = \{0,1\}\cup\{0,1/2,1\}\cup\{0,1/3,2/3,1\}\cup\cdots$$

which is a set of all rational numbers in $[0,1]$. Now, $A_1\subset A_2$ which implies that $A_2 = A_1\cup A_2$. Hence,

$$C_2 = A_2\cup A_3\cup\cdots = \{0,1/2,1\}\cup\{0,1/3,2/3,1\}\cup\cdots = C_1.$$

Note that, $1/2 = 2/4\in A_4$ hence,

$$C_3 = A_3\cup A_4\cdots = \{0,1/3,2/3,1\}\cup\{0,1/4,2/4,3/4,1\}\cup\cdots = C_1.$$

Similarly, $1/3 = 2/6\in A_6, 2/3 = 4/6\in A_6, 1/2 = 2/4\in A_4$ implies that $C_4 = C_1$. In general, any number of the form p/q can be written as $2p/2q\in A_{2q}$ or $3p/3q\in A_{3q}$. Proceeding on these lines we get $C_n = C_1\ \forall\ n\geq 1$. Hence, $\limsup A_n = \bigcap_{n\geq 1} C_n = C_1$. We now find $\liminf A_n = \bigcup_{n\geq 1} B_n$, where $B_n = \bigcap_{k\geq n} A_k$. Observe that $B_1 = \{0,1\}$. Since $1/2\notin A_3$, $B_2 = \{0,1\}$. In general, any number in A_k is not in A_{k+1} as shown below. Note that any number in A_k is of the form $p/k, 0\leq p\leq k$ and any number in A_{k+1} is of the form $r/(k+1), 0\leq r\leq k+1$. If some number in A_k is also in A_{k+1}, then $p/k = r/(k+1)$ for some p such that $0\leq p\leq k$ and for some r such that $0\leq r\leq k+1$, which implies that $r = (p/k)(k+1)$. However, $p \leq k$ and hence r will be an integer if and only if $(k+1)$ is a multiple of k. This is not possible unless $k = 1$. Thus, $A_k \cap A_{k+1} = \{0,1\}\ \forall\ k \geq 1$ and hence $B_n = \{0,1\}\ \forall\ n \geq 1$, which further implies that $\liminf A_n = \{0,1\} \neq \limsup A_n$. Thus $\lim A_n$ does not exist.

(vii) Suppose Q denotes the set of rational numbers. Note that with $C_n = \bigcup_{k\geq n} A_k$,

$$\begin{aligned} C_1 &= (1-1/2, 2) \cap Q, \quad C_2 = (1-1/3, 1+1/2) \cap Q \\ C_3 &= (1-1/4, 1+1/3) \cap Q \\ &\Rightarrow C_n = (1-1/(n+1), 1+1/n) \cap Q, n \geq 1 \\ &\Rightarrow \limsup A_n = \bigcap_{n\geq 1} C_n = \{1\} \cap Q = \{1\}. \end{aligned}$$

Now with $B_n = \bigcap_{k\geq n} A_k$,

$$\begin{aligned} B_1 &= \{1\} \cap Q = \{1\}, \quad B_2 = \{1\} \cap Q = \{1\}, \quad B_3 = \{1\} \\ &\Rightarrow B_n = \{1\}, \ \forall \ n \geq 1 \\ &\Rightarrow \liminf A_n = \bigcup_{n\geq 1} B_n = \{1\}. \end{aligned}$$

Thus, limit of the sequence exists and it is $\{1\}$.

(viii) with $C_n = \bigcup_{k\geq n} A_k$,

$$\begin{aligned} C_1 &= [-1,1] \cup (0,1/2) \cup [-1,1/3] \cup \cdots = [-1,1] \\ C_2 &= (0,1/2) \cup [-1,1/3] \cup \cdots = [-1,1/2) \\ &\Rightarrow C_{2n} = [-1, 1/2n) \ \& \ C_{2n+1} = [-1, 1/(2n+)] \\ &\Rightarrow \limsup A_n = \bigcap_{n\geq 1} C_n = [-1,0]. \end{aligned}$$

Now with $B_n = \bigcap_{k\geq n} A_k$,

$$\begin{aligned} B_1 &= [-1,1] \cap (0,1/2) \cap [-1,1/3] \cap \cdots = \{0\} \\ &\Rightarrow B_n = \{0\}, \ \forall \ n \geq 1 \\ &\Rightarrow \liminf A_n = \bigcup_{n\geq 1} B_n = \{0\}. \end{aligned}$$

Thus, limit of the sequence does not exist.

(ix) Suppose Q denotes the set of rational numbers. Note that with $C_n = \bigcup_{k\geq n} A_k$,

$$\begin{aligned} C_1 &= (0,2) \cap Q, \quad C_2 = (1-1/2, 1+1/2) \cap Q \\ C_3 &= (1-1/3, 1+1/3) \cap Q \\ &\Rightarrow C_n = (1-1/(n+1), 1+1/n) \cap Q, n \geq 1 \\ &\Rightarrow \limsup A_n = \bigcap_{n\geq 1} C_n = \{1\} \cap Q = \{1\}. \end{aligned}$$

Now with $B_n = \bigcap_{k \geq n} A_k$,

$$\begin{aligned} B_1 &= \{1\} \cap Q = \{1\}, \quad B_2 = \{1\} \cap Q = \{1\}, \quad B_3 = \{1\} \\ &\Rightarrow B_n = \{1\}, \ \forall \ n \geq 1 \\ &\Rightarrow \liminf A_n = \bigcup_{n \geq 1} B_n = \{1\}. \end{aligned}$$

Thus, limit of the sequence exists and it is $\{1\}$.

1.6.4 Suppose a sequence $\{A_n, n \geq 1\}$ is defined as follows.

$$A_n = \begin{cases} B, & \text{if } n \text{ is even} \\ C, & \text{if } n \text{ is odd.} \end{cases}$$

Examine whether the sequence $\{A_n, n \geq 1\}$ converges. If yes find the limit. If not, impose a condition on B and C so that it is a convergent sequence.

Solution: By definition, $\liminf A_n = \bigcup_{n \geq 1} \bigcap_{k \geq n} A_k = \bigcup_{n \geq 1} B_n$, where $B_n = \bigcap_{k \geq n} A_k$. If n is even, then $B_n = B \cap C \cap B \cdots = B \cap C$ and if n is odd then $B_n = C \cap B \cap C \cdots = B \cap C$. Hence, $\liminf A_n = B \cap C$. Now, $\limsup A_n = \bigcap_{n \geq 1} \bigcup_{k \geq n} A_k$. Suppose n is even. Then $\bigcup_{k \geq n} A_k = B \cup C \cup B \cdots = B \cup C$. If n is odd then $\bigcup_{k \geq n} A_k = C \cup B \cup C \cdots = B \cup C$. Hence, $\limsup A_n = B \cup C$. Thus, $\lim A_n$ does not exist. It exists if $B \cap C = B \cup C$, that is, if $B = C$.

1.6.5 Examine the limiting behaviour of the following sequence.

$$A_n = \big(-\infty, 0 \big) \cup (1/n, \infty), \quad \forall \ n \in \mathbb{N}.$$

Solution: Observe that $A_n^c = [0, 1/n], n \geq 1$ and it is a decreasing sequence with limit $\bigcap_{n \geq 1} A_n^c = \{0\}$. Hence $\lim_{n \to \infty} A_n = \{0\}^c$.

1.6.6 Give an example of a sequence of sets $\{A_n, n \geq 1\}$ where $\limsup A_n \neq \liminf A_n$.

Solution: Suppose $\{A_n, n \geq 1\}$ is a sequence of sets where,

$$A_n = \begin{cases} (5 - 1/n, 5 + 1/n), & \text{if } n \text{ is odd ,} \\ \emptyset, & \text{if } n \text{ is even.} \end{cases}$$

With $A_n = \emptyset$ for n odd we have,

$$\begin{aligned} \limsup A_n &= \bigcap_{n \geq 1} \bigcup_{k \geq n} A_k = \bigcap_{n \geq 1} C_n = \bigcap_{n \geq 1} (5 - 1/n, 5 + 1/n) \\ &\Rightarrow \limsup A_n = \{5\} \\ \text{Now,} \quad \liminf A_n &= \bigcup_{n \geq 1} \bigcap_{k \geq n} A_k = \bigcup_{n \geq 1} B_n = \emptyset \\ &\Rightarrow \liminf A_n = \emptyset \\ &\Rightarrow \limsup A_n \neq \liminf A_n. \end{aligned}$$

1.6.7 Give an example to show that $\liminf A_n \cup \liminf B_n \neq \liminf (A_n \cup B_n)$..

Solution: Suppose $\{A_n, n \geq 1\}$ and $\{B_n, n \geq 1\}$ are sequences of sets defined as follows. $A_1 = \{1\}, A_{2n} = \{2n\}$ and $A_{2n+1} = \{1, 2, \cdots, 2n + 1\}$ and $B_1 = \{1\}, B_{2n} = \{1, 2, \cdots, 2n\}$ and $B_{2n+1} = \{2n + 1\}$. It is easy to check that $\liminf A_n = \liminf A_n = \emptyset$. Note that $A_n \cup B_n = \{1, 2, \cdots, n\}$ and $\{A_n \cup B_n, n \geq 1\}$ is an increasing sequence. Hence, $\limsup A_n \cup B_n = \lim A_n \cup B_n = \bigcup_{n \geq 1} A_n \cup B_n$. Thus,

$$\limsup A_n \cup B_n = \bigcup_{n \geq 1} \{1, 2, \cdots, n\} = \mathbb{N} \neq \liminf A_n \cup \liminf B_n = \emptyset.$$

1.6.8 Examine whether (i) $\limsup(A_n \bigcap B_n) = \limsup A_n \bigcap \limsup B_n$ and (ii) $\liminf(A_n \bigcup B_n) = \liminf A_n \bigcup \liminf B_n$ if sequences $\{A_n, n \geq 1\}$ and $\{B_n, n \geq 1\}$ are defined as follows. $A_1 = \{1\}$, $A_{2n} = \{2n\}$, $A_{2n+1} = [1, 2n + 1]$ and $B_1 = \{1\}$, $B_{2n} = [1, 2n]$, $B_{2n+1} = \{2n + 1\}$.

Solution: For the given sequences $\{A_n, n \geq 1\}$ and $\{B_n, n \geq 1\}$, $A_n \cap B_n = \{n\}, n \geq 1$. Thus, $\{A_n \cap B_n, n \geq 1\}$ is a sequence of disjoint sets. Hence, $\limsup A_n \cap B_n = \liminf A_n \cap B_n = \emptyset$. Further note that $A_n \cup B_n = [1, n], n \geq 1$. Thus, $\{A_n \cup B_n, n \geq 1\}$ is an increasing sequence of sets with $\limsup(A_n \bigcup B_n) = \liminf(A_n \bigcup B_n) = [1, \infty)$. Further, $\limsup A_n = \limsup B_n = [1, \infty)$ and $\liminf A_n = \liminf B_n = \emptyset$. Hence,

$$\begin{aligned} \limsup(A_n \bigcap B_n) &= \emptyset \neq \limsup A_n \bigcap \limsup B_n = [1, \infty) \\ \& \ \ \liminf(A_n \bigcup B_n) &= [1, \infty) \neq \liminf A_n \bigcup \liminf B_n = \emptyset. \end{aligned}$$

1.6.9 Examine whether

$$\begin{aligned} (i) \liminf A_n \cup \ \liminf B_n &\subseteq \liminf(A_n \cup B_n) \\ &\subseteq \limsup(A_n \cup B_n) \\ &= \limsup A_n \cup \limsup B_n \\ \& \ \ (ii) \liminf A_n \cap \ \liminf B_n &= \liminf(A_n \cap B_n) \\ &\subseteq \limsup(A_n \cap B_n) \\ &\subseteq \limsup A_n \cap \limsup B_n. \end{aligned}$$

Solution: To examine whether $\liminf A_n \cup \ \liminf B_n \subseteq \liminf(A_n \cup B_n)$, observe that

$$\begin{aligned} \omega &\in \liminf A_n \cup \ \liminf B_n \\ &\Rightarrow \omega \in \liminf A_n \text{ and/or } \omega \in \liminf B_n \\ &\Rightarrow \omega \in \bigcup_{n \geq 1} \bigcap_{k \geq n} A_k \text{ and/or } \omega \in \bigcup_{n \geq 1} \bigcap_{k \geq n} B_k \\ &\Rightarrow \omega \in \bigcap_{k \geq n} A_k \text{ for at least one } n, \text{ say, } n_1 \geq 1 \end{aligned}$$

$$
\begin{aligned}
\text{and/or } \omega &\in \bigcap_{k\geq n} B_k \text{ for at least one } n, \text{ say, } n_2 \geq 1 \\
&\Rightarrow \omega \in A_k \ \ \forall \ k \geq n_1 \text{ and/or } \omega \in B_k \ \ \forall \ k \geq n_2 \\
&\Rightarrow \omega \in A_k \ \forall \ k \geq n_0 \text{ and/or } \omega \in B_k \ \forall \ k \geq n_0 \\
&\Rightarrow \omega \in (A_k \cup B_k) \ \forall \ k \geq n_0 \ \text{ for at least one } n_0 \geq 1 \\
&\Rightarrow \omega \in \bigcap_{k\geq n_0} (A_k \cup B_k) \ \text{ for at least one } n_0 \geq 1 \\
&\Rightarrow \omega \in \bigcup_{n\geq 1} \bigcap_{k\geq n} A_k \cup B_k \\
&\Rightarrow \omega \in \liminf(A_n \cup B_n) \\
&\Rightarrow \liminf A_n \cup \ \liminf B_n \subseteq \liminf(A_n \cup B_n),
\end{aligned}
$$

where $n_0 = \max\{n_1, n_2\}$. Further, as proved in Theorem 1.3.6, $\liminf(A_n \cup B_n) \subseteq \limsup(A_n \cup B_n)$. Similarly, it is shown in Example 1.3.17, $\limsup(A_n \cup B_n) = \limsup A_n \cup \limsup B_n$. Thus, (i) is proved. To examine whether $\limsup(A_n \cap B_n) \subseteq \limsup A_n \cap \limsup B_n$, observe that

$$
\begin{aligned}
\omega &\in \limsup(A_n \cap B_n) \Rightarrow \omega \in \bigcap_{n\geq 1} \bigcup_{k\geq n} (A_k \cap B_k) \\
&\Rightarrow \omega \in A_k \cap B_k \text{ for at least one } k \geq n \ \& \ \forall \ n \geq 1 \\
&\Rightarrow \omega \in A_k \text{ for at least one } k \geq n \ \& \ \forall \ n \geq 1 \\
\& \ \omega &\in B_k \text{ for at least one } k \geq n \ \& \ \forall \ n \geq 1 \\
&\Rightarrow \omega \in \bigcap_{n\geq 1} \bigcup_{k\geq n} A_k \ \& \ \omega \in \bigcap_{n\geq 1} \bigcup_{k\geq n} B_k \\
&\Rightarrow \omega \in \limsup A_n \ \& \ \omega \in \limsup B_n \\
&\Rightarrow \omega \in \limsup A_n \cap \limsup B_n \\
&\Rightarrow \limsup(A_n \cap B_n) \subseteq \limsup A_n \cap \limsup B_n.
\end{aligned}
$$

As shown in Example 1.3.17, $\liminf A_n \cap \ \liminf B_n = \liminf(A_n \cap B_n)$. Further, as proved in Theorem 1.3.6, $\liminf(A_n \cap B_n) \subseteq \limsup(A_n \cap B_n)$ and hence (ii) is proved.

1.6.10 Give an example where the sequences $\{A_n, n \geq 1\}$ and $\{B_n, n \geq 1\}$ do not converge, but $\{A_n \cap B_n, n \geq 1\}$ and $\{A_n \cup B_n, n \geq 1\}$ converge.

Solution: Suppose sequences $\{A_n, n \geq 1\}$ and $\{B_n, n \geq 1\}$ are defined as follows.

$$
\begin{aligned}
A_{2n+1} &= \{2n+1\}, \ n \geq 0 \ \ A_{2n} = \{1, 2, \cdots, 2n\}, \ n \geq 1 \\
B_{2n+1} &= \{1, 2, \cdots, 2n+1\}, \ n \geq 0 \quad \& \ \ B_{2n} = \{2n\}, n \geq 1.
\end{aligned}
$$

Note that

$$\begin{aligned}
\limsup A_n &= \bigcap_{n\geq 1}\bigcup_{k\geq n} A_k = \bigcap_{n\geq 1} C_n, \text{ where}\\
C_n &= \bigcup_{k\geq n} A_k = \mathbb{N},\ \forall\ n\geq 1 \Rightarrow \limsup A_n = \mathbb{N}\\
\liminf A_n &= \bigcup_{n\geq 1}\bigcap_{k\geq n} A_k = \bigcup_{n\geq 1} W_n, \text{ where}\\
W_n &= \bigcap_{k\geq n} A_k = \emptyset,\ \ \forall\ n\geq 1 \Rightarrow \liminf A_n = \emptyset.
\end{aligned}$$

Thus, $\lim\limits_{n\to\infty} A_n$ does not exist. Similarly,

$$\begin{aligned}
\limsup B_n &= \bigcap_{n\geq 1}\bigcup_{k\geq n} B_k = \bigcap_{n\geq 1} C_n, \text{ where}\\
C_n &= \bigcup_{k\geq n} B_k = \mathbb{N},\ \forall\ n\geq 1 \Rightarrow \limsup B_n = \mathbb{N}\\
\liminf B_n &= \bigcup_{n\geq 1}\bigcap_{k\geq n} B_k = \bigcup_{n\geq 1} D_n, \text{ where}\\
D_n &= \bigcap_{k\geq n} B_k = \emptyset,\ \ \forall\ n\geq 1 \Rightarrow \liminf B_n = \emptyset.
\end{aligned}$$

Thus, $\lim\limits_{n\to\infty} B_n$ does not exist. Suppose $M_n = A_n\cap B_n,\ n\geq 1$. Observe that $A_1\cap B_1 = \{1\},\ A_2\cap B_2 = \{2\}$ and in general $M_n = A_n\cap B_n = \{n\}$. Hence,

$$\begin{aligned}
\limsup M_n &= \bigcap_{n\geq 1}\bigcup_{k\geq n} M_k = \bigcap_{n\geq 1} N_n, \ \text{ where}\\
N_n &= \bigcup_{k\geq n}\{k\} = \{n, n+1, n+2, \cdots\}\\
&\Rightarrow\ \limsup M_n = \bigcap_{n\geq 1} N_n = \emptyset\\
\liminf M_n &= \bigcup_{n\geq 1}\bigcap_{k\geq n} M_k = \bigcup_{n\geq 1} E_n, \ \text{ where}\\
E_n &= \bigcap_{k\geq n}\{k\} = \emptyset \Rightarrow\ \liminf M_n = \bigcup_{n\geq 1} N_n = \emptyset.
\end{aligned}$$

Hence, $\lim\limits_{n\to\infty}(A_n\cap B_n) = \emptyset$. Suppose $F_n = A_n\cup B_n,\ n\geq 1$. Observe that $A_1\cup B_1 = \{1\},\ A_2\cup B_2 = \{1,2\}$ and in general $F_n = A_n\cup B_n =$

$\{1, 2, \cdots, n\}$. Hence,

$$
\begin{aligned}
\limsup F_n &= \bigcap_{n\geq 1}\bigcup_{k\geq n} F_k = \bigcap_{n\geq 1} G_n, \quad \text{where} \\
G_n &= \bigcup_{k\geq n}\{k\} = \mathbb{N} \Rightarrow \limsup F_n = \bigcap_{n\geq 1} G_n = \mathbb{N} \\
\liminf F_n &= \bigcup_{n\geq 1}\bigcap_{k\geq n} F_k = \bigcup_{n\geq 1} H_n, \quad \text{where} \\
H_n &= \bigcap_{k\geq n}\{k\} = \{1, 2, \cdots, n\} \\
&\Rightarrow \liminf F_n = \bigcup_{n\geq 1} H_n = \mathbb{N} \Rightarrow \lim_{n\to\infty}(A_n \cup B_n) = \mathbb{N}.
\end{aligned}
$$

Thus, the sequences $\{A_n, n \geq 1\}$ and $\{B_n, n \geq 1\}$ do not converge, but $\{A_n \cap B_n, n \geq 1\}$ and $\{A_n \cup B_n, n \geq 1\}$ converge.

1.6.11 Suppose $A = \limsup A_n$ and $B = \liminf A_n$. Then show that $M - A = \liminf(M - A_n)$ and $M - B = \limsup(M - A_n)$ for any $M \in \Omega$.

Solution: Note that using results in Example 1.3.17, we have

$$
\begin{aligned}
\liminf(M - A_n) &= \liminf(M \cap A_n^c) = \liminf M \cap \liminf A_n^c \\
&= M \cap (\limsup A_n)^c = M - \limsup A_n = M - A.
\end{aligned}
$$

By the definition,

$$
\begin{aligned}
\limsup(M - A_n) &= \limsup(M \cap A_n^c) = \bigcap_{n\geq 1}\bigcup_{k\geq n}(M \cap A_k^c) \\
&= \bigcap_{n\geq 1} D_n \quad \text{where} \quad D_n = \bigcup_{k\geq n}(M \cap A_k^c).
\end{aligned}
$$

Note that

$$
\begin{aligned}
\omega \in D_n &\iff \omega \in (M \cap A_k^c) \text{ for at least one } k \geq n \\
&\iff \omega \in M \;\&\; \omega \in A_k^c \text{ for at least one } k \geq n \\
&\iff \omega \in M \;\&\; \omega \in \bigcup_{k\geq n} A_k^c \iff \omega \in M \cap (\bigcup_{k\geq n} A_k^c)
\end{aligned}
$$

Further, $\omega \in \bigcap_{n\geq 1} D_n \iff \omega \in D_n \;\; \forall \;\; n \geq 1$

$$
\begin{aligned}
&\iff \omega \in M \;\&\; \omega \in \bigcup_{k\geq n} A_k^c \;\; \forall \;\; n \geq 1 \\
&\iff \omega \in M \;\&\; \omega \in \bigcap_{n\geq 1}\bigcup_{k\geq n} A_k^c \iff \omega \in M \cap (\limsup A_k^c) \\
&\iff \omega \in M \cap (\liminf A_k)^c \iff \omega \in (M - \liminf A_k) \\
&\iff \omega \in (M - B) \Rightarrow M - B = \limsup(M - A_n).
\end{aligned}
$$

1.6.12 Suppose Ω is a finite set. Examine whether (i) a collection of all finite subsets of Ω is a field and (ii) a collection of all singleton subsets of Ω is a field.

Solution: (i) Suppose $\mathbb{C}$ a collection of all finite subsets of Ω. Note that $A \in \mathbb{C} \Rightarrow A^c \in \mathbb{C}$ as Ω is finite. Further a union of two finite sets is again a finite set and hence $\mathbb{C}$ is closed under finite unions. Thus, $\mathbb{C}$ is a field. (ii) Suppose $\mathbb{C}_1$ a collection of all singleton subsets of Ω. If $A \in \mathbb{C}$ then $A^c \notin \mathbb{C}$ as A^c is not a singleton set, unless Ω has exactly two elements. Further a union of two singleton sets is not a singleton set and hence $\mathbb{C}_1$ is not a field.

1.6.13 Examine whether the following classes of subsets of Ω are field and/or sigma field, where Ω is an uncountable set. Suppose $A \subseteq \Omega$
(i) $\{\Omega, \emptyset\}$ (ii) $\{\Omega, \emptyset, A\}$, (iii) $\{\Omega, \emptyset, A, A^c\}$, (iv) $\mathbb{P}(\Omega)$, (v) the class of all finite subsets of Ω and (vi) the class of all countable subsets of Ω.

Solution: (i) and (iii) are fields and sigma fields, (iv) is sigma field. (ii) is not a field since it is not closed under complementation. In (v) class of all finite subsets will not include Ω and hence cannot be a field or a sigma field. In (vi) class of all countable subsets will not include Ω and hence cannot be a field or a sigma field.

1.6.14 Suppose $\mathbb{F}$ is a sigma field of subsets of Ω. Suppose $E \in \mathbb{F}$ and $E \neq \emptyset$. Show that $\mathbb{F}_E = \{A \cap E | A \in \mathbb{F}\}$ is the sigma field of subsets of E.

Solution: Suppose $A = \Omega$, then $A \cap E = E$. Thus, E plays the role of Ω for the collection $\mathbb{F}_E$. Note that $M = (A \cap E)^c = A^c \cup E^c \in \mathbb{F}$. From the definition of $\mathbb{F}_E$,

$$B \in \mathbb{F}_E \;\Rightarrow\; B = A \cap E \;\Rightarrow\; B^c = E \cap (A \cap E)^c = E \cap M \in \mathbb{F}_E.$$

Thus, $\mathbb{F}_E$ is closed under complementation. Further,

$$\forall\ n \geq 1, B_n \in \mathbb{F}_E \;\Rightarrow\; B_n = A_n \cap E \;\Rightarrow\; \bigcup_{n\geq 1} B_n = \Big(\bigcup_{n\geq 1} A_n\Big) \cap E \in \mathbb{F}_E$$

since $\bigcup_{n\geq 1} A_n \in \mathbb{F}$. Thus, $\mathbb{F}_E$ is closed under complementation with respect to whole space as E and it is also closed under countable union and hence it is a sigma field of subsets of E.

1.6.15 Suppose $\mathbb{C}_2$ is a class of intervals of the type $(a, b], a < b \in \mathbb{R}$ and $\mathbb{C}_6$ is a class of intervals of the type $(-\infty, a], a \in \mathbb{R}$. Then show that the minimal sigma field generated by $\mathbb{C}_2$ is the same as that by $\mathbb{C}_6$.

Solution: Suppose $\sigma(\mathbb{C}_2)$ is the minimal sigma field generated by $\mathbb{C}_2$. Note that,

$$\begin{aligned}
\forall\ n \geq 1,\ (-n, a] \in \mathbb{C}_2 \;&\Rightarrow\; \forall\ n \geq 1,\ (-n, a] \in \sigma(\mathbb{C}_2) \\
&\Rightarrow\; \bigcup_{n\geq 1} (-n, a] = (-\infty, a] \in \sigma(\mathbb{C}_2) \\
&\Rightarrow\; \mathbb{C}_6 \subseteq \sigma(\mathbb{C}_2) \;\Rightarrow\; \sigma(\mathbb{C}_6) \subset \sigma(\mathbb{C}_2).
\end{aligned}$$

Now by the definition of $\mathbb{C}_6$,

$$\begin{aligned}(-\infty, a] \in \mathbb{C}_6 \quad &\Rightarrow \quad (-\infty, a] \in \sigma(\mathbb{C}_6) \\ &\Rightarrow \quad (-\infty, a]^c = (a, \infty) \in \sigma(\mathbb{C}_6) \\ &\Rightarrow \quad \text{for } \ b > a, \ \ (-\infty, b] \cap (a, \infty) = (a, b] \in \sigma(\mathbb{C}_6) \\ &\Rightarrow \quad \mathbb{C}_2 \subseteq \sigma(\mathbb{C}_6) \quad \Rightarrow \quad \sigma(\mathbb{C}_2) \subseteq \sigma(\mathbb{C}_6).\end{aligned}$$

Thus, $\sigma(\mathbb{C}_2) = \sigma(\mathbb{C}_6)$.

1.6.16 Give an example of a class of events, which is closed under finite unions and finite intersections but not under complementation.

Solution: Suppose $\Omega = \mathbb{R}$ and class $\mathbb{A}$ is a class of intervals of the type $(x, \infty), \ x \in \Omega$. With $u = \min\{x, y\} \quad \& \quad v = \max\{x, y\}$,

$$(x, \infty) \cup (y, \infty) = (u, \infty) \in \mathbb{A} \ \ \& \ \ (x, \infty) \cap (y, \infty) = (v, \infty) \in \mathbb{A}$$

but $(x, \infty)' = (-\infty, x] \notin \mathbb{A}$.

1.6.17 Suppose Ω is a finite set and $\mathbb{A}$ is a field of subsets of Ω. Show that is a sigma field.

Solution: Since $\mathbb{A}$ is a field of subsets of Ω, it is closed under complementation and finite union. Further, Ω is a finite set implies that any sequence $\{A_n, n \geq 1\}$ of sets in $\mathbb{A}$ is a finite collection. Hence, a countable union $\bigcup_{n \geq 1} A_n$ is a finite union. Thus, $\mathbb{A}$ is closed under complementation and countable union and hence is a sigma field.

1.6.18 Suppose $\{\mathbb{F}_n, n \geq 1\}$ is a sequence of fields of subsets of Ω, such that $\mathbb{F}_n \subseteq \mathbb{F}_{n+1}, \ \ \forall \ \ n \geq 1$. (i) Prove that $\bigcup_{n=1}^{\infty} \mathbb{F}_n$ is a field. (ii) What can one say if $\{\mathbb{F}_n, n \geq 1\}$ is a sequence of sigma fields? Justify.

Solution: (i) Observe that

$$\begin{aligned}A \in \bigcup_{n=1}^{\infty} \mathbb{F}_n \quad &\Rightarrow \quad A \in \mathbb{F}_n \ \text{ for at least one } n \geq 1 \\ &\Rightarrow \quad A^c \in \mathbb{F}_n \ \text{ for at least one } n \geq 1 \ \Rightarrow \ A^c \in \bigcup_{n=1}^{\infty} \mathbb{F}_n.\end{aligned}$$

Suppose for $m = 1, 2$

$$\begin{aligned}A_m \in \bigcup_{n=1}^{\infty} \mathbb{F}_n &\Rightarrow A_m \in \mathbb{F}_M \ \text{ for some } M_m \geq 1 \\ &\Rightarrow A_m \in \mathbb{F}_n \ \ \forall \, n \geq M_m \ \text{ since } \ \mathbb{F}_n \subseteq \mathbb{F}_{n+1} \\ &\Rightarrow \bigcup_{m=1}^{2} A_m \in \mathbb{F}_n \ \ \forall \, n \geq M_0, \ \text{ where } \ M_0 = \max\{M_1, M_2\} \\ &\Rightarrow \bigcup_{m=1}^{2} A_m \in \bigcup_{n=1}^{\infty} \mathbb{F}_n.\end{aligned}$$

Hence, $\bigcup_{n=1}^{\infty} \mathbb{F}_n$ is a field.
(ii) Suppose $\Omega = \{1, 2, \cdots, \}$ and $\mathbb{F}_n = \sigma(\{A_1, A_2, \cdots, A_n\})$ where $A_n = \{n\}, n \geq 1$. Thus, $\mathbb{F}_n$ consists of all finite and co-finite sets and hence $\bigcup_{n=1}^{\infty} \mathbb{F}_n$ also consists of all finite and co-finite sets. As shown in Example 1.4.6, $\bigcup_{n=1}^{\infty} \mathbb{F}_n$ is a field. Now

$$\forall\ n \geq 1,\ B_n = \{2n\} \in \bigcup_{n=1}^{\infty} \mathbb{F}_n \text{ but } \bigcup_{n=1}^{\infty} B_n = \{2, 4, 6, \cdots, \} \notin \bigcup_{n=1}^{\infty} \mathbb{F}_n,$$

since it is neither finite nor its complement is finite. Hence, $\bigcup_{n=1}^{\infty} \mathbb{F}_n$ is not a sigma field.

1.6.19 Suppose $\Omega = \mathbb{N}$, the set of all natural numbers and $A_n = \{1, 2, \ldots, n\}$, $n \geq 1$. If $\mathbb{A}_n = \sigma(\{A_n\})$, verify whether $\{\mathbb{A}_n, n \geq 1\}$ is an increasing sequence.

Solution: Observe that $\forall\ n \geq 1$,

$$\begin{aligned} A_n &\subseteq A_{n+1} \subset \sigma(\{A_{n+1}\}) = \mathbb{A}_{n+1} \\ &\Rightarrow \mathbb{A}_n = \sigma(\{A_n\}) \subseteq \sigma(A_{n+1}) = \mathbb{A}_{n+1} \\ &\Rightarrow \{\mathbb{A}_n, n \geq 1\} \text{ is an increasing sequence.} \end{aligned}$$

The second statement follows by the definition of the minimal sigma field.

1.6.20 Suppose $\mathbb{P} = \{A_1, A_2, \ldots, \}$ is a countable partition of Ω and $\mathbb{A} = \{E | E = \emptyset \text{ or } E \text{ is a union of sets in } \mathbb{P}\}$. Show that $\mathbb{A}$ is the minimal sigma field containing $\mathbb{P}$.

Solution: Since $\mathbb{P} = \{A_1, A_2, \ldots, \}$ is a partition of Ω, $A_i \cap A_j = \emptyset$ for all $i \neq j$ and $\bigcup_{i \geq 1} A_i = \Omega$. Note that $\emptyset$ & $\Omega \in \mathbb{A}$. Any set $E \in \mathbb{A}$ is of the form $E = \bigcup_{i \in T} A_i$ where $T \subset \mathbb{N}$. Hence, $E^c = \bigcup_{i \in T^c} A_i$ and hence is in $\mathbb{A}$. Thus, $\mathbb{A}$ is closed under complementation. Further, suppose $\{E_n, n \geq 1\}$ is a sequence of sets in $\mathbb{A}$. Then $\forall\ n \geq 1$, $E_n = \bigcup_{i \in T_n} A_i$, $T_n \subseteq \mathbb{N}\ \ \forall\ \ n \geq 1$. As a consequence,

$$\bigcup_{n \geq 1} E_n = \bigcup_{n \geq 1} (\bigcup_{i \in T_n} A_i) = \bigcup_{i \in \bigcup_{n \geq 1} T_n} A_i \in \mathbb{A}.$$

Thus, $\mathbb{A}$ is a sigma field. Further, $\forall\ n \geq 1,\ A_n \in \mathbb{P} \ \Rightarrow A_n \in \mathbb{A}$ Thus, $\mathbb{P} \subset \mathbb{A}$ and $\mathbb{A}$ is a sigma field generated by $\mathbb{P}$. Suppose $\sigma^*(\mathbb{P})$ is any other sigma field generated by $\mathbb{P}$. Note that for any $E \in \mathbb{A}$, E is either $\emptyset$ or is of the form $E = \bigcup_{i \in T} A_i$ where $T \subset \mathbb{N}$ and hence, it is in $\sigma^*(\mathbb{P})$. Thus, $\mathbb{A} \subset \sigma^*(\mathbb{P})$ and hence $\mathbb{A}$ is the minimal sigma field containing $\mathbb{P}$.

1.6.21 Suppose $\mathbb{P}_1 = \{A_i, i = 1, 2, \cdots, m\}$ and $\mathbb{P}_2 = \{B_j, j = 1, 2, \cdots, n\}$ are two partitions of Ω. Show that (i) $\mathbb{P} = \{A_i \cap B_j, i = 1, 2, \cdots, m, j = 1, 2, \cdots, n\}$ is a partition of Ω and (ii) $\sigma(\sigma(\mathbb{P}_1) \cup \sigma(\mathbb{P}_2)) = \sigma(\mathbb{P})$.

Solution: (i) Suppose $C_{ij} = A_i \cap B_j, \quad i = 1, 2, \cdots, m$ and $j = 1, 2, \cdots, n$. Then for $i \neq r$ and $j \neq s$,

$$C_{ij} \cap C_{rs} = (A_i \cap B_j) \cap (A_r \cap B_s) = (A_i \cap A_r) \cap (B_j \cap B_s) = \emptyset.$$

Similarly, even if $i = r$ and $j \neq s$ or $i \neq r$ and $j = s$, then $C_{ij} \cap C_{rs} = \emptyset$. Thus, $C_{ij} = A_i \cap B_j, \quad i = 1, 2, \cdots,$ and $j = 1, 2, \cdots,$ are all disjoint sets. Note that,

$$\bigcup_{i=1}^{k} \bigcup_{j=1}^{l} C_{ij} = \bigcup_{i=1}^{k} \bigcup_{j=1}^{l} (A_i \cap B_j) = \bigcup_{i=1}^{k} \Big(\bigcup_{j=1}^{l} (A_i \cap B_j) \Big) = \bigcup_{i=l}^{k} (A_i \cap \Omega) = \Omega.$$

Further, $A_i \in \mathbb{A}$ and $B_j \in \mathbb{A}$, a sigma field of subsets of Ω. Hence $C_{ij} = A_i \cap B_j \in \mathbb{A}$, $i = 1, 2, \cdots, m$ and $j = 1, 2, \cdots, n$. Thus, $\mathbb{P} = \{C_{ij}, i = 1, 2, \cdots m, j = 1, 2, \cdots, n\}$ forms a measurable partition of Ω.

(ii) As shown in Example 1.4.2, the sigma field generated by $\mathbb{P}_1$ is given by,

$$\sigma(\mathbb{P}_1) = \{A \mid A = \sum_{t \in T} A_t\} \text{ where } T \text{ is the powerset corresponding to}$$

$\{1, 2, \cdots, m\}$, where summation over empty set is taken as the empty set. Similarly,

$$\sigma(\mathbb{P}_2) = \{B \mid B = \sum_{s \in S} B_s\} \text{ where } S \text{ is the powerset corresponding to}$$

$\{1, 2, \cdots, n\}$, where summation over empty set is taken as the empty set. Observe that any set $A_i \in \mathbb{P}_1$ can be expressed as $A_i = \bigcup_{j=1}^{n} (A_i \cap B_j) \in \mathbb{P}$. Similarly, any set $B_j \in \mathbb{P}_2$ can be expressed as $B_j = \bigcup_{i=1}^{m} (A_i \cap B_j) \in \mathbb{P}$. Hence,

$$\begin{aligned}
\mathbb{P}_i &\subseteq \mathbb{P}, \quad i = 1, 2 \\
&\Rightarrow \sigma(\mathbb{P}_i) \subseteq \sigma(\mathbb{P}), \quad i = 1, 2 \text{ by Theorem 1.4.6} \\
&\Rightarrow \sigma(\mathbb{P}_1) \cup \sigma(\mathbb{P}_1) \subseteq \sigma(\mathbb{P}) \\
&\Rightarrow \sigma\Big(\sigma(\mathbb{P}_1) \cup \sigma(\mathbb{P}_1)\Big) \subseteq \sigma(\mathbb{P}),
\end{aligned}$$

the last step follows by the definition of the minimal sigma field. To establish inclusion in the reverse direction, note that if $A \in \sigma(\mathbb{P})$, then either $A \in \mathbb{P}_1$ or $A \in \mathbb{P}_2$ or A is a finite or union or intersection of sets in $\sigma(\mathbb{P}_1)$ and $\sigma(\mathbb{P}_2)$ and hence any set in $\mathbb{P}$ is in $\sigma\Big(\sigma(\mathbb{P}_1) \cup \sigma(\mathbb{P}_1)\Big)$. Thus,

$$\mathbb{P} \subseteq \sigma\Big(\sigma(\mathbb{P}_1) \cup \sigma(\mathbb{P}_1)\Big) \Rightarrow \sigma(\mathbb{P}) \subset \sigma\Big(\sigma(\mathbb{P}_1) \cup \sigma(\mathbb{P}_1)\Big).$$

Hence $\sigma(\sigma(\mathbb{P}_1) \cup \sigma(\mathbb{P}_2)) = \sigma(\mathbb{P})$.

1.6.22 Suppose $(\Omega, \mathbb{A}, P)$ is a probability space. Prove that for $\{A_1, A_2, \cdots, A_n\}$ in $\mathbb{A}$, $P(\bigcap_{i=1}^n A_i) \geq \sum_{i=1}^n P(A_i) - (n-1)$.

Solution: From the property of the probability measure P,

$$\begin{aligned} 1 &\geq P(A_1 \cup A_2) = P(A_1) + P(A_2) - P(A_1 \cap A_2) \\ \Rightarrow\ P(A_1 \cap A_2) &\geq P(A_1) + P(A_2) - 1 = P(A_1) + P(A_2) - (2-1). \end{aligned}$$

Now,

$$\begin{aligned} P(A_1 \cap A_2 \cap A_3) &\geq P(A_1 \cap A_2) + P(A_3) - 1 \\ &\geq P(A_1) + P(A_2) - 1 + P(A_3) - 1 \\ &= P(A_1) + P(A_2) + P(A_3) - 2 \\ &= P(A_1) + P(A_2) + P(A_3) - (3-1). \end{aligned}$$

We assume $P\Big(\bigcap_{i=1}^m A_i\Big) \geq \sum_{i=1}^m P(A_i) - (m-1)$. It is true for $m = 2, 3$. Observe that

$$\begin{aligned} P\Big(\bigcap_{i=1}^{m+1} A_i\Big) &= P\Big(\bigcap_{i=1}^{m} A_i \cap A_{m+1}\Big) \geq P(\bigcap_{i=1}^{m} A_i) + P(A_{m+1}) - 1 \\ &\geq \sum_{i=1}^{m} P(A_i) - (m-1) + P(A_{m+1}) - 1 \\ &= \sum_{i=1}^{m+1} P(A_i) - (m+1-1). \end{aligned}$$

Thus, we get the required result by induction.

1.6.23 Suppose $(\Omega, \mathbb{A}, P)$ is a probability space. Suppose $\{A_1, A_2, A_3\} \in \mathbb{A}$ are such that $P(A_1 \cap A_2 \cap A_3) = P(A_1^c \cap A_2^c \cap A_3^c)$. Then show that each is equal to
$\frac{1}{2}[1-(P(A_1)+P(A_2)+P(A_3))+P(A_1 \cap A_2)+P(A_1 \cap A_3)+P(A_2 \cap A_3)]$.

Solution: We have

$$\begin{aligned} P(A_1^c \cap A_2^c \cap A_3^c) &= 1 - P(A_1 \cup A_2 \cup A_3) \\ &= 1 - P(A_1) - P(A_2) - P(A_3) + P(A_1 \cap A_2) \\ &\quad + P(A_1 \cap A_3) + P(A_2 \cap A_3) - P(A_1 \cap A_2 \cap A_3) \\ &= 1 - P(A_1) - P(A_2) - P(A_3) + P(A_1 \cap A_2) \\ &\quad + P(A_1 \cap A_3) + P(A_2 \cap A_3) - P(A_1^c \cap A_2^c \cap A_3^c) \\ \Rightarrow\ P(A_1^c \cap A_2^c \cap A_3^c) &= \frac{1}{2}[1 - (P(A_1) + P(A_2) + P(A_3)) + P(A_1 \cap A_2) \\ &\quad + P(A_1 \cap A_3) + P(A_2 \cap A_3)]. \end{aligned}$$

1.6.24 Show that if $\sum_{n \geq 1} P(A_n) < \infty$, then $P(\limsup A_n) = 0$. Examine if $P(\liminf A_n) = 0$.

Solution: By definition of limit supremum of a set we have,

$$\limsup A_n = \bigcap_{n\geq 1}\bigcup_{k\geq n} A_k = \bigcap_{n\geq 1} C_n, \quad \text{where} \quad C_n = \bigcup_{k\geq n} A_k.$$

Now,

$$\begin{aligned}
C_n &= \bigcup_{k\geq n} A_k = A_n \bigcup C_{n+1} \quad \Rightarrow \quad C_n \supseteq C_{n+1} \\
&\Rightarrow \quad \{C_n, n\geq 1\} \text{ is a non-increasing sequence of events in } \mathbb{A} \\
&\Rightarrow \quad \lim_{n\to\infty} C_n = \bigcap_{n=1}^{\infty} C_n = \limsup A_n \\
&\Rightarrow \quad P(\limsup A_n) = P(\bigcap_{n=1}^{\infty} C_n) = P(\lim_{n\to\infty} C_n) \\
&= \lim_{n\to\infty} P(C_n) = \lim_{n\to\infty} P(\bigcup_{k\geq n} A_k) \leq \lim_{n\to\infty} \sum_{k\geq n} P(A_k) = 0
\end{aligned}$$

in view of the fact that $\sum_{k\geq n} P(A_k)$ forms a tail of the convergent series $\sum_{n=1}^{\infty} P(A_n)$. Thus, $P(\limsup A_n) \leq 0$, but being probability it cannot be less that 0. Hence, $\sum_{n=1}^{\infty} P(A_n) < \infty \;\; \Rightarrow \;\; P(\limsup A_n) = 0$. Since $P(\liminf A_n) \leq P(\limsup A_n)$, $P(\liminf A_n) = 0$

1.6.25 Suppose $\Omega = \mathbb{W}$, a set of whole numbers and $\mathbb{A} = \mathbb{P}(\Omega)$ and a set function P on $(\Omega, \mathbb{P}(\Omega))$ is defined as $P(A) = \sum_{i\in A} p(1-p)^i$ for $A \in \mathbb{P}(\Omega)$ and $p \in (0,1)$. Examine whether P is a probability measure on $(\Omega, \mathbb{P}(\Omega))$.

Solution: Observe that $P(A) = \sum_{i\in A} p(1-p)^i \geq 0 \;\; \forall \;\; A \in \mathbb{P}(\Omega)$. Further,

$$P(\Omega) = \sum_{i\in\Omega} p(1-p)^i = \sum_{i=0}^{\infty} p(1-p)^i = 1.$$

To examine σ-additivity, suppose $\{A_n, n\geq 1\}$ is a sequence of disjoint sets in $\mathbb{A}$. Then

$$P\Big(\sum_{n\geq 1} A_n\Big) = \sum_{i\in \sum_{n\geq 1} A_n} p(1-p)^i = \sum_{n\geq 1}\sum_{i\in A_n} p(1-p)^i = \sum_{n\geq 1} P(A_n).$$

Hence, we conclude that P is a probability measure on $(\Omega, \mathbb{P}(\Omega))$.

1.6.26 Suppose $(\Omega, \mathbb{P}(\Omega), P)$ is a probability space, where $P(\{\omega\}) = p \in (0,1)$. Show that Ω must be finite. If Ω consists of k elements, then p must be $1/k$.

Solution: Suppose $\Omega = \{1, 2, \cdots, \}$ which is countably infinite. Then

$$P(\Omega) = \sum_{\omega\in\Omega} P(\{\omega\}) = \sum_{\omega\in\Omega} p = \infty,$$

which is a contradiction. Hence, Ω must be finite. If it consists of k elements, then $P(\Omega) = 1 \Rightarrow p = 1/k$.

1.6.27 Suppose $\Omega = \mathbb{N}$, a set of natural numbers and P is a set function defined on $(\Omega, \mathbb{P}(\Omega))$ such that

$$P(A) = \begin{cases} 0, & \text{if } A \text{ is finite} \\ 1, & \text{if } A \text{ is infinite.} \end{cases}$$

Is P a probability measure on $(\Omega, \mathbb{P}(\Omega))$?

Solution: Suppose P is a probability measure. If $A_n = \{2n\}$, $n \geq 1$, then for each $n \geq 1$ A_n is a finite set and hence $P(A_n) = 0$. Further, $\bigcup_{n\geq 1} A_n = \{2, 4, \cdots, \}$ is a set of all even numbers which is infinite, which implies $P(\bigcup_{n\geq 1} A_n) = 1$. But $\sum_{n\geq 1} P(A_n) = 0$. Thus the condition of sigma additivity is not satisfied and hence P cannot be a probability measure.

1.6.28 Suppose $\Omega = \mathbb{W}$, a set of whole numbers, $\mathbb{A}$ is a power set of Ω. Examine whether set functions P defined below in (i), (ii) and (iii) are probability measures on $(\Omega, \mathbb{A})$. (i) P(A) = number of points in A, (ii) for a non-empty set $A \in \mathbb{A}$, $P(A) = \sum_{x\in A} \mu(\{x\})$, where $\mu(\{x\}) = 1/2^x, x = 1, 2, \cdots$, $\mu(\{0\}) = 1/2$ and $\mu(\emptyset) = 0$ and (iii) $\mu(\{x\}) = 1/2^{x+1}, x \in \Omega$.

Solution:(i) Suppose $A = \{1, 2\}$, then $P(A) = 2 > 1$. Hence the set function P defined in (i) cannot be a probability measure.
(ii) Note that $P(\Omega) = \sum_{x\geq 0} \mu(\{x\}) = 1/2 + \sum_{x\geq 1}(1/2^x) = 1/2 + 1 > 1$ and the set function P defined in (ii) cannot be a probability measure.
(ii) With $\mu(\{x\}) = 1/2^{x+1}$, $x \in \Omega$, $P(A) \geq 0$, $P(\Omega) = \sum_{x\geq 0} \mu(\{x\}) = 1$. Suppose $\{A_n, n \geq 1\}$ is a sequence of disjoint events. Then A_n is any subset of Ω and $P(A_n) = \sum_{x\in A_n} \mu(\{x\})$. Hence,

$$P(\sum_{n\geq 1} A_n) = \sum_{x\in \sum_{n\geq 1} A_n} \mu(\{x\}) = \sum_{n\geq 1} \sum_{x\in A_n} \mu(\{x\}) = \sum_{n\geq 1} P(A_n).$$

Thus, P satisfies sigma additivity condition and hence P is a probability measure.

1.6.29 Suppose $(\Omega, \mathbb{A}, P)$ is a probability space and $\{A_n, n \geq 1\}$ and $\{B_n, n \geq 1\}$ are sequences of sets from $\mathbb{A}$. Suppose $P(\limsup A_n) = P(\limsup B_n) = 1$. Examine whether (i) $P(\limsup A_n \cap \limsup B_n) = 1$ and (ii) $P(\limsup A_n \cap B_n) = 1$.

Solution: (i) Note that

$$\begin{aligned} P(\limsup A_n \cup \limsup B_n) &= P(\limsup A_n) + P(\limsup B_n) \\ &- P(\limsup A_n \cap \limsup B_n) \\ &= 2 - P(\limsup A_n \cap \limsup B_n) \leq 1 \\ &\Rightarrow P(\limsup A_n \cap \limsup B_n) \geq 1 \\ &\Rightarrow P(\limsup A_n \cap \limsup B_n) = 1. \end{aligned}$$

(ii) Suppose $\Omega = \mathbb{N}$, $\mathbb{A} = \mathbb{P}(\Omega)$ and the probability measure P is defined as $P(\{x\}) = pq^{x-1}$, $x \in \mathbb{N}$ and $q = 1-p$. As in Example 1.3.18, we define sequences $\{A_n, n \geq 1\}$ and $\{B_n, n \geq 1\}$ of sets in $\mathbb{A}$ as follows.

$$\begin{aligned} A_{2n+1} &= \{2n+1\},\ n \geq 0 \quad A_{2n} = \{1, 2, \cdots, 2n\},\ n \geq 1 \\ B_{2n+1} &= \{1, 2, \cdots, 2n+1\},\ n \geq 0 \quad \& \ B_{2n} = \{2n\}, n \geq 1. \end{aligned}$$

In the same example, it is shown that

$$\begin{aligned} \limsup A_n &= \limsup B_n = \mathbb{N} \\ \Rightarrow \ P(\limsup A_n) &= P(\limsup B_n) = 1 \\ \limsup(A_n \cap B_n) &= \emptyset \ \Rightarrow \ P \limsup((A_n \cap B_n)) = 0. \end{aligned}$$

Thus, $P(\limsup A_n \cap \limsup B_n) = 1$ but $P(\limsup(A_n \cap B_n)) \neq 1$.

1.6.30 Suppose $(\Omega, \mathbb{A})$ is a measurable space and $B \in \mathbb{A}$ is any non empty subset of Ω. Suppose a class is defined as $B \cap \mathbb{A} = \{B \cap A | A \in \mathbb{A}\} = \mathbb{A}_B$ say. Show that $\mathbb{A}_B$ is a sigma field of subsets of B. A function $P_B : (B, \mathbb{A}_B) \to [0,1]$ is defined as $P_B(A) = P(B \cap A)/P(B) \ \forall \ A \in \mathbb{A}$. Show that P_B is a probability measure on $(B, \mathbb{A}_B)$.

Solution: The result that $\mathbb{A}_B$ is a sigma field of subsets of B follows from the solution of Exercise 1.6.22. To examine $P_B(A) = P(B \cap A)/P(A)$ is a probability measure on $(B, \mathbb{A}_B)$, we examine whether Kolmogorov's three axioms are satisfied. Observe that (i) $P_B(A) = P(B \cap A)/P(A) \geq 0$. (ii) $P_B(B) = P(B \cap B)/P(B) = 1$. Suppose $\{A_n, n \geq 1\}$ is a sequence of disjoint events in $\mathbb{A}$. Then, $D_n = B \cap A_n \in \mathbb{A}_B$ and $\{D_n, n \geq 1\}$ is a sequence of disjoint events. Further, $P(\sum_{n \geq 1} A_n) = \sum_{n \geq 1} P(A_n)$. Now,

$$\begin{aligned} P_B(\bigcup_{n \geq 1} D_n) &= P(B \cap \bigcup_{n \geq 1} B \cap A_n)/P(B) \\ &= P(\bigcup_{n \geq 1} B \cap A_n)/P(B) = \sum_{n \geq 1} P_B(D_n) \end{aligned}$$

1.6.31 Suppose $\{B_n, n \geq 1\}$ is a sequence of sets from $\mathbb{A}$ such that $P(B_n) = 1, \ \forall \ n \geq 1$. Then show that $P(\bigcap_{n \geq 1} B_n) = 1$.

Solution: Observe that

$$\begin{aligned} \forall \ n \geq 1, \ P(B_n) = 1 \ &\Rightarrow \ \forall \ n \geq 1, \ P(B_n^c) = 0 \\ &\Rightarrow \ P(\bigcup_{n \geq 1} B_n^c) \leq \sum_{n \geq 1} P(B_n^c) = 0 \\ &\Rightarrow \ P(\bigcap_{n \geq 1} B_n) = 1 - P(\bigcup_{n \geq 1} B_n^c) = 1. \end{aligned}$$

Answers to the multiple choice questions, based on Chapter 1, are given in Table 11.1.

TABLE 11.1
Answer Key to MCQs in Chapter 1

Q.No.	1	2	3	4	5	6	7	8
Ans	b	a	a,b,d	a,b,d	b	a,d	c	c,d
Q.No.	9	10	11	12	13	14	15	16
Ans	a,d	b,c,d	a,b,c,d	d	c	a,b,c,d	b,c	b
Q.No.	17	18	19	20	21	22	23	24
Ans	b	a,c,d	c	b	a,b,c,d	a,b	a,b	

11.2 Chapter 2

2.6.1 Suppose Ω is a sample space corresponding to the experiment of rolling a die once. Suppose $\mathbb{A} = \{\Omega, \emptyset, \{1,3,5\}, \{2,4,6\}\}$. Define two functions on $(\Omega, \mathbb{A})$ such that one is a random variable on $(\Omega, \mathbb{A})$ and the other is not a random variable on $(\Omega, \mathbb{A})$.

Solution: Suppose $A = \{1,3,5\}$ and $B = \{1,2,3,4\}$. Functions X and Y are defined as follows.

$$X(\omega) = \begin{cases} 10, & \text{if} \quad \omega \in A \\ 20, & \text{if} \quad \omega \in A^c \end{cases} \quad \& \quad Y(\omega) = \begin{cases} 10, & \text{if} \quad \omega \in B \\ 20, & \text{if} \quad \omega \in B^c. \end{cases}$$

Then, $X^{-1}(\mathbb{B}) = \{\Omega, \emptyset, A, A^c\} = \mathbb{A}$. Hence, X is a random variable on $(\Omega, \mathbb{A})$. Further, $Y^{-1}(\mathbb{B}) = \{\Omega, \emptyset, B, B^c\}$ and $B \notin \mathbb{A}$. Hence, Y is not a random variable on $(\Omega, \mathbb{A})$.

2.6.2 Suppose $\Omega = (0,1]$ and $\mathbb{A}$ is a minimal sigma field consisting of all Borel sets which are subsets of Ω. Examine whether a function $X(\omega) = 6\omega + 2$ is a random variable on $(\Omega, \mathbb{A})$. (i) Use the economical definition. (ii) Use the result that a Borel function of a random variable is a random variable.

Solution: (i) The range of a function X is $(2,8]$. Note that

$$X^{-1}(-\infty, x]) = [X \leq x] \quad = \quad \begin{cases} \emptyset, & \text{if} \quad x \leq 2 \\ (0, (x-2)/6], & \text{if} \quad 2 < x < 8 \\ \Omega, & \text{if} \quad x \geq 8. \end{cases}$$

Thus, $\forall\ x \in \mathbb{R}, X^{-1}(-\infty, x]) \in \mathbb{A}$. Hence, X is a random variable on $(\Omega, \mathbb{A})$, measurable with respect to $\mathbb{A}$.
(ii) Suppose Y is an identity function defined on $(\Omega, \mathbb{A})$. Then $\forall\ S \in \mathbb{B},\ Y^{-1}(S) = S \in \mathbb{A}$. Hence, Y is $\mathbb{A}$ measurable random variable. Now, $X(\omega) = g(Y(\omega))$ where $g(x) = 6x + 2$. Now g is a continuous function, hence a Borel function, implying that X is a random variable on $(\Omega, \mathbb{A})$.

2.6.3 Suppose X is a random variable, measurable with respect to $\mathbb{A}$. Examine whether (i) $Y_1 = e^X$ and $Y_2 = e^X + X^3 + 5X$ are random variables, measurable with respect to $\mathbb{A}$. (ii) If X is a positive random variable measurable with respect to $\mathbb{A}$, examine whether $Y_3 = \log X$ is a random variable. Is $\underline{Z} = (Y_1, Y_2, Y_3)'$ a random vector measurable with respect to $\mathbb{A}$? Justify.

Solution: (i) Observe that $e^X = g(X)$ where $g(x) = e^x$ is a continuous and hence a Borel function. Hence, e^X is a random variable, measurable with respect to $\mathbb{A}$. Note that X^3 and $5X$, being Borel functions of X, are random variables, measurable with respect to $\mathbb{A}$. A class of random variables is closed under arithmetic operations and hence $e^X + X^3 + 5X$ is a random variable, measurable with respect to $\mathbb{A}$. Alternatively, $e^X + X^3 + 5X = g(X)$, where $g(x) = e^x + x^3 + 5x$. It is a continuous and hence a Borel function. Thus, being Borel function of a random variable, $e^X + X^3 + 5X$ is a random variable, measurable with respect to $\mathbb{A}$.
(ii) Since $X > 0$, $\log X$ is defined. Further $g(x) = \log x, \ x > 0$ is a continuous and hence a Borel function. Thus, being Borel function of a random variable X, $\log X$ is a random variable. Since Y_1, Y_2, Y_3 are random variables, measurable with respect to $\mathbb{A}$, $\underline{Z} = (Y_1, Y_2, Y_3)'$ is a random vector measurable with respect to $\mathbb{A}$.

2.6.4 (i) Suppose A and B are mutually exclusive events. Find the minimal sigma field induced by $I_A + I_B$. (ii) Suppose A and B are not mutually exclusive events. Find the minimal sigma field induced by $I_A + I_B$.

Solution: (i) By the definition of an indicator function, if A and B are mutually exclusive events, then $I_A + I_B$ is given by,

$$(I_A + I_B)(\omega) \quad = \quad \begin{cases} 0, & \text{if} \quad \omega \notin A, \omega \notin B, \ \text{ that is } \ \omega \in A^c \cap B^c, \\ 1 & \text{if} \qquad\qquad \omega \in A \cup B = A + B. \end{cases}$$

Hence, the minimal sigma field induced by $I_A + I_B$ is $\{\Omega, \emptyset, A \cup B, (A \cup B)^c\}$.
(ii) Suppose A and B are not mutually exclusive events. Then $I_A + I_B$ is given by,

$$(I_A + I_B)(\omega) \quad = \quad \begin{cases} 0, & \text{if} \quad \omega \in (A \cup B)^c \\ 1, & \text{if} \quad \omega \in (A \cap B^c) \cup (A^c \cap B) \\ 2, & \text{if} \quad \omega \in A \cap B. \end{cases}$$

Hence, the inverse images of Borel sets are give by

$$(I_A + I_B)^{-1}(S) \; = \; \begin{cases} \emptyset, & \text{if} 0 \notin S, 1 \notin S, 2 \notin S,, \\ (A \cup B)^c, & \text{if} 0 \in S, 1 \notin S, 2 \notin S, \\ (A \cap B^c) \cup (A^c \cap B), & \text{if} 0 \notin S, 1 \in S, 2 \notin S, \\ A \cap B, & \text{if} 0 \notin S, 1 \notin S, 2 \in S, \\ (A \cap B)^c, & \text{if} 0 \in S, 1 \in S, 2 \notin S, \\ (A \cup B)^c \cup (A \cap B), & \text{if} 0 \in S, 1 \notin S, 2 \in S, \\ A \cup B, & \text{if} 0 \notin S, 1 \in S, 2 \in S, \\ \Omega, & \text{if} 0 \in S, 1 \in S, 2 \in S. \end{cases}$$

Hence, the minimal sigma field induced by $I_A + I_B$ is $\{\Omega, \emptyset, A \cup B, (A \cup B)^c, A \cap B, (A \cap B)^c, (A \cap B^c) \cup (A^c \cap B), (A \cup B)^c \cup (A \cap B)\}$.

2.6.5 Suppose $\Omega = \{a, b, c\}$ and $\mathbb{A} = \{\Omega, \emptyset, \{a\}, \{b, c\}\}$. Examine whether $\mathbb{A}$ is a sigma field. If yes, specify any non-constant function on $(\Omega, \mathbb{A})$ which is measurable with respect to $\mathbb{A}$.

Solution: It is easy to verify that $\mathbb{A}$ is closed under complementation and countable union and hence it is a sigma field. Suppose a function on $(\Omega, \mathbb{A})$ is defined as

$$X(\omega) \quad = \quad \begin{cases} 10, & \text{if} \quad \omega = a \\ 30, & \text{if} \quad \omega = b, c. \end{cases}$$

Then $X^{-1}(\mathbb{B}) = \mathbb{A}$ and hence X is measurable with respect to $\mathbb{A}$.

2.6.6 Suppose X and Y are random variables on $(\Omega, \mathbb{A}, P)$ such that $\sigma(X) = \sigma(Y)$. Show that $\sigma(X + Y) \subseteq \sigma(X)$.

Solution: Note that $X + Y$ is a Borel function of $(X, Y)'$ and hence sigma field induced by $X + Y$ is included in the sigma field induced by $(X, Y)'$. Thus,

$$\begin{aligned} \sigma(X + Y) &\subseteq \sigma(X^{-1}(\mathbb{B}) \cup Y^{-1}(\mathbb{B})) = \sigma(X^{-1}(\mathbb{B}) \cup X^{-1}(\mathbb{B})) \\ &= \sigma(X^{-1}(\mathbb{B})) = X^{-1}(\mathbb{B}) = \sigma(X). \end{aligned}$$

2.6.7 Suppose X is a random variable. Show that $\sigma(X^2) \subseteq \sigma(X)$.

Solution: We have a result which states that sigma field induced by a Borel function of a random variable is included in the sigma field induced by the random variable. Note that X^2 is a Borel function of X and hence $\sigma(X^2) \subseteq \sigma(X)$. If X^2 is a one-to-one function of X, then $\sigma(X^2) = \sigma(X)$, otherwise $\sigma(X^2) \subset \sigma(X)$.

2.6.8 Give an example of a random variable such that $\sigma(X^2) \neq \sigma(X)$.

Solution: Suppose X is a random variable defined on $(\Omega, \mathbb{A})$ to $(\mathbb{R}, \mathbb{B})$ as $X(i) = i$, where $\Omega = \{-1, 0, 1\}$. Then we have obtained a sigma field induced by X and $Y = X^2$ in Example 2.2.12. These are given by,

$$X^{-1}(\mathbb{B}) = \{\Omega, \emptyset, \{-1\}, \{0\}, \{1\}, \{-1, 0\}, \{-1, 1\}, \{0, 1\}\}$$

and $Y^{-1}(\mathbb{B}) = \{\Omega, \emptyset, \{0\}, \{-1, 1\}\}$. We note that $\sigma(X^2) \neq \sigma(X)$. Observe that X^2 is not a one-to-one function of X.

2.6.9 Give an example of a random variable such that $\sigma(X^2) = \sigma(X)$.

Solution: Suppose X is a random variable defined on $(\Omega, \mathbb{A})$ to $(\mathbb{R}, \mathbb{B})$ as follows. For $A \in \mathbb{A}$,

$$X(\omega) \quad = \quad \begin{cases} 2, & \text{if} \quad \omega \in A \\ 3, & \text{if} \quad \omega \notin A. \end{cases}$$

Then sigma fields induced by X and $Y = X^2$ are the same and are given by $\{\Omega, \emptyset, A, A^c\}$. Observe that X^2 is a one-to-one function of X.

2.6.10 Suppose $\Omega = \{1, 2, 3, 4\}$, $A = \{1, 2\}$ and $\mathbb{A} = \{\Omega, \emptyset, A, A^c\}$. Functions X, Y and W on $(\Omega, \mathbb{A})$ are defined as follows. $X(\omega) = 4 \quad \forall \;\; \omega \in \Omega$, $Y(\omega) = 1$ if $\omega \in A$ and 0 otherwise and $W(\omega) = 5\omega \;\; \forall \;\; \omega \in \Omega$. Examine whether $\underline{Z} = (X, Y, W)$ is random vector on $(\Omega, \mathbb{A})$. If not, modify $\mathbb{A}$, so that $\underline{Z} = (X, Y, W)$ is random vector on $(\Omega, \mathbb{A})$. Find the sigma field induced by $\underline{Z}$.

Solution: It is known that $\underline{Z} = (X, Y, W)$ is random vector on $(\Omega, \mathbb{A})$ if and only if each of the components is $\mathbb{A}$ measurable random variable. We use this result to examine whether $\underline{Z} = (X, Y, W)$ is random vector on $(\Omega, \mathbb{A})$. Observe that $X(\omega) = 4 \;\forall\;\; \omega \in \Omega$. Hence, $X^{-1}(\mathbb{B}) = \{\Omega, \emptyset\} \subset \mathbb{A}$. Thus, X is $\mathbb{A}$ measurable random variable. $Y(\omega) = 1$ if $\omega \in A$ and 0 otherwise. Hence, $Y^{-1}(\mathbb{B}) = \{\Omega, \emptyset, A, A^c\} = \mathbb{A}$. Thus, Y is $\mathbb{A}$ measurable random variable. Further, $W(\omega) = 5\omega \;\; \forall \;\; \omega \in \Omega$ and $W^{-1}(\mathbb{B}) = \mathbb{P}(\Omega)$. It is not included in $\mathbb{A}$. Thus, W is not $\mathbb{A}$ random variable. Hence, $\underline{Z} = (X, Y, W)$ is not a random vector on $(\Omega, \mathbb{A})$. Suppose $\mathbb{A} = \mathbb{P}(\Omega)$. Then X, Y, W are measurable with respect to $\mathbb{A}$ and hence $\underline{Z} = (X, Y, W)$ is random vector on $(\Omega, \mathbb{A})$. By definition

$$\sigma(\underline{Z}) = \sigma(\sigma(X) \cup \sigma(Y) \cup \sigma(W)) = \sigma(\mathbb{P}(\Omega)) = \mathbb{P}(\Omega).$$

2.6.11 Suppose $\Omega = \mathbb{N}$, the set of natural numbers, and $\mathbb{A} = \mathbb{P}(\mathbb{N})$. Construct one finite and one countably infinite measurable partition of Ω. Give an example of one simple random variable and one elementary random variable defined on $(\mathbb{N}, \mathbb{P}(\mathbb{N}))$.

Solution: Suppose E and O denote the set of even natural numbers and odd natural numbers respectively. Then a finite partition $\mathbb{P}_f$ is defined as $\mathbb{P}_f = \{E, O\}$. Suppose a function X on $(\Omega, \mathbb{A})$ is defined as follows.

$$X(\omega) \quad = \quad \begin{cases} 200, & \text{if} \quad \omega \in E \\ 300, & \text{if} \quad \omega \notin E. \end{cases}$$

$$\text{Then} \quad X^{-1}(\mathbb{B}) = \{\Omega, \emptyset, E, E^c\} \subset \mathbb{A} = \mathbb{P}(\mathbb{N})$$

which implies that X is the simple random variable on $(\Omega, \mathbb{A})$. Suppose a countable partition $\mathbb{P}_c$ is defined as $\mathbb{P}_c\{\{1\}, \{2\}, \{3\}, \cdots, \}$ and a function X on $(\Omega, \mathbb{A})$ is defined as $X(i) = i, i \in \mathbb{N}$, that is $X = \sum_{i \in \mathbb{N}} i I_{\{i\}}$ then the sigma field induced by X is $\mathbb{P}(\mathbb{N})$. Hence, X is the elementary random variable defined on $(\mathbb{N}, \mathbb{P}(\mathbb{N}))$.

2.6.12 Prove or disprove: $X^{-1}(\lim S_n) = \lim X^{-1}(S_n)$, where S_n is a Borel set.

Solution: Suppose $\{S_n, n \geq 1\}$ is a sequence of Borel sets. We assume that $\lim S_n$ exists. Thus, $\lim S_n = \limsup S_n = \liminf S_n$. By definition

$\limsup S_n = \bigcap_{n\geq1} B_n$ where $B_n = \bigcup_{k\geq n} S_k$ and $\liminf S_n = \bigcup_{n\geq1} C_n$ where $C_n = \bigcap_{k\geq n} S_k$. Observe that

$$\begin{aligned} X^{-1}(\limsup S_n) &= X^{-1}(\bigcap_{n\geq1} B_n) = \bigcap_{n\geq1} X^{-1}(B_n) = \bigcap_{n\geq1} X^{-1}(\bigcup_{k\geq n} S_k) \\ &= \bigcap_{n\geq1}\bigcup_{k\geq n} X^{-1}(S_k) = \limsup X^{-1}(S_n). \end{aligned}$$

Using similar steps, it follows that $X^{-1}(\liminf S_n) = \liminf X^{-1}(S_n)$. Now we have, assumed that $\lim S_n$ exists. Thus,

$$\begin{aligned} \lim S_n &= \limsup S_n = \liminf S_n \\ \Rightarrow\ X^{-1}(\lim S_n) &= X^{-1}(\limsup S_n) = X^{-1}(\liminf S_n) \\ \Rightarrow\ X^{-1}(\lim S_n) &= \limsup X^{-1}(S_n) = \liminf X^{-1}(S_n) \\ \Rightarrow\ X^{-1}(\lim S_n) &= \lim X^{-1}(S_n). \end{aligned}$$

2.6.13 Show that a necessary and sufficient condition for a function to be measurable is that its positive and negative parts are measurable.

Solution: Suppose f defined on $(\Omega, \mathbb{A})$ is a measurable function with respect to $\mathbb{A}$, Ω may be $\mathbb{R}$ and $\mathbb{A}$ may be $\mathbb{B}$, that is, f may be a Borel measurable function. Then $f^+ = g(f)$ and $f^- = h(f)$ where g and h are Borel measurable functions. Hence, f^+ and f^- are measurable functions. Conversely, assume that f^+ and f^- are measurable functions. Further, $f = f^+ - f^- = g_2(f^+, f^-)$ where g_2 is a Borel measurable function from $(\mathbb{R}^2, \mathbb{B}^2)$ to $(\mathbb{R}, \mathbb{B})$. Hence f is a measurable function.

2.6.14 Suppose f and g are Borel functions on $(\mathbb{R}, \mathbb{B})$. For $A \in \mathbb{B}$, a function h is defined as $h = f$ on A and $h = g$ on A^c. Show that h is a Borel function.

Solution: It is given that f and g are Borel functions. Since $A \in \mathbb{B}$, I_A and I_{A^c} are Borel measurable functions. A function h can be expressed as $h = fI_A + gI_{A^c} = g_4(f, g, I_A, I_{A^c})$ where g_4 is a Borel measurable function from $(\mathbb{R}^4, \mathbb{B}^4)$ to $(\mathbb{R}, \mathbb{B})$. Hence h is a Borel measurable function.

2.6.15 Suppose $|X|$ is $\mathbb{A}$ measurable. Examine whether X^+ and X^- are also $\mathbb{A}$ measurable.

Solution: Suppose $\Omega = \{H, T\}$, $\mathbb{A} = \{\Omega, \emptyset\}$ and X is defined as $X = 1$ if $\omega = H$ and $X = -1$ if $\omega = T$. Thus $|X| \equiv 1$ and is $\mathbb{A}$ measurable. Note that

$$X^+ = \begin{cases} 1, & \text{if} \quad \omega = H, \\ 0, & \text{if} \quad \omega = T. \end{cases} \quad \& \quad X^- = \begin{cases} 0, & \text{if} \quad \omega = H, \\ 1, & \text{if} \quad \omega = T. \end{cases}$$

Hence, both X^+ and X^- are not $\mathbb{A}$ measurable.

2.6.16 Suppose $(X_1, X_2, \cdots, X_n)'$ is a random vector on $(\Omega, \mathbb{A})$. Show that (i) $\sum_{i=1}^{n} l_i X_i$ is real valued random variable where $l_1, l_2, \cdots, l_n$ are any real numbers, (ii) $(\sin X_1, X_2, X_3, \cdots, e^{X_{n-1}}, |X_n|)$ is a random vector and (iii) $(X_2, X_3, \cdots, X_{n-1})$ is a random vector, all defined on the same probability space $(\Omega, \mathbb{A})$.

Solution: Since $(X_1, X_2, \cdots, X_n)'$ is a random vector on $(\Omega, \mathbb{A}, P)$, each component is a random variable on $(\Omega, \mathbb{A})$.
(i) Note that $\sum_{i=1}^{n} l_i X_i = g((X_1, X_2, \cdots, X_n)')$ where $g(x) = \sum_{i=1}^{n} l_i x_i$ is a Borel function from $(\mathbb{R}^n, \mathbb{B}^n)$ to $(\mathbb{R}, \mathbb{B})$. Hence, it is a random variable.
(ii) Since each X_i is a random variable, being Borel functions, $\sin X_1, e^{X_{n-1}}, |X_n|$ are random variables and hence $(\sin X_1, X_2, X_3, \cdots, e^{X_{n-1}}, |X_n|)$ is a random vector.
(iii) Since each X_i is a random variable, $(X_2, X_3, \cdots, X_{n-1})$ is a $(n-2)$ dimensional random vector.

2.6.17 Suppose a measurable space $(\Omega, \mathbb{A})$ is defined as $\Omega = [0, \infty)$ and $\mathbb{A}$ is a sigma field of subsets of Ω. Suppose a sequence $\{X_n, n \geq 1\}$ of $\mathbb{A}$ measurable random variables and a $\mathbb{A}$ measurable random variable X are defined as follows.

$$X_n(\omega) = \exp(-n\omega), \ n \geq 1 \ \& \ X(\omega) = 0 \quad \forall \ \omega \in \Omega.$$

(i) Examine if $X_n \to X$ on Ω or any subset of Ω. (ii) If not, identify another X such that $X_n \to X$ on Ω.

Solution: Observe that,

$$X_n(\omega) = \exp(-n\omega) \quad \to \quad 0 = X(\omega), \ \ \forall \ \omega \in (0, \infty) \ .$$

But at $\omega = 0, X_n(\omega) = 1 \ \forall \ \ n \geq 1$ hence converges to $1 \neq X(\omega) = 0$. Thus the sequence of random variables $\{X_n, n \geq 1\}$ does not converge point-wise to X. However, note that $X_n(\omega) \to X(\omega)$, except for $\omega = 0$. Thus, if $A = (0, \infty)$, then $X_n \to X$ on A.
(ii) Suppose a random variable X is defined as follows.

$$X(\omega) \quad = \quad \begin{cases} 0, & \text{if} \quad \omega \in (0, \infty) \\ 1, & \text{if} \quad \omega = 0. \end{cases}$$

Then $X_n \to X$ on Ω.

2.6.18 Suppose a random variable X is defined on $(\Omega, \mathbb{A}, P)$ and $X \sim N(0, 1)$ distribution. Suppose $A_n = [X \leq 1/n]$. Find $\lim_{n \to \infty} P(A_n)$.

Solution: Observe that $A_n = [X \leq 1/n] = \{\omega | X(\omega) \leq 1/n\} = X^{-1}(S_n)$, where $S_n = (-\infty, 1/n]$. Note that $S_{n+1} \subset S_n, \quad \forall \ \ n \geq 1$. Hence, $\{S_n, n \geq 1\}$ is a decreasing sequence with limit $S = (-\infty, 0]$. Hence, $\{A_n = X^{-1}(S_n), n \geq 1\}$ is also a decreasing sequence with limit

$A = X^{-1}(S) = X^{-1}(-\infty, 0] = [X \leq 0]$. Now by the continuity theorem for a probability measure,

$$\lim_{n\to\infty} P(A_n) = P(\lim_{n\to\infty} A_n) = P[X \leq 0] = 1/2.$$

2.6.19 Suppose a random variable X is defined on $(\Omega, \mathbb{A}, P)$ and $X \sim \chi^2_5$ and $A_n = [X \leq 1/n]$. Find $\lim_{n\to\infty} P(A_n)$.

Solution: As in the previous example,

$$\lim_{n\to\infty} P(A_n) = P(\lim_{n\to\infty} A_n) = P[X \leq 0] = 0,$$

since $X \sim \chi^2_5$.

2.6.20 Suppose a random variable X is defined on $(\Omega, \mathbb{A}, P)$ and X has exponential distribution with mean 1 and $A_n = [1-1/n \leq X \leq 2+1/n]$. Find $\lim_{n\to\infty} P(A_n)$. Suppose $B_n = [n < X \leq n+1]$. Find $\lim_{n\to\infty} P(B_n)$.

Solution: As in Exercise 2.6.17, note that $\{A_n, n \geq 1\}$ is a decreasing sequence and decreases to $[X \in [1,2]] = A$. Hence,

$$\begin{aligned}\lim_{n\to\infty} P(A_n) &= P(\lim_{n\to\infty} A_n) = P(A) = P[1 \leq X \leq 2] \\ &= e^{-1} - e^{-2} = 0.2325.\end{aligned}$$

Observe that $B_n = [n < X \leq n+1] = [X \in (n, n+1]$ is a sequence of mutually exclusive events and hence its limit is $\emptyset$. Hence, $\lim_{n\to\infty} P(B_n) = P(\lim_{n\to\infty} B_n) = 0$.

2.6.21 Suppose a random variable X is defined on $(\Omega, \mathbb{A}, P)$ and $P[X = n] = a_n, n \in \mathbb{N}, a_n \geq 0$ and $\sum_{n\in\mathbb{N}} a_n = 1$. Find $\lim_{n\to\infty} P[X = n]$.

Solution: Suppose $A_n = [X = n] = X^{-1}(\{n\})$. Then $\{A_n, n \geq 1\}$ is a sequence of mutually exclusive events and hence its limit is $\emptyset$. Hence,

$$\lim_{n\to\infty} P(A_n) = P(\lim_{n\to\infty} A_n) = 0.$$

Answers to the multiple choice questions, based on Chapter 2, are given in Table 11.2.

TABLE 11.2
Answer Key to MCQs in Chapter 2

Q.No.	1	2	3	4	5	6	7	8	9	10
Ans	a	b	b	a,b	a,b	a,b	b	d	b	b
Q.No.	11	12	13	14	15	16	17	18	19	20
Ans	b	c	b	d	b	a,b,c,d	b	d	a,c	a
Q.No.	21	22	23	24	25	26	27	28	29	30
Ans	c	a	a	b	c					

11.3 Chapter 3

3.5.1 Suppose functions F are defined as follows. In each case examine whether F is a distribution function. If yes, find the set of points of discontinuities and points of increase of a distribution function. Decompose the distribution function and decide the type of distribution function F.

$$(i)\ F(x) = \begin{cases} 0, & \text{if} \quad x < 0, \\ x^2/4, & \text{if} \quad 0 \le x < 2, \\ 1, & \text{if} \quad x \ge 2. \end{cases} \quad (ii)\ F(x) = \begin{cases} 0, & \text{if} \quad x < 0, \\ \frac{x}{2}, & \text{if} \quad 0 \le x \le \frac{2}{3}, \\ 1, & \text{if} \quad x > \frac{2}{3}. \end{cases}$$

$$(iii) F(x) = \begin{cases} 0, & \text{if} \quad x \le 1, \\ 1 - \frac{1}{x^2}, & \text{if} \quad x > 1. \end{cases} \quad (iv) F(x) = \begin{cases} 0, & \text{if} \quad x \le 0, \\ \frac{1}{2} + \frac{1}{2}e^{-x}, & \text{if} \quad x > 0. \end{cases}$$

$$(v) F(x) = \begin{cases} 0, & \text{if} \quad x < 0, \\ 1 - \frac{1}{2}e^{-x}, & \text{if} \quad x \ge 0. \end{cases} \quad (vi) F(x) = \begin{cases} 0, & \text{if} \quad x < 0, \\ \sin x, & \text{if} \quad 0 \le x < \frac{\pi}{2}, \\ 1, & \text{if} \quad x \ge \frac{\pi}{2}. \end{cases}$$

(vii) $F(x) = \sin x, \quad 0 < x < \pi/2.$
(viii) $F(x) = \sin x, \quad x \in \mathbb{R}.$

$$(ix) F(x) = \begin{cases} 0, & \text{if} \quad x < 0, \\ x/2, & \text{if} \quad 0 \le x < 1, \\ 1, & \text{if} \quad x \ge 1. \end{cases} \quad (x)\ F(x) = \begin{cases} 0, & \text{if} \quad x < 2, \\ x/3, & \text{if} \quad 2 \le x < 3, \\ 1, & \text{if} \quad x \ge 3. \end{cases}$$

Solution: (i) It satisfies the four requirements of a distribution function and hence is a distribution function. Further, set of points of discontinuities is $\emptyset$. Hence, it is a distribution function of an absolutely continuous random variable. The set of points of increase is $[0, 2]$.

(ii) Here F is not right continuous at $x = 2/3$ and hence it is not a distribution function.

(iii) It satisfies the four requirements of a distribution function and hence is a distribution function. Further, set of points of discontinuities is $\emptyset$. Hence, it is a distribution function of an absolutely continuous random variable. The set of points of increase is $[1, \infty)$.

(iv) For $x > 0$ it is decreasing and as $x \to \infty$ it tends to $1/2$. Further, F is not right continuous at $x = 0$ and hence it is not a distribution function.

(v) It satisfies the four requirements of a distribution function and hence is a distribution function. It can be decomposed as $F = l_1F_1 + l_2F_2$, where $l_1 = l_2 = 1/2$, F_1 is a distribution function of a discrete random variable X_1 which is degenerate at 0 and F_2 is a distribution function of an absolutely continuous random variable X_2 which has exponential distribution with mean 1.

(vi) It satisfies the four requirements of a distribution function and hence is a distribution function. The set of points of discontinuities is $\emptyset$ and the set of points of increase is $[0, \pi/2]$. It is an absolutely continuous distribution function.

(vii) Domain of F is not $\mathbb{R}$ and hence it is not a distribution function.
(viii) F is not non-decreasing on $\mathbb{R}$ and hence it is not a distribution function.
(ix) It satisfies the four requirements of a distribution function and hence is a distribution function. Note that 1 is a point of discontinuity with size 1/2. Hence G and H are given by,

$$G(x) = \begin{cases} 0, & \text{if} \quad x < 1, \\ 1/2, & \text{if} \quad x \geq 1 \end{cases} \quad \& \quad H(x) = \begin{cases} 0, & \text{if} \quad x < 0, \\ x/2, & \text{if} \quad 0 \leq x < 1, \\ 1/2, & \text{if} \quad x \geq 1. \end{cases}$$

Hence, $F_1(x) = G(x)/l_1$ & $F_2(x) = H(x)/l_2$ where $l_1 = l_2 = 1/2$ and $F(x) = l_1F_1(x) + l_2F_2(x)$.
(x) It satisfies the four requirements of a distribution function and hence is a distribution function. Further, 2 is a point of discontinuity with size 2/3. Hence G and H are given by,

$$G(x) = \begin{cases} 0, & \text{if} \quad x < 2, \\ 2/3, & \text{if} \quad x \geq 2 \end{cases} \quad \& \quad H(x) = \begin{cases} 0, & \text{if} \quad x < 2, \\ (x-2)/3, & \text{if} \quad 2 \leq x < 3, \\ 1/3, & \text{if} \quad x \geq 2. \end{cases}$$

Hence, $F_1(x) = G(x)/l_1$ & $F_2(x) = H(x)/l_2$ where $l_1 = 2/3$ & $l_2 = 1/3$ and $F(x) = l_1F_1(x) + l_2F_2(x)$.

3.5.2 If D is the set of points of discontinuity of a distribution function, give an example of each of the following types of distribution functions.
(i) D is empty. (ii) D is singleton. (iii) D is finite, but not singleton. (iv) D is countably infinite. Write the distribution function or probability density function or probability mass function.

Solution: (i) If the set $D = \emptyset$, then the corresponding distribution function is continuous. For example, the distribution function of any absolutely continuous distribution, such as exponential distribution with probability density function $f(x) = \lambda e^{-\lambda x}, x > 0$ and 0 otherwise.
(ii) For a degenerate distribution, D is singleton. The probability mass function of a random variable $X \equiv C$ is $P[X = C] = 1$.
(iii) For a Binomial $B(n, p)$ distribution, D is finite and is given by $D = \{0, 1, \cdots, n\}$. Its probability mass function is $P[X = i] = \binom{n}{i}p^i(1-p)^{n-i}$, $i \in D$. (iv) For a Poisson distribution D is countably infinite, given by $D = W$, the set of whole numbers. Its probability mass function is $P[X = i] = e^{-\lambda}\lambda^i/i!$, $i \in D$.

3.5.3 A function $F : \mathbb{R} \to \mathbb{R}$ is defined as $F(x) = 0$ if $x \leq 0$ and $F(x) = \alpha + ke^{-x^2/2}$ if $x > 0$. Determine values of α and k so that F is a distribution function.

Solution: For F to be a distribution function it has to be a right continuous function on $\mathbb{R}$. Thus, from the right continuity at $x = 0$ we have,

$$\lim_{h \to 0} F(0 + h) = F(0) = 0 \quad \Rightarrow \quad \lim_{h \to 0}(\alpha + ke^{-h^2/2}) = \alpha + k = 0.$$

Further, $\lim_{x\to\infty} F(x) = 1 \Rightarrow \alpha + 0 = 1 \quad \Rightarrow \alpha = 1$ and hence $k = -1$.

3.5.4 Examine whether a convex combination of k distribution functions is a distribution function.

Solution: Suppose $F(x) = \sum_{i=1}^{k} \alpha_i F_i(x),\ x \in \mathbb{R}$ is a convex combination of distribution functions F_i, $\alpha_i \geq 0, i = 1, 2, \cdots, k$ and $\sum_{i=1}^{k} \alpha_i = 1$. We examine whether F satisfies the four requirements of a distribution function.
(i) Observe that $\forall\ x \in \mathbb{R}$,

$$\begin{aligned} 0 \leq F_i(x) \leq 1 \quad &\Rightarrow \quad 0 \leq \alpha_i F_i(x) \leq \alpha_i, \quad \forall\ i = 1, 2, \cdots, k \\ &\Rightarrow \quad 0 \leq \sum_{i=1}^{k} \alpha_i F_i(x) \leq \sum_{i=1}^{k} \alpha_i \quad \Rightarrow \quad 0 \leq F(x) \leq 1. \end{aligned}$$

(ii) Further $\forall\ x_1 < x_2 \in \mathbb{R}$

$$\begin{aligned} F_i(x_1) \leq F_i(x_1) \quad &\Rightarrow \quad \alpha_i F_i(x_1) \leq \alpha_i F_i(x_1) \ \ \forall\ i = 1, 2, \cdots, k \\ &\Rightarrow \quad \sum_{i=1}^{k} \alpha_i F_i(x_1) \leq \sum_{i=1}^{k} \alpha_i F_i(x_2) \\ &\Rightarrow \quad F(x_1) \leq F(x_1) \ \Rightarrow \ F \ \text{ is non-decreasing on } \ \mathbb{R}. \end{aligned}$$

(iii) F_i being a distribution function is right continuous for each i. Thus, for $x \in \mathbb{R}$ and for each i,

$$\begin{aligned} \lim_{h \to 0} F_i(x+h) = F_i(x) \quad &\Rightarrow \quad \lim_{h \to 0} \alpha_i F_i(x+h) = \alpha_i F_i(x) \\ &\Rightarrow \quad \lim_{h \to 0} \sum_{i=1}^{k} \alpha_i F_i(x+h) = \sum_{i=1}^{k} \alpha_i F_i(x) \\ &\Rightarrow \quad \lim_{h \to 0} F(x+h) = F(x) \\ &\Rightarrow \quad F \ \text{ is right continuous on } \ \mathbb{R}. \end{aligned}$$

(iv) Note that for each i

$$\begin{aligned} \lim_{x \to \infty} F_i(x) = 1 \quad &\Rightarrow \quad \lim_{x \to \infty} \alpha_i F_i(x) = \alpha_i \\ &\Rightarrow \quad \lim_{x \to \infty} \sum_{i=1}^{k} \alpha_i F_i(x) = \sum_{i=1}^{k} \alpha_i \ \Rightarrow \ \lim_{x \to \infty} F(x) = 1. \end{aligned}$$

On similar lines, we have $\lim_{x \to -\infty} F(x) = 0$. Thus, a convex combination of k distribution functions is a distribution function.

3.5.5 Does there exist a distribution function F such that $\forall\ x, y \in \mathbb{R}$ $F(x+y) = F(x) + F(y)$? Justify your answer.

Solution: Allow $x, y \to \infty$ in $F(x+y) = F(x) + F(y)$, then we get $1 = 2$. Hence, we conclude that there does not exist a distribution function such that $F(x+y) = F(x) + F(y) \ \ \forall\ x, y \in \mathbb{R}$.

3.5.6 Suppose F and G are distribution functions of X and Y respectively. If $X > Y$ on Ω, show that $F(x) \leq G(x) \ \ \forall\ x \in \mathbb{R}$. Show that the converse is not true.

Solution: Since $X > Y$ on Ω, $\forall\ x \in \mathbb{R}$,

$$X(\omega) \leq x \ \Rightarrow \ Y(\omega) \leq x \ \Rightarrow \ P[X \leq x] \leq P[Y \leq x] \ \Rightarrow \ F(x) \leq G(x).$$

To show that the converse is not true, suppose $X \sim U(0,1)$ and $Y = q(1-X)$ where $0 < q < 1$. Suppose F and G denote the distribution functions of X and Y respectively. From F we find the distribution function G of Y and it is given by $G(x) = 1 - F(1 - x/q)$. Thus, F and G are given by,

$$F(x) = \begin{cases} 0, & \text{if} \quad x < 0, \\ x, & \text{if} \quad 0 \le x < 1, \\ 1, & \text{if} \quad x \ge 1 \end{cases} \quad \& \quad G(x) = \begin{cases} 0, & \text{if} \quad x < 0, \\ x/q, & \text{if} \quad 0 \le x < q, \\ 1, & \text{if} \quad x \ge q. \end{cases}$$

Thus, $Y \sim U(0,q)$. For $x < 0$, $F(x) = G(x) = 0$. For $0 \le x < q < 1$, $F(x) = x < x/q = G(x)$. For $q \le x < 1$, $F(x) = x < G(x) = 1$ and for $x \ge 1$, $F(x) = G(x) = 1$. Thus, $F(x) \le G(x) \;\; \forall \;\; x \in \mathbb{R}$. However, when $X(\omega) = 0, Y(\omega) = q > X(\omega)$. Thus, $X(\omega)$ is not larger than $Y(\omega)$ for all $\omega \in \Omega$.

3.5.7 Suppose a distribution function $F : \mathbb{R} \to [0,1]$ is defined as follows.

$$F(x) = \begin{cases} 0, & \text{if} \quad x < -2, \\ 0.2, & \text{if} \quad -2 \le x < 1, \\ 0.2 + x^2/10, & \text{if} \quad 1 \le x < 2, \\ 0.6 + x/10, & \text{if} \quad 2 \le x < 3, \\ 1, & \text{if} \quad x \ge 3. \end{cases}$$

Decompose F into discrete and continuous components. Identify the distributions in the decomposition.

Solution: We first find a set of points of discontinuities and corresponding size of the discontinuity. Observe that at $F(-2-) = 0 \ne 0.2 = F(-2)$, implying that -2 is a point of discontinuity with size of discontinuity 0.2. Similarly, $F(1-) = 0.2 \ne 0.3 = F(1)$, hence 1 is a point of discontinuitywith size of discontinuity 0.1. Proceeding on similar lines we note that 2 and 3 are also points of discontinuity, with respective sizes 0.2, 0.1. Hence, as in Theorem 3.3.1, we have $G(x) = \sum_{x_n \le x} p_{x_n}, \;\; x \in \mathbb{R}$ and $H(x) = F(x) - G(x), \; x \in \mathbb{R}$. G and H are given by,

$$G(x) = \begin{cases} 0, & \text{if} \quad x < -2, \\ 0.2, & \text{if} \quad -2 \le x < 1, \\ 0.3, & \text{if} \quad 1 \le x < 2, \\ 0.5, & \text{if} \quad 2 \le x < 3, \\ 0.6, & \text{if} \quad x \ge 3 \end{cases} \quad \& \; H(x) = \begin{cases} 0, & \text{if} \quad x < 1, \\ \frac{x^2}{10} - \frac{1}{10}, & \text{if} \quad 1 \le x < 2, \\ \frac{x}{10} + \frac{1}{10}, & \text{if} \quad 2 \le x < 3, \\ 0.4, & \text{if} \quad x \ge 3. \end{cases}$$

Note that $l_1 = G(\infty) = 0.6$ and $H(\infty) = 1 - G(\infty) = l_2 = 0.4$. Suppose F_1 and F_2 are defined as $F_1 = (1/l_1)G$ and $F_2 = (1/l_2)H$ so that $F_1(\infty) = 1$ and $F_2(\infty) = 1$. Thus, F_1 and F_2 are given by,

$$F_1(x) = \begin{cases} 0, & \text{if} \quad x < -2, \\ 2/6, & \text{if} \quad -2 \le x < 1, \\ 3/6, & \text{if} \quad 1 \le x < 2, \\ 5/6, & \text{if} \quad 2 \le x < 3, \\ 1, & \text{if} \quad x \ge 3 \end{cases} \quad \& F_2(x) = \begin{cases} 0, & \text{if} \quad x < 1, \\ \frac{x^2}{4} - \frac{1}{4}, & \text{if} \quad 1 \le x < 2, \\ \frac{x}{4} + \frac{1}{4}, & \text{if} \quad 2 \le x < 3, \\ 1, & \text{if} \quad x \ge 3. \end{cases}$$

It is easy to verify that $F(x) = l_1F_1(x) + l_2F_2(x)$ for all $x \in \mathbb{R}$. F_1 is a discrete distribution function and corresponding probability mass function is given by $\{(-2, 2/6), (1, 1/6), (2, 2/6), (3, 1/6)\}$. F_1 is a continuous distribution function and it is differentiable with corresponding probability density function

$$f_2(x) = \begin{cases} 0, & \text{if} \quad x < 1 \ \ \& \ \ x \geq 3, \\ x/2, & \text{if} \quad 1 \leq x < 2, \\ 1/4, & \text{if} \quad 2 \leq x < 3. \end{cases}$$

3.5.8 Suppose a distribution function $F : \mathbb{R} \to [0, 1]$ is defined as follows.

$$F(x) = \begin{cases} 0, & \text{if} \quad x < 0, \\ (pr + (1-p)x)/5, & \text{if} \quad r \leq x < r+1, r = 0, 1, \cdots, 4, \\ 1, & \text{if} \quad x \geq 5, \end{cases}$$

$0 < p < 1$. Examine whether F is a distribution function. If yes, decompose F into discrete and continuous components. Identify the distributions in the decomposition.

Solution: The given function satisfies the four requirements of the distribution function. Hence, it is a distribution function. It can be expressed as

$$F(x) = \begin{cases} 0, & \text{if} \quad x < 0 \\ (1-p)x/5, & \text{if} \quad 0 \leq x < 1 \\ (p + (1-p)x)/5, & \text{if} \quad 1 \leq x < 2 \\ (2p + (1-p)x)/5, & \text{if} \quad 2 \leq x < 3 \\ (3p + (1-p)x)/5, & \text{if} \quad 3 \leq x < 4 \\ (4p + (1-p)x)/5, & \text{if} \quad 4 \leq x < 5 \\ 1, & \text{if} \quad x \geq 5, \end{cases}$$

Observe that the left hand limit at 0 is 0, $F(0) = 0$ and the right hand limit is also 0. Thus, 0 is a point of continuity. Further, the left hand limit at 1 is $(1-p)/5$, $F(1) = 1/5$ and the right hand limit is also $1/5$. Thus, 1 is a point of discontinuity. Similarly, $2, 3, 4, 5$ are points of discontinuity. Thus, the set D of points of discontinuity is $D = \{1, 2, 3, 4, 5\}$. Note that F can be decomposed as $l_1F_1 + l_2F_2$ where $l_1 = p, l_2 = 1-p$ and F_1, F_2 are as follows.

$$F_1(x) = \begin{cases} 0, & \text{if} \quad x \leq 1 \\ 1/5, & \text{if} \quad 1 \leq x < 2 \\ 2/5, & \text{if} \quad 2 \leq x < 3 \\ 3/5, & \text{if} \quad 3 \leq x < 4 \\ 4/5, & \text{if} \quad 4 \leq x < 5 \\ 1, & \text{if} \quad x \geq 5 \end{cases} \quad \& \quad F_2(x) = \begin{cases} 0, & \text{if} \quad x < 0 \\ x/5, & \text{if} \quad 0 \leq x < 5 \\ 1, & \text{if} \quad x \geq 5, \end{cases}$$

Thus, F_1 is a distribution function of the discrete random variable X_1 with probability mass function $P[X_1 = i] = 1/5, i = 1, 2, 3, 4, 5$, which is the probability mass function of a discrete uniform distribution with support D. Further, F_2 is a distribution function of the absolutely continuous random variable X_2 with probability density function $f(x) = 1/5$,
$x \in (0, 5)$, which is the probability density function of the uniform $U(0, 5)$ distribution.

Answers to the multiple choice questions, based on Chapter 3, are given in Table 11.3.

TABLE 11.3
Answer Key to MCQs in Chapter 3

Q.No.	1	2	3	4	5	6	7	8	9	10
Ans	c	c	c	d	a,b,d	d	b	a	b,c	d

11.4 Chapter 4

4.6.1 Give an example of a discrete random variable with infinite mean.

Solution: Suppose X is a discrete random variable such that $P[X = 2^i] = 2^{-i}$, $i = 1, 2, \cdots$. Observe that

$$P[X < \infty] = \sum_{i \geq 1} P[X = 2^i] = \sum_{i \geq 1} 2^{-i} = 1 \text{ but } E(X) = \sum_{i \geq 1} 2^i 2^{-i} = \infty.$$

Thus, even if X is finite, $E(X)$ is infinite.

4.6.2 Give an example of a discrete random variable with finite mean but with infinite variance.

Solution: Suppose X is a discrete random variable such that $P[X = i] = c/i^3$, $i = 1, 2, \cdots$ and c is a norming constant. Note that

$$\begin{aligned} E(X) &= \sum_{i \geq 1} i(c/i^3) = \sum_{i \geq 1} c/i^2 < \infty \\ \text{but } E(X^2) &= \sum_{i \geq 1} i^2(c/i^3) = \sum_{i \geq 1} c/i = \infty \quad \Rightarrow \quad Var(X) = \infty. \end{aligned}$$

4.6.3 Suppose X and Y are random variables such that $P[|X - Y| \leq M] = 1$ for some $M > 0$. Show that Y has finite mean if X has finite mean.

Solution: It is given that $P[|X - Y| \leq M] = 1$. If $E(X) \geq 0$, note that

$$\begin{aligned} X - M \leq Y \leq X + M \text{ a.s} &\Rightarrow E(X - M) \leq E(Y) \leq E(X + M) \\ &\Rightarrow E(X) - M \leq E(Y) \leq E(X) + M \\ &\Rightarrow -E(X) - M \leq E(Y) \leq E(X) + M \\ &\Rightarrow |E(Y)| \leq E(X) + M \\ &\Rightarrow E(Y) < \infty \text{ if } E(X) < \infty \end{aligned}$$

Similarly, if $E(X) < 0$, note that $E(X) + M < -E(X) + M$. Hence,

$$\begin{aligned} X - M \leq Y \leq X + M \text{ a.s} &\Rightarrow E(X - M) \leq E(Y) \leq E(X + M) \\ &\Rightarrow E(X) - M \leq E(Y) \leq E(X) + M \\ &\Rightarrow E(X) - M \leq E(Y) \leq -E(X) + M \\ &\Rightarrow |E(Y)| \leq -E(X) + M \\ &\Rightarrow E(Y) < \infty \text{ if } E(X) < \infty \end{aligned}$$

4.6.4 Show that for any non-negative random variable X,
(i) $E(X) + E(1/X) \geq 2$ and (ii) $E(\max\{X, 1/X\}) \geq 1$.

Solution: (i) Note that

$$\begin{aligned} X \geq 0 \ \ \& \ \ (X-1)^2 \geq 0 \ \ &\Rightarrow \ \ (X-1)^2/X \geq 0 \Rightarrow X - 2 + 1/X \geq 0 \\ \Rightarrow \ \ X + 1/X \geq 2 \ \ &\Rightarrow \ \ E(X) + E(1/X) \geq 2. \end{aligned}$$

(ii) Observe that

$$\begin{aligned} 0 < X \leq 1 \Rightarrow 1/X \geq 1 \ \ &\Rightarrow \ \ \max\{X, 1/X\} \geq 1 \\ X \geq 1 \Rightarrow 1/X \leq 1 \Rightarrow \max\{X, 1/X\} \geq 1 \ &\Rightarrow \ E(\max\{X, 1/X\}) \geq 1. \end{aligned}$$

4.6.5 Prove that if $Var(X) = 0$, then $X = E(X)$ a.s.

Solution: The result follows from the fact that for a non-negative random variable X, $E(X) = 0$ implies $X = 0$ a.s. Observe that

$$0 = Var(X) = E(X - E(X))^2 \Rightarrow X - E(X) = 0 \text{ a.s.} \Rightarrow X = E(X) \text{ a.s.}$$

4.6.6 Suppose X is an integrable random variable and $A\Delta B$ is a null event. Show that $\int_A X \, dP = \int_B X \, dP$.

Solution: Note that $A\Delta B$ is a null event implies that $P(A\Delta B) = 0$ which further implies that $P(A \cap B^c) = 0$ and $P(A^c \cap B) = 0$. We use the result that if X is an integrable random variable and A is a null event, then $\int_A X \, dP = 0$. Observe that

$$\begin{aligned} \int_A X \, dP &= \int_{A \cap B^c} X \, dP + \int_{A \cap B} X \, dP = \int_{A \cap B} X \, dP \\ &= \int_{A \cap B} X \, dP + \int_{A^c \cap B} X \, dP = \int_B X \, dP \end{aligned}$$

4.6.7 Show that if X is $\mathbb{A}$ measurable and $\int_A X \, dP = 0 \ \forall \ A \in \mathbb{A}$ then $X = 0$ a.s.

Solution: Suppose $\int_A X \, dP = 0 \ \forall \ A \in \mathbb{A}$, but $X \neq 0$ a.s. Thus, the distribution function F of X is not of the form

$$F(x) = \begin{cases} 0, & \text{if } \ x < 0 \\ 1, & \text{if } \ x \geq 0. \end{cases}$$

Thus, $\exists \ a > 0$ such that $0 < F(a) < 1$. Suppose $A = (-\infty, a]$, then $0 < P(A) < 1 \Rightarrow 0 < P(A^c) < 1$. On $A^c, X > a$. Hence, $\int_{A^c} X \, dP > aP(A^c) > 0$ which is a contradiction to the given statement that $\int_A X \, dP = 0 \ \ \forall \ A \in \mathbb{A}$. Hence, the assumption that $X \neq 0$ a.s. is wrong. Thus, $\int_A X \, dP = 0 \ \forall \ A \in \mathbb{A}$ implies that $X = 0$ a.s.

4.6.8 Prove or disprove: $X \equiv 0$ implies $\int_\Omega X dP = 0$, but $\int_\Omega X dP = 0$ may not imply $X \equiv 0$.

Solution: If $X \equiv 0$, that is, $X = 0$ a.s. then it is proved that $E(X) = 0$. However, converse is not true. Suppose X is a discrete random variable with probability mass function$\{(-1, 1/3), (0, 1/3), (1, 1/3)\}$. Then $E(X) = 0$ but $P[X = 0] = 1/3 \neq 1$. Thus, $E(X) = 0$ may not imply $X \equiv 0$.

4.6.9 Suppose $\Omega = \{-2, -1, 3, 7\}$ and $\mathbb{A}$ is a powerset of Ω. A measure μ on $(\Omega, \mathbb{A})$ is defined as $\mu(\{-2\}) = 2, \mu(\{-1\}) = 1, \mu(\{3\}) = 3$ and $\mu(\{7\}) = 7$. Obtain a probability measure P from μ. Suppose a random variable X is defined on $(\Omega, \mathbb{A}, P)$ as follows. Find $\int_\Omega X\ dP$.

$$X(\omega) = \begin{cases} 1, & \text{if} \quad \omega = 3 \text{ or } 7 \\ -1, & \text{if} \quad \omega = -2 \text{ or } -1. \end{cases}$$

Solution: For the given measure $\mu(\Omega) = 13$ and hence the corresponding probability measure P is given by,

$$P(\{-2\}) = 2/13, \ P(\{-1\}) = 1/13, \ P(\{3\}) = 3/13 \ \& \ P(\{7\}) = 7/13.$$

Further, X is a simple random variable. Hence, $\int_\Omega X\ dP = 1(P(\{3\}) + P(\{7\})) - 1(P(\{-2\}) + P(\{-1\})) = 10/13 - 3/13 = 7/13$.

4.6.10 Suppose $(\Omega, \mathbb{A}, P)$ is a probability space where $\mathbb{A}$ is a sigma field generated by a countable partition $\{A_n, n \geq 1\}$ of Ω. Suppose a random variable X is defined on $(\Omega, \mathbb{A}, P)$ as $X = i$ on A_i, $i \geq 1$. Find $\int_\Omega X\ dP$ if the probability measure P is defined as (i) $P(A_i) = c/i^3$, where c is such that $\sum_{i\geq 1} c/i^3 = 1$, (ii) $P(A_i) = c/i^2$, where c is such that $\sum_{i \geq 1} c/i^2 = 1$ and (iii) $P(A_i) = 1/2^i$. In all the three cases, find $\int_\Omega Y\ dP$, when $Y = X^2$ and $Y = e^X$.

Solution: Note that X is an elementary random variable. Hence,

$$\begin{aligned}
(i) \int_\Omega X\ dP &= E(X) = \sum_{i\geq 1} ic/i^3 = c\sum_{i\geq 1} 1/i^2 < \infty \\
(ii) \int_\Omega X\ dP &= E(X) = \sum_{i\geq 1} ic/i^2 = c\sum_{i\geq 1} 1/i = \infty \\
(iii) \int_\Omega X\ dP &= E(X) = \sum_{i\geq 1} i/2^i \\
&= (1/2)(1 + (1/2) + (1/2)^2 + (1/2)^3 + \cdots) \\
&= (1/2)((1/2))^{-2} = 2.
\end{aligned}$$

Suppose $Y = X^2$. Then

$$\begin{aligned}
(i) \int_\Omega Y\, dP &= E(Y) = \sum_{i\geq 1} i^2 c/i^3 = \infty \\
(ii) \int_\Omega Y\, dP &= E(Y) = \sum_{i\geq 1} i^2 c/i^2 = \infty \\
(iii) \int_\Omega X\, dP &= E(Y) = \sum_{i\geq 1} i^2/2^i < \infty.
\end{aligned}$$

since by the ratio test the series $\sum_{i\geq 1} i^2/2^i$ is convergent. Suppose $Y = e^X$. Then

$$\begin{aligned}
(i) \int_\Omega Y\, dP &= E(Y) = \sum_{i\geq 1} e^i c/i^3 = \infty \\
(ii) \int_\Omega Y\, dP &= E(Y) = \sum_{i\geq 1} e^i c/i^2 = \infty \\
(iii) \int_\Omega X\, dP &= E(Y) = \sum_{i\geq 1} e^i/2^i = \infty.
\end{aligned}$$

since in all the three cases, by the ratio test the concerned series are divergent.

4.6.11 Suppose a distribution function $F : \mathbb{R} \to [0, 1]$ is defined as follows.

$$F(x) = \begin{cases} 0, & \text{if} \quad x < 0, \\ 0.2, & \text{if} \quad 0 \leq x < 1, \\ 0.5, & \text{if} \quad 1 \leq x < 2, \\ 0.8, & \text{if} \quad 2 \leq x < 4, \\ 1, & \text{if} \quad x \geq 4. \end{cases}$$

Compute $E(X)$ from $F(x)$, without finding the probability mass function.

Solution: It is clear that the given distribution function F is a discrete distribution function with points of discontinuity as $0, 1, 2, 4$, with respective jump sizes as $0.2, 0.3, 0.3, 0.2$. Thus, we find $E(X)$ using the formula $E(X) = \int_{\mathbb{R}} g(x)dF(x)$ as a Riemann-Stieltjes integral for a step function F. Thus,

$$E(X) = 0 \times 0.2 + 1 \times 0.3 + 2 \times 0.3 + 4 \times 0.2 = 1.7.$$

4.6.12 Suppose a distribution function $F : \mathbb{R} \to [0, 1]$ is defined as follows.

$$F(x) = \begin{cases} 0, & \text{if} \quad x < -2, \\ 1/3, & \text{if} \quad -2 \le x < 0, \\ 1/2, & \text{if} \quad 0 \le x < 5, \\ 1/2 + (x-5)^2/2, & \text{if} \quad 5 \le x < 6, \\ 1, & \text{if} \quad x \ge 6. \end{cases}$$

Find (i) $E(X)$ from $F(x)$. (ii) Decompose F into discrete and continuous components. (iii) Find the probability mass function of a discrete random variable and the probability density function of a continuous random variable involved in the decomposition and hence find $E(X)$.

Solution: It is clear that the given distribution function F is neither discrete nor continuous. Thus, we find $E(X)$ using the formula $E(X) = \int_{\mathbb{R}} g(x)dF(x)$ as a Riemann-Stieltjes integral. Thus,

$$E(X) = -2 \times \frac{1}{3} + 0 \times \frac{1}{6} + \int_5^6 x(x-5)\, dx = -\frac{2}{3} + \frac{17}{6} = \frac{13}{6}.$$

In Example 3.3.1, it is shown that the distribution function F can be expressed as,

$$F(x) = \begin{cases} 0, & \text{if} \quad x < -2, \\ 1/2 \times 2/3, & \text{if} \quad -2 \le x < 0, \\ 1/2 \times 1, & \text{if} \quad 0 \le x < 5, \\ 1/2 + 1/2(x-5)^2, & \text{if} \quad 5 \le x < 6, \\ 1/2 + 1/2, & \text{if} \quad x \ge 6, \end{cases}$$

Suppose

$$F_1(x) = \begin{cases} 0, & \text{if} x < -2, \\ 2/3, & \text{if} -2 \le x < 0, \\ 1, & \text{if} x \ge 0. \end{cases} \;\&\; F_2(x) = \begin{cases} 0, & \text{if} x < 5, \\ (x-5)^2, & \text{if} 5 \le x < 6, \\ 1, & \text{if} x \ge 6. \end{cases}$$

Then $F = (1/2)F_1 + (1/2)F_2$ where F_1 is a distribution function of a discrete random variable with possible values $-2, 0$ with respective probabilities $2/3, 1/3$. Thus its mean is $-4/3$. F_2 is a distribution function of an absolutely continuous random variable with probability density function $f(x) = 2(x-5)$ with support $(5, 6)$. Its mean is $17/3$. Using this approach $E(X) = (1/2)(-4/3) + (1/2)(17/3) = 13/6$.

4.6.13 Suppose a distribution function $F : \mathbb{R} \to [0, 1]$ is defined as follows.

$$F(x) = \begin{cases} 0, & \text{if} \quad x < -2, \\ 0.2, & \text{if} \quad -2 \le x < 1, \\ 0.2 + x^2/10, & \text{if} \quad 1 \le x < 2, \\ 0.6 + x/10, & \text{if} \quad 2 \le x < 3, \\ 1, & \text{if} \quad x \ge 3. \end{cases}$$

Find (i) $E(X)$ from $F(x)$. (ii) Decompose F into discrete and continuous components. (iii) Identify the distributions in the decomposition and hence find $E(X)$.

Solution: In Exercise 7 in Chapter 3, we have noted that $-2, 1, 2, 3$ are points of discontinuities with respective sizes $0.2, 0.1, 0.2, 0.1$. On the intervals $(1, 2)$ and $(2, 3)$, F is differentiable. Hence using the definition of Reimann-Stieltjes integral $E(X)$ is obtained as follows.

$$\begin{aligned} E(X) &= -2 \times 0.2 + 1 \times 0.1 + 2 \times 0.2 + 3 \times 0.1 \\ &+ \int_1^2 x(x/5)\, dx + \int_2^3 x(1/10)\, dx = 2/5 + 7/17 + 1/4 = 67/60. \end{aligned}$$

In the same exercise, we have shown that F can be decomposed as $F(x) = l_1 F_1(x) + l_2 F_2(x)\ \forall\ x \in \mathbb{R}$, where $l_1 = 0.6, l_2 = 0.4$ and F_1 and F_2 are given by,

$$F_1(x) = \begin{cases} 0, & \text{if} \quad x < -2, \\ 2/6, & \text{if} \quad -2 \le x < 1, \\ 3/6, & \text{if} \quad 1 \le x < 2, \\ 5/6, & \text{if} \quad 2 \le x < 3, \\ 1, & \text{if} \quad x \ge 3. \end{cases} \quad \& \ F_2(x) = \begin{cases} 0, & \text{if} \quad x < 1, \\ \frac{x^2}{4} - \frac{1}{4}, & \text{if} \quad 1 \le x < 2, \\ \frac{x}{4} + \frac{1}{4}, & \text{if} \quad 2 \le x < 3, \\ 1, & \text{if} \quad x \ge 3. \end{cases}$$

F_1 is a discrete distribution function and corresponding probability mass function is given by, $\{(-2, 2/6), (1, 1/6), (2, 2/6), (3, 1/6)\}$. Its mean is $2/3$. F_2 is a continuous distribution function and it is differentiable with corresponding probability density function

$$f_2(x) = \begin{cases} 0, & \text{if} \quad x < 1 \ \& \ x \ge 3, \\ x/2, & \text{if} \quad 1 \le x < 2, \\ 1/4, & \text{if} \quad 2 \le x < 3. \end{cases}$$

Mean corresponding to F_2 is $43/24$. Hence, $E(X) = 0.6 \times (2/3) + 0.4 \times (43/24) = 67/60$.

4.6.14 Suppose a distribution function $F : \mathbb{R} \to [0, 1]$ is defined as follows.

$$F(x) = \begin{cases} 0\,, & \text{if} \quad x < 2 \\ (2/3)x - 1\,, & \text{if} \quad 2 \le x < 3 \\ 1\,, & \text{if} \quad x \ge 3. \end{cases}$$

Find (i) $E(X)$ from $F(x)$. (ii) Decompose F into discrete and continuous components. (iii) Identify the distributions in the decomposition and hence find $E(X)$.

Solution: The given distribution function F is neither discrete nor continuous. Thus, we find $E(X)$ using the formula $E(X) = \int_{\mathbb{R}} g(x) dF(x)$ as a Riemann-Stieltjes integral. It is clear that 2 is the only point of discontinuity with jump size 1/3. For $x \in (2,3)$, $F(x)$ is differentiable with derivative $f(x) = 2/3$. Thus,

$$E(X) = 2 \times \frac{1}{3} + \int_2^3 \frac{2}{3} x \, dx = \frac{2}{3} + \frac{5}{3} = \frac{7}{3} .$$

The distribution function F can be expressed as,

$$F(x) \quad = \quad \begin{cases} 0, & \text{if} \quad x < 2, \\ \frac{1}{3}1 + \frac{2}{3}(x-2), & \text{if} \quad 2 \leq x < 3, \\ \frac{1}{3}1 + \frac{1}{3}1, & \text{if} \quad x \geq 3, \end{cases}$$

It is easy to check that $F = (1/3)F_1 + (2/3)F_2$ where

$$F_1(x) = \begin{cases} 0, & \text{if} \quad x < 2, \\ 1, & \text{if} \quad x \geq 2. \end{cases} \quad \& \; F_2(x) = \begin{cases} 0, & \text{if} \quad x < 2, \\ (x-2), & \text{if} \quad 2 \leq x < 3, \\ 1, & \text{if} \quad x \geq 3. \end{cases}$$

Note that F_1 is a distribution function of a degenerate random variable, degenerate at 2. Thus its mean is 2. F_2 is a distribution function of an absolutely continuous random variable with uniform $U(2,3)$ distribution. Its mean is $(5/2)$. Using this approach $E(X) = (1/3)2 + (2/3)(5/2) = 7/3$.

4.6.15 Suppose a distribution function $F : \mathbb{R} \to [0,1]$ is defined as follows.

$$F(x) \quad = \quad \begin{cases} 0\,, & \text{if} \quad x < 0 \\ 1/4\,, & \text{if} \quad 0 \leq x < 1 \\ 1 - (1/2)e^{-(x-1)}, & \text{if} \quad x \geq 1. \end{cases}$$

(i) Find $E(X)$ from $F(x)$. (ii) Decompose F into discrete and continuous components. (iii) Identify the distributions in the decomposition and hence find $E(X)$.

Solution: It is clear that 0 and 1 are the two points of discontinuities with respective sizes 1/4, 1/4. On the interval $(1, \infty)$, F is differentiable with derivative $(1/2)e^{-(x-1)}$. Hence,

$$E(X) = 0 \times (1/4) + 1 \times (1/4) + (1/2) \int_1^\infty x e^{-(x-1)} \, dx = 1/4 + 1 = 5/4.$$

Further, F can be decomposed as $F(x) = l_1 F_1(x) + l_2 F_2(x)$ for all $x \in \mathbb{R}$, where $l_1 = 1/2, l_2 = 1/2$ and F_1 and F_2 are given by,

$$F_1(x) = \begin{cases} 0, & \text{if} \quad x < 0, \\ 1/2, & \text{if} \quad 0 \leq x < 1, \\ 1, & \text{if} \quad x \geq 1. \end{cases} \quad \& \; F_2(x) = \begin{cases} 0, & \text{if} \quad x < 1, \\ 1 - e^{-(x-1)}, & \text{if} \quad x \geq 1. \end{cases}$$

Note that F_1 is a discrete distribution function and corresponding probability mass function is given by, $\{(0, 1/2), (1, 1/2)\}$. Its mean is 1/2. F_2

is a continuous distribution function and it is differentiable with corresponding probability density function

$$f_2(x) = \begin{cases} 0, & \text{if } x < 1, \\ e^{-(x-1)}, & \text{if } x \geq 1. \end{cases}$$

It is the probability density function of a random variable following the exponential distribution with location parameter 1 and scale parameter 1. Mean corresponding to F_2 is 2. Hence, $E(X) = (1/2)(1/2) + (1/2)2 = 5/4$.

4.6.16 Suppose a distribution function $F : \mathbb{R} \to [0, 1]$ is defined as follows.

$$F(x) = \begin{cases} 0, & \text{if} \quad x < 0, \\ (pr + (1-p)x)/k, & \text{if} \quad r \leq x < r+1, r = 0, 1, \cdots, k-1, \\ 1, & \text{if} x \geq k, \end{cases}$$

$0 < p < 1$. (i) Find $E(X)$ from $F(x)$. (ii) Decompose F into discrete and continuous components. (iii) Identify the distributions in the decomposition and hence find $E(X)$.

Solution: In Exercise 8 in Chapter 3, we have discussed such a distribution function with $k = 5$. Proceeding on similar lines we note that the set D of points of discontinuity of F is $D = \{1, 2, \cdots, k\}$ with size of discontinuity to be p/k for each. Further, on the interval $(r, r+1)$, F is continuous and differentiable with derivative $(1-p)/k, r = 0, 1, \cdots, k-1$. Hence,

$$\begin{aligned} E(X) &= (p/k)\sum_{r=1}^{k} r + ((1-p)/k)\sum_{r=0}^{k-1}\int_r^{r+1} x\, dx \\ &= (p/k)(k(k+1)/2) + ((1-p)/2k)\sum_{r=0}^{k-1}[(r+1)^2 - r^2] \\ &= p(k+1)/2 + ((1-p)/2k)\sum_{r=0}^{k-1}(2r+1) \\ &= p(k+1)/2 + ((1-p)/2k)k^2 = p(k+1)/2 + ((1-p)k/2 \\ &= (p+k)/2. \end{aligned}$$

Note that F can be decomposed as $l_1F_1 + l_2F_2$ where $l_1 = p, l_2 = 1-p$ and F_1, F_2 are as follows.

$$F_1(x) = \begin{cases} 0, & \text{if} \quad x \leq 1 \\ \frac{r}{k}, & \text{if} \quad r \leq x < r+1, \\ & \quad r = 0, 1, \cdots, k-1, \\ 1, & \text{if} \quad x \geq k, \end{cases} \quad \& \; F_2(x) = \begin{cases} 0, & \text{if} \quad x < 0 \\ \frac{x}{k}, & \text{if} \quad 0 \leq x < k \\ 1, & \text{if} \quad x \geq k, \end{cases}$$

Thus, F_1 is a distribution function of the discrete random variable X_1 with probability mass function $P[X_1 = i] = 1/k, i = 1, 2, \cdots, k$, which is the probability mass function of a discrete uniform distribution with support D and its mean is $(k+1)/2$. Further, F_2 is a distribution function of the absolutely continuous random variable X_2 with probability density function $f(x) = 1/k, x \in (0,k)$, which is the probability density function of the uniform $U(0,k)$ distribution. Its mean is $k/2$. Hence,

$$E(X) = p(k+1)/2 + (1-p)k/2 = (p+k)/2.$$

4.6.17 Prove that if $X \geq 0$ then $E(X) = 0 \Rightarrow X = 0$ a.s.

Solution: If a non-negative random variable X is a simple random variable, then in Theorem 4.2.2 it is proved that $E(X) = 0$ implies $X = 0$ a.s. Suppose now X is a non-negative random variable. To prove $X = 0$ a.s., we have to prove that $P[X > 0] = 0$, X being a non-negative random variable. Note that

$$P[X > 0] = P[\bigcup_{n\geq 1}[X \geq 1/n]] \leq \sum_{n\geq 1} P[X \geq 1/n] = \sum_{n\geq 1} P(A_n),$$

where $A_n = [X \geq 1/n], n \geq 1$. Note that

$$\begin{aligned}
(1/n)I_{A_n} \leq XI_{A_n} \leq X \quad &\Rightarrow \quad E((1/n)I_{A_n}) \leq E(XI_{A_n}) \leq E(X)\\
&\Rightarrow \quad (1/n)P(A_n) \leq E(XI_{A_n}) \leq E(X) = 0\\
&\Rightarrow \quad P(A_n) = 0 \;\; \forall \;\; n \geq 1\\
&\Rightarrow \quad P[X > 0] = 0 \;\Rightarrow\; P[X = 0] = 1\\
&\Rightarrow \quad X = 0 \;\; \text{a.s.}
\end{aligned}$$

4.6.18 If $X \geq 0$ on A and $\int_A X\,dP = 0$ then show that $X = 0$ a.s. on A.

Solution: Suppose X is a simple random variable defined as $X = \sum_{i=1}^{k} a_i A_i$ where $A_1, A_2, \cdots, A_k$ is a partition of A. It is given that $X \geq 0$ on A. Suppose on $A_1, X = 0$, that is, $a_1 = 0$ and on all other partition sets $X > 0$, that is, $a_i > 0$ for $i = 2, 3, \cdots, k$. Now $\int_A X\,dP = \sum_{i=1}^{k} a_i P(A_i) = 0$. Thus, sum of non-negative numbers is 0, so each term must be 0. Note that

$$\begin{aligned}
0 \quad &= \quad \int_A X\,dP = a_1 P(A_1) + \sum_{i=2}^{k} a_i P(A_i) = \sum_{i=2}^{k} a_i P(A_i)\\
\Rightarrow \; P(A_i) \quad &= \quad 0 \;\; \forall \;\; i = 2, 3, \cdots, k\\
\Rightarrow \; P(A_1) \quad &= \quad P[X = 0] = 1 \;\Rightarrow\; X = 0 \;\; \text{a.s. on} \;\; A.
\end{aligned}$$

Suppose $X \geq 0$. Then $Y = XI_A \geq 0$ and $\int_A X\,dP = 0 \iff E(XI_A) = E(Y) = 0$. Hence the result follows using the similar arguments as in the previous exercise.

4.6.19 If $a \leq X \leq b$ a.s. on A then show that $aP(A) \leq \int_A X\, dP \leq bP(A)$.

Solution: Observe that

$$\begin{aligned} a \leq X \leq b \text{ a.s. on } A &\Rightarrow aI_A \leq XI_A \leq bI_A \\ &\Rightarrow \int_\Omega aI_A \leq \int_\Omega XI_A \leq \int_\Omega bI_A \\ &\Rightarrow aP(A) \leq \int_A X\, dP \leq bP(A). \end{aligned}$$

4.6.20 Examine whether following functions are characteristic functions.
(i) $\phi(t/a)\exp(ibt)$, $a \neq 0, b \in \mathbb{R}$ and ϕ is a characteristic function,
(ii) $\phi(-t)$, where ϕ is a characteristic function,
(iii) $\sum_{i=1}^k a_j\phi_j(t)$, $a_j \geq 0$ and $\sum_{i=1}^k a_j = 1$ and $\phi_j(t)$ are characteristic functions for all $j = 1, 2, \cdots, k$ and (iv) $\phi(t) = 1/(1+t^4)$ where $t \in \mathbb{R}$.

Solution: (i) It is given that ϕ is a characteristic function. Suppose a random variable Y is defined as $Y = X/a + b$. Then

$$\phi_Y(t) = E(e^{itY}) = e^{itb}E(e^{itX/a}) = e^{itb}\phi(t/a).$$

Hence, $\phi(t/a)\exp(ibt)$ is a characteristic function.
(ii) Observe that $E(e^{-itX}) = E(e^{i(-t)X}) = \phi(-t)$. Thus, $\phi(-t)$ is a characteristic function of $-X$ if $\phi(t)$ is a characteristic function of X.
(iii) Suppose a distribution function F of X is a convex combination of k distribution functions given by, $F(x) = \sum_{i=1}^k a_jF_j(x), x \in \mathbb{R}$. Then observe that

$$\begin{aligned} E(e^{itX}) &= \int e^{itx}\, dF(x) = \int e^{itx}\, d\sum_{i=1}^k a_jF_j(x) \\ &= \sum_{i=1}^k a_j \int e^{itx}\, dF_j(x) = \sum_{i=1}^k a_j\phi_j(t). \end{aligned}$$

Hence, $\sum_{i=1}^k a_j\phi_j(t)$ is a characteristic function.
(iv) Suppose $\phi(t) = 1/(1+t^4)$ is a characteristic function. Then it is a differentiable function and derivatives of all order at $t = 0$ exist. Hence, moments of all order exist and can be obtained by taking successive derivatives and substituting $t = 0$. Thus,

$$\frac{d}{dt}\frac{1}{1+t^4} = -\frac{4t^3}{(1+t^4)^2} \quad \& \quad \frac{d^2}{dt^2}\frac{1}{1+t^4} = \frac{-12t^3(1+t^4)^2 + 32t^6(1+t^4)}{(1+t^4)^4}$$
$$\Rightarrow \quad \mu_1' = 0, \quad \mu_2' = 0\ .$$

First two raw moments are 0 implies that $\phi(t)$ is a characteristic function of a random variable X which is degenerate at 0. However, if X is degenerate at 0, then its characteristic function is $E(e^{itX}) = 1$ which is not same as characteristic function $1/(1+t^4)$. Hence our assumption that $\phi(t) = 1/(1+t^4)$ is a characteristic function is wrong. Thus, $\phi(t) = 1/(1+t^4)$ is not a characteristic function.

4.6.21 Suppose a discrete random variable has finite second moment. Show that it has finite mean.

Solution: We have a result which states that if r-th order moments are finite, then all lower order moments are also finite. Alternatively, for a discrete random variable X with probability mass function $\{(i, p_i), i \geq 0\}$, $E(X^2) = \sum_{i\geq 0} i^2 p_i$ and it is given to be finite. Observe that by Cauchy-Schwarz inequality,

$$\begin{aligned}\Big(\sum_{i\geq 0} ip_i\Big)^2 = \sum_{i\geq 0}\sum_{j\geq 0} ip_i jp_j &\leq \Big(\sum_{i\geq 0} i^2 p_i^2 \sum_{j\geq 0} j^2 p_j^2\Big)^{1/2} \\ &\leq \sum_{i\geq 0} i^2 p_i \sum_{j\geq 0} j^2 p_j < \infty.\end{aligned}$$

In the second step, we use the result that if $0 \leq a \leq 1$ then $a^2 \leq a$.

4.6.22 Suppose $L_r = \{Z|E(|Z|^r) < \infty\}$. Prove that the set L_r is closed under addition.

Solution: Follows from the C_r inequality.

4.6.23 Suppose $X \sim U(0,2)$ distribution then without evaluating, show that $E(X\ \log X) \geq 0$.

Solution: Since $X \sim U(0,2)$ distribution, $E(X) = 1$. Suppose $g : \mathbb{R}^+ \to \mathbb{R}$ is defined as $g(x) = x \log x$. Note that

$$g(x) = x\log x \quad \Rightarrow \quad \frac{d}{dx}g(x) = 1 + \log x \quad \Rightarrow \quad \frac{d^2}{dx^2}g(x) = 1/x > 0.$$

Hence g is a convex function and by Jensen's inequality,

$$E(X\ \log X) = E(g(X)) \geq g(E(X)) = g(1) = 1\log 1 = 0.$$

Answers to the multiple choice questions, based on Chapter 4, are given in Table 11.4.

TABLE 11.4
Answer Key to MCQs in Chapter 4

Q.No.	1	2	3	4	5	6	7	8	9	10
Ans	b	d	d	c	d	b,d	d	c	a,c	d
Q.No.	11	12	13	14	15	16	17	18	19	20
Ans	b	a	c	a	c	d	a	b	c,d	a

11.5 Chapter 5

5.5.1 Prove or disprove: A P-null event is independent of itself.

Solution: Suppose N is a P-null event. Hence $P(N) = 0$. Observe that

$$P(N \cap N) = P(N) = 0 = P(N)P(N) \quad \Rightarrow \quad N \text{ is independent of itself.}$$

5.5.2 Suppose N is a P-null event. Examine whether N^c is independent of itself.

Solution: Suppose N is a P-null event. Hence $P(N^c) = 1$. Note that

$$P(N^c \cap N^c) = P(N^c) = 1 = P(N^c)P(N^c) \Rightarrow N^c \text{ is independent of itself.}$$

5.5.3 Suppose $P(C|A \cap B) = P(C|B)$ and $P(A \cap B \cap C) > 0$. Show that $P(A|B \cap C) = P(A|B)$.

Solution: Note that

$$P(A \cap B \cap C) > 0 \Rightarrow P(A \cap B), P(B \cap C) \;\&\; P(B) > 0.$$

Using the definition of conditional probability we have,

$$\begin{aligned} P(C|A \cap B) = P(C|B) \quad &\Rightarrow \quad \frac{P(A \cap B \cap C)}{P(A \cap B)} = \frac{P(B \cap C)}{P(B)} \\ &\Rightarrow \quad \frac{P(A \cap B \cap C)}{P(B \cap C)} = \frac{P(A \cap B)}{P(B)} \\ &\Rightarrow \quad P(A|B \cap C) = P(A|B). \end{aligned}$$

5.5.4 Prove or disprove: (i) Mutually exclusive events are independent. (ii) Independent events are mutually exclusive.

Solution: (i) Suppose A and B are mutually exclusive events, hence $A \cap B = \emptyset \;\Rightarrow\; P(A \cap B) = 0$. Suppose $\Omega = \{1, 2, 3, 4, 5, 6\}$, $A = \{1, 3, 5\}, B = \{2, 4, 6\}$ with $P(A) = P(B) = 1/2$. Here A and B are mutually exclusive events, but $0 = P(A \cap B) \neq P(A)P(B)$. Thus, A and B are not independent.
(ii) Suppose $\Omega = \{a, b, c, d\}, A = \{a, b\}, B = \{b, c\}$ and $\forall \; \omega \in \Omega$, $P(\{\omega\}) = 1/4$. Then $P(A) = P(B) = 1/2$. $A \cap B = \{b\}$, hence $P(A \cap B) = 1/4 = P(A)P(B)$. Thus, A and B are independent. Note that $A \cap B = \{b\} \neq \emptyset$ which implies that A and B are not mutually exclusive events. Thus there is no relation between mutually exclusive events and independent events. Independence is defined in terms of probabilities of events, while definition of mutually exclusive events does not need probabilities associated with these events.

5.5.5 Show that two events A and B are independent if and only if $\sigma(\{A\})$ and $\sigma(\{B\})$ are independent. Show further that it is equivalent to I_A and I_B are independent random variables.

Solution: Note that

$$\sigma(\{A\}) = \{\Omega, \emptyset, A, A^c\} \;\; \& \;\; \sigma(\{B\}) = \{\Omega, \emptyset, B, B^c\}.$$

(i) Suppose A and B are independent. Then as proved in Theorem 5.2.1, A and $\Omega, \emptyset, B, B^c$ are independent. Similarly, A^c and $\Omega, \emptyset, B, B^c$ are independent. Hence, $\sigma(\{A\})$ and $\sigma(\{B\})$ are independent. (ii) Suppose $\sigma(\{A\})$ and $\sigma(\{B\})$ are independent. Hence every event in $\sigma(\{A\})$ is independent of every event in $\sigma(\{B\})$ which implies A and B are independent. I_A and I_B are independent random variables, if the induced sigma fields are independent. Note that

$$\sigma(I_A) = \{\Omega, \emptyset, A, A^c\} \;\; \& \;\; \sigma(I_B) = \{\Omega, \emptyset, B, B^c\}.$$

Thus, I_A and I_B are independent random variables if and only if A and B are independent events.

5.5.6 Give an example of events A_1, A_2, A_3 which are pair-wise independent but not independent.

Solution: Suppose a probability space $(\Omega, \mathbb{A}, P)$ is defined as follows. $\Omega = \{a, b, c, d\}$, $\mathbb{A} = \mathbb{P}(\Omega)$ and $P(\{\omega\}) = 1/4 \; \forall \; \omega \in \Omega$. Suppose events $A_i, \; i = 1, 2, 3$ are defined as follows.

$$A_1 = \{a, b\}, \;\; A_2 = \{a, c\} \;\; \& \;\; A_3 = \{b, c\}.$$

Then it is clear that for $i = 1, 2, 3$,

$$P(A_i) = 1/2, \;\; P(A_i \cap A_j) = 1/4 \;\; i \neq j = 1, 2, 3 \;\; \& \;\; P(A_1 \cap A_2 \cap A_3) = 0.$$

Observe that $\forall \;\; i \neq j = 1, 2, 3, \;\; P(A_i \cap A_j) = P(A_i)P(A_j)$, however $0 = P(A_1 \cap A_2 \cap A_3) \neq P(A_1)P(A_2)P(A_3)$. Thus, A_1, A_2, A_3 are pair-wise independent but not mutually independent.

5.5.7 Suppose $\{A_n, n \geq 1\}$ is a sequence of independent events. Show that $P\left(\bigcup_{r=1}^{k} A_{i_r}\right) = 1 - \prod_{r=1}^{k}(1 - P(A_{i_r}))$, for any combination $i_1, i_2, \cdots, i_k$ of $1, 2, \cdots, k$ and for finite $k \geq 2$.

Solution: Since $\{A_n, n \geq 1\}$ is a sequence of independent events, its every finite sub-collection is a collection of independent events. Thus, for finite $k \geq 2$ and for any combination $i_1, i_2, \cdots, i_k$ of $1, 2, \cdots, k$, $\{A_{i_1}, A_{i_2}, \cdots, A_{i_k}\}$ are independent events. Hence, $\{A_{i_1}^c, A_{i_2}^c, \cdots, A_{i_k}^c\}$

are also independent events. Observe that

$$P\Big(\bigcup_{r=1}^{k} A_{i_r}\Big) = P\Big(\Big(\bigcup_{r=1}^{k} A_{i_r}\Big)^{c}\Big)^{c} = 1 - P\Big(\bigcup_{r=1}^{k} A_{i_r}\Big)^{c} = 1 - P\Big(\bigcap_{r=1}^{k} A_{i_r}^{c}\Big)$$
$$= 1 - \prod_{r=1}^{k} P(A_{i_r}^{c}) = 1 - \prod_{r=1}^{k}(1 - P(A_{i_r})).$$

5.5.8 Suppose X_1 and X_2 are random variables defined on $(\Omega, \mathbb{A}, P)$, where $\Omega = \{a, b, c\}$, $\mathbb{A} = \mathbb{P}(\Omega)$ and P is defined as $\forall\ \omega \in \Omega$, $P(\{\omega\}) = 1/3$.

$$X_1(\omega) = \begin{cases} 0, & \text{if} \quad \omega = a \\ 1, & \text{if} \quad \omega = b, c. \end{cases} \quad \& \ X_2(\omega) = \begin{cases} 2, & \text{if} \quad \omega = b \\ 3, & \text{if} \quad \omega = a, c. \end{cases}$$

(i) Examine whether X_1 and X_2 are independent random variables on $(\Omega, \mathbb{A}, P)$. (ii) Examine whether $5X_1^2 + 4$ and $3X_2^3 + e^{X_2}$ are independent random variables on $(\Omega, \mathbb{A}, P)$.

Solution: (i) To examine whether X_1 and X_2 are independent random variables, we examine whether the induced sigma fields are independent. Note that

$$X_1^{-1}(\mathbb{B}) = \{\Omega, \emptyset, \{a\}, \{b, c\}\} \ \ \& \ X_2^{-1}(\mathbb{B}) = \{\Omega, \emptyset, \{b\}, \{a, c\}\}.$$

Observe that $P(\{a\} \cap \{b\}) = P(\emptyset) = 0 \neq P(\{a\})P(\{b\})$. Hence, X_1 and X_2 are not independent.
(ii) Note that $5X_1^2 + 4$ and $3X_2^3 + e^{X_2}$ are Borel functions of X_1 and X_2 respectively. We have a result which states that if X_1 and X_2 are independent random variables, then any Borel function of X_1 is independent of any Borel function of X_2. However, if X_1 and X_2 are not independent random variables, then we cannot conclude that a Borel function of X_1 is not independent of a Borel function of X_2. Hence, to examine whether $5X_1^2 + 4$ and $3X_2^3 + e^{X_2}$ are independent random variables, we find their induced sigma fields and examine whether these are independent. Observe that $5X_1^2 + 4$ and $3X_2^3 + e^{X_2}$ are one-to-one functions of X_1 and X_2 respectively. Hence the sigma fields induced by $5X_1^2 + 4$ and $3X_2^3 + e^{X_2}$ are the same as those induced by X_1 and X_2 respectively. But from (i) we note that the sigma fields induced by X_1 and X_2 are not independent which implies that the sigma fields induced by $5X_1^2 + 4$ and $3X_2^3 + e^{X_2}$ are not independent. Thus, $5X_1^2 + 4$ and $3X_2^3 + e^{X_2}$ are not independent random variables.

5.5.9 Suppose $(\Omega, \mathbb{A}, P)$ is a probability space, where $\Omega = \{a, b, c, d\}$, $\mathbb{A} = \mathbb{P}(\Omega)$ and $P(\{\omega\}) = 1/4 \ \ \forall\ \omega \in \Omega$. Suppose X_1 and X_2 are random variables defined on $(\Omega, \mathbb{A}, P)$ as follows.

$$X_1(\omega) = \begin{cases} 10, & \text{if} \quad \omega = a, b \\ 20, & \text{if} \quad \omega = c, d. \end{cases} \quad \& \ X_2(\omega) = \begin{cases} 30, & \text{if} \quad \omega = a, c \\ 40, & \text{if} \quad \omega = b, d. \end{cases}$$

Examine whether e^{5X_1-7} and $\log(4X_2+9)$ are independent random variables.

Solution: Note that e^{5X_1-7} and $\log(4X_2+9)$ are Borel functions of X_1 and X_2 respectively. These are independent random variables, if X_1 and X_2 are independent random variables. To examine whether X_1 and X_2 are independent random variables, we examine whether the induced sigma fields are independent. Suppose $A=\{a,b\}$ and $B=\{a,c\}$. Note that

$$X_1^{-1}(\mathbb{B})=\{\Omega,\emptyset,A,A^c\}\ \ \&\ X_2^{-1}(\mathbb{B})=\{\Omega,\emptyset,B,B^c\}.$$

Observe that $P(A\cap B)=P(\{a\})=1/4=P(A)P(B)$. Hence, A and B are independent. As a consequence, $X_1^{-1}(\mathbb{B})$ and $X_2^{-1}(\mathbb{B})$ are independent. Hence, X_1 and X_2 are independent random variables which further implies that e^{5X_1-7} and $\log(4X_2+9)$ are independent random variables.

5.5.10 Suppose $(\Omega,\mathbb{A},P)$ is a probability space and $\{A_1,A_2,A_3\}$ is a measurable partition of Ω. Suppose $B\in\mathbb{A}$. Show that a random variable X defined as $X=\sum_{i=1}^{3} iI_{A_i}$ and I_B are independent if and only if B is independent of each of $\{A_1,A_2,A_3\}$.

Solution: The sigma field induced by I_B is given by $I_B^{-1}(\mathbb{B})=\{\Omega,\emptyset,B,B^c\}$ and the sigma field induced by X is given by

$$X^{-1}(\mathbb{B})=\{\Omega,\emptyset,A_1,A_2,A_3,A_1+A_2,A_1+A_3,A_2+A_3\}.$$

Suppose B is independent of each of $\{A_1,A_2,A_3\}$. Then B is independent of $A_3^c=A_1+A_2, A_2^c=A_1+A_3, A_1^c=A_2+A_3$. Any event is always independent of Ω and $\emptyset$. Similarly, B^c is independent of all sets in $X^{-1}(\mathbb{B})$. Hence, X and I_B are independent random variables. Conversely, suppose X and I_B are independent random variables which implies $X^{-1}(\mathbb{B})$ and $I_B^{-1}(\mathbb{B})$ are independent sigma fields. Thus, every event in $X^{-1}(\mathbb{B})$ is independent of every event in $I_B^{-1}(\mathbb{B})$ and hence B is independent of each of $\{A_1,A_2,A_3\}$.

5.5.11 Suppose $(\Omega,\mathbb{A},P)$ is a probability space and $A,B,C\in\mathbb{A}$ are independent events. Examine whether C is independent of an event obtained from A and B by complementation, union and intersection.

Solution: Using the arguments as in the proof of Theorem 5.2.1, independence of A,B,C implies independence of $A^c,B,C,\ A,B^c,C,$ $A,B,C^c,\ A^c,B^c,C,\ A^c,B,C^c,\ A,B^c,C^c$ and A^c,B^c,C^c. To examine whether, $A\cup B$ is independent of C, observe that

$$\begin{aligned}P((A\cup B)\cap C) &= P((A\cap C)\cup(B\cap C))\\ &= P(A\cap C)+P(B\cap C))-P(A\cap B\cap C)\\ &= P(A)P(C)+P(B)P(C)-P(A)P(B)P(C)\\ &= P(A\cup B)P(C).\end{aligned}$$

Thus, $A \cup B$ is independent of C. Hence, $(A \cup B)^c = A^c \cap B^c$ is independent of C and C^c. Further, $A \cup B$ is independent of C^c. Since A^c, B^c, C are independent, $A^c \cup B^c$ is independent of C. Hence, $(A^c \cup B^c)^c = A \cap B$ is independent of C. It also follows from the definition of independence of A, B, C, since independence of A, B, C implies that all possible pairs are independent and $P(A \cap B \cap C) = P(A)P(B)P(C)$. On similar lines, since A, B^c, C are independent, $A \cup B^c$ is independent of C. Hence, $(A \cup B^c)^c = A^c \cap B$ is independent of C and C^c. Using the same arguments, we can prove that if $A, B, C \in \mathbb{A}$ are independent events, then C is independent of $A^c \cup B, A^c \cap B, A \cap B^c$.

5.5.12 Suppose $(\Omega, \mathbb{A}, P)$ is a probability space and $A, B, C \in \mathbb{A}$ are independent events. Can we claim that $I_A + 3I_B$ and I_C are independent random variables? Justify your answer.

Solution: Since $A, B, C \in \mathbb{A}$ are independent events, I_A, I_B, I_C are independent random variables. Suppose $\underline{Z} = (I_A, I_B)'$. Then the sigma field $\underline{Z}^{-1}(\mathbb{B})$ induced by $\underline{Z}$ is defined as $\underline{Z}^{-1}(\mathbb{B}) = \sigma(I_A^{-1}(\mathbb{B}) \cup I_B^{-1}(\mathbb{B}))$. The sigma fields induced by I_A, I_B, I_C and $\underline{Z}$ are respectively given by,

$$\begin{aligned} I_A^{-1}(\mathbb{B}) &= \{\Omega, \emptyset, A, A^c\}, \quad I_B^{-1}(\mathbb{B}) = \{\Omega, \emptyset, B, B^c\} \\ I_C^{-1}(\mathbb{B}) &= \{\Omega, \emptyset, C, C^c\} \\ \underline{Z}^{-1}(\mathbb{B}) &= \{\Omega, \emptyset, A, A^c, B, B^c, A \cup B, A \cup B^c, \\ &\quad A^c \cup B, A^c \cup B^c, A^c \cap B^c, A^c \cap B, A \cap B^c, A \cap B\}. \end{aligned}$$

To examine whether $\underline{Z} = (I_A, I_B)'$ and I_C are independent, we examine whether $\underline{Z}^{-1}(\mathbb{B})$ and $I_C^{-1}(\mathbb{B})$ are independent. It is known that Ω and $\emptyset$ are independent of any event in $\mathbb{A}$. Further, as shown in the previous exercise, C is independent of all other events in $\underline{Z}^{-1}(\mathbb{B})$. Hence, $\underline{Z} = (I_A, I_B)'$ and I_C are independent. Further, $I_A + 3I_B$ is a Borel function of $(I_A, I_B)'$. Hence by Theorem 5.3.9, $I_A + 3I_B$ and I_C are independent random variables.

5.5.13 Suppose $(\Omega, \mathbb{A}, P)$ is a probability space. Suppose $A, C \in \mathbb{A}$ are independent events and $B, C \in \mathbb{A}$ are independent events. Can we claim that $I_A + I_B$ and I_C are independent random variables? Justify your answer.

Solution: If $A, C \in \mathbb{A}$ are independent events, $B, C \in \mathbb{A}$ are independent events, then $I_A + I_B$ and I_C may or may not be independent random variables. We consider the following probability space $(\Omega, \mathbb{A}, P)$, where $\Omega = \{a, b, c, d\}$, $\mathbb{A} = \mathbb{P}(\Omega)$ and $\forall\ \omega \in \Omega,\ P(\{\omega\}) = 1/4$. Suppose events A, B, C are defined as follows.

$$A = \{a, b\}, \quad B = \{a, c\} \ \ \& \ \ C = \{b, c\}.$$

Observe that

$$(I_A + I_B)(\omega) = \begin{cases} 0, & \text{if} \quad \omega \in A^c \cap B^c = (A \cup B)^c \\ 1, & \text{if} \quad \omega \in (A \cap B^c) \cup (A^c \cap B) \\ 2, & \text{if} \quad \omega \in A \cap B. \end{cases}$$

The sigma fields induced by I_A, I_B, I_C and I_A+I_B are respectively given by,

$$\begin{aligned} I_A^{-1}(\mathbb{B}) &= \{\Omega, \emptyset, A, A^c\}, \quad I_B^{-1}(\mathbb{B}) = \{\Omega, \emptyset, B, B^c\} \\ I_C^{-1}(\mathbb{B}) &= \{\Omega, \emptyset, C, C^c\} \\ (I_A + I_B)^{-1}(\mathbb{B}) &= \{\Omega, \emptyset, (A \cup B)^c, (A \cap B^c) \cup (A^c \cap B), A \cap B, \\ &\quad A \cup B, A^c \cup B^c, (A \cup B)^c \cup (A \cap B)\}. \end{aligned}$$

Note that I_A, I_B, I_C are independent of each other. However,

$$P(A \cap B \cap C) = P(\emptyset) = 0 \neq P(A \cap B)P(C) = (1/4)(1/2)$$

and hence, $(I_A + I_B)^{-1}(\mathbb{B})$ is not independent of $I_C^{-1}(\mathbb{B})$. Thus, $I_A + I_B$ and I_C are not independent random variables. Suppose $\underline{Z}_1 = (I_A, I_B)'$ and $Z_2 = I_C$ are random vectors. Note that $I_A + I_B$ is a Borel function of $\underline{Z}_1$ and it is not independent of I_C. Hence, $\underline{Z}_1$ is also not independent of I_C. Thus, components of $\underline{Z}_1$ are independent of I_C but the random vector $\underline{Z}_1$ is not independent of I_C.

5.5.14 Give two examples in which a random variable is independent of itself.

Solution: Suppose $X \equiv C$ is a degenerate random variable on $(\Omega, \mathbb{A}, P)$. Then the sigma field induced by X is $\{\Omega, \emptyset\}$. Hence, X is independent of itself. Suppose $Y = I_A$ is defined on $(\Omega, \mathbb{A}, P)$. Then the sigma field induced by Y is $\sigma(Y) = \{\Omega, \emptyset, A, A^c\}$. If $P(A) = 0$ then every event in $\sigma(Y)$ is independent of itself and hence Y is independent of itself. Note that if $P(A) = 0$ then $P[Y = 0] = 1$, that is, Y is a degenerate random variable.

5.5.15 Give an example of random variables X_1, X_2, X_3 which are pairwise independent but not independent.

Solution: Random variables X_1, X_2, X_3 are pairwise independent, if and only if all possible pairs are independent. Suppose a probability space $(\Omega, \mathbb{A}, P)$ is defined as follows. $\Omega = \{(1,0,0), (0,1,0), (0,0,1), (1,1,1)$, $\mathbb{A} = \mathbb{P}(\Omega)$ and $P(\{\omega\}) = 1/4 \ \forall \ \omega \in \Omega$. Suppose an event A_i is defined as $A_i = \{\omega | i\text{-th coordinate is } 1\}$, $i = 1,2,3$. Then it is clear that for $i = 1,2,3, P(A_i) = 1/2, P(A_i \cap A_j) = P(A_i)P(A_j) \ \forall \ \ i \neq j = 1,2,3$ and $P(A_1 \cap A_2 \cap A_3) = 1/4$. Observe that $P(A_1 \cap A_2 \cap A_3) \neq P(A_1)P(A_2)P(A_3)$. Thus, $\{A_1, A_2, A_3\}$ is not a collection of independent events. Suppose a random variable X_i is defined as $X_i = 1$ if i-th coordinate is 1 and 0 otherwise, that is, $X_i = I_{A_i}$ for $i = 1,2,3$. Then for $i = 1,2,3, \sigma(X_i) = \{\Omega, \emptyset, A_i, A_i^c\}$. Observe that $\forall \ \ i \neq j = 1,2,3, \sigma(X_i)$ is independent of $\sigma(X_i)$. However, for $A_i \in \sigma(X_i)$, $P(A_1 \cap A_2 \cap A_3) \neq P(A_1)P(A_2)P(A_3)$. Thus, X_1, X_2, X_3 are pair-wise independent but not independent.

5.5.16 Give an example of random variables X_1, X_2, X_3, X_4 which are independent.

Solution: Suppose $(\Omega, \mathbb{A}, P)$ is a probability space and $A_i \in \mathbb{A}$, $i = 1, 2, 3, 4$. If $\{A_1, A_2, A_3, A_4\}$ is a collection of independent events, then $X_i = I_{A_i}, i = 1, 2, 3, 4$ are independent random variables. As an another illustration, suppose $X_i = C_i, i = 1, 2, 3, 4$ are degenerate random variables and hence are independent random variables.

5.5.17 Prove or disprove: If F is a distribution function, then (i) F^4 is also a distribution function, (ii) $2F - F^2$ is also a distribution function.

Solution: Suppose X_1, X_2, X_3, X_4 are independent random variables, each having the same distribution function F.
(i) Suppose $U = \max\{X_1, X_2, X_3, X_4\}$. Then for $x \in \mathbb{R}$,

$$P[U \leq x] = P[\max\{X_1, X_2, X_3, X_4\} \leq x]] = \prod_{i=1}^{4} P[X_i \leq x] = (F(x))^4.$$

Hence, F^4 is a distribution function of U.
(ii) Suppose $V = \min\{X_1, X_2\}$. Then for $x \in \mathbb{R}$,

$$\begin{aligned} P[V \geq x] &= P[\min\{X_1, X_2\} \geq x] = \prod_{i=1}^{2} P[X_i \geq x] = (1 - F(x))^2 \\ \Rightarrow \quad P[V \leq x] &= 1 - (1 - F(x))^2 = 2F - F^2. \end{aligned}$$

Hence, $2F - F^2$ is a distribution function of V.

5.5.18 Suppose $\{X_n, n \geq 1\}$ is a sequence of independent random variables and $\{f_n, n \geq 1\}$ is a sequence of Borel functions. Show that $\{f_n(X_n), n \geq 1\}$ is a sequence of independent random variables.

Solution: Since $\{X_n, n \geq 1\}$ is a sequence of independent random variables, every finite collection $\{X_{i_1}, X_{i_2}, \cdots, X_{i_k}\}$ of $k \geq 2$ random variables is a collection of independent random variables. Further, f_n is a Borel function implies that every finite collection $\{f_{i_1}(X_{i_1}), f_{i_2}(X_{i_2}), \cdots, f_{i_k}(X_{i_k})\}$ of $k \geq 2$ random variables is a collection of independent random variables. Hence, $\{f_n(X_n), n \geq 1\}$ is a sequence of independent random variables.

5.5.19 If X and Y are independent and X is integrable, show that $\int_{[Y \in B]} X \, dP = E(X)P[Y \in B]$ for any Borel set B.

Solution: Since X and Y are independent, $I_{[Y \in B]} = g(Y)$ and X are independent for any Borel set B, g being a Borel function. Observe that

$$\begin{aligned} \int_{[Y \in B]} X \, dP &= \int_{\Omega} X \, I_{[Y \in B]} dP = E(X \, I_{[Y \in B]}) \\ &= E(X)E(I_{[Y \in B]}) = E(X)P[Y \in B]. \end{aligned}$$

5.5.20 Give an example of two random variables X and Y such that X and Y are not independent but the characteristic function of their sum is the product of their characteristic functions.

Solution: Suppose X and Y are identical random variables, each having standard Cauchy distribution. Suppose X and Y are not independent of each other. Note that $\phi_X(t) = \phi_Y(t) = e^{-|t|}$, $t \in \mathbb{R}$. Now

$$\phi_{X+Y}(t) = \phi_{2X}(t) = \phi_X(2t) = e^{-|2t|} = e^{-|t|}e^{-|t|} = \phi_X(t)\phi_Y(t).$$

5.5.21 Suppose X and Y are independent random variables. Show that if X and $X - Y$ are independent, then X must be degenerate.

Solution: Since X and Y are independent, $-X$ and Y are also independent. Hence, $\forall\ t \in \mathbb{R}$,

$$\begin{aligned} \phi_{Y-X}(t) &= \phi_Y(t)\phi_{-X}(t) = \phi_Y(t)\phi_X(-t) \\ \Rightarrow\ \phi_Y(t) &= \phi_{X+(Y-X)}(t) = \phi_X(t)\phi_{Y-X}(t) = \phi_X(t)\phi_Y(t)\phi_X(-t), \end{aligned}$$

the second step follows since X and $X - Y$ are independent. Since $\phi_Y(0) = 1$ and $\phi_Y(t)$ is continuous function, $\phi_Y(t) \neq 0$ in a neighborhood $N_\delta(0)$ of 0. Hence,

$$\begin{aligned} \phi_X(t)\phi_X(-t) &= |\phi_X(t)|^2 = 1\ \forall\ t \in N_\delta(0) \\ \Rightarrow\ (E(\cos(tX))^2 + (E(\sin(tX))^2 &= 1\ \forall\ t \in N_\delta(0) \\ \Rightarrow\ X &= 0 \text{ a.s.} \end{aligned}$$

Thus, X is a degenerate random variable, degenerate at 0.

Answers to the multiple choice questions, based on Chapter 5, are given in Table 11.5.

TABLE 11.5
Answer Key to MCQs in Chapter 5

Q.No.	1	2	3	4	5	6	7	8	9	10
Ans	c,d	d	a	b,c,d	a	a	d	a	b	c
Q.No.	11	12	13	14	15	16	17	18	19	20
Ans	b	c	b	c,d	b					

11.6 Chapter 6

6.5.1 Suppose $\{X_n, n \geq 1\}$ is a sequence of random variables such that $P[X_n = \pm n] = 1/2$. Examine if $X_n \xrightarrow{a.s.} 0$.

Solution: Observe that $P[|X_n| = n] = 1$, hence for $\epsilon < n$,

$$P[|X_n| < \epsilon] = 0 \;\Rightarrow\; X_n \overset{P}{\nrightarrow} X \equiv 0 \;\Rightarrow\; X_n \overset{a.s.}{\nrightarrow} X \equiv 0.$$

6.5.2 Suppose $\{X_n, n \geq 1\}$ is a sequence of random variables defined on the probability space $(\Omega, \mathbb{A}, P)$ where $\Omega = [0, 1]$, $\mathbb{A}$ is a sigma field of subsets of $[0, 1]$ and $P((a, b)) = b - a$ for $(a, b) \in \mathbb{A}$. Suppose $\{X_n\}$ is defined as

$$X_n(\omega) = \begin{cases} 1, & \text{if } 0 \leq \omega < (n+1)/2n \\ 0, & \text{otherwise.} \end{cases}$$

X is a random variable defined as $X(\omega) = 1$ if $0 \leq \omega \leq 1/2$ and 0 otherwise. Examine whether $X_n \overset{a.s.}{\rightarrow} X$.

Solution: Note that X and X_n can be expressed as $X = I_A$ where $A = [0, 1/2]$ and $X_n = I_{A_n}$ where $A_n = [0, 1/2 + 1/2n)$. It is clear that $\{A_n, n \geq 1\}$ is a decreasing sequence of sets and hence is convergent with $\lim_{n\to\infty} A_n = \bigcap_{n\geq 1} A_n = [0, 1/2] = A$. Hence by Theorem 2.3.9

$$I_{A_n} \to I_A \text{ on } \Omega \iff X_n \to X \text{ on } \Omega \;\Rightarrow\; X_n \overset{a.s.}{\rightarrow} X.$$

6.5.3 Suppose $\{X_n, n \geq 1\}$ is a sequence of continuous random variables with support $[-1/n, 1/n]$ and with probability density function given by,

$$f_{X_n}(x) = \begin{cases} n/2, & \text{if } x \in [-1/n, 1/n] \\ 0, & \text{if } x \notin [-1/n, 1/n]. \end{cases}$$

Examine whether the sequence $\{X_n, n \geq 1\}$ converges almost surely.

Solution: From the given probability density function, it follows that $X_n \sim U(-1/n, 1/n)$. Hence,

$$\begin{aligned} E(X_n) &= 0 \;\&\; E(X_n^2) = 1/3n^2 \;\Rightarrow\; \sum_{n=1}^{\infty} E(X_n^2) = \sum_{n=1}^{\infty} 1/3n^2 < \infty \\ &\Rightarrow\; X_n \overset{a.s.}{\rightarrow} 0, \text{ by Theorem 6.4.3.} \end{aligned}$$

6.5.4 Suppose $X_n \sim G(n, n)$. Examine if $X_n \overset{a.s.}{\rightarrow} 1$.

Solution: Since $X_n \sim G(n, n)$, $E(X_n) = 1$ and $Var(X_n) = 1/n$. To examine whether, $X_n \overset{a.s.}{\rightarrow} 1$, by using Theorem 6.4.3, we obtain the fourth central moment of X_n from its cumulant generating function

$C_n(t)$. It is given by,

$$\begin{aligned} C_n(t) &= \log(1-t/n)^{-n} = -n\log(1-t/n) = n\sum_{i\geq 1} t^i/in^i \\ &= \sum_{i\geq 1}(t^i/i!)((i-1)!/n^{i-1}) \;\Rightarrow\; k_2 = 1/n \;\&\; k_4 = 6/n^3 \\ &\Rightarrow\; E(X_n - E(X_n))^4 = E(X_n-1)^4 = k_4 + 3k_2^2 = 6/n^3 + 3/n^2 \\ &\Rightarrow\; \sum_{n\geq 1} E(X_n-1)^4 = \sum_{n\geq 1}(6/n^3+3/n^2) < \infty \;\Rightarrow\; X_n \stackrel{a.s.}{\to} 1. \end{aligned}$$

6.5.5 Suppose $X_n \stackrel{a.s.}{\to} X$. Show that $\liminf X_n$ and $\limsup X_n$ are equivalent random variables.

Solution: By the definition of almost sure convergence,

$$\begin{aligned} X_n \stackrel{a.s.}{\to} X \quad &\Rightarrow\quad P[\lim_{n\to\infty} X_n = X] = 1 \\ &\Rightarrow\quad P[\limsup_{n\to\infty} X_n = \liminf_{n\to\infty} X_n = X] = 1 \\ &\Rightarrow\quad P[\liminf X_n = \limsup X_n] = 1\ . \end{aligned}$$

Hence, $\liminf X_n$ and $\limsup X_n$ are equivalent random variables.

6.5.6 Suppose $(\Omega, \mathbb{A}, P)$ is a probability space, where $\Omega = [0,1]$, $\mathbb{A}$ is a sigma field of subsets of Ω and P is a Lebesgue measure on $(\Omega, \mathbb{A})$. Suppose $A_n = [1/n, 1]$ and $X_n = I_{A_n}, n \geq 1$. Examine whether X_n converges point-wise, almost surely and in probability to 1. Examine whether $\{X_n, n \geq 1\}$ is a monotone sequence of random variables.

Solution: We have $A_n = [1/n, 1]$. It is clear that $\{A_n, n \geq 1\}$ is an increasing sequence of sets and hence is convergent with $\lim_{n\to\infty} A_n = \bigcup_{n\geq 1} A_n = A = (0,1]$ and $P(A) = 1$. Suppose for $n \geq 1$, $X_n = I_{A_n}$ and $X = I_A$, then $X \equiv 1$. Hence by Theorem 2.3.9

$$I_{A_n} \to I_A \;\text{ on }\; \Omega \quad\Longleftrightarrow\quad X_n \to X \equiv 1 \;\text{ on }\; \Omega.$$

Thus, X_n converges to 1 point-wise, hence almost surely and hence in probability. Observe that $\forall\ \epsilon \geq 1$, $P[|X_n - 1| > \epsilon] = 0$. Further $\forall\ \epsilon \in (0,1)$,

$$P[|X_n - 1| > \epsilon] = P[I_{A_n} = 0] = P(A_n^c) = 1/n \to 0 \quad \Rightarrow X_n \stackrel{P}{\to} X \equiv 1\ .$$

It is to be noted that $\{X_n, n \geq 1\}$ is a monotone sequence of random variables. Thus, convergence in probability and almost sure convergence are equivalent.

6.5.7 Suppose $\{X_n, n \geq 1\}$ is a sequence of random variables. If $X_n \stackrel{a.s.}{\to} 0$, show that $\overline{X}_n \stackrel{a.s.}{\to} 0$.

Solution: Suppose $\{a_n, n \geq 1\}$ is a sequence of real numbers and $\{b_n, n \geq 1\}$ is a sequence of positive real numbers. Toeplitz' lemma states that if $a_n \rightarrow a$ and $\sum_{n=1}^{N} b_n \rightarrow \infty$ as $N \rightarrow \infty$, then $\sum_{n=1}^{N} a_n b_n / \sum_{n=1}^{N} b_n \rightarrow a$ as $N \rightarrow \infty$. Now, $X_n \stackrel{a.s.}{\rightarrow} 0$ implies $\forall\, \omega \in N^c$ with $P(N) = 0$, $X_n(\omega) \rightarrow 0$. Hence by Toeplitz' lemma with $b_n = 1$, $\forall\, \omega \in N^c$, $\sum_{n=1}^{N} X_n(\omega)/N \rightarrow 0$ and hence $\overline{X}_N \stackrel{a.s.}{\rightarrow} 0$ as $N \rightarrow \infty$.

6.5.8 Suppose $(\Omega, \mathbb{A}, P)$ is a probability space where $\Omega = \{1, 2, \cdots, \}$ and probability measure P assigns probability $1/(k(k+1))$ to a set $\{k\}, k = 1, 2, \cdots$. (i) Define $\mathbb{A}$ appropriately so that sets $\{k\}$, $k = 1, 2, \cdots$, are in $\mathbb{A}$. If $A_n = \{n, n+1, \cdots\}$, show that it is in $\mathbb{A}$. (ii) Find $P(A_n)$ and $\lim_{n\to\infty} P(A_n)$. (iii) Show that $\{A_n, n \geq 1\}$ is a convergent sequence and find its limit A and $P(A)$. (iv) Verify whether converse of the Borel-Cantelli lemma is true for this sequence of sets. If not, why? Justify your answer.

Solution: (i) With $\mathbb{A} = \mathbb{P}(\Omega)$, sets $\{k\}$, $k = 1, 2, \cdots$, are in $\mathbb{A}$. Since $A_n = \{n, n+1, \cdots\} = \bigcup_{k \geq n}\{k\}$, it is in $\mathbb{A}$.
(ii) Note that

$$P(A_n) = \sum_{k \geq n} P(\{k\}) = \sum_{k \geq n} \frac{1}{k(k+1)} = \sum_{k \geq n} \left[\frac{1}{k} - \frac{1}{k+1}\right] = \frac{1}{n} \rightarrow 0.$$

(iii) Observe that $A_{n+1} \subset A_n, \quad \forall \quad n \geq 1$. Thus $\{A_n, n \geq 1\}$ is a decreasing sequence of events and hence is convergent. Further,

$$\begin{aligned} A_n \quad &\rightarrow \quad A = \bigcap_{n \geq 1} A_n = \emptyset \\ &\Rightarrow \quad P(A) = 0 \quad \Rightarrow \quad P(\lim_{n\to\infty} A_n) = P(\limsup_{n\to\infty} A_n) = 0 \end{aligned}$$

$$\text{But} \quad \sum_{n \geq 1} P(A_n) \quad = \quad \sum_{n \geq 1} 1/n = \infty.$$

Hence, the converse of Borel-Cantelli lemma is not true for this sequence of sets. It is to be noted that

$$P(A_n \cap A_{n+1}) = P(A_{n+1}) = \frac{1}{n+1} \neq \frac{1}{n}\frac{1}{n+1} = P(A_n)P(A_{n+1}).$$

Thus, $\{A_n, n \geq 1\}$ is a sequence of dependent events and hence the converse of the Borel-Cantelli lemma is not true.

6.5.9 If $\{A_n, n \geq 1\}$ is a sequence of independent events converging to A, prove that A is independent of itself.

Solution: It is given that $A_n \rightarrow A$, hence $\liminf A_n = \limsup A_n = A$ which implies that $P(\limsup A_n) = P(A)$. Further, $\{A_n, n \geq 1\}$ is a sequence of independent events, hence by the Borel zero-one law, $P(\limsup A_n) = P(A) = 0$ or 1, depending on convergence or the divergence of the series $\sum_{n\geq 1} P(A_n)$. With $P(A) = 0$ or 1, the identity $P(A \cap A) = P(A)P(A)$ is satisfied. Hence, A is independent of itself.

6.5.10 Prove or disprove: If $\sum_{n\geq 1} P(A_n) < \infty$ then $P(\liminf A_n) = 0$.

Solution: If $\sum_{n\geq 1} P(A_n) < \infty$ then $P(\limsup A_n) = 0$ by the Borel-Cantelli lemma. Now,

$$0 \leq P(\liminf A_n) \leq P(\limsup A_n) = 0 \Rightarrow P(\liminf A_n) = 0 .$$

Alternatively, by the definition of a limit infimum of a sequence of sets we have,

$$\liminf A_n = \bigcup_{n\geq 1}\bigcap_{k\geq n} A_k = \bigcap_{n\geq 1} B_n, \text{ where } B_n = \bigcap_{k\geq n} A_k = A_n \bigcap B_{n+1}$$

$$\Rightarrow B_n \subset B_{n+1} \Rightarrow \lim_{n\to\infty} B_n = \bigcup_{n=1}^{\infty} B_n,$$

$\{B_n, n \geq 1\}$ being a non-decreasing sequence of events in $\mathbb{A}$. Now using the result that a tail of the convergent series converges to 0 we have,

$$\begin{aligned} P(\liminf A_n) &= P(\bigcup_{n=1}^{\infty}\bigcap_{k\geq n} A_k) = P(\bigcup_{n=1}^{\infty} B_n) = P(\lim_{n\to\infty} B_n) = \lim_{n\to\infty} P(B_n) \\ &= \lim_{n\to\infty} P(\bigcap_{k\geq n} A_k) \leq \lim_{n\to\infty} P(\bigcup_{k\geq n} A_k) \\ &\leq \lim_{n\to\infty} \sum_{k\geq n} P(A_k) = 0, \end{aligned}$$

$\sum_{k\geq n} P(A_k)$ being a tail of the convergent series $\sum_{n=1}^{\infty} P(A_n)$. Thus, $\sum_{n=1}^{\infty} P(A_n) < \infty \Rightarrow P(\liminf A_n) = 0$.

6.5.11 Suppose $(\Omega, \mathbb{A}, P)$ is a probability space and $\{A_n, n \geq 1\}$ is a sequence of independent events from $\mathbb{A}$ such that $P(A_n) < 1 \ \forall \ n \geq 1$. Show that (i) $P\Big(\bigcup_{i=1}^n A_i\Big) = 1 - \prod_{i=1}^n (1 - P(A_i))$, (ii) $P\Big(\bigcup_{i=1}^n A_i\Big) \geq 1 - e^{-\sum_{i=1}^n P(A_i)}$,

(iii) $\sum_{n\geq 1} P(A_n) = \infty \Rightarrow P\Big(\bigcup_{n\geq 1} A_n\Big) = 1$ and

(iv) $P(\limsup A_n) = 1 \iff P\Big(\bigcup_{n\geq 1} A_n\Big) = 1$.

Solution: Note that $\{A_n, n \geq 1\}$ is a sequence of independent events implies $\{A_n^c, n \geq 1\}$ is also a sequence of independent events. (i) Hence, noting that $P(A_n) < 1 \ \forall \ n \geq 1$ we have

$$P\Big(\bigcup_{i=1}^n A_i\Big) = 1 - P\Big(\bigcap_{i=1}^n A_i^c\Big) = 1 - \prod_{i=1}^n P(A_i^c) = 1 - \prod_{i=1}^n (1 - P(A_i)).$$

(ii) To prove (ii) we use the result which states that $\forall \ x \in [0, 1]$, $1 - x \leq e^{-x}$. From (i)

$$P\Big(\bigcup_{i=1}^n A_i\Big) = 1 - \prod_{i=1}^n (1 - P(A_i)) \geq 1 - \prod_{i=1}^n e^{-P(A_i)} = 1 - e^{-\sum_{i=1}^n P(A_i)}.$$

(iii) Observe that

$$C_{n+1} = \bigcup_{i=1}^{n+1} A_i = \Big(\bigcup_{i=1}^{n} A_i\Big) \cup A_{n+1} = C_n \cup A_{n+1} \;\Rightarrow \{C_n\} \text{ is } \uparrow \text{ sequence}$$
$$\Rightarrow \lim_{n\to\infty} C_n = \bigcup_{n\geq 1} C_n = \bigcup_{n\geq 1} A_n.$$

Hence,

$$\begin{aligned} P\Big(\bigcup_{n\geq 1} A_n\Big) &= P\Big(\lim_{n\to\infty}\Big(\bigcup_{i=1}^{n} A_i\Big)\Big) = \lim_{n\to\infty} P\Big(\bigcup_{i=1}^{n} A_i\Big) \\ &\geq \lim_{n\to\infty}(1-\exp(-\sum_{i=1}^{n} P(A_i))) \\ &= 1-\exp(-\lim_{n\to\infty}\sum_{i=1}^{n} P(A_i)) = 1-\exp(-\sum_{n\geq 1} P(A_n)) \\ &= 1-\exp(-\infty) = 1 \;\Rightarrow\; P\Big(\bigcup_{n\geq 1} A_n\Big) = 1. \end{aligned}$$

(iv) Since $\{A_n, n \geq 1\}$ is a sequence of independent events, by Borel zero-one law, $P(\limsup A_n) = 1 \iff \sum_{n\geq 1} P(A_n) = \infty$. Thus, by (iii)

$$P(\limsup A_n) = 1 \;\Rightarrow\; \sum_{n\geq 1} P(A_n) = \infty \;\Rightarrow\; P\Big(\bigcup_{n\geq 1} A_n\Big) = 1.$$

In (iii) we have proved that $P\Big(\bigcup_{n\geq 1} A_n\Big) \geq 1 - \exp(-\sum_{n\geq 1} P(A_n))$. Observe that

$$\begin{aligned} 1 &= P\Big(\bigcup_{n\geq 1} A_n\Big) \geq 1-\exp(-\sum_{n\geq 1} P(A_n)) \\ &\Rightarrow \exp(-\sum_{n\geq 1} P(A_n)) \leq 0 \;\Rightarrow\; \exp(-\sum_{n\geq 1} P(A_n)) = 0 \\ &\Rightarrow \sum_{n\geq 1} P(A_n) = \infty \;\Rightarrow\; P(\limsup A_n) = 1. \end{aligned}$$

6.5.12 Suppose $(\Omega, \mathbb{A}, P)$ is a probability space and $\{A_n, n \geq 1\}$ is a sequence of independent events from $\mathbb{A}$. If $\sum_{n\geq 1} P(A \cap A_n) = \infty$ for any $A \in \mathbb{A}$ such that $P(A) > 0$, then show that $P(\limsup A_n) = 1$.

Solution: Note that if $P(A) = 0$, then $P(A \cap A_n) \leq P(A) = 0$ and

$\sum_{n\geq 1} P(A\cap A_n)$ cannot be ∞. Now for $\forall \ n \geq 1$,

$$\begin{aligned} A_n \supset A\cap A_n &\Rightarrow P(A_n) \geq P(A\cap A_n) \\ &\Rightarrow \sum_{n\geq 1} P(A_n) \geq \sum_{n\geq 1} P(A\cap A_n) = \infty \\ &\Rightarrow \sum_{n\geq 1} P(A_n) = \infty \ \Rightarrow \ P(\limsup A_n) = 1. \end{aligned}$$

Answers to the multiple choice questions, based on Chapter 6, are given in Table 11.6.

TABLE 11.6
Answer Key to MCQs in Chapter 6

Q.No.	1	2	3	4	5	6	7	8
Ans	a,c	a,c	b	d	c	c	a,b,cd	a,b,c,d

11.7 Chapter 7

7.5.1 Suppose $\{X_n, n \geq 1\}$ is a sequence of random variables such that $P[X_n = \pm n] = 1/2$. Using the definitions, examine if $X_n \xrightarrow{L} X$, $X_n \xrightarrow{P} 0$ and $X_n \xrightarrow{r} 0$. Verify the known implications among these modes of convergence. Examine if $X_n \xrightarrow{a.s.} X$.

Solution: The distribution function of X_n is given by,

$$F_n(x) = \begin{cases} 0, & \text{if} \quad x < -n \\ \frac{1}{2}, & \text{if} \quad -n \leq x < n \\ 1, & \text{if} \quad x \geq n. \end{cases}$$

Then $\forall \ x \in \mathbb{R}, \ \ F_n(x) \to 1/2 \equiv F(x)$, say. Thus, $F(x)$ satisfies all the properties of a distribution function except that the limits at $-\infty$ and ∞ are not 0 and 1 respectively. Hence, $X_n \overset{L}{\nrightarrow} X$ for any X, but $F_n \xrightarrow{w} F$. Observe that $P[|X_n| = n] = 1$, hence for $\epsilon < n, \ \ P[|X_n| < \epsilon] = 0$ and hence $X_n \overset{P}{\nrightarrow} X \equiv 0$, which further implies that $X_n \overset{a.s.}{\nrightarrow} X \equiv 0$. Similarly $E(|X_n|^r) = n^r \to \infty$ as $n \to \infty$. Hence, $X_n \overset{r}{\nrightarrow} 0$. Thus, we note that

$$X_n \overset{L}{\nrightarrow} 0 \ \Rightarrow \ X_n \overset{P}{\nrightarrow} 0 \ \Rightarrow \ X_n \overset{a.s.}{\nrightarrow} 0 \ \& \ X_n \overset{P}{\nrightarrow} 0 \ \Rightarrow \ X_n \overset{r}{\nrightarrow} 0.$$

7.5.2 Suppose $X_n \equiv 1/n$ and $Y_n \equiv -1/n$. Using the definitions examine whether X_n and Y_n converge in law, in probability, r-th mean and almost surely.

Solution: Suppose $X_n \equiv 1/n$. Then

$$F_{X_n}(x) = \begin{cases} 0, & \text{if} \quad x < 1/n \\ 1, & \text{if} \quad x \geq 1/n \end{cases} \quad \to \quad G(x) = \begin{cases} 0, & \text{if} \quad x \leq 0 \\ 1, & \text{if} \quad x > 0. \end{cases}$$

Thus, $F_{X_n} \to G(x) \ \forall \ x \in \mathbb{R}$, but G is not a distribution function since it is not right continuous at 0. However, we cannot conclude that X_n does not converge in law. Since F_{X_n} may converge to another distribution function, may not be for all $x \in \mathbb{R}$ but for all x in the set of its points of continuity, as is required in the definition. Suppose $X \equiv 0$. Then its distribution function F is given by,

$$F(x) \quad = \quad \begin{cases} 0, & \text{if} \quad x < 0 \\ 1, & \text{if} \quad x \geq 0. \end{cases}$$

Observe that $F_{X_n} \to F(x) \ \ \forall \ \ x \in C(F_X)$. Hence by the definition of convergence in law, $X_n \xrightarrow{L} X \equiv 0$. Since the limit random variable X is degenerate at 0, convergence in law implies convergence in probability. We now examine convergence in probability using the definition. Note that $|X_n| \equiv 1/n$. For given $\epsilon > 0$, $\exists \ n_0$ such that $1/n_0 < \epsilon$ and hence $1/n < \epsilon \ \ \forall \ \ n \geq n_0$. As a consequence

$$\forall \ \epsilon > 1/n, \ P[|X_n| < \epsilon] = 1 \Rightarrow \forall \ n \geq n_0 \ P[|X_n| < \epsilon] = 1 \ \Rightarrow \ X_n \xrightarrow{P} 0.$$

Further for any $r > 0$,

$$E(|X_n|^r) = 1/n^r \to 0 \quad \Rightarrow \quad X_n \xrightarrow{r} 0.$$

To examine almost sure convergence, observe that for a P-null set N

$$\begin{aligned} P[X_n = 1/n] = 1 \quad & \Rightarrow \quad X_n(\omega) = 1/n \ \ \forall \ \ \omega \in N^c = [X_n = 1/n] \\ & \Rightarrow \quad \lim_{n \to \infty} X_n(\omega) = 0 \ \ \forall \ \ \omega \in N^c \ \Rightarrow \ X_n \xrightarrow{a.s.} 0. \end{aligned}$$

Suppose $Y_n \equiv -1/n$. Then

$$F_{Y_n}(x) = \begin{cases} 0, & \text{if} \quad x < -1/n \\ 1, & \text{if} \quad x \geq -1/n \end{cases} \quad \to \quad F(x) = \begin{cases} 0, & \text{if} \quad x < 0 \\ 1, & \text{if} \quad x \geq 0. \end{cases}$$

Observe that $F_{Y_n}(x) \to F(x) \ \ \forall \ \ x \in \mathbb{R}$, where $F(x)$ is a distribution function of $X \equiv 0$. Thus, $F_{Y_n}(x) \to F(x) \ \forall \ x \in C(F_X)$ and hence $Y_n \xrightarrow{L} X \equiv 0$. Since the limit random variable X is degenerate at 0, convergence in law implies convergence in probability. Using the definition, it follows as in the case of $X_n \equiv 1/n$, since $|Y_n| \equiv 1/n$. Further for any $r > 0$,

$$E(|Y_n|^r) = (|-1/n|)^r \to 0 \quad \Rightarrow \quad Y_n \xrightarrow{r} 0.$$

To examine almost sure convergence, observe that for a P-null set N

$$\begin{aligned} P[Y_n = -1/n] = 1 \quad &\Rightarrow \quad Y_n(\omega) = -1/n \ \forall \ \omega \in N^c = [Y_n = -1/n] \\ &\Rightarrow \quad \lim_{n\to\infty} Y_n(\omega) = 0 \ \forall \ \omega \in N^c \ \Rightarrow \ Y_n \stackrel{a.s.}{\to} 0. \end{aligned}$$

It is to be noted that both the sequences converge almost surely, in r-th mean, in probability and in law.

7.5.3 Suppose $\{X_n, n \geq 1\}$ is a sequence of random variables such that $P[X_n = 0] = 1 - 1/n^2$ & $P[X_n = e^n] = 1/n^2$. Using the definitions, examine if $X_n \stackrel{L}{\to} 0$, $X_n \stackrel{P}{\to} 0$ and $X_n \stackrel{r}{\to} 0$. Using a sufficient condition, examine if $X_n \stackrel{a.s.}{\to} 0$. Hence, verify the known implications among these modes of convergence.

Solution: Observe that the distribution function $F_n(x)$ of X_n, given by,

$$F_n(x) = \begin{cases} 0, & \text{if} \quad x < 0 \\ 1 - 1/n^2, & \text{if} \quad 0 \leq x < e^n \\ 1, & \text{if} \quad x \geq e^n \end{cases} \quad \to \quad F(x) = \begin{cases} 0, & \text{if} \quad x < 0 \\ 1, & \text{if} \quad x \geq 0. \end{cases}$$

Note that $F(x)$ is a distribution function of $X \equiv 0$ and $F_n(x) \to F(x)$ for all $x \in \mathbb{R}$, where F is a distribution function of $X \equiv 0$. Thus, $X_n \stackrel{L}{\to} 0$. Since the limit law is degenerate, convergence in law implies convergence in probability. We can verify it independently also. Observe that for $\epsilon > 0$, $P[|X_n| > \epsilon] = 1/n^2 \to 0$ and hence $X_n \stackrel{P}{\to} 0$. Further,

$$\sum_{n\geq 1} P[|X_n| > \epsilon] = \sum_{n\geq 1} 1/n^2 < \infty \quad \Rightarrow \quad X_n \stackrel{a.s.}{\to} 0.$$

$E(|X_n|^r) = e^{nr}/n^2 \to \infty$ as $n \to \infty$. Hence, $X_n \stackrel{r}{\nrightarrow} 0$. It is to be noted that $X_n \stackrel{r}{\nrightarrow} 0$ but $X_n \stackrel{P}{\to} 0$. Further, $X_n \stackrel{a.s.}{\to} 0$ but not in r-th mean.

7.5.4 Suppose a probability mass function of a random variable X_n is given by,

$$P[X_n = \sqrt{n}] = 1/n = 1 - P[X_n = 0].$$

Examine whether $X_n \stackrel{P}{\to} 0$, $X_n \stackrel{L}{\to} 0$ and $X_n \stackrel{q.m.}{\to} 0$.

Solution: From the given probability distribution of X_n,

$$P[|X_n| > \epsilon] = P[|X_n| = \sqrt{n}] = 1/n \quad \to 0, \quad \text{as} \ \ n \to \infty, \ \ \forall \ \epsilon > 0.$$

Thus, $X_n \stackrel{P}{\to} 0$ and hence $X_n \stackrel{L}{\to} 0$. Now,

$$\forall \ n, \ \ E(|X_n - 0|^2) = E(X_n^2) = n(1/n) = 1 \ \ \Rightarrow \ \ X_n \stackrel{q.m.}{\nrightarrow} 0.$$

7.5.5 Suppose a probability mass function of a random variable X_n is given by,

$$P[X_n = 0] = 1/n^2 = 1 - P[X_n = n].$$

Examine whether $X_n \xrightarrow{P} 0$, $X_n \xrightarrow{L} 0$, $X_n \xrightarrow{a.s.} 0$ and $X_n \xrightarrow{q.m.} 0$.

Solution: From the given probability mass function of a random variable X_n, its distribution function $F_{X_n}(x)$ for any $x \in \mathbb{R}$ is given by,

$$F_{X_n}(x) = \begin{cases} 0, & \text{if} \quad x < 0 \\ 1/n^2, & \text{if} \quad 0 \le x < n \\ 1, & \text{if} \quad x \ge n. \end{cases}$$

Observe that

$$F_{X_n}(x) \to 0 \ \forall \ x \in \mathbb{R} \ \Rightarrow \ X_n \stackrel{L}{\nrightarrow} 0 \ \Rightarrow \ X_n \stackrel{P}{\nrightarrow} 0 \ \Rightarrow \ X_n \stackrel{q.m.}{\nrightarrow} 0.$$

Further, $X_n \stackrel{P}{\nrightarrow} 0 \ \Rightarrow \ X_n \stackrel{a.s.}{\nrightarrow} 0$. Note that $E(X_n^2) = n^2(1-1/n^2) \to \infty$ also implies that $X_n \stackrel{q.m.}{\nrightarrow} 0$.

7.5.6 Suppose $\{X_n, n \ge 1\}$ is a sequence of discrete random variables with probability mass function given by,

$$P[X_n = x] = \begin{cases} 1/n, & \text{if} \quad x = n \\ 1 - 1/n, & \text{if} \quad x = 0. \end{cases}$$

Discuss the limiting behaviour of $\{X_n, n \ge 1\}$ in probability, in law and in quadratic mean.

Solution: From the given probability mass function, $E(X_n^2) = n \to \infty$ hence $X_n \stackrel{q.m.}{\nrightarrow} 0$. Now, given $\epsilon > 0, \exists \ n_0 \in \mathbb{N}$ such that $\epsilon \le n_0$. Hence $P[|X_n| > \epsilon] = 0 \ \forall \ n \le n_0$ and $\forall \ n > n_0$, $P[|X_n| > \epsilon] = 1/n \to 0$ as $n \to \infty$. As a consequence, $X_n \xrightarrow{P} 0 \ \Rightarrow \ X_n \xrightarrow{L} 0$.

7.5.7 Suppose $\{X_n, n \ge 1\}$ is a sequence of random variables defined as $X_n = X + Y_n$ where X is a random variable and $\{Y_n, n \ge 1\}$ is a sequence of random variables such that $E(Y_n) = 1/n$ & $Var(Y_n) = \sigma^2/n$, where $\sigma > 0$ is a constant. Verify whether $X_n \xrightarrow{P} X$.

Solution: Observe that

$$\begin{aligned} E(X_n - X)^2 &= E(Y_n)^2 = Var(Y_n) + (E(Y_n^2))^2 = \sigma^2/n + 1/n^2 \to 0 \\ &\Rightarrow \ X_n \xrightarrow{q.m.} X \ \Rightarrow \ X_n \xrightarrow{P} X. \end{aligned}$$

7.5.8 Suppose $\{X_n, n \ge 1\}$ is a sequence of random variables defined on the probability space $(\Omega, \mathbb{A}, P)$ where $\Omega = [0,1]$, $\mathbb{A}$ is a sigma field of subsets of $[0,1]$ and $P((a,b)) = b - a$ for $(a,b) \in \mathbb{A}$. Suppose $\{X_n\}$ is defined as

$$X_n(\omega) = \begin{cases} 1, & \text{if} \quad 0 \le \omega < (n+1)/2n \\ 0, & \text{otherwise.} \end{cases}$$

X is a random variable defined as $X(\omega) = 1$ if $0 \le \omega \le 1/2$ and 0 otherwise. Examine whether $X_n \xrightarrow{a.s.} X$, $X_n \xrightarrow{P} X$, $X_n \xrightarrow{L} X$ and $X_n \xrightarrow{q.m.} X$.

Solution: Note that X and X_n can be expressed as $X = I_A$ where $A = [0, 1/2]$ and $X_n = I_{A_n}$ where $A_n = [0, 1/2 + 1/2n)$. It is clear that $\{A_n, n \ge 1\}$ is a decreasing sequence of sets and hence is convergent with $\lim_{n\to\infty} A_n = \bigcap_{n\ge1} A_n = [0, 1/2] = A$. Hence by Theorem 2.3.9

$$I_{A_n} = X_n \to I_A = X \text{ on } \Omega \;\Rightarrow\; X_n \xrightarrow{a.s.} X \Rightarrow X_n \xrightarrow{P} X \Rightarrow X_n \xrightarrow{L} X.$$

Further observe that $X_n - X = I_{B_n}$ where $B_n = [1/2, 1/2 + 1/2n)$. Hence,

$$E(X_n - X)^r = 1/2n \to 0 \;\Rightarrow\; X_n \xrightarrow{q.m.} X \;\Rightarrow\; X_n \xrightarrow{P} X.$$

7.5.9 Suppose $\{X_n, n \ge 1\}$ is a sequence of discrete random variables with probability mass function given by, $P[X_n = -(n+4)] = 1/(n+4)$, $P[X_n = -1] = 1 - 4/(n+4)$ and $P[X_n = n+4] = 3/(n+4)$. Discuss the limiting behaviour of $\{X_n\}$ in probability, in law and in quadratic mean to $X \equiv -1$ and $X \equiv 0$. Examine whether $E(X_n)$ converges.

Solution: Observe that for $\epsilon > 0$,

$$\begin{aligned} P[|X_n + 1| < \epsilon] &\ge P[X_n = -1] = 1 - 4/(n+4) \to 1 \\ &\Rightarrow X_n \xrightarrow{P} -1 \Rightarrow \; X_n \xrightarrow{L} -1 \\ &\Rightarrow X_n \overset{P}{\nrightarrow} 0 \Rightarrow \; X_n \overset{L}{\nrightarrow} 0. \\ \text{Now } \; E(X_n) &= 4/(n+4) + 1 \to 1. \\ E(X_n^2) &= 1 - 4/(n+4) + 4(n+4) \to \infty \;\Rightarrow\; X_n \overset{q.m.}{\nrightarrow} 0. \end{aligned}$$

Further $E((X_n + 1)^2) \to \infty \;\Rightarrow\; X_n \overset{q.m.}{\nrightarrow} -1$.

7.5.10 Suppose $\{X_n, n \ge 1\}$ is a sequence of discrete random variables with probability mass function given by, $P[X_n = \pm 1/n] = 1/2$. Discuss the limiting behaviour of $\{X_n\}$ in probability, in law, almost surely and in quadratic mean to X. Examine whether $E(X_n) \to E(X)$.

Solution: From the given probability mass function,

$$E|X_n|^r = 1/n^r \to 0 \;\Rightarrow\; X_n \xrightarrow{r} 0 \;\Rightarrow\; X_n \xrightarrow{P} 0 \;\Rightarrow\; X_n \xrightarrow{L} 0.$$

$$\text{Further, } \forall \; r \ge 2, \; \sum_{n\ge1} E|X_n|^r = \sum_{n\ge1} 1/n^r < \infty \;\Rightarrow\; X_n \xrightarrow{a.s.} 0.$$

Observe that $E(X_n) = 0 \;\forall\; n \ge 1$ and hence $E(X_n) \to 0 = E(X)$ where $X \equiv 0$.

7.5.11 Suppose $\{X_n, n \geq 1\}$ is a sequence of continuous random variables with support $[-1/n, 1/n]$ and with the probability density function given by,

$$f_{X_n}(x) = \begin{cases} n/2, & \text{if } x \in [-1/n, 1/n] \\ 0, & \text{if } x \notin [-1/n, 1/n]. \end{cases}$$

Examine whether the sequence $\{X_n, n \geq 1\}$ converges in quadratic mean, in probability, in law and almost surely.

Solution: From the given probability density function, it follows that $X_n \sim U(-1/n, 1/n)$. Hence,

$$\begin{aligned} E(X_n) &= 0 \;\; \& \;\; E(X_n^2) = 1/3n^2 \to 0 \\ &\Rightarrow \;\; X_n \stackrel{q.m.}{\to} 0 \;\; \Rightarrow \;\; X_n \stackrel{P}{\to} 0 \;\; \Rightarrow \;\; X_n \stackrel{L}{\to} 0. \\ \text{Further, } \sum_{n=1}^{\infty} E(X_n^2) &= \sum_{n=1}^{\infty} 1/3n^2 < \infty \;\Rightarrow\; X_n \stackrel{a.s.}{\to} 0, \end{aligned}$$

by Theorem 6.4.3.

7.5.12 Suppose $(\Omega, \mathbb{A}, P)$ is a probability space, where $\Omega = [0, 1]$, $\mathbb{A}$ is a sigma field of subsets of Ω and $P(A_n) = 1/n$ for $A_n = [0, 1/n)$. Suppose $X_n = e^n I_{A_n}, n \geq 1$. Show that $X_n \stackrel{P}{\to} 0$, but $X_n \stackrel{r}{\nrightarrow} 0$.

Solution: For any $\epsilon > 0$, as $n \to \infty$,

$$P[|X_n| > \epsilon] = P[I_{A_n} = 1] = 1/n \to 0 \;\; \Rightarrow \;\; X_n \stackrel{P}{\to} 0.$$

On the other hand, for any $r > 0$,

$$E(|X_n|^r) = e^{nr}/n \to \infty \;\; \text{as} \;\; n \to \infty \;\; \Rightarrow \;\; X_n \stackrel{r}{\nrightarrow} 0.$$

7.5.13 Suppose $\{X_1, X_2, \cdots, X_n\}$ are independent and identically distributed random variables with the probability density function $f(x) = \alpha x^{-\alpha-1}$ for $x > 1, \alpha > 0$. Suppose $Y_n = n^{-1/\alpha} X_{(n)}$. Examine whether Y_n converges in law.

Solution: To find the limit law of Y_n, observe that the distribution function of X_1 is

$$F(x) = \begin{cases} 0, & \text{if } x < 1 \\ 1 - x^{-\alpha}, & \text{if } x \geq 1. \end{cases}$$

Now, $X_{(n)} \geq 1 \;\; \Rightarrow \;\; Y_n \geq n^{-1/\alpha}$. Hence, the distribution function F_n of Y_n given by,

$$F_n(x) = \begin{cases} 0, & \text{if } x < n^{-1/\alpha} \\ (1 - n^{-1}x^{-\alpha})^n, & \text{if } x \geq n^{-1/\alpha} \end{cases}$$

Observe that

$$F_n(x) \;\to\; F(x) = \begin{cases} 0, & \text{if} \quad x < 0 \\ \exp(-x^{-\alpha}), & \text{if} \quad x \geq 0. \end{cases}$$

Note that F is non-decreasing and continuous. Further,

$$\lim_{x\to\infty} \exp(-x^{-\alpha}) = 1 \;\;\&\;\; \lim_{x\to-\infty} F(x) = 0 \;\;\Rightarrow\;\; Y_n \xrightarrow{L} X,$$

where F is the distribution function of X.

7.5.14 Suppose $\{X_n, n \geq 1\}$ is a sequence of independent and identically distributed random variables, such that $X_n = 1$ and 0 with probability 1/2 each. Suppose $Y_n = 1 - X_n$. Show that $X_n \xrightarrow{L} X$ and $Y_n \xrightarrow{L} Y$, where X and Y are identically distributed random variables, with values 1 and 0 with equal probabilities 1/2. Examine whether $X_n + Y_n \xrightarrow{L} X + Y$. Examine whether X_n and Y_n are independent random variables.

Solution: Since $\{X_n, n \geq 1\}$ is a sequence of independent and identically distributed random variables, it follows that $X_n \xrightarrow{L} X$, where $X = 1$ and 0 with probability 1/2 each. Further $Y_n = 1 - X_n$ and hence by Slutsky's theorem $Y_n \xrightarrow{L} Y = 1 - X$. It is to be noted that X and Y are identically distributed random variables, with values 1 and 0 with equal probabilities 1/2. Further, $X_n + Y_n \equiv 1$ and possible values of $X + Y$ are $0, 1, 2$, with probabilities $1/4, 1/2, 1/4$ respectively. It thus clearly follows that $X_n + Y_n \overset{L}{\nrightarrow} X + Y$. Observe that,

$$\begin{aligned} P[X_n = 1, Y_n = 0] &= P[X_n = 1, X_n = 1] \\ &= P[X_n = 1] = 1/2 \neq P[X_n = 1]P[Y_n = 0] = 1/4, \end{aligned}$$

which implies that X_n and Y_n are not independent random variables.

7.5.15 Suppose $\{X_n, n \geq 1\}$ is a sequence of random variables such that X_n follows $N(0, \sigma^2/n)$. Suppose a function $g : \mathbb{R} \to \mathbb{R}$ is defined as

$$g(x) \;=\; \begin{cases} 0, & \text{if} \quad x \leq 0 \\ 1, & \text{if} \quad x > 0. \end{cases}$$

Show that $X_n \xrightarrow{P} X \equiv 0$, but $g(X_n) \overset{P}{\nrightarrow} g(X)$.

Solution: It is clear that g is not a continuous function and 0 is a point of discontinuity. Observe that,

$$E(X_n^2) = \sigma^2/n \to 0 \;\;\Rightarrow\;\; X_n \xrightarrow{q.m.} X \equiv 0 \;\;\Rightarrow\;\; X_n \xrightarrow{P} X \equiv 0.$$

Since the distribution X_n is symmetric around 0, the distribution of $g(X_n)$ is as given below.

$$g(X_n) \;=\; \begin{cases} 0, & \text{with probability} \quad 1/2 \\ 1, & \text{with probability} \quad 1/2. \end{cases}$$

However, $X \equiv 0$ which implies that $g(X) \equiv 0$. Hence,

$$|g(X_n) - g(X)| = \begin{cases} 0, & \text{with probability } 1/2 \\ 1, & \text{with probability } 1/2. \end{cases}$$

Thus, for

$$\epsilon \in (0,1), \quad P[|g(X_n) - g(X)| < \epsilon] = 1/2 \nrightarrow 1 \quad \Rightarrow g(X_n) \stackrel{P}{\nrightarrow} g(X).$$

7.5.16 Suppose $X_n \sim G(n, n)$. Examine if $X_n \stackrel{r}{\to} 1$ for $r > 1$ and $X_n \stackrel{a.s.}{\to} 1$.

Solution: In Example 7.3.24, it is shown that if $X_n \sim G(n, n)$, then $X_n \stackrel{q.m.}{\to} 1$, $X_n \stackrel{P}{\to} 1$ and $X_n \stackrel{L}{\to} 1$. To examine whether, $X_n \stackrel{r}{\to} 1$ for $r > 2$, we need moments of X_n of order r. We obtain these from the cumulant generating function $C_n(t)$ of X_n. It is given by,

$$\begin{aligned} C_n(t) &= \log(1 - t/n)^{-n} = -n \log(1 - t/n) \\ &= n\Big(\sum_{i \geq 1} t^i / in^i\Big) = \Big(\sum_{i \geq 1} (t^i/i!)((i-1)!/n^{i-1})\Big). \end{aligned}$$

Thus, i-th cumulant k_i is given by $k_i = (i-1)!/n^{i-1}$, $i \geq 1$ and hence cumulants of all order larger than 1 converge to 0 as $n \to \infty$, which implies that $E(X_n - 1)^r \to 0$. It is to be noted that $E|X_n - 1|^{2k} = E(X_n - 1)^{2k}$ and $E|X_n - 1|^{2k+1} = E(X_n - 1)^{2k+1}$, when $X_n - 1 > 0$ and $E|X_n - 1|^{2k+1} = -E(X_n - 1)^{2k+1}$, when $X_n - 1 \leq 0$. Hence, $E(X_n - 1)^r \to 0$ implies that $X_n \stackrel{r}{\to} 1$ for $r > 2$. To examine almost sure convergence, observe that

$$\begin{aligned} k_2 = 1/n \ \&\ k_4 = 6/n^3 &\Rightarrow E(X_n - E(X_n))^4 = k_4 + 3k_2^2 \\ &\Rightarrow E(X_n - 1)^4 = 6/n^3 + 3/n^2 \\ &\Rightarrow \sum_{n \geq 1} E(X_n - 1)^4 = \sum_{n \geq 1} (6/n^3 + 3/n^2) < \infty \\ &\Rightarrow X_n \stackrel{a.s.}{\to} 1 \end{aligned}$$

7.5.17 Suppose $\{X_n, n \geq 1\}$ is a sequence of random variables such that X_n has Cauchy distribution with probability density function given by $f(x, n) = (n/\pi)(1/(1 + n^2x^2))$, $x \in \mathbb{R}$. Examine whether $X_n \stackrel{P}{\to} 0$, $X_n \stackrel{r}{\to} 0$, $r \geq 1$ and $X_n \stackrel{L}{\to} 0$.

Solution: It is given that X_n has Cauchy distribution with location parameter 0 and scale parameter $1/n$. Hence, its distribution function is given by $F_n(x) = (1/2) + (1/\pi)\tan^{-1}(nx)$, $x \in \mathbb{R}$. Hence $\forall \ \epsilon > 0$, as $n \to \infty$

$$\begin{aligned} P[|X_n| < \epsilon] &= F_n(\epsilon) - F_n(-\epsilon) = (2/\pi)\tan^{-1}(n\epsilon) \to (2/\pi)(\pi/2) = 1 \\ &\Rightarrow X_n \stackrel{P}{\to} 0 \ \Rightarrow \ X_n \stackrel{L}{\to} 0. \end{aligned}$$

Convergence in law can be verified independently using the definition as follows. Observe that

$$F_n(x) = \begin{cases} 1/2 + (1/\pi)\tan^{-1}(nx), & \text{if } x < 0 \\ 1/2 & \text{if } x = 0 \\ 1/2 + (1/\pi)\tan^{-1}(nx), & \text{if } x > 0 \end{cases} \to \begin{cases} 0, & \text{if } x < 0 \\ 1/2 & \text{if } x = 0 \\ 1, & \text{if } x > 0. \end{cases}$$

If F_X denotes the distribution function of $X \equiv 0$. Then $F_n(x) \to F_X(x)$ $\forall\ x \in C(F_X) = \mathbb{R} - \{0\}$ and hence $X_n \xrightarrow{L} 0$. It is to be noted that for all $r \geq 1$, $E(|X_n|^r) = 2\int_0^\infty \frac{n}{\pi}\frac{x^r}{1+n^2x^2}\,dx$ and it is a divergent integral. Hence, we cannot define convergence in r-th mean.

7.5.18 Suppose $\{X_n, n \geq 1\}$ is a sequence of random variables such that X_n has Cauchy distribution with probability density function given by $f(x,n) = (n/\pi)(1/(n^2 + x^2))$, $x \in \mathbb{R}$. Examine whether $X_n \xrightarrow{P} 0$, $X_n \xrightarrow{r} 0$, $r \geq 1$, $X_n \xrightarrow{L} 0$ and $X_n \xrightarrow{a.s.} 0$.

Solution: It is given that X_n has Cauchy distribution with location parameter 0 and scale parameter n. Hence its distribution function is given by $F_n(x) = 1/2 + (1/\pi)\tan^{-1}(x/n)$, $x \in \mathbb{R}$. Hence $\forall\ \epsilon > 0$, as $n \to \infty$,

$$P[|X_n| < \epsilon] = F_n(\epsilon) - F_n(-\epsilon) = (2/\pi)\tan^{-1}(\epsilon/n) \to 0 \ \Rightarrow \ X_n \overset{P}{\nrightarrow} 0.$$

To verify convergence in law, observe that for all x, $F_n(x) \to 1/2$. Hence, X_n does not converge in law. Convergence in r-th mean is not defined as $X_n \notin L_r$ for any $r \geq 1$. Since $X_n \overset{P}{\nrightarrow} 0$, $X_n \overset{a.s}{\nrightarrow} 0$.

7.5.19 Suppose X is a real random variable defined on a probability space on $(\Omega, \mathbb{A}, P)$ and $\{A_n, n \geq 1\}$ is a sequence of sets in $\mathbb{A}$ such that $P(A_n) \to 0$. Then show that $XI_{A_n} \xrightarrow{P} 0$.

Solution: For $\epsilon > 0$ observe that

$$P[|I_{A_n}| > \epsilon] = P[I_{A_n} = 1] = P(A_n) \to 0 \ \Rightarrow \ I_{A_n} \xrightarrow{P} 0\,.$$

Further X is a real random variable and hence is bounded in probability. Thus, $XI_{A_n} \xrightarrow{P} 0$.

7.5.20 Show that equivalent random variables are identically distributed.

Solution: Suppose X and Y are equivalent random variables, that is, $P[X = Y] = P(N^c) = 1$, where $N^c = \{\omega | X(\omega) = Y(\omega)\}$. Observe that for any $x \in \mathbb{R}$

$$\begin{aligned} F_X(x) &= P[X \leq x] = P[X \leq x, N^c] + P[X \leq x, N] \\ &= P[X \leq x, N^c] \ \text{ as } \ P[X \leq x, N] \leq P(N) = 0 \\ &= P[Y \leq x, N^c] \ \text{ as on } \ N^c, X(\omega) = Y(\omega) \\ &= P[Y \leq x, N^c] + P[Y \leq x, N] = P[Y \leq x] = F_Y(x). \end{aligned}$$

Thus, X and Y are identically distributed random variables.

7.5.21 Suppose $X_n = \min\{Y_1, Y_2, \cdots, Y_n\}$, where $\{Y_n, n \geq 1\}$ is a sequence of independent and identically distributed random variables, each having an exponential distribution with location parameter θ and scale parameter 1. (i) Show that $X_n \xrightarrow{P} \theta$ and $X_n \xrightarrow{a.s.} \theta$. (ii) Suppose $U_n = n(X_n - \theta)$, then show that $U_n \xrightarrow{L} U$, where U has exponential distribution with location parameter 0 and scale parameter 1. (iii) From the limit law of U_n, find the limit law of $V_n = \sqrt{n}(X_n - \theta)$. (iv) From the limit law of U_n, find the limit law of $\exp(-U_n)$.

Solution: A random variable Y_1 has exponential distribution with location parameter θ and scale parameter 1. Hence, its distribution function $F(y)$ is given by,

$$F(y) = \begin{cases} 0, & \text{if } y < \theta, \\ 1 - e^{-(y-\theta)}, & \text{if } y \geq \theta. \end{cases}$$

Thus, the distribution function $F_n(y)$ of $X_n = \min\{Y_1, Y_2, \cdots, Y_n\}$ is,

$$F_n(y) = 1 - (1 - F(y))^n = \begin{cases} 0, & \text{if } y < \theta \\ 1 - e^{-n(y-\theta)}, & \text{if } y \geq \theta. \end{cases}$$

Hence, for $\epsilon > 0$, $P[|X_n - \theta| > \epsilon]$ is obtained as follows.

$$\begin{aligned} P[|X_n - \theta| > \epsilon] &= 1 - P[\theta - \epsilon < X_n < \theta + \epsilon] \\ &= 1 - P[\theta < X_n < \theta + \epsilon] \\ &= 1 - P[X_n < \theta + \epsilon] + P[X_n < \theta] \\ &= 1 - (1 - e^{-n(\theta+\epsilon-\theta)}) = e^{-n\epsilon} \\ &\to 0 \text{ as } n \to \infty \ \forall \ \epsilon > 0. \end{aligned}$$

Consequently, $X_n \xrightarrow{P} \theta$. To examine almost sure convergence, observe that $\forall \ \epsilon > 0, \ e^{-\epsilon} < 1$. Hence,

$$\sum_{n=1}^{\infty} P[|X_n - \theta| > \epsilon] = \sum_{n=1}^{\infty} e^{-n\epsilon} = \sum_{n=1}^{\infty} (e^{-\epsilon})^n < \infty \quad \Rightarrow \quad X_n \xrightarrow{a.s.} \theta.$$

(ii) From the distribution function F_n of X_n, it follows that for each n, X_n has an exponential distribution with location parameter θ and scale parameter n, which further implies that for each n, $U_n = n(X_n - \theta)$ has the exponential distribution with location parameter 0 and scale parameter 1. Hence, $U_n \xrightarrow{L} U$, where U follows the exponential distribution with location parameter 0 and scale parameter 1.
(iii) Observe that V_n can be written as, $V_n = \sqrt{n}(X_n - \theta) = (1/\sqrt{n})U_n$. Now $U_n \xrightarrow{L} U$ implies that the sequence $\{U_n, n \geq 1\}$ is bounded in probability. Hence, $V_n = (1/\sqrt{n})U_n \xrightarrow{P} 0$, which further implies that

$\sqrt{n}(X_n - \theta) \xrightarrow{L} 0$. Thus, the limit law of $\sqrt{n}(X_n - \theta)$ is degenerate at 0.
(iv) By the continuous mapping theorem, $U_n \xrightarrow{L} U \Rightarrow \exp(-U_n) \xrightarrow{L} \exp(-U)$. Since U follows the exponential distribution with location parameter 0 and scale parameter 1, the distribution of $\exp(-U)$ is uniform $U(0,1)$.

7.5.22 Suppose $X_n \xrightarrow{L} X$ and $Y_n \xrightarrow{L} C$, where $C \neq 0$ is a degenerate random variable. Using the definition of convergence in law, show that (i) $X_nC \xrightarrow{L} XC$ and (ii) $X_n/C \xrightarrow{L} X/C$, provided these are defined. Hence show that $X_nY_n \xrightarrow{L} XC$, and $X_n/Y_n \xrightarrow{L} X/C$, provided these are defined.

Solution: (i) Suppose $C > 0$ and x/C is a point of continuity of F_X. Note that for $C > 0$,

$$\begin{aligned} F_{X_nC}(x) &= P[X_nC \leq x] = P[X_n \leq x/C] = F_n(x/C) \\ &\rightarrow F_X(x/C) = P[X \leq x/C] = F_{XC}(x). \end{aligned}$$

Now, $X_nC \xrightarrow{L} XC$ if x is a point of continuity of F_{XC}. To prove it, observe that since x/C is a point of continuity of F_X,

$$\begin{aligned} & \lim_{h\to 0} F_X(x/C - h) = F_X(x/C) \\ \iff & \lim_{h\to 0} P[X \leq x/C - h] = P[X \leq x/C] \\ \iff & \lim_{h\to 0} P[XC \leq x - Ch] = P[XC \leq x] \\ \iff & \lim_{h\to 0} F_{XC}(x - Ch) = F_{XC}(x). \end{aligned}$$

Thus, x is a point of continuity of F_{XC}. Hence for $C > 0$, $X_nC \xrightarrow{L} XC$. Suppose $C < 0$ and $C = -D$. Then

$$X_n(-D) \xrightarrow{L} X(-D) \Rightarrow -X_nC \xrightarrow{L} -XC \Rightarrow X_nC \xrightarrow{L} XC,$$

by continuous mapping theorem. To find the limit law of X_nY_n, note that by Theorem 7.3.7,

$$(X_nY_n) - (X_nC) = X_n(Y_n - C) \xrightarrow{P} 0.$$

Hence by Theorem 7.3.5, limit law of (X_nY_n) is the same as that of (X_nC).
(ii) Suppose xC is a point of continuity of F_X. For $C > 0$,

$$\begin{aligned} F_{X_n/C}(x) &= P[X_n/C \leq x] = P[X_n \leq xC] = F_n(xC) \\ &\rightarrow F_X(xC) = P[X \leq xC] = F_{X/C}(x). \end{aligned}$$

Now, $X_n/C \xrightarrow{L} X/C$ if x is a point of continuity of $F_{X/C}$. Since xC is a point of continuity of F_X, we have

$$\begin{aligned} \lim_{h\to 0} F_X(xC-h) &= F_X(xC) \\ \iff \lim_{h\to 0} P[X \le xC-h] &= P[X\le xC] \\ \iff \lim_{h\to 0} P[X/C \le x-h/C] &= P[X/C\le x] \\ \iff \lim_{h\to 0} F_{X/C}(x-h/C) &= F_{X/C}(x). \end{aligned}$$

Thus, x is a point of continuity of $F_{X/C}$. Hence, $X_n/C \xrightarrow{L} X/C$. If $C<0$, with $C=-D$ we have

$$X_n/(-D) \xrightarrow{L} X/(-D) \;\Rightarrow\; -X_n/C \xrightarrow{L} -X/C \;\Rightarrow\; X_n/C \xrightarrow{L} X/C,$$

by continuous mapping theorem. Observe that

$$(X_n/Y_n)-(X_n/C) = X_n(1/Y_n - 1/C) \xrightarrow{P} 0$$

by Theorem 7.3.7. Hence, the limit law of (X_n/Y_n) is the same as that of (X_n/C).

7.5.23 Suppose $X_n \sim \chi^2_n$, $n\ge 1$ and $Y_n = (X_n-n)/\sqrt{2n}$. Examine whether (i) $X_n/n \xrightarrow{P} 1$, $X_n/n \xrightarrow{q.m.} 1$ and (ii) $Y_n \xrightarrow{L} Z \sim N(0,1)$. What is the distribution of Y_n^2 as $n\to\infty$?

Solution: Since $X_n \sim \chi^2_n$, $E(X_n)=n$ and $Var(X_n)=2n$. Hence,

$$\begin{aligned} E(X_n/n-1)^2 &= E(X_n-n)^2/n^2 = 2n/n^2 \to 0 \;\Rightarrow\; X_n/n \xrightarrow{q.m.} 1 \\ &\Rightarrow\; X_n/n \xrightarrow{P} 1. \end{aligned}$$

(ii) Since $X_n \sim \chi^2_n$, its characteristic function is $\phi_{X_n}(t) = (1-2it)^{-n/2}$. Suppose $\phi_n(t)$ denotes the characteristic function of Y_n. Then $\forall\ t\in\mathbb{R}$

$$\begin{aligned} \phi_n(t) = E(e^{itY_n}) &= E(e^{it(X_n-n)/\sqrt{2n}}) = e^{-it\sqrt{n}/\sqrt{2}}E(e^{itX_n/\sqrt{2n}}) \\ &= e^{-it\sqrt{n}/\sqrt{2}}(1-2it/\sqrt{2n})^{-n/2} \\ \Rightarrow \log\phi_n(t) &= -it\sqrt{n}/\sqrt{2} - (n/2)\log(1-\sqrt{2}it/\sqrt{n}) \\ &= -it\sqrt{n}/\sqrt{2} + (n/2)(\sqrt{2}it/\sqrt{n} - t^2/n - it^3 2^{3/2}/3n^{3/2} \\ &\quad + \cdots) \\ &= -t^2/2 + \delta_n \;\to\; -t^2/2 = \phi(t), \quad \text{say} \end{aligned}$$

since $\delta_n \to 0$ as $n\to\infty$. Thus, a sequence $\{\phi_n(t), n\ge 1\}$ of characteristic functions converges to $\phi(t) = \exp(-t^2/2)$, which is continuous at $t=0$. Hence, by the continuity theorem for characteristic functions,

$Y_n \xrightarrow{L} Z$, where $\phi(t) = \exp(-t^2/2)$ is a characteristic function of Z. It is known that $\phi(t) = \exp(-t^2/2)$ is a characteristic function of the standard normal distribution hence $Z \sim N(0,1)$.

By the continuous mapping theorem, $Y_n \xrightarrow{L} Z \Rightarrow Y_n^2 \xrightarrow{L} Y \sim \chi_1^2$.

7.5.24 Suppose $X_n \sim P(\lambda_n)$ and $Y_n = (X_n - \lambda_n)/\sqrt{\lambda_n}$. Show that as $\lambda_n \to \infty$, $Y_n \xrightarrow{L} Z \sim N(0,1)$. What is the distribution of Y_n^2 as $\lambda_n \to \infty$?

Solution: Suppose $X_n \sim P(\lambda_n)$. Then $E(X_n) = \lambda_n$, $Var(X_n) = \lambda_n$ and its characteristic function is given by $\phi_{X_n}(t) = \exp\{\lambda_n(e^{it} - 1)\}$. Suppose $\phi_n(t)$ denotes the characteristic function of Y_n. Then $\forall \ t \in \mathbb{R}$

$$\begin{aligned}
\phi_n(t) = E(e^{itY_n}) &= E(e^{it\frac{(X_n-\lambda_n)}{\sqrt{\lambda_n}}}) = e^{-it\sqrt{\lambda_n}}E(e^{i\frac{t}{\sqrt{\lambda_n}}X_n}) \\
&= e^{-it\sqrt{\lambda_n}}\exp\{\lambda_n(e^{i\frac{t}{\sqrt{\lambda_n}}} - 1)\} \\
\Rightarrow \log\phi_n(t) &= -it\sqrt{\lambda_n} + \lambda_n(e^{i\frac{t}{\sqrt{\lambda_n}}} - 1) \\
&= -it\sqrt{\lambda_n} + \lambda_n(1 + it/\sqrt{\lambda_n} - t^2/2\lambda_n - it^3/6(\lambda_n)^{3/2} \\
&\quad + \cdots - 1) = -t^2/2 + \delta_n \ \to \ -t^2/2 = \phi(t), \ \text{say}
\end{aligned}$$

since $\delta_n \to 0$ as $\lambda_n \to \infty$. Thus, a sequence $\{\phi_n(t), n \geq 1\}$ of characteristic functions converges to $\phi(t) = \exp(-t^2/2)$, which is a continuous function at $t = 0$. Further $\phi(t) = \exp(-t^2/2)$ is a characteristic function of Z which follows $N(0,1)$ distribution. Hence, by the continuity theorem and the uniqueness theorem of the characteristic functions $Y_n \xrightarrow{L} Z \sim N(0,1)$. By the continuous mapping theorem,
$Y_n \xrightarrow{L} Z \Rightarrow Y_n^2 \xrightarrow{L} Y \sim \chi_1^2$.

7.5.25 Suppose $\{X_n, n \geq 1\}$ is a sequence of independent random variables each having uniform $U(0,1)$ distribution. Suppose $X_{(1)} = \min\{X_1, X_2, \cdots, X_n\}$ and $X_{(n)} = \max\{X_1, X_2, \cdots, X_n\}$. Show that (i) $nX_{(1)} \xrightarrow{L} X$, where X has exponential distribution with mean 1, (ii) $1 - (X_{(1)} + X_{(n)}) \xrightarrow{P} 0$ and (iii) $n(1 - X_{(n)})X_{(1)} \xrightarrow{P} 0$.

Solution: (i) If $X_1 \sim U(0,1)$, then as shown in Example 7.2.5, the distribution function $F_{X_{(1)}}(x)$ of $X_{(1)}$ is given by,

$$F_{X_{(1)}}(x) \quad = \quad \begin{cases} 0, & \text{if} \quad x < 0 \\ 1 - (1-x)^n, & \text{if} \quad 0 \leq x < 1 \\ 1, & \text{if} \quad x \geq 1. \end{cases}$$

From the distribution function of $X_{(1)}$, the distribution function of $nX_{(1)}$ is given by,

$$F_{nX_{(1)}}(x) \quad = \quad P[X_{(1)} \leq x/n] = \begin{cases} 0, & \text{if} \quad x < 0 \\ 1 - (1-x/n)^n, & \text{if} \quad 0 \leq x < n \\ 1, & \text{if} \quad x \geq n \end{cases}$$

It then follows that

$$F_{nX_{(1)}}(x) \;\to\; \begin{cases} 0, & \text{if } \; x < 0 \\ 1 - e^{-x}, & \text{if } \; x \geq 0. \end{cases}$$

But the limit is the distribution function of the exponential distribution with mean 1, which is continuous for all $x \in \mathbb{R}$ and hence $nX_{(1)} \xrightarrow{L} X$, where X has exponential distribution with mean 1.
(ii) As shown in Example 7.2.5,

$$X_{(1)} \xrightarrow{P} 0 \;\&\; X_{(n)} \xrightarrow{P} 1 \;\Rightarrow\; 1-(X_{(1)}+X_{(n)}) = -X_{(1)}+(1-X_{(n)}) \xrightarrow{P} 0.$$

(iii) Observe that $n(1 - X_{(n)})X_{(1)} = (nX_{(1)})(1 - X_{(n)})$. Now by (i) $nX_{(1)} \xrightarrow{L} X$ and hence the first factor $nX_{(1)}$ is bounded in probability and the second factor $(1 - X_{(n)}) \xrightarrow{P} 0$ which implies that $n(1 - X_{(n)})X_{(1)} \xrightarrow{P} 0$. Alternatively as shown in Example 7.3.3, $n(1 - X_{(n)})$ converges in law and hence is bounded in probability, $X_{(1)} \xrightarrow{P} 0$ which implies that $n(1 - X_{(n)})X_{(1)} \xrightarrow{P} 0$.

7.5.26 Suppose $\{X_n, n \geq 1\}$ is a sequence of independent and identically distributed random variables, each having $U(0,1)$ distribution. Find the limiting distribution of $n(1 - X_{(n-1)})$.

Solution: As shown in Example 7.3.3, $n(1 - X_{(n)}) \xrightarrow{L} X$, where X follows exponential distribution with mean 1. To find the limiting distribution of $n(1-X_{(n-1)})$, we examine whether $n(1-X_{(n-1)})-n(1-X_{(n)}) = n(X_{(n-1)} - X_{(n)}) \xrightarrow{P} 0$. Observe that for $\epsilon > 0$

$$\begin{aligned} P[|n(X_{(n-1)} - X_{(n)})| < \epsilon] &\geq P[X_{(n-1)} = X_{(n)}] = P[X_{(n-1)} \geq X_n] \\ &= E(P[X_{(n-1)} \geq X_n | X_n]) = E(1 - X_n^{n-1}) \\ &= \int_0^1 (1 - u^{n-1})\, du = 1 - 1/n \;\to\; 1 \\ &\Rightarrow n(X_{(n-1)} - X_{(n)}) \xrightarrow{P} 0 \\ &\Rightarrow n(1 - X_{(n-1)}) - n(1 - X_{(n)}) \xrightarrow{P} 0. \end{aligned}$$

Hence, $n(1-X_{(n-1)})$ and $n(1-X_{(n)})$ converge in law to the same random variable. However, $n(1 - X_{(n)}) \xrightarrow{L} X$ and hence $n(1 - X_{(n-1)}) \xrightarrow{L} X$, where X follows exponential distribution with mean 1.

7.5.27 Suppose $\{X_1, X_2, \cdots, X_n\}$ is a random sample from a normal $N(\theta, 1)$ distribution and $\theta \in [0, \infty)$. The maximum likelihood estimator $\hat{\theta}_n$ of θ is given by,

$$\hat{\theta}_n \;=\; \begin{cases} \overline{X}_n, & \text{if } \; \overline{X}_n \geq 0 \\ 0, & \text{if } \; \overline{X}_n < 0, \end{cases}$$

Examine whether $\hat{\theta}_n \xrightarrow{P} \theta$. Find the limit law of $Y_n = \sqrt{n}(\hat{\theta}_n - \theta)$. Identify the limiting distribution at $\theta = 0$.

Solution: To verify $\hat{\theta}_n \xrightarrow{P} \theta$, we note that $\overline{X}_n \xrightarrow{q.m.} \theta$ and hence $\overline{X}_n \xrightarrow{P_\theta} \theta$, for all $\theta \geq 0$. Now for $\epsilon > 0$,

$$P_\theta[|\hat{\theta}_n - \overline{X}_n| < \epsilon] \geq P_\theta[\hat{\theta}_n = \overline{X}_n] = P_\theta[\overline{X}_n \geq 0] = 1 - \Phi(-\sqrt{n}\theta) \to \; 1$$

if $\theta > 0$. As a consequence, if $\theta > 0$ then $\hat{\theta}_n - \overline{X}_n \xrightarrow{P_\theta} 0$ and $\overline{X}_n \xrightarrow{P_\theta} \theta$ implies that $\hat{\theta}_n \xrightarrow{P_\theta} \theta$ when $\theta > 0$. Suppose $\theta = 0$. Then using the fact that $\hat{\theta}_n \geq 0$, for $\epsilon > 0$, as $n \to \infty$,

$$P_0[|\hat{\theta}_n| > \epsilon] = P_0[\hat{\theta}_n > \epsilon] = P_0[\overline{X}_n > \epsilon] = 1 - \Phi(\sqrt{n}\epsilon) \;\to\; 0 \;\Rightarrow\; \hat{\theta}_n \xrightarrow{P_0} 0\,.$$

Thus, $\hat{\theta}_n \xrightarrow{P_\theta} \theta$ for all $\theta \geq 0$.
(ii) Since $X \sim N(\theta, 1)$, $\sqrt{n}(\overline{X}_n - \theta) \sim N(0, 1)$ for each n and hence in limit. Further, $\sqrt{n}(\overline{X}_n - \theta) - \sqrt{n}(\hat{\theta}_n - \theta) = \sqrt{n}(\overline{X}_n - \hat{\theta}_n)$. Observe that for $\theta > 0$ and for $\epsilon > 0$,

$$\begin{aligned} P_\theta[|\sqrt{n}(\overline{X}_n - \hat{\theta}_n)| < \epsilon] &\geq P_\theta[\overline{X}_n = \hat{\theta}_n] = P_\theta[\overline{X}_n \geq 0] \\ &= 1 - \Phi(-\sqrt{n}\theta) \;\to\; 1 \quad \text{if} \quad \theta > 0. \end{aligned}$$

Thus, for $\theta > 0$, $\sqrt{n}(\overline{X}_n - \theta) - \sqrt{n}(\hat{\theta}_n - \theta) \xrightarrow{P_\theta} 0$, hence $\sqrt{n}(\overline{X}_n - \theta)$ and $\sqrt{n}(\hat{\theta}_n - \theta)$ have the same limit law. But $\sqrt{n}(\overline{X}_n - \theta) \xrightarrow{L} Z \sim N(0, 1)$ and hence $\sqrt{n}(\hat{\theta}_n - \theta) \xrightarrow{L} Z \sim N(0, 1)$, for $\theta > 0$. Suppose now $\theta = 0$.

$$\text{If} \quad x < 0, \quad P_0[\sqrt{n}(\hat{\theta}_n - 0) \leq x] = 0 \quad \text{as } \hat{\theta}_n \geq 0.$$

For $x = 0$, $P_0[\sqrt{n}\hat{\theta}_n \leq 0] = P_0[\sqrt{n}\hat{\theta}_n = 0] = P_0[\overline{X}_n < 0] = \Phi(0) = 1/2$.

$$\begin{aligned} \text{For } x > 0, \; P_0[\sqrt{n}\hat{\theta}_n \leq x] &= P_0[\sqrt{n}\hat{\theta}_n \leq 0] + P_0[0 < \sqrt{n}\hat{\theta}_n \leq x] \\ &= 1/2 + P_0[0 < \sqrt{n}(\overline{X}_n) \leq x] \\ &= 1/2 + \Phi(x) - \Phi(0) = \Phi(x). \end{aligned}$$

$$\text{Thus } P_0[\sqrt{n}\hat{\theta}_n \leq x] = \begin{cases} 0, & \text{if} \quad x < 0 \\ 1/2, & \text{if} \quad x = 0, \\ \Phi(x), & \text{if} \quad x > 0, \end{cases}$$

which shows that at $\theta = 0$, the asymptotic distribution of $\sqrt{n}\hat{\theta}_n$ is not normal and 0 is a point of discontinuity. Suppose U_1 is a random variable with a distribution degenerate at 0. Then its distribution function is given by,

$$F_{U_1}(x) = \begin{cases} 0, & \text{if} \quad x < 0 \\ 1, & \text{if} \quad x \geq 0 \end{cases}$$

Suppose a random variable U_2 is defined as $U_2 = |U|$ where $U \sim N(0,1)$. Then $P[U_2 \leq x] = 0$ if $x < 0$. Suppose $x \geq 0$, then

$$P[U_2 \leq x] = P[|U| \leq x] = P[-x \leq U \leq x] = \Phi(x) - \Phi(-x) = 2\Phi(x) - 1.$$

Thus, the distribution function of U_2 is given by,

$$F_{U_2}(x) = \begin{cases} 0, & \text{if } x < 0 \\ 2\Phi(x) - 1, & \text{if } x \geq 0 \end{cases}$$

Thus, $P_0[\sqrt{n}(\hat{\theta}_n - 0) \leq x] \to (1/2)F_{U_1}(x) + (1/2)F_{U_2}(x)$.

7.5.28 Suppose $X_n \xrightarrow{P} X$. Show that $|X_n| \xrightarrow{L} |X|$.

Solution: A function $g(x) = |x|$, $x \in \mathbb{R}$ is a continuous function, hence by invariance of convergence in probability under continuous transforms, $X_n \xrightarrow{P} X \Rightarrow |X_n| \xrightarrow{P} |X|$. Further, convergence in probability implies convergence in law. Hence, $|X_n| \xrightarrow{L} |X|$. Alternatively,

$$X_n \xrightarrow{P} X \Rightarrow X_n \xrightarrow{L} X \Rightarrow |X_n| \xrightarrow{L} |X|,$$

by continuous mapping theorem.

7.5.29 Suppose $X_n \xrightarrow{L} X$ where X is a continuous random variable. Using the definition of convergence in law, show that $|X_n| \xrightarrow{L} |X|$.

Solution: $X_n \xrightarrow{L} X \Rightarrow F_{X_n}(x) \to F(x) \ \forall \ x \in C(F_X) = \mathbb{R}$ as X is a continuous random variable. Now, $|X_n|$ being a non-negative random variable, $F_{|X_n|}(x) = 0 \ \forall \ x < 0$. For $x \geq 0$

$$F_{|X_n|}(x) = P[|X_n| \leq x] = P[-x \leq X_n \leq x] = F_{X_n}(x) - F_{X_n}(-x-)$$

as $F_{X_n}(x)$ may not be continuous at $-x$. Since $X_n \xrightarrow{L} X$, $F_{X_n}(x) \to F_X(x)$. To find $\lim_{n\to\infty} F_{X_n}(-x-0)$ observe that,

$$\begin{aligned} \lim_{n\to\infty} F_{X_n}(-x-) &= \lim_{n\to\infty} \lim_{h\to 0} F_{X_n}(-x-h) \\ &= \lim_{h\to 0} \lim_{n\to\infty} F_{X_n}(-x-h) \\ &= \lim_{h\to 0} F_X(-x-h) \ \text{ as } \ -x-h \in C(F_X) = \mathbb{R} \\ &= F(-x) \quad \text{as } X \text{ is a continuous random variable} \end{aligned}$$

In the second step, we can interchange the limits as F_{X_n} is a monotone and bounded function. Thus, $\forall \ x \geq 0$

$$F_{|X_n|}(x) = F_{X_n}(x) - F_{X_n}(-x-0) \to F_X(x) - F_X(-x) = P[|X| \leq x].$$

Hence, $|X_n| \xrightarrow{L} |X|$.

7.5.30 Suppose $(\Omega, \mathbb{A}, P)$ is a probability space, where $\Omega = [0,1]$, $\mathbb{A}$ is a sigma field of subsets of Ω and $P(A_n) = 1/n$ for $A_n = [0, 1/n)$. Suppose the random variable X and X_n, $n \geq 1$ are defined on this space as follows.

$$X(\omega) = \begin{cases} 0, & \text{if } 0 \leq \omega \leq 1/2 \\ 1, & \text{if } 1/2 < \omega \leq 1 \end{cases} \quad \& \ X_n(\omega) = \begin{cases} 1, & \text{if } 0 \leq \omega \leq 1/2 \\ 0, & \text{if } 1/2 < \omega \leq 1. \end{cases}$$

Examine whether $X_n \xrightarrow{L} X$ and $X_n \xrightarrow{P} X$.

Solution: From the definitions of X_n and X, observe that for any $x \in \mathbb{R}$,

$$F_{X_n}(x) = F_X(x) \quad = \quad \begin{cases} 0, & \text{if } \quad x < 0 \\ 1/2, & \text{if } \ 0 \leq x < 1 \\ 1, & \text{if } \quad x \geq 1. \end{cases}$$

Thus, $F_{X_n}(x) = F_X(x) \ \forall \ x \in \mathbb{R}$. Hence, $X_n \xrightarrow{L} X$. Now $\forall \ \omega \in \Omega$,

$$|X_n(\omega) - X(\omega)| = 1 \ \Rightarrow \ \forall \ \epsilon \in (0,1) \ P[|X_n - X| > \epsilon] = 1 \ \Rightarrow \ X_n \overset{P}{\nrightarrow} X.$$

7.5.31 Suppose $\{X_n, n \geq 1\}$ is a sequence of independent and identically distributed random variables, each having the exponential distribution with scale parameter 1. Examine whether $X_{(n)} - \log n \xrightarrow{L} X$, where the distribution function of X is $F_X(x) = \exp(-e^{-x})$, $\ -\infty < x < \infty$.

Solution: The distribution function of X_n for each n is given by,

$$F(x) \quad = \quad \begin{cases} 0, & \text{if } \ x < 0 \\ 1 - e^{-x}, & \text{if } \ x \geq 0. \end{cases}$$

Hence, The distribution function of $X_{(n)}$ is given by,

$$F_{X_{(n)}}(x) \quad = \quad \begin{cases} 0, & \text{if } \ x < 0 \\ (1 - e^{-x})^n, & \text{if } \ x \geq 0. \end{cases}$$

Observe that

$$\begin{aligned} P[X_{(n)} - \log n \leq x] &= P[X_{(n)} \leq (x + \log n)] = F_{X_{(n)}}(x + \log n) \\ &= 0 \ \text{if } \ x + \log n < 0 \quad \Longleftrightarrow \quad x < -\log n. \end{aligned}$$

If $x + \log n \geq 0$, that is, if $x \geq -\log n$ then

$$\begin{aligned} P[X_{(n)} - \log n \leq x] &= F_{X_{(n)}}(x + \log n) \\ &= (1 - e^{-(x + \log n)})^n = (1 - e^{-x}/n)^n \\ &\to \exp(-e^{-x}), \quad -\infty < x < \infty \\ &\Rightarrow X_{(n)} - \log n \xrightarrow{L} X, \end{aligned}$$

where $F_X(x) = \exp(-e^{-x})$, $\ -\infty < x < \infty$.

7.5.32 Suppose m_n is a median of a random variable $X_n,\ n \geq 1$. Show that if $X_n \xrightarrow{L} X$, then any limit point of m_n is a median of X.

Solution: Suppose m is any limit point of m_n. For $\epsilon > 0$ such that $m+\epsilon$ and $m-\epsilon$ are continuity points of the distribution of X, we have $m - \epsilon < m_n < m + \epsilon$ for sufficiently large n. By the definition of the median we get,

$$\frac{1}{2} \leq P[X_n \leq m_n] \leq P[X_n \leq m+\epsilon] \ \& \ \frac{1}{2} \leq P[X_n \geq m_n] \leq P[X_n \geq m-\epsilon].$$

Allowing $n \to \infty$, we obtain

$$1/2 \leq P[X \leq m + \epsilon] \ \ \& \ \ 1/2 \leq P[X \geq m - \epsilon].$$

Now allowing $\epsilon \to 0$, we get

$$1/2 \leq P[X \leq m] \ \ \& \ \ 1/2 \leq P[X \geq m]$$

and hence m is a median of X.

7.5.33 If $\{X_n, n \geq 1\}$ is Cauchy in r-th mean, show that it is Cauchy in probability.

Solution: The sequence $\{X_n, n \geq 1\}$ is Cauchy in r-th mean implies that there exists a random variable X such that $X_n \xrightarrow{r} X$. Hence, $X_n \xrightarrow{P} X$ and hence it is Cauchy in probability.

7.5.34 If $X_n \xrightarrow{r} X$ and $Y_n \xrightarrow{r} Y$, examine whether (i) $X_n \pm Y_n \xrightarrow{r} X \pm Y$ and (ii) $X_nY_n \xrightarrow{r} XY$.

Solution: It is given that $X_n \xrightarrow{r} X$ and $Y_n \xrightarrow{r} Y$. Thus, it is assumed that X_n, X, Y and Y_n belong to $\mathbb{L}_r$. By C_r inequality

$$\begin{aligned} E(|(X_n \pm Y_n) - (X \pm Y)|^r) &= E(|(X_n - X) \pm (Y_n - Y)|^r) \\ &\leq C_r(E(|X_n - X|^r) + E(|Y_n - Y|^r)) \to 0 \\ &\Rightarrow X_n \pm Y_n \xrightarrow{r} X \pm Y. \end{aligned}$$

To examine whether $X_nY_n \xrightarrow{r} XY$, we again use C_r inequality and also use Holder's inequality proved in Section 4.5. It states that for $s > 1$ and $1/s + 1/l = 1$, $E(|XY|) \leq E^{1/s}(|X|^s)E^{1/l}(|Y|^l)$. Thus,

$$\begin{aligned} E(|(X_nY_n) - (XY)|^r) &= E(|X_nY_n - X_nY + X_nY - XY|^r) \\ &= E(|X_n(Y_n - Y) + Y(X_n - X)|^r) \\ &\leq C_r(E(|Y_n - Y|^r|X_n|^r) + E(|X_n - X|^r|Y|^r)) \\ &\leq C_r(E^{1/l}(|Y_n - Y|^{rl})E^{1/s}(|X_n|^{rs}) \\ &+ C_r(E^{1/l}(|X_n - X|^{rl})E^{1/s}(|Y|^{rs})) \ \to \ 0, \end{aligned}$$

assuming that $X_n \xrightarrow{rl} X$ and $Y_n \xrightarrow{rl} Y$, $E(|X_n|^{rs}) < \infty$ and $E(|Y|^{rs}) < \infty$, where $1/l + 1/s = 1$. It then follows that $X_nY_n \xrightarrow{r} XY$.

7.5.35 If $X_n \xrightarrow{r} X$ and $Y_n \xrightarrow{r} Y$, examine whether $E(X_nY_n) \to E(XY)$.

Solution: It is given that $X_n \xrightarrow{r} X$ and $Y_n \xrightarrow{r} Y$. Thus, it is assumed that X_n, X, Y and Y_n are belong to $\mathbb{L}_r$ and hence in $\mathbb{L}_s$ for any $s < r$. We use Holder's inequality to examine whether $E(X_nY_n) \to E(XY)$. Observe that $1/r+1/s = 1 \;\Rightarrow\; s = r/(r-1) < r$. Hence, $E^{1/s}(|X_n|^s) < \infty$ and $E^{1/s}(|Y|^s) < \infty$. Now using Holder's inequality we have,

$$\begin{aligned}|E(X_nY_n) - E(XY)| &= |E(X_nY_n - X_nY + X_nY - XY)| \\ &\le E(|Y_n - Y||X_n|) + E(|X_n - X||Y|) \\ &\le E^{1/r}(|Y_n - Y|^r)E^{1/s}(|X_n|^s) \\ &+ E^{1/r}(|X_n - X|^r)E^{1/s}(|Y|^s) \;\to\; 0,\end{aligned}$$

as $X_n \xrightarrow{r} X$ and $Y_n \xrightarrow{r} Y$. Thus, $E(X_nY_n) \to E(XY)$.

7.5.36 Suppose $\{X_n, n \ge 1\}$ is a sequence of independent random variables such that $\sum_{i=1}^n (X_i - E(X_i))/s_n \xrightarrow{L} Z \sim N(0,1)$, where $s_n^2 = \sum_{i=1}^n Var(X_i)$. Show that $\sum_{i=1}^n (X_i - E(X_i))/n \xrightarrow{P} 0$ if and only if $s_n/n \to 0$ as $n \to \infty$.

Solution: Since $\sum_{i=1}^n (X_i - E(X_i))/s_n \xrightarrow{L} Z \sim N(0,1)$, it is bounded in probability. Suppose $s_n/n \to 0$ as $n \to \infty$. Then

$$\frac{1}{n}\sum_{i=1}^n (X_i - E(X_i)) = \frac{s_n}{n}\left(\frac{1}{s_n}\sum_{i=1}^n (X_i - E(X_i))\right) \xrightarrow{P} 0.$$

Suppose now that $\sum_{i=1}^n (X_i - E(X_i))/n \xrightarrow{P} 0$ but $s_n/n \to c > 0$. Then

$$\begin{aligned}\frac{1}{s_n}\sum_{i=1}^n (X_i - E(X_i)) &= \frac{n}{s_n}\left(\frac{1}{n}\sum_{i=1}^n (X_i - E(X_i))\right) \\ &\xrightarrow{P} 0 \;\Rightarrow\; \frac{1}{s_n}\sum_{i=1}^n (X_i - E(X_i)) \xrightarrow{L} 0\end{aligned}$$

which is a contradiction to $\sum_{i=1}^n (X_i - E(X_i))/s_n \xrightarrow{L} Z \sim N(0,1)$. Thus, $s_n/n \to 0$ as $n \to \infty$.

7.5.37 Suppose $\{X_1, X_2, \cdots, X_n\}$ is a random sample from a normal $N(\theta, 1)$ distribution, where $\theta \in \{0, 1\}$. The maximum likelihood estimator $\hat{\theta}_n$ of θ is given by,

$$\hat{\theta}_n = \begin{cases} 1, & \text{if } \overline{X}_n > 1/2 \\ 0, & \text{if } \overline{X}_n \le 1/2. \end{cases}$$

Examine whether (i) $\hat{\theta}_n \xrightarrow{P} \theta$. (ii) Suppose $Y_n = \sqrt{n}(\hat{\theta}_n - \theta)$. Examine the convergence in law of Y_n. (iii) Examine whether $\hat{\theta}_n \xrightarrow{a.s.} \theta$. (iv) Examine whether for large n, $\hat{\theta}_n = \theta$ almost surely.

Solution: (i) To verify $\hat{\theta}_n \xrightarrow{P} \theta$ we have to check whether $\hat{\theta}_n \xrightarrow{P_0} 0$ and $\hat{\theta}_n \xrightarrow{P_1} 1$. Observe that for all $\epsilon \leq 1$,

$$P_0[|\hat{\theta}_n - 0| < \epsilon] \geq P_0[\hat{\theta}_n = 0] = P_0\left[\overline{X}_n \leq 1/2\right] = \Phi\left(\sqrt{n}/2\right) \to 1$$

as $n \to \infty$. On similar lines for all $\epsilon \leq 1$,

$$P_1[|\hat{\theta}_n - 1| < \epsilon] \geq P_1[\hat{\theta}_n = 1] = P_1\left[\overline{X}_n > 1/2\right] = 1 - \Phi\left(-\sqrt{n}/2\right) \to 1$$

as $n \to \infty$. For $\epsilon > 1, P_\theta[|\hat{\theta}_n - \theta| < \epsilon] = 1$ for $\theta = 0$ and $\theta = 1$. Thus, $\hat{\theta}_n \xrightarrow{P} \theta$.
(ii) Suppose $Y_n = \sqrt{n}(\hat{\theta}_n - \theta)$. We find the limit law of $\sqrt{n}(\hat{\theta}_n - \theta)$ for $\theta = 0$ and $\theta = 1$. For $\theta = 0$,

$$P_0[\sqrt{n}(\hat{\theta}_n - 0) \leq x] = P_0[\hat{\theta}_n \leq x/\sqrt{n}] = 0 \ \text{ if } \ x < 0 \quad \text{as } \ \hat{\theta}_n = 0 \ \text{ or } \ 1.$$

For $x = 0$, $P_0[\sqrt{n}(\hat{\theta}_n - 0) \leq 0] = P_0[\hat{\theta}_n = 0] = P_0[\overline{X}_n \leq 1/2] = \Phi(\sqrt{n}/2)$.
For $x > 0$,

$$P_0[\hat{\theta}_n \leq x/\sqrt{n}] = \begin{cases} P_0[\hat{\theta}_n = 0] = \Phi(\sqrt{n}/2), & \text{if } \ 0 < x/\sqrt{n} < 1 \\ 1, & \text{if } \ x/\sqrt{n} \geq 1 \,. \end{cases}$$

$$\text{Thus,} \quad P_0[\sqrt{n}(\hat{\theta}_n - 0) \leq x] = \begin{cases} 0, & \text{if } \ x < 0 \\ \Phi(\sqrt{n}/2), & \text{if } \ x = 0 \\ \Phi(\sqrt{n}/2), & \text{if } \ 0 < x < \sqrt{n} \\ 1, & \text{if } \ x \geq \sqrt{n} \,. \end{cases}$$

Consequently as $n \to \infty$,

$$P_0[\sqrt{n}(\hat{\theta}_n - 0) \leq x] \to \begin{cases} 0, & \text{if } \ x < 0 \\ 1, & \text{if } \ x \geq 0 \,. \end{cases}$$

Thus, for $\theta = 0$, $\sqrt{n}(\hat{\theta}_n - 0) \xrightarrow{L} U \equiv 0$. Suppose $\theta = 1$,

$$\begin{aligned} P_1[\sqrt{n}(\hat{\theta}_n - 1) \leq x] &= P_1[\hat{\theta}_n \leq 1 + x/\sqrt{n}] = 0 \\ &\quad \text{if } \ 1 + x/\sqrt{n} < 0 \quad \iff \quad x < -\sqrt{n}. \end{aligned}$$

Suppose $-\sqrt{n} < x < 0 \quad \iff \quad 1 + x/\sqrt{n} < 1$. Then

$$\begin{aligned} P_1[\hat{\theta}_n \leq 1 + x/\sqrt{n}] &= P_1[\hat{\theta}_n = 0] = P_1[\overline{X}_n \leq 1/2] \\ &= P_1[\sqrt{n}(\overline{X}_n - 1) \leq \sqrt{n}(1/2 - 1)] = \Phi(-\sqrt{n}/2). \end{aligned}$$

$$\text{For } x = 0, \quad P_1[\sqrt{n}(\hat{\theta}_n - 1) \leq 0] = P_1[\hat{\theta}_n \leq 1] = 1.$$

For $x > 0 \iff 1 + x/\sqrt{n} > 1$, then $P_1[\hat{\theta}_n \leq 1 + x/\sqrt{n}] = 1$. Thus,

$$P_1[\sqrt{n}(\hat{\theta}_n - 1) \leq x] = \begin{cases} 0, & \text{if} \quad x < -\sqrt{n} \\ \Phi(-\sqrt{n}/2), & \text{if} \quad -\sqrt{n} < x < 0 \\ 1, & \text{if} \quad x \geq 0\,. \end{cases}$$

Hence, as $n \to \infty$,

$$P_1[\sqrt{n}(\hat{\theta}_n - 1) \leq x] \to \begin{cases} 0, & \text{if} \quad x < 0 \\ 1, & \text{if} \quad x \geq 0\,. \end{cases}$$

Thus, for $\theta = 1$ also, $\sqrt{n}(\hat{\theta}_n - 1) \xrightarrow{L} U \equiv 0$. Hence, the limit law of $\sqrt{n}(\hat{\theta}_n - \theta)$ is degenerate at 0 for $\theta \in \{0, 1\}$.

(iii) It is noted in (i) that for $\epsilon > 1, P_\theta[|\hat{\theta}_n - \theta| > \epsilon] = 0$ for $\theta = 0$ and $\theta = 1$. For large n and $0 < \epsilon \leq 1$,

$$\begin{aligned} P_\theta[|\hat{\theta}_n - \theta| \geq \epsilon] &= 1 - \Phi\left(\frac{\sqrt{n}}{2}\right) = \Phi\left(-\frac{\sqrt{n}}{2}\right) \approx \phi\left(\frac{\sqrt{n}}{2}\right)\frac{2}{\sqrt{n}} \\ &= \frac{2}{\sqrt{2\pi}}\frac{e^{-n/8}}{\sqrt{n}}, \end{aligned}$$

where $\phi(\cdot)$ is the probability density function of the standard normal distribution. By the ratio test for convergence of series,

$$\begin{aligned} \frac{e^{-(n+1)/8}}{\sqrt{n+1}}\frac{\sqrt{n}}{e^{-n/8}} &= \frac{\sqrt{n}}{\sqrt{n+1}}e^{-1/8} \to e^{-1/8} < 1 \\ \Rightarrow \sum_{n \geq 1} P_\theta[|\hat{\theta}_n - \theta| > \epsilon] &= \frac{2}{\sqrt{2\pi}}\sum_{n \geq 1}\frac{e^{-n/8}}{\sqrt{n}} < \infty \\ \Rightarrow \hat{\theta}_n &\xrightarrow{a.s.} \theta, \end{aligned}$$

by the sufficient condition of almost sure convergence. Further we have,

$$\begin{aligned} P_0[\hat{\theta}_n = 0] &= P_1[\hat{\theta}_n = 1] = \Phi\left(\sqrt{n}/2\right) \\ &\iff P_\theta[\hat{\theta}_n = \theta] = \Phi\left(\sqrt{n}/2\right) \ \ \forall \ \theta \\ \Rightarrow P_\theta[\hat{\theta}_n \neq \theta] &= 1 - \Phi\left(\sqrt{n}/2\right) = \Phi\left(-\sqrt{n}/2\right) \\ &= \frac{2}{\sqrt{2\pi}}\frac{e^{-n/8}}{\sqrt{n}} \ \ \& \ \ \frac{2}{\sqrt{2\pi}}\sum_{n \geq 1}\frac{e^{-n/8}}{\sqrt{n}} < \infty. \end{aligned}$$

Suppose an event A_n is defined as $A_n = \{\omega | \hat{\theta}_n(\omega) \neq \theta\}$, then by the Borel-Cantelli lemma,

$$\begin{aligned} \sum_{n \geq 1} P_\theta[\hat{\theta}_n \neq \theta] < \infty &\Rightarrow \sum_{n \geq 1} P(A_n) < \infty \\ &\Rightarrow P(\limsup A_n) = 0 \iff P(\liminf A_n^c) = 1 \\ &\Rightarrow P_\theta\{\omega | \hat{\theta}_n(\omega) = \theta, \ \ \forall \ \ n \geq n_0(\omega)\} = 1. \end{aligned}$$

TABLE 11.7
Answer Key to MCQs in Chapter 7

Q.No.	1	2	3	4	5	6	7	8
Ans	a,d	a,b	a,b,c	a,b,c	a,b,d	c	a,b,c,d	a,b,c
Q.No.	9	10	11	12	13	14	15	16
Ans	a,b,c,d	a,b,c,d	a,b,c,d	a,b,c,d	a	c	d	a,c
Q.No.	17	18	19	20	21	22	23	24
Ans	d	a,b,c,d	a,b,c,d	c	a,b,d	a,b,c	d	d
Q.No.	25	26	27	28	29	30	31	32
Ans	a,b,d	c	a,d	b,d	a,b,c,d	a,c	b	d
Q.No.	33	34	35	36	37	38	39	40
Ans	a,b,c,d	a,b,c,d	c	c	b	b	c	b
Q.No.	41	42	43	44	45	46	47	48
Ans	d	d	d	d	c	d	a,b,c,d	

Hence we conclude that for large n, $\hat{\theta}_n = \theta$ almost surely.

Answers to the multiple choice questions, based on Chapter 7, are given in Table 11.7.

11.8 Chapter 8

8.4.1 Suppose a sequence $\{(X_n, Y), n \geq 1\}$ of random variables is such that the conditional distribution of X_n given Y is uniform $U(Y - 1/n, Y + 1/n)$, $n \geq 1$ and the marginal distribution of Y is exponential with location parameter 1 and scale parameter 1. Examine whether the sequence $\{X_n, n \geq 1\}$ converges in quadratic mean, probability, law and almost surely to Y. Examine whether a sequence $\{E(X_n), n \geq 1\}$ converges to $E(Y)$. Check whether the conditions of the dominated convergence theorem are satisfied.

Solution: From the conditional distribution of X_n given Y and the marginal distribution of Y, we have $\forall\ n \geq 1$,

$$\begin{aligned} E(Y) = 2,\ E(X_n|Y) = Y \ &\Rightarrow \quad E(X_n) = E(Y) = 2, \\ E((X_n - Y)^2|Y) &= E((X_n - E(X_n|Y))^2|Y) = Var(X_n|Y) = 1/3n^2 \\ &\Rightarrow E(X_n - Y)^2) = E(E((X_n - Y)^2|Y)) = 1/3n^2 \to 0 \\ &\Rightarrow X_n \stackrel{q.m.}{\to} Y \ \Rightarrow \ X_n \stackrel{P}{\to} Y \ \Rightarrow \ X_n \stackrel{L}{\to} Y. \end{aligned}$$

$$\text{Further, } \sum_{n\geq 1} E(X_n - Y)^2 \;=\; \sum_{n\geq 1} 1/3n^2 < \infty \;\Rightarrow\; X_n \overset{a.s}{\to} Y.$$

Thus, the sequence converges in quadratic mean, probability, law and almost surely. Further, $E(Y) = E(X_n) = 2 \;\forall\; n \geq 1$, Thus, $\{E(X_n), n \geq 1\}$ is a sequence of constants 2 and hence has limit 2. Observe that $|X_n| = X_n \leq Y+1$ and $E(Y+1) = 3$. Thus, X_n is dominated by an integrable random variable and $X_n \overset{P}{\to} Y$. Thus, the requirements of the Lebesgue dominated probability convergence theorem are satisfied. It is to be noted that since $\{E(X_n), n \geq 1\}$ is a sequence of constants, we do not need the dominated convergence theorem to find its limit.

8.4.2 Suppose $\{X_n, n \geq 1\}$ is a sequence of random variables such that X_n follows exponential distribution with mean θ^n, $0 < \theta < 1$. Show that $E(\sum_{n\geq 1} X_n) = \theta/(1-\theta)$. State the results that you may use.

Solution: Note that $\{X_n, n \geq 1\}$ is a sequence of non-negative random variables. Hence by Corollary 8.2.4,

$$E\Big(\sum_{n\geq 1} X_n\Big) \;=\; \sum_{n\geq 1} E(X_n) = \sum_{n\geq 1} \theta^n = \theta \sum_{n\geq 1} \theta^{n-1} = \theta/(1-\theta).$$

8.4.3 Suppose $\{X_n, n \geq 1\}$ is a sequence of identically distributed random variables such that $E(|X_1|) < \infty$. Suppose $Y_n = (1/n)\max_{1\leq i\leq n} |X_i|$. Show that $\lim_{n\to\infty} E(Y_n) = 0$.

Solution: Note that $U_n = \max_{1\leq i\leq n} |X_i|$ is a non-negative random variable. Hence,

$$E(Y_n) = (1/n)E(U_n) = (1/n)\int_0^\infty P[U_n > x]\; dx.$$

Suppose $f_n(x) = (1/n)P[U_n > x]$. Then $\lim_{n\to\infty} f_n(x) = 0$ for any $x > 0$. We use the following lemma to prove the required result.
Lemma: Suppose $h_n(x) \to h(x)$ and $|h_n(x)| \leq g(x)$ where $\int g(x)\; dx$ is finite. Then

$$\lim_{n\to\infty} \int h_n(x)\; dx = \int \lim_{n\to\infty} h_n(x)\; dx = \int h(x)\; dx.$$

Note that this lemma is a version of Lebesgue dominated almost sure convergence theorem. Observe that

$$\begin{aligned}
P[U_n > x] &= P[\bigcup_{i=1}^{n}[|X_i| > x]] \leq \sum_{i=1}^{n} P[|X_i| > x] \\
\Rightarrow \ 0 &\leq f_n(x) \leq (1/n)\sum_{i=1}^{n} P[|X_i| > x] = P[|X_1| > x] \\
\& \int_0^\infty P[|X_1| > x] &= E(|X_1|) < \infty \\
\Rightarrow \lim_{n\to\infty} E(Y_n) &= \lim_{n\to\infty}\int_0^\infty f_n(x)\, dx = \int_0^\infty \lim_{n\to\infty} f_n(x) = 0,
\end{aligned}$$

by the above lemma.

8.4.4 Show that X is an integrable random variable if and only if $\sum_{n\geq 1} P[|X| \geq cn] < \infty$, for some $c > 0$. In particular, deduce that if the above series converges for some $c > 0$ then it converges for all $c > 0$.

Solution: Suppose a random variable Y is defined as X/c. Then $P[|X| \geq cn] = P[|Y| \geq n]$ and hence the result follows from Lemma 8.2.2. Further, if the result is true for any $c > 0$ it is true for all $c > 0$.

8.4.5 Suppose X is a non-negative random variable. Show that (i) $\lim_{n\to\infty} nE(I_{[X>n]}/X) = 0$ and (ii) $\lim_{n\to\infty}(1/n)E(I_{[X>1/n]}/X) = 0$.

Solution: (i) Observe that $I_{[X>n]}/X = 0$ if $X \leq n$. Note that

$$X > n \ \Rightarrow \ 1/X < 1/n \ \Rightarrow \ n/X < 1 \ \Rightarrow \ (n/X)I_{[X>n]} < 1.$$

Suppose $A_n = [X > n]$, then $A_{n+1} \subset A_n \ \forall \ n \geq 1$. Thus, $\{A_n, n \geq 1\}$ is a decreasing sequence with limit $\bigcap_{n\geq 1} A_n = \emptyset$. Hence, $I_{[X>n]} \stackrel{a.s.}{\to} 0$. Observe that $n/X < 1$ implies that $(n/X)I_{[X>n]} \stackrel{a.s.}{\to} 0$. Hence, by the Lebesgue dominated almost sure convergence theorem,

$$\lim_{n\to\infty} nE(I_{[X>n]}/X) = E(\lim_{n\to\infty}(n/X)I_{[X>n]}) = 0.$$

(ii) Note that $I_{[X>1/n]}/X = 0$ if $X \leq 1/n$. Further,

$$X > 1/n \ \Rightarrow \ 1/nX < 1 \ \Rightarrow \ (1/nX)I_{[X>1/n]} < 1.$$

Suppose $B_n = [X > 1/n]$, then $B_{n+1} \subset B_n \ \forall \ n \geq 1$. Thus, $\{B_n, n \geq 1\}$ is an increasing sequence with limit $\bigcup_{n\geq 1} A_n = \Omega$. Hence, $I_{[X>1/n]} \stackrel{a.s.}{\to} 1$ and hence in probability. Observe that X being a real random variable, $1/X$ is bounded in probability. Hence, $(1/n)(1/X)I_{[X>1/n]} \stackrel{P}{\to} 0$. Hence, by the Lebesgue dominated probability convergence theorem,

$$\lim_{n\to\infty}(1/n)E(I_{[X>1/n]}/X) = E(\lim_{n\to\infty}(1/n)E(I_{[X>1/n]}/X) = 0.$$

8.4.6 Suppose $\{X_n, n \geq 1\}$ is a sequence of random variables defined on $(\Omega, \mathbb{A}, P)$ and Y is an integrable random variables such that $|X_n| \leq Y$ a.s. Prove that if $X_n \xrightarrow{L} 0$, then $\lim_{n\to\infty} E(X_n) = 0$.

Solution: Note that if $X_n \xrightarrow{L} 0$, then $X_n \xrightarrow{P} 0$ and the result follows from case (i) of Lebesgue dominated probability convergence theorem.

Answers to the multiple choice questions, based on Chapter 8, are given in Table 11.8.

TABLE 11.8
Answer Key to MCQs in Chapter 8

Q.No.	1	2	3	4	5	6	7
Ans	a,b	a	c	b,c	a,b,c	d	a,b,c,d

11.9 Chapter 9

9.4.1 Suppose $\{X_n, n \geq 1\}$ is a sequence of independent random variables such that $P[X_n = 2^n] = 1/2^n,\ n \geq 1$. Examine whether the sequence $\{X_n, n \geq 1\}$ obeys the WLLN. Find more that one sequences $\{a_n, n \geq 1\}$ and $\{b_n, n \geq 1\}$ with which WLLN holds.

Solution: We have

$$E(X_n) = 1 \ \ \& \ \ Var(X_n) = 2^n - 1$$

$$\Rightarrow Var(S_n) = 2(2^n - 1) - n \Rightarrow Var(S_n/b_n) = \frac{2^{n+1}}{b_n^2} - \frac{n+2}{b_n^2}.$$

Hence, if we select $b_n^2 = (2+\delta)^{n+1}$, $\delta > 0$, then $b_n > 0$, $b_n \to \infty$ and $Var(S_n/b_n) \to 0$. Hence by Chebyshev's inequality, it follows that $(S_n - a_n)/b_n \xrightarrow{P} 0$, with $a_n = n$ and with $b_n^2 = (2+\delta)^{n+1}, \delta > 0$. With $\delta > 0$, we have many options for b_n. Suppose $a_n = 0$.Then $\forall\ \epsilon > 0$

$$P[|S_n/b_n| > \epsilon] \leq E(S_n^2)/b_n^2\epsilon^2 = (2(2^n - 1) - n + n^2)/b_n^2\epsilon^2 \to 0.$$

It also follows from $S_n/b_n = (S_n - n)/b_n + n/b_n \xrightarrow{P} 0$. Thus, the sequence $\{X_n, n \geq 1\}$ obeys the WLLN with $a_n = n$ and $b_n = (2+\delta)^{(n+1)/2}$ as well as with $a_n = 0$ and $b_n = (2+\delta)^{(n+1)/2}$. Note that $(S_n - a_n)/b_n = S_n/b_n - a_n/b_n \xrightarrow{P} 0$ for any sequence $\{a_n, n \geq 1\}$ such that $a_n/b_n \to 0$.

9.4.2 Suppose $\{X_n, n \geq 1\}$ is a sequence of independent random variables with $E(X_n) = 0$ and $V(X_n) = n^2$. Examine whether the sequence $\{X_n, n \geq 1\}$ obeys the WLLN.

Solution: We have $E(X_n) = 0,\ Var(X_n) = n^2$. Hence,

$$Var(S_n) = \frac{n(n+1)(2n+1)}{6} \Rightarrow Var(S_n/b_n) = \frac{n(n+1)(2n+1)}{6b_n^2}.$$

The expression of $Var(S_n/b_n)$ suggests that if we select $6b_n^2 = n^{3+\delta}, \delta > 0$, then $b_n > 0,\ b_n \to \infty$ and $Var(S_n/b_n) \to 0$. Hence by Chebyshev's inequality, it follows that $(S_n - a_n)/b_n \xrightarrow{P} 0$, with $a_n = 0$ and with $b_n^2 = n^{3+\delta}/6$. Thus, the sequence $\{X_n, n \geq 1\}$ follows WLLN.

9.4.3 Suppose $\{X_n, n \geq 1\}$ is a sequence of independent random variables with $E(X_n) = 0$ and $V(X_n) = n$. Examine whether the sequence $\{X_n, n \geq 1\}$ obeys the WLLN.

Solution: We have $E(X_n) = 0$ and $Var(X_n) = n$. Hence,

$$Var(S_n) = \frac{n(n+1)}{2} \Rightarrow Var(S_n/b_n) = \frac{n(n+1)}{2b_n^2}.$$

The expression of $Var(S_n/b_n)$ suggests that if we select $2b_n^2 = n^{2+\delta}, \delta > 0$, then $b_n > 0,\ b_n \to \infty$ and $Var(S_n/b_n) \to 0$. Hence by Chebyshev's inequality, it follows that $(S_n - a_n)/b_n \xrightarrow{P} 0$, with $a_n = 0$ and with $b_n^2 = n^{2+\delta}/2$. Thus, the sequence $\{X_n, n \geq 1\}$ follows WLLN.

9.4.4 Suppose $\{X_n, n \geq 1\}$ is a sequence of independent random variables with $P[X_n = 2^n] = P[X_n = -2^n] = 2^{-2(n+1)}$ and $P[X_n = 0] = 1 - 2^{-2n-1}$. Examine whether the sequence $\{X_n, n \geq 1\}$ obeys the WLLN.

Solution: From the given probability distribution of X_n, we have

$$E(X_n) = 0, \quad Var(X_n) = \frac{22^{2n}}{2^{2n+2}} = \frac{1}{2} \quad \& \quad Var(S_n/b_n) = \frac{n}{2b_n^2}.$$

Thus, if we select $2b_n^2 = n^{1+\delta}, \delta > 0$, then $b_n > 0,\ b_n \to \infty$ and $Var(S_n/b_n) \to 0$. Hence by Chebyshev's inequality, it follows that $(S_n - a_n)/b_n \xrightarrow{P} 0$, with $a_n = 0$ and with $b_n^2 = (1/2)n^{1+\delta}$.

9.4.5 Suppose $\{X_n, n \geq 1\}$ is a sequence of independent random variables with $P[X_n = n] = P[X_n = -n] = 1/(2n^\lambda)$ and $P[X_n = 0] = 1 - 1/n^\lambda,\ \lambda > 0$. Examine whether the sequence $\{X_n, n \geq 1\}$ obeys the WLLN.

Solution: From the given probability distribution of X_n, we have $E(X_n) = 0$ and $Var(X_n) = n^{2-\lambda}$. Hence,

$$Var(S_n/b_n) = \frac{1}{b_n^2}\sum_{i=1}^{n} i^{2-\lambda} \quad < \quad \frac{1}{b_n^2}\sum_{i=1}^{n} i^2 = \frac{n(n+1)(2n+1)}{6b_n^2}.$$

If we select $b_n^2 = n^{3+\delta},\ \delta > 0$, then $b_n > 0,\ b_n \to \infty$ and $Var(S_n/b_n) \to 0$ and the sequence $\{X_n, n \geq 1\}$ follows WLLN with $a_n = 0$ and $b_n^2 = n^{3+\delta}$.

9.4.6 Suppose $\{X_n, n \geq 1\}$ is a sequence of independent random variables with $P[X_n = n^\alpha] = P[X_n = -n^\alpha] = 1/2n^2$ and $P[X_n = 0] = 1 - 1/n^2$, where $\alpha < 3/2$. Examine whether the sequence $\{X_n, n \geq 1\}$ obeys the WLLN.

Solution: We have $E(X_n) = 0$ and $Var(X_n) = n^{2\alpha-2}$. It is to be noted that $\alpha < 3/2 \;\Rightarrow\; 2\alpha - 2 < 3 - 2 = 1$. Hence,

$$Var(S_n/b_n) = \frac{1}{b_n^2}\sum_{i=1}^{n} i^{2\alpha-2} \;<\; \frac{1}{b_n^2}\sum_{i=1}^{n} i = \frac{n(n+1)}{2b_n^2}.$$

If we select $b_n^2 = n^{2+\delta}$, $\delta > 0$, then $b_n > 0$, $b_n \to \infty$ and $Var(S_n/b_n) \to 0$ and the sequence $\{X_n, n \geq 1\}$ follows WLLN with $a_n = 0$ and $b_n^2 = n^{2+\delta}$.

9.4.7 Suppose $\{X_n, n \geq 1\}$ is a sequence of independent random variables such that X_1 is degenerate at 0 and $\forall\; n \geq 2$,

$$P[X_n = \pm n/\log n] = \log n/(2n) \;\;\&\;\; P[X_n = 0] = 1 - (\log n)/n.$$

Examine whether the sequence $\{X_n, n \geq 1\}$ obeys the WLLN.

Solution: We have $E(X_n) = 0$ and $Var(X_n) = n/\log n, n \geq 2$. Hence,

$$Var(S_n/b_n) = \frac{1}{b_n^2}\sum_{i=2}^{n} \frac{i}{\log i} \;<\; \frac{1}{b_n^2 \log 2}\sum_{i=2}^{n} i = \frac{1}{b_n^2 \log 2}\left(\frac{n(n+1)}{2} - 1\right).$$

If we select $b_n^2 = n^{2+\delta}$, $\delta > 0$, then $b_n > 0$, $b_n \to \infty$ and $Var(S_n/b_n) \to 0$ and the sequence $\{X_n, n \geq 1\}$ follows WLLN with $a_n = 0$ and $b_n^2 = n^{2+\delta}$.

9.4.8 Suppose $\{X_n, n \geq 1\}$ is a sequence of independent and identically distributed random variables with characteristic function

$$\phi(t) \;=\; \begin{cases} 1 - |t|, & \text{if } \;|t| \leq 1 \\ 0, & \text{if } \;|t| > 1. \end{cases}$$

Show that $S_n/n \xrightarrow{L} X$, where $X \sim C(0,1)$, Cauchy distribution with location parameter 0 and scale parameter 1 and hence WLLN does not hold.

Solution: Note that S_n is a sum of n independent and identically distributed random variables, hence its characteristic function is given by, $\phi_{\frac{S_n}{n}}(t) = (\phi(t/n))^n$. Since $\phi(t) = 0$ for $|t| > 1$, $\phi(t/n) = 0$ if $n < |t|$ and $\forall\; n \geq |t|$, it is $1 - |t/n|$. Thus $\forall\; n \geq |t|$,

$$\phi_{\frac{S_n}{n}}(t) = (\phi(t/n))^n = (1 - |t/n|)^n \to e^{-|t|} \;\;\forall\;\; t \in \mathbb{R}.$$

However, $e^{-|t|}$ is a characteristic function Cauchy $C(0,1)$ distribution.

Hence, $S_n/n \xrightarrow{L} X \sim C(0,1)$ distribution which further implies that S_n/n is bounded in probability. To show that WLLN does not hold, we assume the contrary. Suppose $\{X_n, n \geq 1\}$ satisfies the WLLN, that is, there exist sequences $\{a_n, n \geq 1\}$ of real numbers and $\{b_n, n \geq 1\}$ of positive real numbers with $b_n \to \infty$ such that $(S_n - a_n)/b_n \xrightarrow{P} 0$ as $n \to \infty$. Then using the result that the sequence $\{(S_n - a_n)/b_n, n \geq 1\}$ is bounded in probability, we have

$$\frac{(S_n - a_n)}{n} = \frac{(S_n - a_n)}{b_n}\frac{b_n}{n} \xrightarrow{P} \begin{cases} 0, & \text{if} \quad n \to \infty \text{ faster than } b_n \\ \infty, & \text{if} \quad b_n \to \infty \text{ faster than } n. \end{cases}$$

If $b_n \to \infty$ at the same rate as that of n, then also $(S_n - a_n)/n \xrightarrow{P} 0$, since $(S_n - a_n)/b_n \xrightarrow{P} 0$. Observe that if $b_n \to \infty$ faster than n, then $(S_n - a_n)/n \stackrel{P}{\nrightarrow} 0$ and hence the WLLN does not hold with such sequences $\{b_n, n \geq 1\}$. Hence we focus on sequences $\{b_n, n \geq 1\}$ which $\to \infty$ at a rate slower than or equal to n. Since convergence in probability implies convergence in law, for all such sequences $\{b_n, n \geq 1\}$, $(S_n - a_n)/n \xrightarrow{L} 0$. We have proved that $S_n/n \xrightarrow{L} X \sim C(0,1)$ distribution. Now we make following three cases. (i) If the sequence $\{a_n, n \geq 1\}$ of real numbers is such that $a_n/n \to 0$, then by Slusky's theorem $(S_n - a_n)/n$ also converges in law to Cauchy $C(0,1)$ distribution. (ii) If the sequence $\{a_n, n \geq 1\}$ of real numbers is such that $a_n/n \to a$, then by Slusky's theorem $(S_n - a_n)/n \xrightarrow{L} X + a$. (iii) If the sequence $\{a_n, n \geq 1\}$ of real numbers is such that $a_n/n \to \infty$, then for $(S_n - a_n)/n$ to converge in probability or in law to 0, S_n/n must $\to \infty$. It is a contradiction to the result that S_n/n is bounded in probability. Hence we discard all the sequences $\{a_n, n \geq 1\}$ as considered in case (iii). For sequences in cases (i) and (ii) $(S_n - a_n)/n \xrightarrow{L} X$ or $X + a$. It is known that the limit random variables in convergence in law are identically distributed. This result gets violated when we assume that $\{X_n, n \geq 1\}$ satisfies the WLLN, since in that case $(S_n - a_n)/n \xrightarrow{L} 0$. Hence, we claim that $\{X_n, n \geq 1\}$ does not satisfy the WLLN.

9.4.9 Suppose $\{X_n, n \geq 1\}$ is a sequence of independent random variables such that $P[X_n = \sqrt{n}] = P[X_n = -\sqrt{n}] = 1/2$. (i) Show that the characteristic function $\phi_n(t)$ of S_n/n is $\prod_{k=1}^{n} \cos\left(t\sqrt{k}/n\right)$ and converges to $\exp(-t^2/4)$ as $n \to \infty$. (ii) Hence, show that $S_n/n \stackrel{P}{\nrightarrow} 0$. (iii) Find a sequence $\{a_n, n \geq 1\}$ and $\{b_n, n \geq 1\}$ such that $(S_n - a_n)/b_n \xrightarrow{P} 0$ as $n \to \infty$.

Solution: From the probability distribution of X_n, $E(X_n) = 0$ and $Var(X_n) = n$. Further, the characteristic function $\phi_k(t)$ of X_k is $\phi_k(t) = E(e^{itX_k}) = (1/2)(e^{it\sqrt{k}} + e^{-it\sqrt{k}}) = \cos(t\sqrt{k})$. Hence the characteristic function of S_n/n is given by, $\phi_n(t) = \phi_{S_n/n}(t) =$

$\prod_{k=1}^{n}\cos\left(t\sqrt{k}/n\right)$, $\forall\ t\in\mathbb{R}$. To find its limit as $n\to\infty$, observe that

$$\begin{aligned}\log\phi_n(t) &= \log\left(\prod_{k=1}^{n}\cos\left(t\sqrt{k}/n\right)\right)=\sum_{k=1}^{n}\log\cos\left(t\sqrt{k}/n\right)\\ &= \sum_{k=1}^{n}\log\left(1-\frac{t^2k}{2n^2}+\frac{t^4k^2}{4n^4}-\cdots\right)\\ &= -\sum_{k=1}^{n}\left(\frac{t^2k}{2n^2}-\frac{t^4k^2}{4n^4}-\cdots\right)-\sum_{k=1}^{n}\frac{1}{2}\left(\frac{t^2k}{2n^2}-\cdots\right)^2-\cdots\\ &= -\frac{t^2}{2n^2}\frac{n(n+1)}{2}+\frac{t^4}{4n^4}\frac{n(n+1)(2n+1)}{6}-\cdots.\end{aligned}$$

Hence,

$$\log\phi_n(t)=-\frac{t^2}{4}\frac{n(n+1)}{n^2}+\Delta_n\ \to\ -t^2/4,$$

since Δ_n consists of all the terms which converge to 0 as $n\to\infty$. Hence, $\phi_n(t)\to\exp(-t^2/4)$ which is the characteristic function of Z which has normal $N(0,1/2)$ distribution. Thus, $S_n/n=\overline{X}_n\xrightarrow{L}Z$, without any norming.

(ii) Suppose $S_n/n\xrightarrow{P}0$. Then $S_n/n\xrightarrow{L}0$. But we have proved that $S_n/n\xrightarrow{L}Z\sim N(0,1/2)$. Thus the result that the limit random variables in convergence in law are identically distributed gets violated. Hence, we conclude that $S_n/n\not\xrightarrow{P}0$.

(iii) Observe that $Var(S_n/b_n)=n(n+1)/2b_n^2$. Hence, if we take $b_n=n^{1+\delta}$ with $\delta>0$ then $Var(S_n/b_n)\to0$ and the WLLN holds with $b_n=n^{1+\delta}$.

9.4.10 Suppose $\{X_n,n\geq1\}$ is a sequence of random variables with $E(X_n)=0$, $Var(X_n)=\sigma^2$ and $Cov(X_n,X_m)=\sigma^2\alpha^{|m-n|}$, $|\alpha|<1$. Verify whether WLLN holds.

Solution: Since $E(X_n)=0$, $Var(X_n)=\sigma^2$ and $Cov(X_n,X_m)=\sigma^2\alpha^{|m-n|}$,

$$\begin{aligned}Var(S_n) &= Var\Big(\sum_{i=1}^{n}X_i\Big)=n\sigma^2+2\sum_{i=1}^{n}\sum_{j=i+1}^{n}Cov(X_i,X_j)\\ &= n\sigma^2+2\sigma^2\sum_{i=1}^{n}\sum_{j=i+1}^{n}\alpha^{|j-i|}\\ &= n\sigma^2+2\sigma^2((n-1)\alpha+(n-2)\alpha^2+\cdots+\alpha^{n-1})\\ \Rightarrow\ Var(S_n/n) &= \frac{\sigma^2}{n}+\frac{2\sigma^2}{n^2}((n-1)\alpha+(n-2)\alpha^2+\cdots+\alpha^{n-1})\\ &\to 0\ \text{ as }\ n\to\infty.\end{aligned}$$

Hence if we take $a_n=0$ and $b_n=n$, then $Var((S_n-a_n)/b_n)\to0$ and the WLLN holds.

TABLE 11.9
Answer Key to MCQs in Chapter 9

Q.No.	1	2	3	4	5	6	7	8	9
Ans	b	c	a,b,c,d	a,b,d	c	a,b,d	b	a,b,c,d	a

9.4.11 Suppose $\{X_n, n \geq 1\}$ is a sequence of independent and identically distributed random variables with $Var(X_1) < \infty$. Show that $(2/n(n+1))\sum_{i=1}^{n} iX_i \xrightarrow{P} E(X_1)$.

Solution: It is clear that

$$E\left(\frac{2}{n(n+1)}\sum_{i=1}^{n} iX_i\right) = \frac{2}{n(n+1)}\sum_{i=1}^{n} iE(X_i) = E(X_1)$$

$$Var\left(\frac{2}{n(n+1)}\sum_{i=1}^{n} iX_i\right) = \frac{4}{n^2(n+1)^2}\sum_{i=1}^{n} i^2 Var(X_i)$$

$$= \frac{4}{n^2(n+1)^2} Var(X_1)\frac{n(n+1)(2n+1)}{6} \to 0,$$

as $n \to \infty$. The required result follows from Chebyshev's inequality.

Answers to the multiple choice questions, based on Chapter 9, are given in Table 11.9.

11.10 Chapter 10

10.4.1 Suppose $\{X_n, n \geq 1\}$ is a sequence of independent and identically distributed random variables, each having $P(\lambda)$ distribution. Suppose $S_n = \sum_{i=1}^{n} X_i$. Find $\lim_{n\to\infty} P[S_n \leq n\lambda + \sqrt{n\lambda}]$.

Solution: Since $X_n \sim P(\lambda)$ distribution, $E(X_n) = Var(X_n) = \lambda < \infty$. Hence by the CLT, $(S_n - n\lambda)/\sqrt{n\lambda} \xrightarrow{L} Z \sim N(0,1)$. Hence,

$$P[S_n \leq n\lambda + \sqrt{n\lambda}] = P[(S_n - n\lambda)/\sqrt{n\lambda} \leq 1] \to \Phi(1) = 0.8413.$$

10.4.2 Suppose $\{X_n, n \geq 1\}$ is a sequence of independent and identically distributed random variables, each having Pareto-type distribution, with probability density function $f(x) = c/(|x|^3(\log|X|)^2)$ for $|x| > 2$ and 0 otherwise, where c is a normalizing constant. Examine whether the sequence satisfies CLT. Examine whether one can verify Lyapounov's condition.

Solution: Observe that $f(x) = f(-x)\ \forall\ x \in \mathbb{R}$. Thus the distribution is symmetric around 0. Hence $E(X_n) = 0$ and $Var(X_n) = E(X_n^2)$.

With $\log x = y$, $E(X_n^2)$ is obtained as follows.

$$\begin{aligned} E(X_n^2) &= \int_{-\infty}^{-2} x^2 f(x)\, dx + \int_{2}^{\infty} x^2 f(x)\, dx \\ &= 2\int_{2}^{\infty} x^2 f(x)\, dx = 2c\int_{2}^{\infty} x^2/(x^3(\log x)^2)\, dx \\ &= 2c\int_{2}^{\infty} 1/(x(\log x)^2)\, dx = 2c\int_{\log 2}^{\infty} (1/y^2)\, dy = \frac{2c}{\log 2} < \infty. \end{aligned}$$

Hence, the sequence satisfies CLT. To examine whether Lyapounov's condition is satisfied, we find $E(|X_n|^{2+\delta})$, $\delta > 0$ as follows.

$$\begin{aligned} E(|X_n|^{2+\delta}) &= 2\int_{2}^{\infty} x^{2+\delta} f(x)\, dx = 2c\int_{2}^{\infty} x^{2+\delta-3}/(\log x)^2)\, dx \\ &= 2c\int_{\log 2}^{\infty} e^{y\delta}/(y^2)\, dy \quad \text{with} \quad \log x = y \\ &> 2c\int_{\log 2}^{\infty} y^3\delta^3/(6y^2)\, dy = (2c/6)\delta^3\int_{\log 2}^{\infty} y\, dy = \infty. \end{aligned}$$

Hence, one cannot verify Lyapounov's condition.

10.4.3 Prove that $\lim_{n\to\infty} e^{-n}\sum_{i=0}^{n} n^i/i! = 1/2$.

Solution: Suppose $\{X_n, n \geq 1\}$ is a sequence of independent and identically distributed random variables, each following $P(1)$ distribution. Thus, $E(X_n) = Var(X_n) = 1\ \forall\ n \geq 1$. Hence by the additive property of Poisson distribution, $Y_n = \sum_{i=1}^{n} X_i \sim P(n)$ distribution and by the CLT, $\sum_{i=1}^{n}(X_i - 1)/\sqrt{n} = (Y_n - n)/\sqrt{n} \xrightarrow{L} Z \sim N(0,1)$. Note that

$$e^{-n}\sum_{i=0}^{n} n^i/i! = P[Y_n \leq n] = P[(Y_n - n)/\sqrt{n} \leq 0] \to P[Z \leq 0] = 1/2.$$

10.4.4 Evaluate $\lim_{n\to\infty}\sum_{i=0}^{[nq/p]}\binom{i+k-1}{i}p^k q^{i-k}$, where $0 < p < 1$ and $q = 1-p$.

Solution: Note that $E_n = \sum_{i=0}^{[nq/p]}\binom{i+k-1}{i}p^k q^{i-k} = P[X_n \leq nq/p]$, where X_n follows negative binomial distribution with parameters n and p. Hence $X_n \stackrel{d}{=} \sum_{i=1}^{n} Y_i$, where Y_i follows geometric distribution with parameter p and support as set W of whole numbers. Further, $E(Y_i) = q/p$ and $Var(Y_i) = q/p^2$. Hence by the CLT, $(\sum_{i=1}^{n} Y_i - nq/p)/\sqrt{nq/p^2} \xrightarrow{L} Z \sim N(0,1)$ distribution. Thus,

$$\begin{aligned} E_n &= P[X_n \leq nq/p] = P\Big[\frac{(X_n - nq/p)}{\sqrt{nq/p^2}} \leq \frac{(nq/p - nq/p)}{\sqrt{nq/p^2}}\Big] \\ &= P\Big[(\sum_{i=1}^{n} Y_i - nq/p)/\sqrt{nq/p^2} \leq 0\Big] \to P[Z \leq 0] = 1/2. \end{aligned}$$

10.4.5 Examine whether $(X_n - E(X_n))/\sqrt{V(X_n)} \xrightarrow{L} Z \sim N(0,1)$ as $n \to \infty$, if $X_n \sim G(\alpha, n)$, where n is a natural number.

Solution: Observe that $X_n \sim G(\alpha, n) \quad \Rightarrow \quad X_n \stackrel{d}{=} \sum_{i=1}^n Y_i$, where $Y_i \sim G(\alpha, 1)$, $i = 1, 2, \cdots, n$. Hence, $E(Y_i) = 1/\alpha$ and $Var(Y_i) = 1/\alpha^2$. Hence by the CLT,

$$(X_n - E(X_n))/\sqrt{V(X_n)} \stackrel{d}{=} \Big(\sum_{i=1}^n Y_i - n(1/\alpha)\Big)/\sqrt{n/\alpha^2} \xrightarrow{L} Z \sim N(0,1).$$

10.4.6 Evaluate $\lim_{n\to\infty} \int_0^{n/\alpha+2\sqrt{n}/\alpha} \frac{\alpha^n}{\Gamma(n)} x^{n-1} e^{-\alpha x}\, dx$.

Solution: Observe that $I_n = \int_0^{n/\alpha+2\sqrt{n}/\alpha} \frac{(\alpha)^n}{\Gamma(n)} x^{n-1} e^{-\alpha x}\, dx$ is the same as $P[X_n \le n/\alpha + 2\sqrt{n}/\alpha]$, where $X_n \sim G(\alpha, n)$. Note that $X_n \stackrel{d}{=} \sum_{i=1}^n Y_i$, where $Y_i \sim G(\alpha, 1)$ with $E(Y_i) = 1/\alpha$ and $Var(Y_i) = 1/\alpha^2$. Hence by the CLT, $(\sum_{i=1}^n Y_i - n/\alpha)/\sqrt{n/\alpha^2} \xrightarrow{L} Z \sim N(0,1)$ distribution. Thus,

$$\begin{aligned} I_n &= P[X_n \le n/\alpha + 2\sqrt{n}/\alpha] = P\Big[\frac{(X_n - n/\alpha)}{\sqrt{n/\alpha^2}} \le \frac{2\sqrt{n}/\alpha}{\sqrt{n/\alpha^2}}\Big] \\ &= P\Big[\sum_{i=1}^n (Y_i - n/\alpha)/\sqrt{n/\alpha^2} \le 2\Big] \to \Phi(2) = 0.9773. \end{aligned}$$

10.4.7 Suppose $\{X_n, n \ge 1\}$ is a sequence of independent and identically distributed random variables each having standard normal distribution. Suppose $Y_n = \sum_{i=1}^n X_i^2$. Show that $(Y_n - n)/\sqrt{(2n)} \xrightarrow{L} Z \sim N(0,1)$. What is the distribution of $(Y_n - n)^2/(2n)$ for large n? Justify.

Solution: Since $X_i \sim N(0,1)$ distribution, $\{X_1^2, X_2^2, \cdots, X_n^2\}$ are independent and identically distributed random variables, each having χ_1^2 distribution with mean 1 and variance 2. Hence by the CLT,

$$(Y_n - n)/\sqrt{(2n)} = \sum_{i=1}^n (X_i^2 - E(X_i^2))/\sqrt{(nVar(X_i^2))} \xrightarrow{L} Z \sim N(0,1).$$

By the continuous mapping theorem, it follows that $(Y_n - n)^2/(2n) \xrightarrow{L} U \sim \chi_1^2$ distribution.

10.4.8 Suppose $X \sim P(\lambda)$, $\lambda > 0$. Prove that for large λ, distribution of $(X - \lambda)^2/\lambda$ can be approximated by the chi-square distribution.

Solution: The characteristic function ϕ of $X \sim P(\lambda)$ is $\phi(t) = \exp\{\lambda(e^{it} - 1)\}$. Suppose $Y_\lambda = (X - \lambda)/\sqrt{\lambda}$. Then its characteristic function is given by

$$\begin{aligned}
\phi_{Y_\lambda}(t) &= E(\exp(itY_\lambda) = E(\exp(it(X-\lambda)/\sqrt{\lambda}) \\
&= e^{-it\sqrt{\lambda}}\exp\{\lambda(e^{it/\sqrt{\lambda}} - 1)\} \\
\Rightarrow \quad \log\phi(t)_{Y_\lambda} &= -it\sqrt{\lambda} + \lambda(e^{it/\sqrt{\lambda}} - 1) \\
&= -it\sqrt{\lambda} + \lambda(it/\sqrt{\lambda} - (1/2)t^2/\lambda - (1/6)it^3/\lambda^{3/2} + \cdots) \\
&= -t^2/2 + \Delta_\lambda \to -t^2/2 \quad \Rightarrow \quad \phi_{Y_\lambda}(t) \to \exp(-t^2/2),
\end{aligned}$$

as $\lambda \to \infty$, since $\Delta_\lambda \to 0$ as $\lambda \to \infty$. Hence, by the continuity theorem and the uniqueness theorem of characteristic functions $Y_\lambda \xrightarrow{L} Z \sim N(0,1)$. Further by the continuous mapping theorem, $(X-\lambda)^2/\lambda \xrightarrow{L} U \sim \chi^2_1$ distribution. Thus, for large λ, distribution of $(X-\lambda)^2/\lambda$ can be approximated by the chi-square distribution.

10.4.9 Suppose $\{X_n, n \geq 1\}$ is a sequence of independent and identically distributed random variables, each following $U(0,\theta)$ distribution, $\theta > 0$. Suppose $S_n = \left(\prod_{i=1}^n X_i\right)^{1/n}$. Find the limiting distribution of $\sqrt{n}(S_n - \theta e^{-1})$.

Solution: Suppose $T_n = -\log S_n = (1/n)\sum_{i=1}^n(-\log X_i) = (1/n)\sum_{i=1}^n Y_i$ where $Y_i = -\log X_i$, $i = 1, 2, \cdots, n$. If $X \sim U(0,\theta)$ and $Y = -\log X$, then the probability density function $f_Y(y,\theta)$ of Y is given by,

$$f_Y(y,\theta) = (1/\theta)e^{-y}, \quad -\log\theta < y < \infty.$$

Using the method of integration by parts, we have

$$\begin{aligned}
E(Y) &= \frac{1}{\theta}\int_{-\log\theta}^{\infty} ye^{-y} = \frac{1}{\theta}\left[\int_{-\log\theta}^{\infty} y\frac{d}{dy}(-e^{-y})\right] = 1 - \log\theta \\
E(Y^2) &= \frac{1}{\theta}\int_{-\log\theta}^{\infty} y^2e^{-y} = \frac{1}{\theta}\left[\int_{-\log\theta}^{\infty} y^2\frac{d}{dy}(-e^{-y})\right] \\
&= \frac{1}{\theta}\left[(-y^2e^{-y})_{-\log\theta}^{\infty} + 2\int_{-\log\theta}^{\infty} ye^{-y}\right] \\
&= (\log\theta)^2 + 2 - 2\log\theta = 1 + (1-\log\theta)^2 .
\end{aligned}$$

Thus, $Var(Y) = 1$ and $\{Y_1, Y_2, \cdots, Y_n\}$ are independent and identically distributed random variables with finite mean $1 - \log\theta = \phi$ and variance 1. Hence by the Lindeberg-Levy CLT, $\sqrt{n}(T_n - \phi) \xrightarrow{L} Z \sim N(0,1)$. Now to find the limiting distribution of $\sqrt{n}(S_n - \theta e^{-1})$, we consider a transformation $g(x) = e^{-x}$, it is a differentiable function with

$g'(x) = -e^{-x} \neq 0$. Hence, $\sqrt{n}(g(T_n) - g(\phi)) \xrightarrow{L} Z_1 \sim N(0, (g'(\phi))^2)$ by Theorem 7.3.9. Note that

$$\begin{aligned} g(T_n) &= e^{-T_n} = S_n = \Big(\prod_{i=1}^{n} X_i\Big)^{1/n}, \quad g(\phi) = e^{-1+\log\theta} = \theta e^{-1} \\ \& \ (g'(\phi))^2) &= (-e^{-1+\log\theta})^2 = \theta^2 e^{-2}. \end{aligned}$$

Thus, the limiting distribution of $\sqrt{n}(S_n - \theta e^{-1})$ is normal $N(0, \theta^2 e^{-2})$. From Lemma 7.3.2, we also conclude that $S_n \xrightarrow{P} \theta e^{-1}$. In the terminology of large sample inference, we say that S_n is a consistent and asymptotically normal estimator of θe^{-1} with approximate variance $\theta^2 e^{-2}/n$.

10.4.10 Suppose $\{X_n, n \geq 1\}$ is a sequence of independent and identically distributed positive random variables. Suppose $S_n = (\prod_{i=1}^n X_i)^{1/n}$. Find the non-degenerate limiting distribution of suitably normalized S_n.

Solution: Suppose $T_n = \log S_n = (\sum_{i=1}^n \log X_i)/n$. Since $\{X_n, n \geq 1\}$ is a sequence of independent and identically distributed positive random variables, $\{\log X_n, n \geq 1\}$ is a sequence of independent and identically distributed random variables. Suppose $\mu = E(\log X)$ and $\sigma^2 = Var(\log X) < \infty$. Then $\sqrt{n}(T_n - \mu)/\sigma \xrightarrow{L} Z \sim N(0,1)$, by the Lindeberg-Levy CLT. Suppose a function g is defined as $g(x) = e^x$, $x > 0$. Then it is differentiable and $g'(x) = e^x \neq 0 \ \forall \ x > 0$. Hence by the δ method, $\sqrt{n}(g(T_n) - g(\mu))/\sigma \xrightarrow{L} Z \sim N(0, (g'(\mu))^2)$ which implies that $\sqrt{n}(S_n - e^{\mu})/ \xrightarrow{L} Z_1 \sim N(0, \sigma^2 e^{2\mu})$ distribution.

10.4.11 Suppose $\{X_n, n \geq 1\}$ is a sequence of independent and identically distributed random variables, each following a Poisson $P(\theta)$ distribution. Find the non-degenerate limiting distribution of suitably normalized sample variance.

Solution: Since $X \sim P(\theta)$, $E(X) = Var(X) = \theta < \infty$. The sample variance S_n^2 is given by $S_n^2 = (1/n)\sum_{i=1}^n (X_i - \overline{X}_n)^2$. To find its limiting distribution, observe that

$$\begin{aligned} S_n^2 &= \frac{1}{n}\sum_{i=1}^{n}(X_i - \overline{X}_n)^2 = \frac{1}{n}\sum_{i=1}^{n}((X_i - \theta) - (\overline{X}_n - \theta))^2 \\ &= \frac{1}{n}\sum_{i=1}^{n}(Y_i - \overline{Y}_n)^2 = \frac{1}{n}\sum_{i=1}^{n} Y_i^2 - \overline{Y}_n^2, \end{aligned}$$

where $Y_i = X_i - \theta$. Thus, $E(Y_i) = 0$ and $E(Y_i^2) = Var(Y_i) = \theta$. Hence by the WLLN, $\overline{Y}_n \xrightarrow{P} E(Y_i) = 0$. By the CLT $\sqrt{n}(\overline{Y}_n) \xrightarrow{L} Z_1 \sim N(0, \theta)$. Hence, it is bounded in probability, as a consequence, $\sqrt{n}(\overline{Y}_n)(\overline{Y}_n) \xrightarrow{P} 0$. Further, by the CLT applied to $\{Y_1^2, Y_2^2, \cdots, Y_n^2\}$,

$$\sqrt{n}((1/n)\sum_{i=1}^{n} Y_i^2 - \theta) \xrightarrow{L} Z_2 \sim N(0, v(\theta)), \text{ where}$$

$$\begin{aligned} v(\theta) &= Var(Y_i^2) = E(X_i - \theta)^4 - (E(X_i - \theta)^2)^2 \\ \Rightarrow \quad \sqrt{n}(S_n^2 - \theta) &= \sqrt{n}\Big(\frac{1}{n}\sum_{i=1}^{n} Y_i^2 - \overline{Y}_n^2 - \theta\Big) \\ &= \sqrt{n}\Big(\frac{1}{n}\sum_{i=1}^{n} Y_i^2 - \theta\Big) - \sqrt{n}(\overline{Y}_n)(\overline{Y}_n) \xrightarrow{L} Z_2 \end{aligned}$$

where $Z_2 \sim N(0, v(\theta))$. To find $v(\theta)$ we need to find the fourth central moment of $Poi(\theta)$ distribution. We obtain it from its cumulant generating function $C_X(t)$. The moment generating function $M_X(t)$ of X is $M_X(t) = \exp(\theta\{e^t - 1\})$ and hence $C_X(t) = \theta(e^t - 1) = \theta(t + t^2/2! + t^3/3! + t^4/4! + \cdots)$. Using the relation between cumulants k_i and central moments μ_i we have $k_i = \mu_i = \theta, i = 2, 3$ and $\mu_4 = k_4 + 3k_2^2 = \theta + 3\theta^2$. Thus, $v(\theta) = E(X_i - \theta)^4 - (E(X_i - \theta)^2)^2 = \mu_4(\theta) - \theta^2 = \theta + 2\theta^2$.

10.4.12 Suppose $X_n \sim B(n, p)$, $0 < p < 1$, $n \geq 1$ and $Y_n = \log(X_n/n)$ if $X_n \neq 0$ and $Y_n = 1$ if $X_n = 0$. Show that $\sqrt{n}(Y_n - \log p) \xrightarrow{L} Z_1 \sim N(0, q/p)$, where $q = 1 - p$.

Solution: Note that Y_n can be expressed as

$$Y_n = \log(X_n/n)I_{[X_n \neq 0]} + 1 \times I_{[X_n = 0]} = \log(X_n/n)(1 - U_n) + U_n,$$

where $U_n = I_{[X_n=0]}$. Observe that,

$$\begin{aligned} \sqrt{n}(Y_n - \log p) &= \sqrt{n}\Big(\log(X_n/n)(1 - U_n) + U_n - \log p\Big) \\ &= \sqrt{n}\,(\log(X_n/n) - \log p) - \sqrt{n}U_n \log(X_n/n) + \sqrt{n}U_n. \end{aligned}$$

We discuss the limiting behaviour of these terms. For $\epsilon > 0$,

$$\begin{aligned} P[\sqrt{n}|U_n| > \epsilon] &= P[U_n = 1] = P[X_n = 0] = q^n \\ \Rightarrow \sum_{n\geq 1} P[\sqrt{n}|U_n| > \epsilon] &= \sum_{n\geq 1} q^n < \infty \Rightarrow \sqrt{n}U_n \xrightarrow{a.s.} 0. \quad (11.1) \end{aligned}$$

by Corollary 6.3.1. Hence, $\sqrt{n}U_n \xrightarrow{a.s.} 0$. Further, if $X_n \sim B(n, p)$, then $X_n \stackrel{d}{=} \sum_{i=1}^{n} W_i$ where $\{W_1, W_2, \cdots, W_n\}$ are independent and identically distributed random variables each having Bernoulli $B(1, p)$ distribution with mean p and variance pq, which is positive and finite.

Hence by the Lindeberg-Levy CLT,

$$\begin{aligned}
\Big(\sum_{i=1}^{n} W_i - np\Big)/\sqrt{npq} &\stackrel{L}{\to} Z \sim N(0,1)\\
\Rightarrow \quad \sqrt{n}\Big((X_n/n) - p\Big) &\stackrel{L}{\to} Z_1 \sim N(0, pq)\\
\Rightarrow \quad \sqrt{n}\Big(\log(X_n/n) - \log p\Big) &\stackrel{L}{\to} Z_1 \sim N(0, q/p),
\end{aligned}$$

by the δ method. Observe that as in Example 6.3.8,

$$\begin{aligned}
(X_n/n) &\stackrel{a.s.}{\to} p \Rightarrow \log(X_n/n) \stackrel{a.s.}{\to} \log p\\
\sqrt{n}U_n &\stackrel{a.s.}{\to} 0 \text{ by Equation (11.1)}\\
\Rightarrow \sqrt{n}U_n \log(X_n/n) &\stackrel{a.s.}{\to} 0 \Rightarrow \sqrt{n}U_n \log(X_n/n) \stackrel{P}{\to} 0.
\end{aligned}$$

Hence,

$$\begin{aligned}
\sqrt{n}(Y_n - \log p) &= \sqrt{n}\,(\log(X_n/n) - \log p) - \sqrt{n}U_n \log(X_n/n) + \sqrt{n}U_n\\
&\stackrel{L}{\to} Z_1 \sim N(0, q/p) \text{ by Slutsky's theorem.}
\end{aligned}$$

10.4.13 Suppose $\{X_n, n \geq 1\}$ is a sequence of independent random variables with $P[X_n = \pm n] = n^{-\lambda}/2$ and $P[X_n = 0] = 1 - n^{-\lambda}$, $\lambda > 0$. Show that Lyapounov's condition for the central limit theorem holds for this sequence for some λ.

Solution: We verify the condition for $\delta = 1$. From the given distribution of X_k,

$$\begin{aligned}
&E(X_k) = 0, \; Var(X_k) = k^{2-\lambda} \;\; \& \;\; E(|X_k|^3) = k^{3-\lambda}\\
\Rightarrow \quad & s_n^2 = \sum_{k=1}^{n} Var(X_k) = \sum_{k=1}^{n} k^{2-\lambda} \approx \int_0^n x^{2-\lambda}\,dx = n^{3-\lambda}/(3-\lambda)\\
\& \quad & u_n = \sum_{k=1}^{n} E(|X_k|^3) = n^{4-\lambda}/(4-\lambda)\\
\Rightarrow \quad & u_n/s_n^3 = \frac{n^{4-\lambda}}{4-\lambda}\Big(\frac{3-\lambda}{n^{3-\lambda}}\Big)^{3/2} = \frac{(3-\lambda)^{3/2}}{4-\lambda}\frac{1}{n^{(1-\lambda)/2}} \to 0
\end{aligned}$$

if $\lambda < 1$. Hence with $\delta = 1$, Lyapounov's condition for the central limit theorem holds for this sequence when $\lambda < 1$. Hence, Lindeberg's condition also holds. Thus, the CLT holds for this sequence for $\lambda < 1$. Note that

$$\max_{1 \leq k \leq n} \sigma_k/s_n^2 = n^{2-\lambda}(3-\lambda)/n^{3-\lambda} = 1/n \to 0.$$

Hence Feller's condition is also satisfied.

10.4.14 Suppose $\{X_n, n \geq 1\}$ is a sequence of independent random variables with $P[X_n = \pm n^{\lambda}] = 1/2$. Examine whether Lyapounov's condition, Lindeberg's condition and Feller's condition are satisfied. Examine if CLT holds for some values of λ.

Solution: From the given distribution of X_k,

$$\begin{aligned} E(X_k) &= 0,\ Var(X_k) = k^{2\lambda} \ \ \&\ \ E(|X_k|^3) = k^{3\lambda} \\ \Rightarrow\ s_n^2 &= \sum_{k=1}^{n} Var(X_k) = \sum_{k=1}^{n} k^{2\lambda} \approx \int_0^n x^{2\lambda}\, dx = n^{2\lambda+1}/(2\lambda+1). \end{aligned}$$

Note that $s_n^2 > 0$, if $2\lambda + 1 > 0 \iff \lambda > -1/2$. Further,

$$\begin{aligned} u_n &= \sum_{k=1}^{n} E(|X_k|^3) = \sum_{k=1}^{n} k^{3\lambda} \approx \int_0^n x^{3\lambda}\, dx = n^{3\lambda+1}/(3\lambda+1) \\ \Rightarrow\ u_n/s_n^3 &= \frac{n^{3\lambda+1}}{3\lambda+1}\left(\frac{2\lambda+1}{n^{2\lambda+1}}\right)^{3/2} = \frac{(2\lambda+1)^{3/2}}{3\lambda+1}\frac{1}{n^{1/2}} \to 0. \end{aligned}$$

Thus, if $\lambda > -1/2$ with $\delta = 1$, Lyapounov's condition for the central limit theorem holds for this sequence and hence Lindeberg's condition also holds. Thus, the CLT holds for this sequence for $\lambda > -1/2$. Note that for $\lambda > 0$,

$$\max_{1\leq k\leq n} \sigma_k^2/s_n^2 = n^{2\lambda}(2\lambda+1)/n^{2\lambda+1} = (2\lambda+1)/n \to 0.$$

If $\lambda \in (-1/2, 0)$, then

$$\max_{1\leq k\leq n} \sigma_k^2/s_n^2 = (2\lambda+1)/n^{2\lambda+1} \to 0.$$

Hence Feller's condition is also satisfied. We now verify Lindeberg's condition separately. Note that $n^{\lambda}/s_n = 1/(n^{1/2}(2\lambda+1)) \to 0$, for all $\lambda \geq -1/2$. Hence given $\epsilon > 0$, $\exists\ n_0$ such that $\forall\ n \geq n_0$, $n^{\lambda}/s_n < \epsilon \iff \epsilon s_n > n^{\lambda}$. Hence, $\forall\ k = 1, 2, \cdots, n$, $[|X_k| > \epsilon s_n] \subseteq [|X_k| > n^{\lambda}]$.

Case (i) Suppose $\lambda > 0$. Then $\forall\ k = 1, 2, \cdots, n$

$$\begin{aligned} k \leq n \ &\Rightarrow\ |X_k| = k^{\lambda} \leq n^{\lambda} = |X_n| \ \Rightarrow\ P[|X_k| > n^{\lambda}] = 0 \\ &\Rightarrow\ P[|X_k| > \epsilon s_n] \leq P[|X_k| > n^{\lambda}] = 0 \\ &\Rightarrow\ g_n(\epsilon) = (1/s_n^2)\sum_{k=1}^{n}\int_{[|X_k|\geq \epsilon s_n]} x_k^2 dF_k(x_k) \to 0. \end{aligned}$$

Case (ii) Suppose $-1/2 < \lambda \leq 0$. Note that $1/s_n = \sqrt{(2\lambda+1)/n^{(2\lambda+1)/2}} \to 0$. Hence given $\epsilon > 0$, $\exists\ n_1$ such that $\forall\ n \geq n_1$, $1/s_n < \epsilon$, that is, $\epsilon s_n > 1$. Further, for $-1/2 < \lambda \leq 0$, $|X_k| = k^{\lambda} \leq 1$. Thus, $\forall\ n \geq n_1$, $\epsilon s_n > 1 > |X_k|$ which implies that $P[|X_k| > \epsilon s_n] = 0\ \forall\ k = 1, 2, \cdots, n$. Hence $g_n(\epsilon) \to 0$ and Lindeberg's condition holds.

10.4.15 Suppose $\{X_n, n \geq 1\}$ is a sequence of independent random variables such that for $\alpha \geq 1/2$,

$$P[X_n = \pm n^{\alpha}] = n^{1-2\alpha}/2 \;\; \& \;\; P[X_n = 0] = 1 - n^{1-2\alpha}, \;\; n \geq 1.$$

Show that $\forall \;\; \alpha \in [1/2, 1), \; S_n/s_n \xrightarrow{L} Z \sim N(0,1)$.

Solution: Observe that $\alpha \geq 1/2 \;\Rightarrow\; 1 - 2\alpha \leq 0$. Hence, $1 - n^{1-2\alpha} \leq 1$ and $n^{1-2\alpha}/2 \leq 1$. From the given probability distribution we have

$$\begin{aligned} E(X_k) &= 0, \;\; Var(X_k) = E(X_k^2) = k \;\Rightarrow\; s_n^2 = n(n+1)/2 \\ E(|X_k|^3) &= k^{1+\alpha} \;\Rightarrow\; \sum_{k=1}^{n} E(|X_k|^3) \approx \int_0^n x^{1+\alpha}\, dx = n^{2+\alpha}/(2+\alpha) \\ \Rightarrow \frac{\sum_{k=1}^{n} E(|X_k|^3)}{s_n^3} &= \frac{2^{3/2}\; n^{2+\alpha}}{(n(n+1))^{3/2}(2+\alpha)} \\ &= \frac{2^{3/2}}{(2+\alpha)} \frac{1}{(1+1/n)^{3/2} n^{1-\alpha}} \to 0, \quad \text{for} \;\; \alpha < 1. \end{aligned}$$

Hence by Lyapounov's CLT, $S_n/s_n \xrightarrow{L} Z \sim N(0,1)$ distribution for $\alpha < 1$.

10.4.16 Suppose $\{X_n, n \geq 1\}$ is a sequence of independent random variables such that $X_k \sim U(0,k)$ distribution. Suppose $Y_k = X_k/n^{1/2}$, $k = 1, 2, \cdots, n, \;\; n \geq 1$. Show that $\{Y_k, k \geq 1\}$ satisfies Lyapounov's, Lindeberg's and Feller's conditions and hence CLT is valid for $S_n = \sum_{k=1}^{n}(Y_k - E(Y_k))$.

Solution: Since $X_k \sim U(0,k)$

$$E(X_k) = k/2 \;\; Var(X_k) = k^2/12 \;\&\; E(X_k^3) = k^3/4$$

$$E|X_k - E(X_k)|^3 = \frac{1}{k}\int_0^k |x - k/2|^3 \, dx = k^3/32$$

$$\begin{aligned} Y_k = X_k/n^{1/2} &\Rightarrow E(Y_k) = k/2n^{1/2} \;\&\; Var(Y_k) = k^2/12n \\ &\Rightarrow s_n^2 = \sum_{k=1}^{n} Var(Y_k) = n(n+1)(2n+1)/72n \to \infty \\ &\Rightarrow E|Y_k - E(Y_k)|^3 = k^3/32n^{3/2} \\ &\Rightarrow \sum_{k=1}^{n} E|Y_k - E(Y_k)|^3 = n^2(n+1)^2/128n^{3/2}. \end{aligned}$$

TABLE 11.10
Answer Key to MCQs in Chapter 10

Q.No.	1	2	3	4	5	6	7	8	9	10
Ans	d	a,b,c,d	a,d	d	a,b,c,d	a,c,d	b	c,d	a,c,d	b
Q.No.	11	12	13	14	15	16	17	18	19	20
Ans	c	a	b	c	c	b	b			

Observe that $\max_{1\leq k\leq n} Var(Y_k)/s_n^2 = \frac{n^2/12n}{n(n+1)(2n+1)/72n} \to 0$ which implies that Feller's condition is satisfied. To examine whether Lyapounov's condition is satisfied, note that with $\delta = 1$,

$$\frac{\sum_{k=1}^{n} E|Y_k - E(Y_k)|^3}{s_n^{3/2}} = \frac{n^2(n+1)^2}{128n^{3/2}} \frac{(72n)^{3/2}}{(n(n+1)(2n+1))^{3/2}} \to 0$$

which implies that Lyapounov's condition is satisfied. Lyapounov's condition implies Lindeberg's condition and hence it is also satisfied. Hence, CLT is valid for $S_n = \sum_{k=1}^{n}(Y_k - E(Y_k))$.

Answers to the multiple choice questions, based on Chapter 10, are given in Table 11.10.

Bibliography

[1] **Apostol T.** (1967). *Calculus*, Vol I, second edition, John Wiley, New York.

[2] **Ash, R. B.** (1970) *Probability and Measure Theory*, second edition, Academic Press, USA

[3] **Athreya, K. B. and Lahiri, S. N.** (2006). *Measure Theory and Probability Theory*, Springer, New York.

[4] **Bhat, B. R.** (2007). *Modern Probability Theory*, New Age International Publishers, New Delhi.

[5] **Billingsley, P.** (1986). *Probability and Measure*, John Wiley, London.

[6] **Breiman, L.** (1968) *Probability*, Addison Wesley, London.

[7] **Chow, Y. S. and Teicher, H.** (1979). *Probability Theory*, Springer, New York.

[8] **Chung, K. L.** (2001). *A Course in Probability Theory*, third edition, Academic Press, London.

[9] **Crawley, M. J.** (2007). *The R Book*, John Wiley, London.

[10] **Dalgaard, P.** (2008). *Introductory Statistics with R*, second edition, Springer, New York.

[11] **Deshmukh, S. R. and Kulkarni, M. G.** (2021). *Asymptotic Statistical Inference: A Basic Course Using R*, Springer, New York.

[12] **Grimmett, G. R. and Stirzaker, D. R.** (1982) *Probability and Stochastic Processes.* Clarendon Press, Oxford.

[13] **Gut, A.** (2005). *Probability: A Graduate Course*, Springer, New York.

[14] **Johnson, N. L., Kemp, A. W. and Kotz, S.** (2005). *Univariate Discrete Distributions*, third edition, Wiley, New Jersey.

[15] **Kumar, A. and Kumaresan, S.** (2014) *A Basic Course in Real Analysis*, Chapman and Hall/CRC, USA.

[16] **Loeve, M.** (1977). *Probability Theory I*, fourth edition, Springer, New York.

[17] **Purohit, S. G., Gore, S. D. and Deshmukh, S. R.** (2008). *Statistics using R*, second edition Narosa Publishing House, New Delhi.

[18] **R Core Team** (2019). *R: A language and environment for statistical computing*, R Foundation for Statistical Computing, Vienna, Austria. URL- https://www.R-project.org/.

[19] **Rao, C. R.** (1978). *Linear Statistical Inference and its Applications*, Wiley, New York.

[20] **Serfling, R. J.** (1980). *Approximation Theorems of Mathematical Statistics*, John Wiley, New York.

[21] **Shao, J.** (2003). *Mathematical Statistics*, second edition, Springer, New York.

[22] **Sivaprasad, M. and Deshmukh, S. R.** (2003). *Introduction to Stochastic Processes Using R*, Springer, New York.

[23] **Stoyanov, J.** (1996). *Counter Examples in Probability*, John Wiley, New York.

[24] **van der Vaart, A.** (1998). *Asymptotic Statistics*, Cambridge University Press, Cambridge.

[25] **Verzani, J.** (2005). *Using R for Introductory Statistics*, Chapman and Hall/CRC Press, New York.

Index